## Basic to Advanced Computer Aided Design Using

# NX 8.5

## Modeling, Drafting, and Assemblies

### A Project Oriented Learning Manual

By:

Stephen M. Samuel PE

DESIGN VISIONARIES

Superior Vision Yields
Optimal Products

Basic to Advanced Computer Aided Design Using NX 8.5



**ISBN:** 978-1935951049

**Published by:**

Design Visionaries

7034 Calcaterra Drive

San Jose, CA 95120

info@designviz.com

www.designviz.com

www.nxtutorials.com

Local Phone: (408) 997 6323

Fax: (408) 997 9667

DESIGN VISIONARIES

Superior Vision Yields Optimal Products

**Printed in the United States of America**

**Published June 2013**

# Dedication

We dedicate this book to those of us who truly poses a social consciousness – those brave enough to believe in the Einsteinian quote "Only a life lived for others is worth living".

## About the main Author

Stephen M. Samuel PE, Founder and President of Design Visionaries, has over 20 years' experience in developing and using high-end CAD tools and mentoring its users. During a ten-year career at Pratt & Whitney Aircraft, he was responsible for implementing advanced CAD/CAM technology in a design/manufacturing environment. He has trained thousands of engineers in Unigraphics, written self-paced courses in UG Advanced Modeling and Best Practices, and performed design work for numerous Fortune 500 companies. Stephen is the author of distinctive publications on Nastran, UGNX CAD, Sold Edge, SolidWorks, and Teamcenter Engineering PLM. Stephen holds several US patents and enjoys a life of creativity and intellectual challenge in the city of San Jose, CA. He happily shares his life with three amazing children, his wife and an 81 year old powerhouse of a mother that lives in a home right next door.

## Editors

Huge thanks to Landon Ritchie. This book would not be what it is without your invaluable help and knowledge of the intricacies of NX.

## Acknowledgements

We would like to thank the following people for their tireless efforts. Without the contributions from each of you, this book would be a mere shadow of what it has become:

Jennine Scott, Landon Ritchie.

## What readers have to say about our previous books:

### Bryan McDonald, Product Design Manager at Apple Computers

"Bottom line: this is an excellent book. If anyone wants to learn how to use Unigraphics quickly, efficiently and practically, this is the way to do it."

### Fred Dyen, Director of St. Louis University's Aviation Maintenance Institute (AMI)

"Practical Unigraphics NX Modeling for Engineers was extremely effective and much better than [other] textbooks. I would highly recommend this book to other professors and students alike."

### Dr. Pat Spicer, Professor at Western Illinois University

"The UG NX textbook is well organized. Its tutorial style of learning is easy for students to utilize. The practice exercises are essential. From my experience in teaching students to use UG software, I have found that this is the best textbook currently on the market for teaching UG NX."

## Preface

Dear reader,

Thank you for purchasing our NX 8.5 Basic to Advanced book, the newly revised version of our previous CAD training text books. Design Visionaries is an engineering consulting firm that performs many design projects great and small, including industrial design, product design and engineering analysis. Our customers entrust us with the design of medical devices, aerospace components, heavy machinery, consumer products, etc.

The methods outlined in this book go beyond an academic use of the software. They are tricks of the trade that come from thousands of hours of actual use of the software to design some of the most difficult products in the world. In addition, Design Visionaries offers world class on-site training which enables us to develop and evolve our training material to provide maximum benefit. Please enjoy this text, and we invite you to log on to our websites – www.designviz.com and nxtutorials.com, where you can download the part files pack that accompanies this book. There are also additional free materials, other advanced materials, products, and goodies.

Thank you,

Stephen Samuel

June 2013.

# CONTENTS

COPYRIGHT © 2013 BY DESIGN VISIONARIES, INC ..... I
EDITORS ..... III
HUGE THANKS TO KATHERINE ROBBINS AND MANSI. THIS BOOK WOULD NOT BE WHAT IT IS WITHOUT YOUR INVALUABLE HELP AND KNOWLEDGE OF THE INTRICACIES OF NX. ..... IV
ACKNOWLEDGEMENTS ..... IV
**CONTENTS ..... 7**
**1. AN OVERVIEW OF NX8.5 ..... 13**
1.1 THE PHILOSOPHY OF THIS BOOK ..... 13
1.2 BASIC CONCEPTS OF NX 8.5 ..... 13
1.3 A BASIC MODELING PROCEDURE ..... 15
**2. THE NX 8.5 INTERFACE ..... 17**
2.1 INTRODUCTION ..... 17
2.2 USING THE THREE-BUTTON MOUSE FUNCTIONALITY ..... 17
2.3 THE NX 8.5 GRAPHICS WINDOW ..... 18
2.4 SELECTING GEOMETRY ..... 23
2.5 COMMAND FINDER ..... 26
2.6 THE WCS ..... 27
2.7 FILE MANAGEMENT ..... 29
**3. FUNDAMENTAL CURVES ..... 32**
3.1 INTRODUCTION ..... 32
3.2 BASIC CURVES ..... 33
3.3 BASIC CURVES WITH GEOMETRIC RELATIONS TO OTHER CURVES ..... 37
3.4 BASIC CURVES WITH SPECIFIC DIMENSIONS ..... 40
3.5 BASIC CURVE TRIM OPERATIONS ..... 43
3.6 ASSOCIATIVE LINES AND ARCS ..... 47
3.7 HELIX ..... 50
3.8 SELF-TEST ..... 52
**4. EDITING ENTITIES ..... 53**
4.1 INTRODUCTION ..... 53
4.2 DELETING ENTITIES ..... 53
4.3 HIDING ENTITIES ..... 54
4.4 CHANGING AN ENTITY'S APPEARANCE ..... 57
4.5 SELF-TEST ..... 59
**5. SKETCHES ..... 60**
5.1 INTRODUCTION ..... 60
5.2 CREATING A SKETCH ..... 64
5.3 DEFINING GEOMETRIC CONSTRAINTS ..... 77
DETAIL: THE REFERENCE SKETCH FEATURES ..... 87
5.4 MIRROR CURVE OPERATION ..... 89
5.5 PLACING A SKETCH AT THE END OF A CURVE ..... 91
5.6 THE PERIMETER CONSTRAINT ..... 93
5.7 THE STUDIO SPLINE ..... 95
5.8 DIRECT SKETCHING ..... 98

5.9 Self-Test ... 103
**6. DESIGN FEATURES ... 104**
6.1 Introduction ... 104
6.2 Extruded Features ... 106
Detail: Extruding Intersecting Curves ... 113
6.3 Text Curves ... 116
6.4 Revolved Features ... 121
6.5 Primitive Features ... 124
6.6 Practice: The Pool Rack ... 128
6.8 Practice: The Hinge ... 150
6.9 Practice: Emboss ... 152
6.10 Self-Test ... 155
**7. DETAIL FEATURES ... 156**
7.1 Introduction ... 156
7.2 Edge and Face Blend Features ... 157
7.3 Drafts ... 174
7.4 Chamfer Features ... 180
7.5 Bridge Surfaces ... 184
7.6 Practice: The Pool Rack, Part 2 ... 187
7.7 Self-Test ... 188
**8. REFERENCE ENTITIES ... 189**
8.1 Introduction ... 189
8.2 Datum Plane ... 190
8.3 Datum Axis Features ... 199
8.4 Point ... 200
8.5 Self-Test ... 202
**9. OTHER BASIC TECHNIQUES ... 203**
9.1 Introduction ... 203
9.2 Sketching on a Solid Face ... 203
9.3 Shell Feature ... 206
9.4 The Trim Body Feature and the Split Operation ... 209
9.5 Offset Face Features ... 213
9.6 Thicken Sheet ... 215
9.7 Tube Features ... 217
9.8 Pattern Feature ... 220
9.9 Extract ... 245
9.10 Practice: Attachment Arm ... 247
9.11 Practice: The Clip ... 249
9.12 Self-Test ... 251
**10. INTERMEDIATE CURVES ... 252**
10.1 Introduction ... 252
10.2 Spline Curves ... 252
10.3 Offset Curves ... 256
10.4 Bridge Curves ... 258
10.5 Join Curves ... 261

10.6 Project Curves ..... 264
10.7 Isoparametric Curves ..... 267
10.8 Practice: Creating Curve Geometry ..... 270
10.9 Self-Test ..... 271
**11. COMBINE BODIES ..... 272**
11.1 Introduction ..... 272
11.2 Unite and Subtract ..... 273
11.3 Intersection ..... 278
11.4 Inferred Boolean ..... 282
11.5 Sew Operation ..... 286
11.6 Unsew ..... 289
11.7 Emboss Body ..... 295
11.8 Practice: The Stepped Plug ..... 297
11.9 Self-Test ..... 299
**12. EXPRESSIONS ..... 300**
12.1 Introduction ..... 300
12.2 Pre-Determined Expressions ..... 301
12.3 Creating Expressions during Modeling ..... 306
12.4 Renaming Existing Variables ..... 312
12.5 Practice: The Door Latch ..... 315
12.6 Self-Test ..... 317
**13. EDIT FEATURE OPERATIONS ..... 318**
13.1 Introduction ..... 318
13.2 The Part Navigator ..... 318
13.3 Editing Feature Parameters ..... 323
13.4 Deleting, Suppressing, and Reordering Features ..... 324
13.5 Self-Test ..... 329
**14. SYNCHRONOUS MODELING ..... 330**
14.1 Introduction: A Double-Edged Sword ..... 330
14.2 Move Face ..... 331
14.3 Resize Face ..... 336
14.4 Offset Region ..... 338
14.5 Replace Face ..... 340
14.6 Pull Face ..... 343
14.7 Replace Face with Offset ..... 344
14.8 Linear Dimension ..... 346
14.9 Angular Dimension ..... 349
14.10 Radial Dimension ..... 352
14.11 Co-Axial Constraint ..... 354
14.12 Make Tangent ..... 356
14.13 Pattern Face ..... 358
14.14 - Resize & Label Chamfer ..... 362
14.15 - Make Offset ..... 368
14.16 Self-Test ..... 371
**15. SURFACE FEATURES ..... 372**

15.1 Introduction ... 372
15.2 Surface Feature – Bounded Plane ... 373
15.3 Surface Feature - Through Curves ... 375
15.4 Sweep Along Guide ... 377
15.5 Swept Surface Features ... 380
15.6 Through Curve Mesh ... 384
15.7 N-Sided Surface Modeling Technique ... 392
15.8 More About Surface Techniques ... 395
15.9 Self-Test ... 401
**16. BASIC ASSEMBLIES ... 402**
16.1 Introduction ... 402
16.2 Bottom-Up Assembly Design ... 402
16.3 Top-Down Assembly Design ... 411
16.4 Editing Assembly Components ... 414
16.5 Modifying Assembly Components ... 420
16.6 Exploded Views in an Assembly ... 429
16.7 Assembly Constraints ... 432
16.8 Make Unique ... 446
16.9 Self-Test ... 449
**17. BASIC DRAFTING ... 450**
17.1 Introduction ... 450
17.2 Creating Drawings ... 450
17.3 Creating Drawing Borders ... 454
17.4 Adding Basic Drawing Views ... 465
17.5 Editing Drawings and Views ... 471
17.6 Advanced Drawing Views ... 476
17.7 Creating Dimensions ... 486
17.8 Creating and Editing Notes ... 498
17.9 Drawing Symbols ... 505
17.10 Self-Test ... 513
**18. MASTER MODELING ... 514**
18.1 What is a "Master Model?" ... 514
18.2 How to make an Associative Assembly using a Master Model ... 516
18.3 Wave Geometry Linker ... 523
18.4 Self-Test ... 528
**19. SHEET METAL FEATURES ... 529**
19.1 Introduction ... 529
19.2 Sheet Metal Flange Features ... 529
19.3 Inset Flange ... 532
19.4 Sheet Metal from Solid ... 534
19.5 Contour Flange ... 536
19.6 Bridge Bend ... 539
19.7 Bead ... 541
19.8 Drawn Cutout ... 543
19.9 Corner ... 545

19.10 Unbend/Rebend ....... 547
19.11 Relief ....... 548
19.12 Solid Punch ....... 550
19.13 Bend ....... 552
19.14 Self-Test ....... 554
**20. INJECTION MOLDED PARTS AND CASTINGS ....... 555**
20.1 Introduction ....... 555
20.2 Draft from Edges ....... 555
20.3 Body Draft ....... 557
20.4 Isocline Curves ....... 560
20.5 Divide Face ....... 565
20.6 Sewing Surfaces ....... 568
20.7 Patching Surfaces and Solids ....... 572
20.8 Ribs ....... 575
20.9 Reducing Surface Patch Count using Join Curves ....... 584
20.10 Lips ....... 586
20.11 Heat Stakes ....... 590
20.12 Snap-Fit ....... 596
20.13 Plastic Rules of Thumb ....... 600
20.14 Final Project 'PCB Component Assembly' ....... 604
20.15 Self-Test ....... 615
**21. HOW TO MAKE NX 8.5 RUN GREAT! ....... 616**
21.1 Introduction ....... 616
21.2 Customized Keyboard shortcuts ....... 617
21.3 Full menus ....... 618
21.4 Save Layout ....... 619
21.6 View Animation Speed ....... 622
21.7 Shaded Face Edges ....... 623
21.8 Assigning Feature Names in a Part ....... 624
21.9 Using Grids ....... 625
21.10 Section Tool ....... 626
21.11 True Shading Editor ....... 629
21.12 True Studio Task ....... 634
**22. PRACTICE PROJECTS ....... 638**
22.1 Introduction ....... 638
22.2 Free Form Solid Using Swept ....... 639
22.3 The Product Housing ....... 642
22.4 Advanced Face Blend Features ....... 646
22.5 Face Blending - Tangency Controlled ....... 650
22.6 A Bridge Surface using the Wave Geometry Linker ....... 652
22.7 Applying Shrinkage with Non-Uniform Scale ....... 656
22.8 Coil Springs ....... 658
22.9 NX 8.5 Final Project ....... 663
**23. GOODIES ....... 671**
23.1 Introduction ....... 671

23.2 Creative 3D Models....671
23.3 Cool Stuff to Build....674
23.4 Haven't Had Enough?....675
**GLOSSARY....676**
**INDEX....692**
**TEST ANSWERS....699**

# 1. An Overview of NX8.5

## 1.1 The Philosophy of this Book

This book has been written with a certain underlying philosophy that comes from years of engineering design which we would like to share with you. Engineers are pretty bright in general, so we've written this book to take advantage of that fact. Our book begins with the basics and examples explained to every last detail. As the book progresses, more and more is left to the reader. We believe this enables faster learning as you won't have to sift through copious and superfluous instructions. We hope you enjoy this material that we've truly poured our hearts into.

Using NX 8.5 is like playing a piano. In the same way that chords are as important as individual notes, NX commands are far more powerful when used in concert with others. Our book makes an effort to show not only the details of the most important commands, but the powerful combinations that we have used to bring about excellent designs.

All projects and practice exercises can be downloaded from www.designviz.com. Use them for your reference or as an example of how each model/project should look upon completion.

For even more tutorials, visit nxtutorials.com. There are some great tutorials on some very useful but esoteric NX functions. They even have some awesome projects with plans already available if you need a quick break from NX and want to get your hands on some real building!

## 1.2 Basic Concepts of NX 8.5

Siemens PLM Software NX 8.5™ (commonly referred to as "**Unigraphics**", "**UG**", or "**NX**") is a 3D Computer-Aided Design (CAD) software that is used by companies all over the world to design state-of-the-art products. This manual teaches you the basics of modeling, assemblies, and drafting in NX 8.5. Each of these functions is organized into a discrete "**Application**". This book concentrates on what is typically the most complicated task in NX 8.5: 3D geometric modeling.

Before continuing the discussion of how to use the NX 8.5 modeling tools, it is useful to distinguish between '**Parametric**', '**Non-Parametric**', and '**Hybrid**' models.

### *Parametric Models:*

In addition to the geometric information, a parametric model has variables (parameters) that control its geometry. The shape and size of a model may be changed by modifying these parameters. Parametric models are built by constructing a sequence of geometric operations known as "Features". Parametric models are therefore 'smart' models that remember how they were built and can 'rethink' themselves as their parameters or features are changed.

## ***Non-parametric Models:***

A non-parametric model has none of the parameters or feature-driven geometry; it is sometimes referred to as a 'dumb' model. To modify a non-parametric model, it is often necessary to rebuild a lot of the geometry.

## ***Hybrid Models:***

A Hybrid Model is neither completely parametric nor non-parametric; rather it is comprised of both types of geometry. Portions of the geometry may be so complicated that it is more efficient to make it non-parametric, while some of the geometry is still feature-driven. This is a very flexible and powerful modeling technique.

Building a completely parametric model takes more planning and foresight; however, the extra effort can pay off if you are building multiple versions of a model, or if you can anticipate potential design modifications. The experienced NX 8.5 user will determine which modeling approach is appropriate for each design project.

## ***Curves:***

Curves can be created in three-dimensional space. Basic curves are typically non parametric, but NX 8.5 has the capability to create some associative curves like lines, curves, circles, and helices where you can control certain parameters; plus, unlike the basic curves, they appear in the modeling tree. Each curve has a set of control points associated with it. For example, lines have control points at both ends and at the midpoint, while circles have a basic control point at their center. These control points are useful for a variety of tasks, such as selecting features, building new geometry, and making measurements of existing geometry.

Curves can be placed in many ways, such as at *x, y,* or *z* coordinates, or they can be connected to another curve at control points. They can also be placed in relation to other curves (parallel, perpendicular, tangent, normal, etc.).

> **Note:** NX 8.5 distinguishes between selecting a curve itself and selecting one of its control points. This can be a source of confusion to a new user. If you are having difficulty selecting geometry, pay close attention to your selection technique. To avoid the ambiguity between selecting a curve and its control points, you can zoom in closer to the area in question or rotate the model into a better orientation. You can also use the snap point tool bar to control the selection.

## ***Sketches:***

Sketches are by definition, planar (2D) parametric features. Sketches are always defined upon a planar entity, with a reference that orients the sketch horizontally or vertically. As with curves, sketch entities also have control points.

The general approach in creating a sketch is to define its orientation, quickly create the curves that define it (do not worry about dimensions when creating geometry), and constrain it by placing dimensions or geometric constraints. At any point after the sketch is constrained, the dimensions that define the sketch can be changed and the model will then update.

Once a sketch is completed, it may be used to create other geometry. A typical example is to extrude a sketch profile for a specified distance, creating a solid body. Sketches may also be used to define complex geometry, such as free-form surfaces.

### ***Reference Features:***

Reference Features include the Datum Plane, Datum Axis and Datum Coordinate features, which are used as construction entities and may be thought of as parametric reference geometry. A common use for Reference Features is to position sketches in order to create meaningful geometric relationships.

### ***Design Features:***

Design Features are used to create solids. They are typically based on either curves or sketches (i.e. the sweep entity). Although Design Features are parametric in nature, they are only fully parametric if they are created upon parametric geometry, (i.e. solids created from sketches).

### ***Surfaces:***

Simple or free-form surfaces can be created using different Surface or Mesh Surface menu options. Like Design Features, Surfaces are also based upon curves or sketches. They are the most powerful and mathematically complex of the geometric tools in NX and are often used when the desired geometry is too complicated to create using Design Features.

### ***Feature Operations:***

Feature Operations are used to add detail to the existing geometry. These are great tools and save a lot of time while creating complex geometry. The most commonly used Feature Operations are Blend, Draft, Chamfer, Instance, Hollow, Trim, and Boolean Operations (Unite, Subtract, and Intersect).

## 1.3 A Basic Modeling Procedure

NX 8.5 allows the user a great deal of freedom to determine the method in which a model is built. The basic approach to building a model is outlined below.

1. **THINK.** Step 1 is to carefully consider the geometry that you wish to create and to develop a strategy for building it. NX 8.5 works best if you have a good idea of which operations you need to perform and in what sequence they need to be applied. Although you may not know the final configuration of what you are building until it is done, you will find that there are certain universal techniques that can be applied to almost any modeling situation.

2. **SKETCH.** The next step is to sketch the basic two-dimensional shape, or outline, of what you are building. Generally just a neat freehand sketch is adequate. Remember

that the idea for this step is to save time – you do not need to sketch the perfect finished shape just yet. The geometry will take its accurate form later when you apply dimensions.

> **Note:** Sometimes people like to use primitive features (block, cylinder, cone, or sphere) for this step, but in most cases, the use of a good sketch is a superior modeling approach. Defining the shape using non-parametric curves often does not work as well either, since defining them is harder than sketching and modifying the shape.

3. **CONSTRAIN.** Put all of the dimensional and geometric constraints on the sketch. Use enough to capture the design intent and constrain the sketch.

> **Note:** If the sketch geometry changes color, you either have an over-constrained sketch or you have created conflicting constraints. There will be more on that later.

4. **ENUMERATE.** Once you have a fully constrained sketch, input the actual numerical values of each dimension. At this time, you can also relate dimensions to one another (by referring to their variable names) in order to further capture design intent.

5. **POSITION.** You can position your sketch using the sketchers standard dimensioning and constraining tools. Each strategy comes with its own pros and cons.

6. **PERFORM SOLID OPERATIONS.** Once the first sketch is complete, you may extrude or revolve it to create the basic building block of your model.

7. **ADD FEATURES.** Now that you have the first solid object, you may proceed by adding detail features such as bosses, pads, holes, or even more sketches (to be used for uniting, subtracting, or intersecting).

8. **REPEAT THE ABOVE.** A model may consist of many sketches or form features, each created in the above sequence. Mastering these modeling steps will make you far more productive.

9. **DRAFT, ANALYZE, AND MACHINE.** Once you have performed the steps above to a substantial degree, you may now start the associated functions.

# 2. The NX 8.5 Interface

## 2.1 Introduction

The look and feel of NX 8.5 is similar to that of many Windows™ based products. Commands are arranged at the top of the screen in a menu bar, and selections are made primarily with the left mouse button. Using the menu bar, different tool palettes can be activated. These tool palettes contain the icons from which most geometry is created.

You will find that NX 8.5 file management functions are very similar to most typical Windows-based products, such as Microsoft Word™. Creating new files, saving, and closing are done using pull-down menus and dialog boxes.

## 2.2 Using the Three-Button Mouse Functionality

The NX 8.5 user interface is primarily icon-based. Most commands are made by using the mouse to navigate between menus and to pick geometry. **NX 8.5 requires the use of a three-button mouse to maximize productivity**. The primary functions of the mouse buttons are illustrated below.

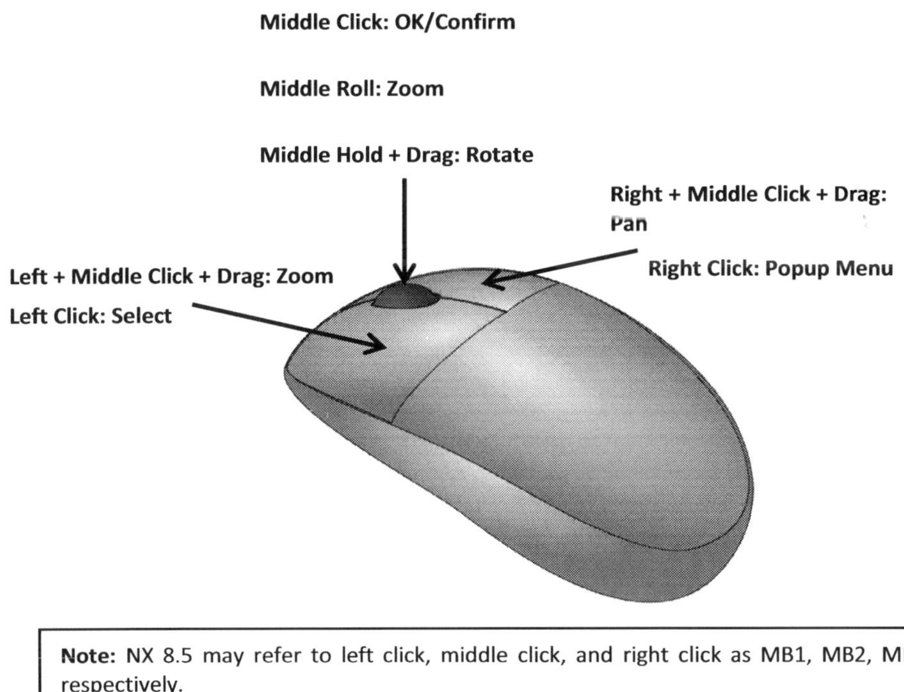


> **Note:** NX 8.5 may refer to left click, middle click, and right click as MB1, MB2, MB3, respectively.

## ***Detail: The Mouse Button Functions***

**Rotate**:

Press and hold the **middle mouse button** and move the cursor. The rotate icon will appear. Drag in the direction of the rotation you want.

To rotate the model around a single axis, put the cursor near the screen edge. Assuming the screen to be the x-y plane and z facing towards you as in top view:

Place it on the left or right side of the model for x-axis rotation.

Place it on the bottom for y-axis rotation.

Place it on the top for z-axis rotation.

**Pan**:

Drag while pressing and holding the **middle and right mouse buttons** simultaneously.

**OR:**

Press and hold **<Shift>** and drag using the **middle mouse button**.

**Zoom**:

Drag while pressing and holding the **left and middle mouse buttons** simultaneously.

**OR:**

Press and hold down <**Ctrl**> and drag using the **middle mouse button**.

In Windows, if you have a wheel mouse, each click will zoom in or out by 25%.

**Orient**:

To orient the view, press **F8**. This moves the model to the closest pre-defined view (top, front, isometric, etc.) to what you have on screen.

## ***Detail: Refreshing the View***

NX 8.5 will often leave traces of previous modeling operations in the Graphics Window. To avoid confusion, you can use **View / Refresh** or **F5** to remove extraneous display items.

> **Note:** This functionality is identical to the navigation menus that are available from the toolbar and view popup menus except that you do not need to enter a special mode.

## 2.3 The NX 8.5 Graphics Window

### ***Detail: Roles***

The NX 8.5 Graphics Window display, toolbars, and menu options depend on what role you select. Roles can be found in the resource bar as shown in the figure below. We will be using the **Advanced with full menus** role throughout the book. The toolbars may be re-arranged vertically or horizontally using drag-and-drop techniques. To save a personal role, right-click the background in the roles palette, select **New User Role** and customize it for your needs.

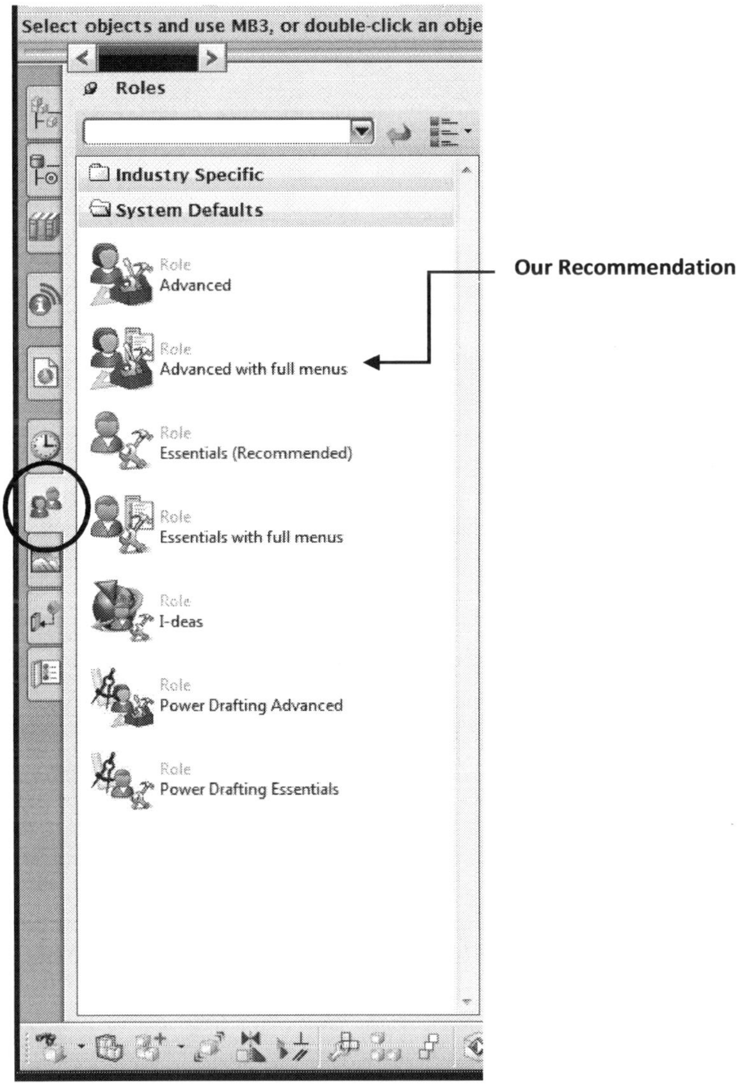

## ***Detail: The Basic Display and some useful commands***

The most commonly used functions on the **Menu Bar** are:

- **FILE**: performing file management tasks
- **EDIT**: deleting, hiding, changing entity attributes, etc.
- **VIEW**: manipulating the display of geometry in the Graphics Window
- **INSERT**: tools for creating geometry (Curves, Sketches, Detail Features, and Surfaces) are located here
- **START**: starting different applications such as Modeling and Drafting

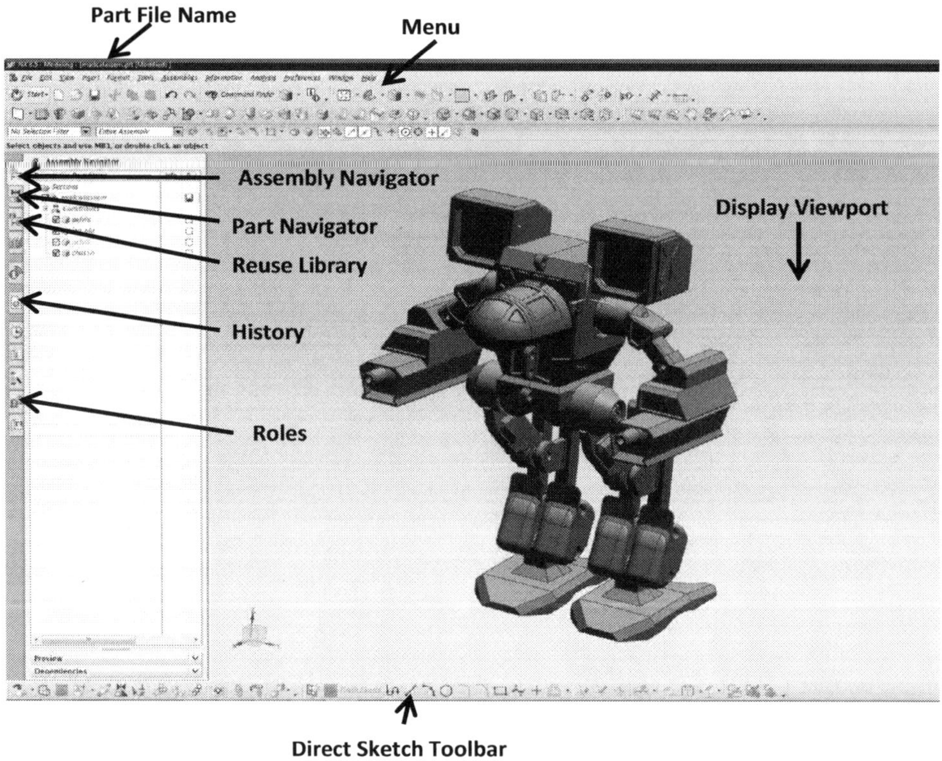


Commands like Open, Save, etc. can be accessed from the menus; however, it is faster to access the commands from a Toolbar icon or using keyboard shortcuts. You can display any toolbars you need by selecting them from the Toolbar Selection Menu, as shown above. You can right click in the toolbar area to get the Toolbar Selection Menu.

The **FILE NAME** section of the screen displays the active part, i.e. the part that you are currently working on. If you are working with assemblies, it shows both a Displayed Part and a Work Part. You will learn more about this concept later.

All geometry is displayed in the main **GRAPHICS WINDOW**.

The **CUE DISPLAY** contains important information about what NX is expecting from you. If you are performing an operation and are uncertain what to do next, READ THE CUE! (Most of the time it is actually helpful!)

The **STATUS DISPLAY** often gives a dialog of what operations are being done. This display is sometimes a good supplement to the cue, and it can be particularly helpful when picking geometry in the Graphics Window. If you are ever uncertain about what geometry you just picked, look at the status.

## *Detail: The View Pop-Up Menu*

Clicking the right mouse button activates the View Pop-Up menu (shown in the figure below) when the cursor is in the Graphics Window. This menu allows you to change the view options easily. The Shortcut Toolbar is a new feature added in NX 8.5

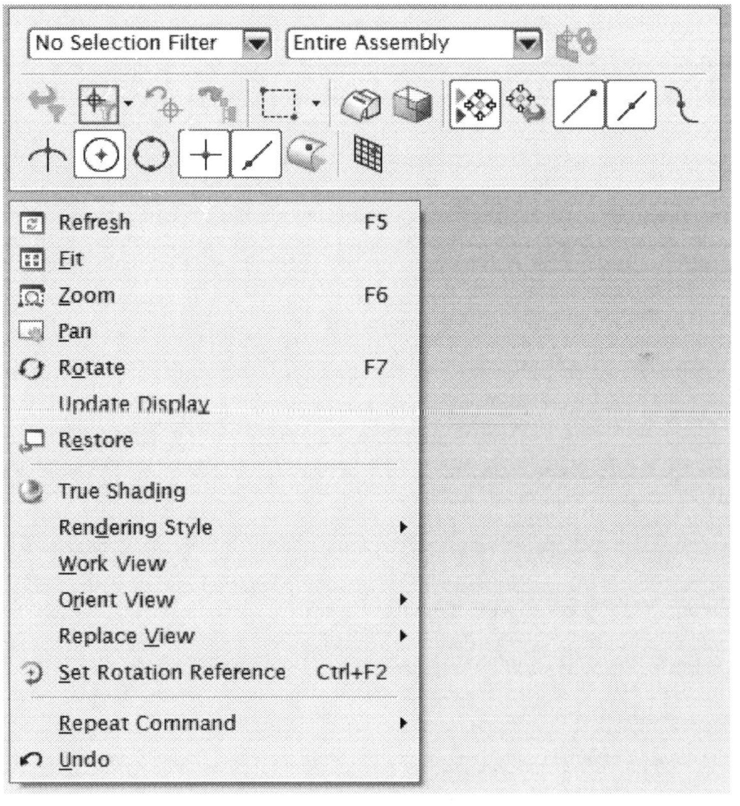


The following is a list of the most commonly used View Pop-Up menu functions:

- **REFRESH:** Cleans up any out-of-date items displayed in the Graphics Window.
- **FIT:** Resizes the model to fill the entire Graphics Window.
- **ZOOM:** Turns the cursor into a magnifying glass. Click and drag to define a rectangular zoom region. To exit zoom mode or leave any of the following modes where the cursor changes appearance, click the middle mouse button.
- **ORIENT VIEW:** Allows you to reorient the model to another standard view, such as Top, Front, or Right, but does not replace the current work view.
- **REPLACE VIEW:** Displays a submenu that allows you to replace the current work view with another standard view, such as Top, Front, or Right. By default, NX 8 starts with the model oriented in the Top view.
- **SET ROTATE POINT:** Allows you to set a point that acts as the center of rotation.

**Note:** When rotating an object after setting a rotate point, NX 8.5 displays the rotate point with a green cross that is visible during the object's rotation. To remove the user-defined point, choose **CLEAR ROTATE POINT**.

- **UNDO:** Undoes the previous operation.

**Note:** NX 8.5 saves a 'snapshot' of the model prior to performing an operation - such as creating a curve or deleting an entity. The "**Undo**" command essentially returns the model to the saved snapshot. Several actions, such as saving a model or invoking another Application will purge the Undo List, making the Undo command no longer available.

- **F4:** Re-does the last command.

## ***Detail: The Resource Bar***

The Resource Bar contains several "fly-out" icons such as the Model Navigator, the Assembly Navigator shown below, and the Recent Part History. Each has a "Push Pin" that will keep the fly-out open. They may also be "undocked" by clicking the pin again.

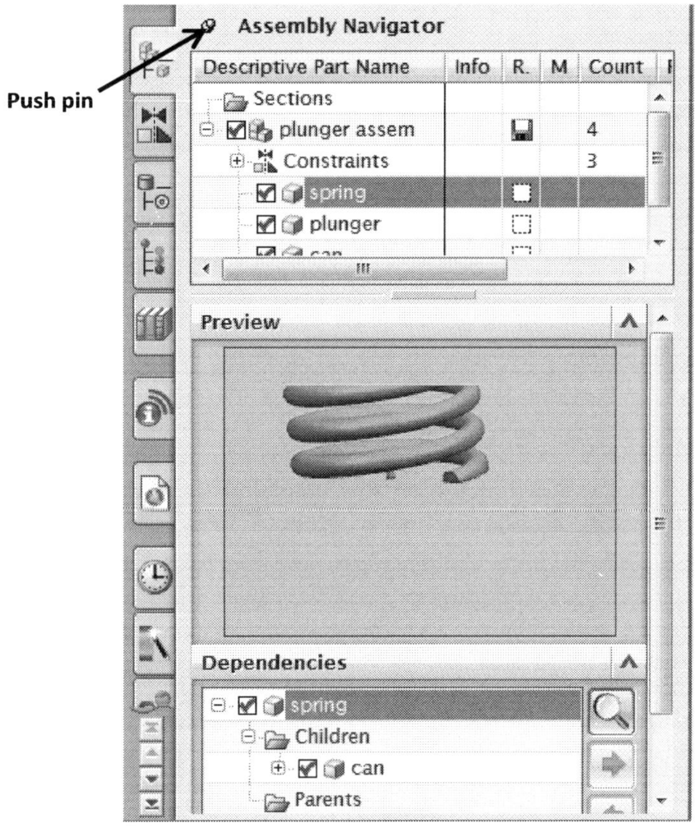


## 2.4 Selecting Geometry

In NX, all menus and commands are selected with the left mouse button. In general, geometry is selected with the left mouse button, either in the Graphics Window or when using one of the NX Navigators.

### *Selection Filters:*

In NX 8.5, the 'Selection Toolbar' seen below affects which entities may be selected. There are many Selection Filter options available, including Select General Type Objects, Select Features, Select Layers, and Select Components. Pay attention to the setting of the Selection filter to ensure that you are able to choose appropriate items.

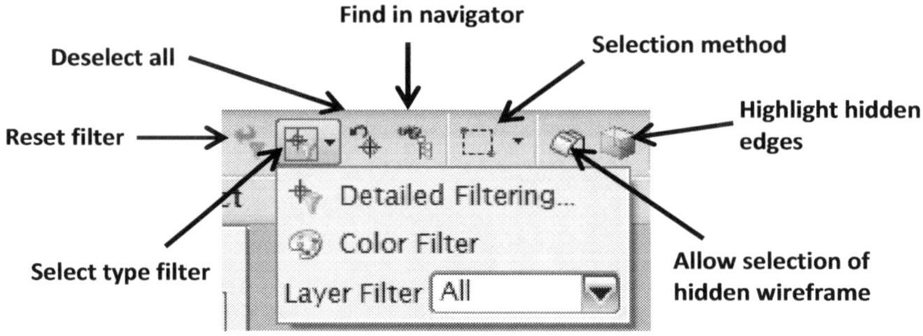


You can unselect items that have already been selected by holding down the 'Shift' key and selecting the item again.

## *The Selection Ball:*

When selecting geometry, the cursor display will change into a 'Selection Ball' as it is moved into the Graphics Window, as shown below. Any geometry that falls within the Selection Ball is a candidate for selection, assuming that it is a valid entity for the active dialog box.

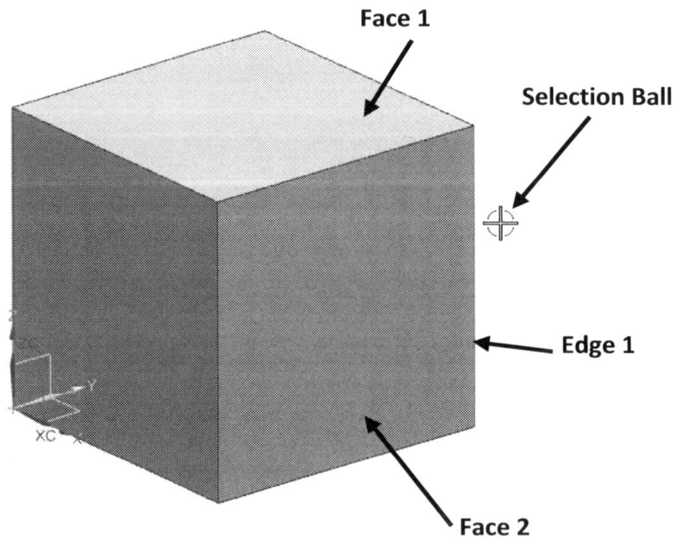


You can adjust the size and behavior of the Selection Ball using the options found in the ***Preferences / Selection*** Dialog box.

## ***Quick Pick:***

Sometimes it is not possible to position the Selection ball such that the geometric entity you are trying to select is unambiguous. If you hold the cursor still for a second then click the 'Quick Pick' dialog box. This allows you to choose among multiple entities within the Selection Ball as shown in the figure below. You can control the delay time for quick pick from the ***Preferences / Selection*** menu.

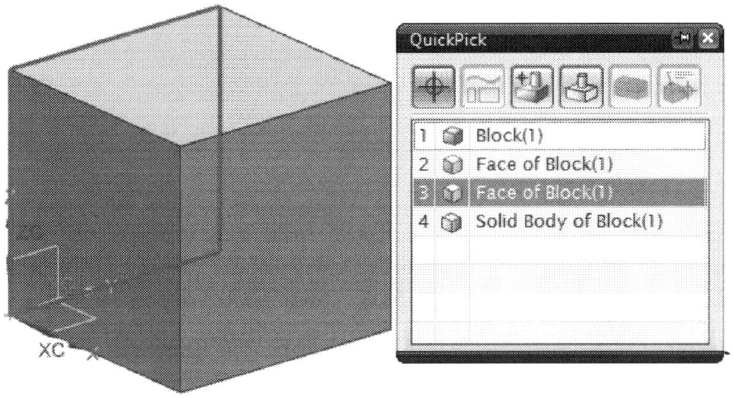


How is the selection order determined in NX 8.5? The simplest answer is to visualize a light beam of the same diameter as the selection ball shot directly into the computer screen. The first item that this light beam pierces is the first item that NX 8.5 highlights, the next item pierced is the next item highlighted, and so on. If there is still ambiguity, edges typically are highlighted before faces, and faces before solid bodies. If the thing that you are trying to select is the first thing on the list in the Quick Pick menu, just left click again and it will be accepted.

Another new addition to NX 8.5 is the shortcut toolbar, which will display relevant commands which can be used on the geometry selected.

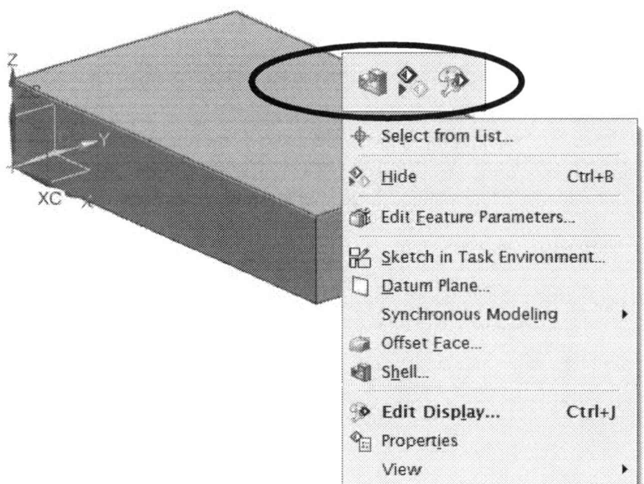

## 2.5 Command Finder

The command finder is a godsend: use it to find commands that you cannot locate in the menus or toolbars. Unfortunately, there is no finder for the command finder so you will have to look for it yourself!

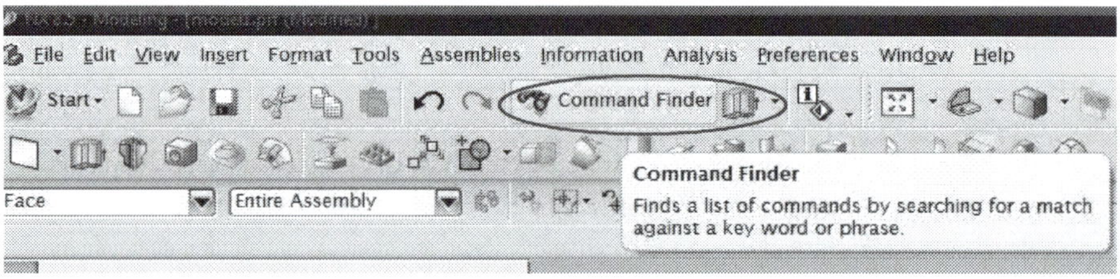


You can even type in commands that exist in other programs and let NX find it for you. For example, type "hollow," a command from some other program, into the command finder, and the Command Finder will locate the NX equivalent, "Shell."

After the search results come back, hover the cursor over the correct result and it will show you exactly where it is located, as shown in the following figure. If it is not in a toolbar, it will show you the path. To use the tool instantly, you can just click on it in the command finder and it will give you the tool operation window.

Sometimes a feature you are looking for may not be displayed. Instead of spending forever trying to get around a tool you wish was there, try customizing a menu too see if the tool you want is just hiding from you!

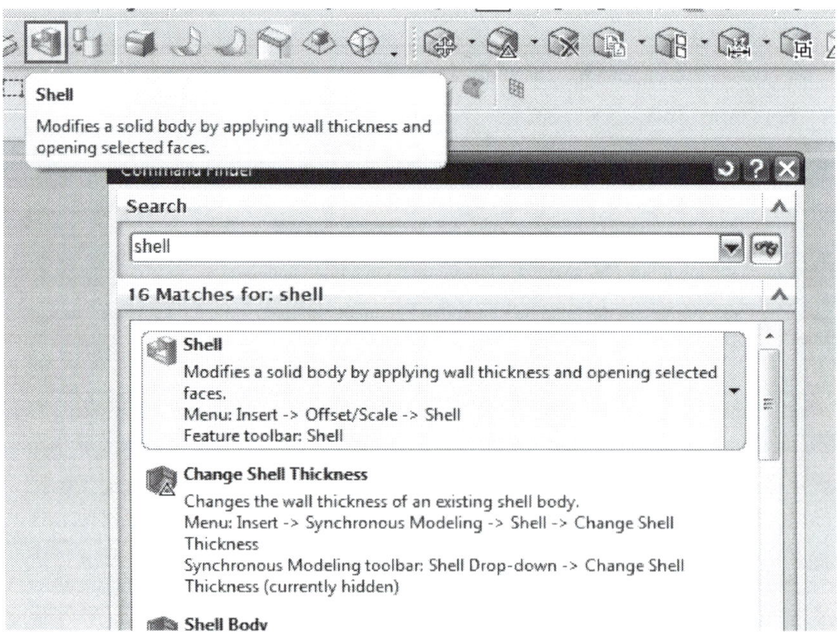


Again - this is a fantastic tool for quickly finding features that are hidden in NX's vast library of capabilities! So no more excuses!

## 2.6 The WCS

The WCS (Work Coordinate System) is an important part of NX 8.5. The WCS is a local coordinate system that you can reposition into convenient locations while performing modeling operations. NX 8.5 performs many operations with respect to the XC-YC plane of the WCS, which is sometimes referred to as the Work Plane.

**WCS**

**Dynamic WCS**

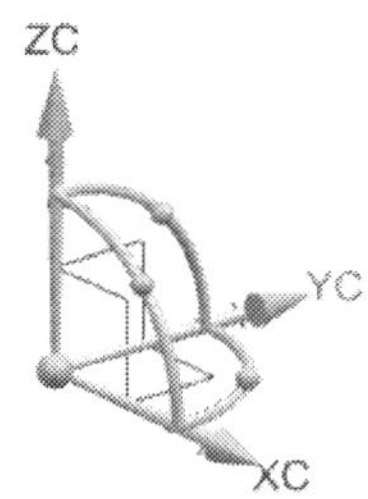


### *The WCS and the Absolute Coordinate System:*

By default, the WCS is positioned at the Absolute Coordinate System origin when you create a new part. The Absolute Coordinate System is the (0, 0, 0) origin of the model. When working with assemblies, it is often convenient to create all components such that they share a common Absolute Coordinate System. The WCS may be placed at the Absolute Coordinate System origin at any time by choosing: ***Format / WCS / Orient,*** selecting ***Absolute CSYS*** from the Type drop-down menu, then clicking on ***OK***.

### *The Dynamic WCS:*

The WCS may be moved using the ***Format / WCS / Dynamics*** option. This allows you to move the WCS interactively by dragging it to new locations in the Graphics Window. The Dynamic WCS has several handles, as shown below.

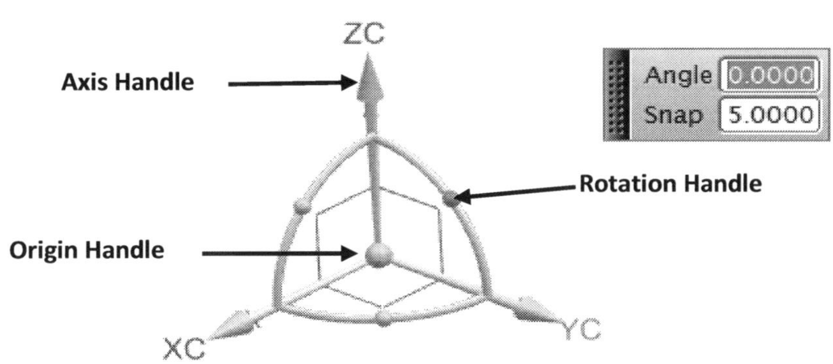

The Origin Handle allows you to drag the WCS to a new location. The Axis Handles constrain the WCS motion along the corresponding axis, and the Rotation Handles constrain the WCS rotation about the corresponding axis. The Dynamic Input box allows you to enter specific offset values. To exit Dynamic WCS mode, you can hit the **Escape** key or click the **middle mouse button**.

## ***Detail: "Object/ Action" Menus***

In NX 8.5, the "Object/Action" paradigm has been implemented for many functions. What this means is that you can often select an object in the Graphics Window and click the right mouse button to access a group of operations associated to the object.

For example, when you select a feature and click the right mouse button, a pop-up menu appears, as shown below.

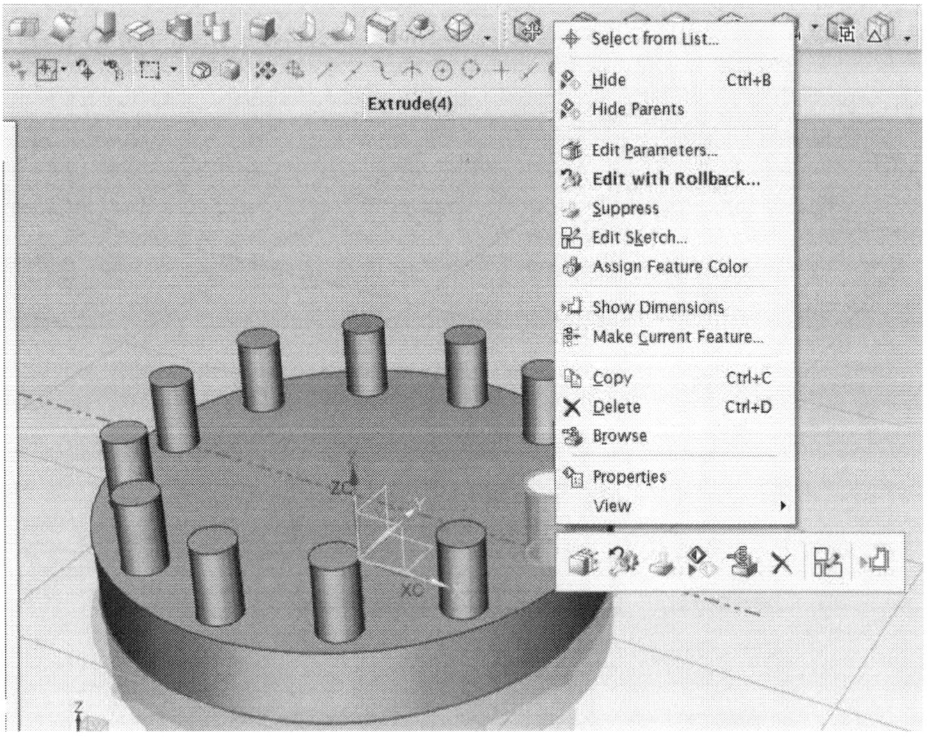

## 2.7 File Management

File management operations are accessed through the 'File' heading on the menu bar. This is where you can create new files, save files, open existing files, and close files. The file manipulation functions work in a familiar fashion to anyone with experience in Windows-based applications.

### *Detail: Creating New Files*

Before any geometry is created, a new part database must be created. The part/filename and units must be specified at this time.

Project 2.7.1

Here you will create a new part file called **training01_*<your initials>*.**prt.

- Select ***File / New***. The 'New File' dialog box will appear, as shown below.
- Select the desired units from the Units dropdown menu near the top middle of the dialog box. Remember that once chosen, units can be difficult to change after the part has been created and saved.
- Select the **Model** template. This is the default template which we will be using throughout this book.
- Use the Folder selection field of the dialog box to either type in or navigate to the directory where you would like the new file to be created.
- Click in the file Name box and type **training01_*<your initials>.***
- Choose **OK**.

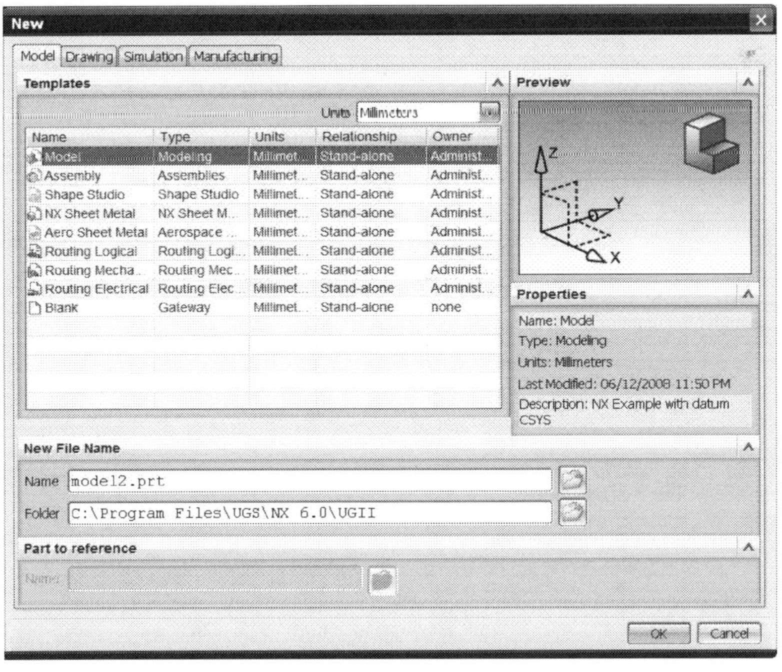

> **Note:** The dialog box in the above figure offers the user a variety of named templates to choose from. These templates will automatically open up applicable toolbars and application packages relevant to the discipline and are customizable by the user.

The WCS will appear and the file name will be shown at the top of the Graphics Window.

***End of Project***

## ***Detail: Saving Files***

Although you created a new file and entered its name, it will not be saved to disk until you explicitly save the part database. You can save a part with its existing name, or you can save it with a new name.

> **Note:** NX 8.5 automatically saves all part databases with a **".prt"** extension. When you save a file, it overwrites the version that is on disk. If you want to keep a copy of an earlier version of a part, you should manually archive it in a different folder.

## *Project 2.7.2*

Create two copies of the same file while getting around the 'Save As' function.

With the file 'training01_<*your initials*>' open, select File / Save.
The file will be saved to disk.

- When you save a model with no specified file name, you will be prompted to give it a name (eg. 'block' instead of 'model1').

***End of Project***

> **Note:** Use the hot key listings next to some of the menu selections to make your CADing more efficient.

## *Detail: Opening Existing Files*

Select ***File / Open*** to open existing part files.

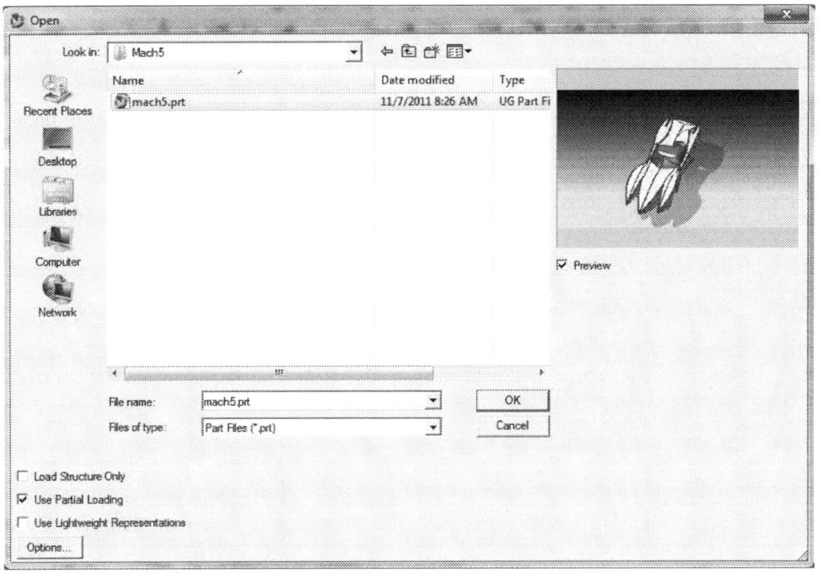

To open a file, either **double-click** on the file name or single click on the file name and then click on **OK**. Before opening the file you can also select the load options as shown above. You can also specify if you do not want to load the components of an assembly by checking the 'Do Not Load Component" check box.

## *Detail: Working with Multiple Parts*

NX 8.5 allows you to simultaneously work on many parts. The name of the active part is always displayed at the top of the Graphics Window. To change to another already opened part, select ***Window / another_part_name***, and that part becomes active. If the part name is not displayed in the Window menu, you can select ***Window/ More...*** to get a full listing of open parts.

## *Detail: Closing Files and Exiting NX 8.5*

The ***File/ Close*** commands can be used to close individual files or all the files that are currently loaded in a session. Selecting ***File / Close / Selected Parts*** will display a 'Close Part' window that lists all the loaded parts which you can select individually to close. Selecting ***File / Close / All Parts*** will close all loaded parts. You will be given the opportunity to abort the close operation if you are closing a modified part that has not been saved.

The File menu also allows you to exit an NX 8.5 session using ***File / Exit***. If you do this, NX 8.5 will prompt you for confirmation to exit if there are any open files that have been modified.

# 3. Fundamental Curves

## 3.1 Introduction

In NX 8.5, the definition of 'curve' includes lines, arcs, circles, ellipses, and splines, as well as other non-solid, non-sketch geometry (e.g. text can also be treated as a set of curves). Curves can either be parametric or 'dumb.' Parametric curves are defined in space by specific, numeric parameters which can be modified under ***Tools / Expression***. All parametric curves are listed in the Part Navigator. 'Dumb' curves are non-quantified entities which do not have parameters defining their position. A 'dumb' curve will not be listed in the Part Navigator.

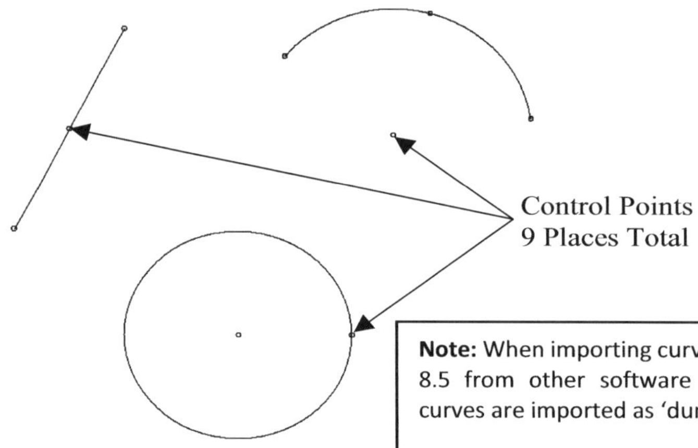


**Note:** When importing curves into NX 8.5 from other software packages, curves are imported as 'dumb.'

**CONTROL POINTS:** Every curve has a set of control points that are associated with it, as shown above. Lines have control points at their ends and at the midpoint. Arcs have control points at their ends, midpoint, and center. Circles have a control point at the center and at a 'start' point.

Control points are useful for a variety of tasks, such as building new geometry or making measurements of existing geometry. NX 8.5 is sensitive to how you pick curve entities and may provide different results based on whether you select the curve itself or its control points.

**Important Note**: When this book asks you to select a curve, pick anywhere on the curve, but **not** at the control points. When you are asked to select a center point, an end point, or a middle point, you should pick on the control point itself, not the curve. In NX 8.5, a control point is selected if it is inside the selection ball when you pick. You can also set your snap points to control what you pick as shown below.

Curves can be defined in many ways. They can be placed at screen picks, at *X, Y, Z* coordinates, or they can be connected to another curve at control points. They can also be placed by relations to other curves, such as parallel, perpendicular, tangent, or normal.

In this chapter, we will specifically discuss 2D basic curves (non-parametric lines, arcs and circles), parametric lines and arcs, and helix curves. Text will be covered in Chapter 6.

## 3.2 Basic Curves

The easiest way to create curve geometry is by making screen picks.

### *Detail: Basic Curve from Screen Picks*

Time to create some random curve geometry!

### *Project 3.2.1*

- Create a new file called ***curve_<your initials>.prt***
- When you choose the Model template, you are automatically placed in the modeling application, which we will use for all projects and exercises until Chapter 16. If you are within a different template (i.e. Assembly or Blank), then you will need to select ***Start / Modeling.***
- Select Insert / Curve / Basic Curves.

> **Note:** The system will display the 'Basic Curves' dialog box, as shown below, and a separate Tracking Bar. To see a brief description of an icon, hold the cursor over it for a moment.

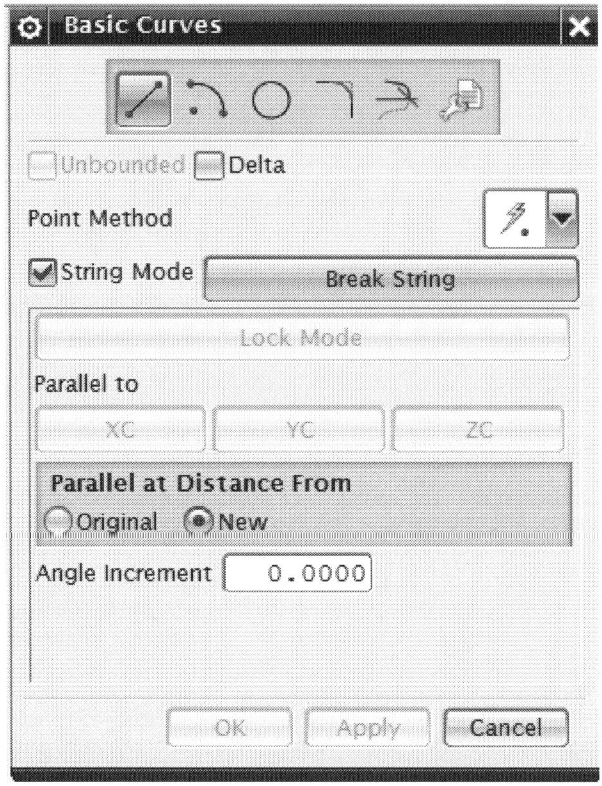


- Select the **Line** icon from the 'Basic Curves' dialog box shown above. The system will highlight the active icon.
- Set the point method to **Inferred Point** . This tells NX 8.5 to 'guess' how you want to create curves based upon where you pick in the Graphics Window – it does a pretty good job once you get used to it.

- Make sure that the **'String Mode'** box is checked. When this option is enabled, NX 8.5 automatically starts defining a new entity at the end point of the last entity that you created.

> ***Note:*** *In this detail, all screen picks are automatically placed in the XC-YC plane of the WCS. This is always the case for* <u>*screen picks*</u>*.*

- Pick the screen locations 1, 2, and 3 in the Graphics Window to create the lines shown in the figure below. NX 8.5 will interpret the screen picks as the line endpoints. Remember to read the Cue; it will prompt you for a start and end point for each line.

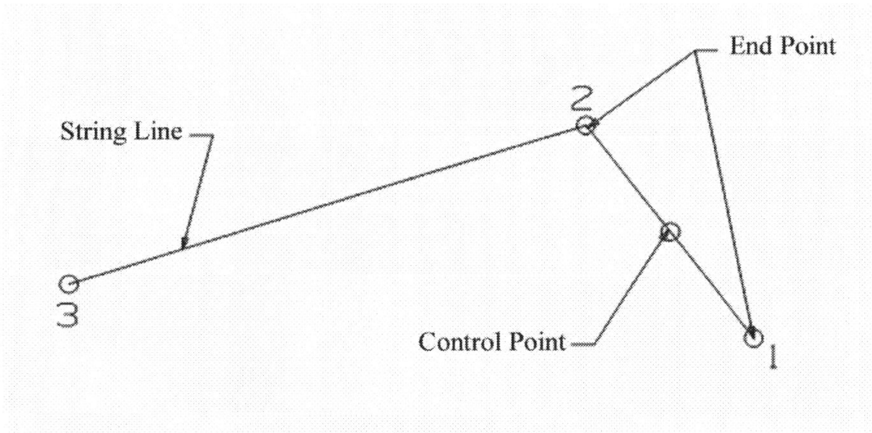


- Select **Break String** from the Basic Curves dialog box to begin a line with a new end point.

> **Note:** Clicking the middle mouse button also interrupts String Mode.

- Uncheck the **String Mode** radio button.

> **Note:** Although string mode can be very useful, it is often desirable to disable it temporarily.

- Create two or three more lines – *Have a ball!*
- Select the ***Circle*** icon  .
- Click in the Graphics Window at the approximate location of the circle center as in the following figure.

Click where you want the circumference of the circle to be. *Compare your results to the figure below.*

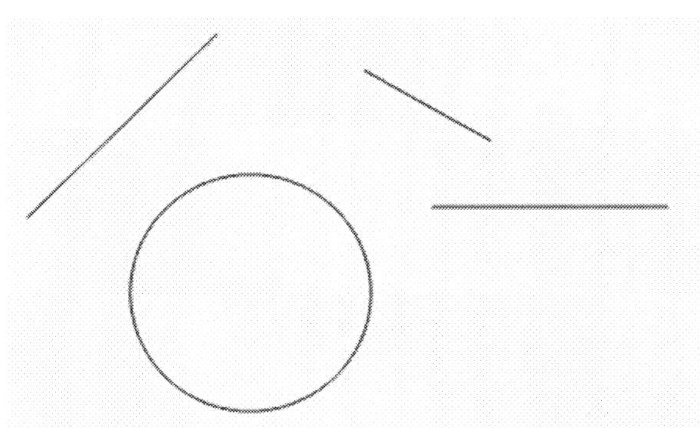

- Select the ***Arc*** icon .
- Select the Start, End, Point on Arc button.
- Pick the first endpoint of the arc.
- Pick the second endpoint of the arc.

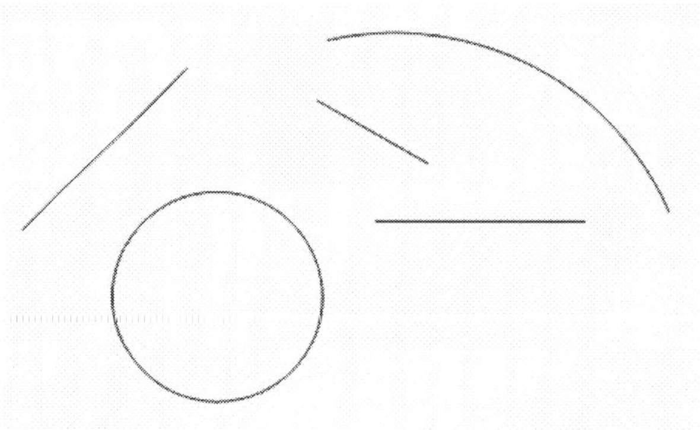

- Select a third point to define the arc itself. Compare your results to the figure shown above.

> **Note:** The arc will drag more intuitively if the cursor is kept roughly equidistant from the endpoints.

***End of Project***

## 3.3 Basic Curves with Geometric Relations to Other Curves

It is often necessary to create curves with certain relations to other curves, such as parallel, perpendicular, tangent, etc. The approach for creating such geometry is to pick an existing curve on which the new geometry will be based and then pick some second defining restraint. For this topic, make sure that the point method is set to 'Infer'.

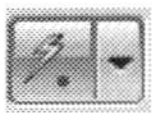

### *Detail: Tangent Lines*

To create tangent lines, simply pick a circle or an arc and choose an end point for the line. In the following project you will create a line tangent to a circle and to a point, as in the following figure.

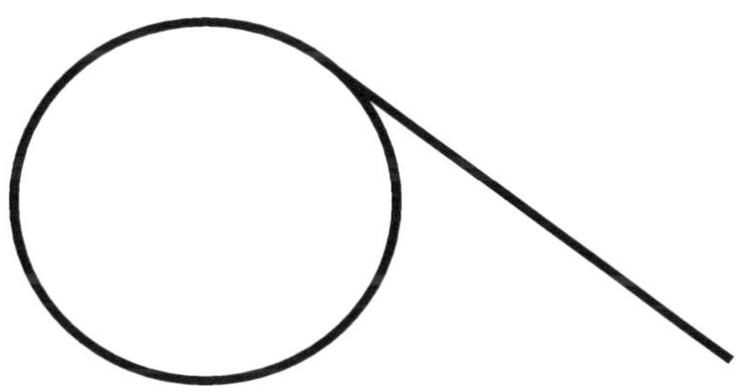

### *Project 3.3.1*

- Create a new part file named ***curve2_<your initials>.***
- Create a circle of any dimension.
- Select ***Insert / Datum/Point / Point***  .
- Select the ***Associative Point*** icon from the dialog box as shown below.

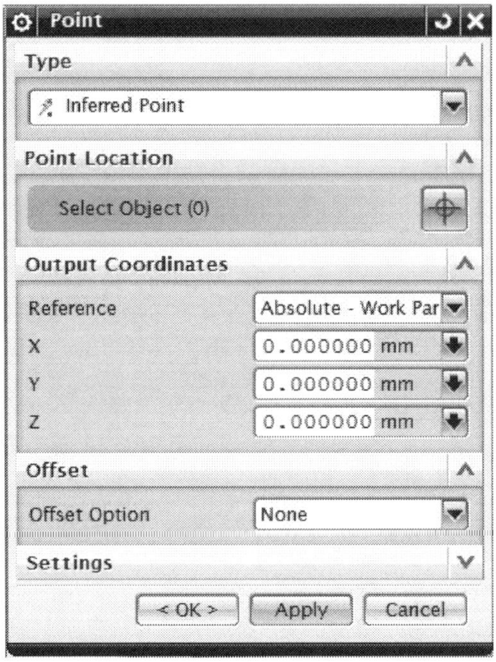


> **Note:** The order of your picks is not significant here. For example, you could have picked the point first and the circle second and gotten the same resulting lines.

- Pick a point anywhere on the screen. Recall that this is done with the Quick View Pop-Up menu activated by the right mouse button.
- To create Line 1, select Insert / Curve / Basic Curves / Line.
- Pick the circle. Do not pick a control point. You will see a tangent line that can be dragged to an endpoint.
- Pick the point.
- Create Line 2 just as you did Line 1. If the tangent line comes from the wrong 'side' of the circle, simply pass the cursor through the inside of the circle and the tangency point will be corrected.

***End of Project***

## Detail: Parallel and Perpendicular Lines

Parallel and perpendicular lines are created in much the same fashion as tangent lines. In this project you will create the geometry shown below.

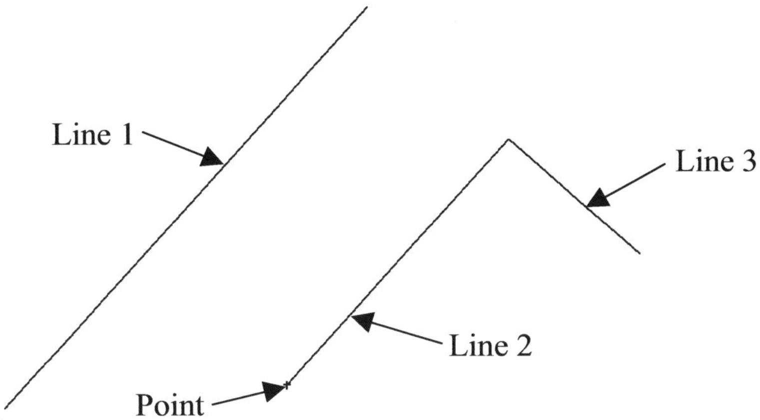


## *Project 3.3.2*

- Create a new part file.
- Create Line 1 and the Point in the approximate positions shown.
- Select Insert / Curve / Basic Curves / Line.
- Pick Line 1. Be careful not to pick a control point.
- Pick the Point. You will now be able to drag a line that is either parallel or perpendicular to line 1, depending upon where the cursor is in the Graphics Window.
- Drag a line to the approximate length and position of Line 2 and click with the left mouse button.
- Create Line 3 in the same fashion as Line 2. Instead of picking the point, however, pick the other endpoint of line 2.

***End of Project***

**Note:** The order of your picks is not critical in this project - you could have picked the point first.

In the last two projects, you may have found that picking on a control point would simply connect a curve to that control point. Selecting on a curve itself, however, tells NX 8.5 that you wish to create the new curve based on some geometric relation to the selected curve. This interface may take just a bit of practice, but it can become a very handy tool.

## 3.4 Basic Curves with Specific Dimensions

The easiest way to create a curve with specific dimensions is to use the inputs in the 'Tracking Bar' window as illustrated in the figure below. The *X, Y,* and *Z* curve coordinates are displayed on the left side and are based upon the WCS location. On the right side, the length, angle, and offset are displayed. If you are creating an arc or circle, the radius and diameter inputs would be displayed instead.

The coordinate input would be used, for example, when the cue is prompting you for a line start point, a line end point, a circle center, etc. In this situation, instead of making a screen pick, you would double click in an input box to activate it and then type in a value (**Tab** and **Shift-Tab** will work to toggle between boxes). When you are done typing in values, hit the **Enter** key.

The radius and diameter inputs can be used in a similar fashion. In addition, length and angle inputs can be used when prompted for the second point during the definition of a line.

## ***Detail: Lines***

In the next project you will create the two lines shown below.

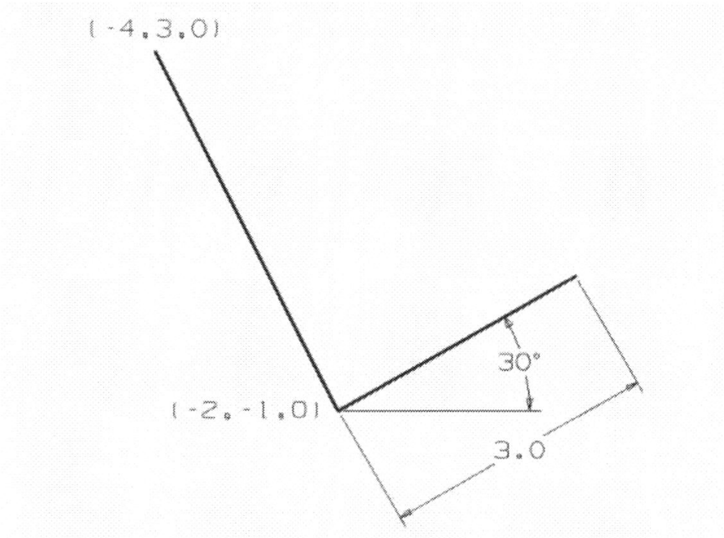


> **Note:** When using the input boxes, it is important to <u>not</u> put your cursor in the Graphics Window. If you do so, the value of the input boxes will change to values that correspond to the cursor location – and the numbers you have typed in will be lost. Keep the cursor in the Tracking Bar window.

### *Project 3.4.1*

- Select Insert / Curve / Basic Curves / Line.
- Uncheck the 'String Mode' box to disable String Mode for this project.
- Double-click on the XC box and type **-2.**
- Double click on the YC box (or use the **Tab** key) and type **-1.**
- Hit **Enter**.
- Double click on the XC box and type **-4.**
- Double click on the YC box and type **3.**
- Hit **Enter**.
- Input a line start point of XC= **-2**, YC= **-1.**
- Hit **Enter**.
- Input length = **3** and angle = **30.**

Be careful to keep your cursor out of the Graphics Window.

- Hit **Enter**. Compare your results to the previous figure. You will not see the point data or dimensions.

***End of Project***

## ***Detail: Circles***

In the next project you will use the graphics window inputs to create the geometry shown below.

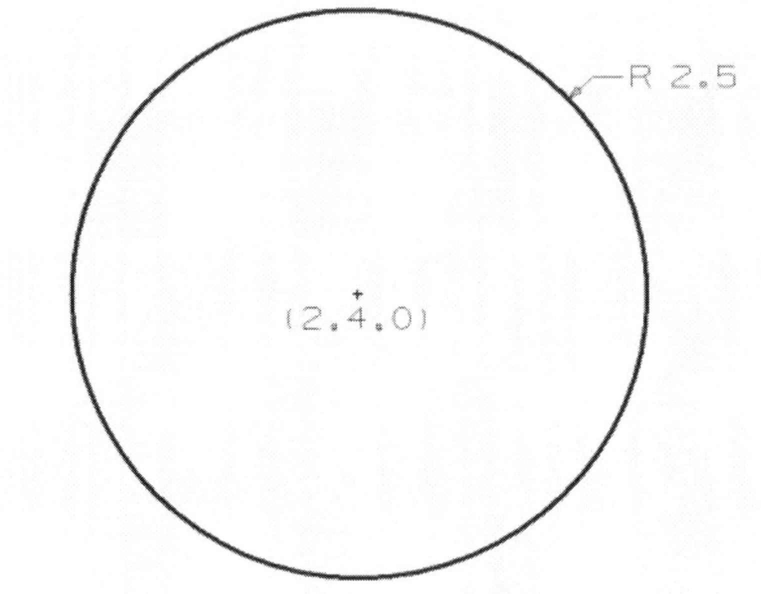


### ***Project 3.4.2***

- Select Insert / Curve / Basic Curves / Circle
- The cue should prompt you for the location of the circle's center.

- Input XC = **2.**
- Input YC = **4** and hit **Enter**.
- Type **2.5** in the input box with the radius symbol and hit **Enter**.

***End of Project***

## 3.5 Basic Curve Trim Operations

There are many situations where existing curves must be extended to some boundary, or a perhaps curves may overlap and require trimming. In NX 8.5, this is accomplished using the 'Trim' operations.

### *Detail: Using Trim to Shorten*

The figure below shows a horizontal line that must be 'cut' at the boundary of two other lines.

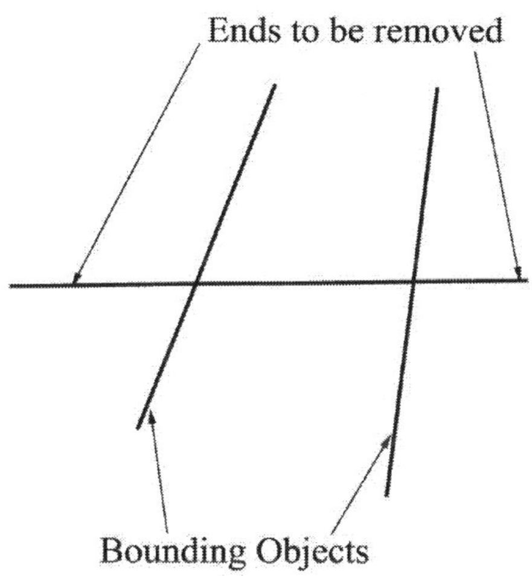


### *Project 3.5.1*

- Create the three lines approximately as shown above the **Basic Curve** function.
- Select ***Edit / Curve / Trim*** or select the **Trim** 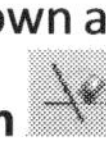 button from the Basic Curves menu.

> **Note:** Feel free to experiment with the setting shown next. Notice that when the associative toggle is activated and the input curves are set to "keep" you will see evidence of your original curves after the trim operation.

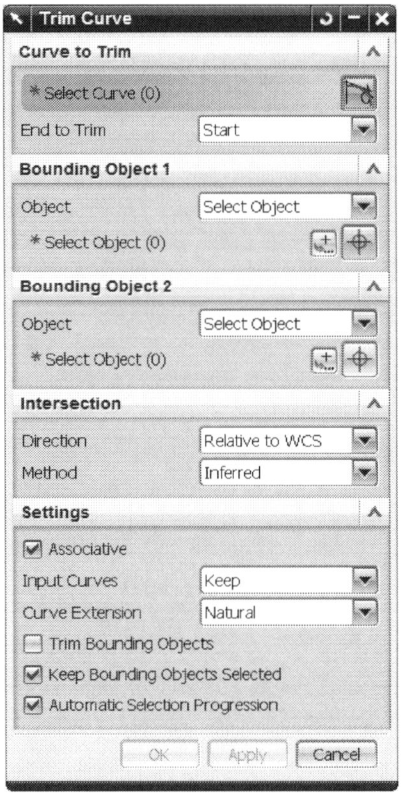


**Note:** For most actions, a dialog box will appear containing various needed parameters. The orange shading highlights the current unsatisfied parameter. By following the cue and completing the action, usually by selecting a line or object, it will become satisfied. The completed parameter turns green, and the orange highlight will progress to the next parameter, and so on. You will become accustomed to what parameters each action requires, and before you know it, you won't even need to look at the dialog box.

- Pick either end of the horizontal line that is to be removed.
- Pick one of the lines as shown in the previous geometry figure to define Bounding Object 1.
- Pick the other line to define Bounding Object 2.

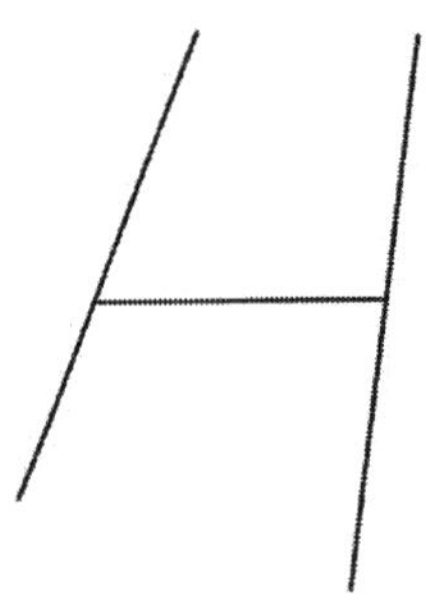

**Note:** When trimming to shorten, always pick the portion of the curve to be removed

- Hit the **middle mouse button** for NX 8.5 to accept this step. Compare your result to the figure above.

***End of Project***

## Detail: Using Trim to Extend

The Trim function may also be used to extend a curve, as shown below.

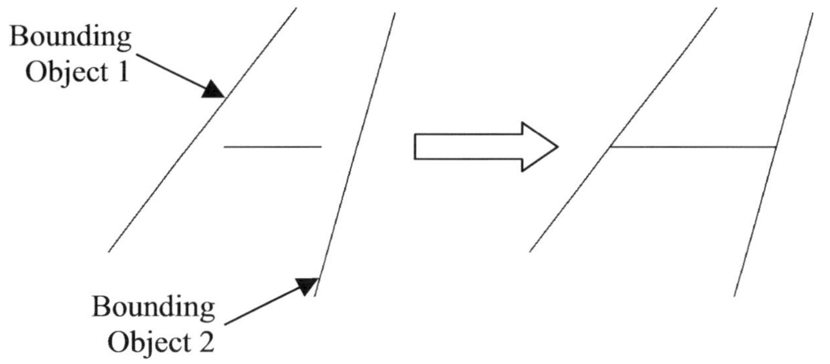


### Project 3.5.2

- Create the three curves approximately as shown in the left hand side of the figure above.
- Select Edit / Curve / Trim.
- Pick the horizontal line to be extended and middle click.
- Pick bounding object 1.
- Pick bounding object 2 and **middle click** to finish.

***End of Project***

## ***Detail: Trimming with Only One Bounding Object***

Often there may be only one bounding object, as shown below. The Trim function can accomplish this by skipping selection of the 2nd bounding object.

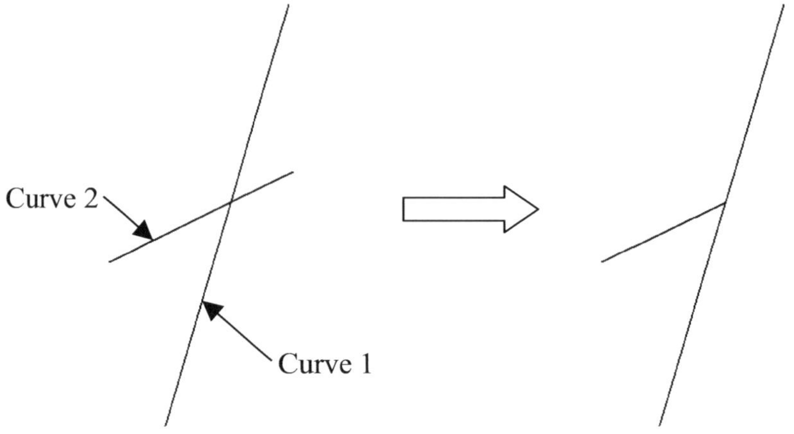


### *Project 3.5.3*

- Create Curve 1 and Curve 2 as shown in the left hand side above.
- Select Edit / Curve / Trim.
- Pick the part of Curve 2 that is to be removed.
- Pick Curve 1 as the first bounding object.

*The cue is prompting you for a second bounding object, which is an 'Optional step'. Since there is none, middle clicking tells NX to 'move on.'*

- **Middle click** and compare your results to the figure above.

***End of Project***

## 3.6 Associative Lines and Arcs

NX 8.5 has functions that allow you to create parametric lines, arcs and circles quickly. Until this point in the chapter, we have discussed basic lines and curves that NX 8.5 treats as 'dumb' curves. In order to assign and control the relationships associated with a curve, associative curves with parametric qualities need to be created. Once they have been placed in the model, their parameters can be viewed and modified in **Tools / Expression**. All variables associated with a given curve will be listed under **Expression**.

### *Detail: Parametric Line*

The Line and Arc commands can be used to create associative lines and arcs with parametric values.

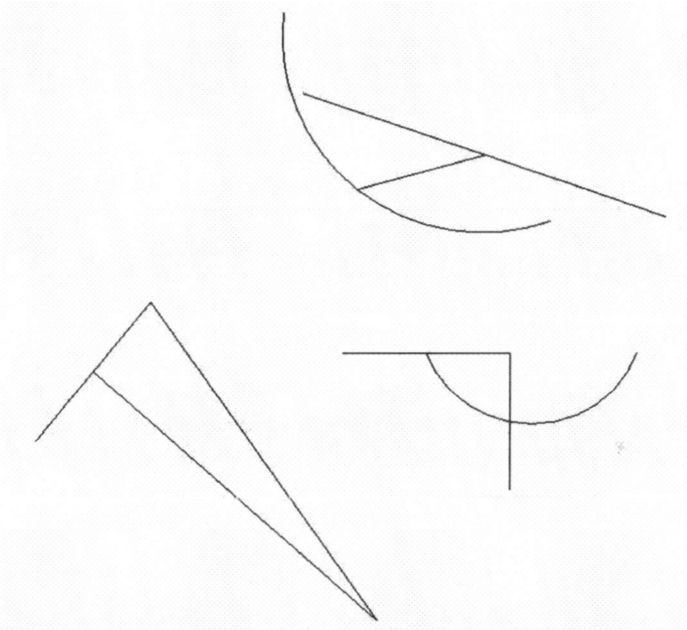

### *Project 3.6.1*

- Create a new part.
- Select the **Line** icon  from the toolbar or select ***Insert / Curve / Line***.

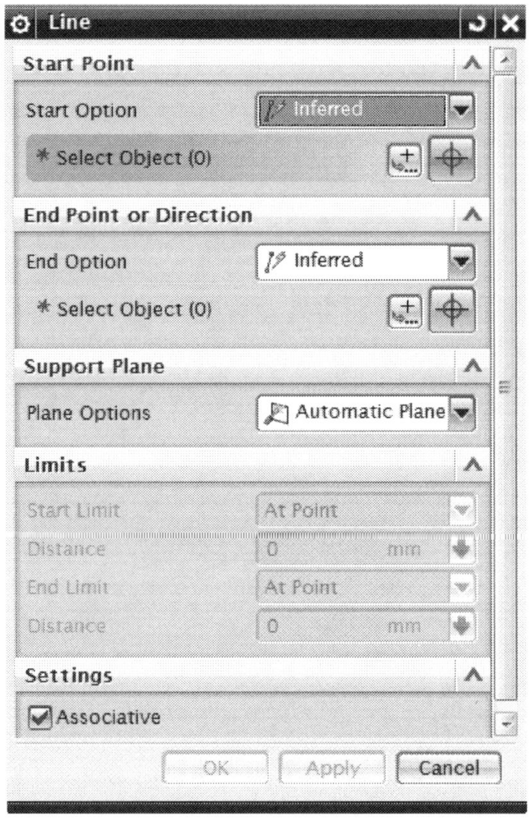


- Make sure that the 'Associative' check box is checked as in the above figure. When this option is enabled, NX 8.5 *automatically creates parametric lines* with associations to other existing parametric curves.

- Click anywhere on the screen and create lines similar to those in the preceding geometry figure. **Middle click** after every line you create. Notice NX 8.5 has no default control points for Associative lines and arcs. Instead, Snap Points must be assigned to allow you to choose End Point, Mid Point, Point on Curve, etc. The figure below shows the Snap Point window.

> ***Note:*** *The Snap Point Toolbar is available in any command where it is required.*

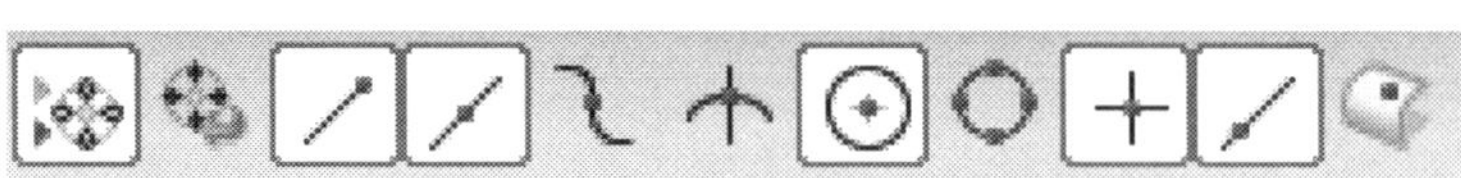

- Select the **Arc / Circle** icon  or select ***Insert / Curve / Arc/Circle***.

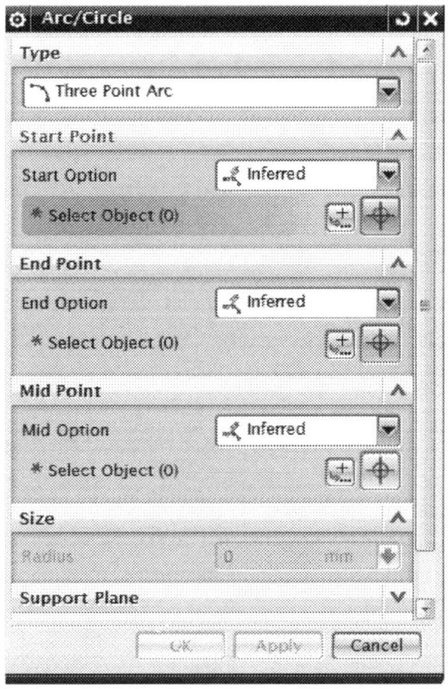


- Using the **Line** and **Arc** command with **End Point** and **Mid Point** Snap Points on, create lines and arcs similar to those in the previous geometry figure.
- Notice that by **right-clicking** on any of the lines or arcs and selecting **Edit Parameters**, an edit window appears as in the figure below. You can modify the parametric values associated with the curve – a feature not possible with Basic Curves.

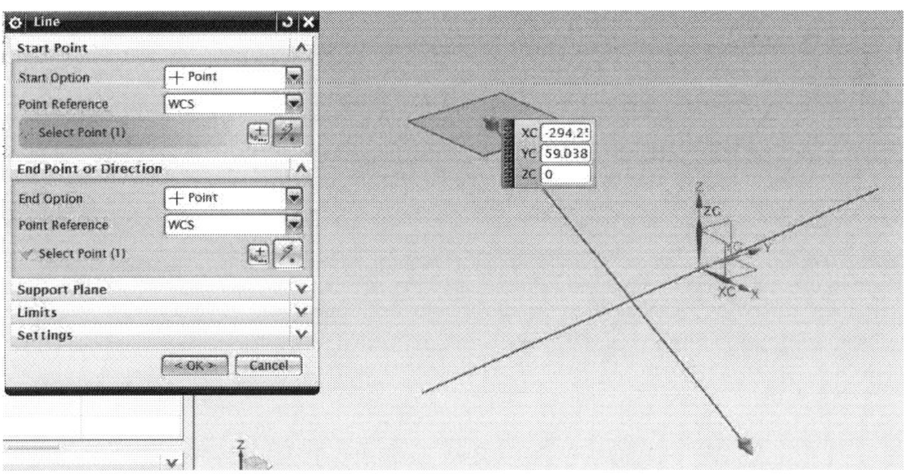


> **Note:** Click on the **Part Navigator** icon located on the left side of the screen. Notice how associative lines and arcs drawn are listed, whereas basic curves are <u>not</u> listed.

***End of Project***

## 3.7 Helix

A helix can be created for use as a guide string. This is especially useful for designing springs like the one seen below. The helix likes to assume that the ZC-axis will be right up the middle.

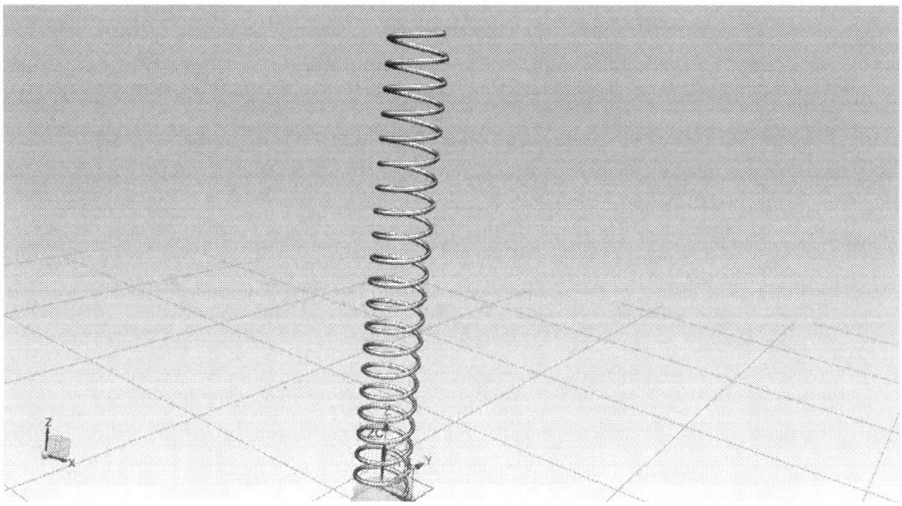

### *Detail: Parametric Helix*

Parametric helices can be made quickly using the Helix command. The 'Use Law' switch allows you to vary the radius of the turns as you go. You will see a variety of choices, the most common of which is a linear law.

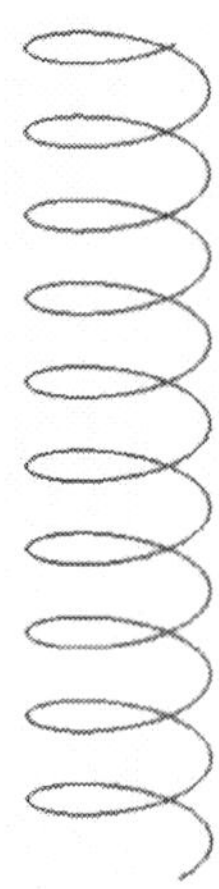

## *Project 3.7.1*

- Select Insert / Curve / Helix
- Enter **20** in the 'End Limit' dialog box
- Enter **2** in the 'Pitch' dialog box
- Enter **2** in the 'Radius' value dialog box, as shown on the next page.

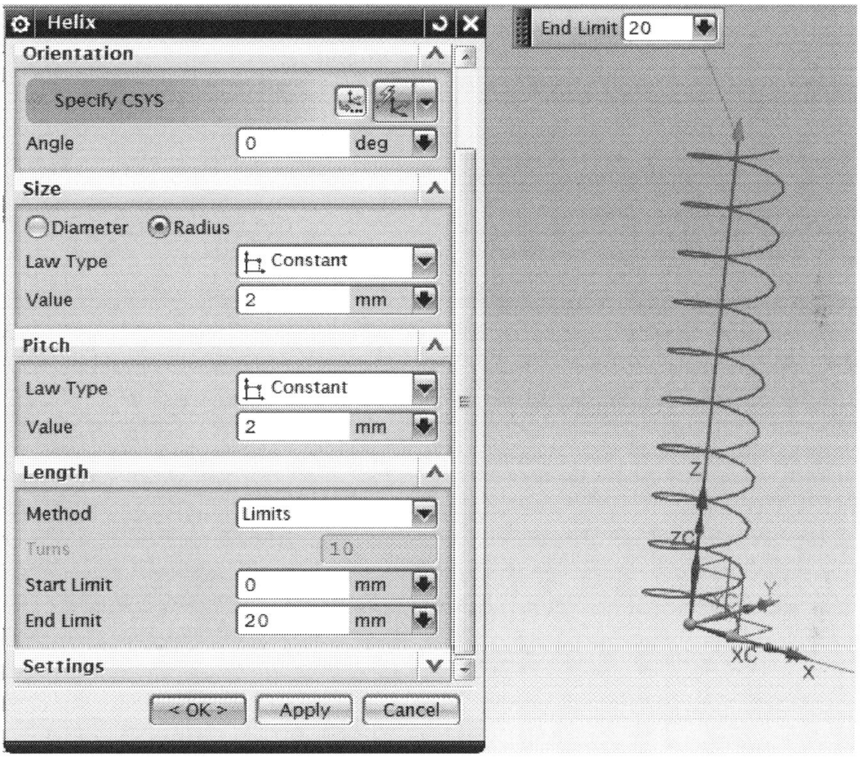


- Choose **OK** *or* **middle click**. Compare your results with our previous geometry.

***End of Project***

## 3.8 Self-Test

(All questions are true/false. Answers are at the back of the book.)

1. If you are creating curves with String Mode turned on, you can break the string with a middle click.

2. When trimming a curve to shorten it, select on the part of the curve that you want to remain.

3. The offset operation cannot produce parametric output.

# 4. Editing Entities

## 4.1 Introduction

In this chapter we will focus on the most common editing functions including deleting, hiding, transforming and changing the appearance of on object. All of these commands are found under the 'Edit' menu in the top toolbar.

## 4.2 Deleting Entities

Deleting is one of the most commonly used operations. In NX 8.5, simply choose ***Edit / Delete***, select the items to be deleted, and choose OK to complete the operation.

### *Detail: Deleting Curves*

You will use the delete command to remove some curves.

### *Project 4.2.1*

Create a shape with basic lines similar to the one shown below.

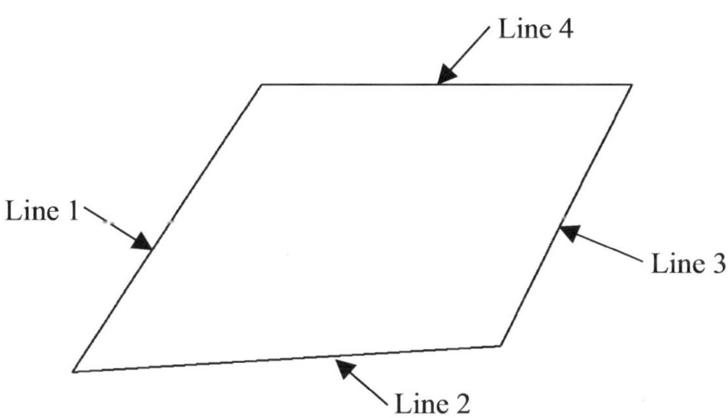


- Select ***Edit / Delete***. The Class Selection menu window will appear, as shown in the next figure.

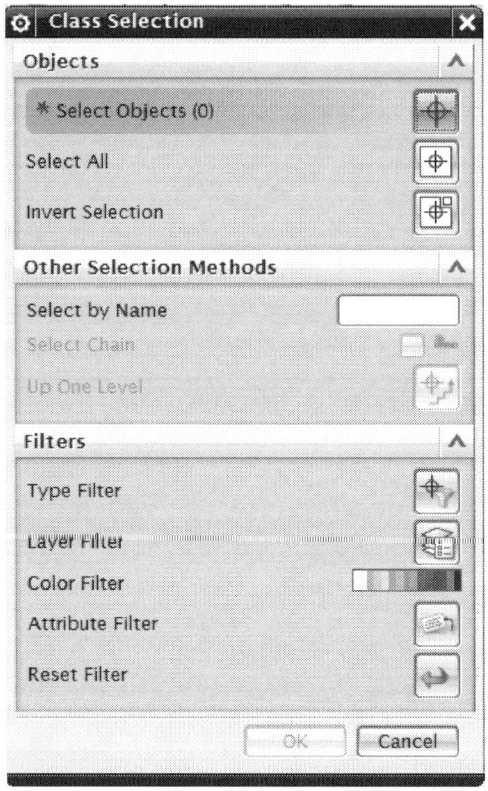


- Pick Line 1.
- Pick Line 2.
- Both lines will become highlighted. Choose **OK**. *Line 1 and Line 2 will be deleted.*

***End of Project***

## 4.3 Hiding Entities

Hiding is often used when the graphics window is too cluttered and you wish to simplify the display. The steps for hiding an entity are very similar to deleting – but do not confuse the two functions!

## ***Detail: Hiding Curves***

You will form a model containing circles and points, hide all the entities, and then recall them.

## ***Project 4.3.1***

- Select ***Insert / Curve / Basic Curves*** to create several circles. Select ***Insert / Datum/Point / Point*** to create several non-associative points. The results should look similar to the following figure.

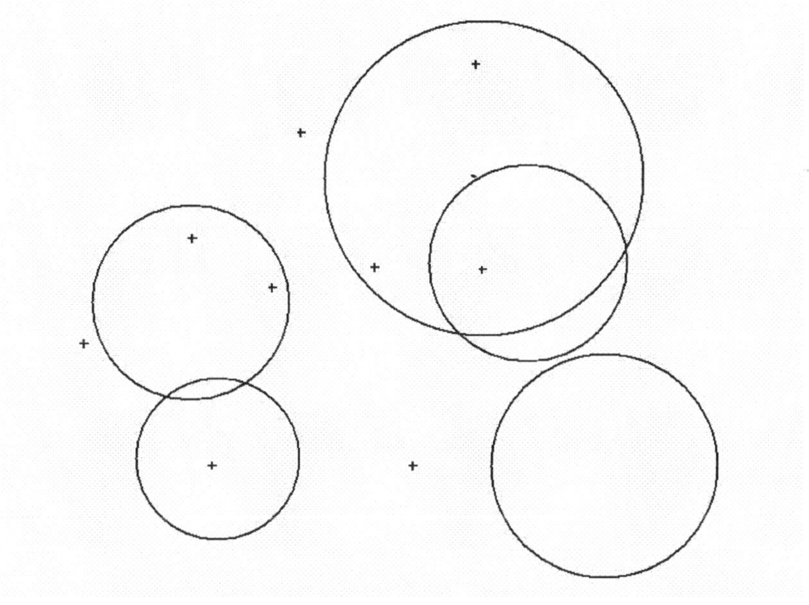

- Select Edit / Show and Hide / Hide.
- **Hint:** Ctrl+B is the keyboard shortcut.
- The **Class Selection** menu window will appear, as shown on the following page. You may use it to apply selection filters, or bypass it by selecting from the Graphics Window.

- A Select by Type menu will appear, with items already highlighted.
- From the menu, select the **Type** filter method.
- Select **Point.**
- Choose **OK.**
- Choose **Select All**. Because you have applied a filter, only the points will highlight. If you had not applied the filter, you would have had to pick each point individually.
- Choose **OK**. The points will have disappeared from the view.
- Select **Edit / Show and Hide / Show All**. Hint: Ctrl+Shift+U is the keyboard shortcut. The points will reappear.

***End of Project***

> **Note:** Under the **Edit / Show and Hide** menu are several different ways to 'unhide' objects. Showing can be done at any time, and the hidden geometry will reappear. By using the options in this submenu, you can be selective about which types of entities are shown (unhidden). It is left to the reader to experiment with these options.

Applying selection filters with the Class Selection menu can be useful when there are many entities in the Graphics Window. For example, if there were many different colored lines and you want to select only those that are colored green, select the **Color** Filter box, select **GREEN**, and Choose OK. The Class Selection menu is still visible, but now you will only be able to select green geometry.

The 'Select All' button is also a useful function of the Class Selection menu. Its function should be self-explanatory.

## 4.4 Changing an Entity's Appearance

It is often useful to change the appearance of an entity when there are many objects displayed in the graphics window. This is accomplished using the ***Edit / Object Display*** functionality.

### ***Detail: Changing the Appearance of Curves***

You will modify the display of lines and several circles, changing the font to thin, dashed, and magenta for the circles only.

### *Project 4.4.1*

Randomly create two lines and a number of circles, as shown below.

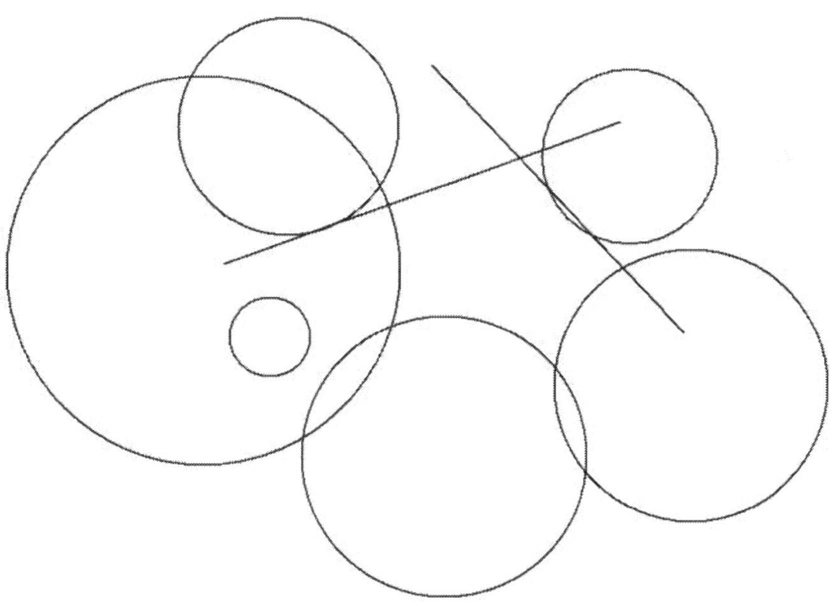

- Select **Edit / Object Display**. Ctrl+J is the keyboard shortcut.
- Pick the two lines.
- Choose **Invert Selection** from the 'Class Selection' menu. All the circles should highlight, and the lines should become de-selected. You could have directly picked all the circles, but this method is slightly faster.
- Choose **OK.**

The Edit Object Display window will open. From the menu shown on the next page, select the 'Color' Magenta, the 'Line Font' *Dashed*, and the Width '*Thin Width.*' *Line Font and Width are shown graphically.*

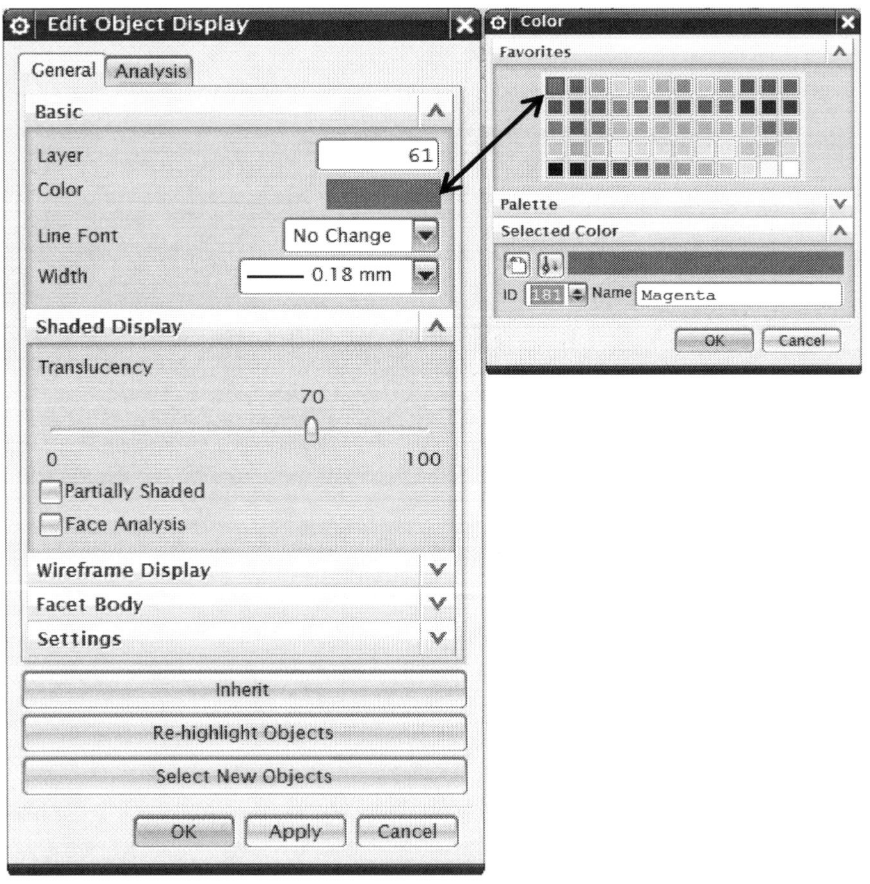


- Choose **OK**.

> **Note:** The width of a curve is typically not displayed in the Graphics Window; it is only seen when plotted. If you would like to see the line width displayed in the Graphics Window, activate Preferences / Visualization. Choose the Line tab and check Show Widths

***End of Project***

## 4.5 Self-Test

(All questions are true/false. Answers are at the back of the book.)

1. You can undelete an object at any time.
2. You can show an object at any time.
3. When the Class Selection menu appears, you must apply a filter before proceeding.

# 5. Sketches

## 5.1 Introduction

Sketching is one of the most powerful tools available in NX 8.5. The basic idea behind sketching is that it helps create a 2D shape that is controlled by numerical values and geometric constraints.

The latest version of NX offers two different sketch methods. You can access them through the toolbars or under the Insert menu as shown in the following figure.

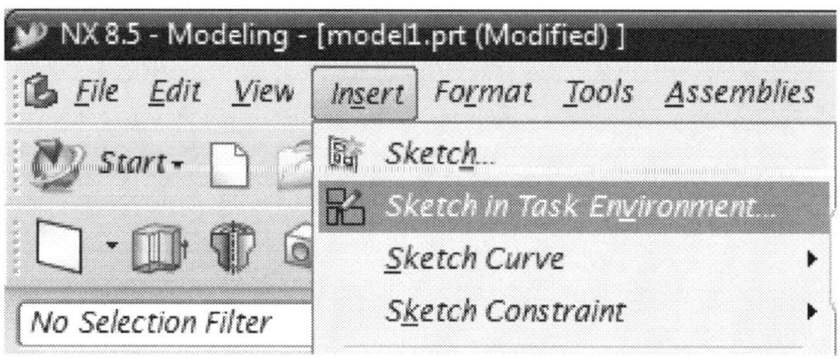


**Sketch in Task Environment**: This option is the original method that existing NX users know and love. To start a new sketch, simply choose ***Insert / Sketch in Task Environment***. Sketches are *always* defined using a sketch plane and a reference direction. Once in sketch mode, the program allows for the creation of entities that capture the shape of a geometric feature. If the user does not choose a sketch plane, the computer will automatically choose a default one. There is even a menu choice that allows the user to select the normal direction along a curve to create a sketch plane called 'Sketch On Path.'

**Direct Sketching**: This option promotes a faster and more fluid design experience. As of NX 8.5, it is possible to create a sketch without entering the 'Sketch Environment.' Using this method limits the number of button clicks needed to complete the sketch before it is possible to start creating the sketch's profile. The program also has a 'Direct Sketch' toolbar, as shown below (the same tools found in the normal sketch environment), where the user may simply click on one of the tools and begin sketching at any location.

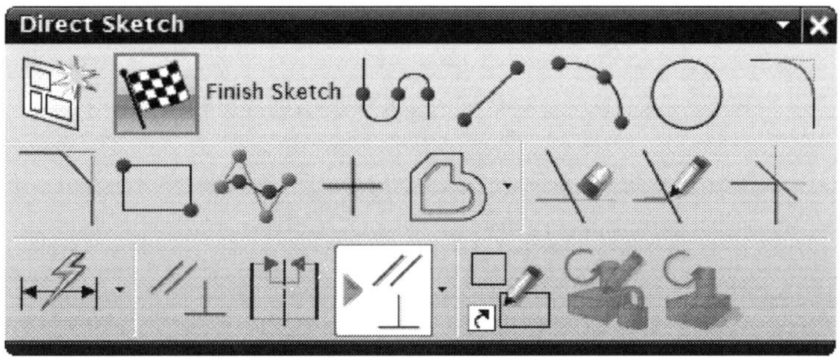


**The method for using 'Direct Sketch' is simple:**

- Select a tool that you wish to draw your sketch with from the 'Direct Sketch' toolbar, for example, the profile tool.
- Selecting where to create your sketch. You can either hover you cursor over the plane or face that you wish to sketch on, or left mouse button click. In either case the face will immediately highlight and remain highlighted unless another plane is selected.
- Start drawing your profile.

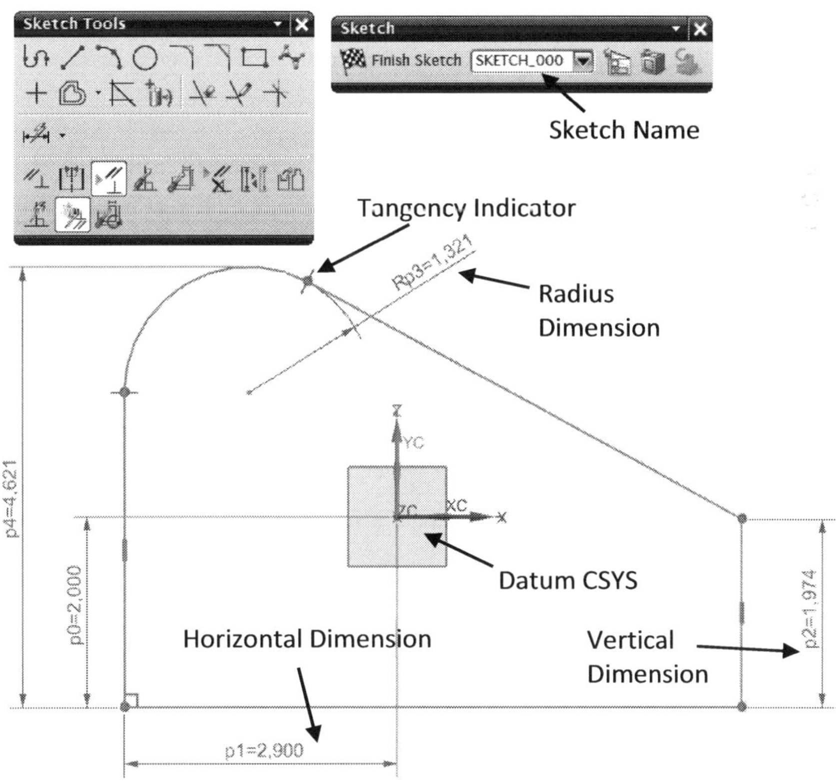

- Click on **Finish Sketch** in the 'Direct Sketch' toolbar.

The following figure shows all the main features of the sketch environment.

Another feature found in both sketch modes is the **'Continuous Auto Dimensioning'** tool.  If the tool is activated, as the profile of your sketch is drawn, dimensions will appear. If this tool is used in combination with the already existing (and fantastic) 'Inferred Constraints' tool, the sketch will be fully constrained by the time the profile is finished. This tool is auto-activated, so you need to turn it off before you start sketching, if you so prefer.

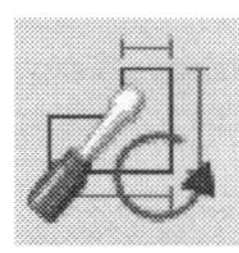

> **Note:** Usually the user will want dimensions other than the inferred dimensions. Any new dimensions created will override the auto dimension. This means that the user will not need to delete any auto-dimensions.

Once the sketch curves are drawn, they can be further defined using manually applied constraints and dimensions. You can change these dimensions and constraints until you get the geometry desired.

The final step of the sketch process is to finish the sketch and exit sketch mode. A sketch may then be used to define subsequent geometric features. The figure below shows how a sketch can be used to create an extruded solid.

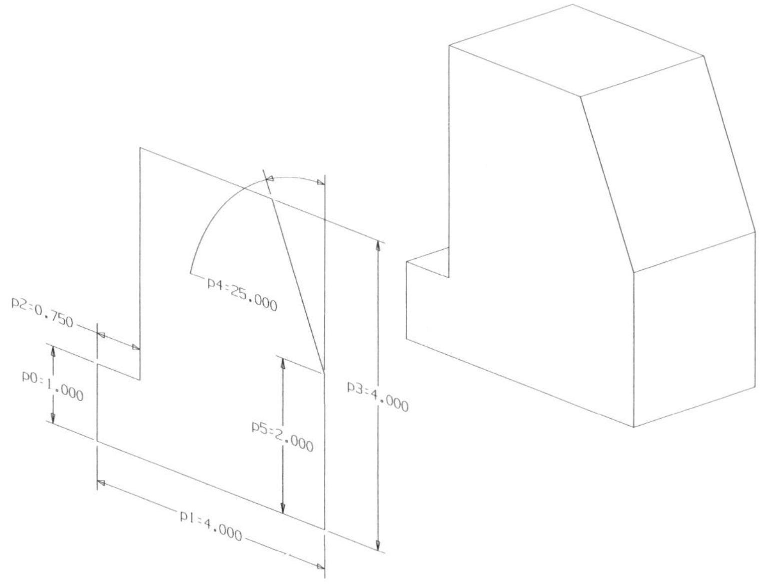


The shape of the solid will change as you modify the sketch, as shown in the following figure.

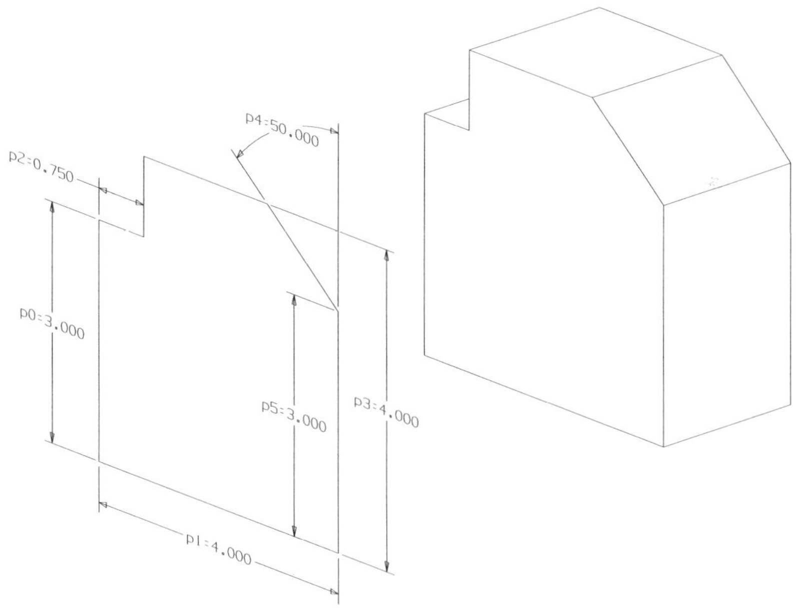

## 5.2 Creating a Sketch

- To create a sketch using the original method, choose ***Insert / Sketch in task environment***.

If you are working in a part file that contains no solid geometry, NX 8.5 will automatically create a datum CSYS that defines the sketch plane. By default, a sketch is created on the XC-YC plane of the WCS (Working Coordinate System). The datum axes are infinitely long lines that are coincident with the X and Y axes. The datum entities are displayed as shown below.

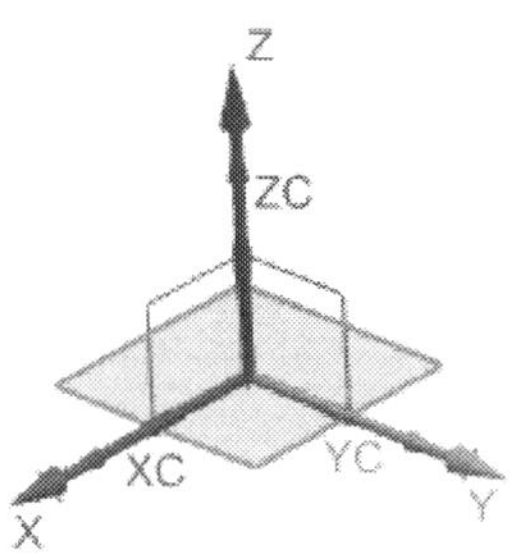

- The next step in the sketching process is to create curves that capture the desired shape. The following figure illustrates some of the tools available for sketching geometry.

> **Note:** You can also use the Trim and Extend Sketcher functions to edit sketch curves.

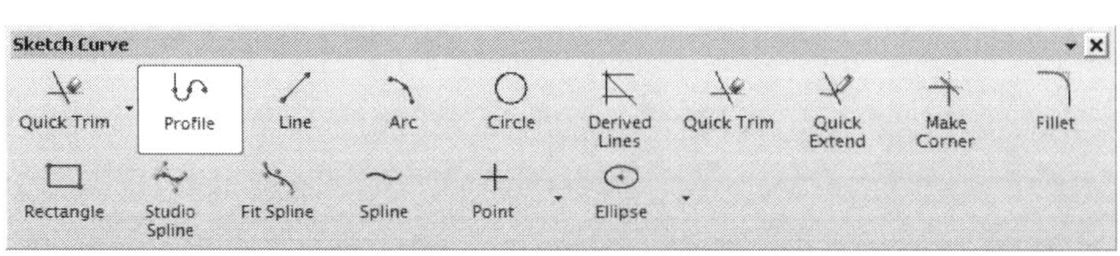

- The next step in the sketching process is to dimension and constrain the sketch. The next figure illustrates some commonly used tools for dimensioning and constraining sketch geometry.

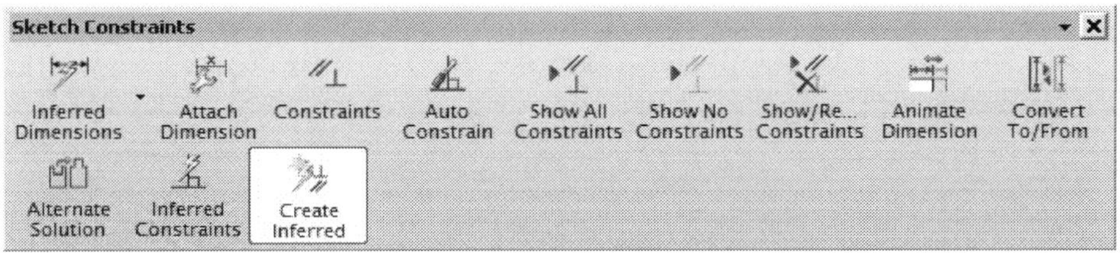


- The final step in the sketching process is to choose the 'Finish Sketch' icon located in the upper left corner of the Graphics Window when you are in Sketch Mode. This tells NX that you are finished sketching and are ready for bigger and better things!

Imagine that you want to sketch a pitcher. The figure below illustrates the initial steps of the sketching process. The datum entities and the shape of the desired geometry have been created in three separate sketches.

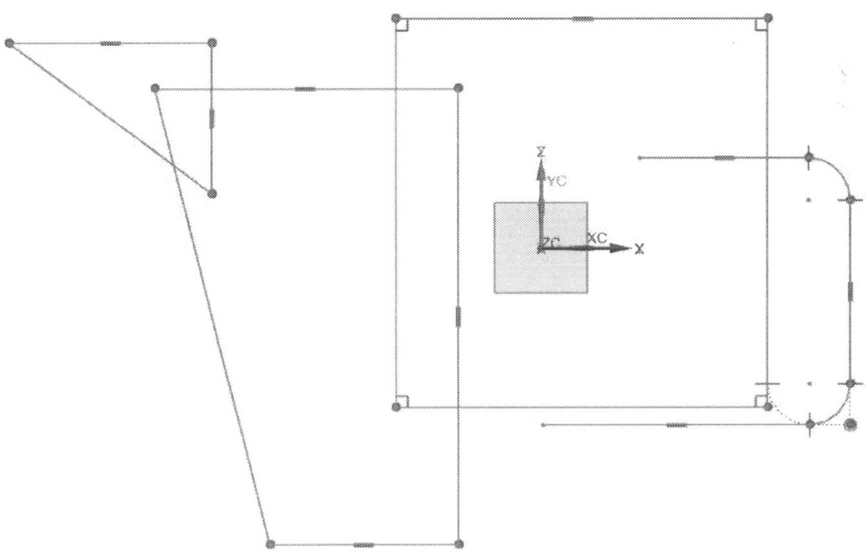

Once the basic shape is sketched, add the exact dimensions and constraints. In the next figure, you can see that the pitcher has been sized and all the geometry is ready to create the solid model.

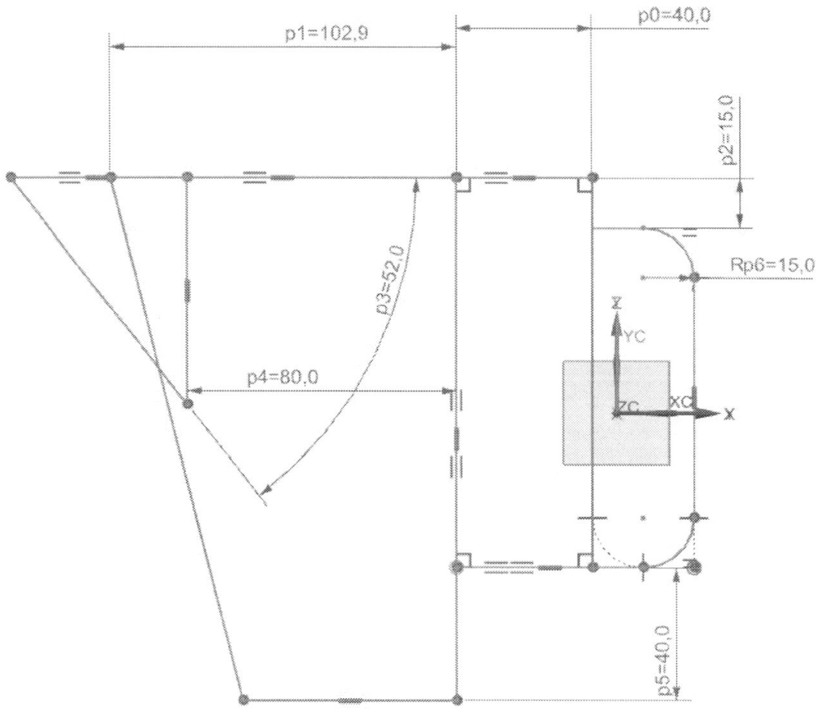


> ***Note:*** *In the previous figure, there are only five types of dimensions displayed: Horizontal, Vertical, Perpendicular, Radial, and Angular.*

## ***Detail: Dynamic Input Boxes***

Dynamic input boxes are used to quickly view and modify input data related to a current action or feature. The input fields display labeled values you can change by clicking in the value field and typing a new value.

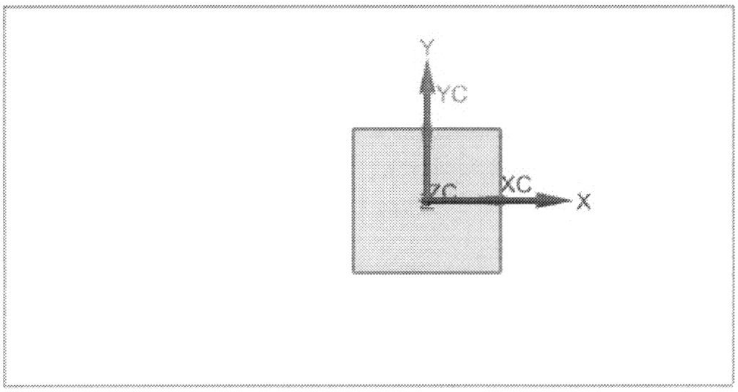


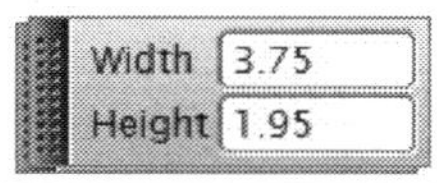


---

## ***Detail: Adding Dimensions to a Sketch***

Eight dimension types may be placed on sketch entities using the 'Dimensions' dialog box shown below.

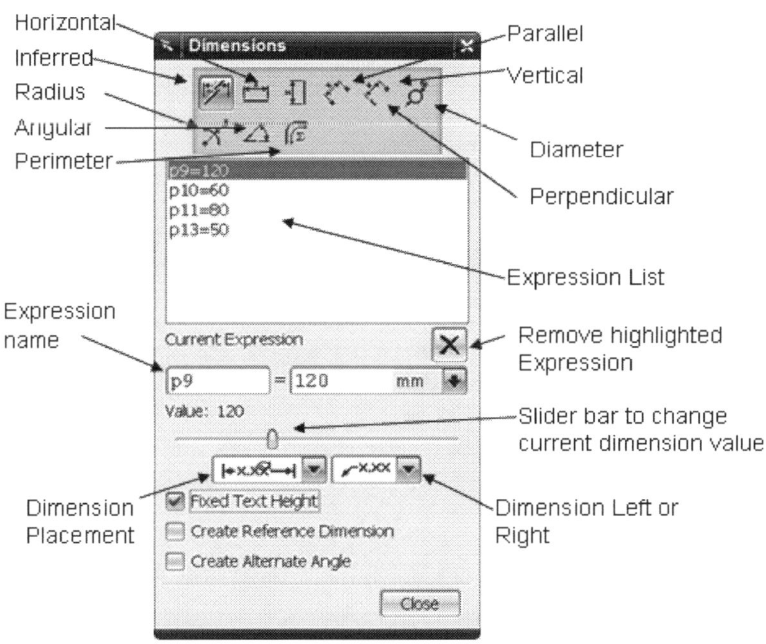

By default, NX 8.5 uses the 'Inferred' dimension tool. To use this option, select the entity or entities that you want to dimension, and NX 8.5 determines the dimension type automatically, based upon the cursor position.

## *Project 5.2.1*

- Create a new part file.
- Select Insert / Sketch in task environment.
- Choose a plane and select **OK**.

A datum plane coincident with the x-y plane, has been created along with horizontal and vertical datum axes, as shown in the next figure.

- The **Profile** tool should automatically open, but if it does not, select the **Profile** icon  .
- Using left hand mouse button clicks, create the shape shown below. The red dimensions are "**Continuous Auto Dimensions**". By default they appear to ensure that the shape you create is fully dimensioned.

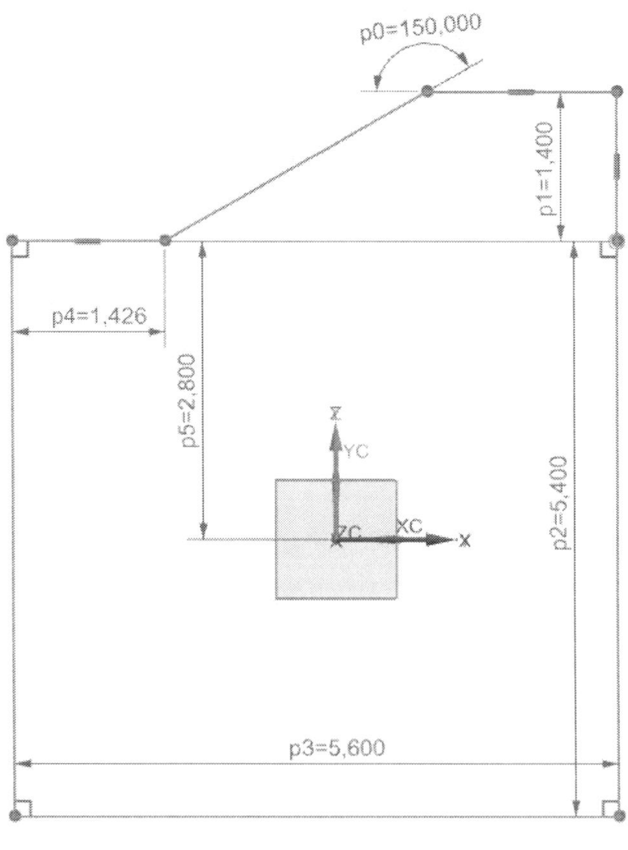

- As you select screen locations, lines will be drawn between each sequential click. Make sure that when you draw the last line, you place the capture circle on top of the very first location. This will create a closed loop of curves as shown above.
- Select the **Inferred** icon from the Sketch Constraints Toolbar.

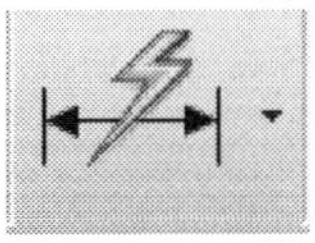

To use the 'Inferred' tool, click once on the entity that you want to dimension and select a screen location near the entity to place the dimension. If you rotate your sketch for any reason, you can orient it back to the sketch by clicking the 'Orient View to Sketch' icon.

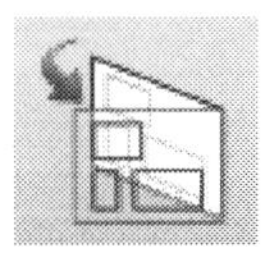

- Create dimensions similar to those shown below.

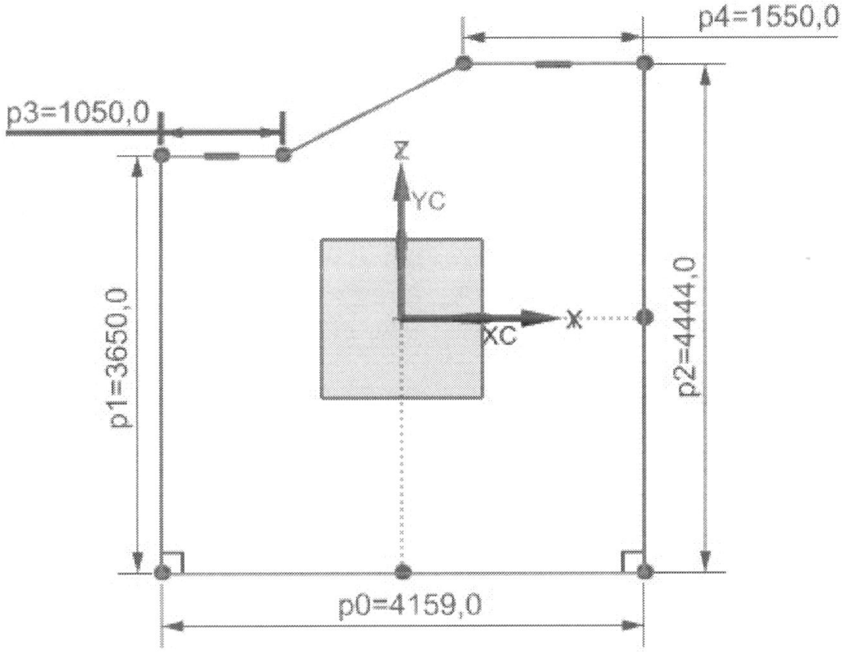


**Note:** NX's default decimal point character may be a comma (i.e. 4,232) rather than a period (4.232). This can be changed in Preferences / Annotation / Units.

- While in the "Inferred Dimensions" tool and as you select the variuous entities to create your own dimensions, the red dimensions disapper. They are "weak" dimensions and are superceeded by the dimensions that you create. When you click directly on the red dimensions, they become "strong" dimensions. This is another method of fully constraining a sketch and giving it the dimension scheme that most suits your geometry. Some folks would rather the inferred dimension never get created in the first place. This can be accomplished by turning off the "Continuous Auto Dimensioning" function. You'll find it in the sketch menu as shown in the following figure.

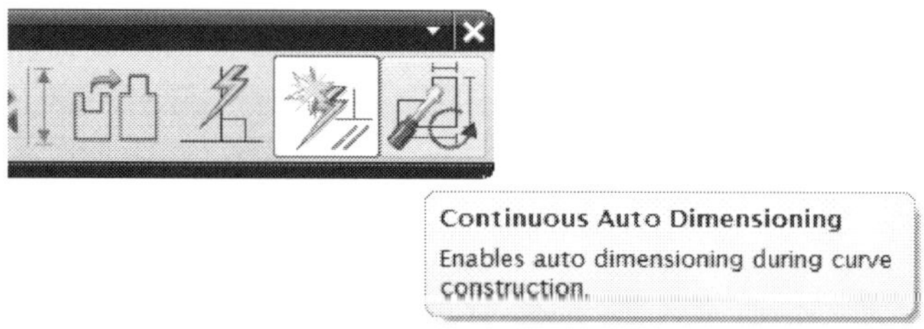


- Once you have created all the dimensions, select each of them and input the correct dimension value. Each time you type in a new value hit 'Enter' on the keyboard to apply the change.
- You can reposition the dimensions by clicking and dragging them. When you are finished, the sketch should appear similar to the figure below.

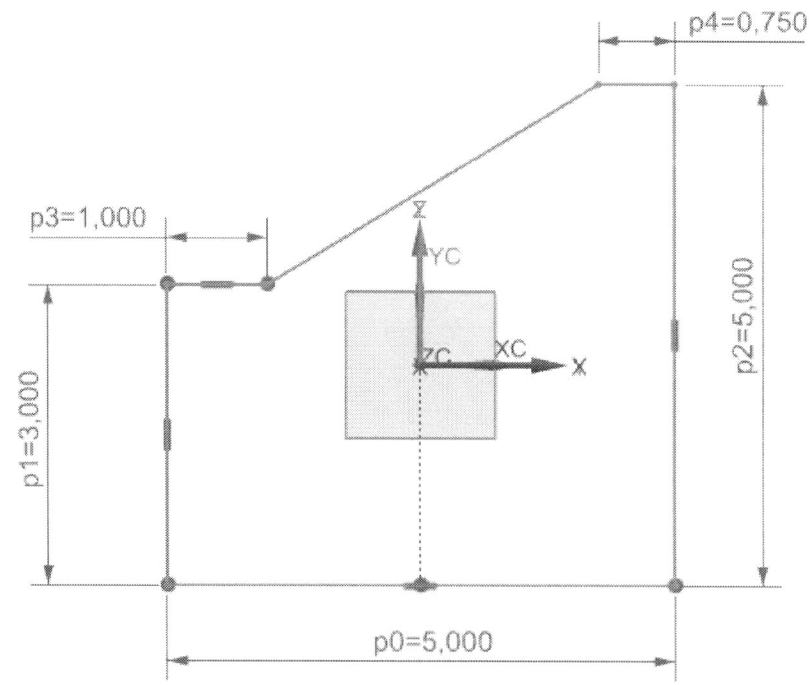

- Exit the sketcher by choosing the **Finish Sketch** icon .

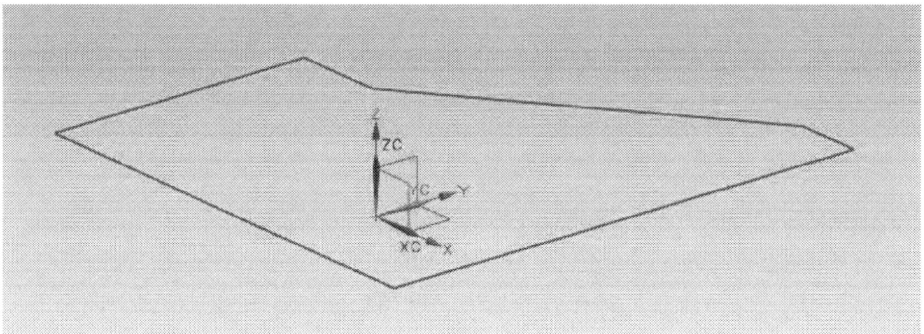


Behold the finished product. This contour can now be used to create solid geometry using many of the solid creation commands.

***End of Project***

## ***Detail: Angular Sketch Dimensions***

Four different angles can be used to dimension any two intersecting lines.

## ***Project 5.2.2***

Create the sketch shown in the following figure.

- Create a new part file.
- Select Insert / Sketch in task environment.
- Select the **Profile** icon and create a sketch as shown below.

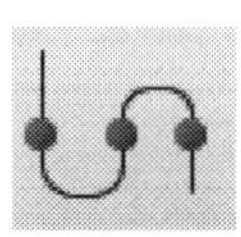

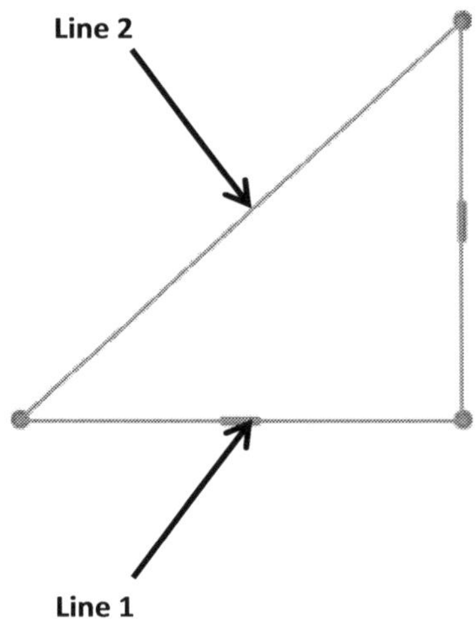


- Select the **Inferred** icon .
- Create an angular dimension by selecting Line 1 and Line 2 as shown above. Place any one of the available dimensions by picking in the desired quadrant.

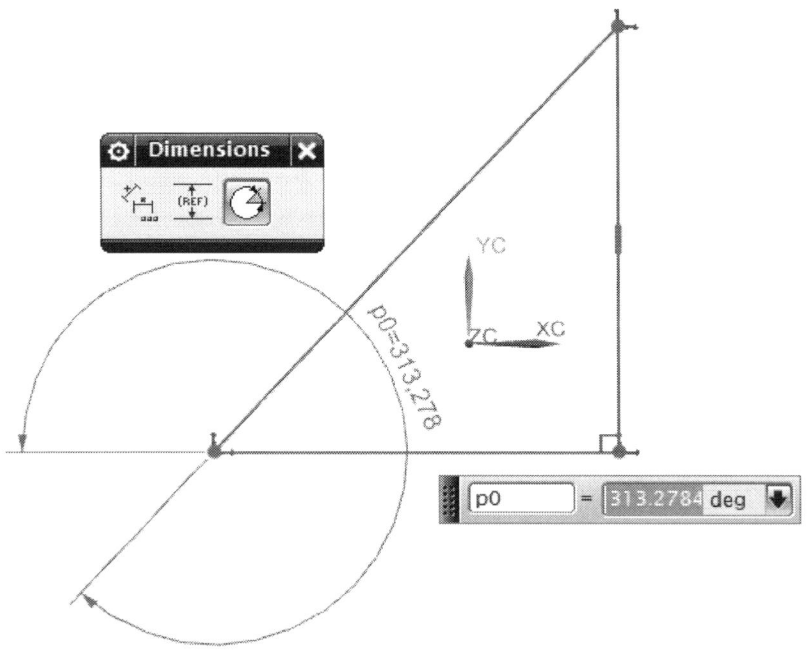


- Exit the sketcher by choosing the **Finish Sketch** icon.

***End of Project***

## ***Detail: Perpendicular and Parallel Sketch Dimensions***

**Perpendicular Dimension**: Dimension that is placed perpendicular to a line. The first entity chosen when creating a perpendicular dimension *must* be a line.

**Parallel Dimension**: Dimension that runs parallel to a line drawn between two points.

## *Project 5.2.3*

Create a sketch that resembles the one shown below.

- Create a new part file.
- Select Insert / Sketch in task environment. Hit OK.
- Create a sketch with the Profile tool as shown in the figure below.

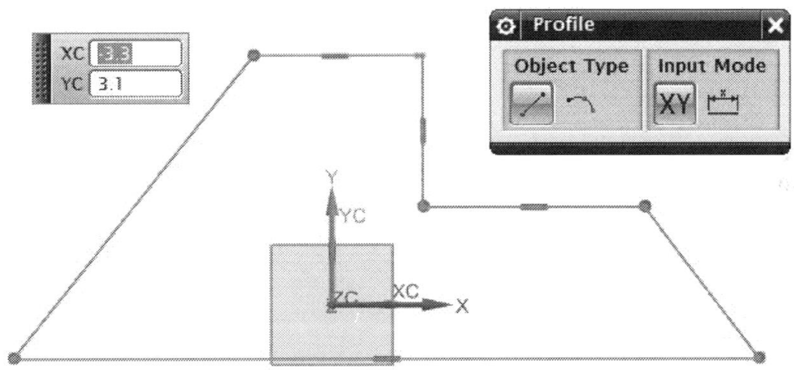


- Select the **Inferred** icon  from the Sketch Toolbar.
- Select the **Sketch Dimensions Dialog** icon  from the Dimensions dialog box. Here you can manually choose which kind of dimension you wish to place.
- Create the dimension as shown in the following figure. NX 8.5's Inferred Dimension should generally choose the most obvious dimension type you are aiming for, but if it doesn't, you can choose manually. Click the Perpendicular dimension button. *Note: Pick locations in order.*

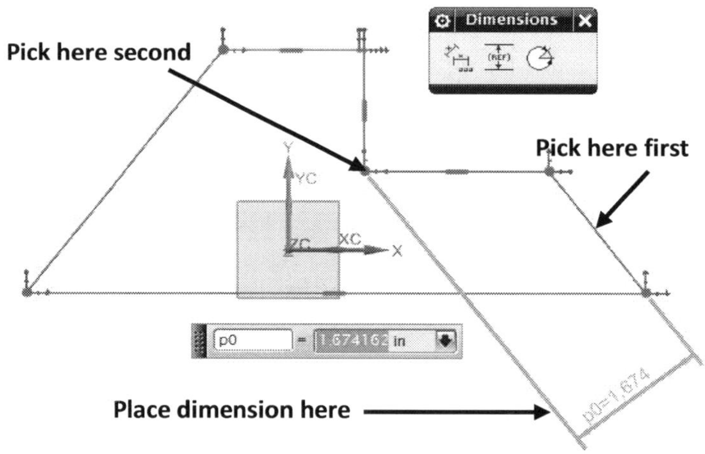


- Create the dimension as shown below. Use the parallel dimension in this case.

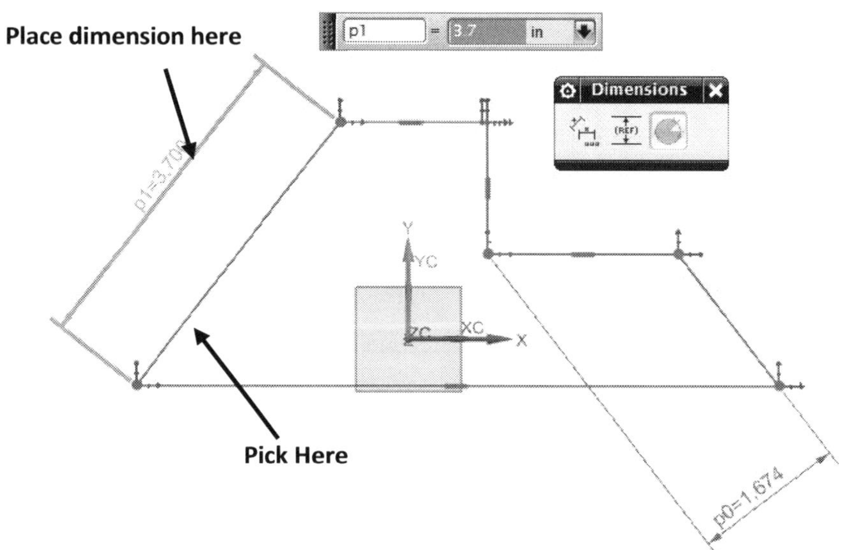


- Exit the sketch by choosing the **Finish Sketch** icon.

> **Note:** Do not apply conflicting or redundant constraints. This will confuse NX 8.5.

***End of Project***

## *Detail: Delete or Undo Dimensions on a Sketch*

***Edit / Delete*** may be used when you have placed a dimension on a sketch that is not required. You may also use the ***Edit / Undo*** command; however, in some cases NX may undo more than you bargained for. *Be careful, the Redo command is limited.*

## *Detail: Radius and Diameter Sketch Dimensions*

You may apply either a radius or a diameter dimension to sketched arcs and circles. The inferred tool creates a diameter dimension on circles and a radius dimension on arcs (see below).

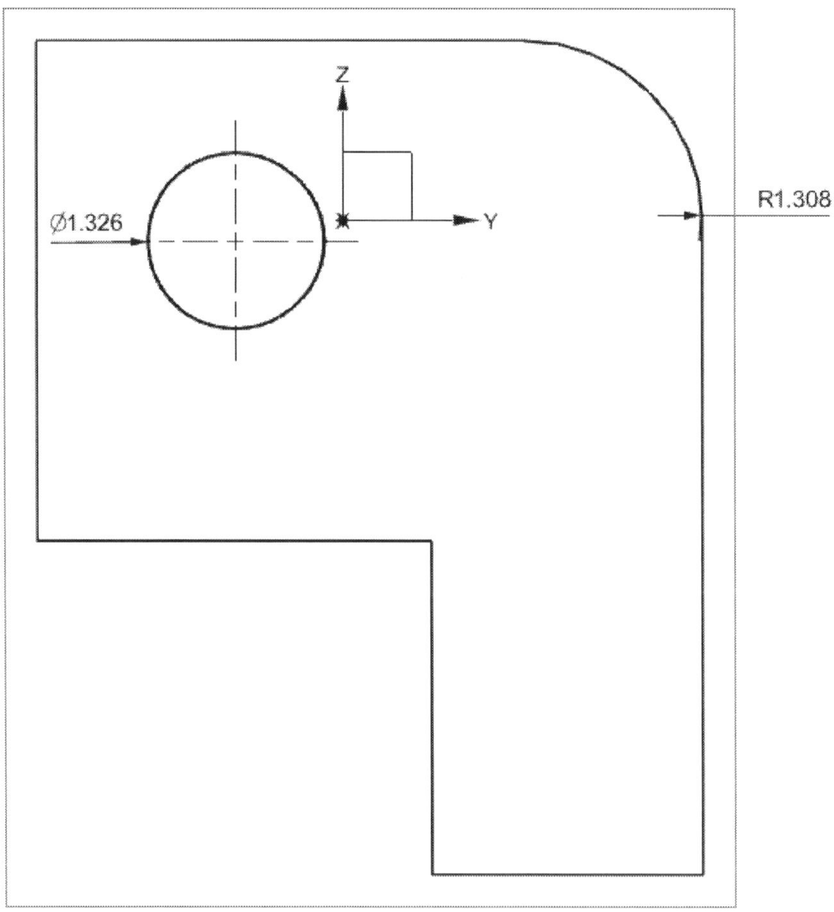

## ***Detail: Moving a Previously Placed Dimension***

Dimensions created with the Auto Placement option are automatically placed in a "good" location by NX 8.5. For horizontal and vertical dimensions, the placement is right in the middle of the span of the dimension. For diameter dimensions, the placement is usually radial to the arc.

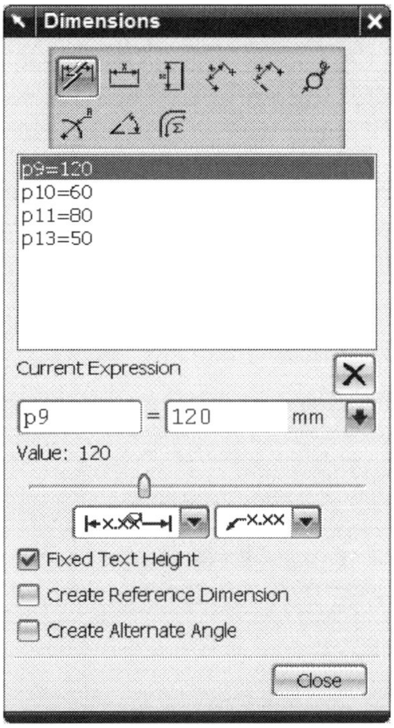


To change the position of a Sketch dimension, click and hold on a dimension with the left mouse button and drag it to a new location. If a dimension was created using 'Auto Placement,' the system will limit its movement. To convert it to manual placement, select the dimension, select the Sketch Dimension dialog icon, and choose one of the 'Manual Placement' options (see previous figure).

## 5.3 Defining Geometric Constraints 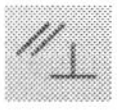

Creating geometric constraints is analogous to specifying a set of rules that sketch geometry must follow, such as forcing the radii of all circles to be equal or making two lines parallel. Geometric Constraints are applied using the Create Constraints option, accessed from the Sketch Constraints toolbar.

## ***Detail: The Degree Of Freedom Indicators***

When applying dimensions or constraints, you will notice arrows are displayed on some of the vertices of sketched lines, or perhaps at the middle of sketched arcs. These are the degree of freedom indicators indicating which entities and vertices have not been fully constrained. As you place more and more dimensions and geometric constraints on a sketch, the degree of freedom indicators disappear one by one. To constrain a sketch fully, you must apply enough dimensions and geometric constraints to remove all of the degree of freedom indicators. You will also need to fix the sketch with respect to previously existing geometry. The figure below shows a simple sketch of a triangle and its degree of freedom indicators (red arrows on all three vertexes).

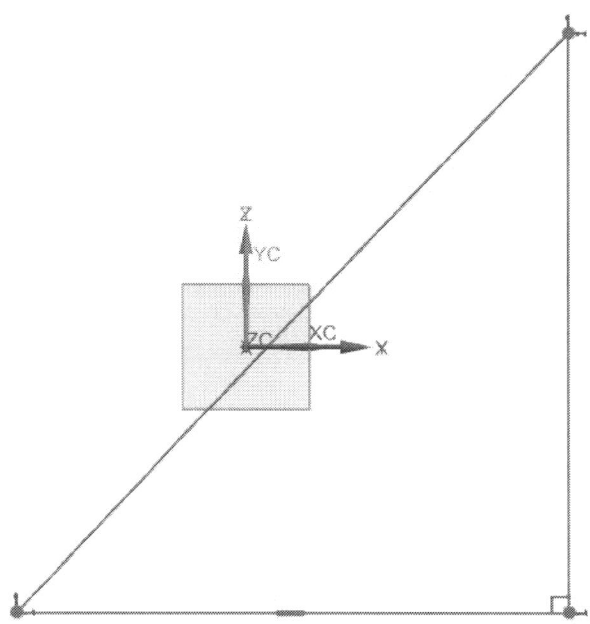


> **Note:** It is not always necessary to fully constrain each sketch, but doing so makes a sketch behave more predictably.

## ***Detail: The Fix Constraint*** 

One of the geometric constraints is the 'Fix' constraint. It is used to anchor one point or line on the sketch. To ensure that it has a "zero point," the Fix Constraint does not fix a point in space – it fixes a point with respect to all other geometry in the sketch.

## ***Detail: The Coincident Constraint*** 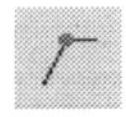

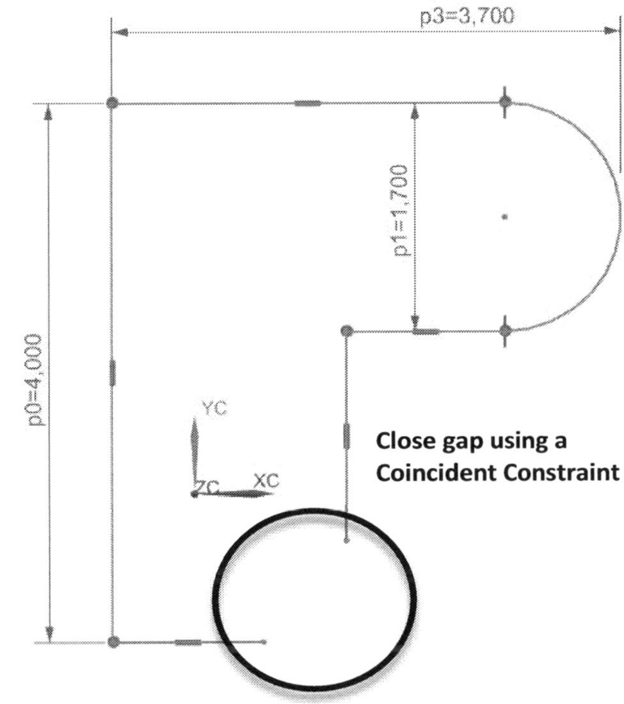


The 'Coincident' constraint is used to place a vertex (e.g. a circle center or endpoint of a line) onto another vertex (see figure below).

To use this constraint, select the desired vertices, then select the Coincident constraint from the list of available constraints.

### *Project 5.3.1*

Place a **Fix** and a **Coincident** Geometric Constraint onto a sketch.

- Construct a sketch that resembles the following figure. Make sure it is *not* a closed loop.

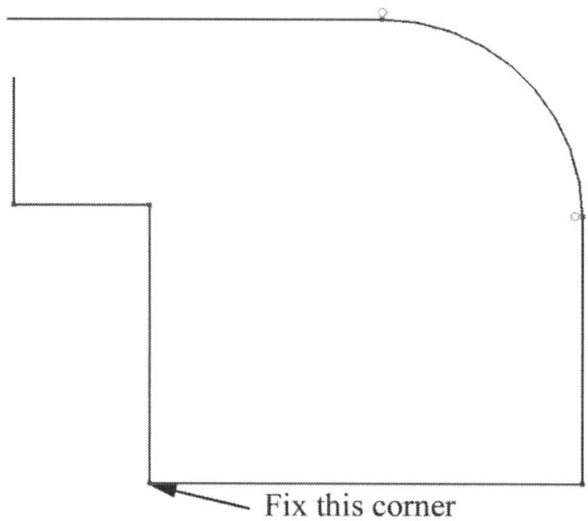


- Select the **Constraints** icon 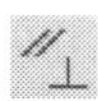 from the Sketch Constraints Toolbar.
- Select the lower left corner of the sketch as shown. Either the horizontal or vertical line can be highlighted.
- Select the **Fix** constraint icon 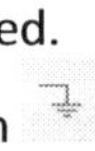 .

Notice how the fix constraint removes the degree of freedom indicators.

- Select the two points as shown in the next figure.

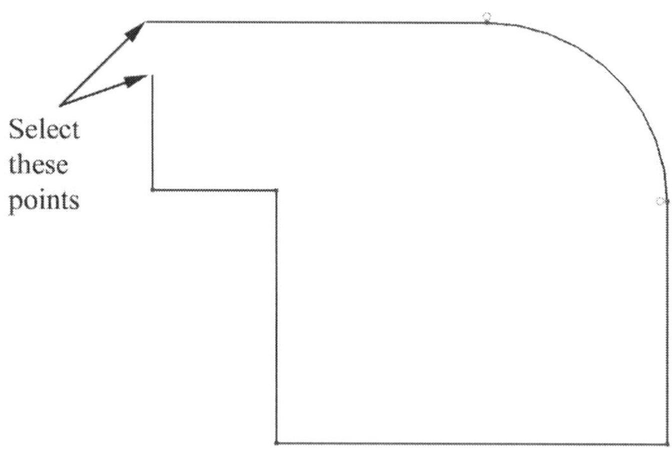


- Select the **Coincident** constraint icon  .
- Exit the sketcher.

***End of Project***

## ***Detail: Horizontal and Vertical Geometric Constraints***

When you sketch a line that is nearly horizontal or vertical, NX 8.5 will often automatically apply a horizontal or vertical geometric constraint to them. You may also place these constraints manually if this does not occur.

## *Project 5.3.2*

Create several Horizontal and Vertical Constraints.

- Sketch a profile similar to the one shown below.

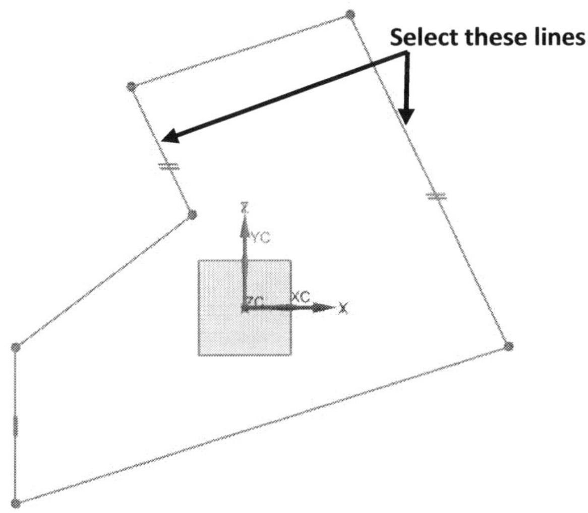


- Select **Constraints** from the Sketch Constraints Toolbar.
- Select the lines as shown.
- Select the **Vertical** geometric constraint . *The sketch should look similar to the following figure.*

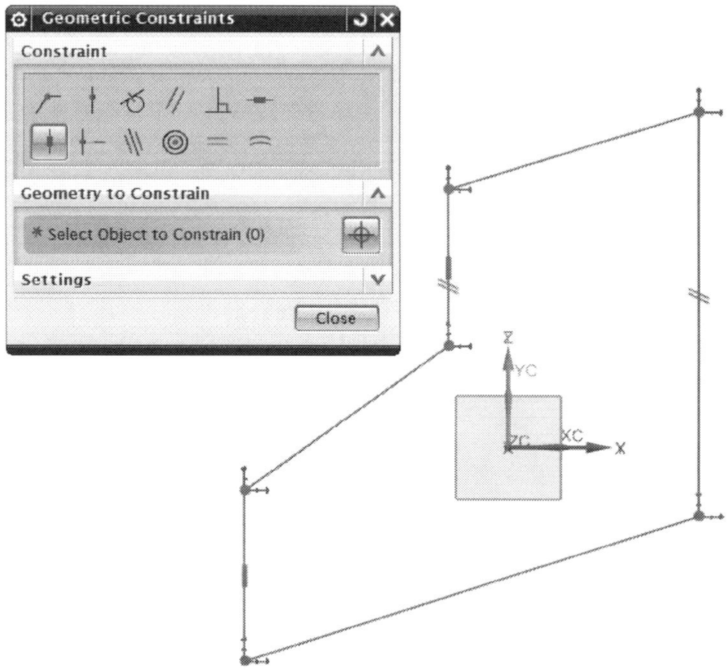


- Select the other lines and apply a **Horizontal** constraint .

Your sketch should look like the figure below.

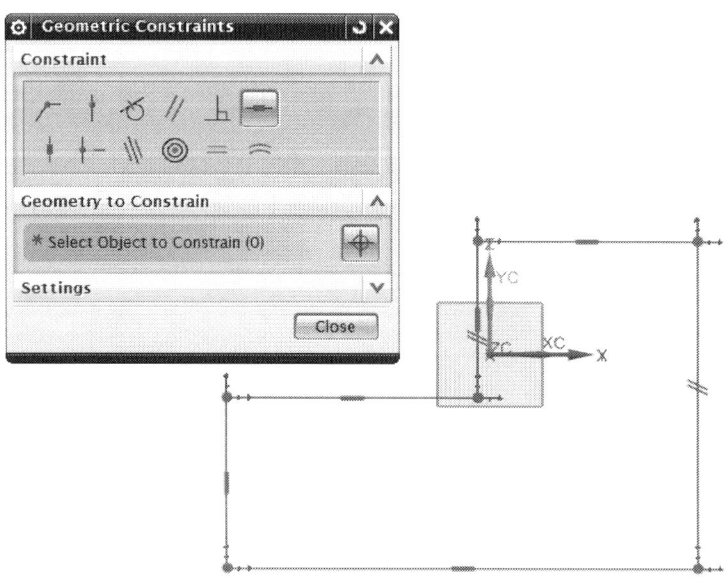


- Exit the sketcher.

***End of Project***

## ***Detail: The Point on Curve Constraint*** 

'Point on Curve' is a very powerful constraint option. The name is a bit misleading– it should really be thought of as a point constrained to be on an infinite extension of a curve. It may be used to align points, endpoints of curves, or arc center points to lines or other curves. It is especially useful for creating geometry where elements must be aligned to each other. For example, if you had a series of holes that lined up with a slot or edge, you may use the 'Point on Curve' constraint.

The following figure contains a good example of the application for a Point on Curve constraint. The holes on the right hand portion of the shape shown in the following figure should be aligned with the angled (p1 in our sketch) edge on the left-hand side.

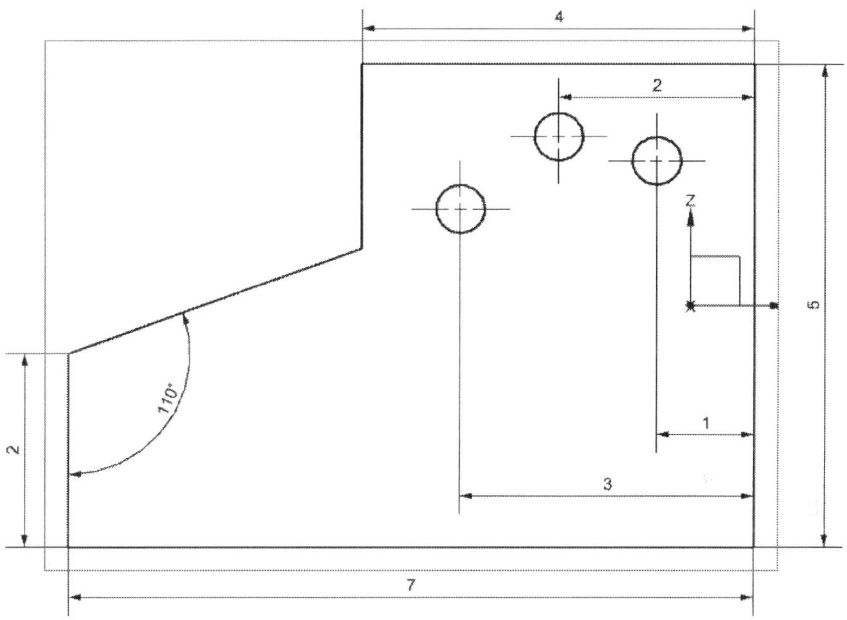


In the next figure, the point on curve constraint has been applied. Each center point of the circles has been aligned with the angled edge on the left-hand side of the shape. You can see that as the angle is changed, the location of the holes follows.

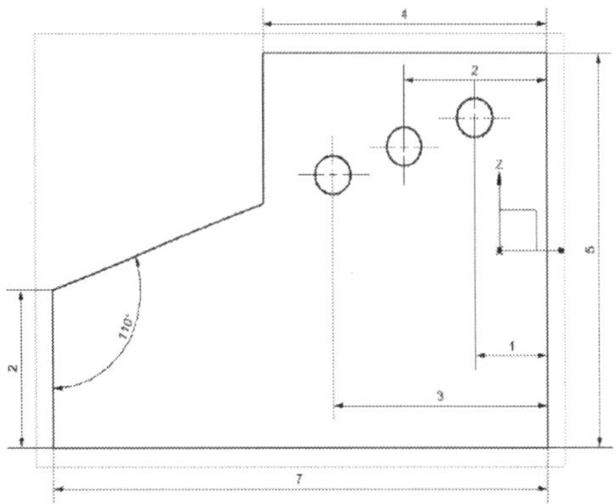

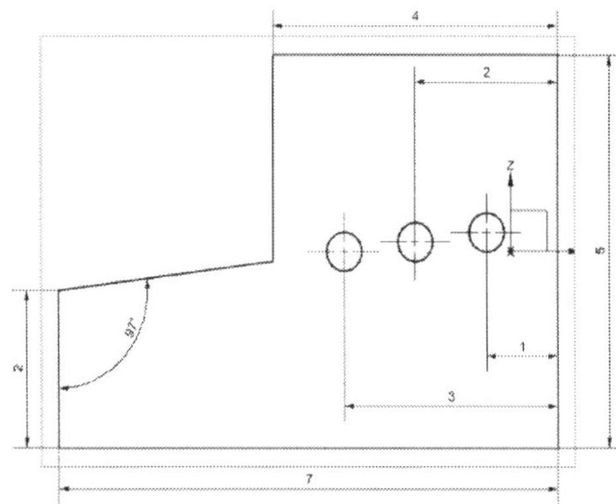

Now it's your turn.

The following project illustrates the use of the Point on Curve constraint.

## *Project 5.3.3*

Create the sketch depicted below.

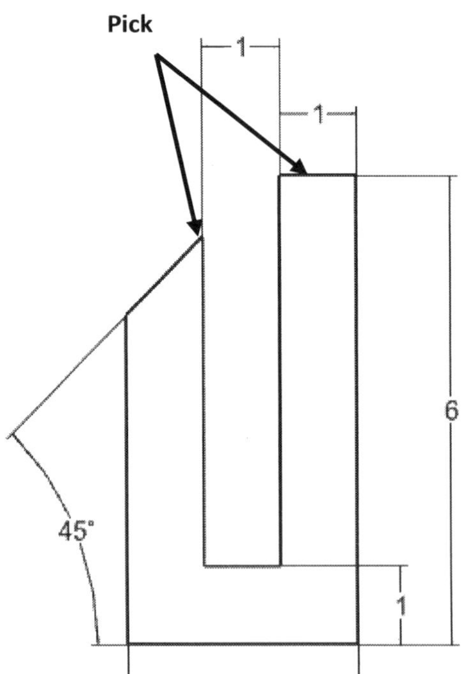

- Select **Constraints** from the 'Sketch Constraints' Toolbar.
- Select the two entities labeled 'Pick' in the diagram above. A list of constraints will appear, including the 'Point on Curve' constraint. It does not matter which entity is selected first.
- Select the **Point on Curve** constraint  .
- The result is shown below.

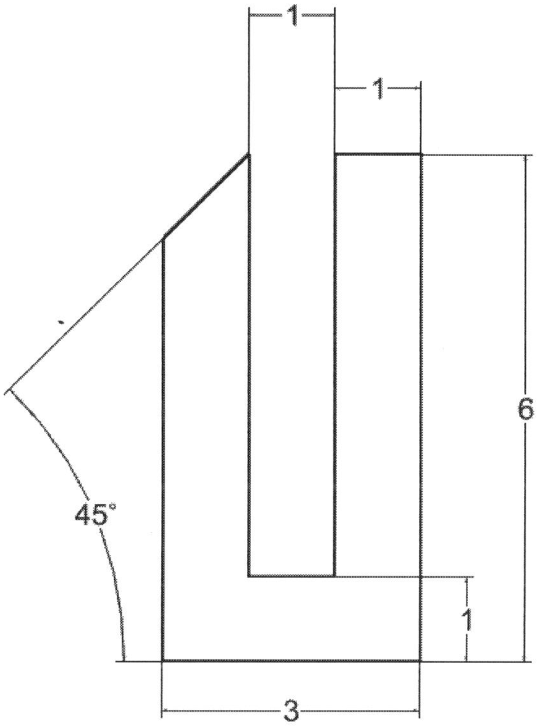


- Exit the sketcher.

***End of Project***

## ***Detail: The Collinear Line Constraint*** 

The 'Collinear' geometric constraint is similar to the Point on Curve constraint. The Collinear constraint aligns two or more lines with each other, as shown in the following figure.

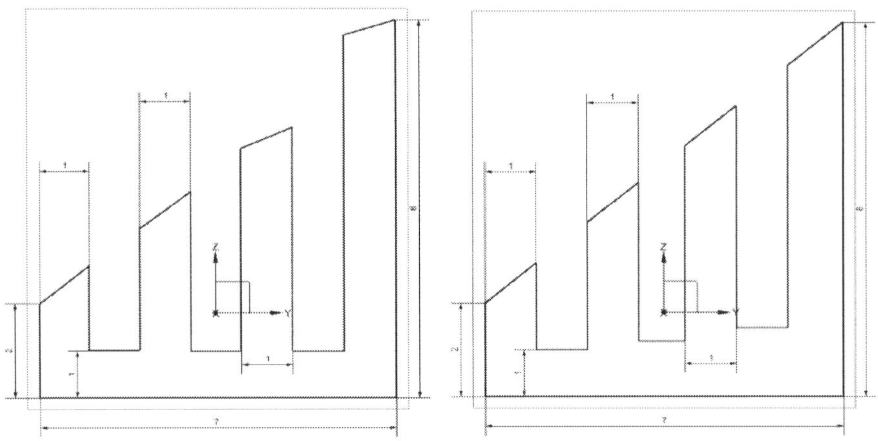

To apply the collinear geometric constraint, select the appropriate lines, then select the collinear geometric constraint from the list of available constraints.

## ***Detail: The Tangent Constraint*** 

The 'Tangent' geometric constraint applies a tangency condition to the selected curves (lines or arcs), as shown below. To apply this constraint, select the two entities near the desired tangency location, then select the 'Tangent' constraint.

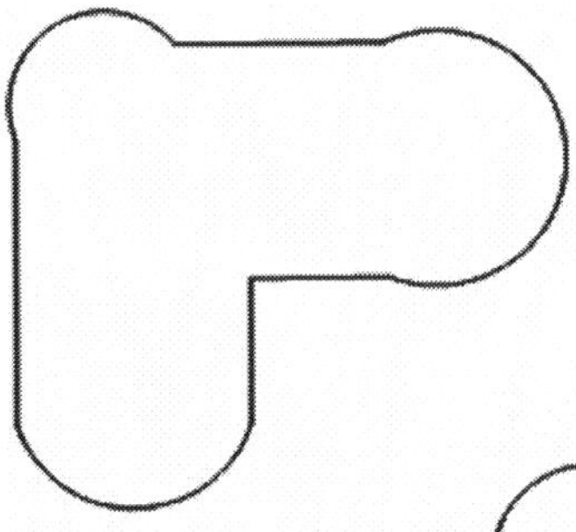

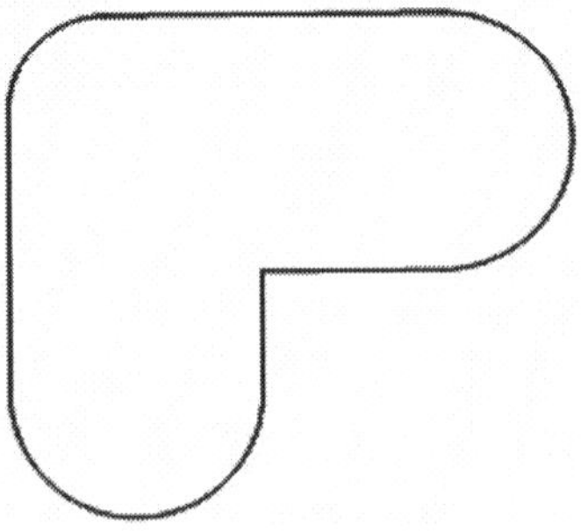

> **Note:** To fully constrain a sketch, NX 8.5 requires you also to fully constrain all Reference Features located in that sketch.

## Detail: The Reference Sketch Features

The 'Convert To/From Reference' sketch tool allows the creation of Reference features within a sketch plane. Reference features can be constrained and dimensioned just as regular sketch features. All non-reference features are considered 'Active' in the sketch. Reference lines show up as dotted lines in your sketch. They remain once you exit the Sketcher, but can only be used as reference locations for features.

### *Project 5.3.4*

Create a sketch with a line as shown in the following figure. Constrain it so that the base is at the origin and the dimensions are specified as shown.

- Select the **Convert To/From Reference** icon.
- Select the line.

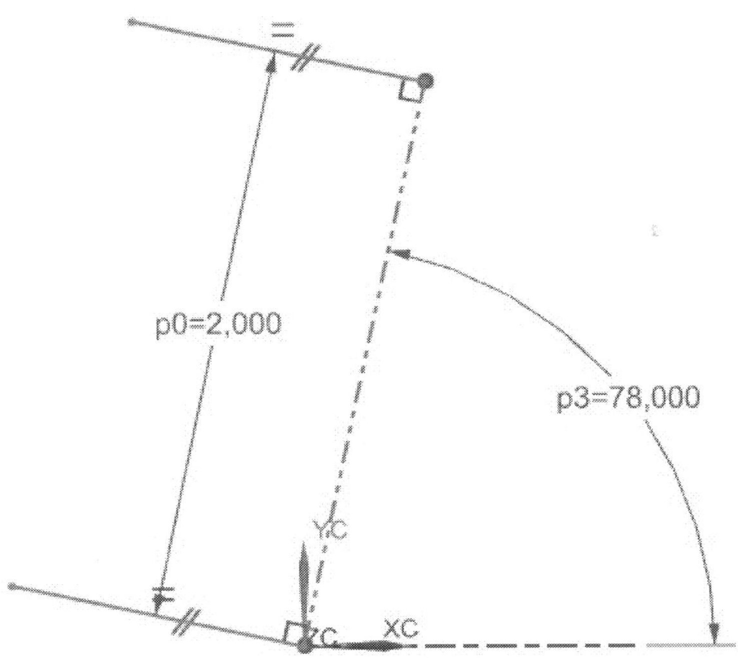


- Select **OK**. The dotted line indicates it is now a Reference line.
- Create three circles with Radius = **0.25in** equidistant across the line as shown below.

**Hint:** Select the two endpoints and the midpoint of the line as the center points of each of the circles.

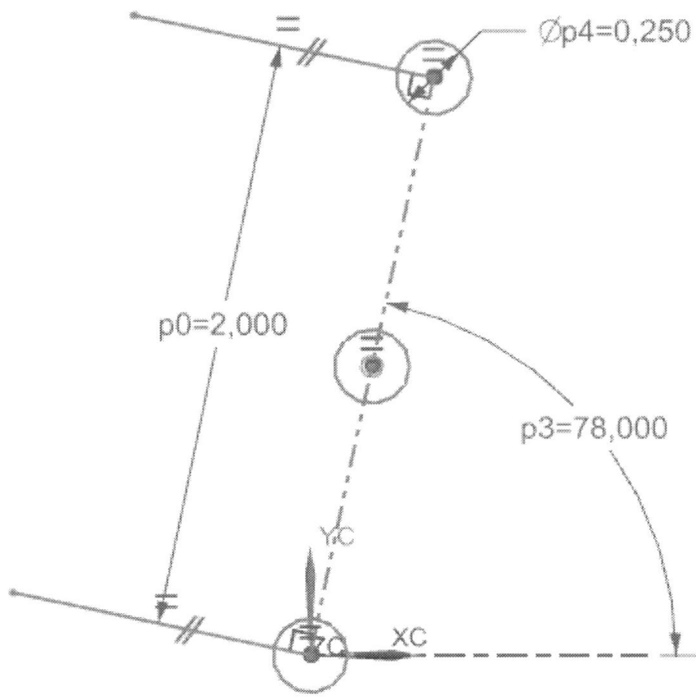


- Exit the sketcher by choosing the **Finish Sketch** button. Compare your results to the figure above.

**End of Project**

## 5.4 Mirror Curve Operation

Sketching surfaces often requires several identical features to be placed on the same sketch plane. NX 8.5 offers a mirror curve operation that allows you to replicate your drawing. The resulting sketch has the same properties as the original, and if either the original or mirrored sketch is modified, the entire sketch updates accordingly.

### ***Detail: Mirror Curve***

The Sketch Tools toolbar is shown below. If it is not a default option when you enter the Sketcher, right click over the toolbar area and select Sketch Tools.

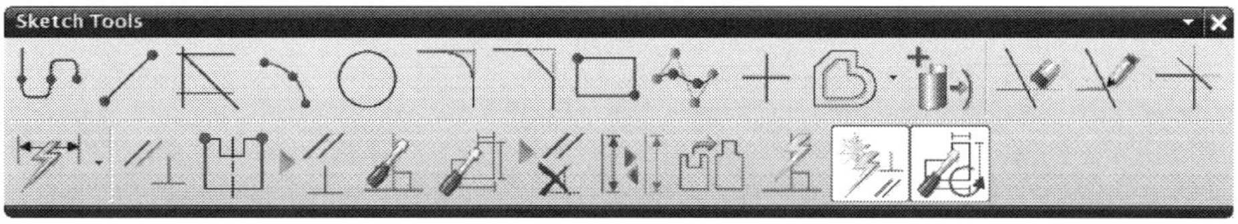


### ***Project 5.4.1***

Enter the sketcher and sketch *anything* in the upper right (+XC, +YC) quadrant of the sketch plane.

- Draw a curve as shown next.

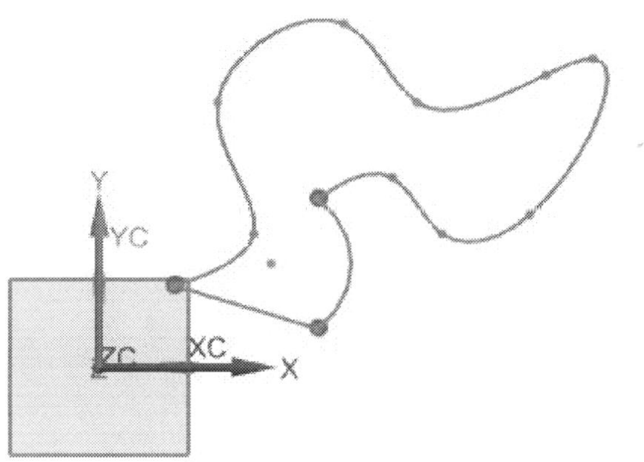


- Select the Mirror Curve sketch icon or select ***Insert / Curve from Curves / Mirror Curve***.

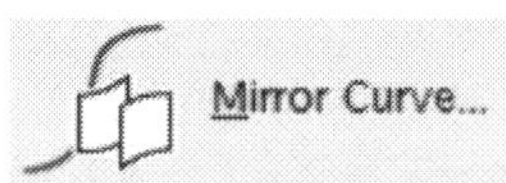


Read the cue line – it is asking you to select the pattern of objects to mirror.

- Select all the lines and arcs in your first quadrant sketch. Select them individually or left click in a blank portion of the sketch and create a large selection box to include all lines and arcs.
- Middle click.

Read the cue line – it is asking you to select the centerline.

- Select the **Y-axis**. Note: NX 8.5 automatically lets you select any axis lines as mirror lines.
- Select **Apply**. The mirrored geometry will appear to the left of the Y-axis.
- Select your original sketch again as well as the mirrored geometry.
- Select the X-axis as your centerline about which to mirror.
- Select **OK**.
- The following figure illustrates the result.
- Exit the sketcher.

**Note:** After the mirror command is complete, the mirror line is automatically converted into a Reference line as shown next. You do not need to create a reference line to use for the mirror function.

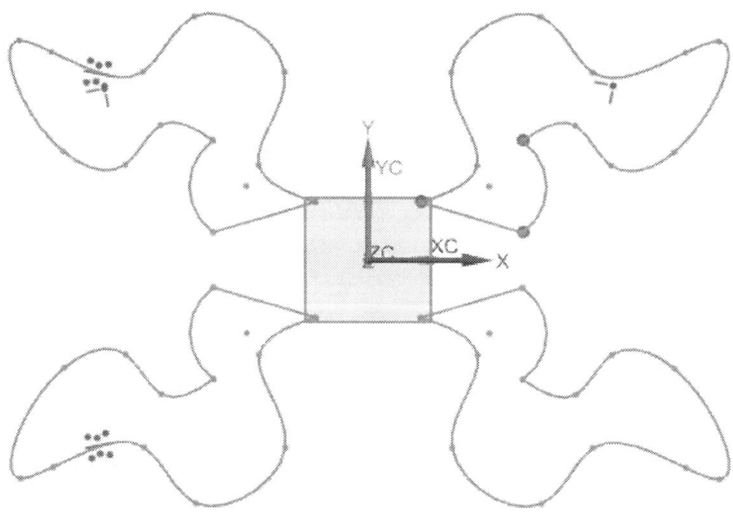


***End of Project***

## 5.5 Placing a sketch at the end of a curve

For many types of geometry the best construction method is to sweep a section curve along a drive curve. Good examples are moldings, railings, and bezels. In cases like these it will be best to use the "On Path" choice in the sketch command dialogue.

### *Detail: Sketch on path*

### *Project 5.5.1*

- Create a new sketch, and select the "On path" choice by clicking on the dropdown button as shown below. Select **OK**, and then select a curve.

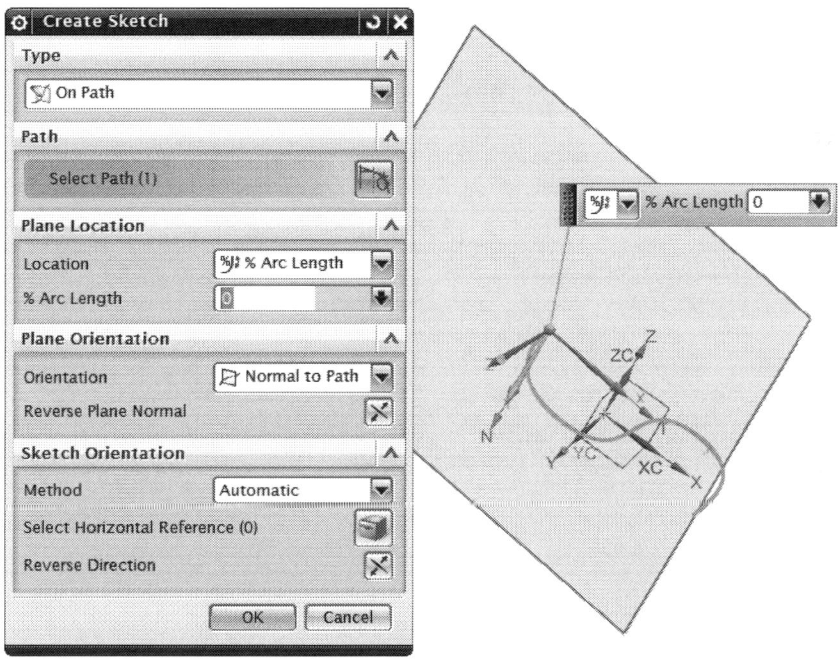


- When the curve is selected, drag the plane that appears by its center point to the end of the curve until the "% Arc Length" box reads 0. The plane will move to the end with a point at the intersection of the sketch plane and the curve. A datum axis for both the horizontal and vertical directions will also be created.
- Now select **OK**.

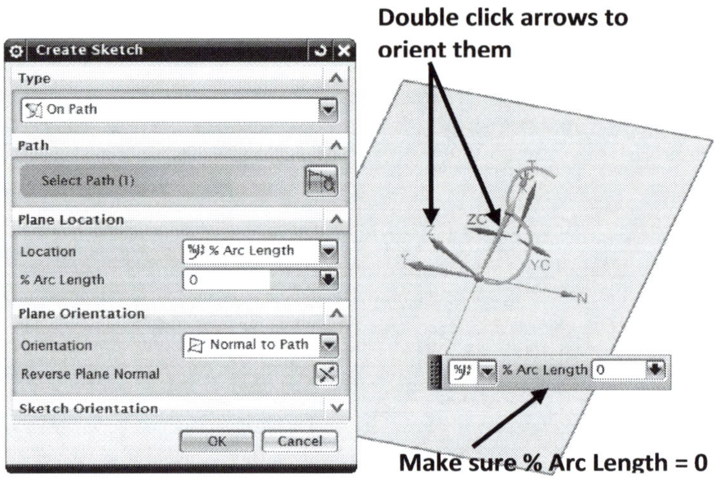


- Create your sketch geometry in the plane and dimension to the center point.
- **Notice:**
    - The center point of the plane is the end of the drive curve.

- Use ***Insert \ Sweep \ Sweep along Guide***, select the section, select the guide, leave the offsets 0, and click **OK**.

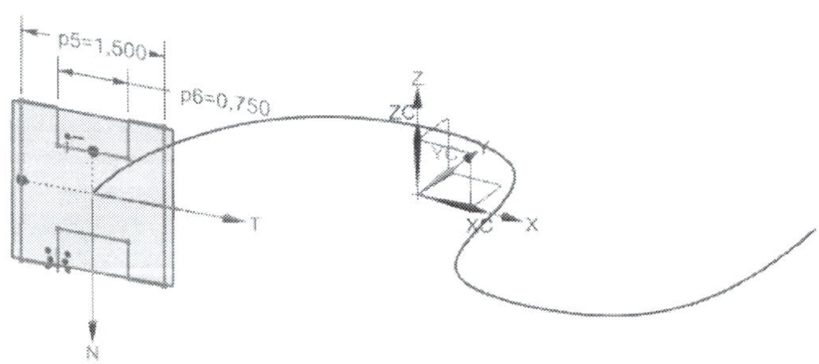


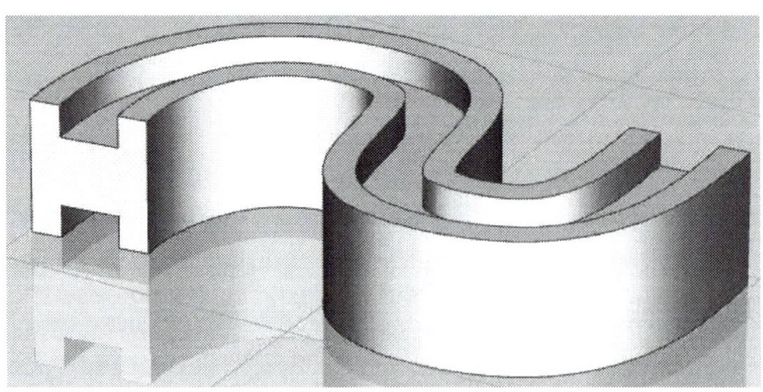

***End of Project***

## 5.6 The Perimeter constraint

Whenever you need to model a molding or some kind of belt that has a constant length, use the perimeter constraint. The perimeter constraint allows you to assign a length value to all the entities of a contour added together. For instance, the shape below is unconstrained +because there is no dimension between centers. If you place a perimeter dimension on the shape, it will be forced to be fully constrained.

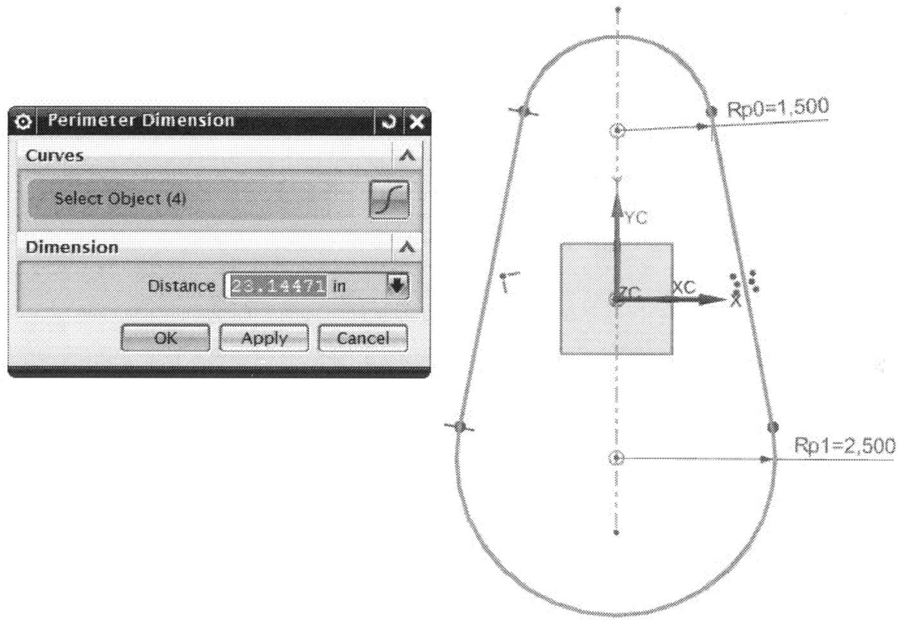


### *Detail: The application of the perimeter constraint:*

This goal is easiest to accomplish if you begin in the sketch dimensions dialog, by selecting Insert / Dimension / Inferred, and click the Sketch Dimensions Dialog icon.

- Select the perimeter icon from the menu .
- Click all the entities that you want to be considered in the perimeter constraint.
- Click the middle mouse button.
- Finally, you can enter your desired value in the appropriate field and hit 'Enter' on your keyboard.

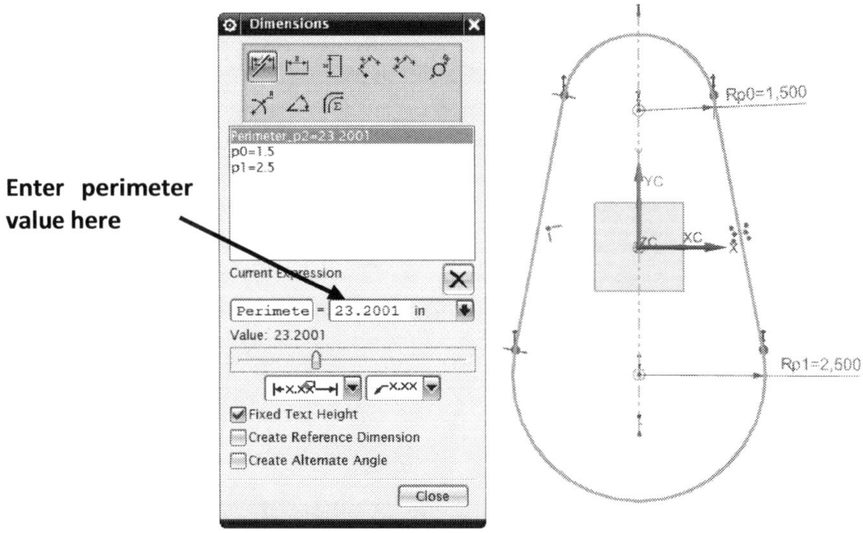


With a perimeter of 20, a radius of 2.5 on the bottom, and a radius of 1.5 on the top, the example sketch appears as shown below.

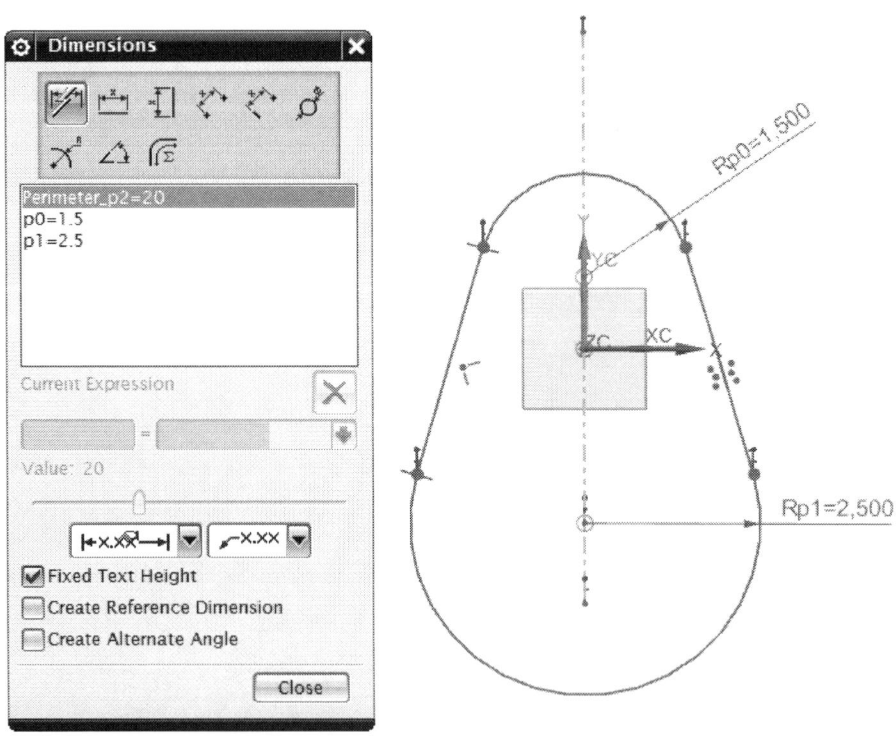

## 5.7 The Studio Spline

Use a studio spline to model a shape you cannot represent with simple arcs and lines. The studio spline is an amazing tool because you get to define the points that control the shape, and as you continue, you will see the effect that each of the points has on the overall spline. You can then begin to drag the points around to get the exact shape you desire. It is great for defining shapes that must be artistic in nature.

### *Detail: The Studio Spline*

Begin by selecting the studio spline icon from the menu 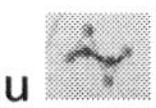 . The menu shown below will appear and you can begin clicking point locations for your spline.

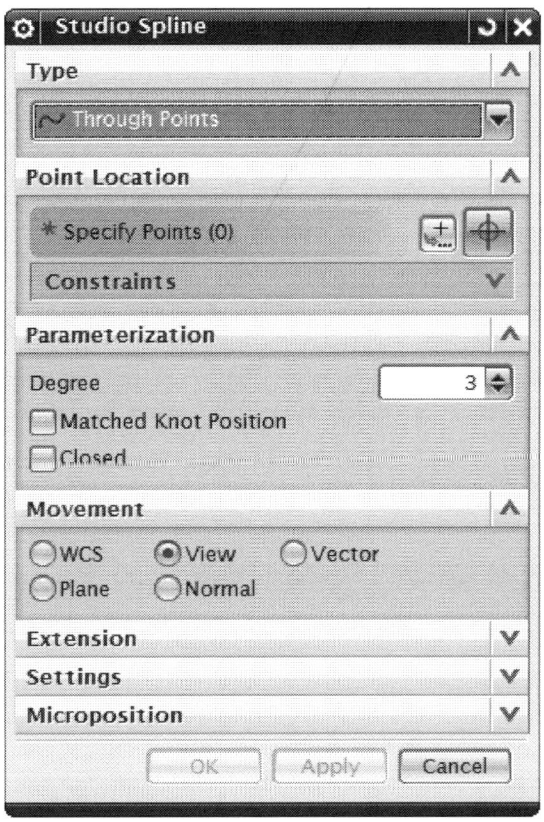


Notice in the following figure that the shape of the spline changes. In the first frame, there are only two points, so the spline is a straight line. In the second frame, the spline takes a turn to meet the new point 3. In the third frame, point 4 is added and the spline changes further. Finally, in the fourth frame, the last point is added to complete the spline.

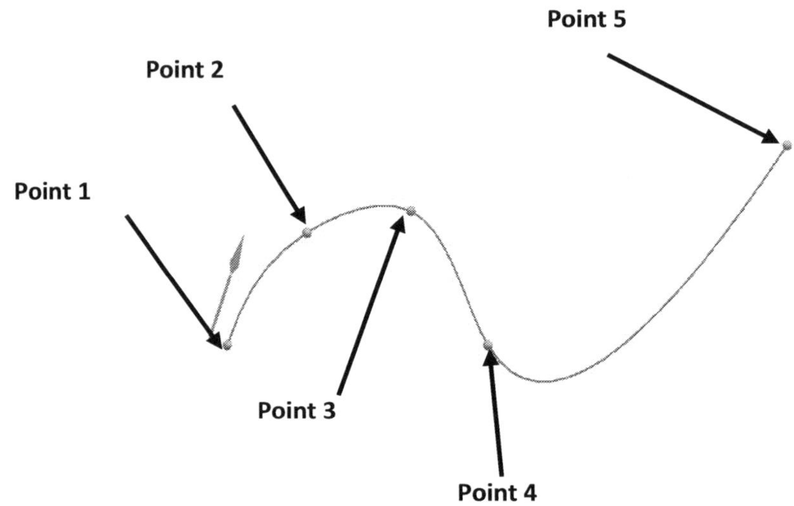


To change the shape of the spline, left click and drag any of the existing points. In the following figure this is done to point 4.

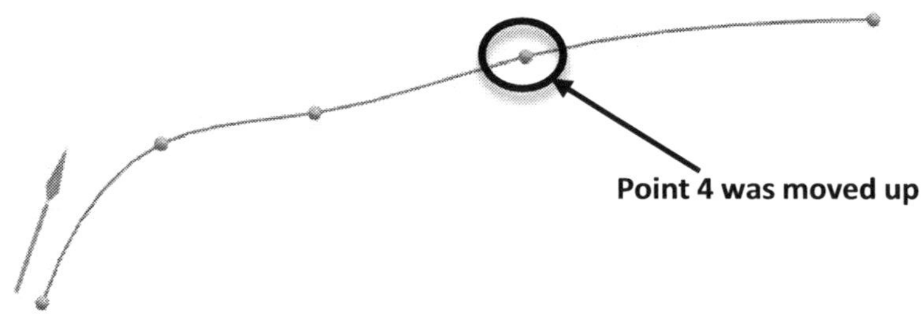

## ***Detail: Closing a Studio Spline***

To make a continuous closed shape, use the "closed" button choice in the studio spline menu.

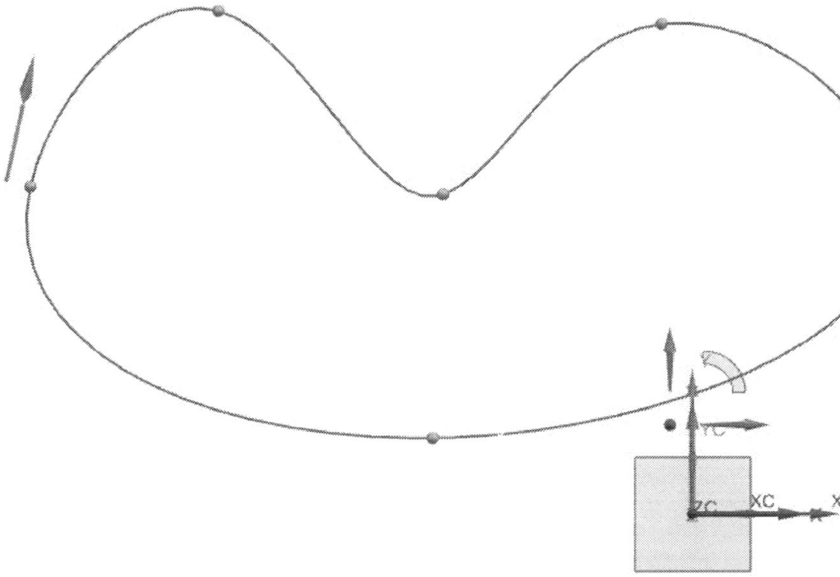


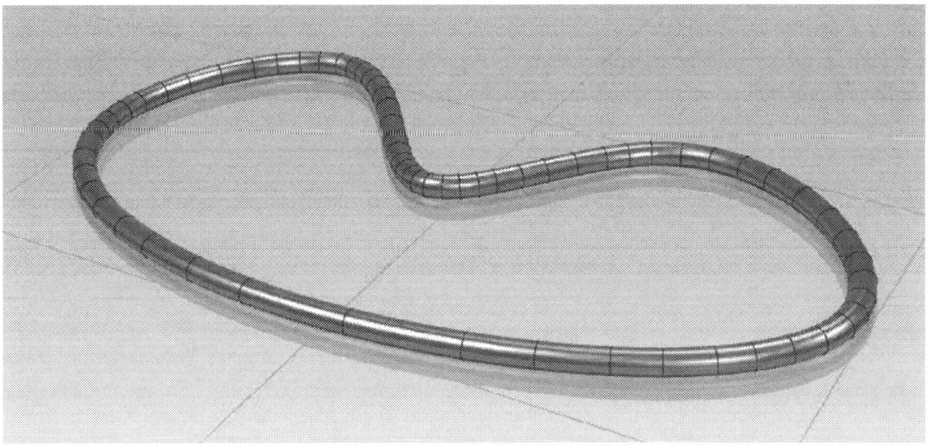

## 5.8 Direct Sketching

This example walks through how to use the new Direct Sketching method described at the beginning of this chapter.

### *Project 5.8.1*

- Open a new modeling part file.
- The modeling environment is now open with a datum CSYS at the center. The datum CSYS will be used for planes on which the sketch is placed. *Notice the 'Direct Sketch' toolbar at the bottom of the screen.*

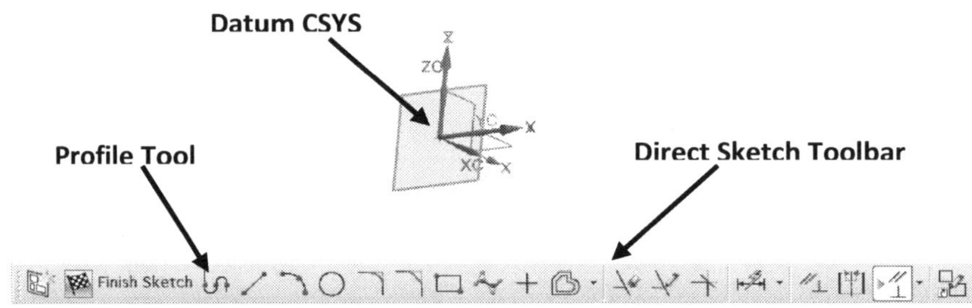


- Click on the **Profile** tool in the Direct Sketch toolbar shown above.
- Click on the XY plane of the datum CSYS as shown in the following figure.

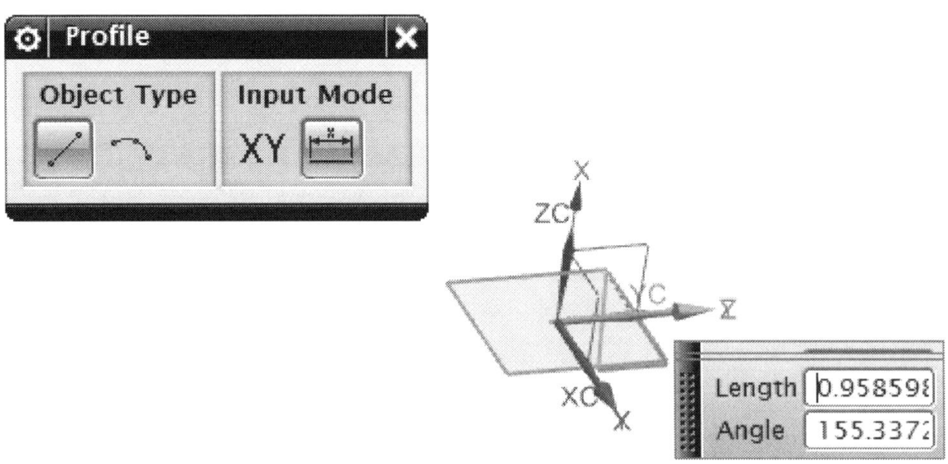

If familiar with the original method of sketching, notice that, when you click the plane to sketch on, the view is NOT oriented to view the sketch plane that you selected. Orienting the screen to the 'Sketch View' tends to give the user a better perspective of the profile they are sketching and an easier ability to sketch horizontal, vertical and perpendicular lines. To view the sketch plane select **View / Orient View to Sketch.**

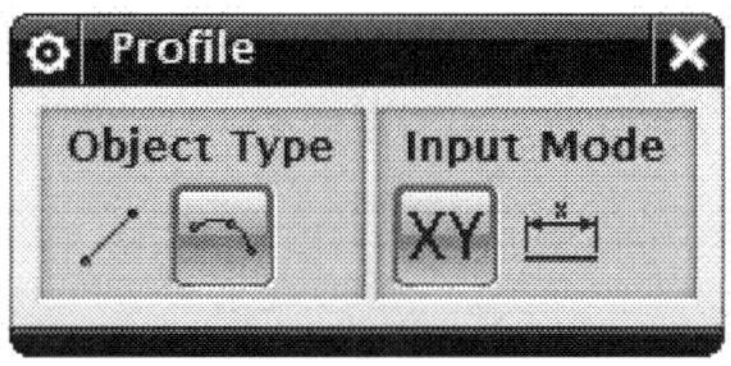


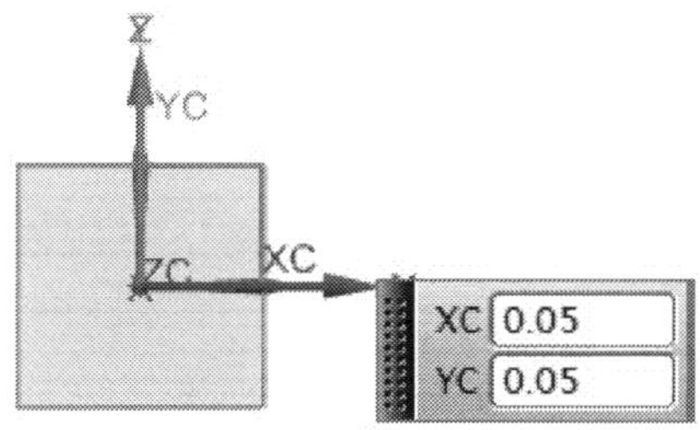


- Draw a line. Notice: NX automatically creates a dimension for you. This only happens when the '**Continuous Auto Dimensioning**' tool is ON, as shown in the following figure.

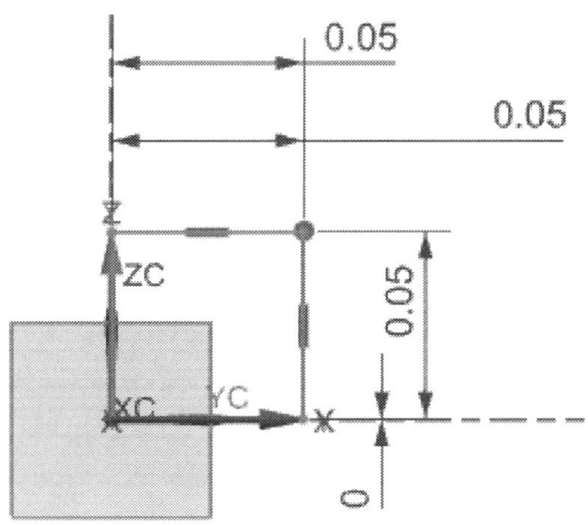


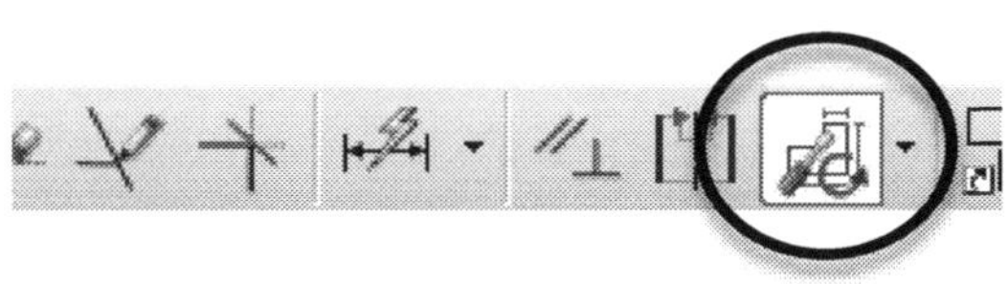

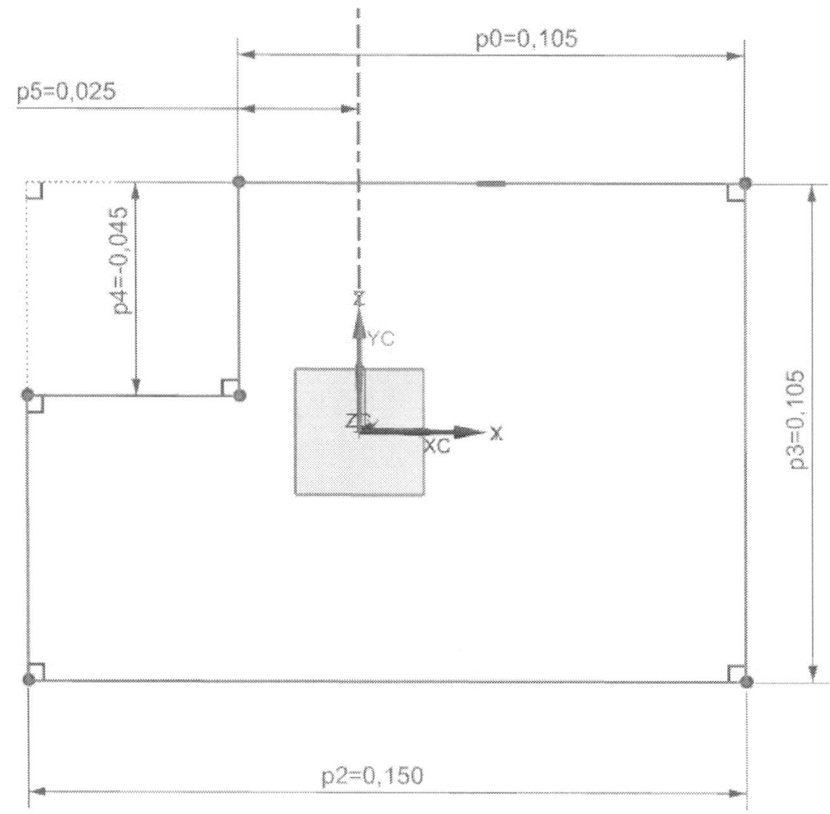


> **Note:** the 'Auto Dimensioning' and 'Infer Constraints' tools will fully constrain the sketch as described at the beginning of the chapter.

- Complete the following profile.
- Now click on the constraint tool in the **Direct Sketch** Tool bar.
- Select the lower line of the sketch and the center point of the Datum CSYS as shown in the following figure, and give them a midpoint constraint.

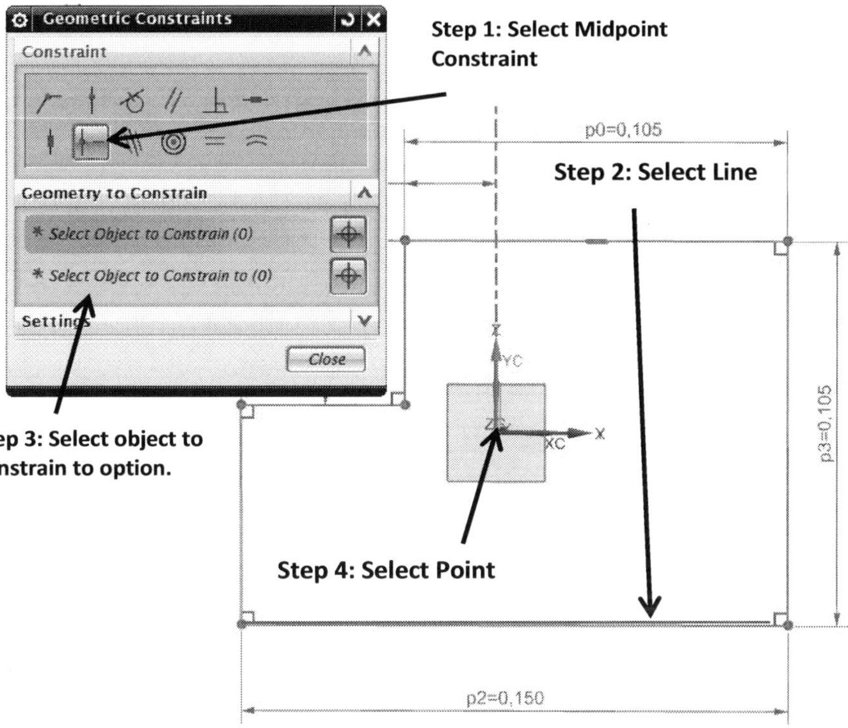


- Repeat this process with the vertical line on the right and the center point of the CSYS.

- Now change the dimensions to match the following figure.

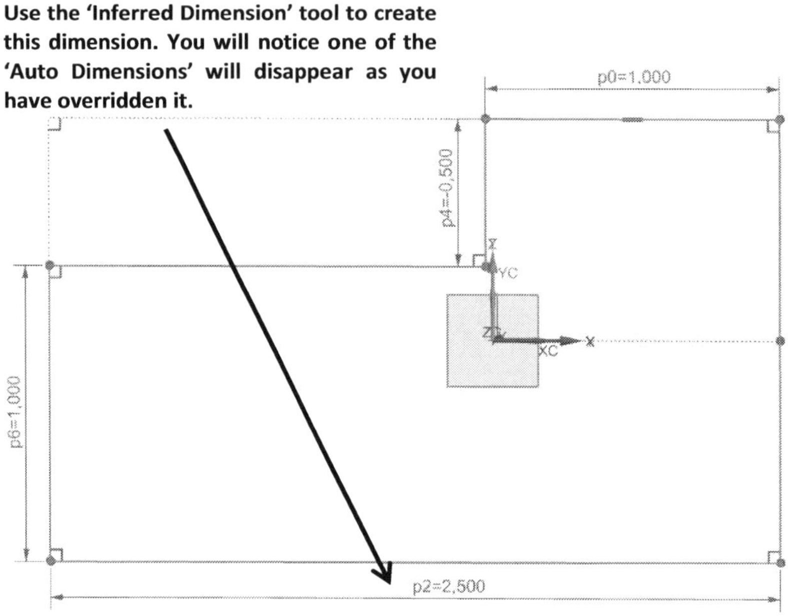


- Click on **Finish Sketch** as highlighted in the following figure.

**End of Project**

## 5.9 Self-Test

(All questions are true/false. Answers are at the back of the book.)

1. A sketch must be planar.
2. The orange degree of freedom indicators appear only when you are working in the constraints menu.
3. If you place a redundant dimension or a constraint on a sketch, NX 8 will automatically convert it to a reference entity.
4. The "On path" choice allows you to orient a sketch.

# 6. Design Features

## 6.1 Introduction

Design features are essential for creating solid geometry. Unlike sketches and curves, design features allow you to create solid geometry directly. Using design features, you can create holes, slots, and other geometry that are commonly used in manufacturing and machining operations. Form features are also parametric. Many people who are accustomed to creating geometry starting with blocks and cylinders call their modeling technique "chunky solids." It's amazing how a complex model can be made using just the most rudimentary items in a combination of subtracts, unites, and intersects.

Design features are most immediately available in the ***Insert / Design Features*** menu or the "Feature" toolbar. The 'Feature' Toolbar is one of the default modeling Toolbars. You might not see all the buttons shown here. To customize your toolbar, click the small arrow on the toolbar and use 'Add or Remove Buttons.' There is also an option to place the labels below each icon as in the figure below.

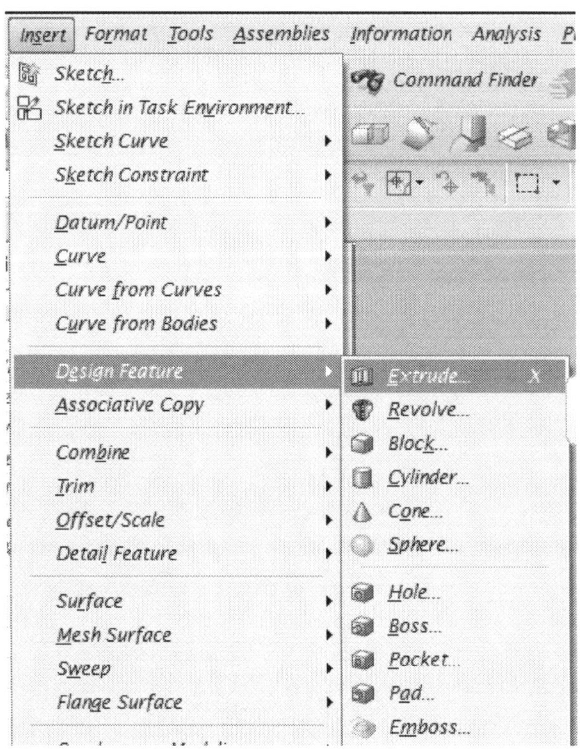

The icons in the ‘Feature’ Toolbar are used to extrude, revolve, or sweep 2D curves, sketches, or solid faces. You will notice that many icons on the ‘Feature’ Toolbar are not part of the Design Feature Menu.

The four icons at the bottom of the figure on the previous page are used to create ‘primitive’ shapes (blocks, cylinders, cones, spheres). Primitives may be used to create geometric models very quickly without using curves or sketches.

Another amazing part of NX is the Design Feature Tree, or the Model Tree. As features are created, they automatically add themselves to the Tree. You can click and drag part features through the design tree in order to reorder them.

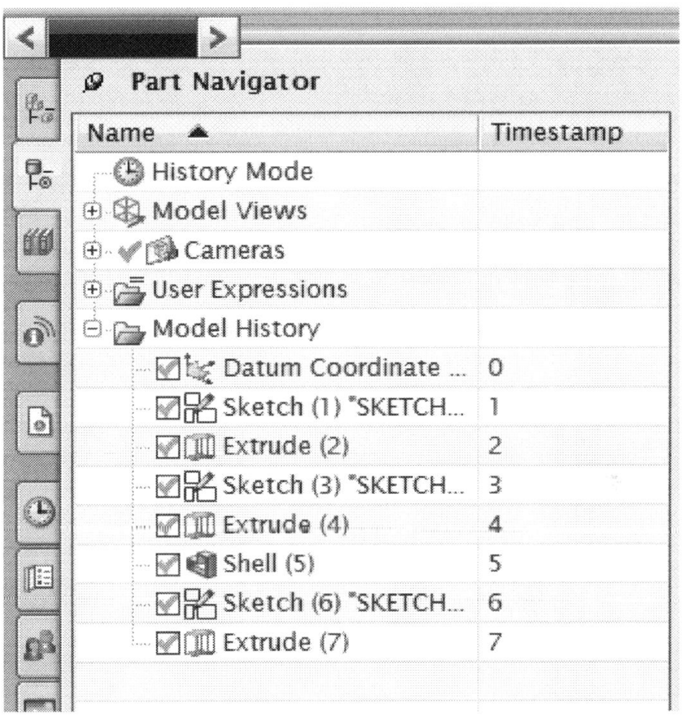

## 6.2 Extruded Features

The 'Extrude' function takes a string of curves (no gaps between entities) and sweeps them along a vector direction to create a 3D body (assuming that you have body type "solid" selected). If you extrude a string of curves that is not closed, the result will be a sheet body (see below).

### *Detail: Extruded Solid Feature*

Extruding a closed string of curves is one of the simplest ways to create a solid.

### *Project 6.2.1*

Using the sketcher, create the geometry shown in the figure below. *Note: To exit the Sketch mode, click the* **Finish Sketch** *icon on the top left corner.*

**Note:** All lines are equal length and all arcs are equal radius.

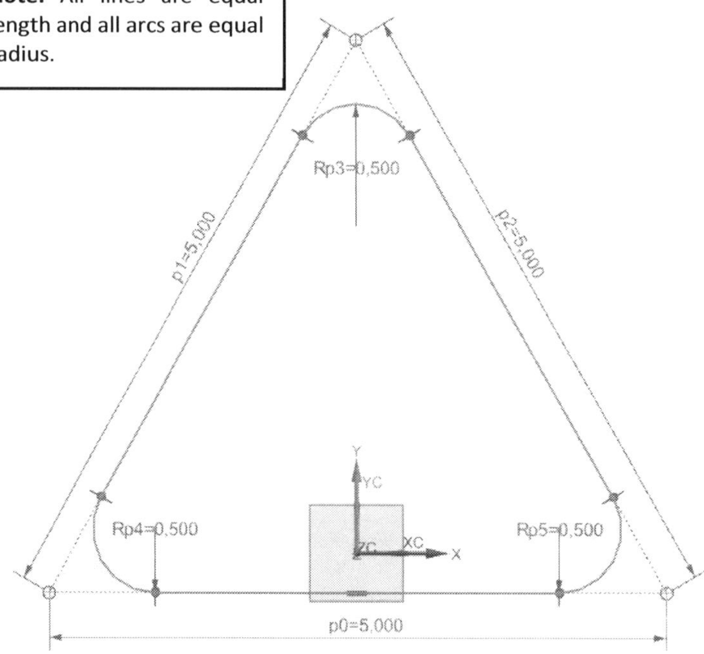


- Select Insert / Design Feature / Extrude .
- The ***"Select Curve"*** icon is already highlighted by default. Select any of the sketch curves. When performing feature operations on a sketch, the entire sketch will be

selected if ***Feature Curves*** or ***Connected Curves*** is selected in the 'Selection Intent' menu for curves. See below.

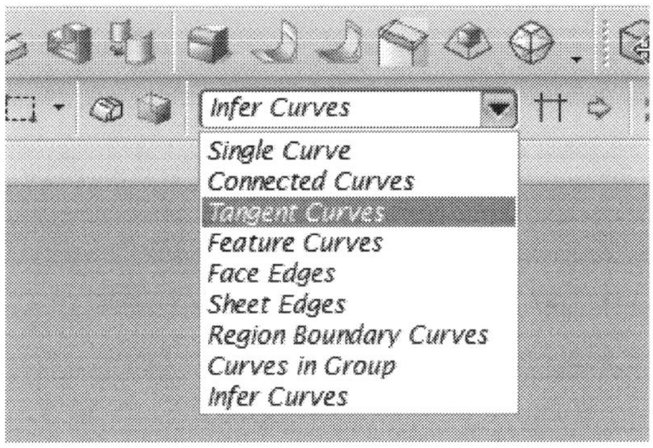


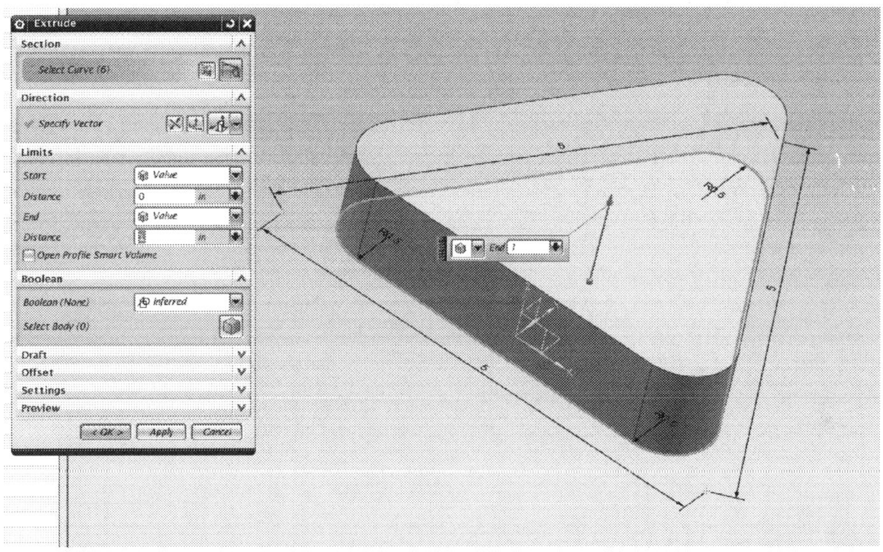


- NX 8.5 will shade the body to be extruded and provide an End dialog box for distance of extrusion, as shown in the previous figure.
- Accept the default direction, or use the 'Reverse Direction Button' to flip the direction in order to get the solid body "preview" in the +Z axis direction.

> **Note:** NX 8.5 will often let you select multiple geometric entities. Choosing OK signifies that you are ready to proceed. Also make sure that you select the right Boolean operation every time you use Extrude; by default, NX 8.5 will create a new body every time.

- You may choose the Direction button which will open a 'Vector Constructor' menu as follows. The default extrude direction is shown as a dashed arrow and is normal to the sketch plane.

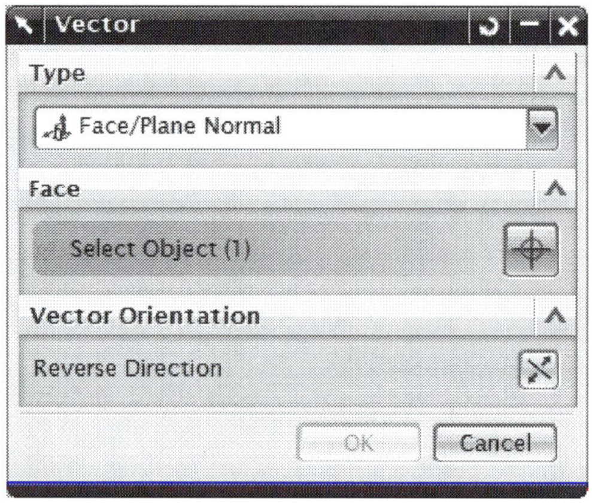


- Choose **OK**.
- Enter a **Start Value** of **1** and an **End Value** of **3**.

Choose **OK.** (Results are shown below).

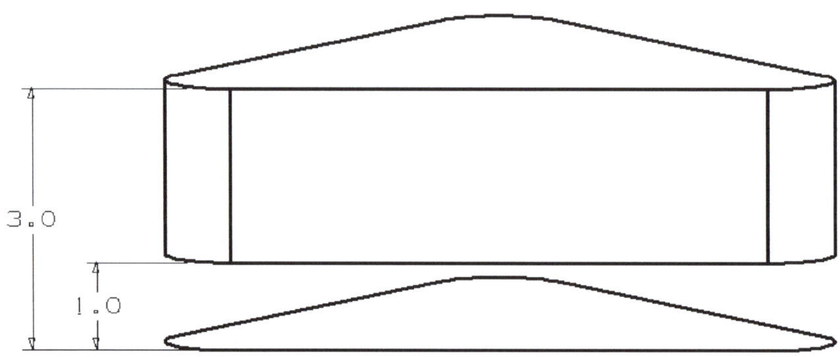


***End of Project***

The start value of 1 specifies that the extrusion should begin 1in from the section string. The end value of 3 specifies that the extrusion should end a total distance of 3in away from the start curves. Thus, the thickness of the extrusion is actually 2in.

If you want to extrude in a direction opposite to the default direction, you can enter negative values for the start and end distances of the extrusion, click "**reverse direction**," or use the Vector Constructor menu. You can also use this menu to define extrusion directions that are not normal to the plane of the section string!

NX 8.5 also has the option to create a sketch if you do not have any predefined curves. You can use the ***"Sketch Section"*** icon to start a new sketch and use it for extrude.

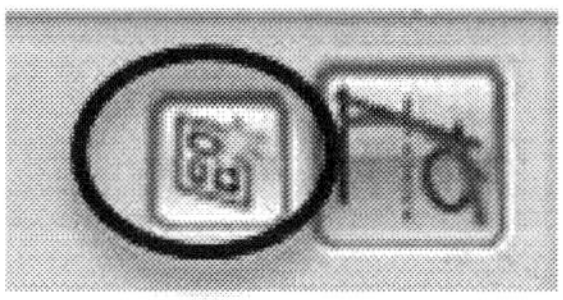

> ***Note:*** *If the extrusion does not require an offset, NX 8 allows you to click and drag the dashed direction arrow to the specified extrusion value, both in the positive and negative direction. This is an easy way to create a quick extrusion. Changing your view to trimetric or isometric may make this easier to see.*

## ***Detail: Extruding With an Offset***

In the last example, you learned that the positive extrusion direction is indicated by a dashed arrow. NX 8.5 also lets you define the extrusion direction. By using the offset parameters in the Extrude dialog box, you can create a solid with sidewalls. Taking advantage of this feature often allows you to make simpler sketches.

## *Project 6.2.2*

- Undo, by using ***Edit / Undo*** (or delete) the Extrude feature that you performed in the last project.
- Create another 'Extrude' feature. This time, click on the 'Offset' dropdown and select **Two-Sided.** Then, enter **0.5** for the **Start** and **0.75** for the **End** as shown below. The figure on the following page shows the offset arrow pointing outward. If your offset arrow is pointing in the opposite direction, you should enter -0.5 and -0.75.

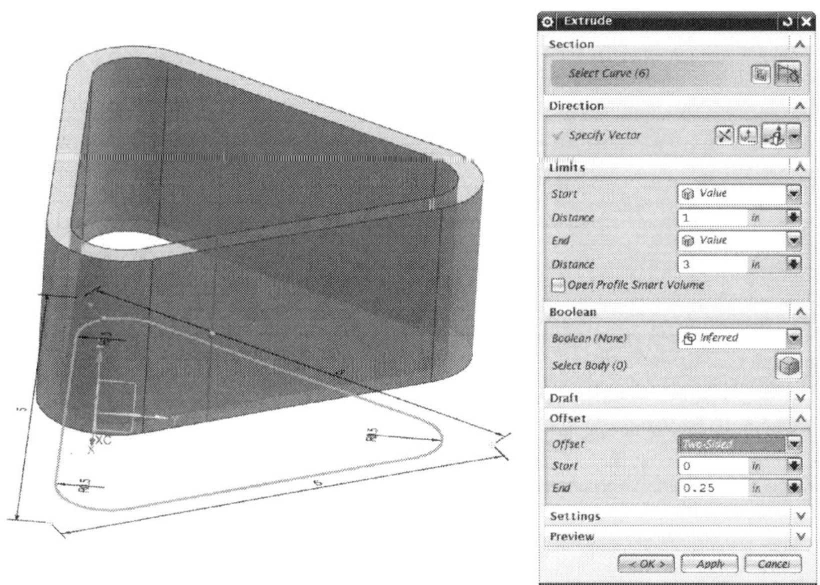


***End of Project***

As seen from a top view, this model will look like the figure below. The inner surface of the solid is offset **0.5in** from the section string in the direction of the offset arrow. The outer surface of the solid is **0.75in** from the section string. This gives a total solid thickness of **.25in**. *You can enter these offset values in either order – i.e. 0.75 could have been entered for the first offset or for the second offset, and the results would have been the same.*

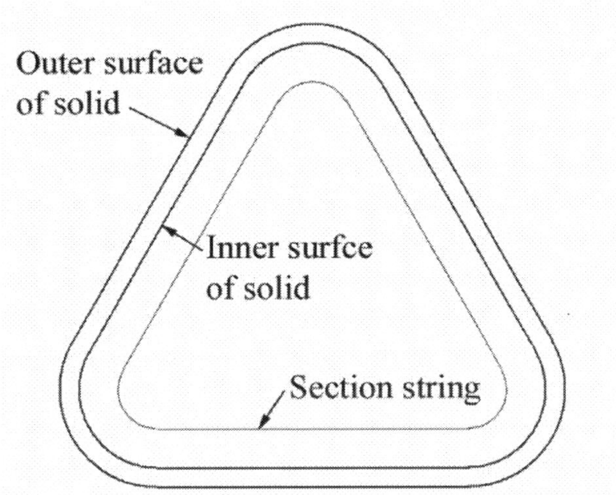


## *Project 6.2.3*

- **Undo** (or delete) the 'Extrude' feature that you created in the last project.
- Create another extrusion, and make one of the offsets **0** and the other offset **1.25**. As seen from the top, your model should look like the following figure.

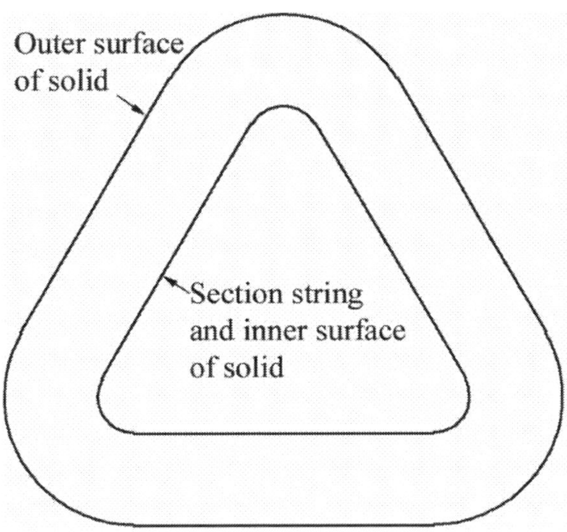

In this case, the inner surface is coincident with the section string because an offset of 0 was entered. The outer surface is 1.25in from the section string.

***End of Project***

## ***Detail: Extruding with a Draft***

You can also specify a draft angle that is applied to the extruded feature. The sidewalls of all 'Extrude' features are assigned a draft angle upon creation. If this is not zero, the sidewalls will taper in or out by the specified value.

## *Project 6.2.4*

- **Undo** (or delete) the 'Extrude' feature that you created in the last project.
- Extrude the original sketch again. This time, however, extrude from **0** to **1**, leave the offsets **0**, and click on the **Draft** dropdown (**From Start Limit**) and type in a value of **20** degrees. You should see the geometry shown in the following figure.

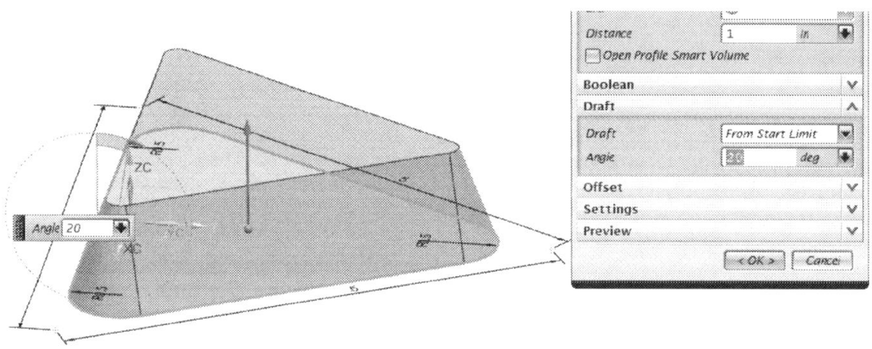


***End of Project***

In this model, the solid faces are angled inward 20 degrees. If a negative value were entered for the draft, the faces would have angled outward.

> **Note:** The top radii at the triangle's corners get smaller or larger when the part is drafted, depending on the angle.

## Detail: Extruding Intersecting Curves

NX makes it possible to select interesting curves and extrude them all at once. This functionality is great for creating features like ribs.

### *Project 6.2.5*

- Create a sketch of the following figure and extrude **20mm** with a 5 degree draft.

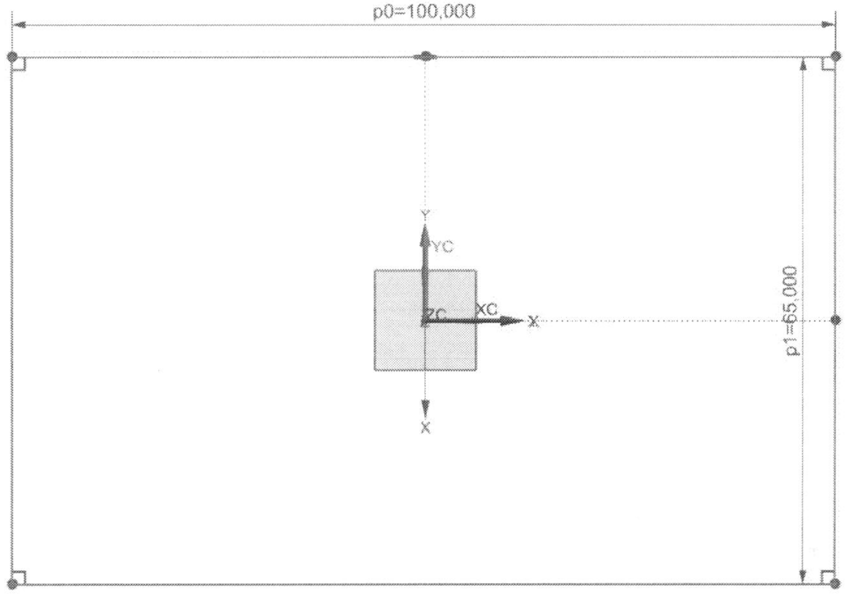


- Use **Insert / Detail Feature / Edge Blend** to apply 5mm blends to all the corners shown in the following figure.
- Use **Insert / Offset Scale / Shell**, select the top face and give the shell a thickness of 2mm.
- Your model should now look like the following figure.

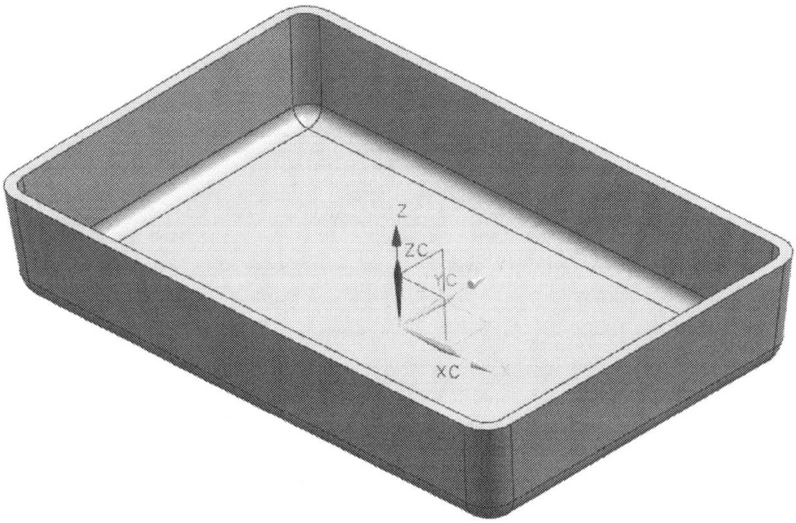


Click on **Insert / Sketch in task environment** and create a sketch on the top face of the part as shown in the following figure. Make sure the lines meet the inside edges of your shell.

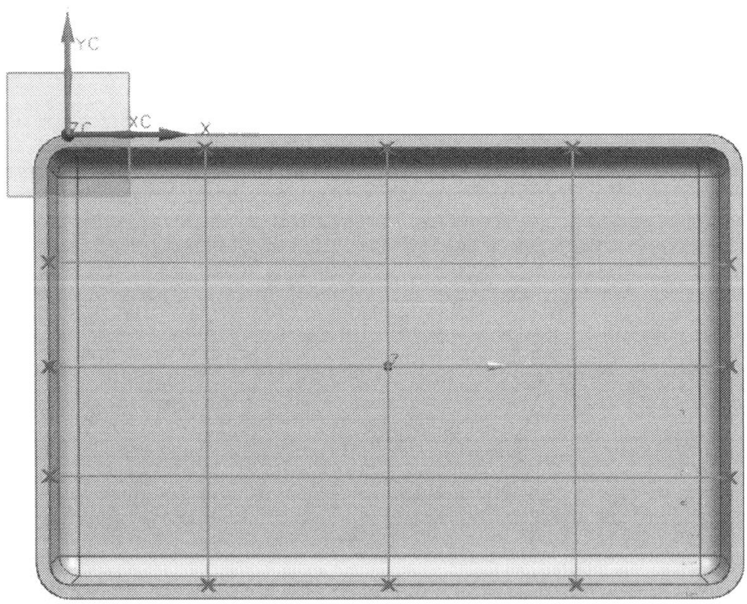


- Now click on the **Extrude** command, and select all the intersecting curves that you created in your sketch and **Extrude**. Use the **parameters** for **offset** and **limits** shown in the extrude window in the following figure. Select **Unite** as your Boolean, then click **OK**.

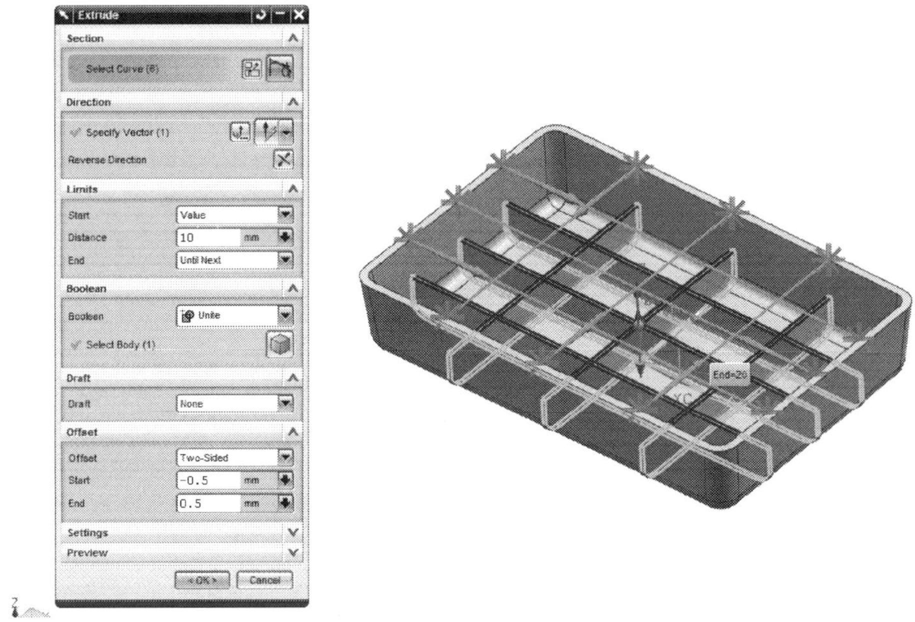


- Your model will now look like the following figure, something that could not have been done previous to NX 8.5 with such ease.

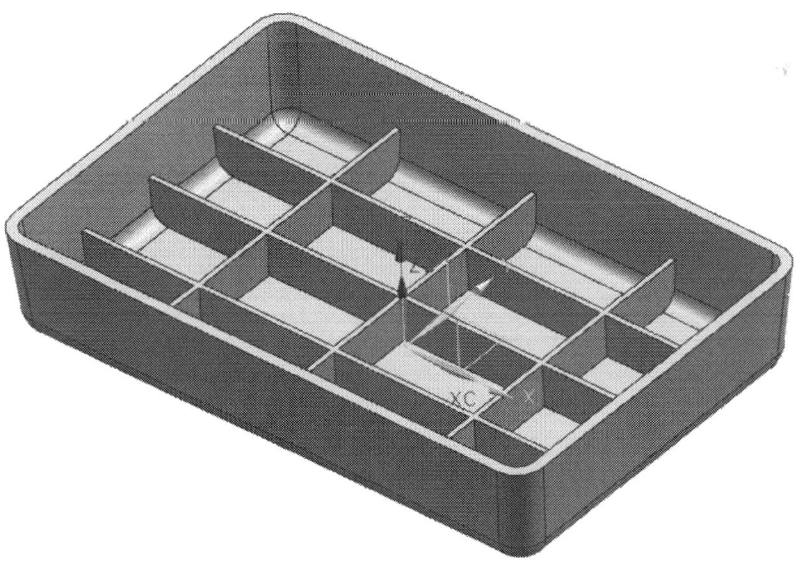

***End of Project***

## 6.3 Text Curves

Text curves allow you to make solid models of actual text. For example, when creating a casting model with a brand name etched onto one side, text curves can be used to extrude letters of the brand name onto the part. But you need to know how to create text before you can extrude it.

### *Detail: Text Curves*

Written text curves can be created quickly using the **Text** command.

### *Project 6.3.1*

- Select Insert / Curve / Text.
- Select text type ***Planar.***
- Enter any text with the font you want in the Text box shown below.

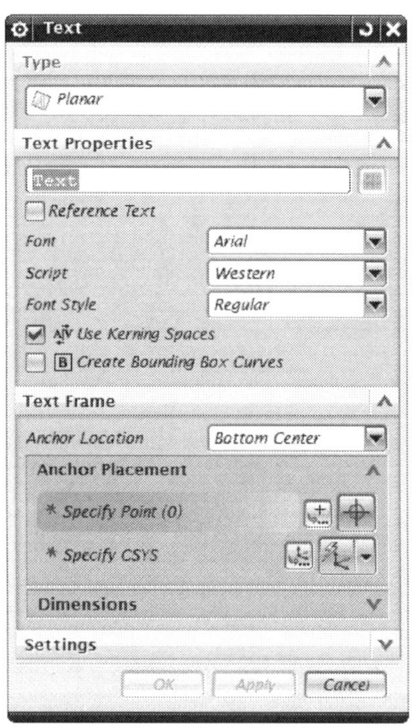

- **Left click** anywhere on the modeling screen to place the text and hit **OK**. Compare your results to the figure above.

***End of Project***

You can also choose to create the text with a bounding box by clicking the 'Create Bounding Box Curves' button shown above.

## ***Detail: Text Curves Using Modifier Handles***

Placement and size of the text can be critical for your design. Text curves can be created very precisely using the dynamic handles.

## *Project 6.3.2*

Follow the instructions in the previous project and place the text on the screen or double click the text to edit it.

- Notice the Dynamic Handles around the text.
- Try to hold and drag the handles one at a time and see how you can stretch and rotate the text with the Dynamic Handles (see below). *Have some fun stretching the text around.*

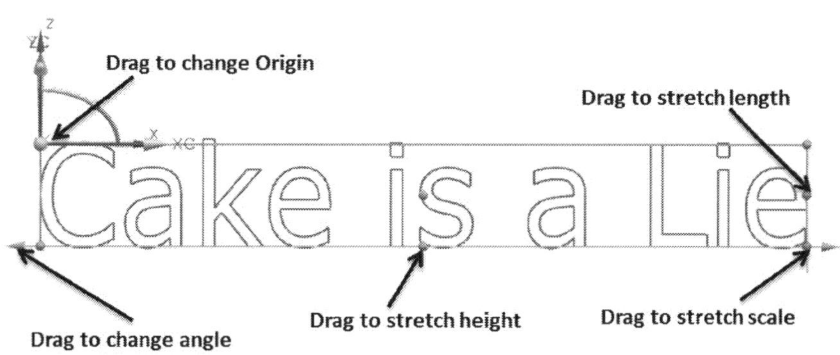


Notice the Dynamic Input Boxes that appear as you drag the Dynamic Handles. It is best practice to use these boxes to get precise results.

- To make the text **2.5in** long, **0.5in** high, select one of the handles at the corner and type the values in the input box as shown below.

- Now, select the rotation handle as shown in the next figure and type the value **15°** in the dynamic input box and hit OK.

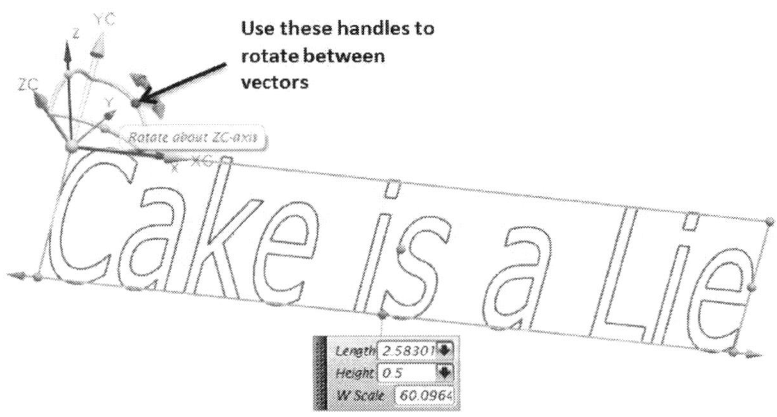


***End of Project***

## ***Detail: Text on Curve***

Text curves can be created using a guide curve that determines the form of text.

## ***Project 6.3.3***

- Create a Cylinder and draw an arc on one end of the cylinder.
- Select Insert / Curve / Text.
- Select text type ***On Curve.***
- Type the text using the ***Natural*** vertical direction as shown below.
- Read the Cue: Select curves for text to follow.
- Select the curve.
- Adjust height and length using either the Dynamic or input boxes (see below).

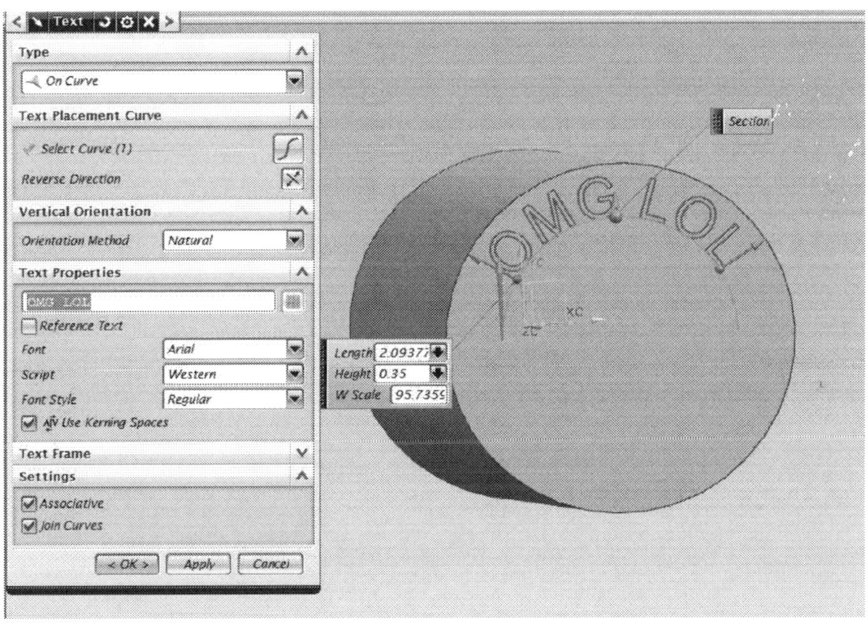


***End of Project***

## ***Detail: Text on Face***

Text curves can also be created on a face using a guide curve.

### ***Project 6.3.4***

- Create a Cylinder.
- Select Insert / Curve / Text.
- Select text type ***On Face.***
- Type the text as shown in the previous figure.
- Read the Cue, select the cylindrical face to place the text, and middle click.
- Select the edge curve and middle click.
- Adjust the text using handles and input boxes as shown in the next figure.

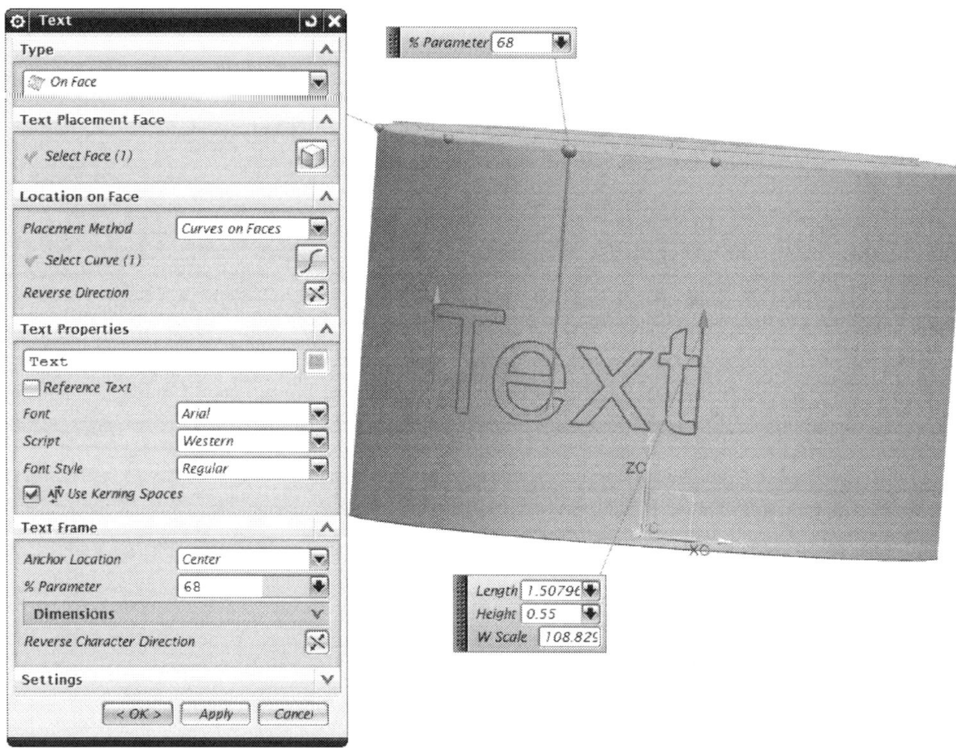


***End of Project***

## 6.4 Revolved Features

Revolving curves about a centerline is another way to create solid geometry. The 'Revolved' feature can be very useful since many common parts are cylindrical in nature. For example, the wheel shown in the figure below could be created very easily by sketching a section and revolving that section about the wheel's center. Just as with the 'Extruded Body' function, you can create either sheet solids or solid bodies by revolving, depending upon whether the section string is open or closed.

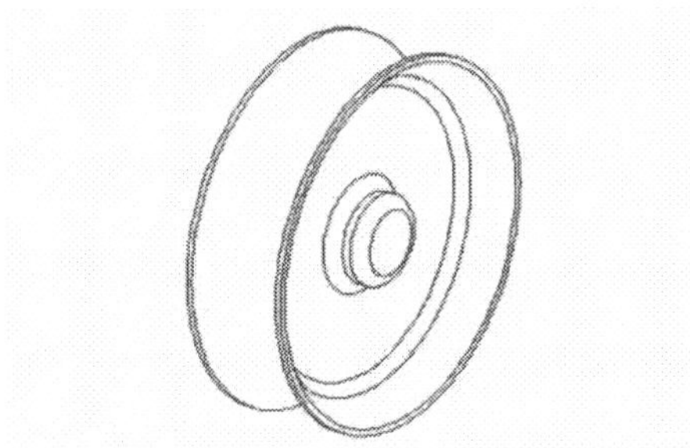

### ***Detail: Revolving About an Axis***

### *Project 6.4.1*

Using the sketcher, create the following shape.

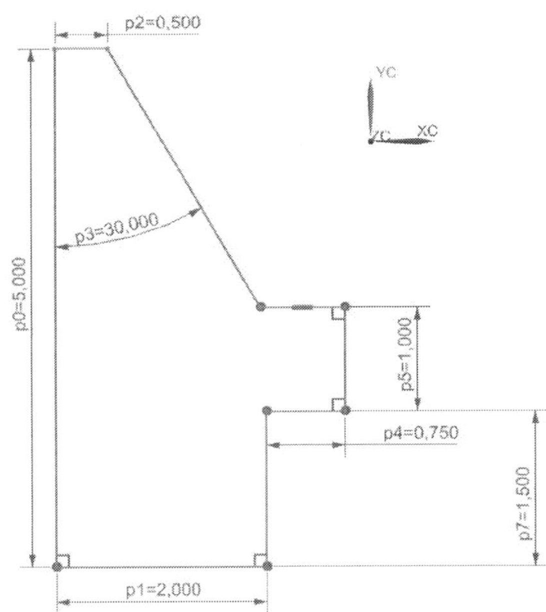


- Select Insert / Design Feature / Revolve or click  .

- Select the sketch as your section geometry and **middle click**.

Read the cue - it is asking to “Select an object to infer the vector.”

- Select the **5in** vertical line shown above.
- Choose **OK** *or* middle click.
- Enter a Start Angle of **0** and an End Angle of **270** in the ‘Revolved Body’ menu, as shown in below.

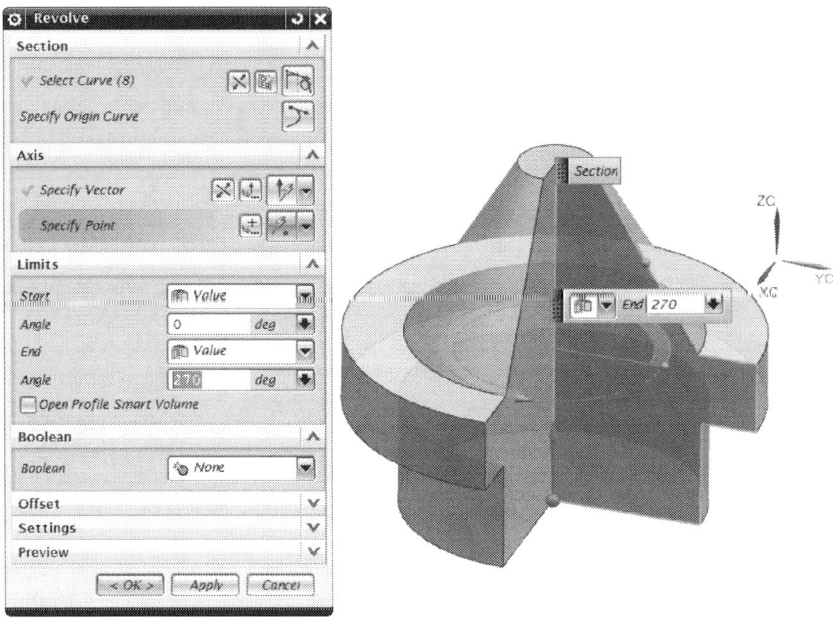


- Choose **OK** to create the geometry.

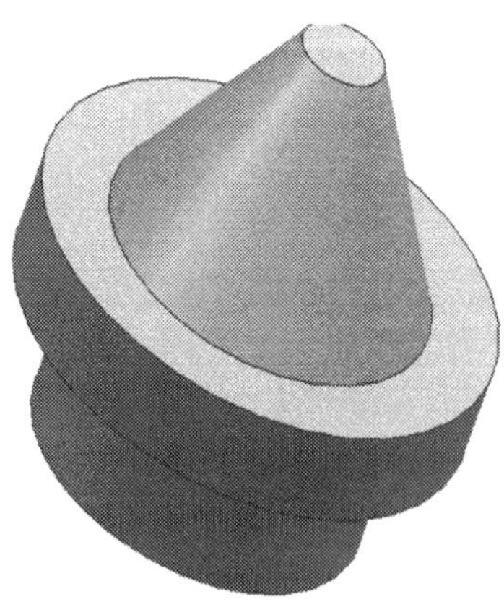

The figure below illustrates the resultant model for End Angle = 360 degrees.

***End of Project***

You can also specify a different string or an axis that is not part of the section string, as shown in the following figure.

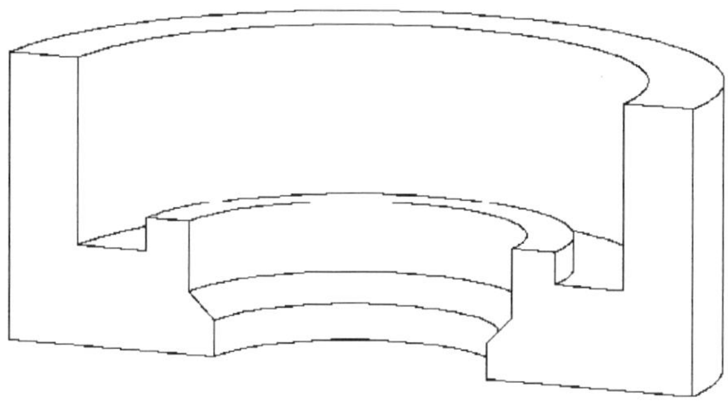

The sketch used to create the model shown in the figure above contains a reference line at a distance from the section string (see below). The reference line is used to specify the vector for the revolution axis.

> **Note:** The Reference Line is a simple line drawn in the sketcher. NX 8.5 does not differentiate between reference lines and sketch boundary lines in the sketcher. **Right click** on a line and select "Convert to Reference" to, well, convert it to reference.

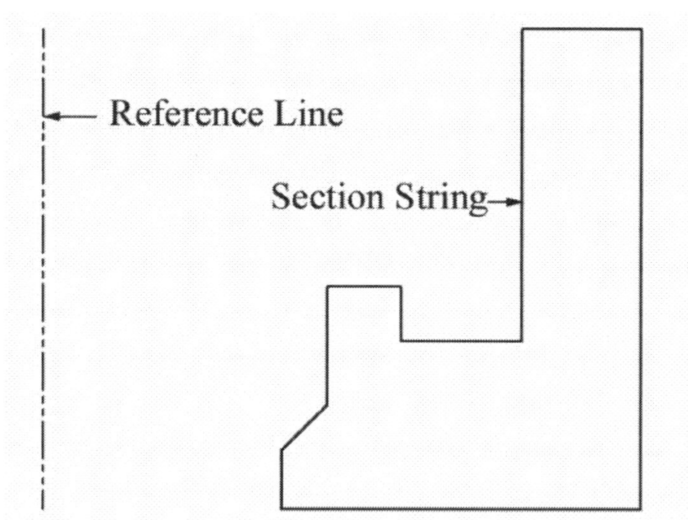

## 6.5 Primitive Features

To create elementary solid shapes, you may define 'Primitives.' The four primitive icons are found in the middle of the Feature menu. They are used to create solid blocks, cylinders, cones, or spheres.

The size of a primitive may be changed parametrically; however, changing its position is ***not*** parametric. *Because of this, some users avoid primitives completely.*

Primitive features are placed and sized according to the location of the WCS.

### *Project 6.5.1*

In this project, you will create two blocks, both 2in by 3in by 6in in size.

- Select Insert / Design Feature / Block  .
- Select **Origin and Edge Lengths** from the 'Block' menu shown below.

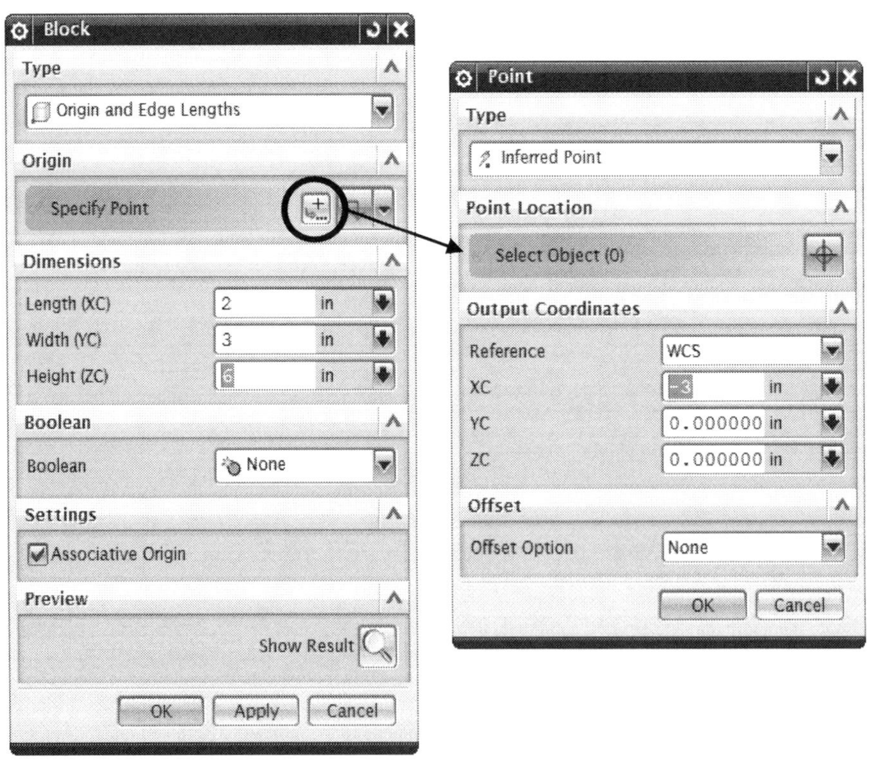

- Using the point constructor menu (from the 'Snap Point' toolbar), input an origin of (**-3, 0, 0**).
- Choose **OK**.
- Enter an X length of **2**, a Y length of **3**, and a Z length of **6** in the box previously shown.
- Choose **OK**.
- Select Insert / Design Feature / Block.
- Select **Two Points and Height** from the Block menu.

*Remember to read the cue. Point 1 will be one of the diagonal points on the base of the block.*

- Using the point constructor menu, enter point 1 coordinates of (**3, 0, 0**).
- Choose **OK**.

Notice the Selection Step icon is now highlighting Point 2 – NX 8.5 is asking for the second diagonal point.

- Using the point constructor menu, Enter point 2 coordinates of (**5, 3, 0**).
- Choose **OK**.
- Input a height of **6.**
- Choose **OK**.
- Your model should resemble the figure below.

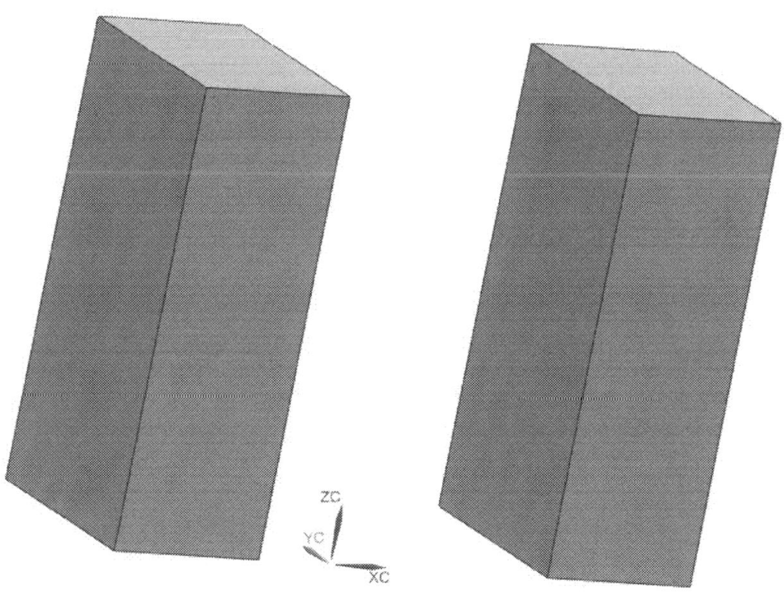


| **Note:** By Default, NX 8.5 will always create a new solid body, but you can choose a Boolean operation such as Unite, Subtract or Intersect depending on your project need. This is located at the bottom right corner of the block menu, as previously shown. |
|---|

***End of Project***

## *Project 6.5.2*

In this project, you will create a cylinder of diameter 0.88in and a height of 2.5in.

- Select Insert / Design Feature / Cylinder 
- Select Axis, Diameter and Height from the Cylinder menu.
- Select the **+ZC** axis. *This defines the center axis of the cylinder.*

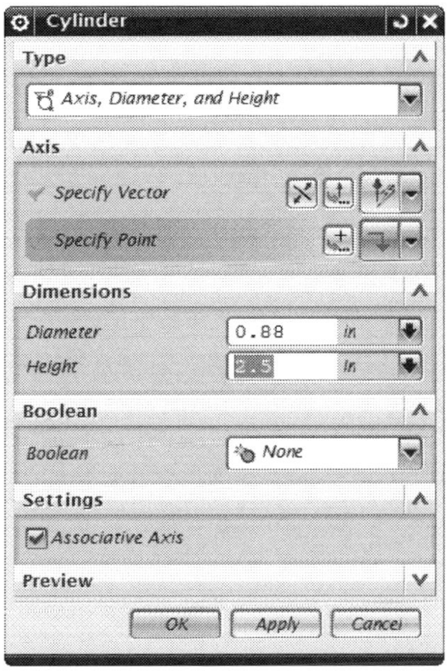


- Enter a diameter of **0.88in** and a height of **2.5in.**
- Choose **OK** if the default base point is (**0, 0, 0**). *If not, click **Reset** and then choose **OK**.*
- Compare your results to the cylinder shown in the following figure.
- Choose **Cancel**. If your part file contains solid geometry, NX 8 will display the Boolean menu. If this is the case, select **Create** before canceling.

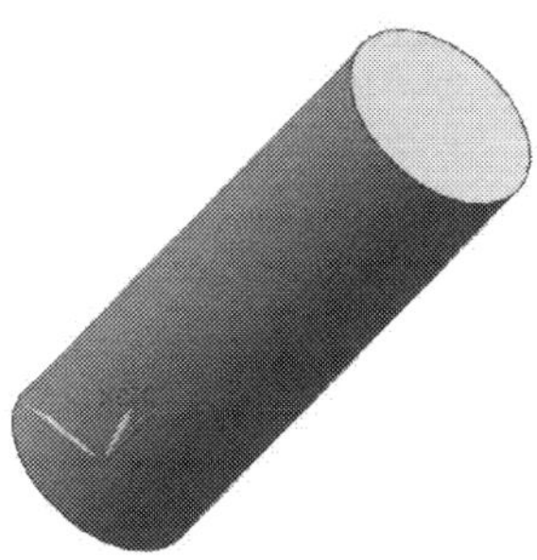

***End of Project***

## *Project 6.5.3*

You can also create the same cylinder using an existing circle.

- Using basic curves (***Insert / Curve / Circle***), create a circle of diameter **0.88in** with its center at (**0, 3, 0**).
- Select **Insert / Design Feature / Cylinder**.
- Select **Arc and Height** in the drop down menu.
- Input a height of **2.5in.**
- Pick the circle located at (0, 3, 0). An arrow will appear to indicate the direction in which the cylinder will be created.
- Choose **OK**.

***End of Project***

> ***Note:*** *The circle that generates the cylinder can be created either with basic curves or with a sketch*, but the cylinder *will not be associated to the sketch geometry. In other words, if you create a primitive cylinder from a sketch and then change the sketch later, the primitive cylinder* **will not** *update along with the sketch. To accomplish this you should use the Extruded Body function, not the Cylinder function.*

## 6.6 Practice: The Pool Rack

Building the solid geometry of a basic pool rack shown below requires two extruded body features. Both can be defined using a single sketch, and will then be united together. This model will be used in a later chapter, so please remember to save the part file as **'pool_rack_<*your initials*>.prt'.**

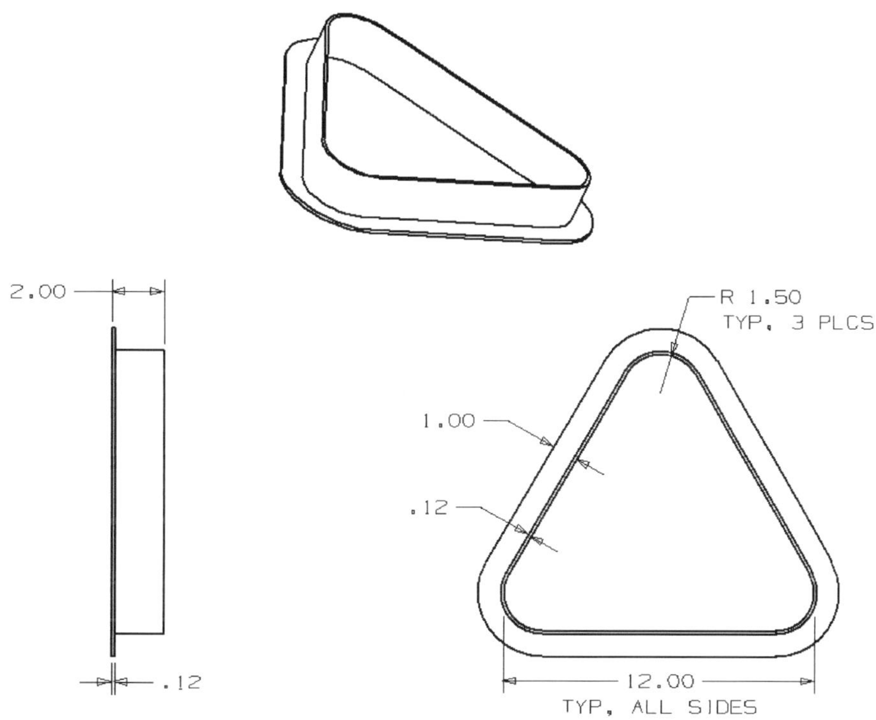

## *Practice Outline*

- Sketch a triangle as shown below. *If you'd like, you can use the sketch made in **Detail: Simple Extrusion** (see the beginning of this chapter), and simply modify the dimensions to match the diagram.*

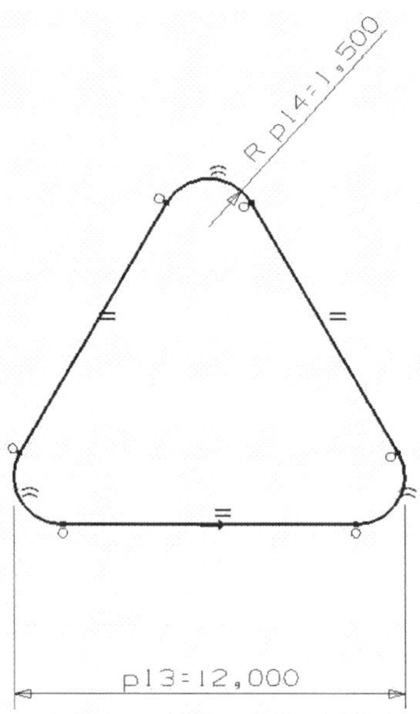


  - Insert a **2in** thick extrusion with an outward offset of **0.12in**.
  - Insert a **0.12in** thick extrusion with an outward offset of **1** with the 'Unite' Boolean operation selected.

Your part should look like this.

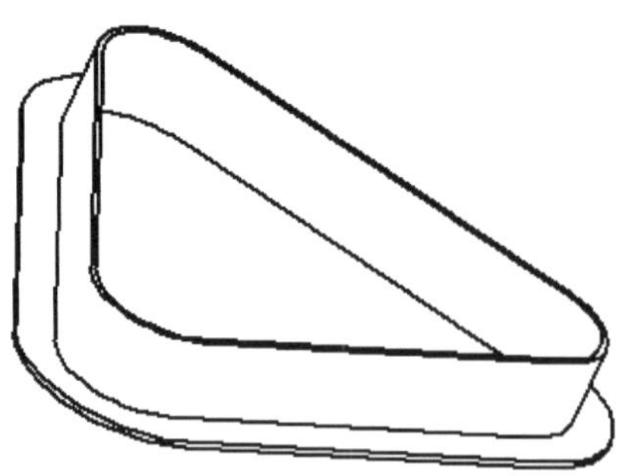

***End of Practice***

## 6.7 Advanced Design Features

'Advanced Design Features' include features such as holes, bosses, pockets, pads, slots, and grooves. They are a great way to add detail to your solid. The base geometry for this model is simply an extruded rectangle. All additional detail in this model is made by using 'Design Features.'

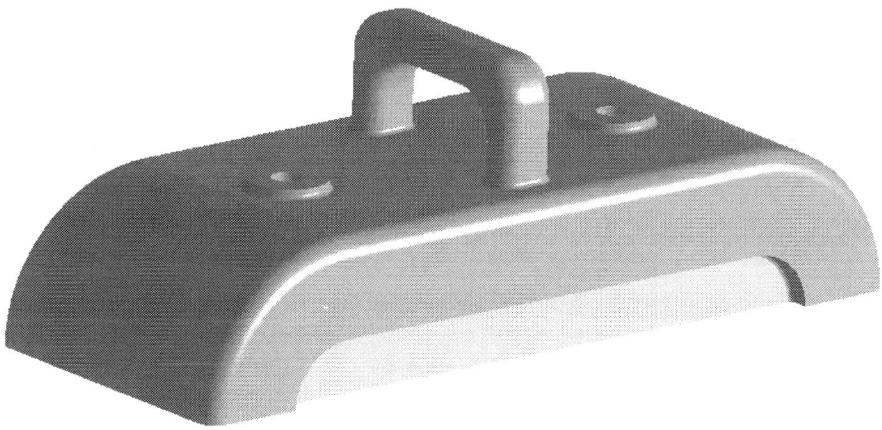

### *Detail: Hole Features*

Although you can create a hole simply by subtracting a cylinder from a solid, using the 'Hole' feature gives you many more options and is much easier to use. You can use the 'Hole' feature to create counterbored, countersunk, threaded, and clearance holes, and NX provides all the information you need. In the following exercise we are going to look at the great functionality of the 'Hole' feature.

## *Project 6.7.1*

- Let's start with a **block** that is **100mm x 80mm x 10mm**, as specified in the following menu.

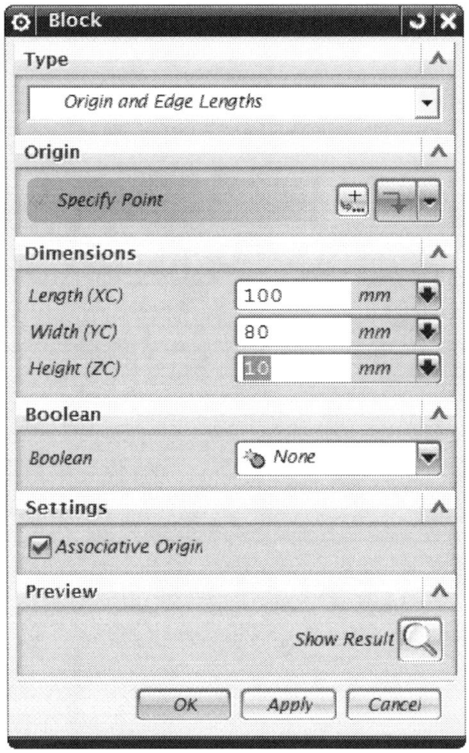


- Next, select the 'Hole' feature through **Insert / Design Feature / Hole**. Under **Type** you have the option of creating a General hole, Drill size hole, screw clearance hole, threaded hole or hole series. Under **Form and Dimensions** you can define whether the hole is a simple hole, a counter-bored hole or a counter sunk hole. It's all very self-explanatory; however, we will go through several of the different types in the following exercise.

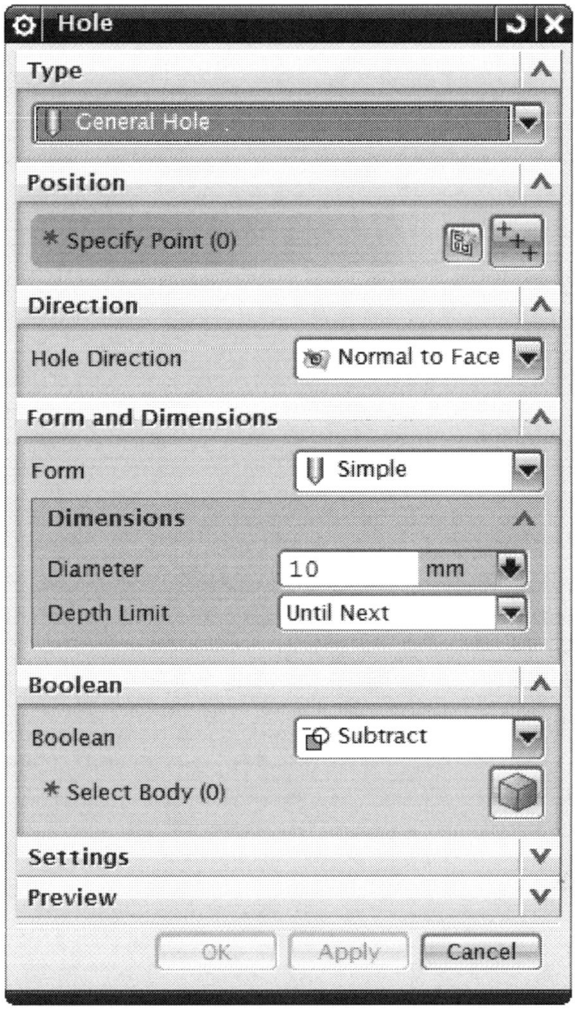


- First, select the **general hole** with a diameter of **10mm**.
- Select the top face of the block that you previously created and you will see the new way to position the hole. As soon as you pick the face, it throws you into the sketcher. It also automatically opens the Point feature, which you use to position your hole, as shown in the following figure.

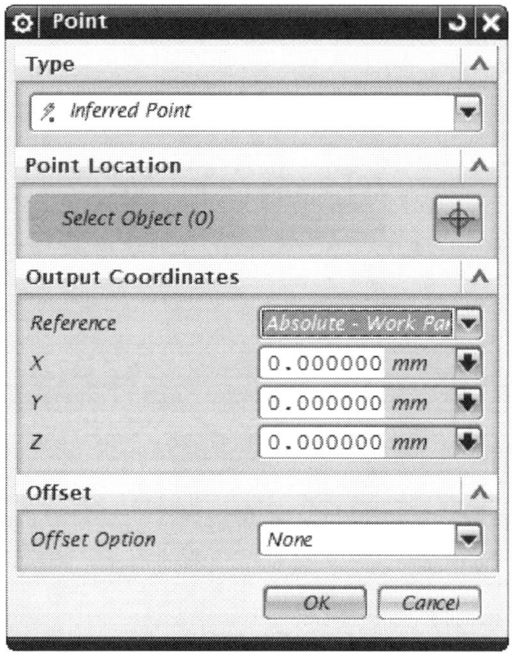


- Click anywhere on the face, then hit cancel to stop creating points. *If you hit **OK,** you will create another point on top of the one you just made.*
- Then position your hole using the dimension tool in the sketcher, as shown below.
- The point is positioned 30mm from both edges.

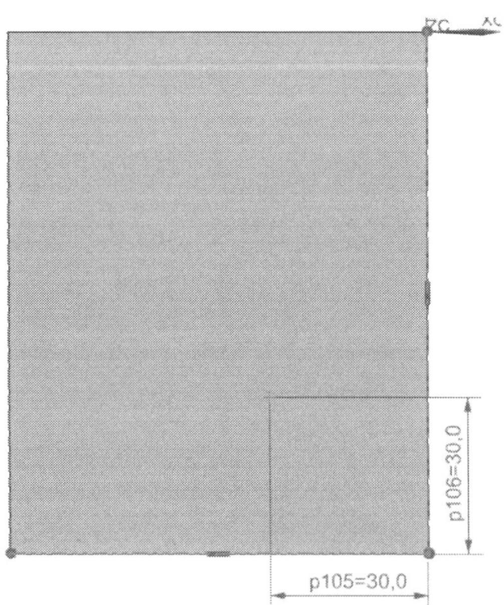


> ***Note:*** *You can create more than one point at a time, making it possible to create multiple holes off of one hole feature using the **feature points** selection method.*

- Once you have positioned your point, click **Finish Sketch**. This will take you back to specifying your hole.

- You will now be looking at a preview of your hole, as shown in the following figure. *If you can't see this, be sure that the preview button is checked at the bottom of the 'Hole' feature window.*

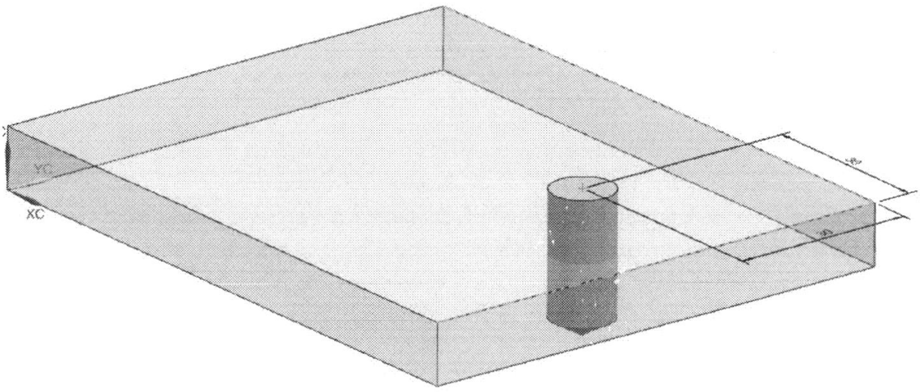


Next we will specify the depth. There are several amazing options here. For this exercise, use **Through Body**. *This way if you change the size of your block the hole will update to always go through it.*

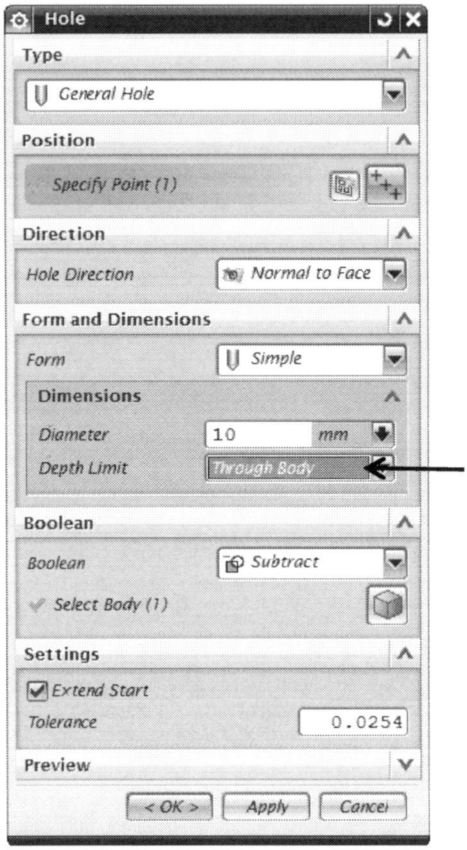


- Click **OK** to create your hole.

Now we will create a threaded hole. Using this hole feature can save you loads of time!

- Select **Insert / Design Feature / Hole**.
- The steps are just the same as before, so select the **face** you want your hole to be on.
- Once you're in the sketcher, place a point on the face and position it wherever you want.
- Click on **Finish Sketch**.
- Now specify that you want a **Threaded Hole** with an **M6 thread size** and a depth limit of **Through Body**, as shown in the following figure.

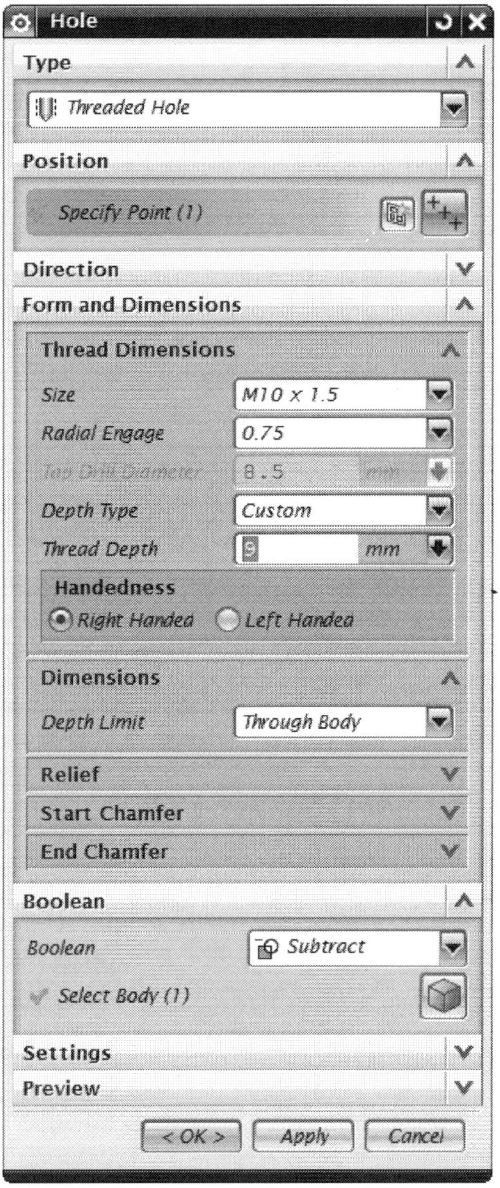


- *Notice all the other specifications that are possible for making holes.*

- Click **OK** to create the threaded hole, as shown in the following figure. *You will notice that the threaded hole is only symbolic. Not showing the threads saves memory and loading time.*

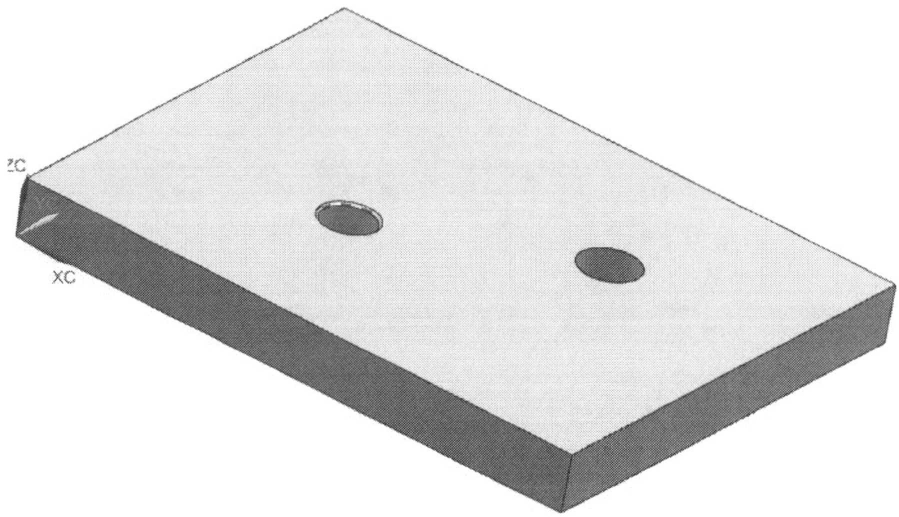


Finally let's quickly create a countersunk screw clearance hole.

- You should be great at this by now, but again select **Insert / Design Feature / Hole**.
- In the sketcher, place a point of the face.
- Choose **screw clearance hole** as the type. Make it a **countersunk M10** hole, with a **through body** depth limit, as shown in the following figure.

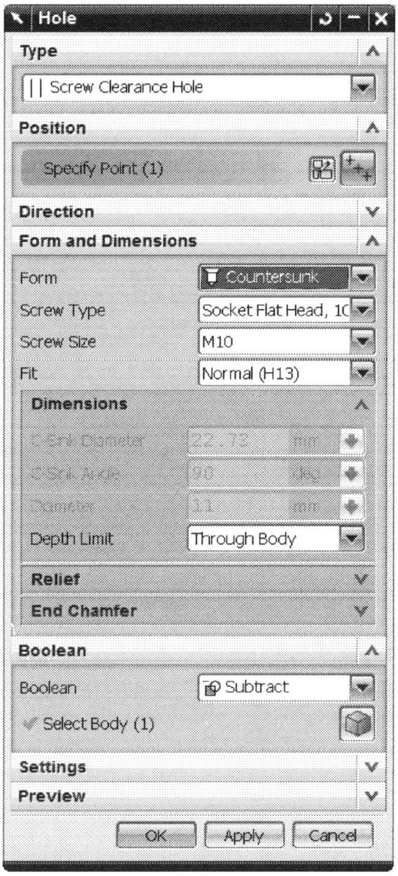


- Click **OK** to finish. Now you have a screw clearance hole!

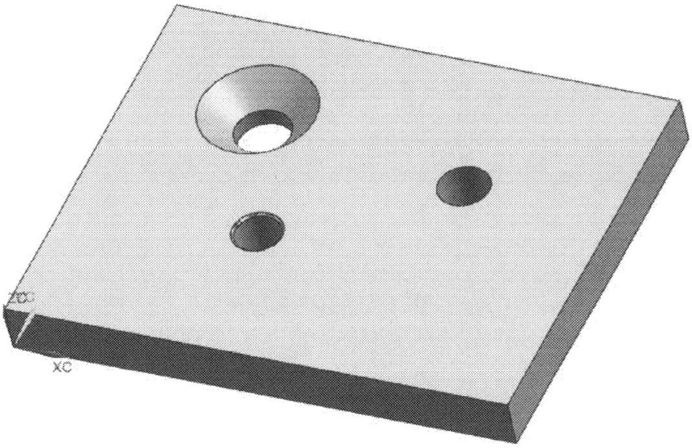

***End of Project***

## ***Detail: Tapered Hole***

The tapered hole is really easy to use, but let's go over it quickly anyway.

## ***Project 6.7.2***

- Start with a simple block.
- Create a 'Hole' feature on the top surface.
- The type of hole has to be **General Hole**.
- Under **Form and Dimension** you can set the form to **Tapered**.

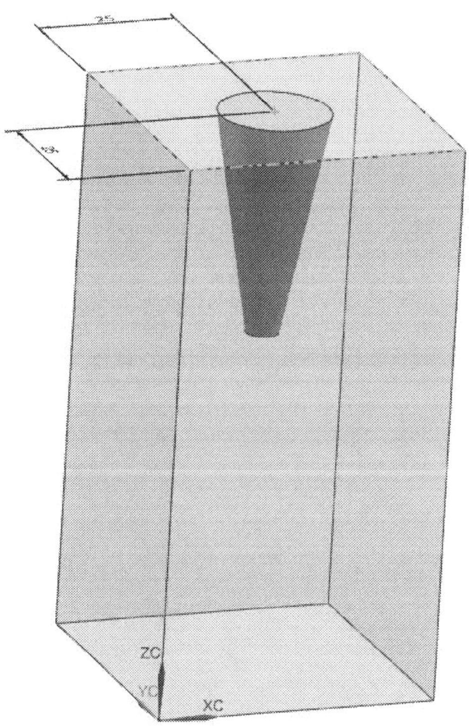


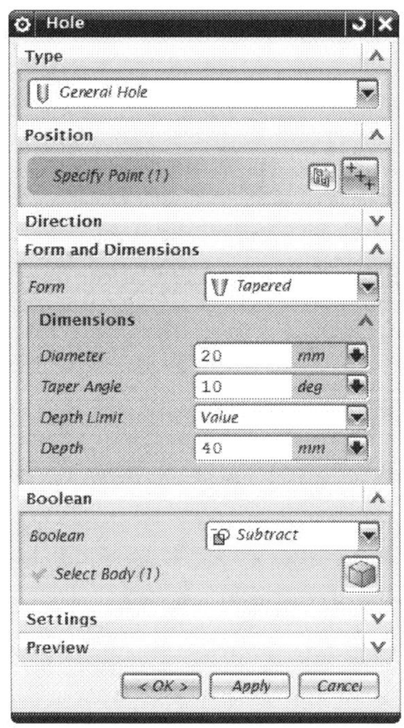


This is very useful for creating holes for self-tapping screws.

***End of Project***

## ***Detail: Hole Depth Limit***

Depth limits are a great parametric feature used to control your Booleans. Just to get you well-adjusted to the differences between all the depth limit options, we will go through them now.

## ***Project 6.7.3***

- Make a simple block.
- Shell one face of it as shown in the following figure.
- Then create a hole on the top surface with the **depth limit** set to **Value**, also shown in the following figure.

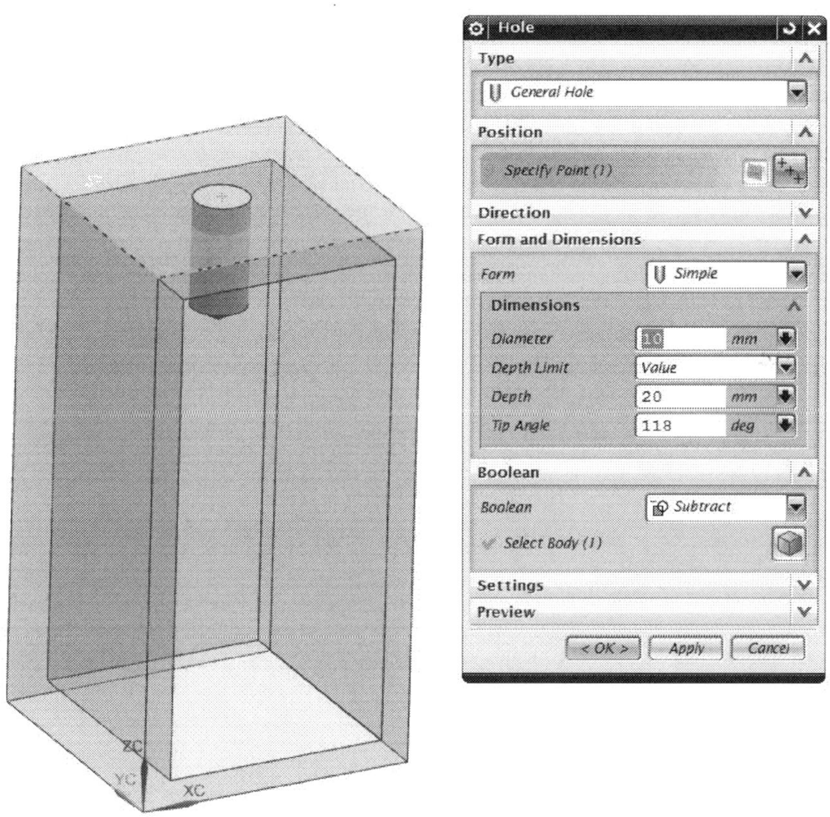


- Now change the depth limit to Until Selected. Select the inner face of the shell and then you will see that the hole goes to that face as in the following figure.

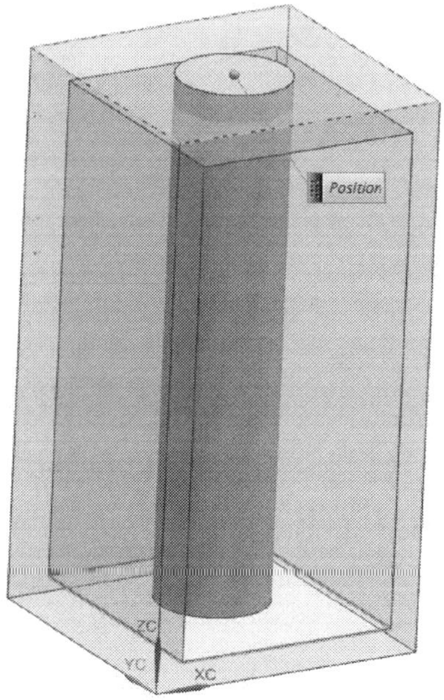


- The **Through Body** depth limit will go through the entire body as shown in the following figure.

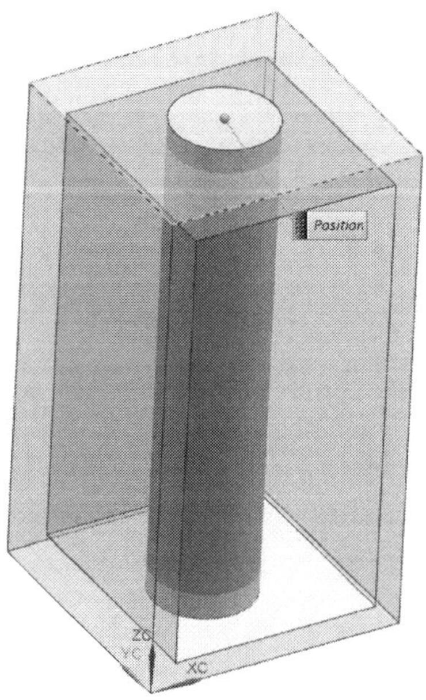


- Finally, try **Until Next**. This will extend the hole to the next face it reaches, as shown in the following figure.

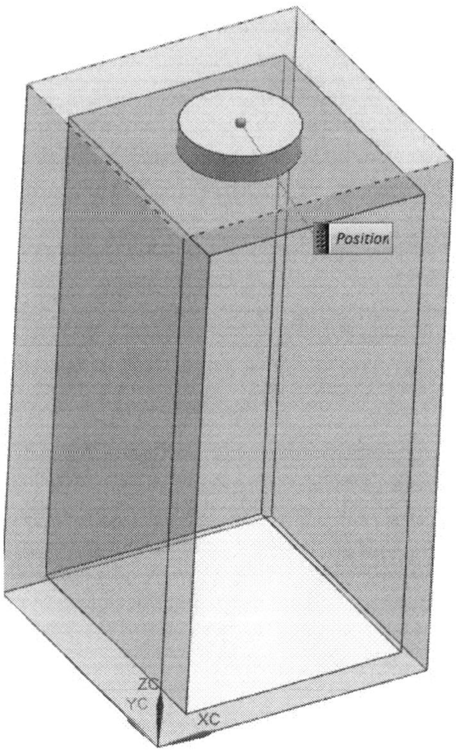


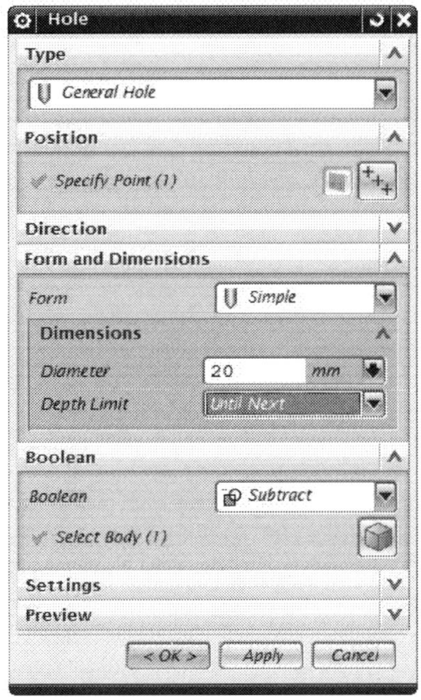


- It's not exactly exciting, but it sure is handy!

***End of Project***

## ***Detail: Rectangular Pocket Features*** 

'Rectangular Pockets' are an easy way to create a rectangular-shaped cut in a solid body (shown below). Like holes, they require a placement plane. The **slot** and **pad** features are very similar to the **pocket** feature – if you understand how to use one, you should be able to employ them all.

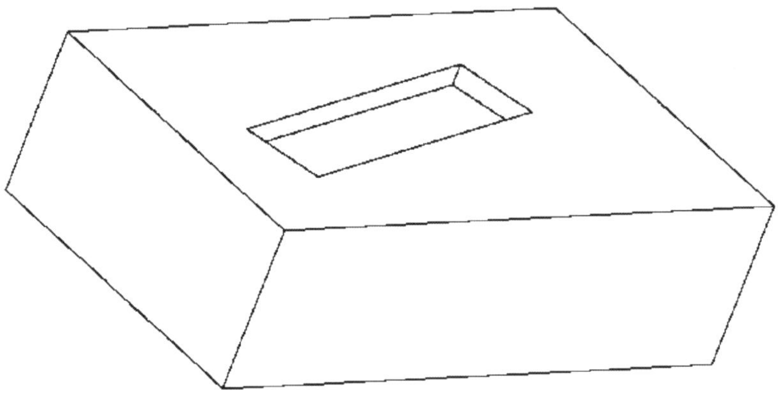

### ***Project 6.7.4***

Create a '**Rectangular Pocket'** feature.

Create a block with dimensions **4in** x **7in** x **2in** as shown below.

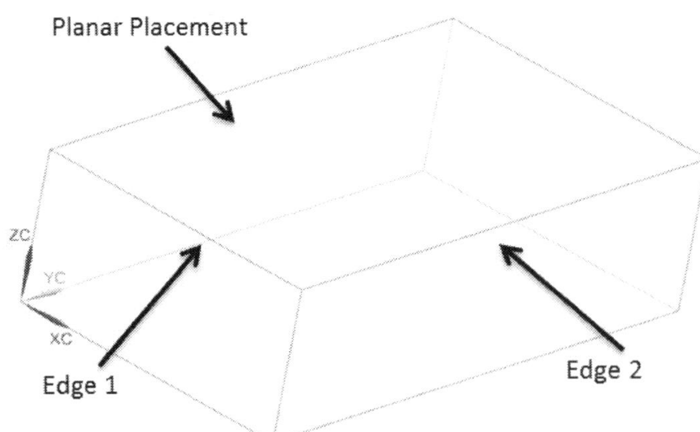


- Select Insert / Design Feature / Pocket.

- Select Rectangular.
- Pick the planar placement face as shown above.
- Pick Edge 1 as the vertical reference.
- Enter the pocket parameters as shown next. The X Length is in the direction of the horizontal reference, and the Y Length is perpendicular to the X Length (in the plane of the placement face). The Z Length is the pocket depth.

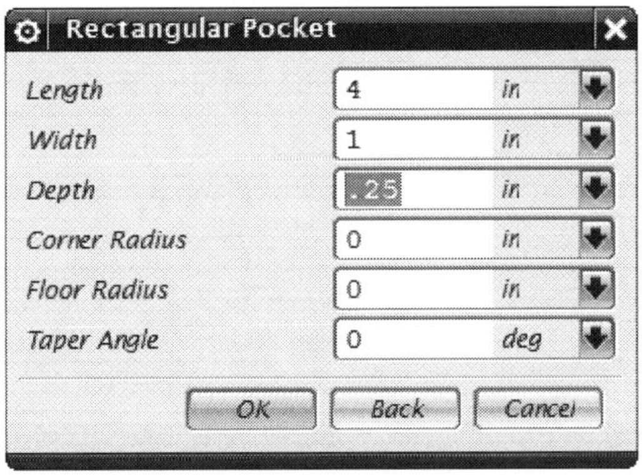


- Choose **OK**. Compare your results to the figure below. The pocket is shown at its initial placement location, and the WCS is shown in a temporary orientation, based upon the horizontal reference selection.

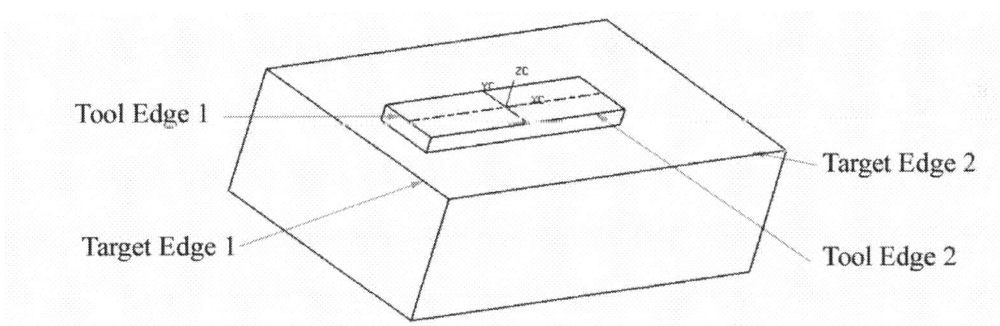


- Select ***Perpendicular*** from the 'Positioning' menu.
- Pick Target Edge 1.
- Pick the –Y half of Tool Edge 1. When positioning pockets, you may specify if the dimensions are relative to the edges or to the centerlines.
- Enter a dimension value of **1.5** and click **OK**.
- Select ***Perpendicular*** from the Positioning menu.
- Pick Target Edge 2.
- Pick the –X half of Tool Edge 2.

- Enter a dimension value of **1.0** and click **OK**.
- Select the ***Angular*** icon from the 'Positioning Method' menu.
- Pick the +X half of Target Edge 2. This is the right half of the Target edge 2 as shown in Figure 38.
- Pick the +X half of Tool Edge 2.
- Enter a dimension value of **20.0** and click **OK**.
- Choose **Cancel**. Compare your results to the figure below (Note: your part may not have dimensions).

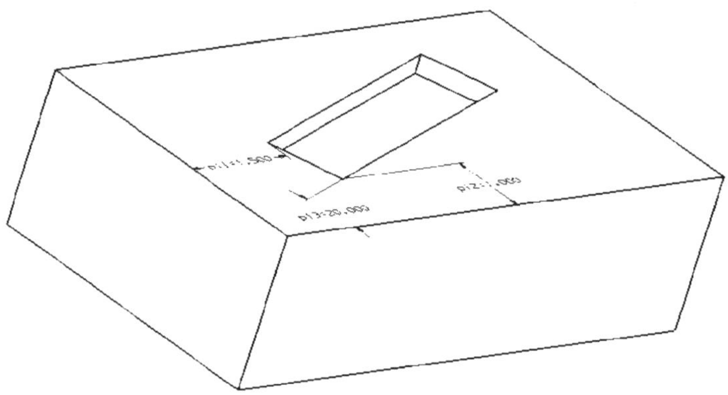

***End of Project***

While creating angular dimensions, it is important to pay attention to which end of the target and tool edges are being selected. Angles are constructed by two vectors whose directions are defined by the "positive" direction of the selected edges. If you find that you often wind up with dimensions that don't seem to be placed sensibly, re-try the angular dimension while paying close attention to how you are picking the edges.

Another common mistake is canceling from the Positioning menu. Selecting Cancel in this menu at any time will throw you completely out of the 'Form Feature' menus. Select **OK** when you are satisfied with the position of a form feature and the feature will be inserted with the dimensions you have applied.

> ***Note:*** *The* **Insert / Design Feature / Pad** *command is very similar to the 'Pocket' feature, but rather than removing material, it adds it. To create a 'Pad', follow the instructions provided for the 'Slot,' including selection of a 'Horizontal Reference.' A 'Corner Radius' and 'Taper Angle' may also be specified for a 'Rectangular Pad.'*

## ***<u>Detail: Groove and Boss Features</u>***

Grooves are an automated way to subtract an annulus (see below) of material from a solid body. They may be placed only on cylindrical or conical faces.

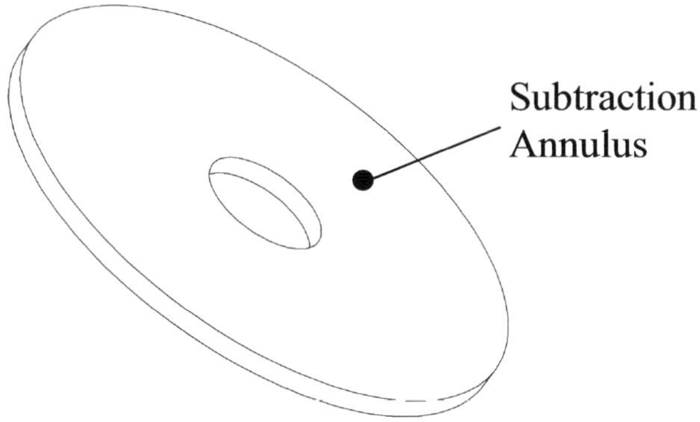


A typical application of a **Groove** feature is to place it on the cylindrical face of a **Boss**. The figure below shows the finished model for this project. Recall that a Boss feature may be constructed similarly to a **Hole** feature.

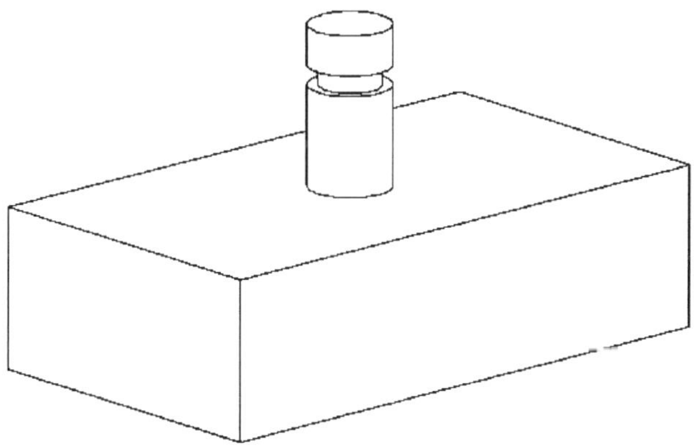

## *Project 6.7.5*

- Create another **4in** x **7in** x **2in** block.
- Create a boss as shown above using ***Insert / Design Feature / Boss*** .

- Use the creation parameters shown in the following figure. The boss should be positioned at the center of the block.

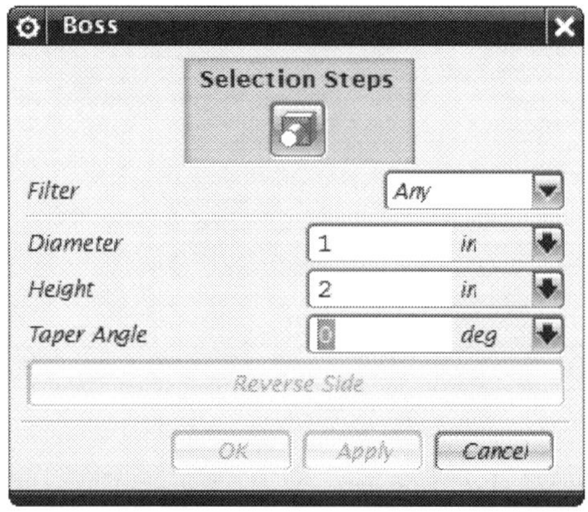


- Select Insert / Design Feature / Groove .
- Select Rectangular.
- Pick the cylindrical face of the boss. *Don't pick the planar end face!*
- In the menu shown below, input a Groove Diameter of **.75** and a Width of **.25**. The groove diameter is the I.D. of the subtraction annulus. The width is the thickness of this annulus.

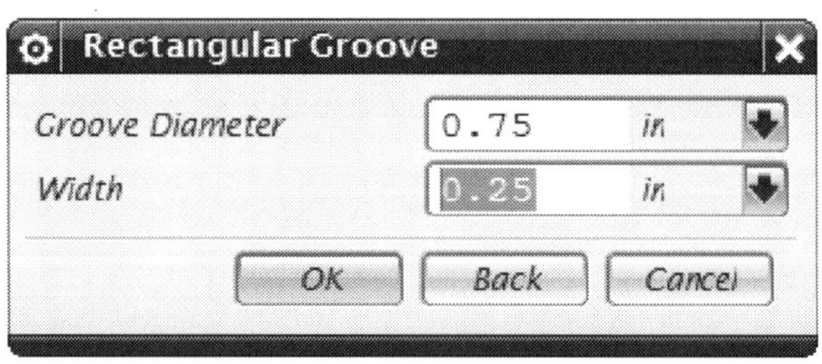


- Choose **OK**. The subtraction annulus will appear as shown below.

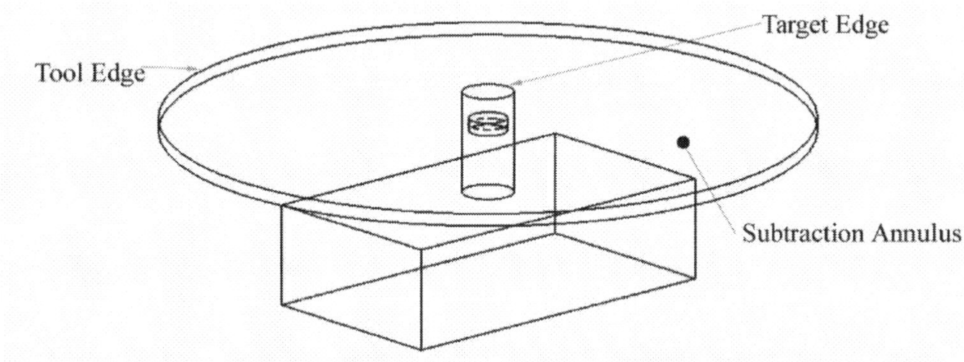


- Pick the Target Edge as shown.
- Pick the Tool Edge as shown. The only meaningful dimension here is along the axis of the cylinder. Therefore, the Positioning menu is not displayed, and you are instead prompted for a target and tool edge.
- Enter a dimension value of **0.5** and click **OK**.

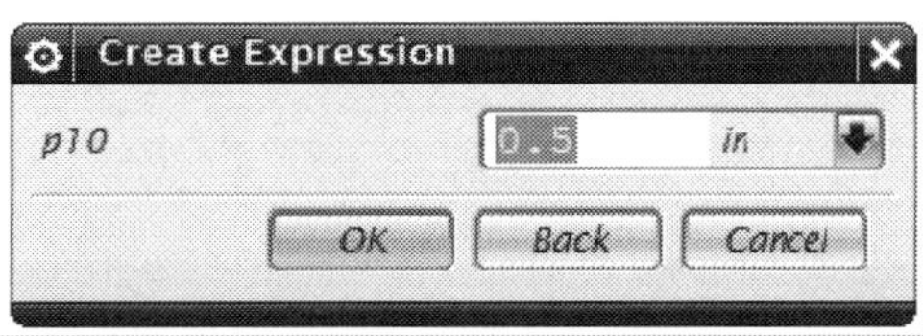


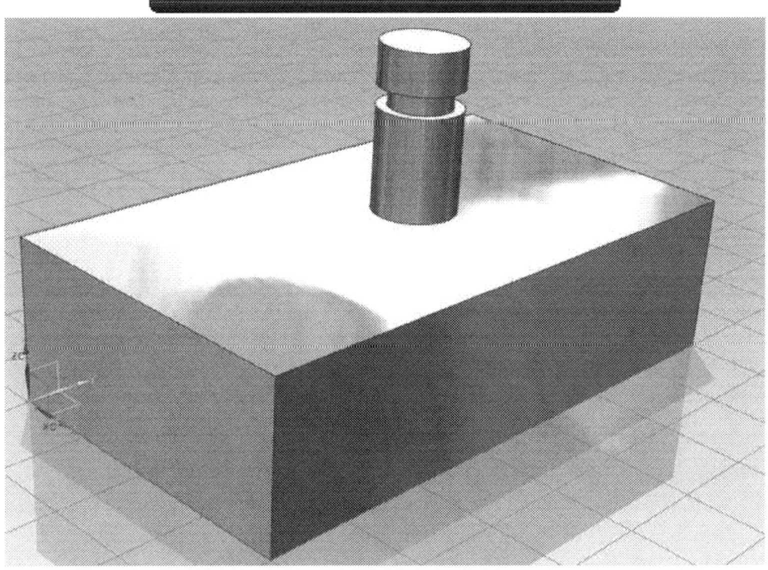

- Choose **Cancel**.

***End of Project***

## 6.8 Practice: The Hinge

The following figure shows the geometry of a basic hinge. Building the solid model will require knowledge of the extruded body and hole features. This model will also be used in a later chapter, so please remember to save both part files as '**hinge1_<*your initials*>.prt**' and '**hinge2<*your initials*>.*prt***'.

### *Practice Outline*

Open a new part in NX 8.5. Make sure you select 'Inches' as your Units.

- Sketch a rectangle and a circle as shown below. Constrain and apply dimensions to make your sketch identical to the diagram.

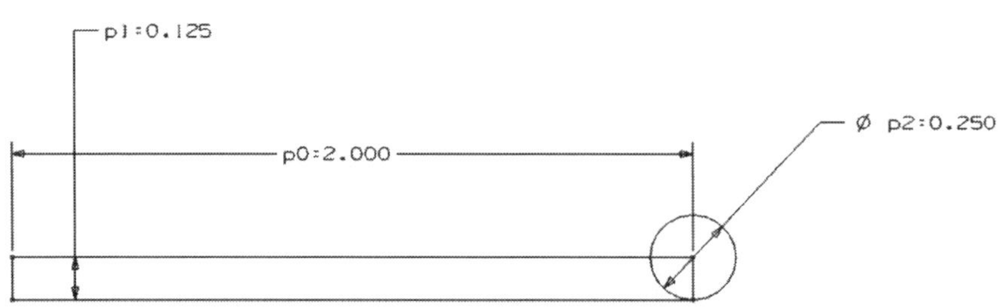


- Select Insert / Design Feature / Extrude.
- Choose **Connected Curves** from the 'Selection Intent' pull-down menu.
- Select the rectangle and extrude **0** to **5.** Select **OK.**
- Select **Extrude**. *Note: Be sure* **Single Curve** *is chosen for Selection Intent.*
- Select the circle and extrude it **0** to **5.**
- Select **Unite** in the Boolean scroll down menu. Select **OK.**

- Select **Hole / Form and Dimensions / Counterbore** and place three thru holes as shown next. Enter counterbore diameter **0.5in**, depth **0.05in** and hole diameter **0.25in**. Position two holes **0.5in** from both edges and the third hole **2.5in** from the smaller edge and **1.3in** from the outer edge.
- Select **Hole /Form and Dimensions/ Simple**. Enter diameter = **0.125in**. Select one of the extruded faces as the placement face and the opposite as the thru face.
- Select **Apply**.
- Apply a **Point on Point** geometric constraint to align the hole to the cylinder (**arc center**). Hit **Cancel**.
- Select ***File / Save*** using the original name (hinge1_*<your initials>*).
- Also, Select ***File / Save*** and then rename using hinge2_*<your initials>.*

*For the following steps, utilize the original sketch.*

- Open hinge1_*<your initials>.*
- **Extrude** the circle from a Start Value of **1** to an End Value of **2.**
- Choose **Subtract** from the Boolean operation menu.
- Select Apply.
- Select the circle and extrude it again **3** to **4.**
- Choose **Subtract** from the Boolean operation menu. Select **Apply.**
- Select Cancel when finished.
- Save hinge1_*<your initials>*. Your finished Hinge1 model should resemble Figure 51.
- Open hinge2_*<your initials>*.
- **Extrude** the circle **0** to **1.**
- Choose **Subtract** from the Boolean operation menu. Select **Apply**.
- Repeat with start and end values of **2** to **3**, and again for **4** to **5**.
- Choose **Cancel** when finished.

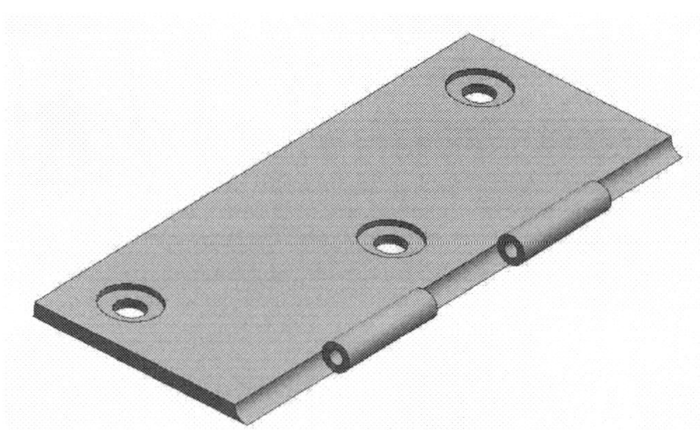

Save Hinge2_*<your initials>*. Your finished Hinge2 model should resemble the figure above.

***End of Project***

## 6.9 Practice: Emboss

The emboss command is used to create geometry that might be punched into a thin sheet of metal or plastic. To use it, you start with a sheet solid and some contour curves such as a sketch or, in many cases, letters. Then the emboss command gives you the opportunity to use various options to create the geometry. You can apply draft angles and "end options" that allow you to control the way the top of the new surface will look.

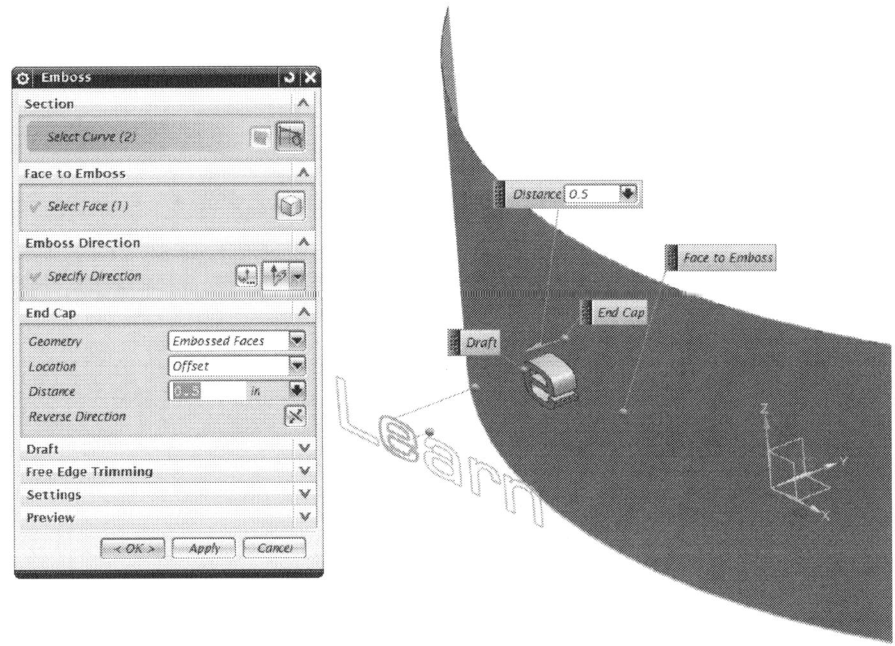


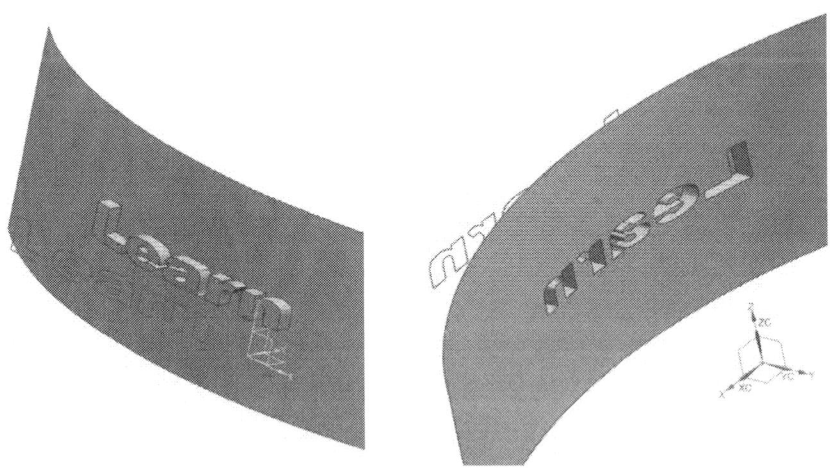

### ***Detail: Emboss limitation***

Although the "Emboss" command allows only one closed profile at a time, it allows you to apply a draft to the geometry as well as a uniform offset from the surface you are embossing. This can be a bit confusing but we will walk you through it in the following example.

## ***Practice Outline***

Open a new part in NX 8. Make sure you select 'Inches' as your Default Unit.

- Create a simple surface and a key hole shaped sketch on an offset plane. Make the offset space about 2 inches away from the surface as shown below:

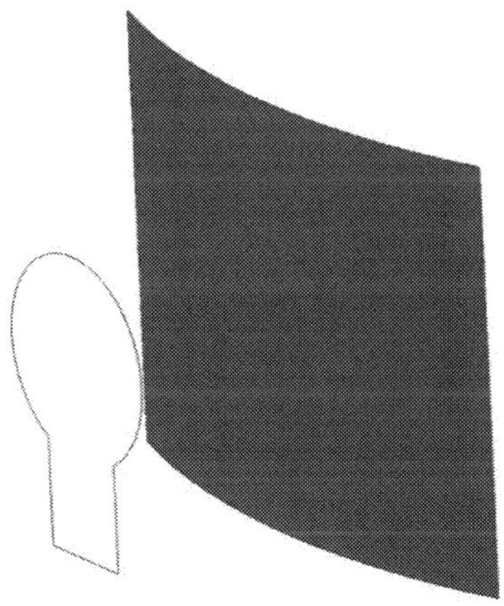

Isometric View

Head on view

- Select Insert / Design Feature / Emboss .
- Select the "**Feature Curves**" from the 'Selection Intent' pull-down menu.
- Select the sketch section curves and **Middle Click.**
- Select the sheet body and **Middle Click.**
- Set the "**End Cap**" in the following way.
- Geometry = Selected Faces
- Location = Translate
- Distance = -0.2
- Set the "Draft" options to:
- Draft = From End Cap
- Angle 1 = 3°
- Check 'Set All to Same Value.'
- Set all other options as default.

- Select **OK**.

The result is shown below:

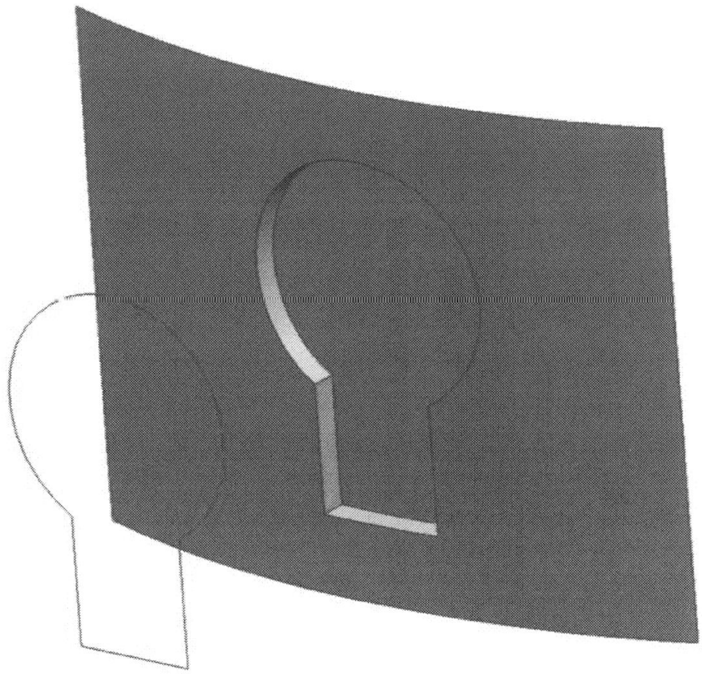

***End of Practice***

## 6.10 Self-Test

(All questions are true/false. Answers are at the back of the book.)

1. An extrusion's start and end distance can be defined by positive or negative values.
2. A positive taper value will angle the sides of an extrusion outward.
3. You can only revolve a section string about an axis that is on the section string itself.
4. A primitive cylinder that is based on a sketch will be associative to that sketch.
5. When creating a rectangular pocket, you must always specify a horizontal or vertical reference.
6. Text curves are parametric.

# 7. Detail Features

## 7.1 Introduction

The 'Detail Features' operation can be used to create curvature, smoothness and rounded qualities on existing surfaces and between surfaces and solids. Like Design Features, Detail Features can be parametrically controlled, allowing you to modify values later.

Detail features are accessed from the ***Insert / Detail Features*** menu as shown next. These commands can also be accessed from the feature operation toolbar. This chapter discusses the functionality of the 'Edge' and 'Face Blends,' 'Draft,' 'Chamfer,' and 'Bridge' commands. Continuing on this foundation, other commands located in the 'Feature Operation Toolbar' are discussed in a later chapter.

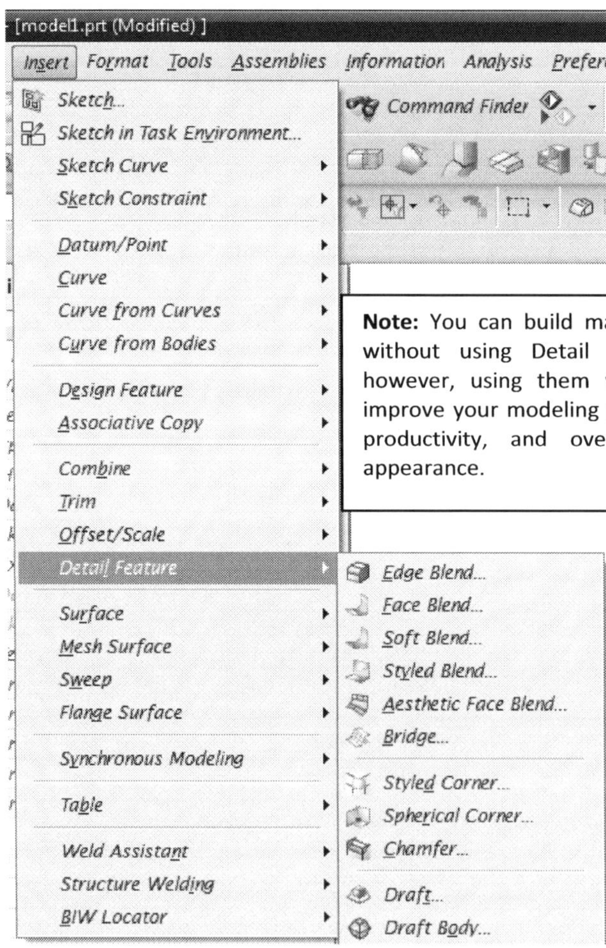


**Note:** You can build many models without using Detail operations; however, using them will greatly improve your modeling proficiency, productivity, and overall model appearance.

## 7.2 Edge and Face Blend Features

The ‘Edge Blend’ and ‘Face Blend’ features are an extremely powerful and flexible function in NX 8.5. Some of the more common applications of blends are discussed in the following text. For most simple blends, either ‘Edge Blend’ or ‘Face Blend’ can be used. For variable radius blends, blends with tangency constraints, and law controlled radii, specific blend operations may be appropriate.

### *Detail: Edge Blend*

An edge blend creates a smooth curve between two faces that join in a sharp corner. The simplest way to use an edge blend is to specify a radius that will be created in the corner. A good way to predict the result is to think of a ball with the specified radius being rolled across the surfaces, blending the faces as in the figure below. Wherever the ball touches the corner is where the blend surface will be created.

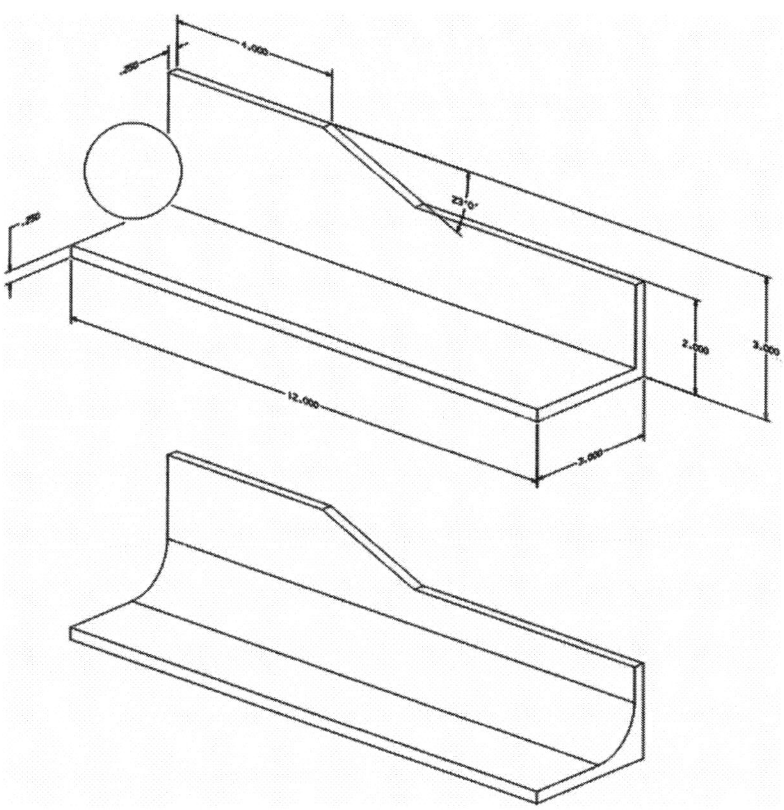

## *Project 7.2.1*

- Sketch the geometry shown below and **Extrude** it **2in**.

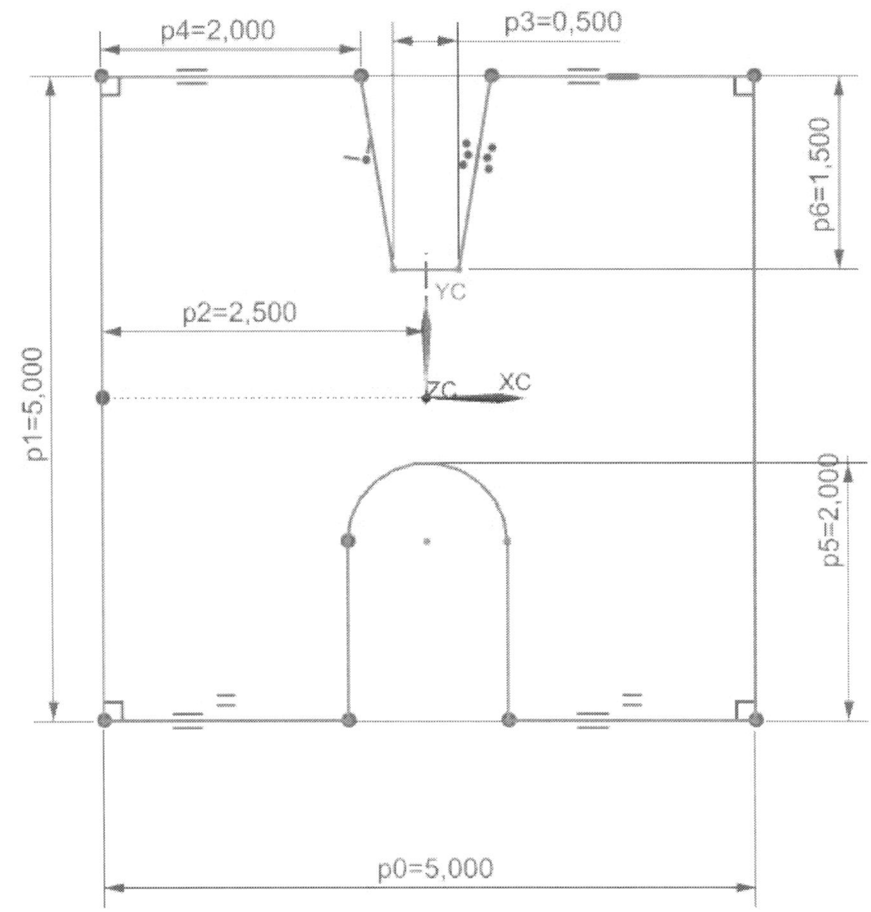


- Select Insert / Detail Feature / Edge Blend .
- Select the two opposite sides with no grooves shown above and enter the radius value **0.5**.
- Select Add New Set.

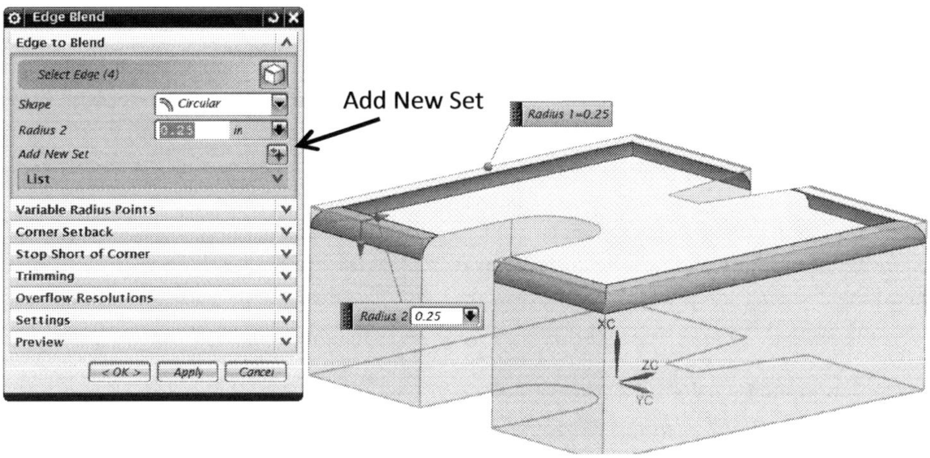


- Select edges for set 2 (four edges near the grooves) as shown above and enter value **0.25.**
- Click **OK.**
- Select Insert / Detail Feature / Edge Blend.
- Select **Tangent Curves** from the **Selection Intent Menu** as shown on the previous page.

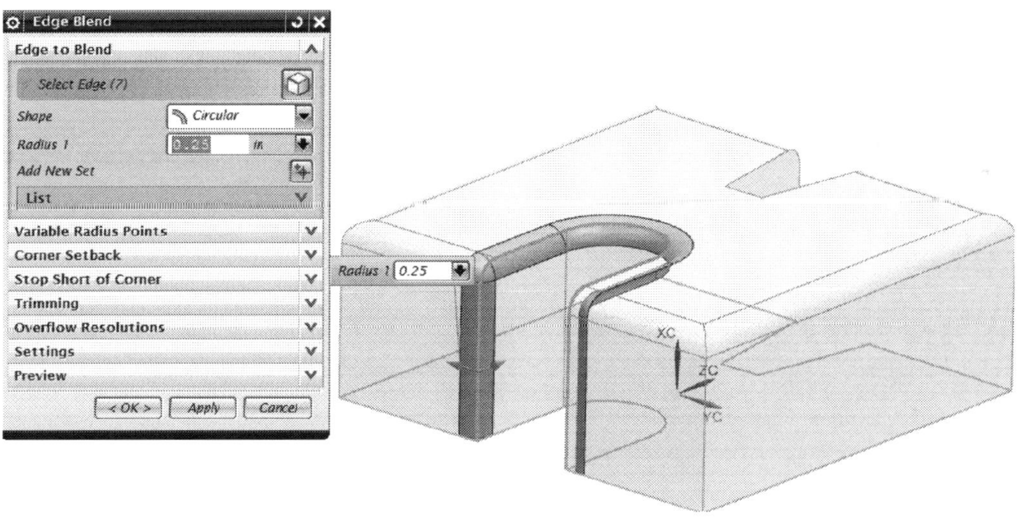


- Select the edge shown above and enter the value **0.25.**
- Click **OK.**

***End of Project***

You can work much faster by using the selection intent menu correctly. Try using other options from the Selection Intent Menu like ***Face Edges*** and ***Vertex Edges.***

## ***Detail: Variable Radius Edge Blend***

The 'Edge Blend' function can also be used to vary the radius of the blend as it traverses the distance of the edge.

## ***Project 7.2.2***

Use the geometry from the previous project.

- Select Insert / Detail Feature / Edge Blend.
- Select the edge shown below.

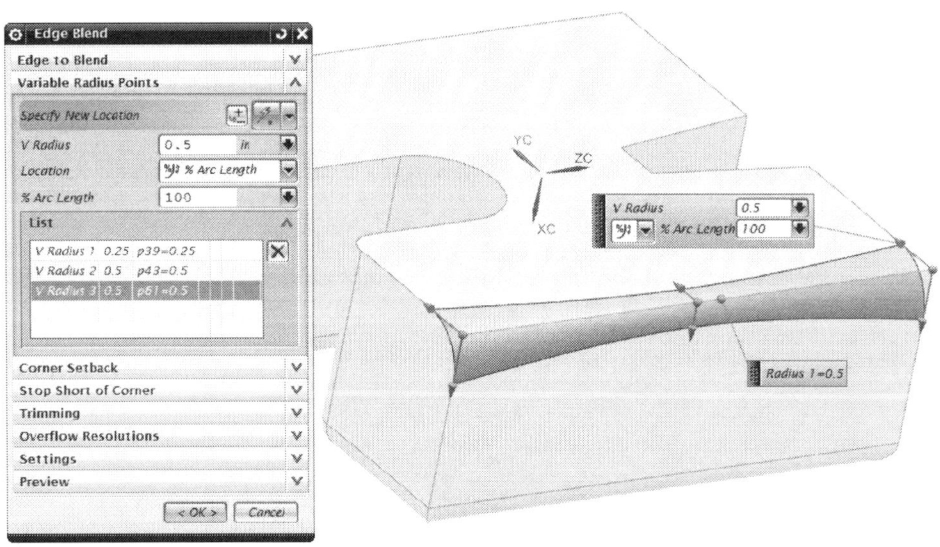


- Under the Variable Radius Points Section, click **Specify New Location**.
- Select the midpoint of the edge.
- Select point 1 and enter the value **0.5**.
- Select point 2 (midpoint) and enter the value **0.25**.
- Select point 3 and enter the value **1.**
- Select **OK.**

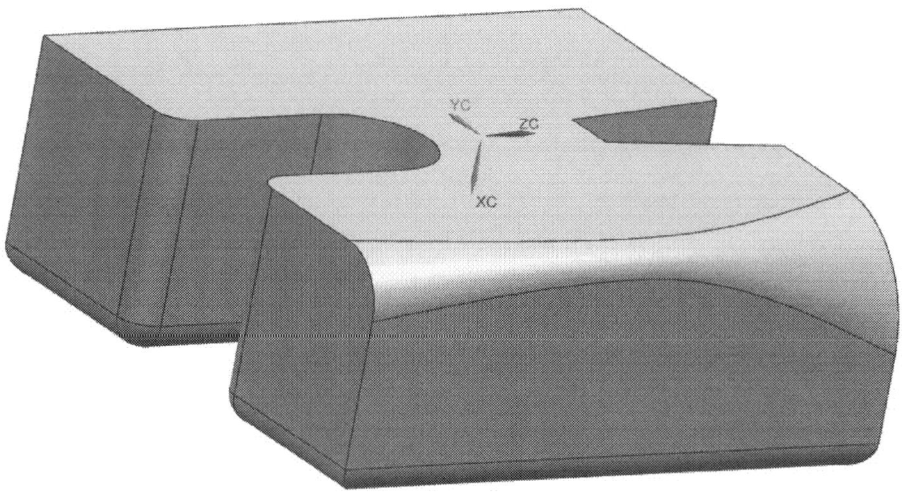


Compare your results to the figure above.

***End of Project***

## ***Detail: Face Blend***

A 'Face Blend' can be created between two sets of faces of solid or sheet bodies that are not necessarily adjacent. There are two main face blends: **Rolling Ball** and **Swept Section**. Rolling Ball is the simpler and more commonly used face blend. Swept Section is an advanced face blend in which the blend surface is controlled by a cross section swept along a spine curve. There are also three different options to define radii: 'Constant', 'Law Controlled' and 'Tangency Controlled.' We will discuss the Rolling Ball type face blend in this chapter. 'Swept Section' is not discussed in this book.

## ***Project 7.2.3***

Use the geometry from the previous project.

- Select Insert / Detail Feature / Face Blend.
- Specify the Type as Two Defining Face Chains.
- Specify the Orientation as **Rolling Ball**
- Select the **First Face Chain** as shown below. Enter a radius value of **0.4**. Make sure that the arrow is pointing towards the other angled face. If it is not, use the **Reverse Normal** button to reverse it.

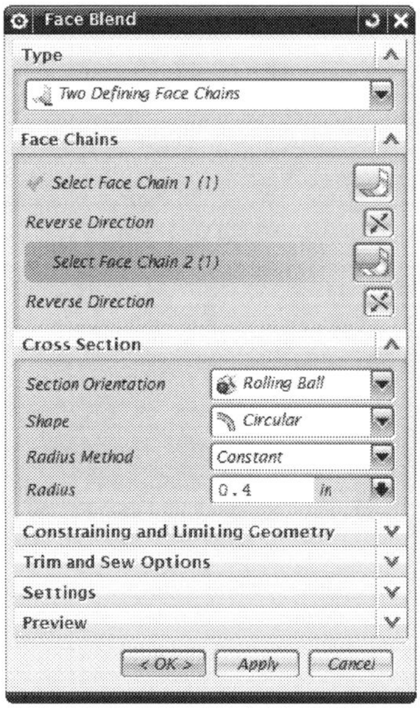


**Note:** Face Blend even allows you to select multiple faces in a set, if you need it.

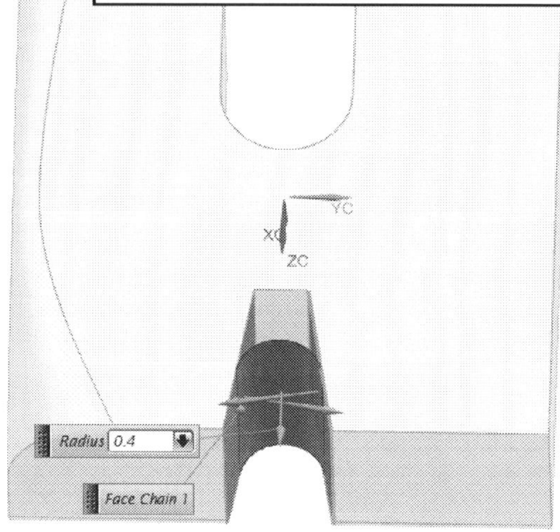


- Middle Click to OK.
- Select the **Second Face Chain** (opposite face). Again make sure that the arrow is pointing inward.
- Hit **OK**.

- Compare your results with the picture above.

***End of Project***

Notice how the blend is created between faces that are not adjacent to each other. This is good functionality in that it lets you skip between surface/surfaces and create clean blended geometry.

## ***Detail: Coincident Edge Blend Features***

When you need to create a blend that does not run out tangent to a face, you can use a 'Coincident Edge' blend. This is an option found under the 'Face Blend' command. Typically, the coincident edge blend is needed in situations where one or more faces are not large enough to accommodate the blend. To create this blend, you must select the edge that will not be tangent.

## *Project 7.2.4*

Create the sketch shown below and extrude it **5in**.

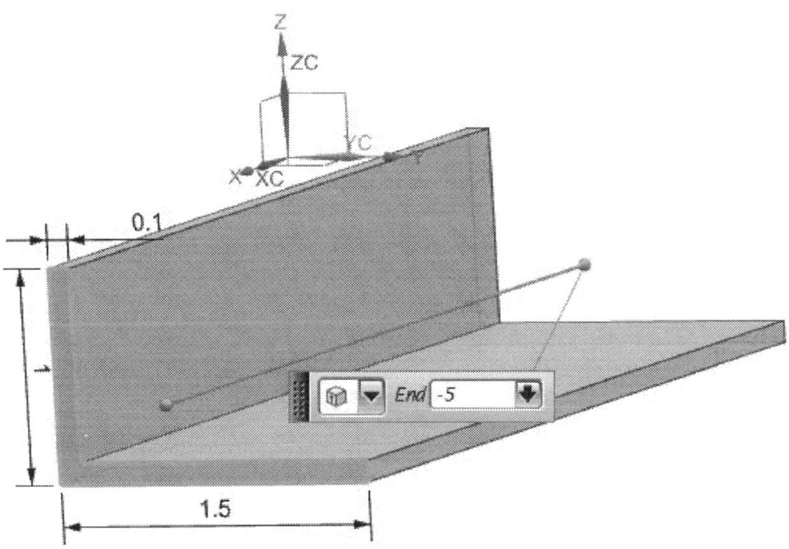


- Select Insert / Detail Feature / Face Blend.
- Select the **First** and **Second Face Chain Sets**. Make sure that the arrows are pointing correctly towards the centerpoint.
- Select the **Coincident Curves** as shown below and type in the radius value **2**.

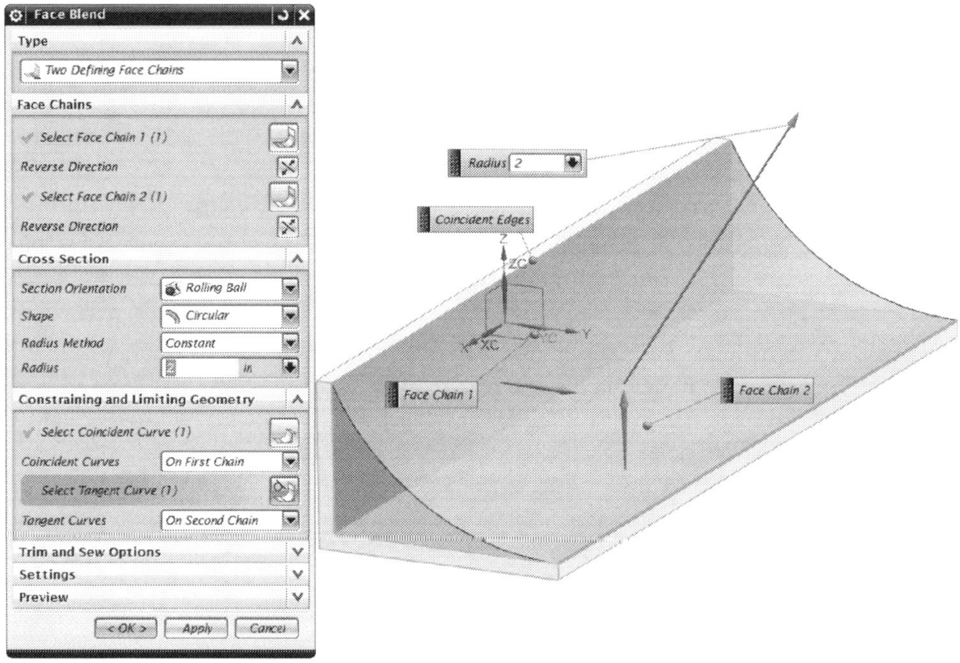


- Hit **Apply** and compare to the following figure.

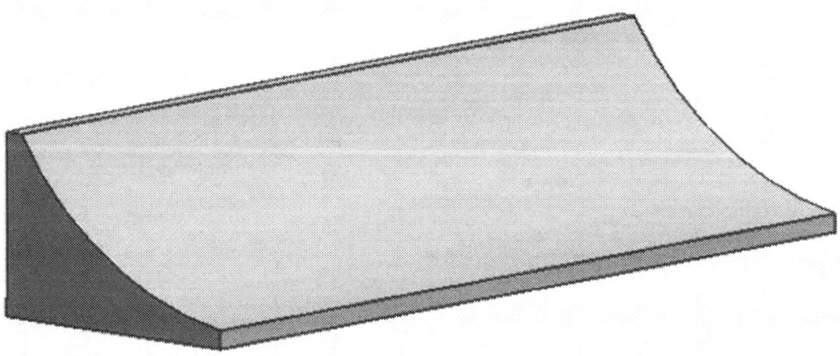

***End of Project***

## ***Detail: Face Blend Using Tangent Curves***

The 'Tangent-Curves' are used when you have a blend whose radius is dictated by the length of one of the sides of the object to be blended.

## ***Project 7.2.5***

Use the initial geometry from the previous project or create a new part with the same geometry.

Create a sketch and cut off the geometry approximately as shown below.

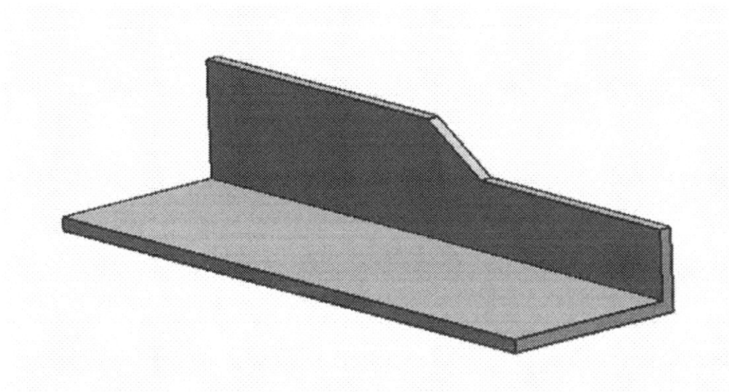

- Select Insert / Detail Feature / Face Blend.
- Select the **First** and **Second Face Chains** as shown below. Make sure that the arrows are pointing correctly towards the centerpoint.

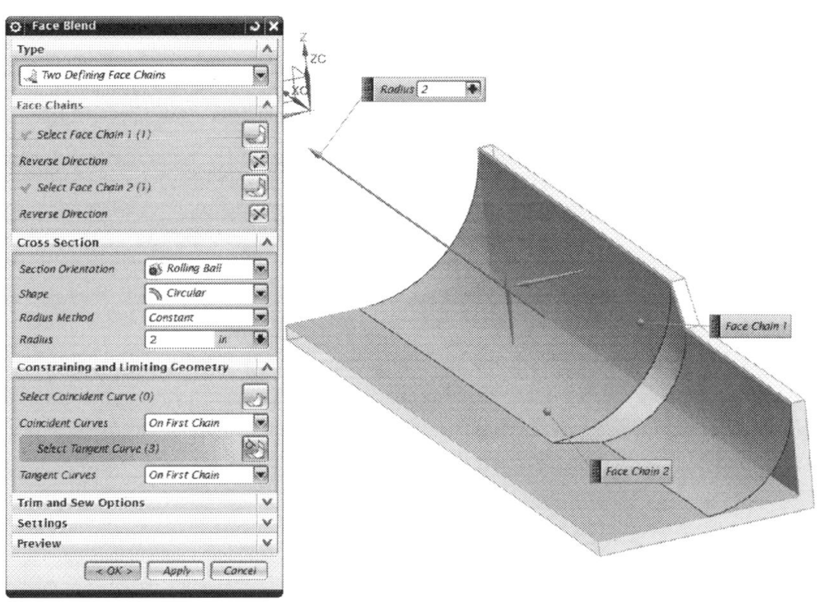

- Skip the 'Radius' field and 'Coincident Edge' option.
- Select the **Tangent Curves** as shown above.
- Set the radius to 2 and hit **OK**.

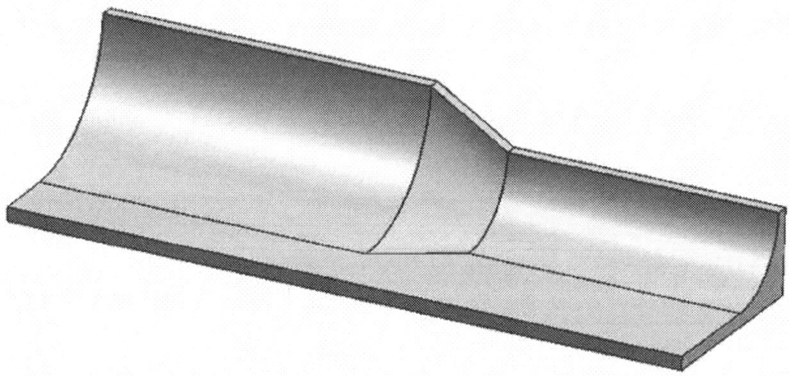

***End of Project***

The order in which you apply blends can make a big difference on complicated models that often have intersecting blends. In general, you should apply the largest radius and most critical blends first and move on to the smaller radius and less critical blends last. However, experience teaches that some experimentation is often required to complete the blending of a solid model. Often times it pays off to do the blending after the part design is completed.

> **Note:** If a solid model is used only to create a drawing, you should be cautious about spending too much time blending. Blends do not always show up correctly in a drawing and will sometimes require a "fudged" approach in drawing views - regardless of whether or not they are modeled correctly.

## ***Detail: Inferred Blend***

The Inferred Blend is a feature that infers a radius when you select 3 adjacent faces. To show you this great feature, we are first going to create a plastic casing for a product and use the 3 face blend option to create blends on our stiffening ribs.

### ***Project 7.2.6***

- Create the following **sketch** on the **XY datum plane**.

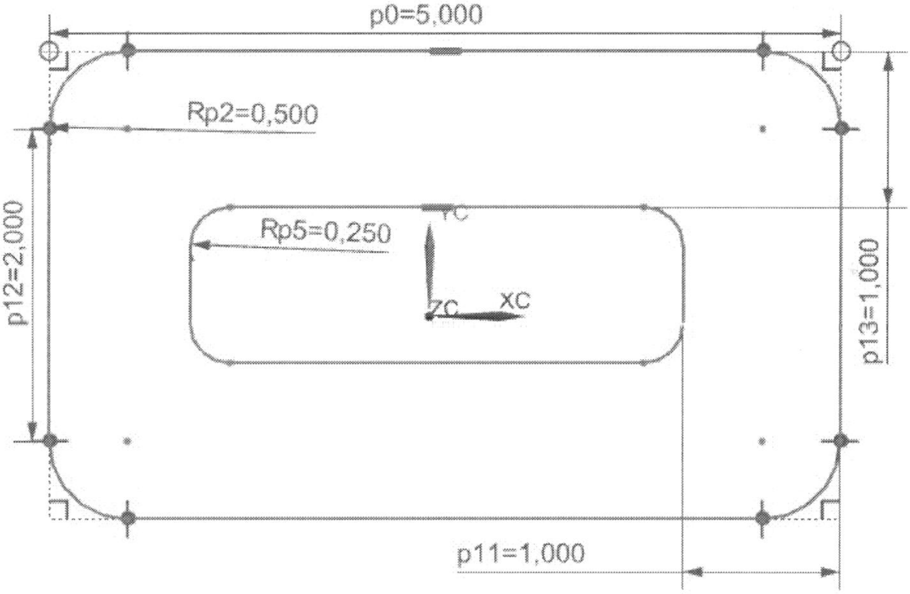


- **Extrude** the outside curves by 0.75in with a **draft angle** of 2.5 degrees, as shown in the following figure.

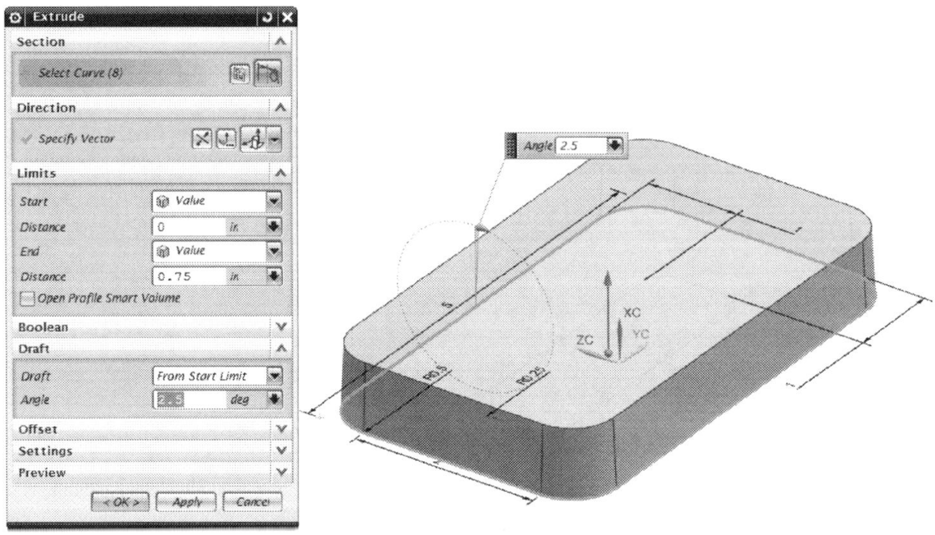


- Select **Insert / Detail Feature / Edge Blend** and give the edge that touches the original sketch a radius of 0.15in. Your model will now look like the following figure.

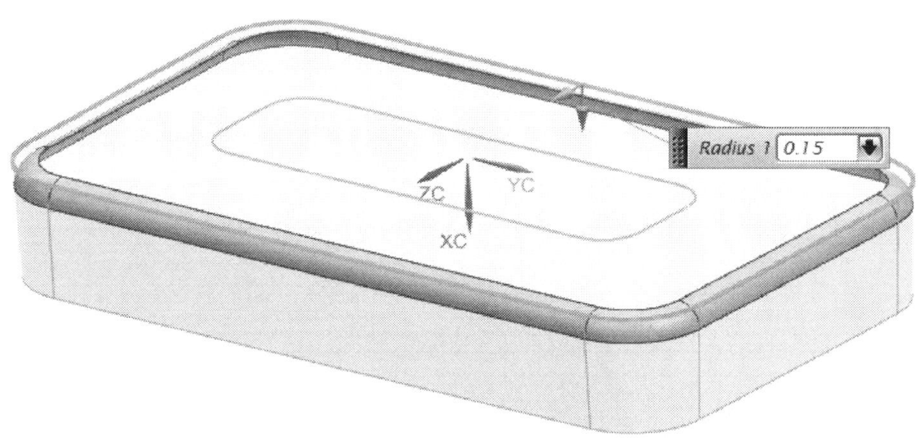


- Now **Extrude** the inside curves of the sketch – through the entire body – and **subtract** from the existing geometry. Your model will look like the following figure.

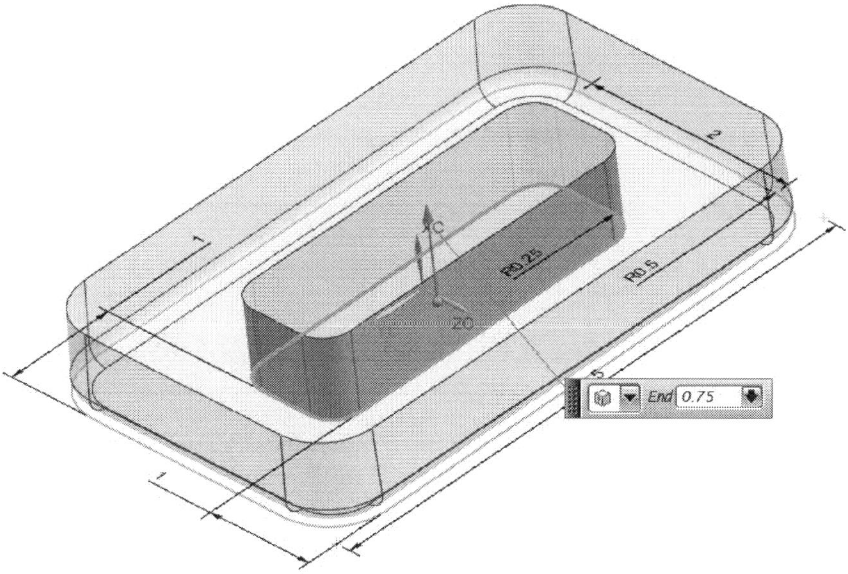


- Select **Insert / Offset Scale / Shell**. Pick the top face and type in 0.1in for the thickness. Your model will now look like the following figure.

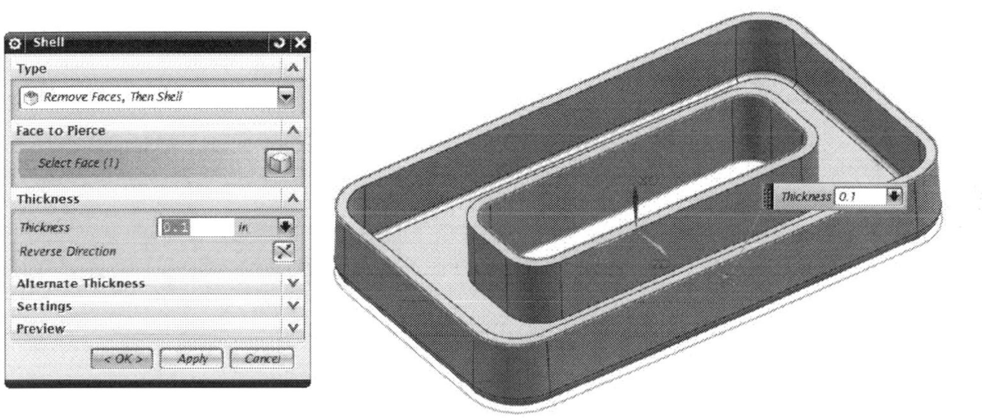


- Select **Insert / Synchronous modeling / Move Face**. We are going to lower the inside lip of the product using this very fast modeling technique. Once you are in the **Move face** command, just click on the face (shown below) and type in a distance of 0.25in.

**Note:** We want the face to move towards the original sketch, so you may need to reverse the direction of the 'move face'.

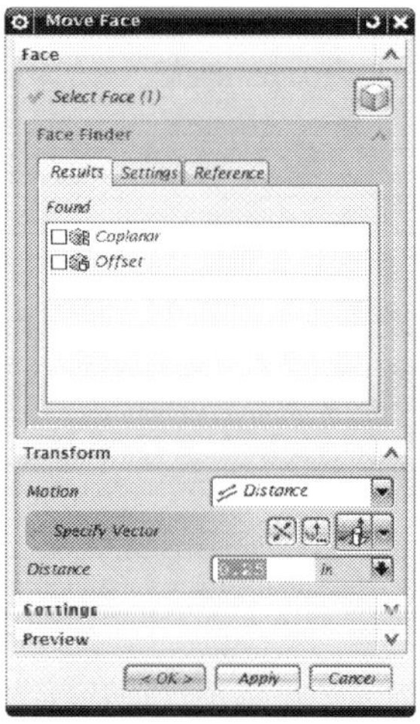


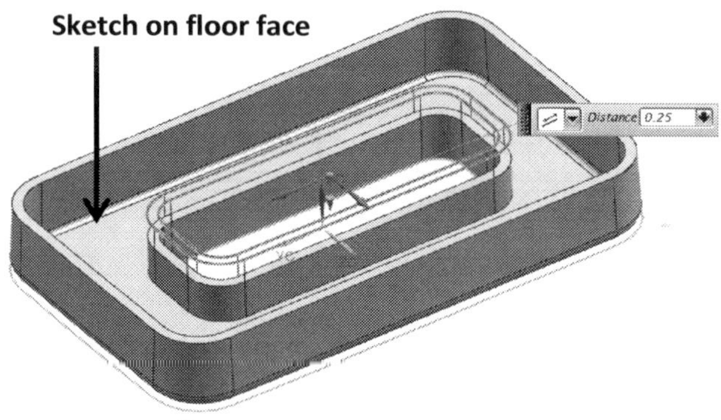


- Now let's create the supporting ribs: create a **sketch** on the floor face (shown on previous figure). Copy the sketch shown below.

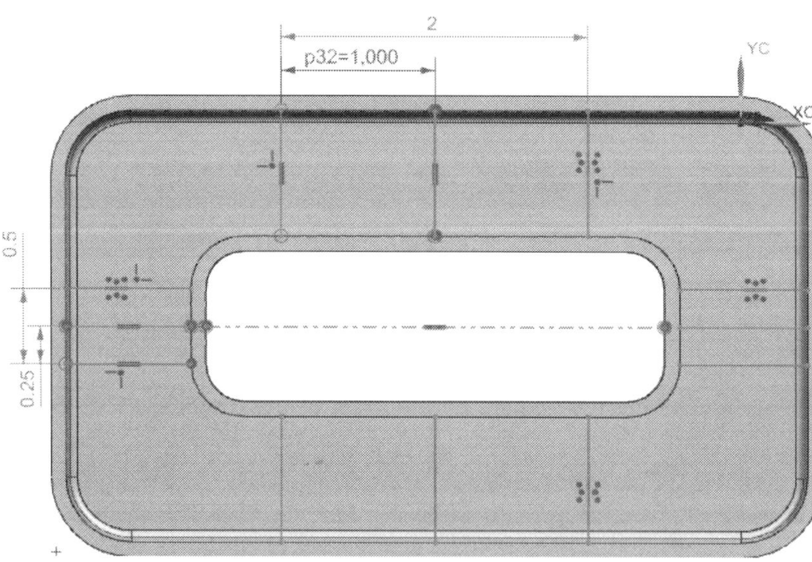

- **Extrude** the sketches of the lines to a distance of 0.35in with a symmetric offset of 0.040625

> **Note:** The symmetric offset calculates to 65% of the existing solids wall thickness – you have to design ribs or bosses with a smaller wall thickness to reduce the occurrence of shrinkage or sinking in the plastic. Add a draft angle of 2 degrees and use the Boolean option **Unite**.

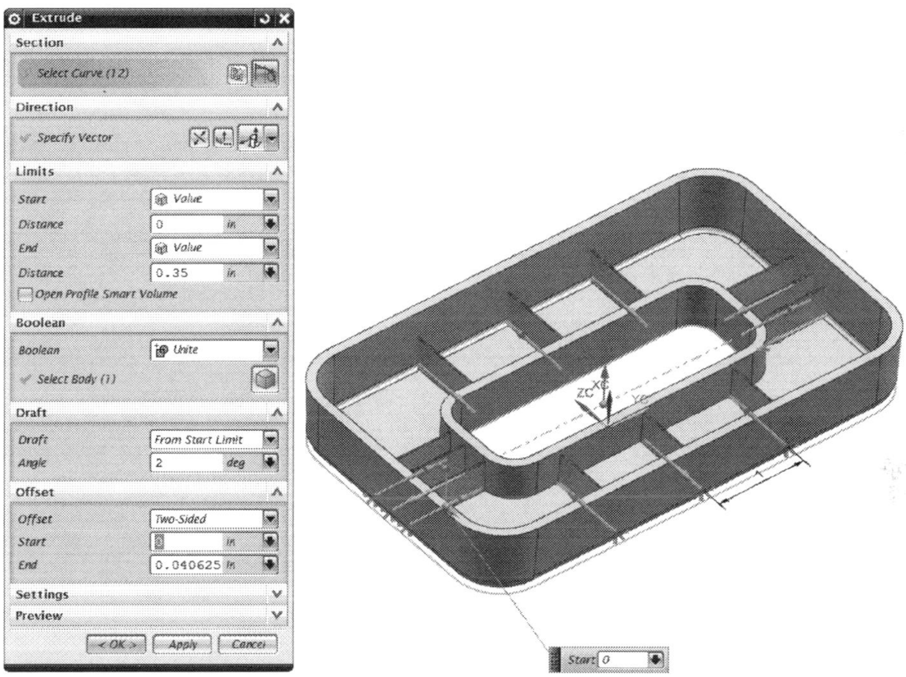

- Your model should now look like the following figure.

- Now time for the star of the show - the 3 Face Blend. Zoom up on one of your ribs and select **Insert / Detail Feature / Face Blend**. Choose **'Three Defining Face Chains'** from the **Type** drop down menu, then select the first side face of your rib, as shown in the following figure.

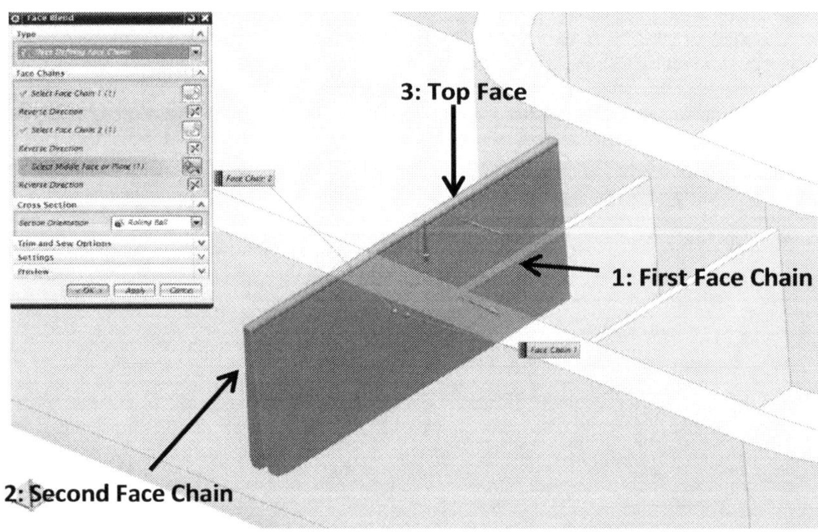


- Move on to the next step of the process in the blend window. Now select the **second face** which is the one opposite the original selection.
- Move on to the final step of the process and pick the **top face** as shown in the previous figure. Once you have all of your selections picked, click **OK** to create your fantastic inferred blend.

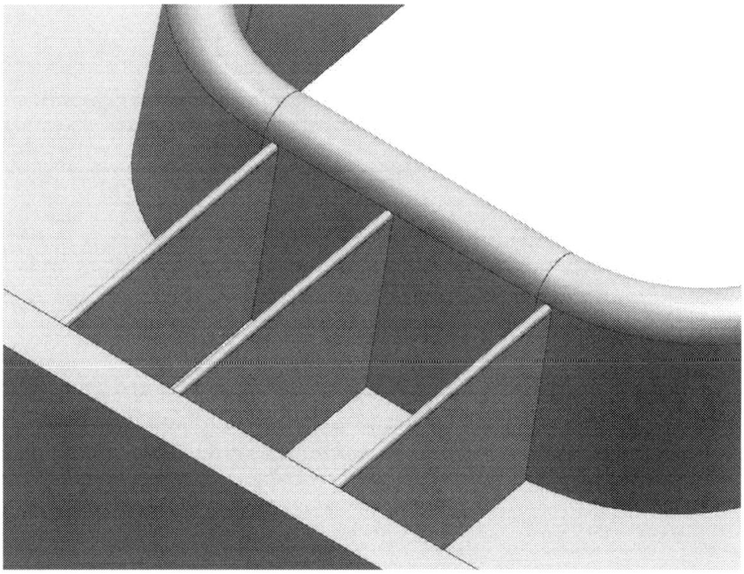

- Repeat the process for the rest of the blends until you have the following figure. You can even try and do the center-loop.

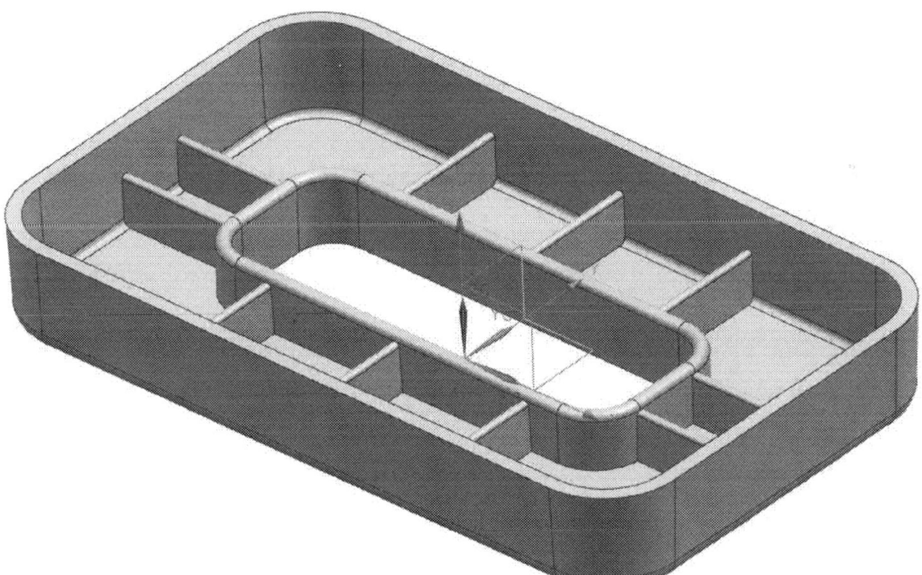

***End of Project***

## 7.3 Drafts

'Draft' is used to apply an angle to the faces of pre-existing solid geometry. You have already learned that a draft angle can be applied to an extruded body at the same time that it is created; however, with this approach, all of the side faces of the extrusion will be drafted. This is often required when creating complex models, such as cast or molded parts. There are four draft features available in NX 8.5: 'Draft from Plane,' 'Draft from Edges,' 'Draft Tangent to Faces,' and 'Draft to Parting Edges.'

### ***Detail: Draft from Stationary Plane***

This feature allows you to specify certain faces of a solid to be drafted by a specified amount. The applied draft angle is measured relative to a specified 'Stationary Plane' (the plane where dimensions remain constant). The 'Stationary Plane' is defined by selecting a Reference point and a 'Draw Direction' (Reference Vector Direction).

Consider the simple shape shown below. Suppose that faces 1 and 2 must be drafted inward by 20 degrees and face 3 must be drafted inward by 30 degrees.

In this case, suppose that the underside of the solid is defined as the Stationary Plane and the 'Draw Direction' is normal to this plane. Upon the application of the draft features, the lower edges will remain fixed, while the faces will become angled inward.

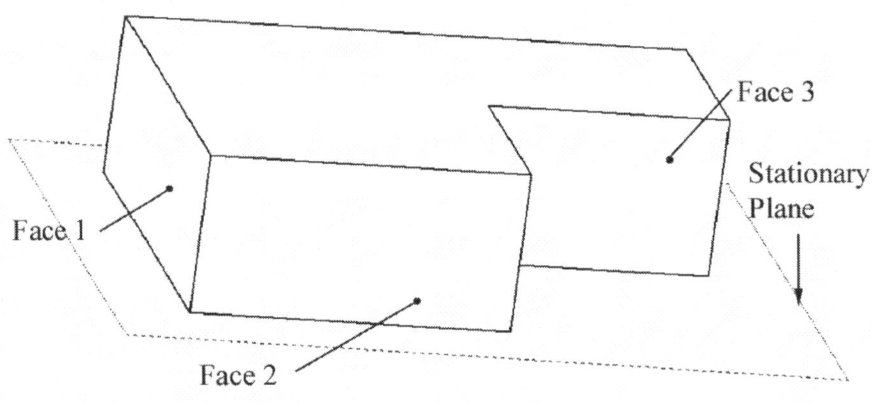


### ***Project 7.3.1***

Sketch the geometry shown below and extrude it **2in**.

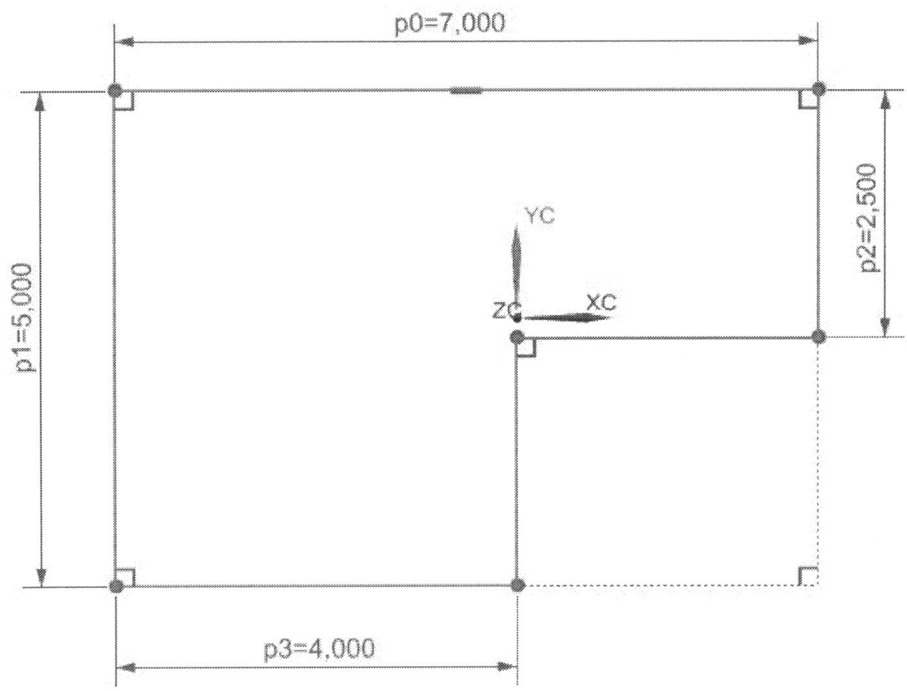


- Select Insert / Detail Feature / Draft.

The Draft menu (shown on following page) will appear.

- Select 'From Plane' for the Draft Type.
- Select Draft direction **ZC.**
- Select the bottom face as the stationary plane.
- Input Angle = **20** degrees.
- Pick Faces 1 and 2 as shown in the following three-dimensional figure for Set 1.
- **Middle click** and select face 3 for Set 2.
- Input Angle = **30** degrees.

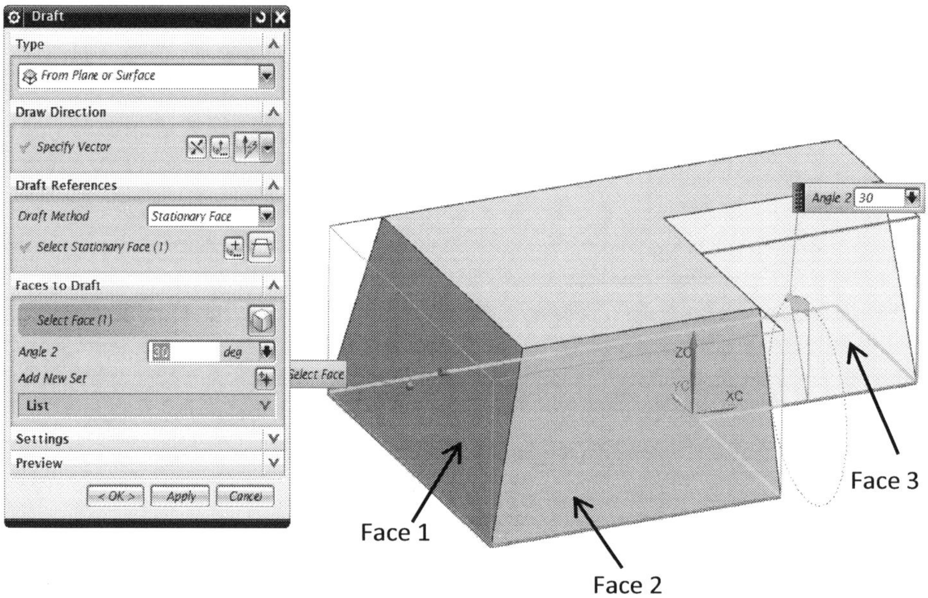


**Note:** Don't forget to read the cue. Cue reading is the easiest way to find out what NX is expecting from you.

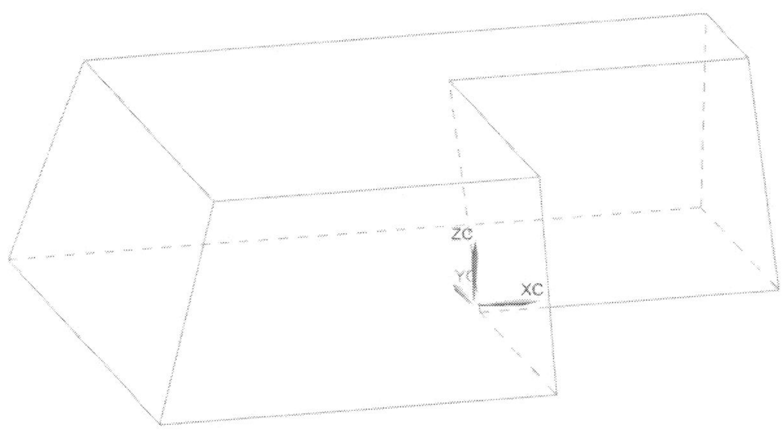


Choose **OK**. Compare your results to the figure above.

***End of Project***

## Detail: Draft from Stationary Edges

Consider the solid body shown below. This geometry is typical of a cast part with a non-planar parting line, which in this case is the bottom surface. You will need to apply drafts to the left, right, front, and rear surfaces. Applying drafts to the left and right surfaces can be accomplished using the Stationary Plane method. Applying a draft to the front and rear surfaces requires a different technique, since the parting line is not planar.

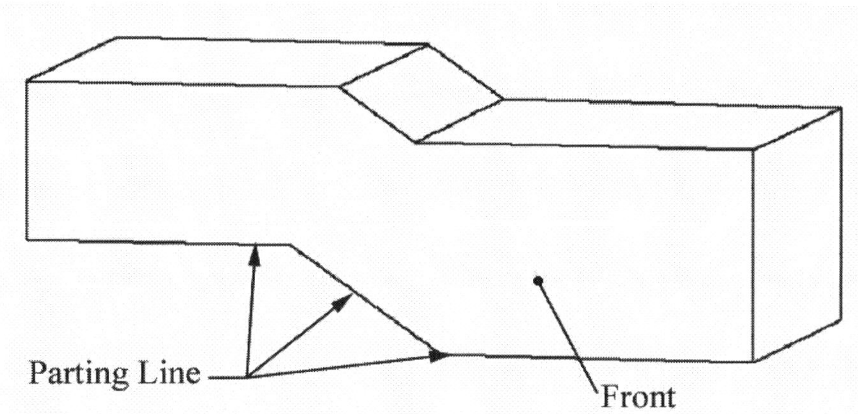


This type of geometry can be created using 'From Edges.' It is similar to the 'From Plane,' but instead of specifying a plane you select edges that remain fixed when the draft is applied.

## *Project 7.3.2*

In this example, a 'From Edges Draft' is applied to the front and rear faces of the solid shown on the previous page.

Create the sketch shown below.

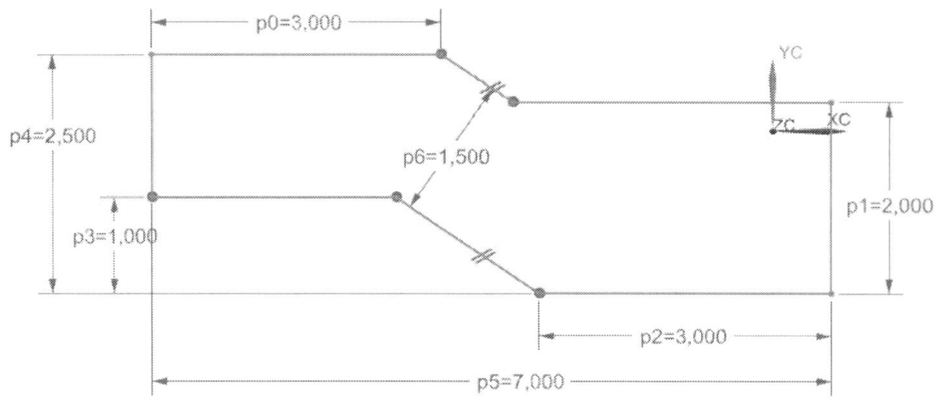


- Extrude it **4in**.
- Select Insert / Detail Feature / Draft.
- Select 'From Edges' for the Draft Type.
- Pick Edge 4 for draw direction. Just as before, this edge defines the Draw Direction vector – an orange arrow will appear to indicate its direction. In this case, it should point upwards. To get the arrow oriented properly, pick the half of the edge that you want the arrow to point toward – i.e. pick the upper half of Edge 4.
- Pick Edges 1, 2, and 3 as shown below to designate the 'Stationary Edges.'

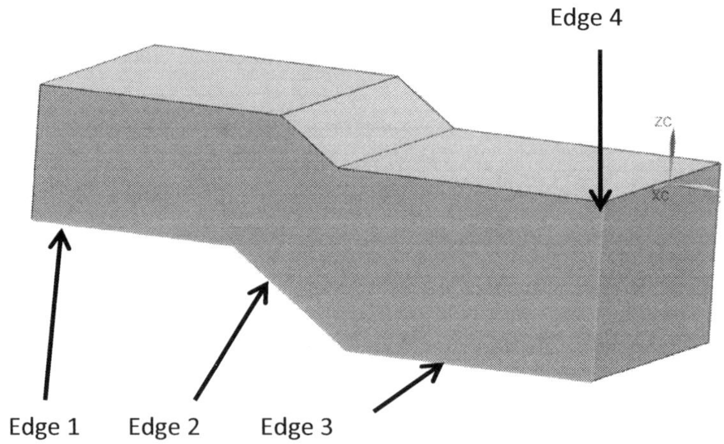

- Input Angle = **30**.
- **Middle click** and select the stationary edges for the rear face. (You can change the angle or keep it the same.)
- Choose **OK**. Compare your result to the Figure below.

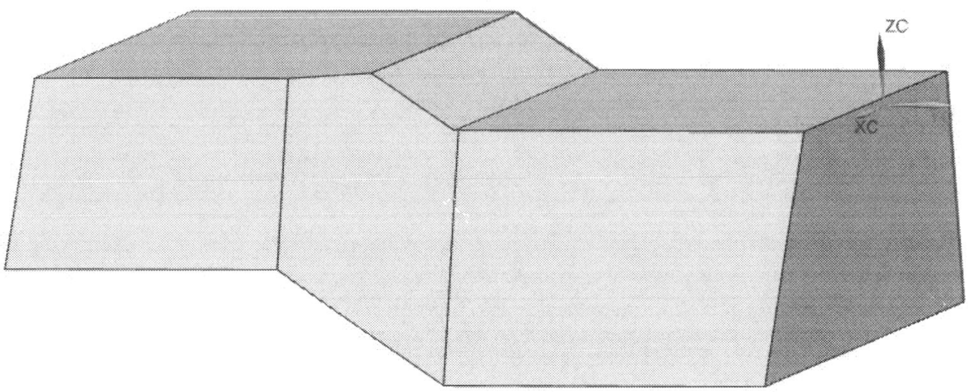


***End of Project***

## 7.4 Chamfer Features

'Chamfer,' like the 'Blend' feature, is used to add manufacturing details to solid bodies. They are both typically attached to model edges and may either add or remove material. Because they are parametric, their size is easily modifiable. A chamfer is a flat surface made by cutting off an edge or corner or by adding a flat surface to a corner at an angle.

### *Detail: Chamfer*

Consider the solid body shown below. It is tempting to use the sketcher to create the entire top view shape as shown, including the radius and chamfer details. The disadvantage of this approach is that it will result in a more complicated sketch. By using the 'Blend' and 'Chamfer' features to add details (instead of the sketcher) the model will usually be easier to create and will respond to modifications in a more robust fashion.

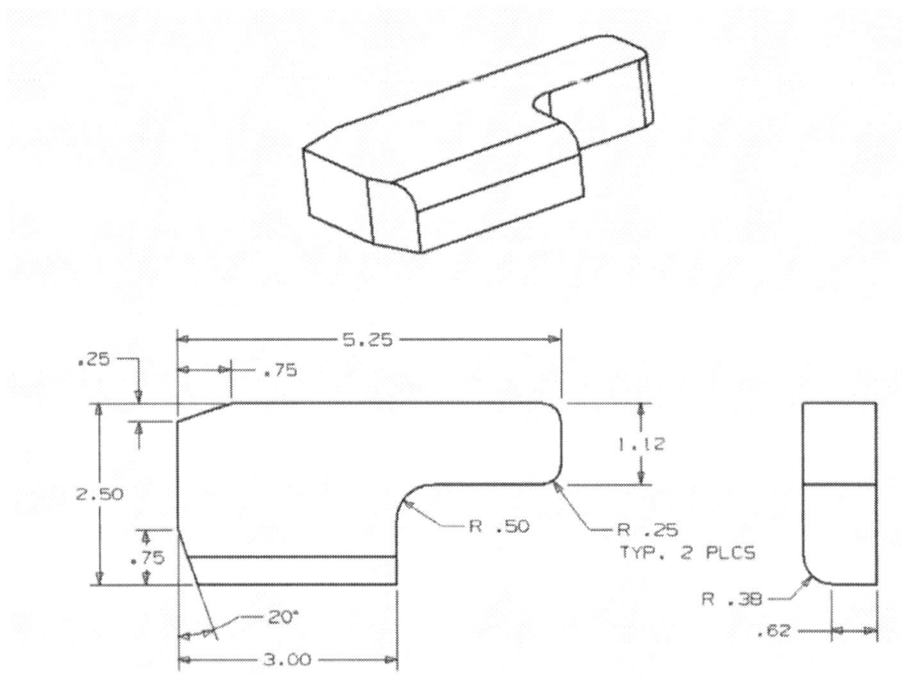


***Project 7.4.1***

Create the sketch shown below.

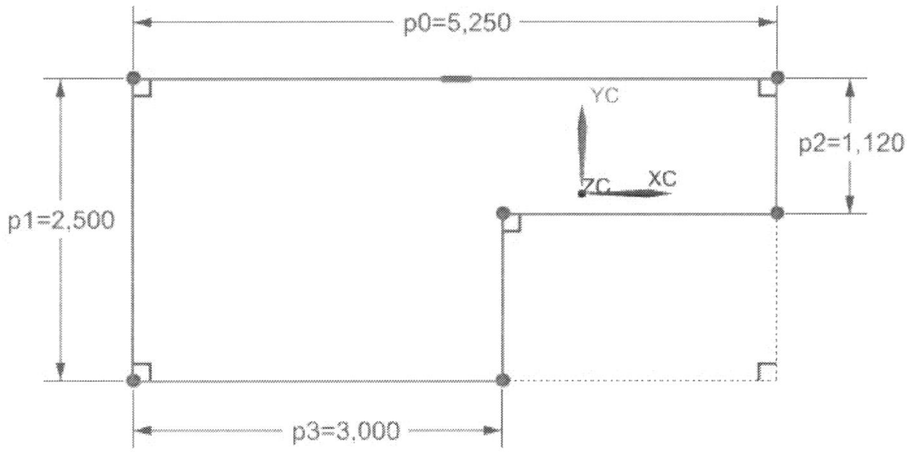


- Extrude the sketch from **0** to **1**.
- Select Insert / Detail Feature / Chamfer  .
- Select **Asymmetric** from the 'Offset' menu shown below.

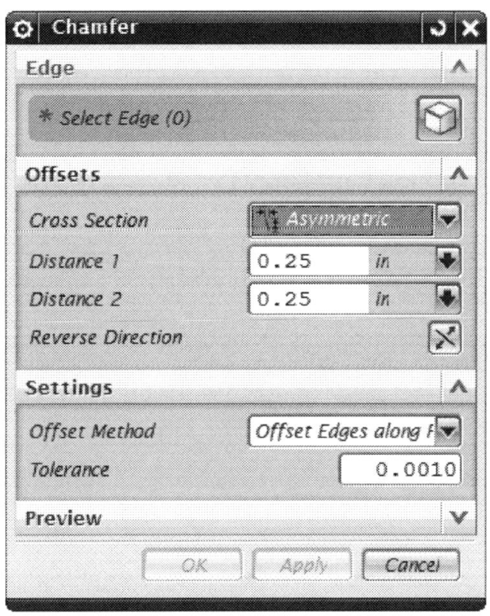


- Pick Edge 1 as shown in the following figure.

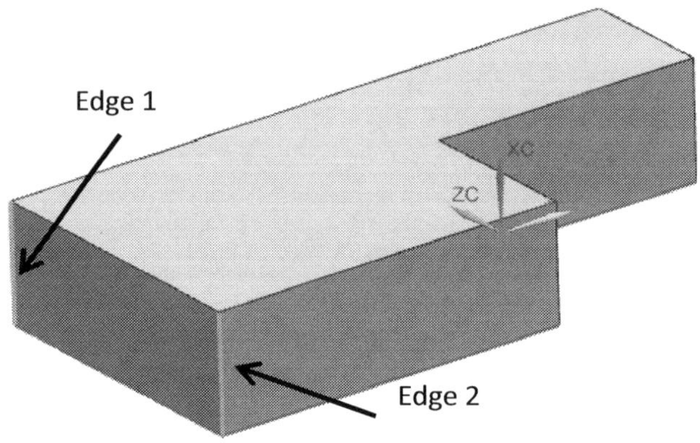


- Choose **OK.**
- Enter the values: Distance 1 = **0.75** and Distance 2 = **0.25.**
- Choose **OK**. You may use the 'Reverse Offsets' icon on the top right corner of the Chamfer menu to flip the offset sides.
- Select Insert / Detail Feature / Chamfer.
- Select Offset and Angle.
- Pick Edge 2 as shown above.
- Enter the values: Offset = **0.75** and Angle = **20.0.**
- Choose **OK**.
- Select Insert / Detail Feature / Edge Blend.
- Pick Edge 3 as shown below.
- 

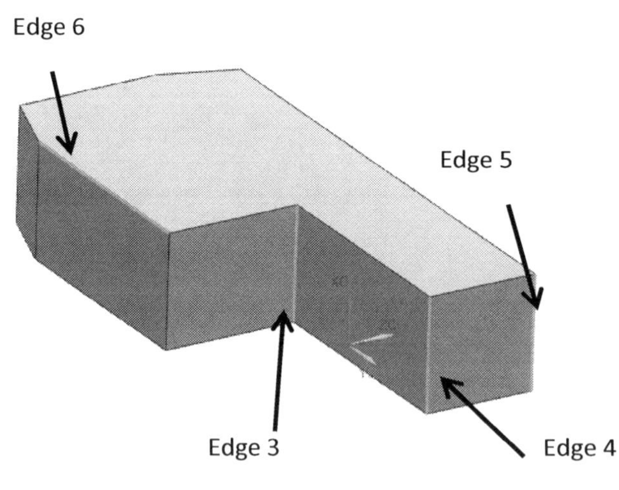

- Enter Radius = 0.5 and **middle click** to create Set 1.
- Pick Edges 4 and 5 as shown in Figure 31.
- Enter Radius = 0.25 and **middle click** to create Set 2.
- Pick Edge 6.
- Enter Radius = **0.38** and **middle click** to create Set 3.
- **Middle click** to complete the 'Edge Blend' feature. Compare to the figure just before the beginning of this project.

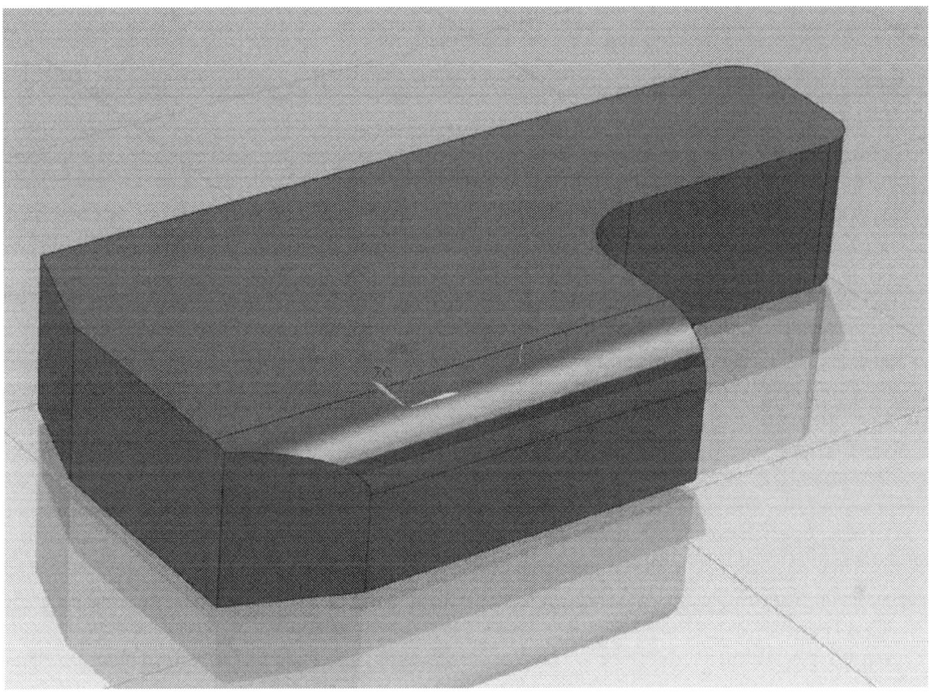

***End of Project***

## 7.5 Bridge Surfaces

You can create a bridge between two faces or fill a gap by using the 'Bridge Surface' command. You can choose either the tangent or curvature continuity type between the bridge surface and the faces to be bridged.

> **Note:** The 'Bridge' command applies to surfaces, which is discussed in more detail later in this book. Therefore, the 'Bridge' project detailed below may seem a little advanced to new users. Fear not, and have fun!

### ***Detail: Creating a Bridge Surface with Primary Faces***

### *Project 7.5.1*

- In the Sketcher, draw two curves similar in shape and size to those below. **Hint:** Use a simple **arc** for your first curve and a **spline** with four points for the second.

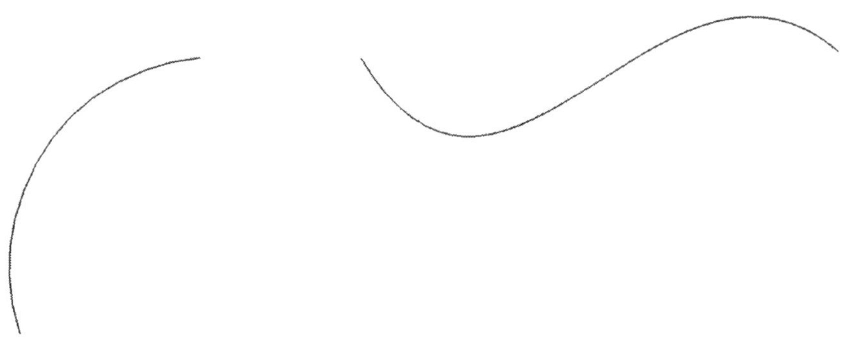

- Finish your sketch and select *Insert / Design Feature / Extrude.*
- **Extrude** the first curve as shown on the following page to create a surface.
- **Extrude** the second curve with a **5** degree **draft** as shown next to create a second surface.

**Note:** Be sure to extrude both curves the same distance.

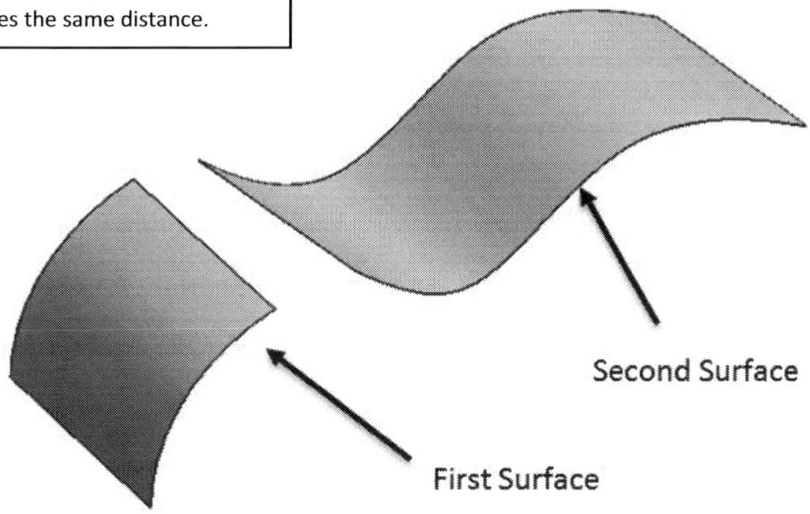


- Select Insert / Detail Feature / Bridge.
- Select Continuity type **Tangent** and select both sheets as shown below.

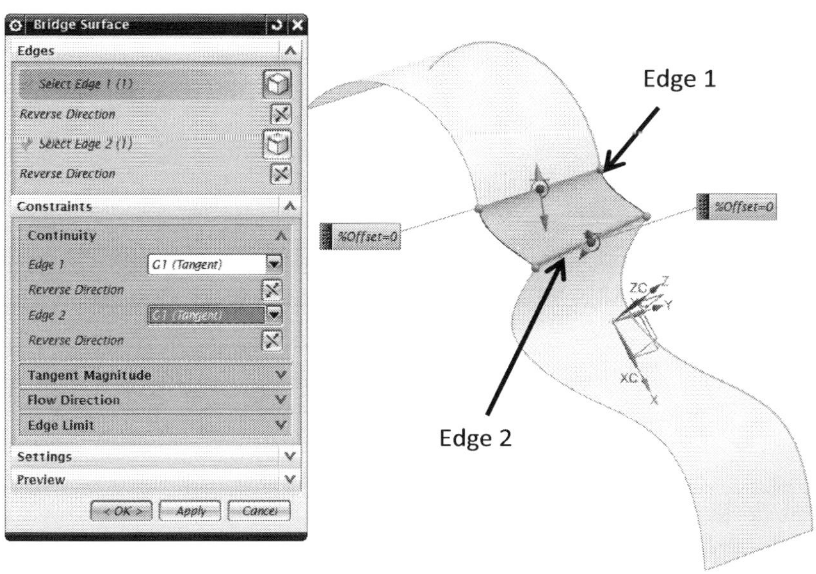

Be sure the arrows on the faces you selected are from the same side as shown. If they are not, select Cancel and repeat the procedure. It helps to click the same spot on each edge.

- Select **Apply** and compare with the figure below.

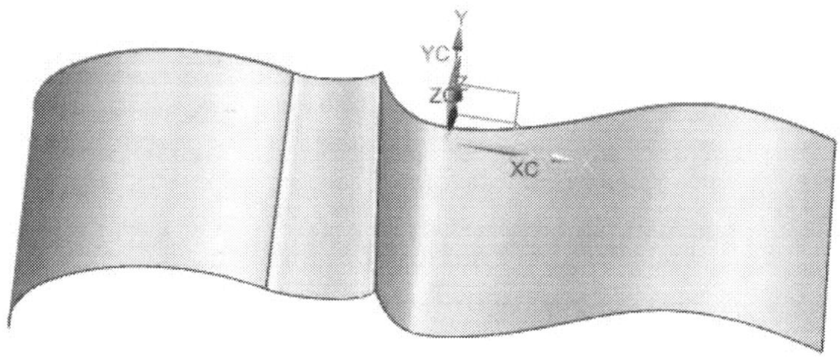


- This operation creates a bridge surface, but if you want to make modifications, Drag the Magnitude sliders.
- The Flow Direction can be changed to alter the course of the drag.

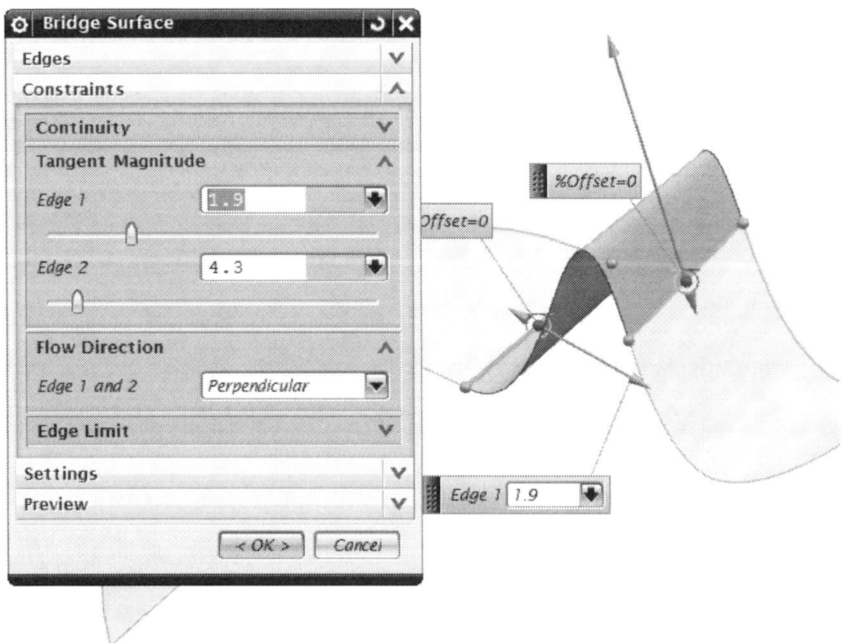


- Choose **OK** and then **Cancel**.

***End of Project***

## 7.6 Practice: The Pool Rack, Part 2

Use the 'Hole,' 'Pocket' and 'Blend' features to add detail to the pool rack model that you made in an earlier project. The finished pool rack is shown below.

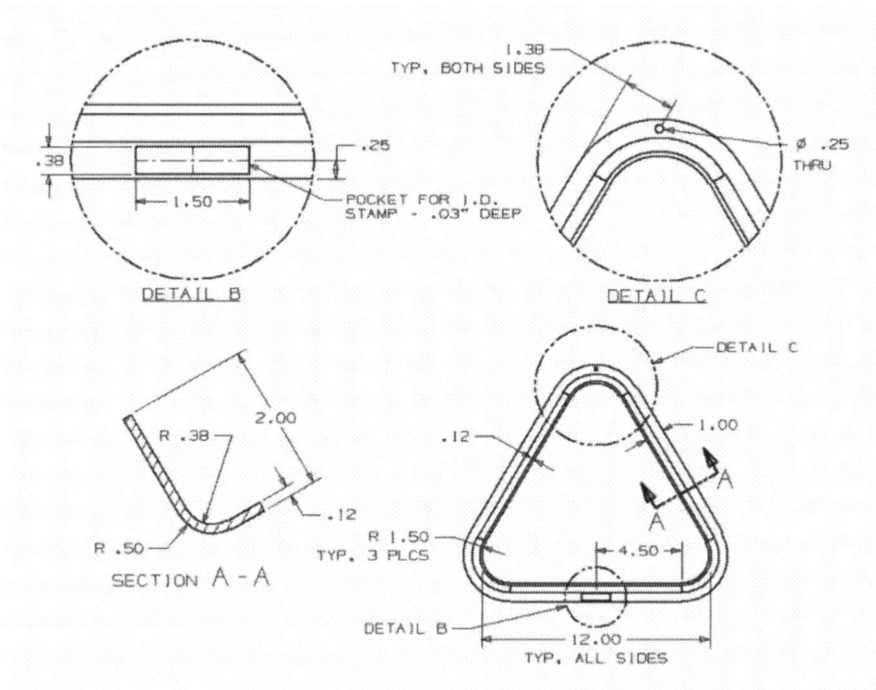


The new details in the pool rack are the 0.38" and 0.5" blends, a pocket in which to place the part number, and a hole for hanging the rack on the wall.

> **Note:** The edges to be blended are a closed loop comprised of several individual edges. For each blend, all adjoining edges must be selected before selecting OK. To make selecting easier, use the 'Tangent Curves' option in the 'Selection Intent' menu. This option will chain all the tangent edges of any edge pick.

## *Practice 7.6.1*

- The pocket has a horizontal positioning dimension of 4.5in. If you pick one of the rack's horizontal edges for a target dimension edge, picking near the right endpoint will place a dimension as shown in the figure above. You could also pick one of the rack's curved edges for a target edge. If that is the case, you will get a menu that allows you to dimension to the arc end point, tangent point, or center point.
- The hole is a simple through hole.

**End of Project**

## 7.7 Self-Test

(All questions are true/false. Answers are at the back of the book.)

1. To vary the radius on a blend along an edge, use the 'Variable Radius' option under the 'Face Blend' command.
2. The 'From Edges' is similar to the 'From Plane,' but instead of specifying a plane, you select edges that remain fixed when the draft is applied.
3. The resultant of a 'Bridge Surface' command is a zero thickness sheet.

# 8. Reference Entities

## 8.1 Introduction

Modeling challenging geometry often requires defining multiple sketches at various positions. To perform these operations parametrically often requires the use of 'Reference Entities' such as **'Datum Planes,' 'Datum Axes,'** and '**Datum Points.'** These can be found under the ***Insert / Datum/Point*** menu shown below.

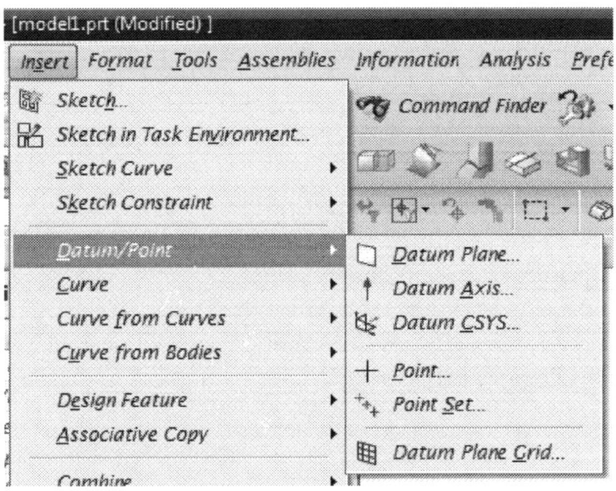


'Reference Entities' are also useful for many other modeling operations such as mirroring, circular instances, and positioning holes.

Often, defining a desired 'Reference Entity' requires a combination of model geometry and Reference features. For example, in the figure below the 'Datum Plane' is oriented at an angle to the top face. Creating an angled 'Datum Plane' requires a straight edge or 'Datum Axis' to define the rotation axis. Since the model in the figure below has no straight edges, a datum axis was created to accomplish this.

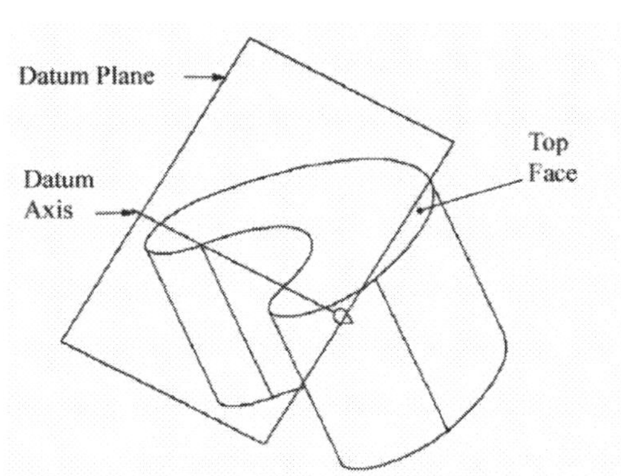

## 8.2 Datum Plane

A 'Datum Plane' is a Reference feature that represents an infinite plane. Datum Planes are very useful for creating sketches, constraining and positioning features, and defining cross-sections for solid models. Datum Planes are defined by specifying a combination of geometric entities and constraints. You may use solid geometric entities (such as faces, vertices, and control points) or other Reference features when defining Datum Planes.

Thus far, when you have created Datum Planes for sketches, you have used the WCS to locate them. However, Datum Planes can also be defined using existing solid geometry.

### ***Detail: Offset Datum Planes***

Datum planes can be offset from a planar solid face, or from another datum plane. Using the solid shown below, create datum planes A and B.

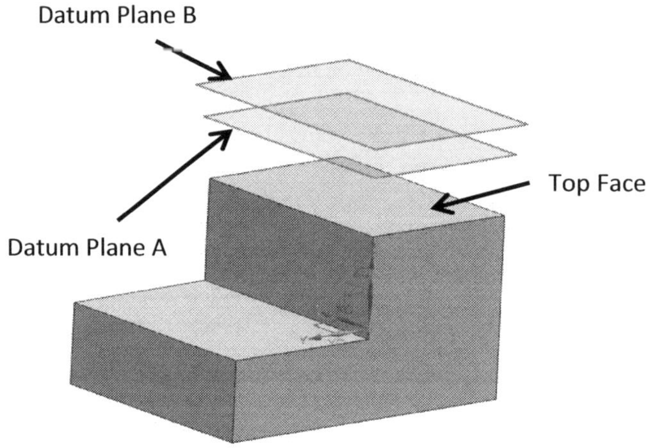


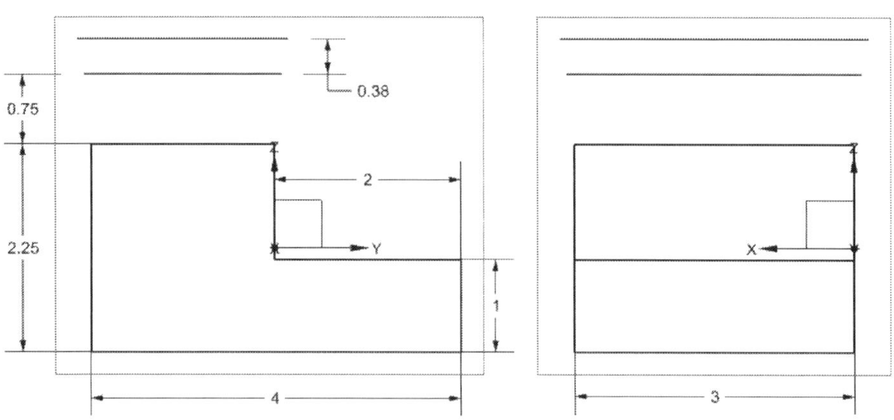


### ***Project 8.2.1***

Create the solid geometry shown in the previous figure, using any method you choose.

- Select Insert / Datum/Point / Datum Plane  .

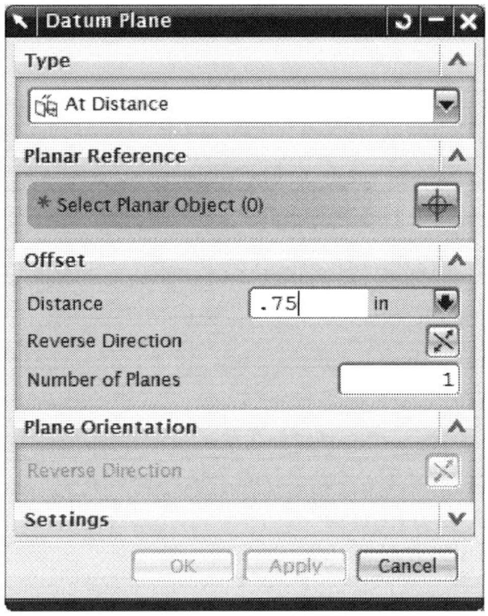


- Set the Type to **At Distance**.
- Select the top face of the solid. The 'Datum Plane' menu (shown in the figure above) will have fields in which to enter the offset value and the number of planes to create.
- A preview of the Datum Plane, along with an arrow appears in the Graphics Window, as shown in the following figure. The arrow indicates the positive offset dircction.

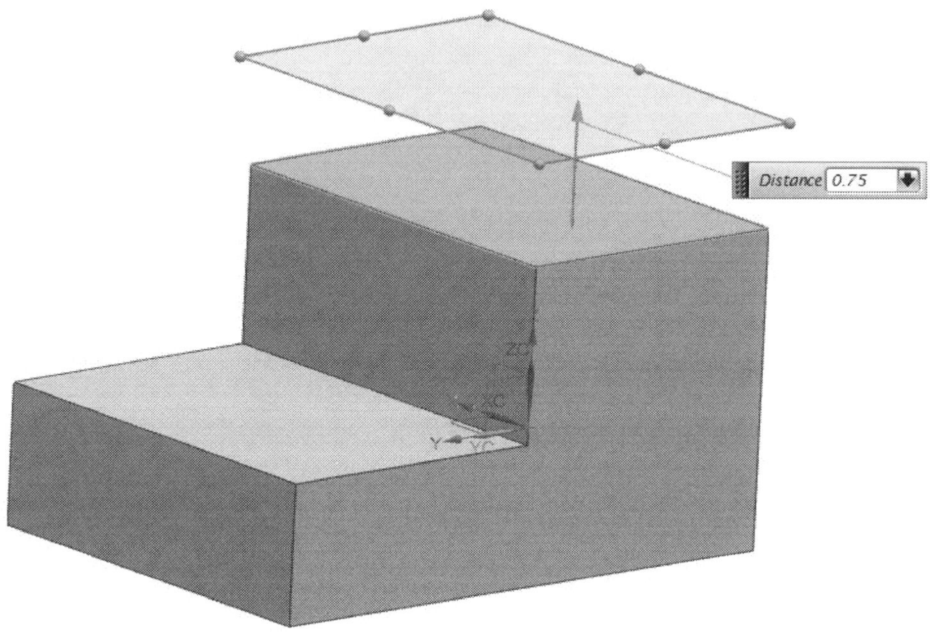


- Enter offset value **0.75** and number of planes **1.**
- Choose **OK**.
- Repeat the process and this time select Datum Plane A (the plane that you just created).
- Type the Offset value **.38**.
- Hit **OK.**

***End of Project***

## *Detail: Centered Datum Planes*

The datum plane shown below is centered between Face 1 and Face 2.

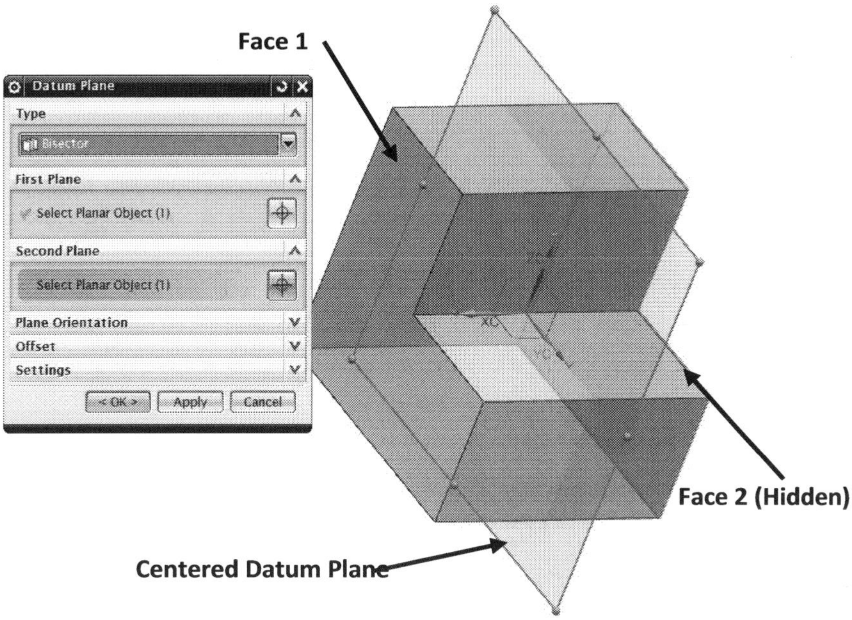


## *Project 8.2.2*

- Either **Undo** or Delete the datum planes you created in the previous project.
- Select Insert / Datum/Point / Datum Plane .
- Set the Type to **Bisector.**
- Select Face 1.
- Rotate the model and select Face 2. NX 8.5 *automatically creates a center plane.*
- Choose **OK** or hit **MB2.**

***End of Project***

## ***Detail: Planes Though Edges***

The datum plane in the figure below is placed though Edge 1 and Edge 2. If you pre-select geometric entities prior to invoking the Datum Plane menu, NX 8.5 will use these entities to infer the datum plane's constraints.

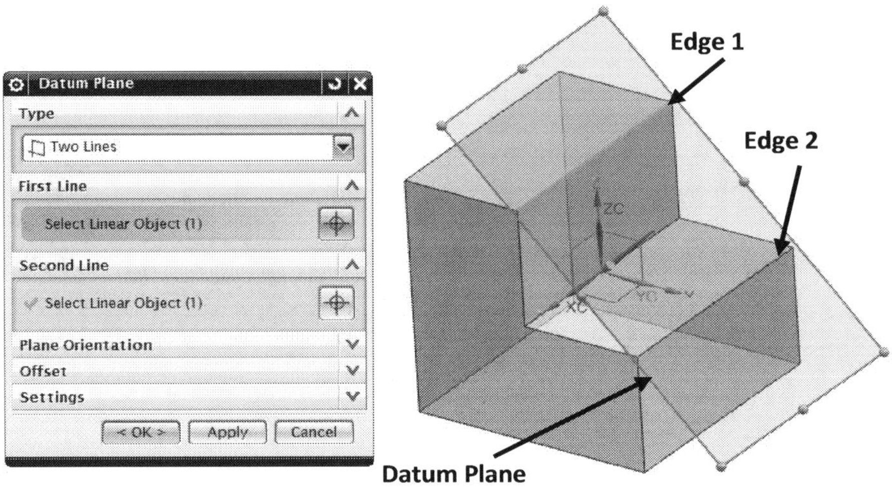


> **Note:** In this project you learned that you can pre-select the objects to create a Datum Plane.

## *Project 8.2.3*

- Either **Undo** or Delete any datum planes that you created in the previous project.
- Select Insert / Datum/Point / Datum Plane.
- Set the Type to **Two Lines.**
- Select Edge 1.
- Select Edge 2. NX 8.5 will display a preview of the Datum Plane in the Graphics Window.
- Choose **OK.**

***End of Project***

## ***Detail: Angled Datum Planes***

Datum planes may be constrained at an angle to a planar face or a datum plane. In order to accomplish this, it must pass through a linear model edge or a Datum Axis. The figure below illustrates a typical example of an angled datum plane.

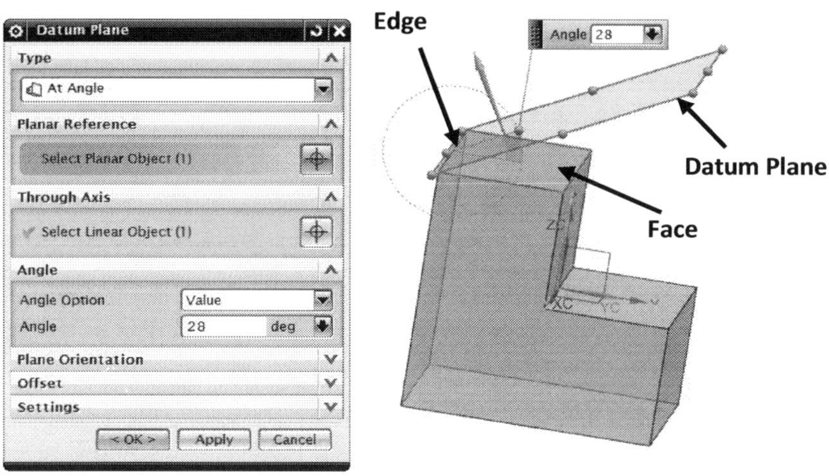


### ***Project 8.2.4***

Create the geometry shown in the figure above.

- Either **Undo** or Delete any datum planes that you created in the previous project.
- Select Insert / Datum/Point / Datum Plane.
- Set the Type to **At Angle**.
- Select the planar face as shown in the figure above. A preview of an offset datum plane appears with an arrow that indicates the positive direction of offset. If you wanted an offset plane, you can accept this result.
- Select the edge as shown in the figure above.
- In the Angle value field input **30** (or **–30**, depending upon which direction works).
- Choose **OK**.

***End of Project***

## Detail: Datum Plane through Three Points

A datum plane can be constrained by three points, as shown in the figure below. To accomplish this, you may select either the vertices or the control points of solid edges.

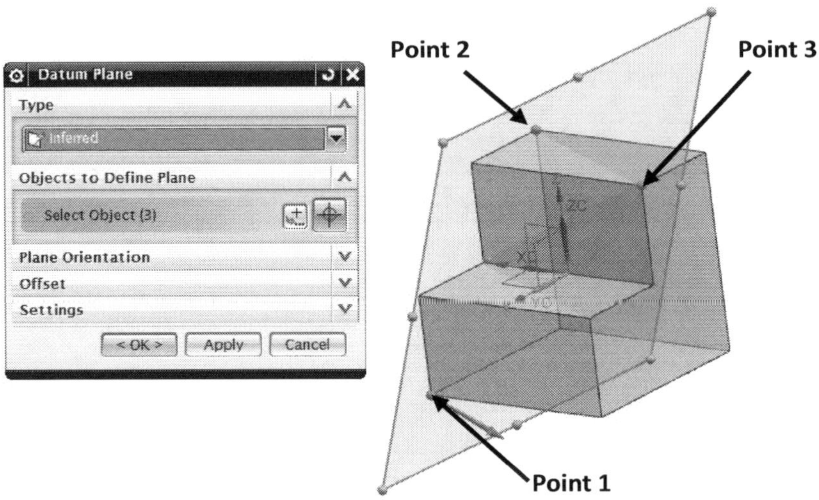


### Project 8.2.5

- Use the solid from the previous project
- Either **Undo** or Delete any datum planes that you created in the previous project.
- Select Insert / Datum/Point / Datum Plane.
- Set the Type to **Inferred**.
- Activate snap point - **End Point**.
- Select point 1.
- Select point 2.
- Select point 3.
- Choose **OK.**

***End of Project***

## ***Detail: Datum Planes defined by Cylinders***

Datum Planes can be constrained to be either tangent to cylindrical faces or through the center axis. To explore these options, construct the geometry shown below.

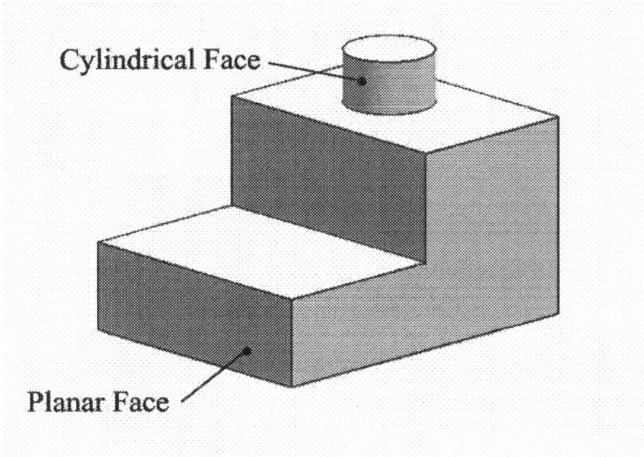


## *Project 8.2.6*

- Use the solid from the previous Project.
- Either **Undo** or Delete any datum planes that you created in the previous project.
- Add a boss with Diameter = **1.0** and Height = **0.6** as shown above.
- Select Insert / Datum/Point / Datum Plane.
- Set the Type to **Tangent**.
- Select the cylindrical face, as shown above. A datum plane preview will appear.
- Select the planar face. The datum plane preview will change.
- Use the **'Alternate Solution'** icon 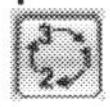 to toggle through the various possible solutions to these constraints. You may wish to rotate the model to examine the results more closely.
- Choose **OK** when the datum plane that is parallel to the planar face appears.

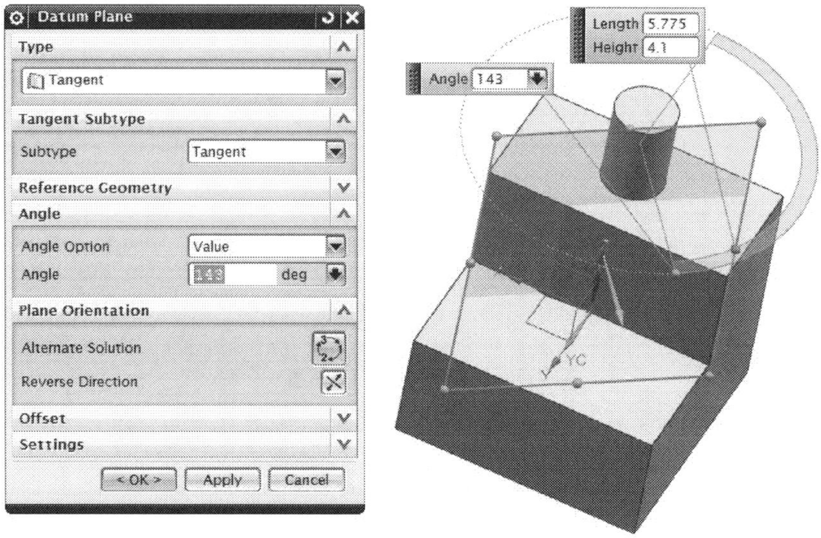


- Select Insert / Datum/Point / Datum Plane.
- Select the axis of the cylindrical face. A datum plane preview that passes through the axis of the cylinder will appear.
- Select any vertical planar face. The datum plane preview will re-orient and change to an angle prompt.
- Enter a value of **45**.
- Choose **OK**.

***End of Project***

| **Note:** For most types of Datum Planes the Inferred type will yield the correct results. |
|---|

## 8.3 Datum Axis Features

A 'Datum Axis' is a Reference feature that defines an infinite line (i.e. vector). They are very useful in many circumstances, such as creating angled datum planes and angular arrays, or for defining the axis for a body of revolution. They are defined by specifying constraints and solid geometric entities, just as for Datum Plane features.

Typical methods to define Datum Axis features include through an edge, through two points on a solid body, through the axis of a cylinder, or at the intersection of two planes.

### *Detail: Datum Axis at Intersections and Through Edges*

You will create the two Datum Axis features shown below.

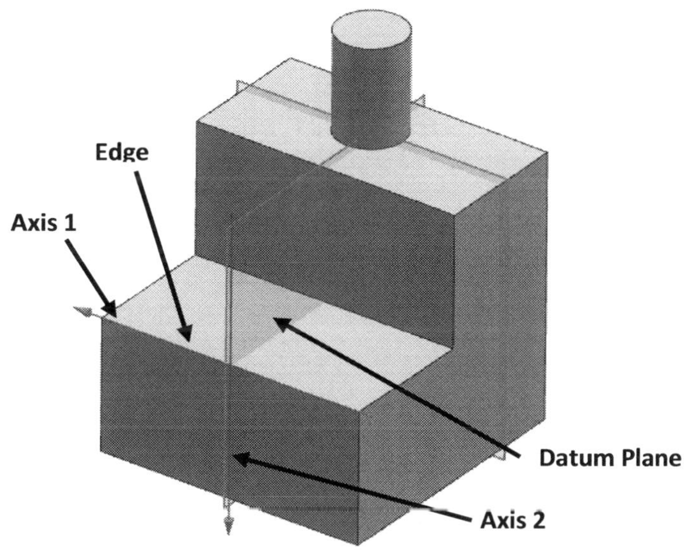


### *Project 8.3.1*

Continue working on the same part you used in the previous Project.

- Select Insert / Datum/Point / Datum Axis 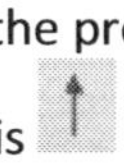 .
- Select the edge as shown above. You should see a preview of Axis 1 appear.
- Choose **OK**.
- Select Insert / Datum/Point / Datum Axis.
- Select the datum plane as shown above.
- Select the face as shown above. You should see a preview of Axis 2 appear.
- Choose **OK**.

***End of Project***

## 8.4 Point

The 'Point' feature creates a point entity that is fully associative and parametric, meaning that design features can be moved around according to the point to which they are set.

### ***Detail: Creating Parametric Point Entities***

The following exercise shows how to create a point entity and demonstrates its parametric capability.

### ***Project 8.4.1***

Create the block shown next with the dimensions X = 5in, Y = 3in and Z = 1in.

- Go to **Insert / Datum/Point / Point** .
- The following figure shows the 'Point' box to control the placement of points. Points can either be placed by the cursor or by typing in the coordinates of exact locations.

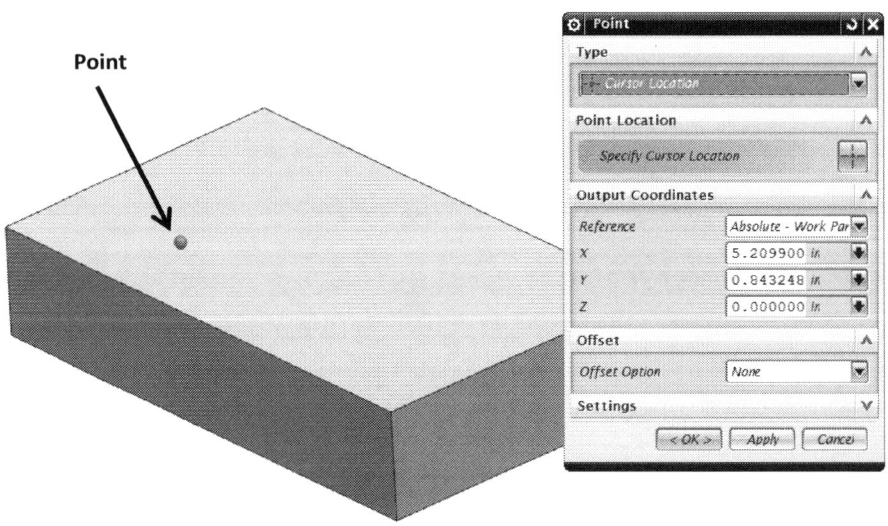


- Create a point by inserting coordinates into the box shown above, making X, Y and Z all 1in. A point will be created as shown above.
- Create a boss feature, *Insert / Design Feature / Boss*, with a height of 1in and a diameter of 0.5in. Place the Boss on the top surface and then use the 'Point on Point' positioning function to position the Boss onto the newly created 'Point' entity, as shown on the following page.

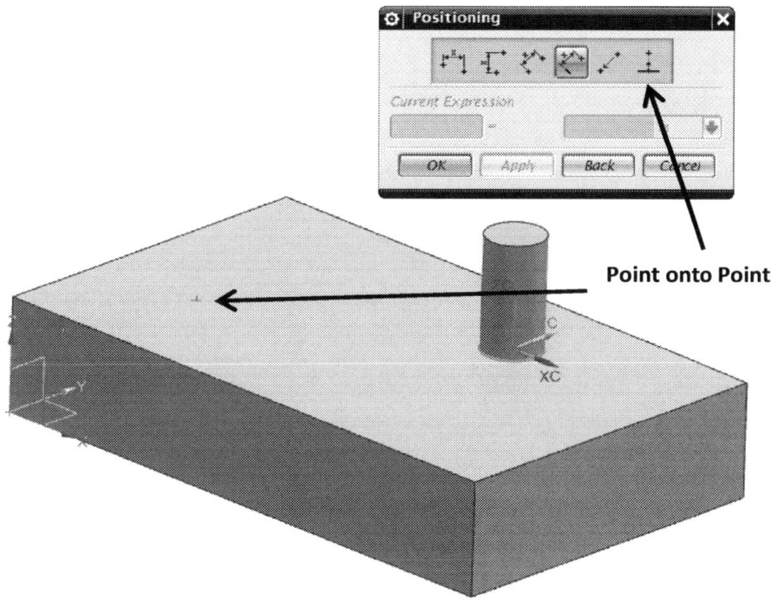


- The model should now look like the figure below.

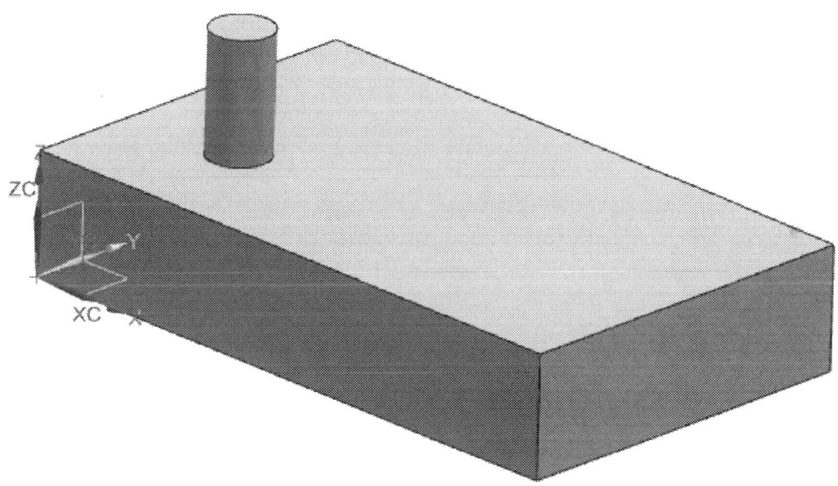

- Go into the part navigator and Right Click on the point that you just created and select 'Edit Parameters'. *This enables you to change the location of the point.*
- Change the coordinates to X = 3 and Y = 2. The boss will change position with the point.

***End of Project***

## 8.5 Self-Test

(All questions are true/false. Answers are at the back of the book.)

1. You can rotate datum planes in either positive or negative angles about a solid edge.

2. If you do not have an appropriate edge to rotate a datum plane about, you can use a datum axis instead.

3. Sketches can be placed on datum planes that you have manually created.

# 9. Other Basic Techniques

## 9.1 Introduction

This chapter will focus on a very useful sketch command: 'Sketching on a Solid Face.' We will also discuss operations found at various locations within the 'Insert' menu and discuss the functionality of the 'Shell,' 'Trim (and Split),' 'Instance (rectangular, circular, and mirror),' 'Offset Face,' 'Thicken Sheet,' 'Tube,' 'Flange,' and 'Extract' features.

With the exception of some types of 'Instances' and 'Split Body,' all 'Feature Operations' result in parametric geometry.

## 9.2 Sketching on a Solid Face

NX 8.5 allows you to sketch on any flat face. Notice that three distinct sketch planes on three non-parallel surfaces needed to be used to create this geometry.

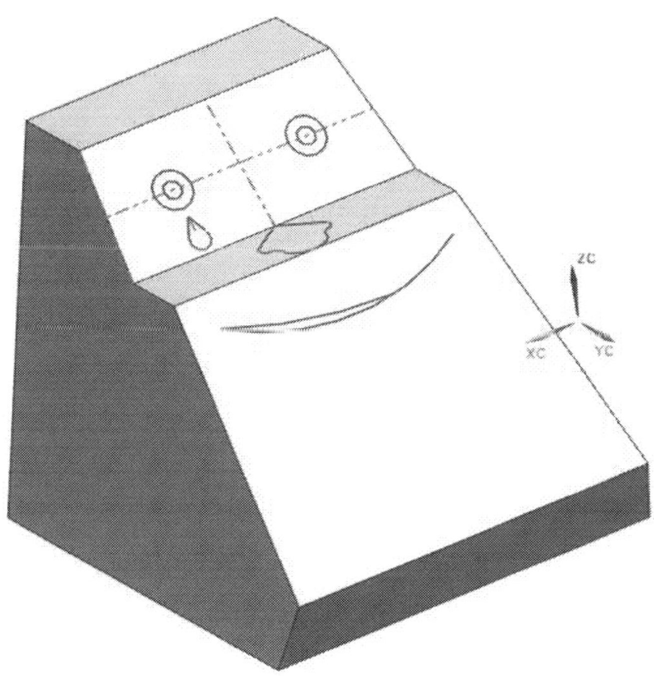


***Project 9.2.1***

- Create the approximate geometry shown next using a basic sketch and extrude it **3** inches. Be sure Faces 1, 2, and 3 are all at different angles to the base of the model.

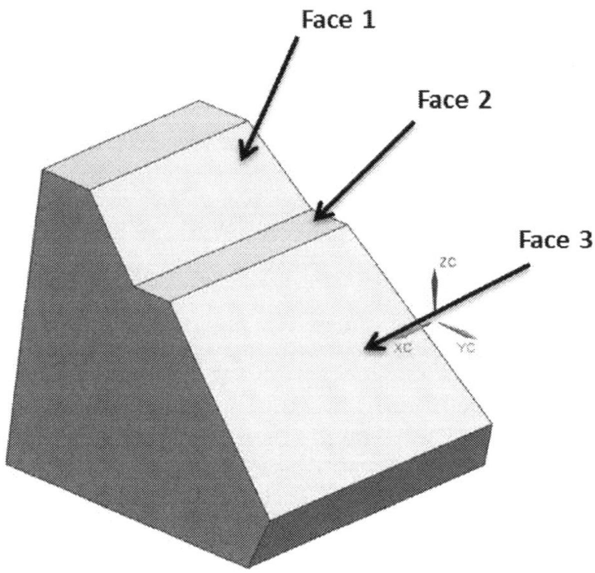


- Insert a regular **Sketch** and select Face 1. (NX 8.5 will highlight the selected face.)
- Choose **OK.**
- NX 8.5 will assume a horizontal sketch axis and re-orient the model so you are now viewing the new sketch plane from above (in the +ZC direction). *Note: you can select an object to align the horizontal sketch axis. You can also define the vertical sketch axis by changing the sketch orientation reference to vertical and then selecting an object to align it.*
- Sketch two pairs of arcs as shown in the figure on the following page.

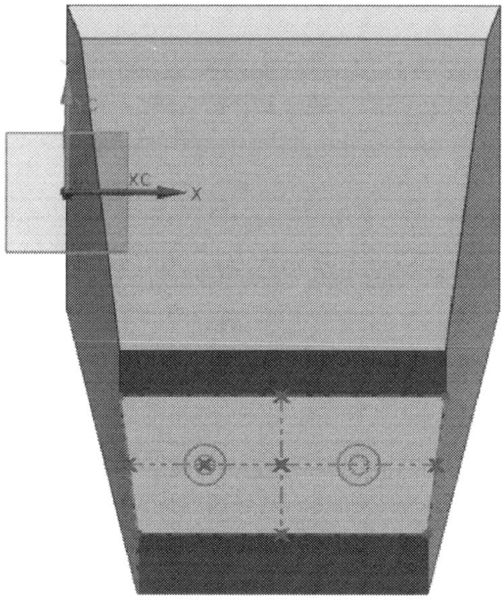


- Select **Finish Sketch** and repeat the procedure to create similar sketches on Faces 2 and 3 as shown below.

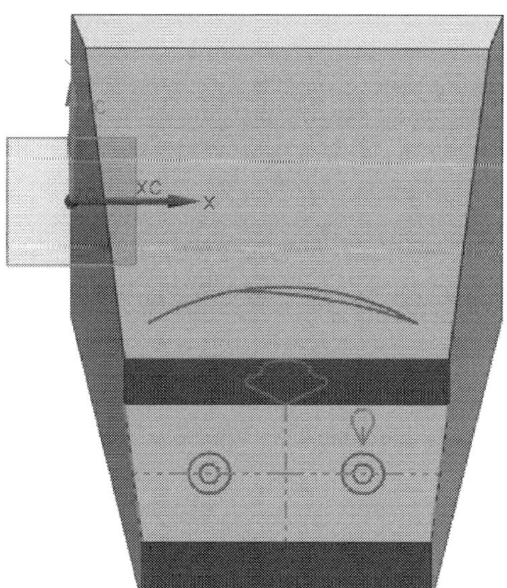


Compare your results to the figure at the beginning of this project.

***End of Project***

## 9.3 Shell Feature

The 'Shell' feature is a method to remove material from an existing solid body, resulting in a constant wall-thickness body. During the definition of a 'Shell' feature, you specify faces that are removed by the feature. The Figure below shows two solids that are shelled with a 0.25in wall thickness. The solid on the left had the top face removed, while the right solid had the top, left, and right faces removed.

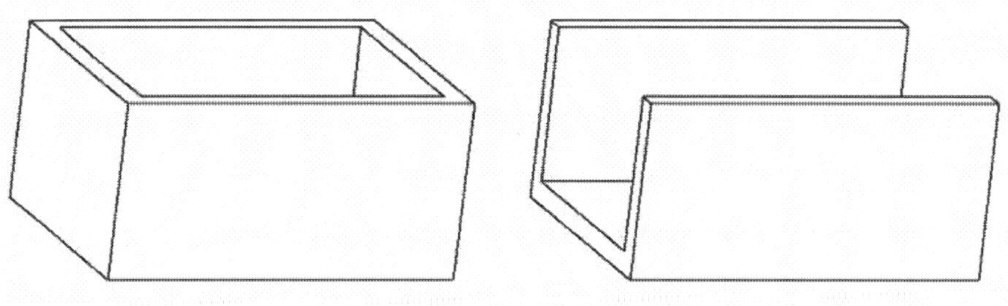

### ***Detail: Shell Feature***

Here you will create the flanged bushing shown on the following page by shelling an extrusion.

### *Project 9.3.1*

- Sketch a circle with a diameter of **3.0in**.
- **Extrude** the circle from **0** to **4** and offset the extrusion **0.75in** outward to create an annular cylinder, as shown on the following page.
- Select Insert / Offset/Scale / Shell .
- Input a thickness of **0.38** in the menu. This will be the wall thickness of the shelled solid.

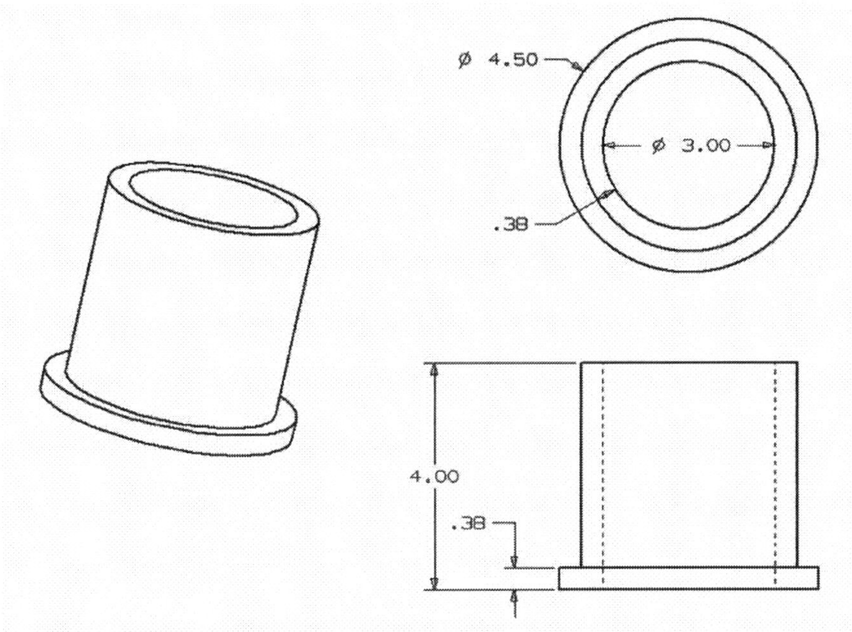


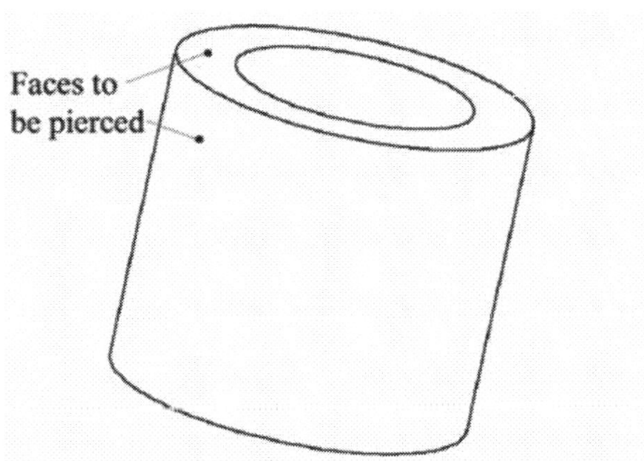


- Read the cue line – it is asking you to select a face to remove.
- Pick the cylindrical face as shown above. This is the first face to be removed.
- Pick the planar face as shown above. This is the second face to be removed.
- Choose **OK.**
- Your model should look like the figure on the following page.

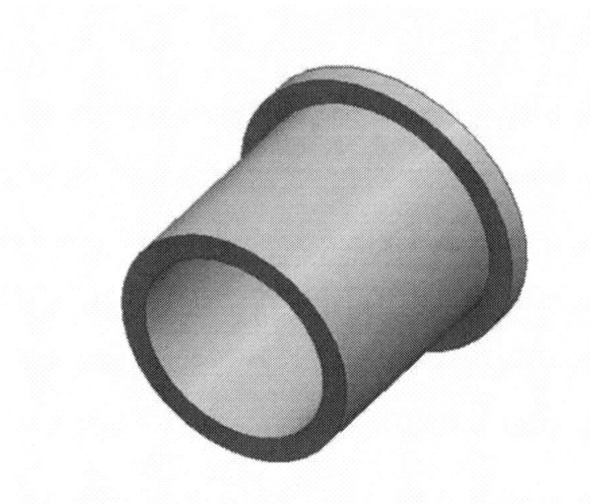

- Save the model and rename it as ***'bushing_<your initials>.prt'*** .

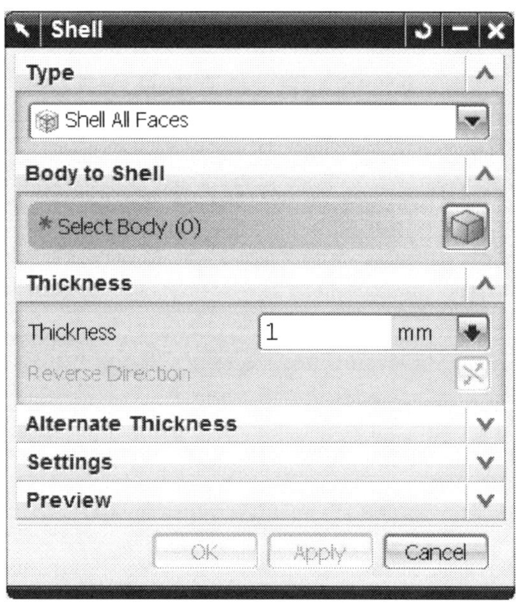


**Note:** If you simply want to give an entire body a wall thickness, i.e. to hollow it, use the shell all faces option as shown in the left figure.

***End of Project***

## 9.4 The Trim Body Feature and the Split Operation

The 'Trim Body' feature is used to remove part of a solid body using either a datum plane or a contoured surface. You can specify which part of the body you wish to remove by changing the direction of the red trimming arrows.

> **Note:** Trim body features fail if the operation splits the body into multiple pieces.

The 'Split' operation is a non-parametric operation that divides one solid body into two or more pieces. Once the Split operation has been performed it cannot be reversed (unless Undo is available), so it should be used carefully. The 'Split' operation has no restriction on the number of bodies created, but the results will *always* be non-parametric.

### ***Detail: The Trim Body Feature***

In these examples you will create a solid body and trim it using a datum plane and a face.

### *Project 9.4.1*

Create a **4in** x **4in** x **8in** block.

- Create a datum plane at the center of the 8in length.
- Select Insert / Trim / Trim Body.
- Select the block as the target body for the Trim operation.
- Choose **OK.**

The 'Trim Body' menu appears, as shown next.

- After selecting the block as the target and the datum plane as the tool, a direction arrow appears. You can choose to reverse the trimming direction using the button in the lower right hand corner of the menu shown above.
- Choose **OK** to accept the default direction, which indicates which portion will be trimmed.
- Choose **Cancel**. Compare your results to the figure below.

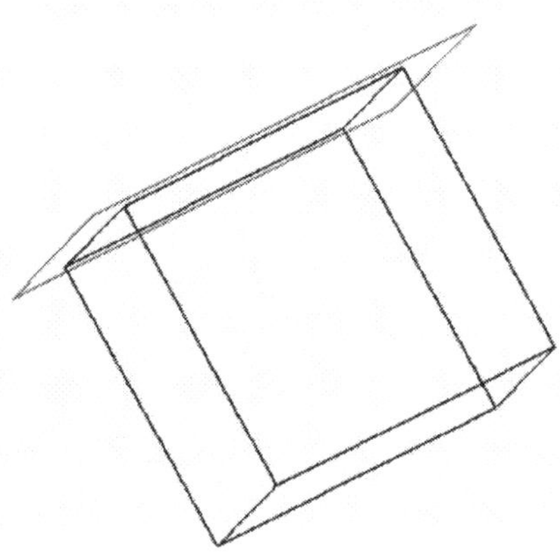

***End of Project***

## *Project 9.4.2*

Undo the trim operation from the last project.

- Create a curve and extrude it to create a sheet body, as shown below. Make sure that all edges of the sheet body extend beyond the block.

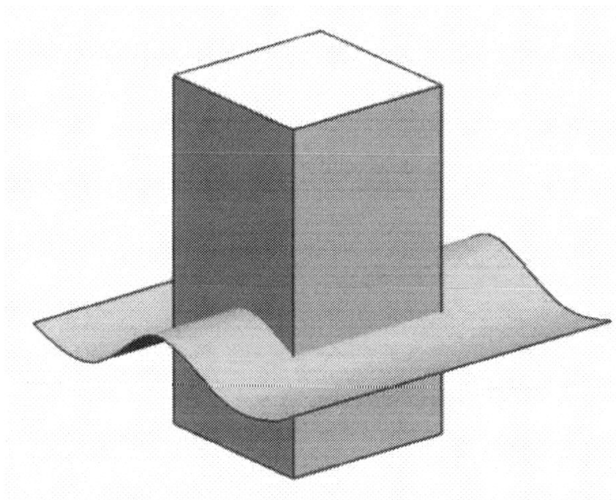

- Select Insert / Trim / Trim Body.
- Select the block as the target body for the 'Trim' operation.
- Choose **OK.**
- Select the sheet body as the tool.
- Choose **OK** to accept the default direction.
- Hide the extruded sheet body.
- Compare your results to the figure below.

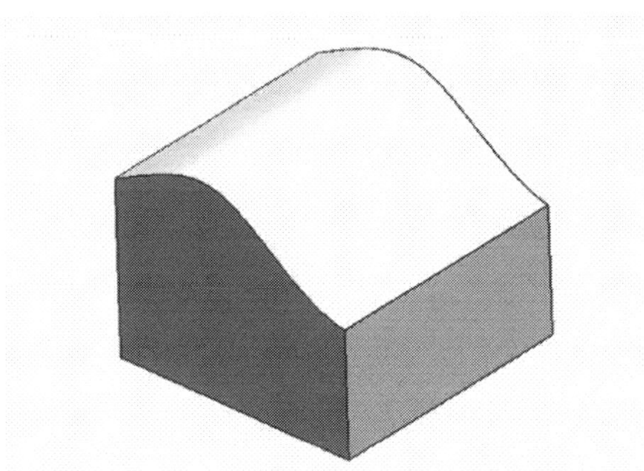

*Choose Cancel to exit.*

***End of Project***

## *Detail: Split Features*

In this example, use the extruded sheet body to split the block, instead of trimming it.

## *Project 9.4.3*

- Undo the trim operation from last project.
- Select Insert / Trim / Split Body.

NX 8.5 may warn you that the Split feature removes parameters from all of the resultant bodies. This warning will appear only once per session. Choose **OK.**

- Select the block as the target body for the Split operation.
- Choose **OK.**
- Select the extruded sheet body as the splitting entity. NX 8.5 now informs you that the parameters have been removed from all of the resultant bodies.
- Compare your results with the figure below. You should have two bodies, not one.

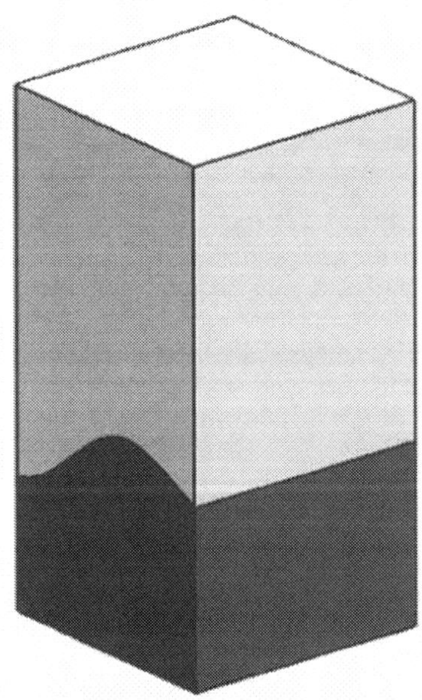

**Note:** The 'Split' operation consumed the splitting sheet body, whereas the 'Trim' feature that you created in the previous project retained it.

- Choose **OK.**
- Choose **Cancel.**

***End of Project***

## 9.5 Offset Face Features

The 'Offset Face' feature offsets one or more solids along their normal vector. The 'Offset Face' feature is very useful when making small changes to existing models and can be applied to both parametric and non-parametric models.

### *Detail: Offset Faces*

In this section you will use the bushing from the previous section and offset the upper face of the flange such that it is only 0.18in thick.

### *Project 9.5.1*

- Open the part file 'bushing_your_initials.prt' .
- Select Insert / Offset/Scale / Offset Face.

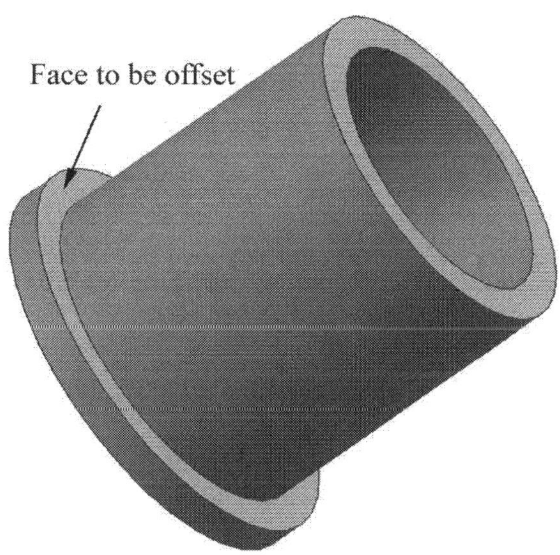


- 

The flange is 0.38in thick; the goal is to decrease the thickness to 0.18in. Thus, this face must be offset by -0.2in.

- Input an Offset Value of **-0.2**.
- Pick the face as shown below.
- Choose **OK**

After selecting OK you will see the thickness of the flange decrease, as shown below.

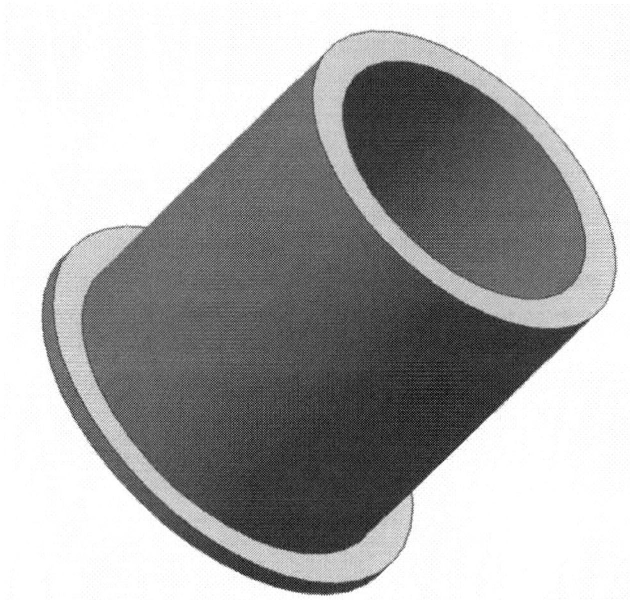

Save your part file.

***End of Project***

## <u>9.6 Thicken Sheet</u>

'Thicken Sheet' is a useful feature when importing surfaces from external software packages. The command allows you to create a solid body from a sheet (surface) body.

### ***<u>Detail: Thicken curve</u>***

'Thicken' can be used to quickly create a solid body from a sheet, as shown below. In this project, you will thicken a sheet.

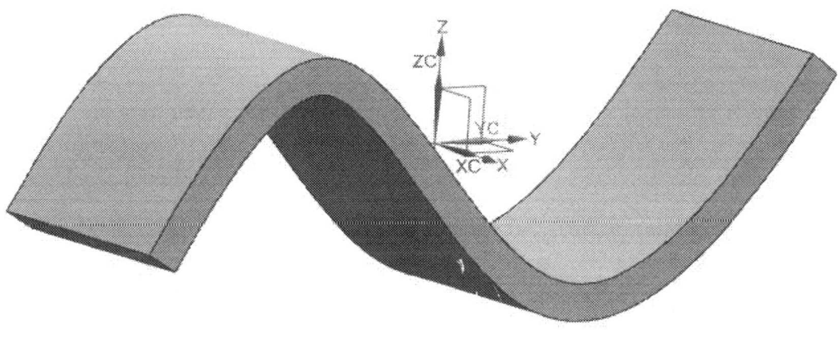


### *Project 9.6.1*

Using splines, create a spline curve similar to the following figure. Extrude the curve to **1** inch.

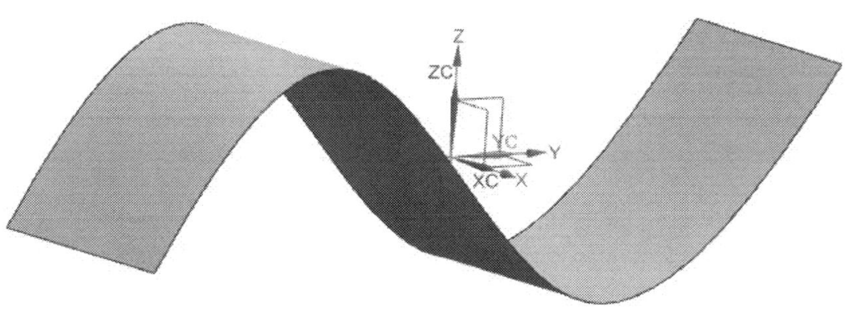


- Select Insert / Offset/Scale / Thicken .

- In the 'Thicken Sheet' window, enter the First Offset = **0.2** as shown below.

**Note:** the direction of the arrow. Input a negative value or click the Reverse Direction button if you would like to reverse the direction.

Read the cue line – NX 8.5 is asking you to select a sheet body to thicken.

- Select the curve sheet.
- Select **OK**.
- Compare your results to the figure at the beginning of this project.

***End of Project***

## 9.7 Tube Features

The 'Tube' feature is a special case of the 'Sweep along Guide' feature, located under ***Insert / Sweep***. It creates solid geometry by sweeping a circular section along an arbitrary guide curve.

The figure below shows a tube feature that has been created using a helical guide curve.

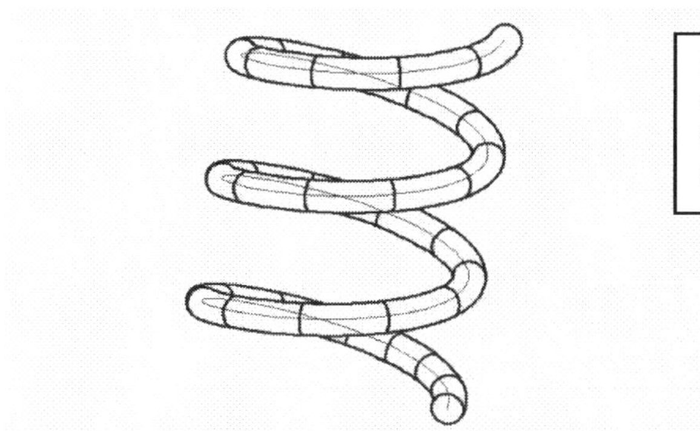

**Note:** This is a common way to create a spring. Use the **Insert / Curve / Helix** command as your guide string.

### ***Detail: Tube Features*** 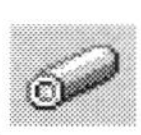

Tube features can also be defined using solid edges as the guide curve. The following figure shows a cylinder with a tube cutout at the end – perhaps the shaft needs to mate against an O-ring.

## *Project 9.7.1*

Create a **2**in diameter cylinder that is **3in** long. *Use any method you choose.*

- Select ***Insert / Sweep / Tube***  .
- Input an Outer Diameter of **0.5** and an Inner Diameter of **0** in the 'Tube' menu (shown below). *Non-zero values for the inner diameter result in a hollow tube.*

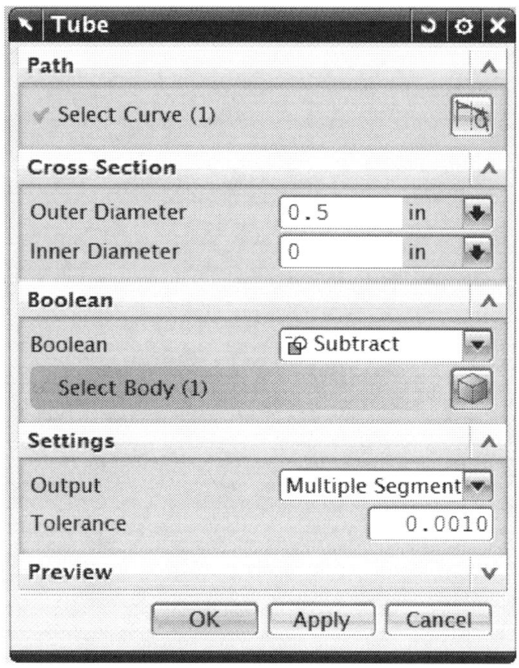


- Choose **OK.**
- Pick the top edge of the cylinder as shown below.

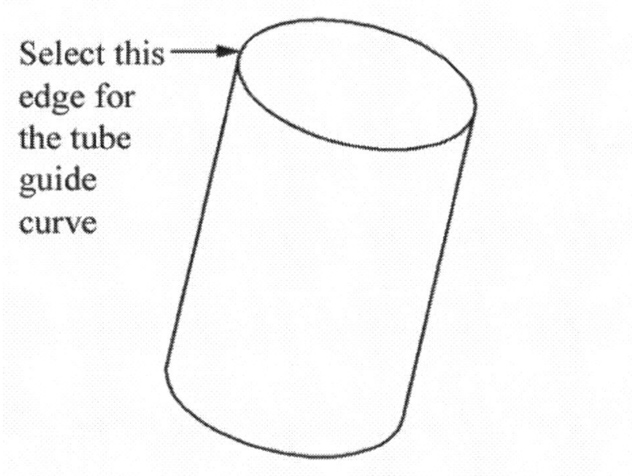


- Choose **OK.**
- Select **Subtract** from the Boolean menu. *Compare your results to the figure at the beginning of the project.*

***End of Project***

## 9.8 Pattern Feature 

The Pattern Feature capability is a powerful method of creating an array of identical features in an associative manner meaning the copies will all take after the original, like cool little children. When you make a parametric change to the mother, the children all do the same. The variety of entities that you can pattern is endless. Imagine a speaker gill, gear teeth, the knurl on a tool handle or any such repeated geometry and you should immediately think “Pattern Feature”. Due to the variety of ways you can pattern geometry the menu is huge:

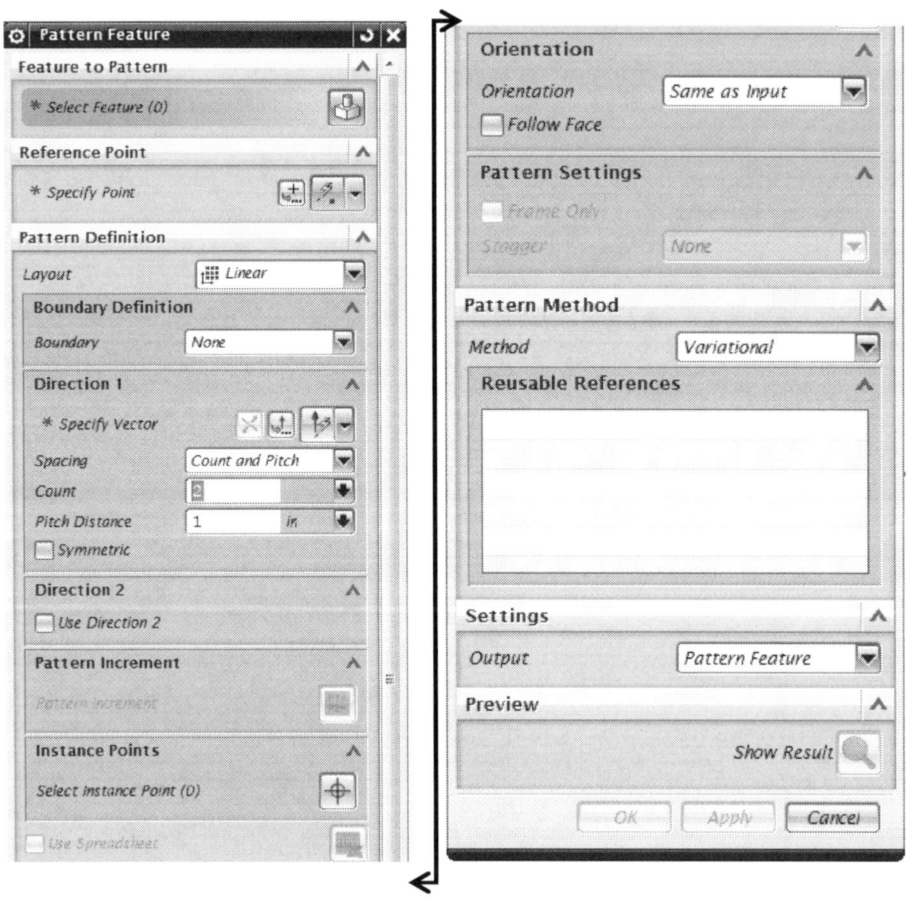

## ***Detail: Linear***

The Linear option of the "layout" choice within the Pattern Geometry command allows you to pattern features in a rectangular fashion, such as shown below.

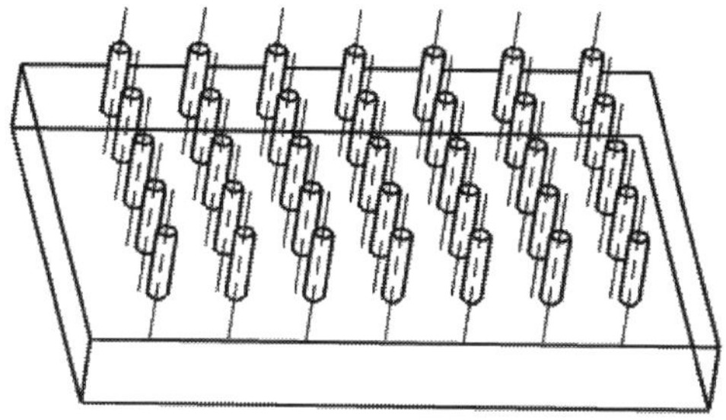

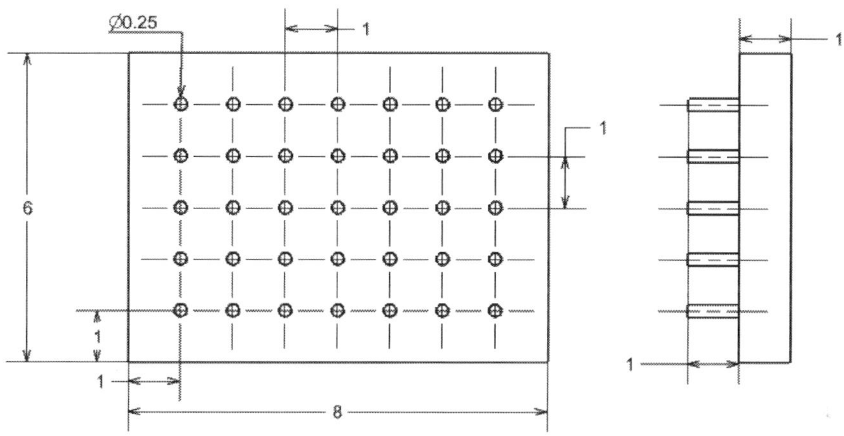


### *Project 9.8.1*

- Create an **8in** x **6in** x **1in** block using any method that you wish. (Orient the 8in dimension along the XC axis and 6in along YC.)
- Create the initial boss, using the dimensions shown above.
- Select Insert / Associative Copy / Pattern Feature  .
- Select the boss and a small yellow cube will appear. This is the feature reference point. It helps you understand what the pattern will look like before you actually have the system calculate the entire array.
- Select **Linear** from the **Layout** choice within the **Pattern** Definition pull down.

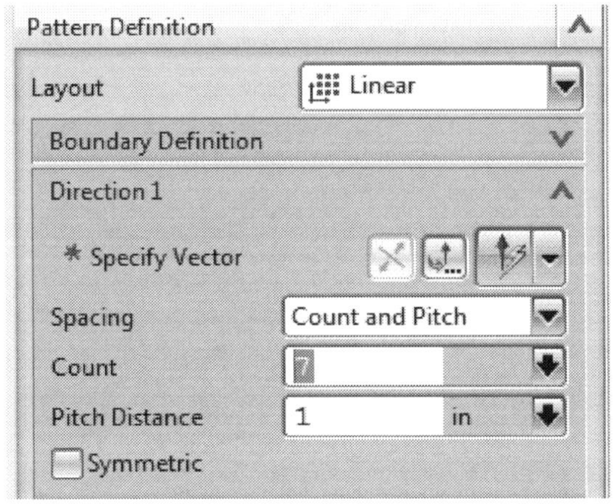


For this project make sure the **Spacing** method is "**Count and Pitch**" Input the **Count** of **7** and the **Pitch Distance** of **1.** To specify the vector you may select one of the long edges of the block. Select the edge in a location that suggests the direction that will fill the block.

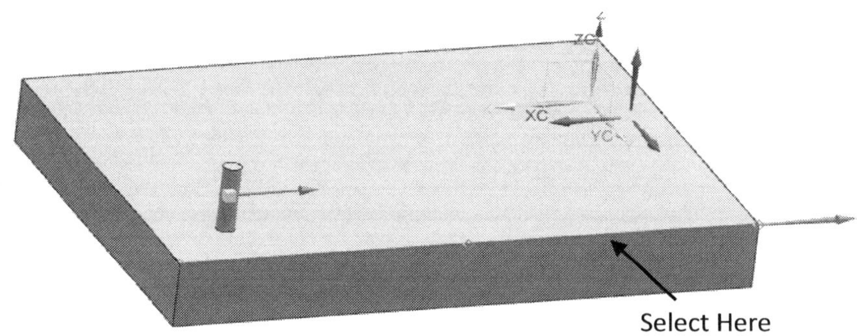


- Next you must click in the box labeled "**Use Direction 2**".
- A similar portion of the dialog box will appear where you can input a **Count** of **5** and a **Pitch Distance** of **1** and select the edge of the block that is 90 degrees away from the first one.

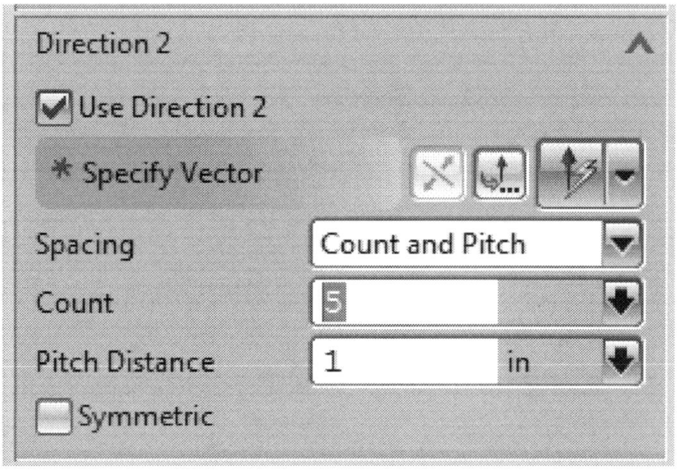


You will get a preview of the positions of the bosses before you **OK** it.

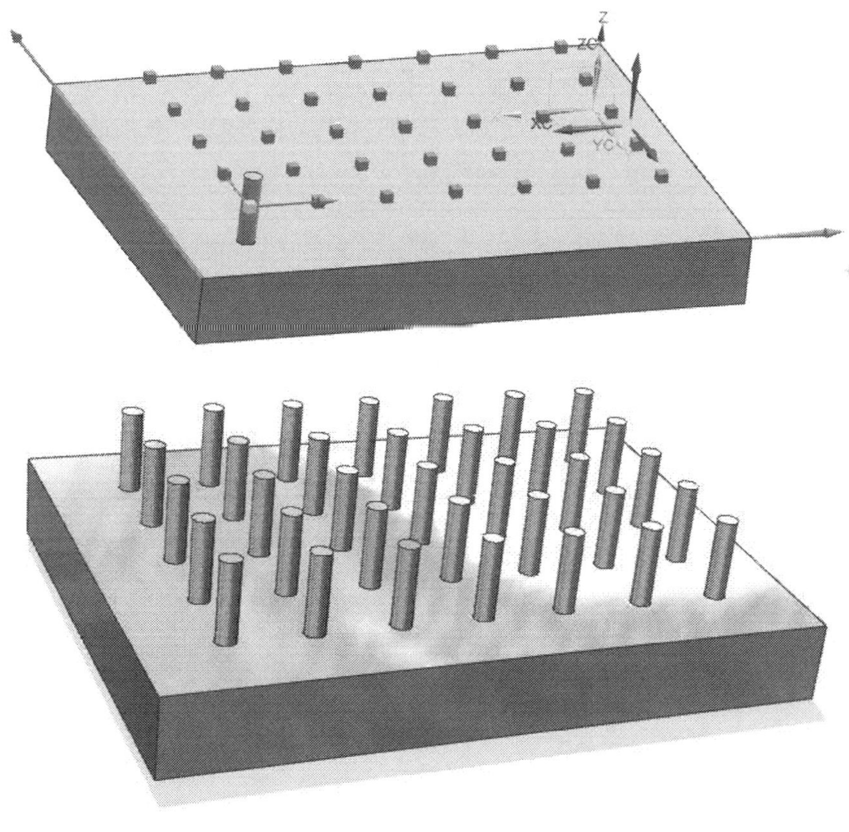

***End of Project***

## ***Detail: Circular Instance Arrays***

You can also use the **Pattern Feature** command in a Circular, Polygon or Spiral mode. The most common is circular

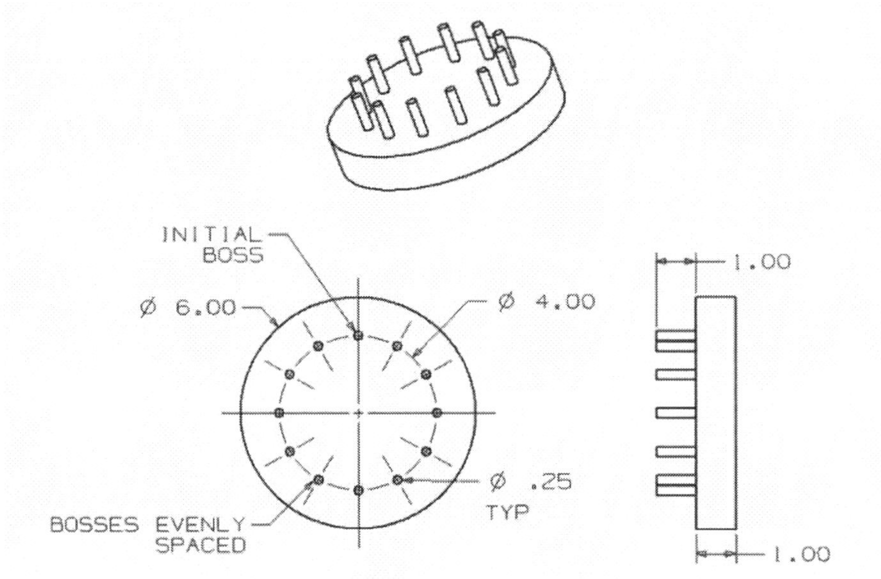


## ***Project 9.8.2***

In addition to learning how to create a circular array, this project requires a new technique for positioning the boss – by dimensioning to datum planes.

- Create Datum Csys at the WCS origin *(Insert / Datum/Point / Datum CSYS / OK)* if you don't have one by default in a new part file.
- Sketch a **6in** diameter circle. Position its center on the x-y datum plane at X=**0**, Y=**0**, as shown in the following figure.

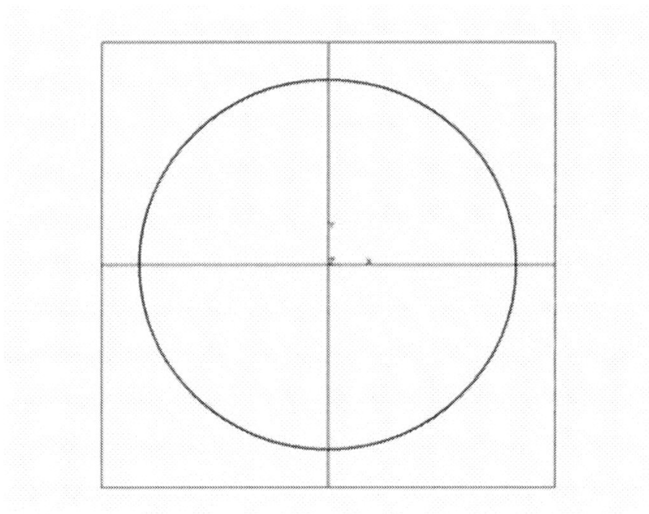

- Extrude the circle **1in.**
- Create the initial boss and position it relative to the datum planes as shown below. (Use the **Point onto Line** command within the positioning window to center the boss onto the Y axis and the **Perpendicula**r command to create the 2 inch dimension away from the x axis

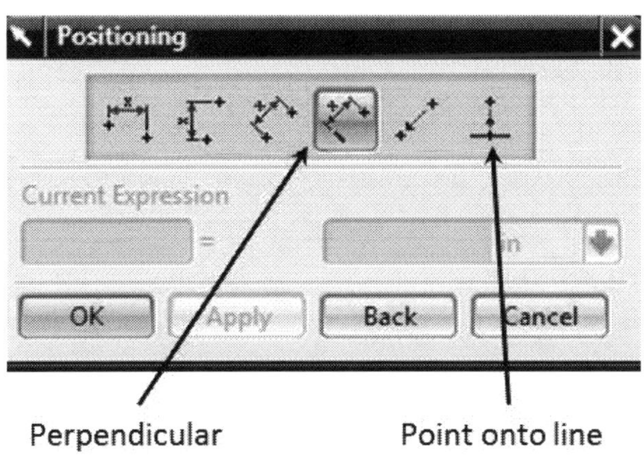


> **Note:** You can achieve the same geometry using the cylinder command. The choice is yours.

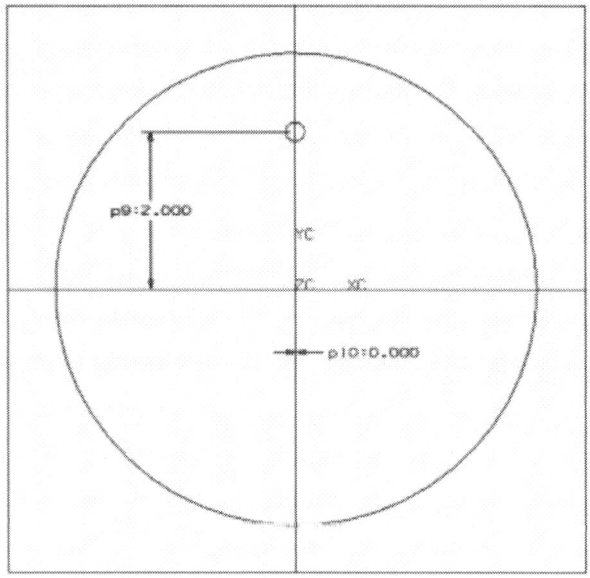


**Note:** The Datum CSYS comes with a datum axis that you will use to create the circular pattern.

- Once you've created the Datum CSYS and the boss on the cylinder select ***Insert / Associative Copy / Pattern Feature*** again.
- Set the Pattern Definition / Layout to **Circular** and select the boss feature
- Set the Spacing to Count and Pitch and select the datum axis that is aligned with the z axis.
- Enter the values shown below: Count = **12** and Pitch Angle = **360/12**. These values determine the number of members in the Pattern and the angular spacing between each instance. Notice that you can let NX evaluate mathematical expressions. Here we have specified a full 360-degree set of pattern instances

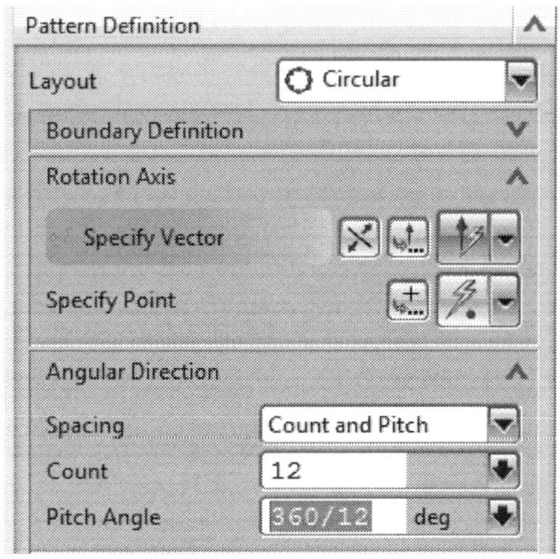


**Note:** when you select the vector, as long as the "Specify Vector" switch is set to Inferred, you will not need to "Specify Point".

- Enter the values shown above: Count = **12** and Pitch Angle = **360/12**. These values determine the number of members in the Pattern and the angular spacing between each instance. Notice that you can let NX evaluate mathematical expressions. Here we have specified a full 360-degree set of pattern instances.

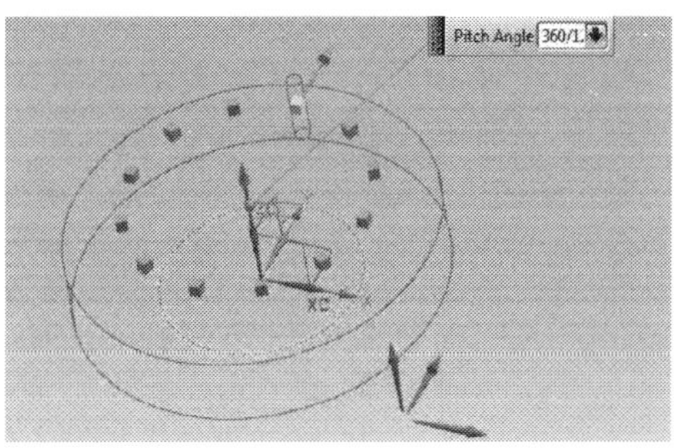


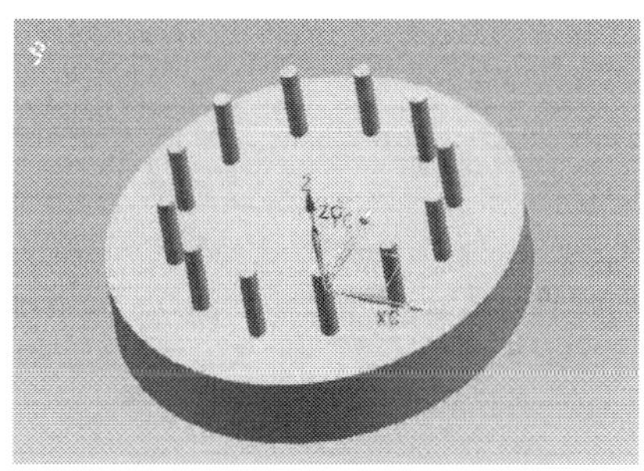

The preview and the finished product are shown above.

- Next come to understand the radiate command. Just edit the pattern feature that you just created and open the **"Radiate"** sub-menu.
- Set the buttons thus: Create Concentric Members = checked on / Include First Circle = checked on / Spacing = Count and Pitch Count = 3 Pitch distance = .4

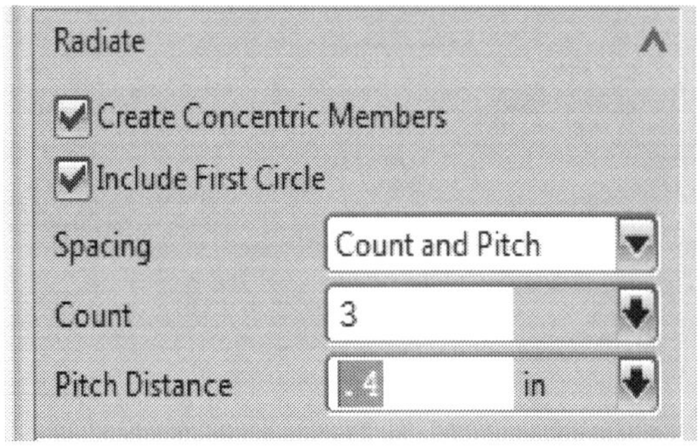


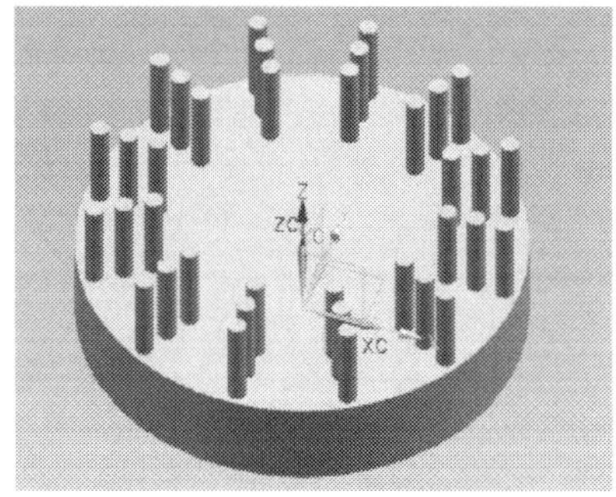

Next come to understand the meaning of **Stagger** within the Pattern Settings dialog. Set the **Stagger** choice to **Angle**. The result is shown below:

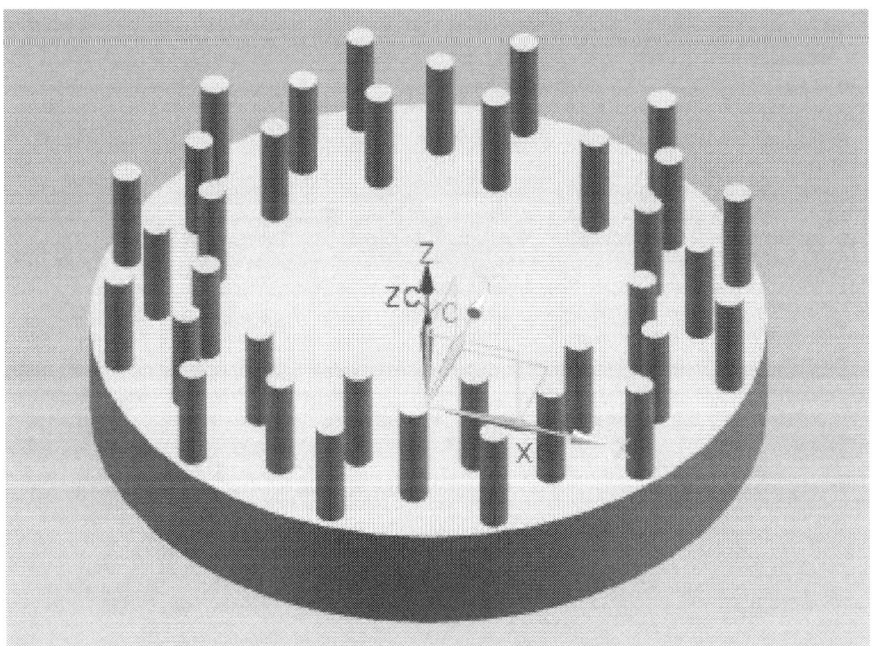

**End of Project**

## ***Detail: "Follow Face"***

It was very difficult before – but not any more

In the old days, when you wanted to repeat a feature on some pattern and you wanted that feature to remain normal to a surface, it was difficult and tedious. The technique is useful when you want to model anything that looks like a hair brush, a heat exchanger, or the texture on anything that is cloth or rubber.

In NX8.5 there is powerful new functionality in the **Pattern Feature** command. It's awesome. In one step you are now able to create a pattern that "**radiates**" according to a "**count**" and a "**pitch angle**", and will also "**follow face**" if you want. This means each associative copy of the original feature will maintain the same relationship relative to a normal vector and selected surface as the original. The next figure shows how a pattern of spines are distributed along a face.

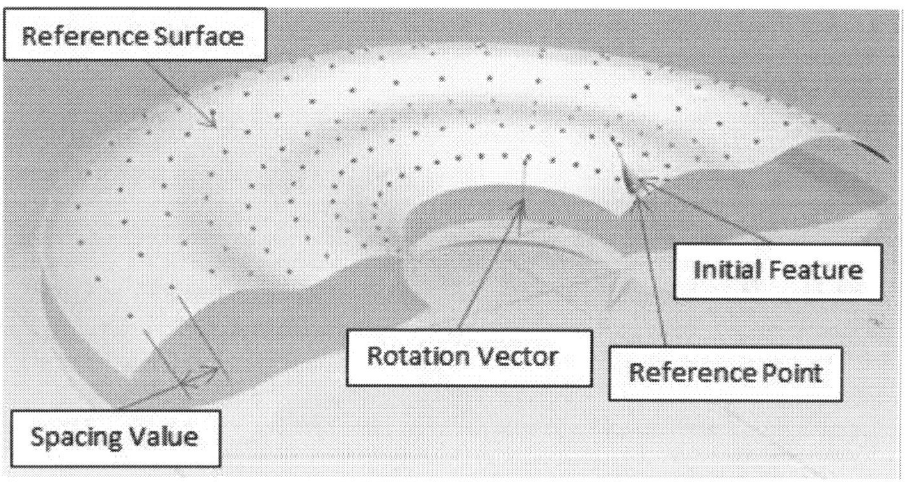


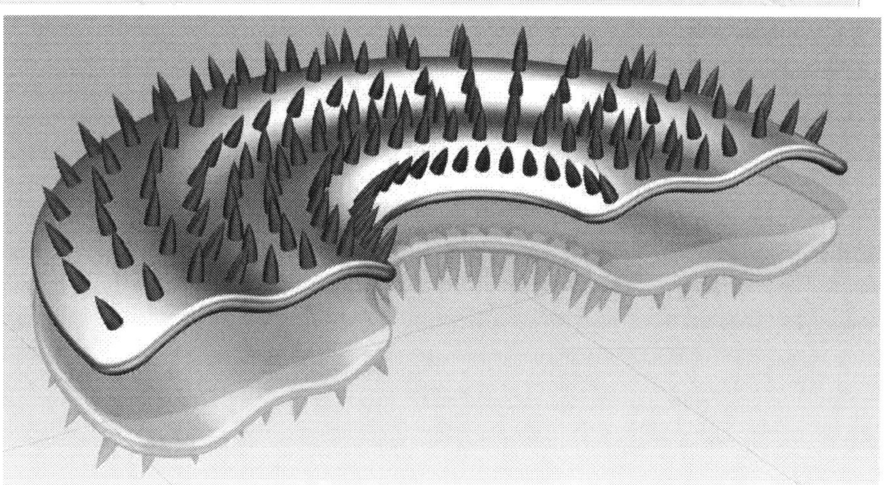

## ***Detail: Pattern Increment***

In addition to the stager and follow face options, there is another truly amazing ability in the **Pattern Face** feature. It's called "**Pattern Increment**". It gives you the ability to add an incremental amount onto one or more of the defining parameters of the feature you are patterning. For example, if you have a series of bosses on a solid block and you want them to get taller and taller as they progress as a pattern, you can use the Increment Option and add a value to the expression that controls it's height.

## *Project 9.8.3*

This project will teach you how to use the Pattern Increment option in the Pattern Feature command. You will also learn about the Along command

- Create a cylinder that is 1 inch in diameter by 6 inches high

- Next offset the top and bottom by .25 inches. This way the shape will be a little bit lower than the wcs and a little bit higher. And build a helix with a radius of .5 a pitch = 1 and number of turns equal to 6.

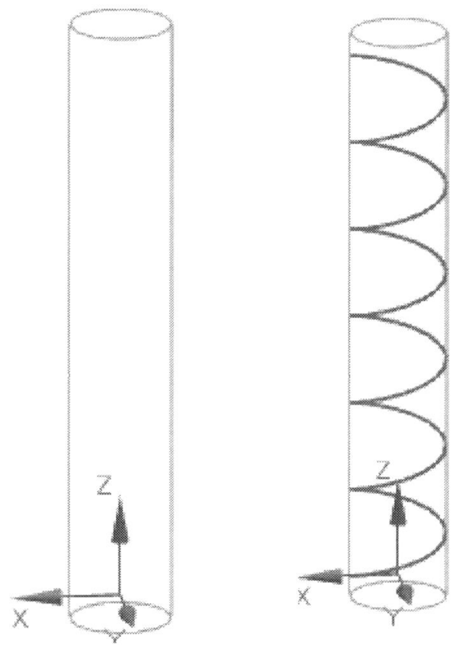


- Using the Sphere command, place a .1 diameter sphere right at the beginning of the helix and unit it on as shown below:

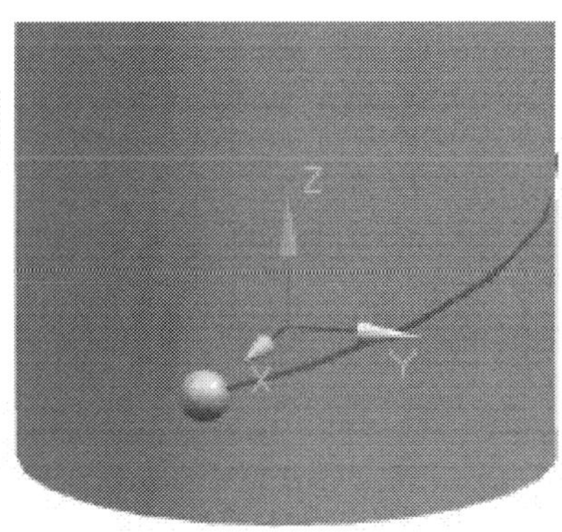


- Next use the Pattern Feature command with **Pattern Definition** set to **Layout = Along** and all other parameters as shown below. You will select the sphere as the feature to be patterned, select the helix as the curve to follow along. Also set the Pattern Increment to .01 on the sphere diameter. This means for every new member of the array, there will be a sphere that is .01 larger than the last one.

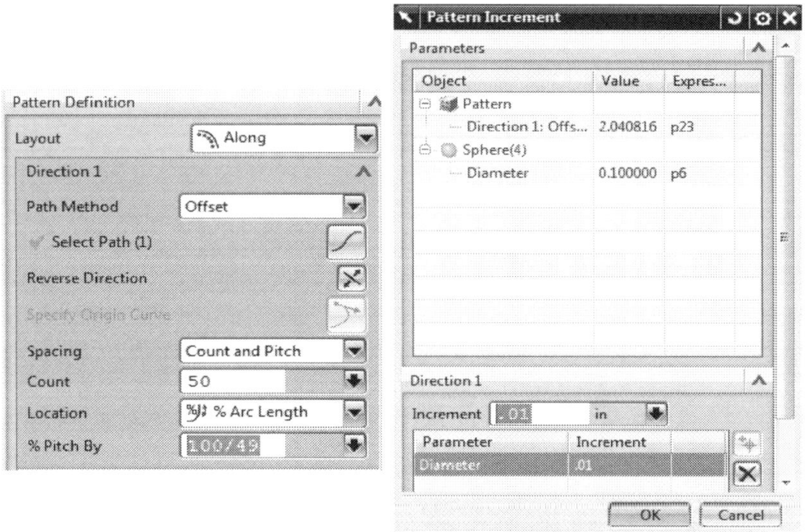


- The result will be a shape that resembles a cudgel as shown below.

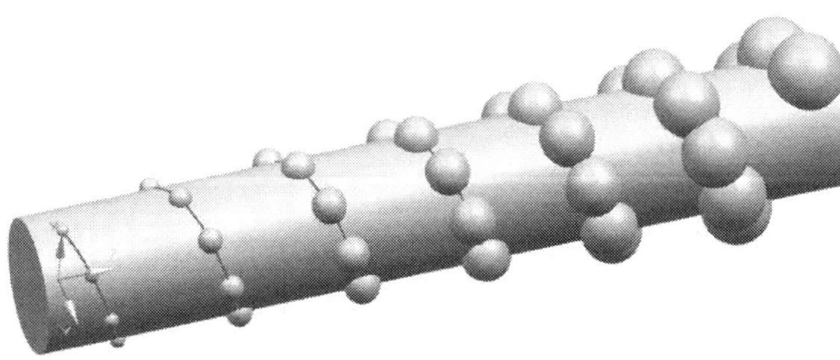

***End of Project***

## ***Detail: Mirror Feature and Mirror Body***

The 'Mirror Body' tool mirrors an entire solid body about a datum plane. It is useful for simplifying the construction of symmetric models. It can be found under the **Insert / Associative Copy** menu.

'Mirror Feature' is similar but mirrors individual geometric features that have been united to a base piece of geometry. When using 'Mirror Feature,' it is important to make sure that the resulting mirrored geometry remains on its original base geometry.

There are many ways to create the geometry shown below. One alternative is to build half of the entire solid and then use the Mirror Body tool. Another alternative would be to create the entire solid geometry with only one of the hole features and then use the Mirror Feature tool. In any case, the use of Mirror Body or Mirror Feature can save you time.

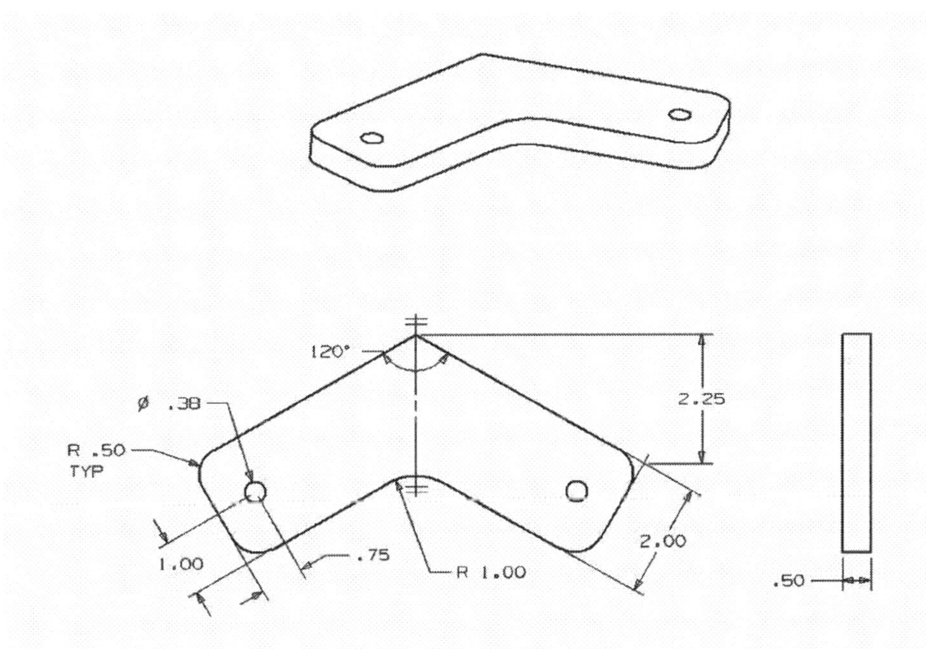

## *Project 9.8.4*

- Create a datum **CSYS**.
- Sketch the geometry shown below and position it using the Datum **CSYS** datum planes.

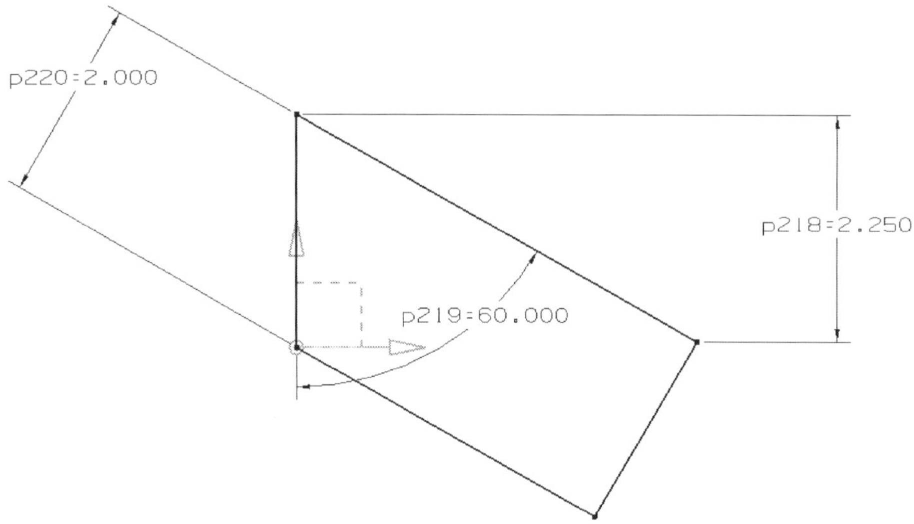


- Extrude the sketch **0.5in.**
- Create the **0.38in** diameter hole and the two **0.5in** blends. *You should now have the model shown below.*

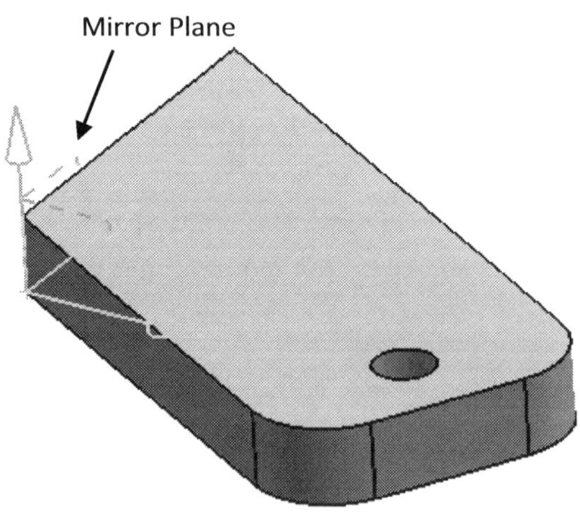

- Select Insert / Associative Copy / Mirror Body 
- Pick the solid body.
- Middle Click to **OK**.
- Pick the mirror plane shown below for the mirror operation. *You should now have two solid bodies, as shown below.*

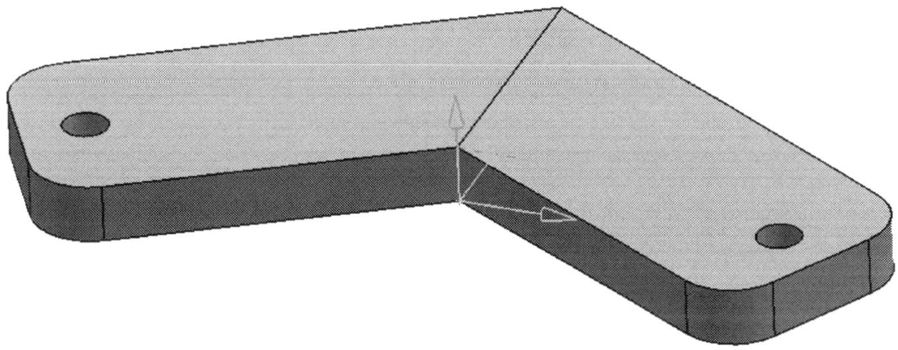

***End of Project***

| **Note:** Adding the 1 inch blend between the two bodies will be covered in Chapter 10. |
| --- |

## ***Detail: Instance Geometry along a Path***

The Instance Geometry toolset adds even more power to the idea of reusing already existing geometry in an associative manner. With Instance Geometry you can create associative patterns

In addition to allowing more types of geometry, Instance Geometry also provides the ability to accomplish more complex patterns than just mirror, rectangular, or circular. The geometry below shows a product that would be easier to design if there were some way to pattern geometry along a randomly shaped path.

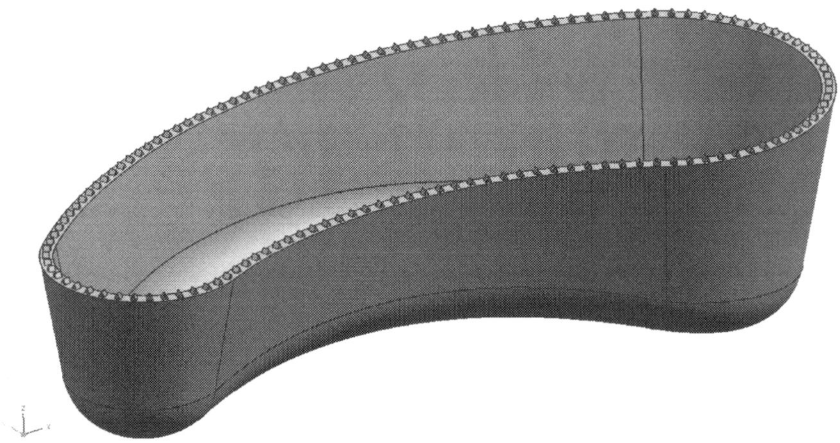

The good news is, the Instance Geometry toolset provides a function for this exact purpose. Appropriately named “Along a Path”, this function is one of the new patterns that can be very useful.

## *Project 9.8.5*

- Create a datum **CSYS**.
- Create a sketch to represent the top shape of the part.

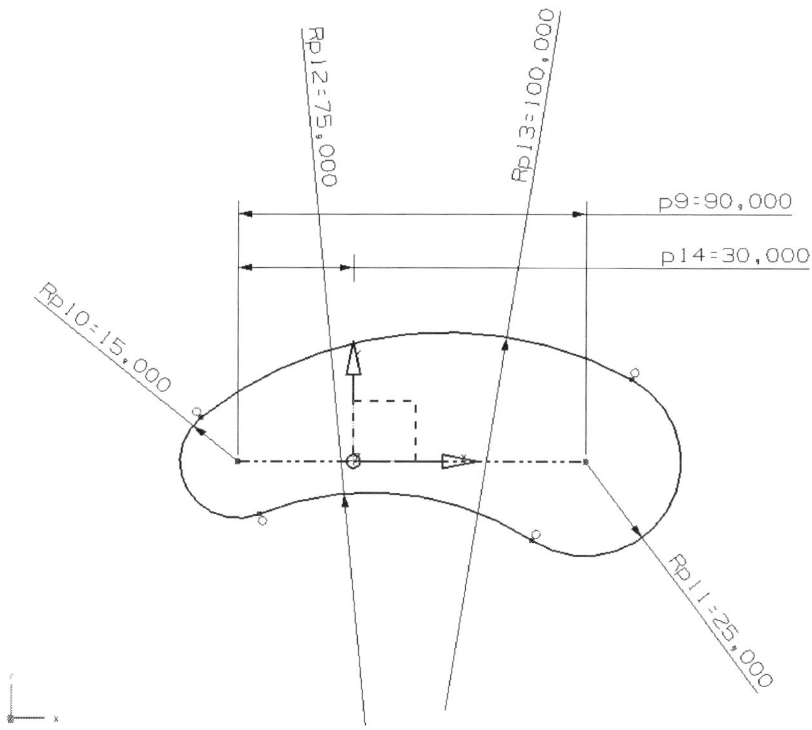


- Create a sketch on the appropriate orthogonal plane that will be used to modify the bottom of our extruded top shape.

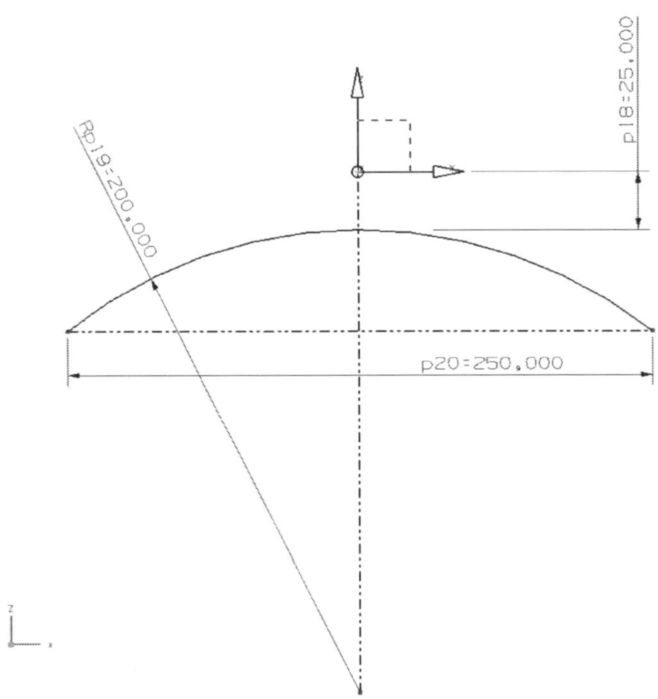


- Extrude the first sketch with a 10 degree draft. Choose a distance great enough that the resulting solid body is fully through the top curve of the second sketch.

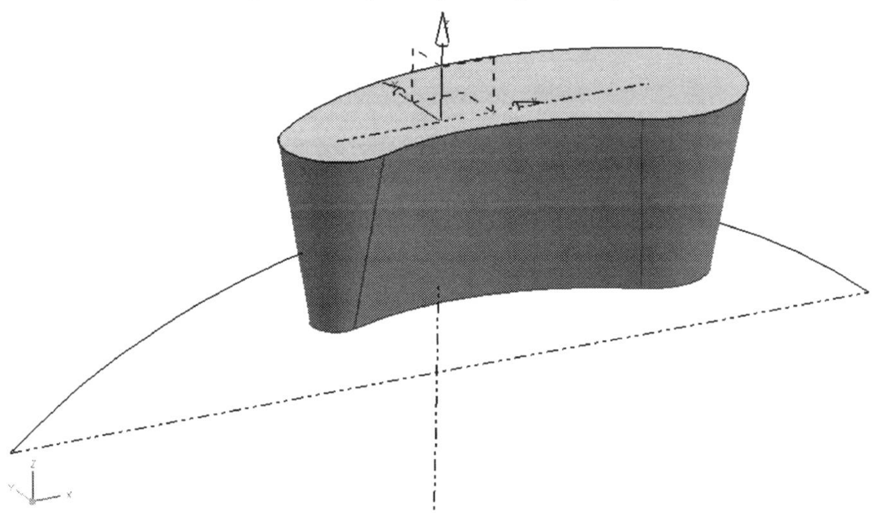

- Symmetrically extrude the top curve of the second sketch to create a sheet that can be used to trim the solid body.

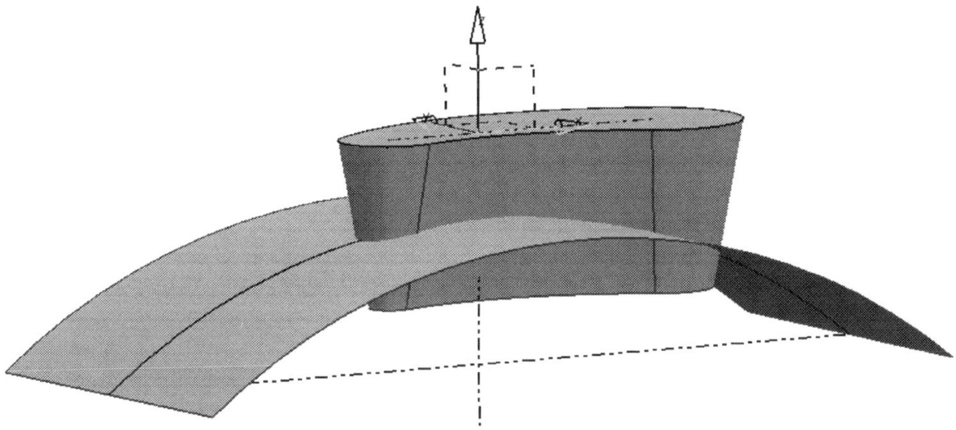

- Trim the body using Insert / Trim / Trim Body.

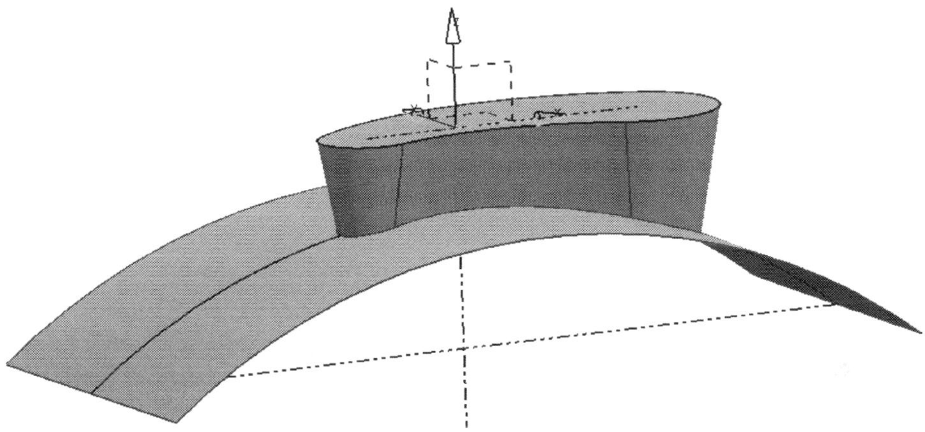

- Clean up your display by hiding the sketches and the sheet: Use CTRL+B as the keyboard shortcut or **Edit / Show and Hide / Hide** as the menu path.

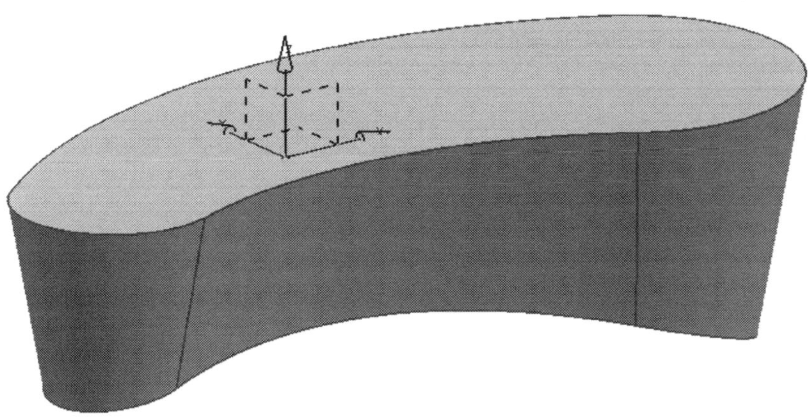

- Add a 10mm edge blend to the bottom of the part using **Insert / Detail Feature / Edge Blend.**

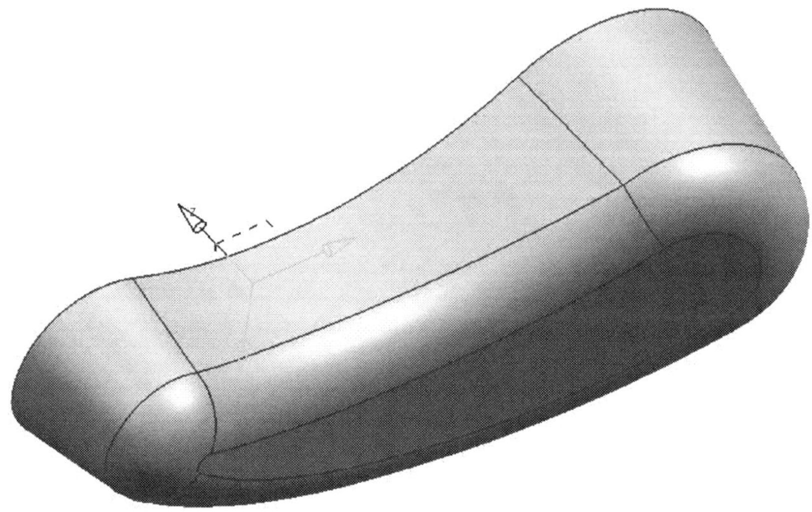

- Shell the part to a wall thickness of 2mm.

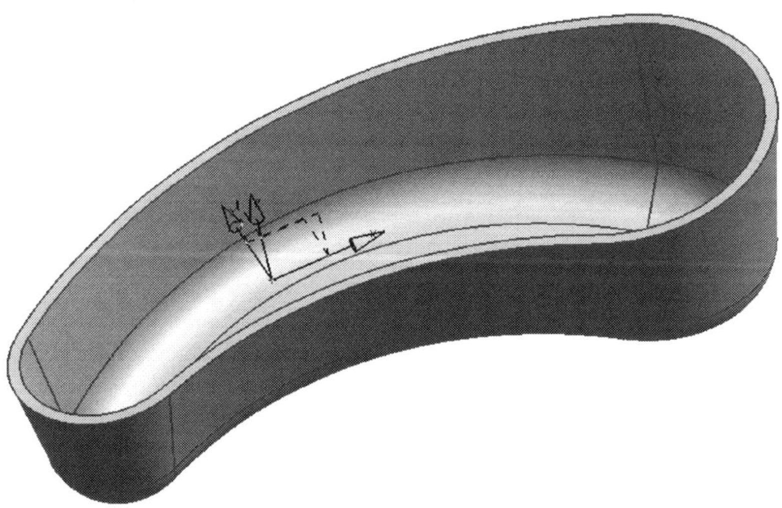

- Now, begin the creation of the first triangular feature. There are many easy strategies available. One is to create a sketch with two single lines superimposed over each other. The longer line acts to provide the position for the feature and is constrained to the endpoints of the inner and outer wall edge curves. The shorter line acts as the guide during a subsequent use of the Sweep along a Guide feature. This is one of many easy strategies to create a feature that has uniform spacing from the outer and inner edges.

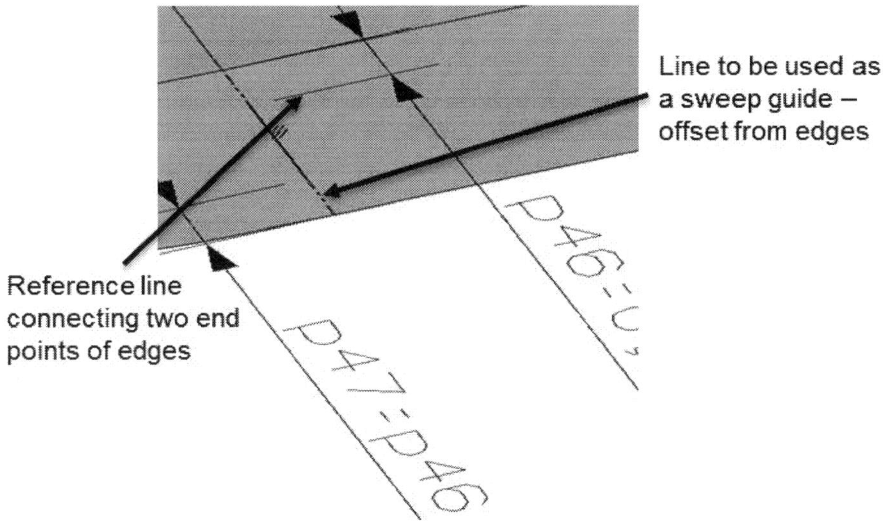


- Create this sketch on the top face at any one of the intersections of radii.

- Take advantage of the new sketch type called "On Path". Create a sketch as shown at either 0% or 100% of the short line from the previous sketch.

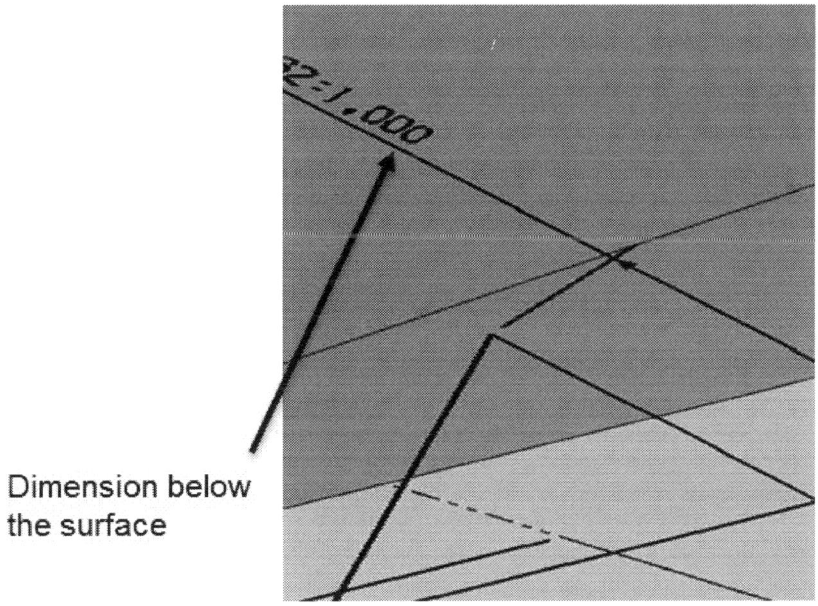


- Create the triangular feature using the Sweep Along Guide tool found in: **Insert / Sweep / Sweep Along Guide** . 

> **Note:** This step is very similar to that used in Section 8.8 for creating a 'tube.' The difference here is that instead of the default circular cross-section utilized in the '**Insert / Sweep / Tube'** option, you will use the triangular sketch as your 'swept' cross-section and the shorter line as the 'guide.' This tool is covered in more detail in Chapter 16.

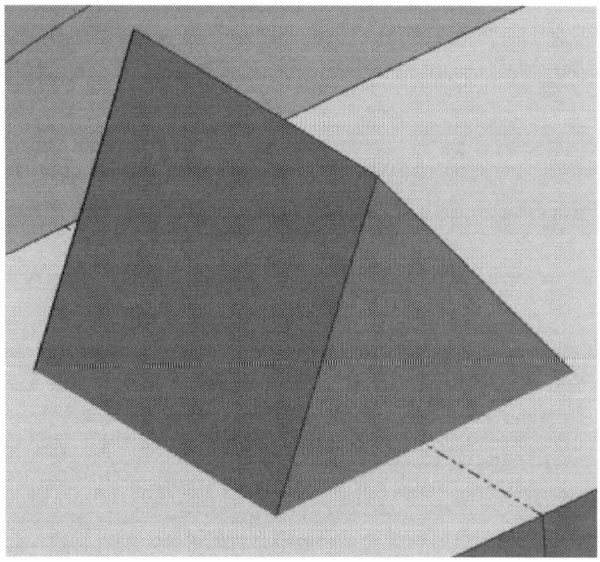

- The goal is to have 100 triangular features distributed evenly around the perimeter of this product. Unlike the previously discussed patterning tools, Instance Geometry 'Along Path' does not ask you for the total desired number; instead, it asks you for the number of copies. In this case we want 99. Begin creating your associative pattern by accessing the Instance Geometry tool through **Insert / Associative Copy / Instance Geometry** and setting the type to Along Path. Fill out the form as seen below.

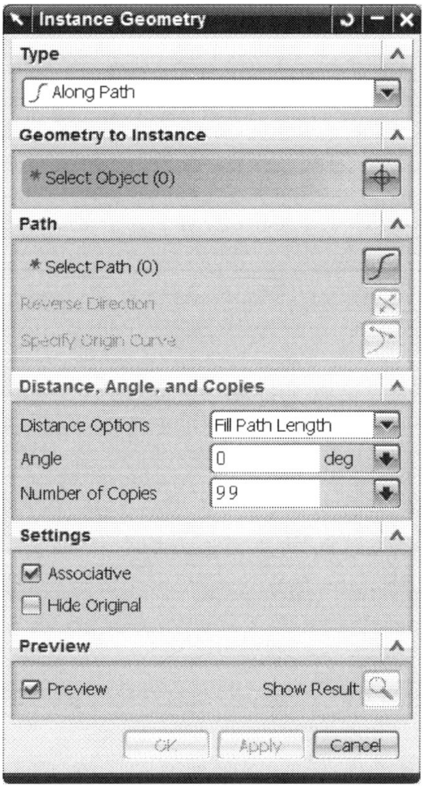


- Select the 'Geometry to Instance,' then select the path (either the inside or outside edge of the product is suitable for the path).

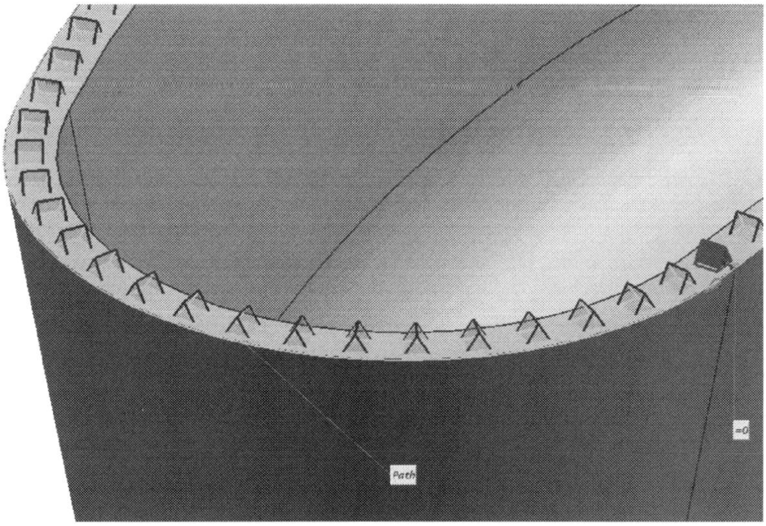

- All of the individual pieces of geometry can be united using the standard **Insert / Combine / Unite** tool. Identify the shelled large piece of geometry as the target and drag a window around all of the little pieces of geometry to identify them as the tools. Combine Bodies is covered in more detail in Chapter 10.

The finished piece of geometry is fully associative; making this is an extremely powerful design feature.

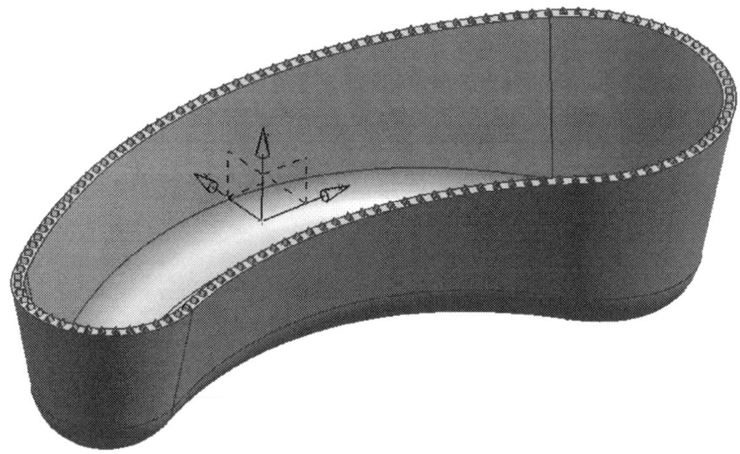

***End of Project***

## 9.9 Extract

The 'Extract' function is very useful when creating geometry that requires a mating surface. The 'Extract' tool allows you to extract existing geometry in the form of a curve, face, region, or body. The original geometry can then be hidden leaving you with only the face or region you desire. New geometry can then be created using your extracted feature, ensuring a perfect match.

### *Detail: Extract Features*

The following project is a very basic example of how this command is used.

### *Project 9.9.1*

Create a model similar to the figure below.

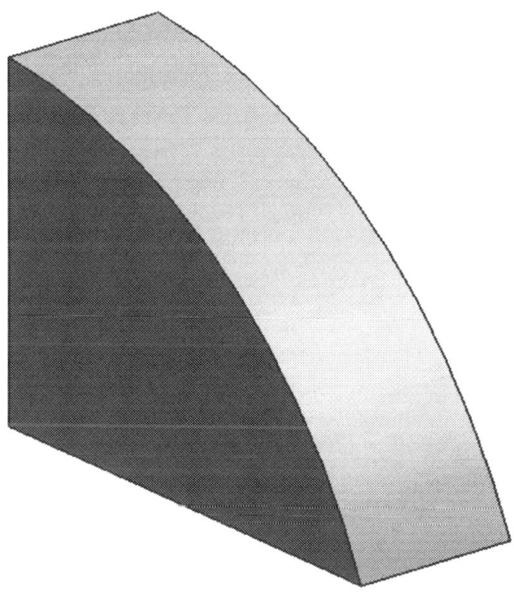

- Select Insert / Associative Copy / Extract Body .
- Choose the **Face** option as shown on the next page and be sure 'Hide Original' is **not** selected.

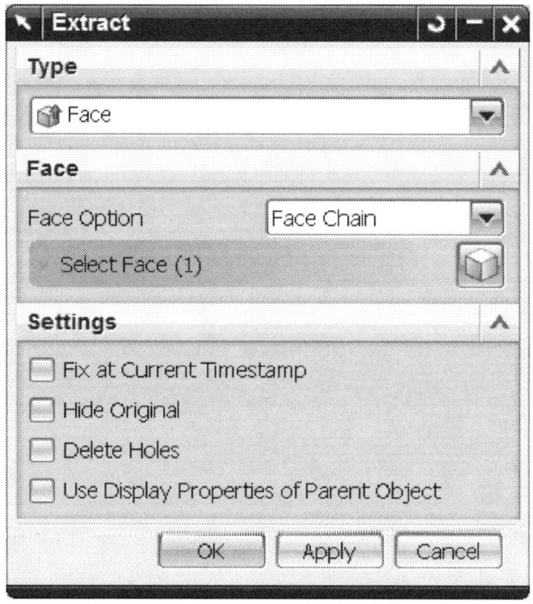


- Read the cue line – Select the curved surface as your face to extract.
- Select **Apply.**
- Now check the **Hide Original** check box.
- Select the vertical surface as your next face to extract.

**Note:** NX 8.5 has now created a curved surface identical to the one you selected. It is located in the same position as the original feature.

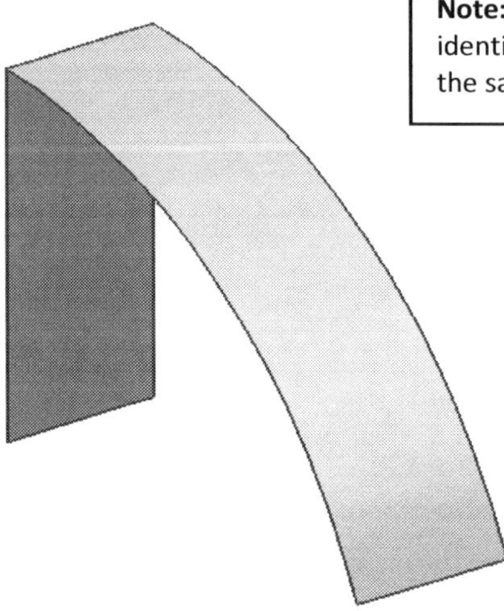

- Choose **OK**. Compare your results to the figure on the previous page.

***End of Project***

## 9.10 Practice: Attachment Arm

A simple approach to building an attachment arm for a robotic fixture, as shown below, involves a combination of the 'Shell,' 'Offset Face,' and 'Tube' features.

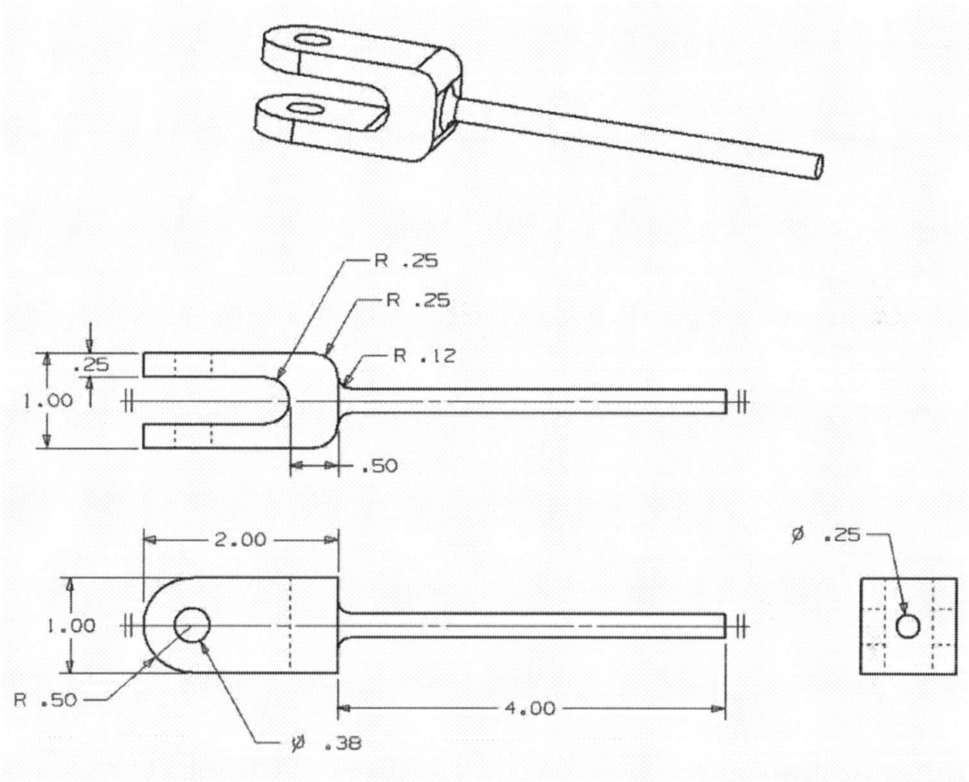


### *Practice Outline*

- Create the sketch as shown below. The shaft and the head will be created separately and then United.

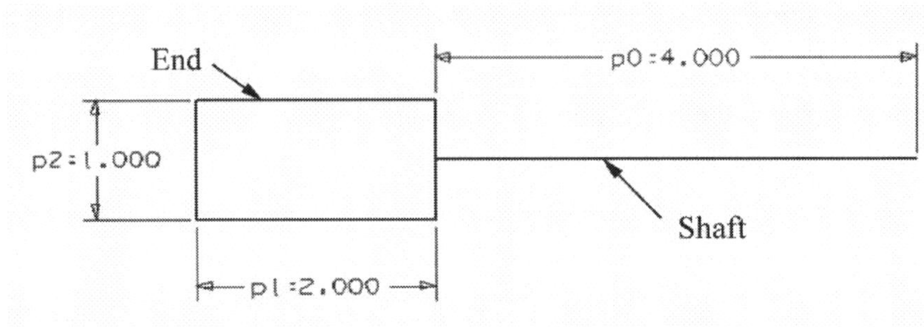


- Extrude the end from **+0.5** to **–0.5**. Do not extrude only in one direction, as the end extrusion must be centered on the sketch plane so that the shaft is located properly.
- Place the two **0.5in** radius blends on the corners of the end extrusion. They will appear as a full radius blend.
- Shell the end by piercing three faces with a **0.25in** wall thickness. Compare to the figure below.

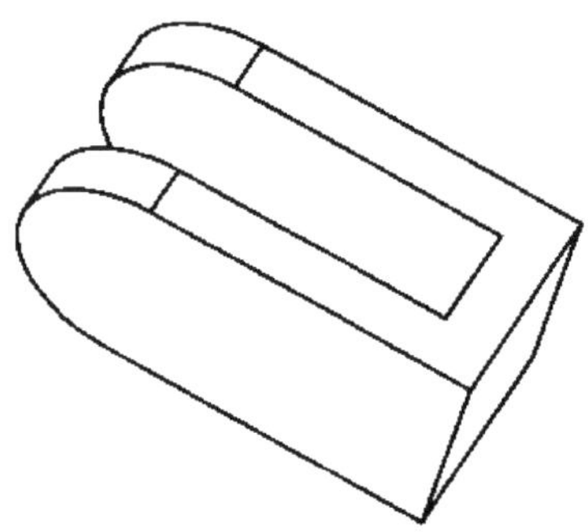

- Now, suppose a design evaluation has resulted in the necessity of making the base 0.5in thick versus the current 0.25in thickness.
- Use the 'Offset Face' feature to achieve the correct thickness.
- Place the hole. Use a 'Point onto Point' dimension to locate the hole at the center of the 'full radius blend'.
- Add the remaining Blend features to the end extrusion.
- Add a 'Tube' feature to create the shaft.
- Unite the shaft to the end extrusion and add the final 0.12in radius blend.
- Compare your model to the figure at the start of this exercise.

***End of Practice***

## 9.11 Practice: The Clip

A good strategy to create the clip model shown below involves using the 'Mirror Body,' 'Draft Faces,' and the 'Boss' features.

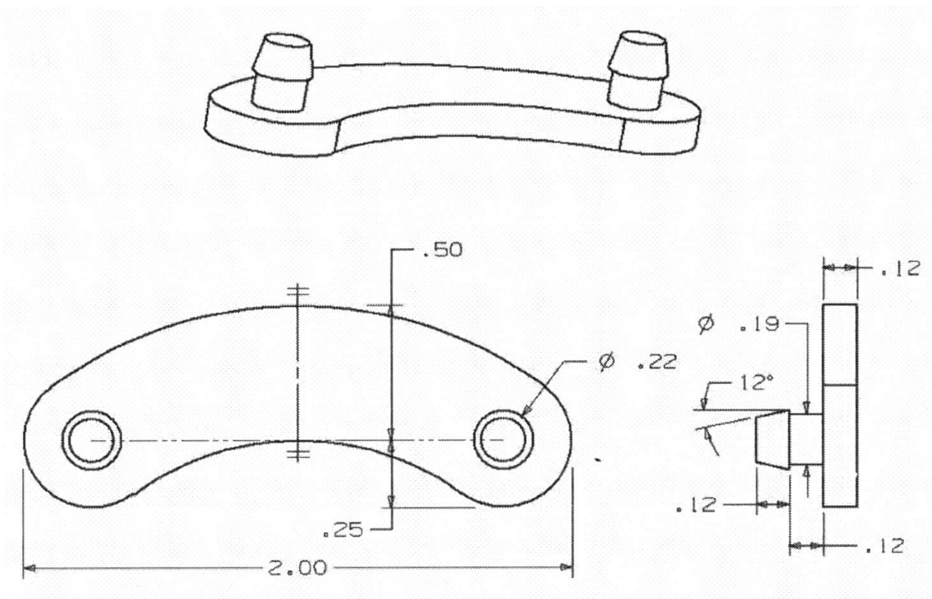


### *Practice Outline*

Create the sketch shown on the following page. The center of Arc 2 is positioned on the horizontal reference line, and the center of Arcs 1 and 3 should be aligned to the vertical line. Make Arc1 and Arc 3 concentric. (Be careful not to over-constrain the sketch – it's pretty easy to do.)

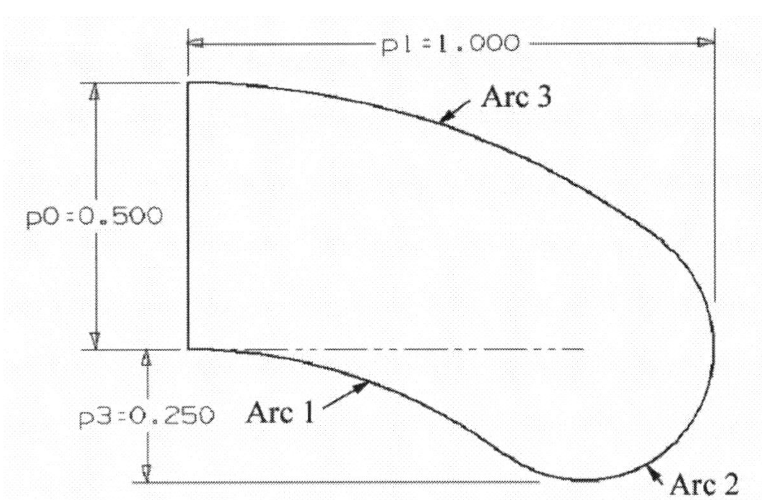

- Extrude the sketch **0.12in** to build the base of the clip.
- Position a boss at the center of Arc 2.
- Position a second boss centered on top of the first boss.
- Draft the second boss with a **12** degree angle. The taper plane is at the base of the second boss. Use the 'Arc Center  Point Method' to pick the reference point.
- Define the taper direction using the 'Face Normal 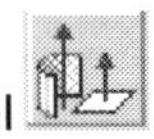 Axis Method,' and pick the cylindrical surface of the boss (see figure below).

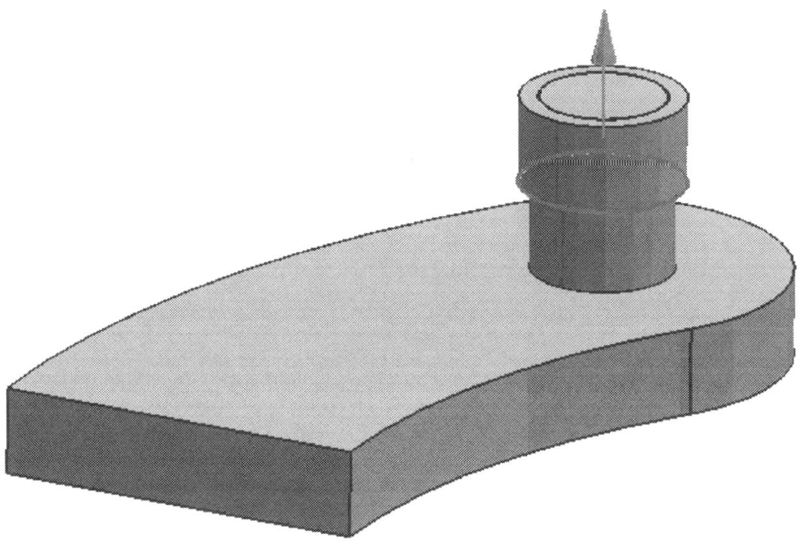

- Create a datum and 'Mirror' the solid.
- 'Unite' the two bodies.

***End of Practice***

## 9.12 Self-Test

(All questions are true/false. Answers are at the back of the book.)

1. The 'Shell' feature creates a uniform wall thickness solid.
2. 'Offset Faces' can only add material to a solid, i.e. to offset outward.
3. The 'Mirror Body' feature requires a datum plane.
4. You can 'Trim' an object into many pieces.
5. Regular dimensions are only available in the Sketcher mode.

# 10. Intermediate Curves

## 10.1 Introduction

Chapter 3 provided an introduction in how to create common curves used often in NX 8.5, but in order to create complex geometry, it is important to step up and master the building blocks – intermediate curves. Some of these are 'Offset,' 'Bridge,' 'Join,' 'Project' and 'Spline' curves. This chapter contains examples of how these curves are created.

## 10.2 Spline Curves

The 'Studio Spline' function allows you to create a closed-loop, non-parametric General Spline curve. 'Spline' curves are extremely useful when creating geometry that is of a free-flowing, organic, or smooth nature (e.g. prosthetic devices and highly designed consumer products).

### ***Detail: Spline***

The figure below is a 'Spline' curve made from 5 points.

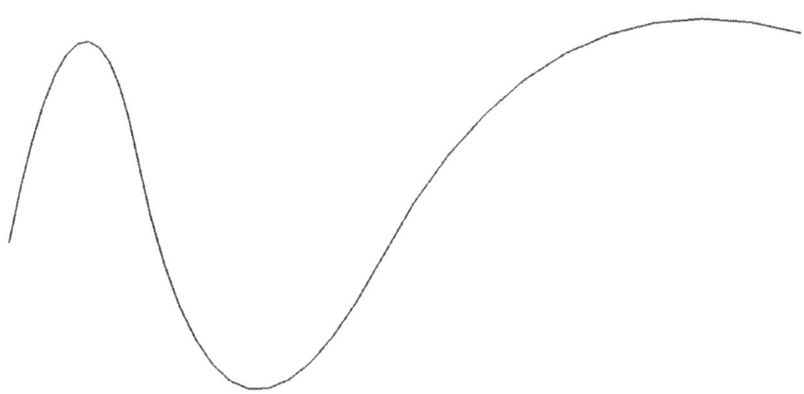

## *Project 10.2.1*

- Select Insert / Curve / Studio Spline 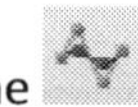 .

The following dialog box is shown.

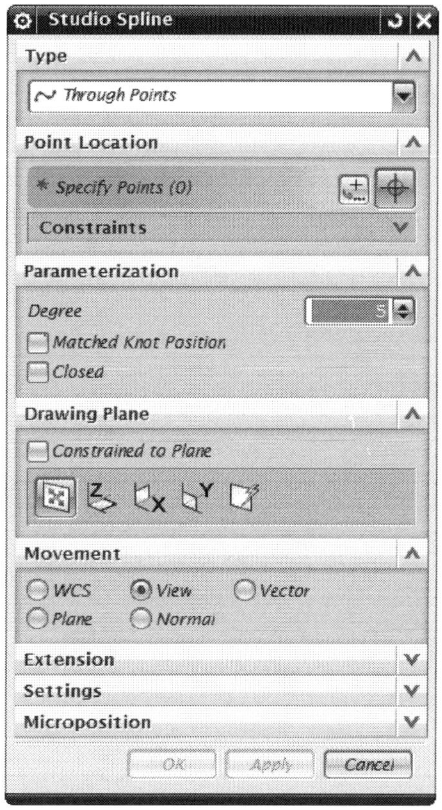


- Click **Through Points** for the Spline creation method.
- Type **2** in the 'Curve Degree Spline' parameters dialog box as shown below.

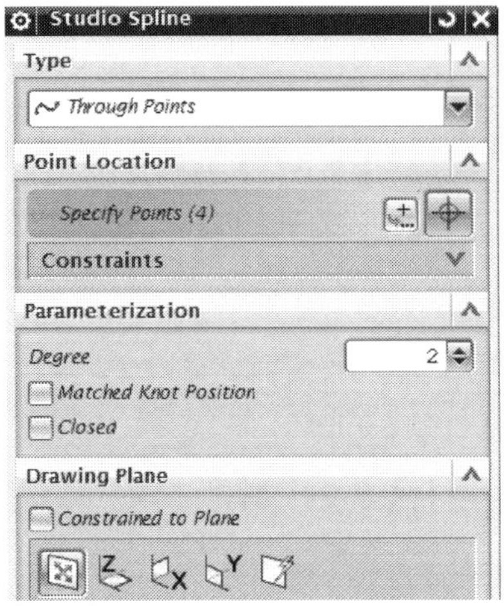


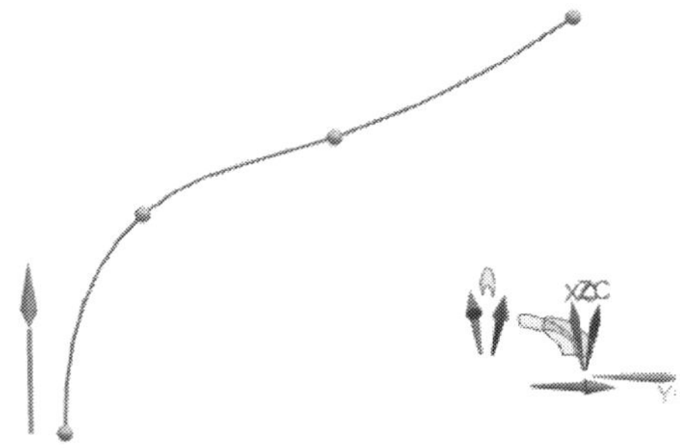

- Click **OK**.
- Click **Point Constructor**  or click in the viewport to create points for the Spline to travel across.

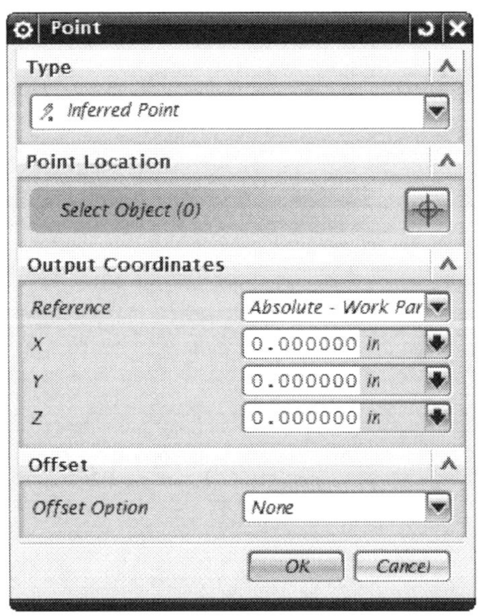


- Click on the X box and type **-5.**
- Click on the Y box (or hit the **Tab** key) and type **1.**
- Leave Z at 0
- Press **OK**.

- Repeat the last three steps to add the following coordinates: (**-4,3,0**), (**-2,-1,0**), (**1,2,0**), (**4,2,0**) .
- After all five coordinates have been entered, click **OK**.
- Compare your results to the figure below

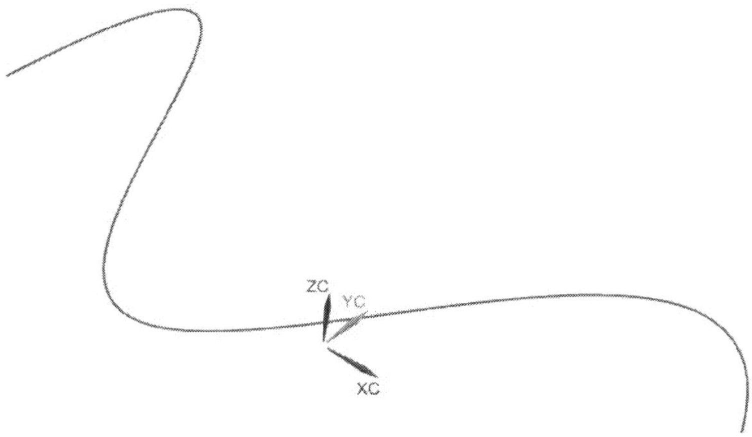


***End of Project***

## 10.3 Offset Curves

The 'Offset' function allows you to select a string of coplanar curves and offset them a specified direction and distance. This function can create either parametric or non-parametric curves.

### *Detail: Offset*

The geometric figure below shows two sets of curves. String #1 is made with basic curves, and String #2 is made using the 'Offset' command.

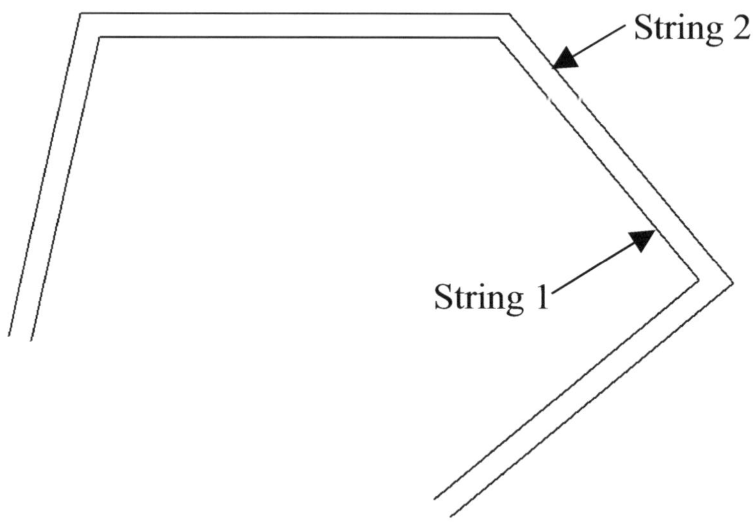


### *Project 10.3.1*

- Create 4 lines (using 'String Mode') that resemble String 1 above. The offset function requires that all input curves share endpoints.

- Select Insert / Curve from Curves / Offset .
- Pick the lines in String 1.
- Choose OK.
- Enter .25 in the distance field of the 'Offset Curve' dialog box, as shown below.

**Note:** An arrow will appear at the end of the string that defines a positive offset direction. If you do not want to offset in the direction of the arrow, you may either type a negative value in the 'Distance' field or select **Reverse Direction**.

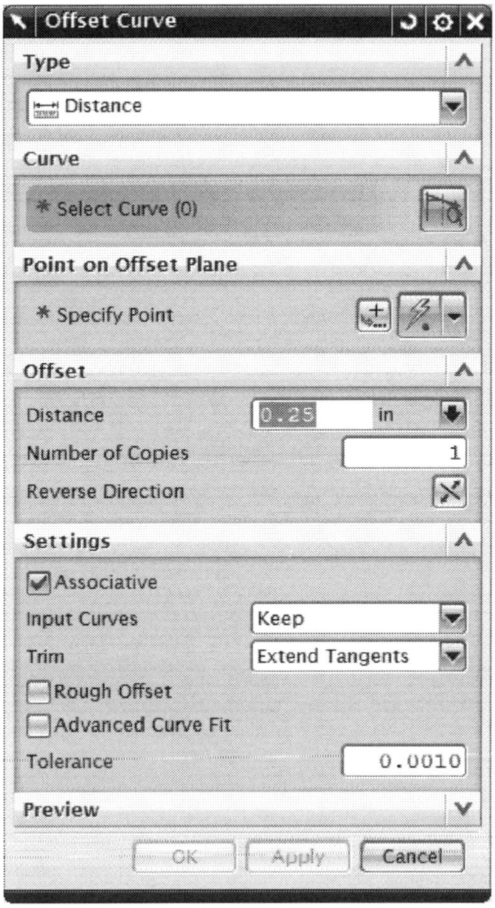


**Note:** This operation creates parametric curves if the **Associative** button is selected. In this case, you will not be able to delete the original curves after offsetting. Turn this radio button off if you do not want the offset curves to be parametric.

Choose **OK**. Compare your results to the previous geometric figure.

***End of Project***

These curves were offset within the plane that they defined. If you want to offset them along the normal to this plane, change the 'Offset By' pop-up from Distance to **Draft**. NX 8.5 will also accept a draft angle.

## 10.4 Bridge Curves 

The 'Bridge' function allows you to select a pair of curves and connect them to provide one continuous, smooth curve. This function can create either non-parametric or parametric curves. The bridge curve is excellent for shapes that have to blend together smoothly, such as the geometry needed to transition from a rectangular opening to an oval shaped opening as shown in the following figure.

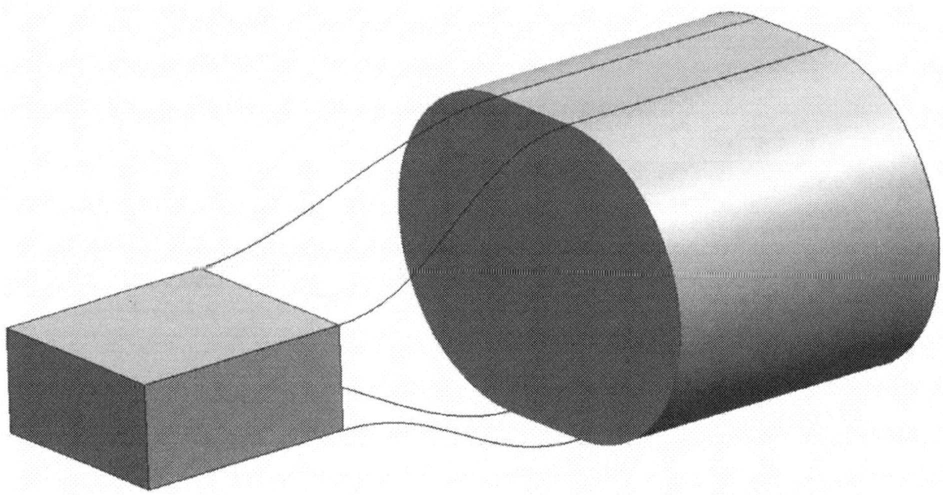

### *Detail: Bridge*

The next figure shows three curves, each comprised of a pair of lines and bridge strings.

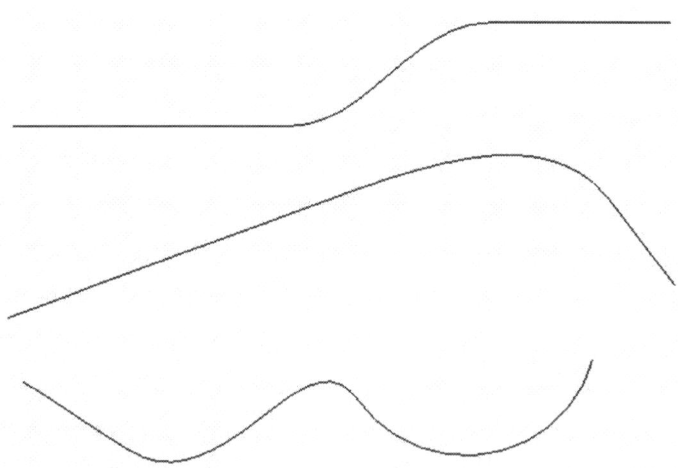

### *Project 10.4.1*

In a new part file, sketch five lines and one arc that resemble the figure below.

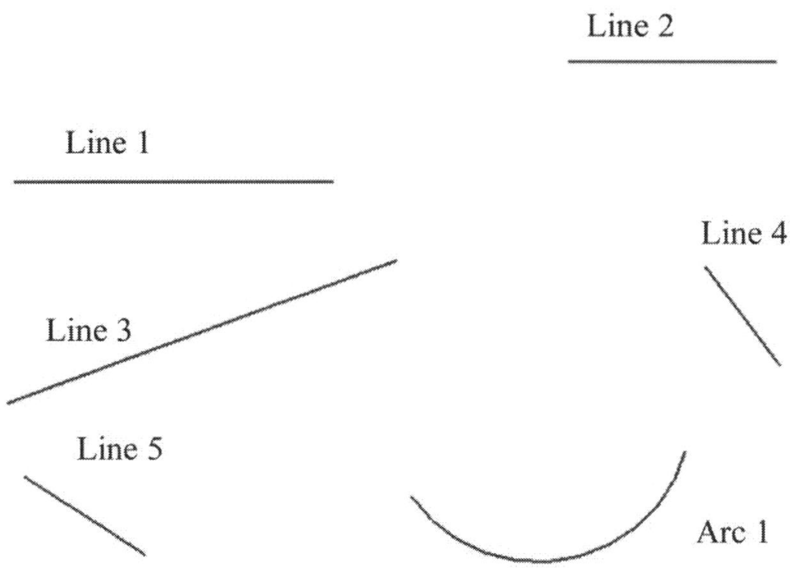


- Click Exit Sketch
- Select Insert / Curve from Curves / Bridge .
- Select Line 1 as prompted by the Cue.
- Select Line 2 as prompted by the Cue.

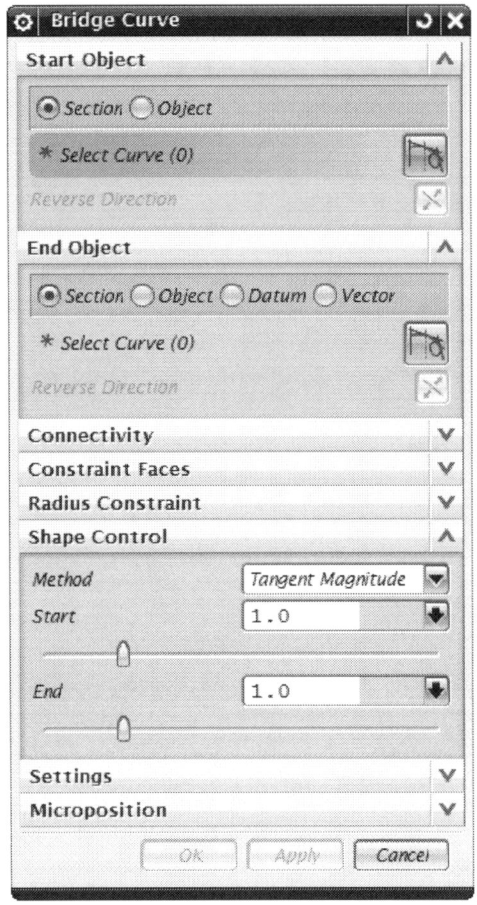


Notice how the 'Bridge Curve' function automatically connects the two curves with a smooth line.

- Click **Apply** and Select lines 3 and 4.
- Click **Apply** and finally select line 5 and arc 1.
- Choose **OK**. Compare your results to the figure at the beginning of this exercise.

**Note:** Be careful when clicking on the curves to bridge. Clicking near the endpoint from which you intend to connect the curve will ensure the 'Bridge' command will operate smoothly. If you click on a curve nearer the other endpoint, the 'Bridge' command will attempt to create a smooth curve from that end instead.

***End of Project***

## 10.5 Join Curves

The 'Join' function allows you to select distinct curves and connect them to provide one continuous curve. This function can create either non-parametric or parametric curves.

Below are shown three curves in 3D space. One curve was created on the XY plane and two curves (one line and one arc) were created in the YZ plane.

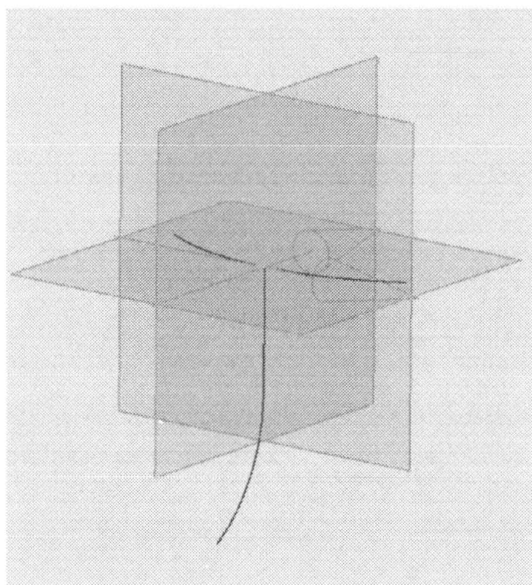

The following figure illustrates the difference between swept features created with and without the 'Join' command. Notice the line separating the swept feature in the first model illustrating two separate lines. The second feature does not have this line and thus illustrates how the Join command was used to connect the two lines shown in the previous figure.

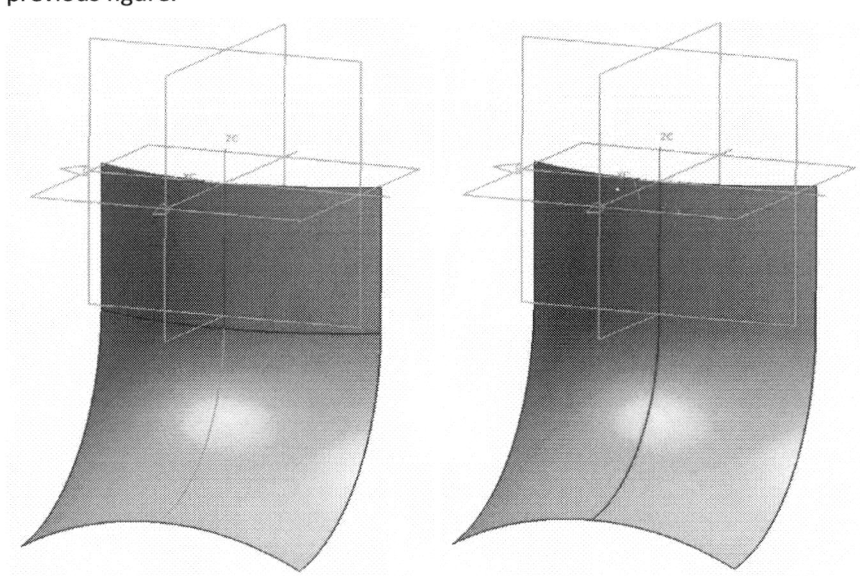

## ***Detail: Join***

The curve below illustrates how the 'Join' Function can be used to unite separate curves and create a single, continuous entity made from two or more individual curves.

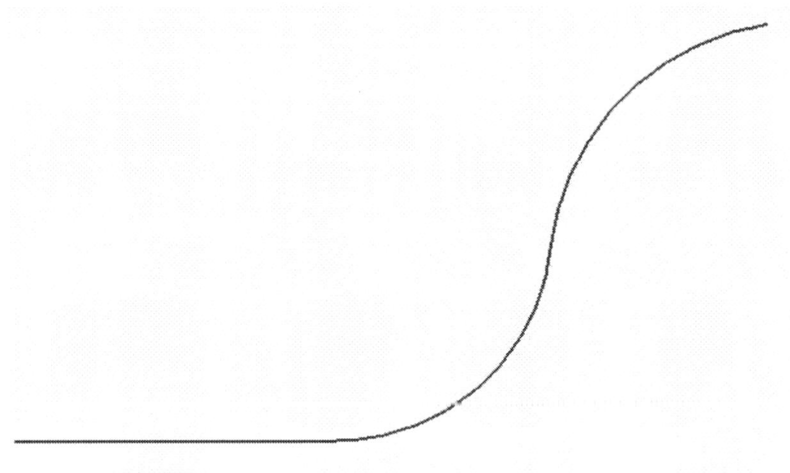

### ***Project 10.5.1***

- Create three basic curves, similar to those shown in the following figure.

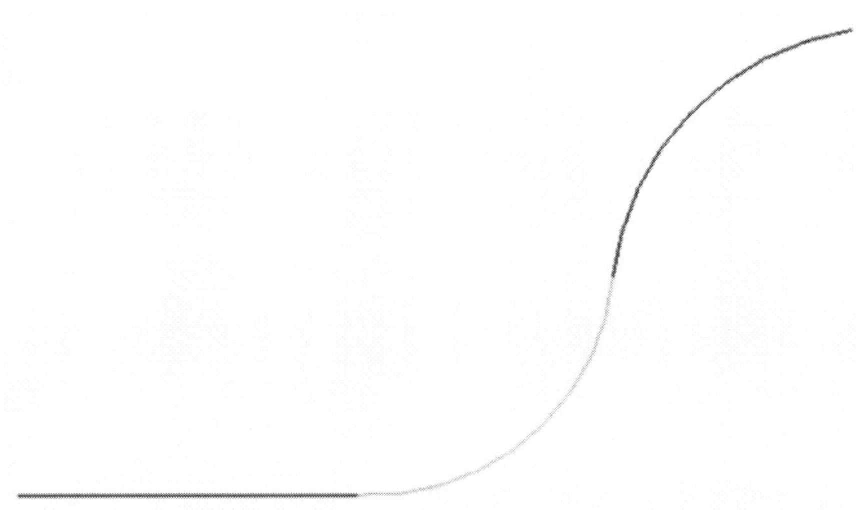

Notice when your cursor is placed over each of the segments they are highlighted individually. NX 8.5 recognizes the three curves as separate entities.

- Select Insert / Curve from Curves / Join .

- Select all three lines as prompted by the Cue.
- Choose **OK**.
- The Join Curves dialog box will open as shown in the next figure.

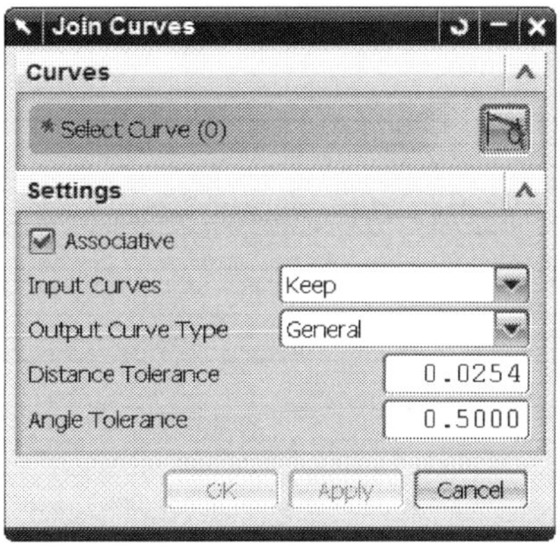


**Note:** Be sure **Input Curves** is toggled to 'Hide' to erase the original three curves.

- Choose **OK**.

**Note:** You may get a warning message that 'the Join curve produced corners.' Select **Yes** to continue.

- Compare your results to the following figure.

Notice now how placing your cursor over any of the three segments highlights the entire curve. NX 8.5 now recognizes one joint curve.

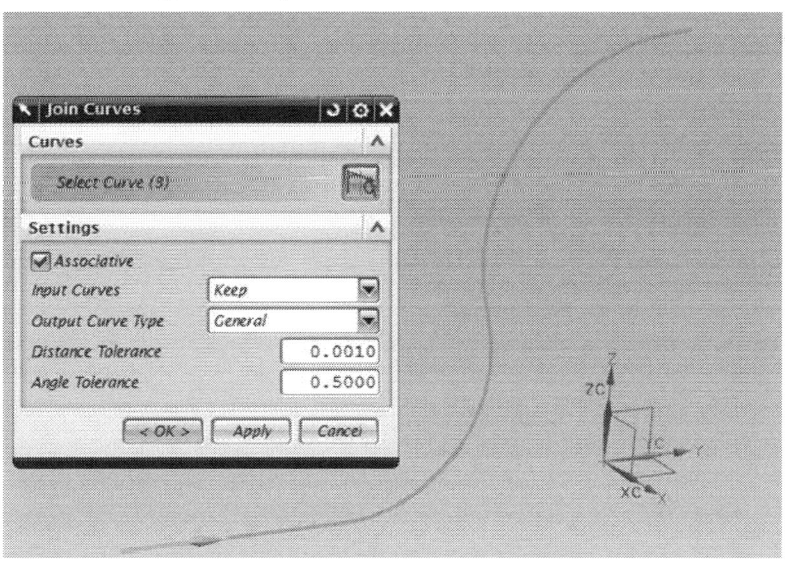


***End of Project***

## 10.6 Project Curves

The 'Project' function allows you to select a curve and a specified plane and project the original curve onto the plane, as shown below. This function can create either non-parametric or parametric curves.

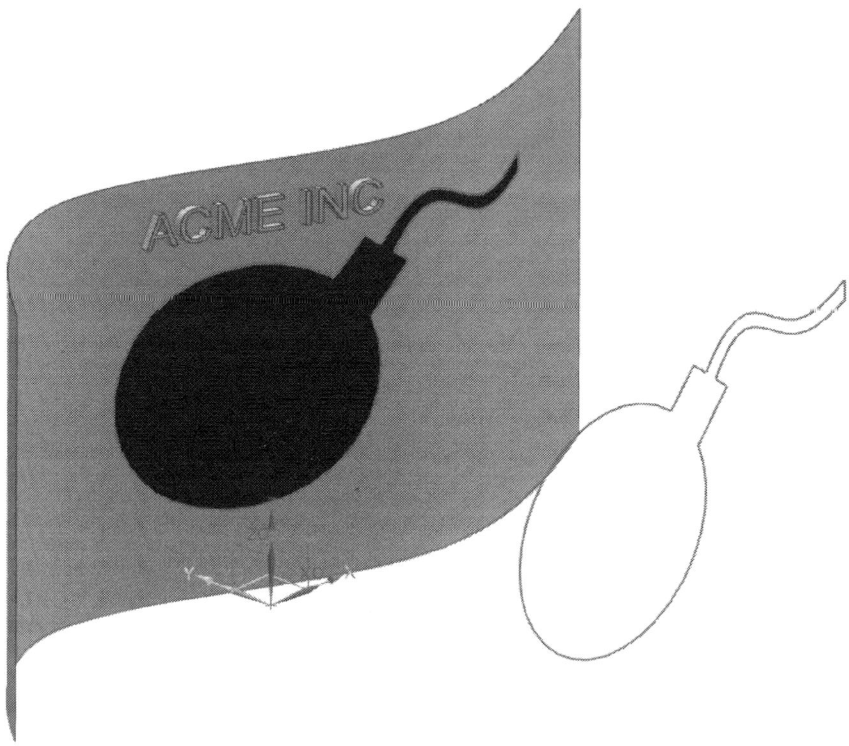


### *Project 10.6.1*

- Create a 'Primitive' cylinder.
- Select Diameter, Height.
- Place the cylinder in the **+XC** direction in the 'Vector Constructor' menu.
- Input the following values: Diameter = **1in**, Height = **8in.**
- Enter cylinder origin values of (**-4, 0.5, -2**). Choose **OK**.
- Select **Cancel**.
- Create a sketch on an XY datum plane.
- Draw a pattern that looks like the figure on the next page.

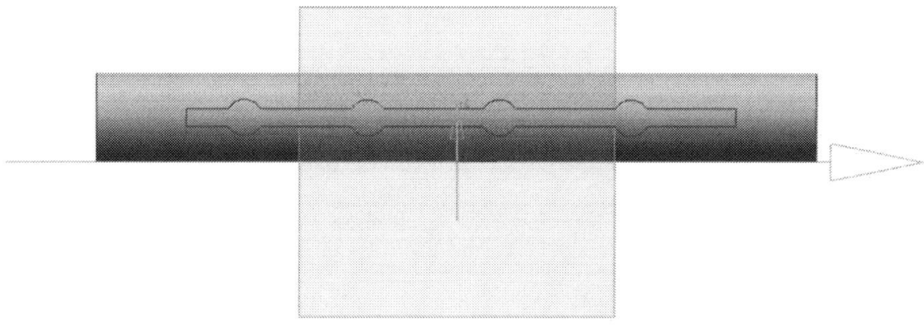

- Select Insert / Curve from Curves / Project  .
- The dialog box in the following figure will open.
- Select Direction Method *'Along Face Normal.'*

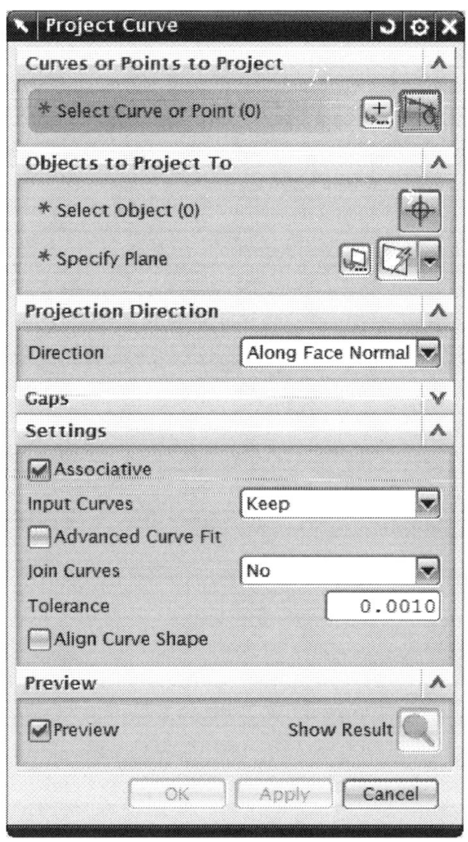


- Read the Cue Line – NX 8.5 is asking you to select the geometry to project.
- Select the pattern curves as prompted by the Cue.

- **Middle click** or hit the **Select Object** radio button under 'Objects to Project To.'
- Read the Cue Line – NX 8 is asking you to select where to project the geometry.
- Click anywhere on the round surface of the cylinder.
- Middle Click.
- Compare your results with the next figure.

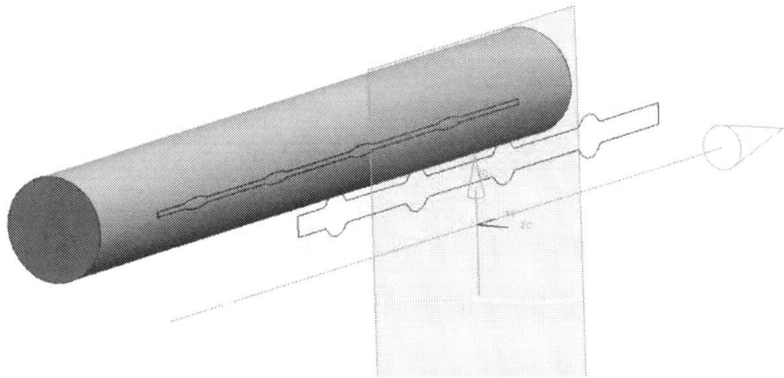

- Now try the same projection using direction method 'Along Vector' and project the curves along vector –ZC.
- Due to the nature of 'Along Vector' and the face of the cylinder being one entity, the curve will project onto both sides of the cylinder it crosses.
- Compare with the following figure.

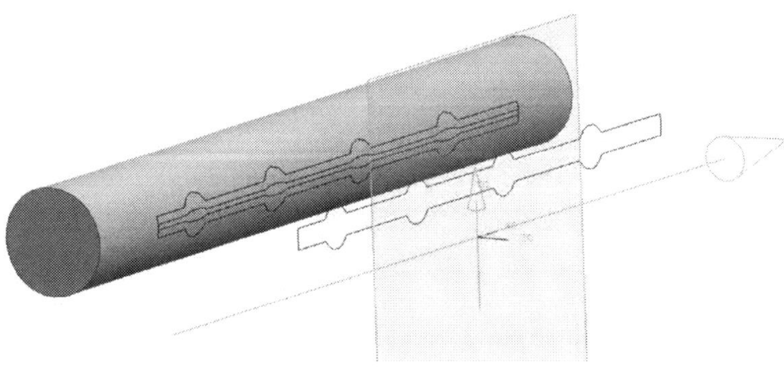

> **Note:** Because the 'Associative' option was checked as the Copy Method setting, if you modify your original text, NX 8.5 will update the projected geometry as well. Selecting the 'Delete' or 'Replace' option would remove the original geometry.

***End of Project***

## 10.7 Isoparametric Curves

Iso parametric curves are incredibly useful. They've been available in NX for many years but now they are parametric entities that update. You can use them to distribute curves along a surface in a rectangularized pattern. Once you have the curves, you can use them as the underlying construction geometry for almost anything. In the example below a surface is created and 2 sets of isoparametric curves are distributed on it. The surface is sewn into a solid and the curves are used to create solid ribs.

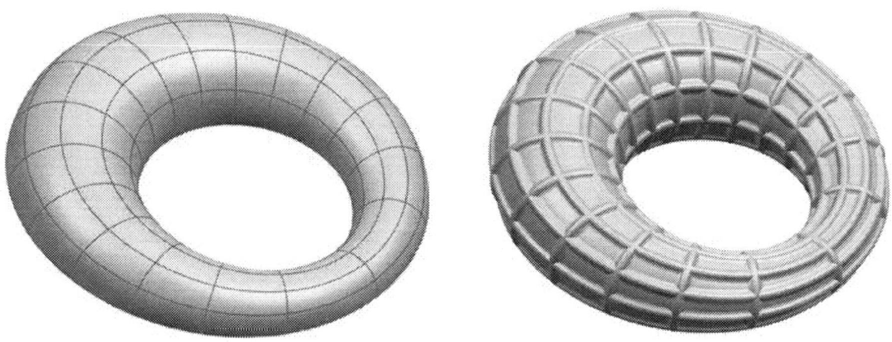

*You see every surface has a U and a V direction. The isoparametirc curve creation tool just lets you access it and lets you make curves on it.*

### *Project 10.7.1*

- Create some sort of spline curve on the X-Y plane as shown below.

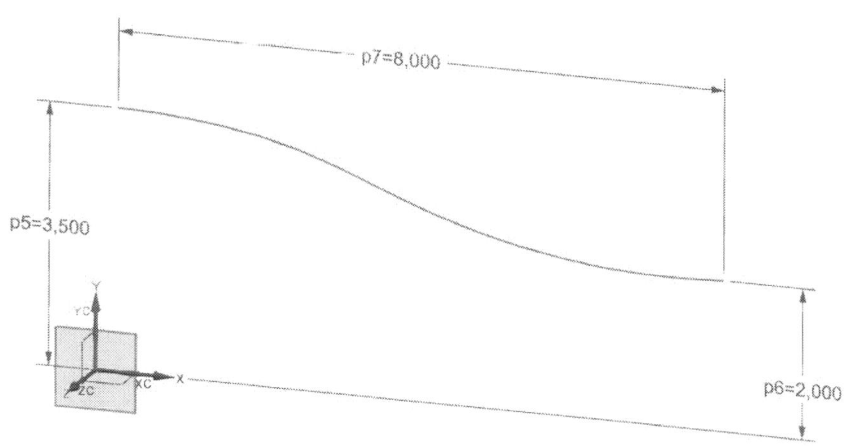

- Now revolve it plus and minus 40 degrees to create a revolved surface resembling the one shown below.

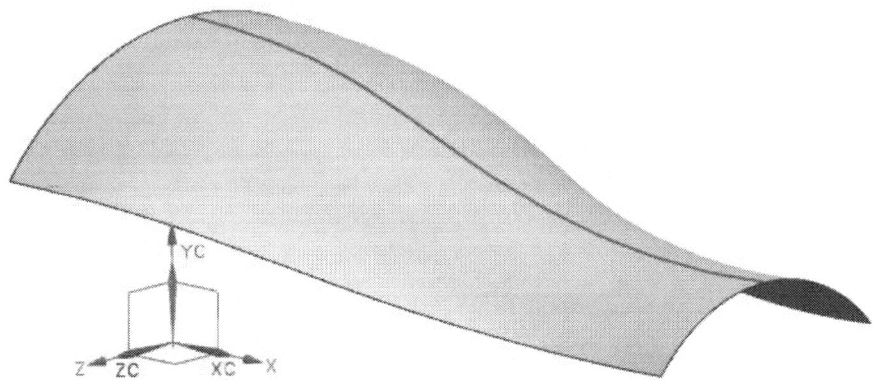


- Now select Insert / Curve from Bodies / IsoParametric Curve... Set the Direction to U and V, Location to Uniform, and Number to 7. Select OK.

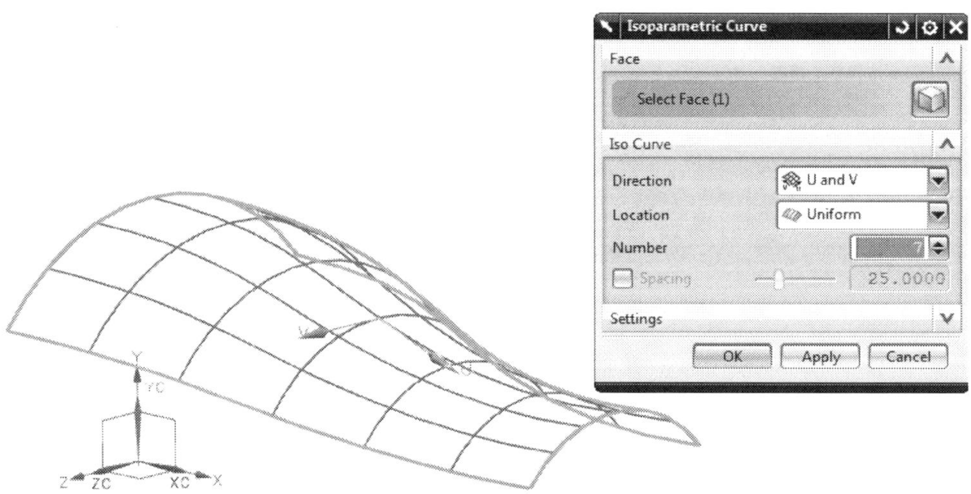


- Using a series of tubes, extrudes and the unite command create the shape shown below:

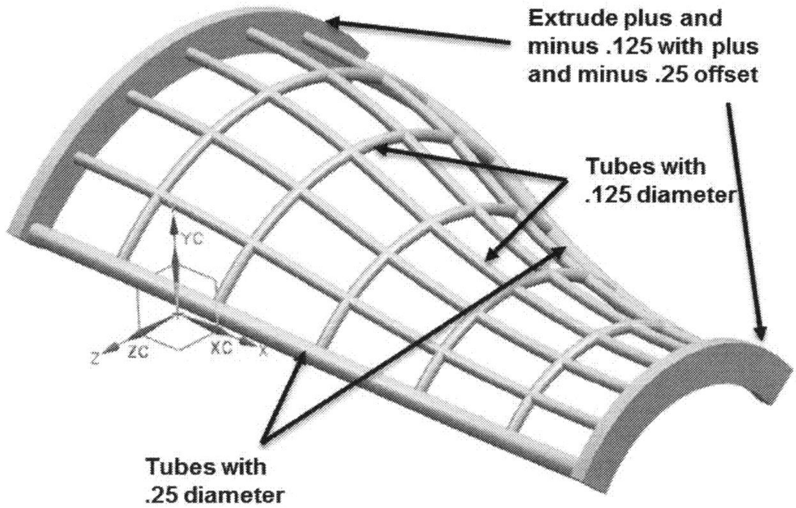


Enter cylinder origin values of (**-4, 0.5, -2**). Choose **OK**.
Select **Cancel**.
Create a sketch on an XY datum plane.
Draw a pattern that looks like the figure on the next page.

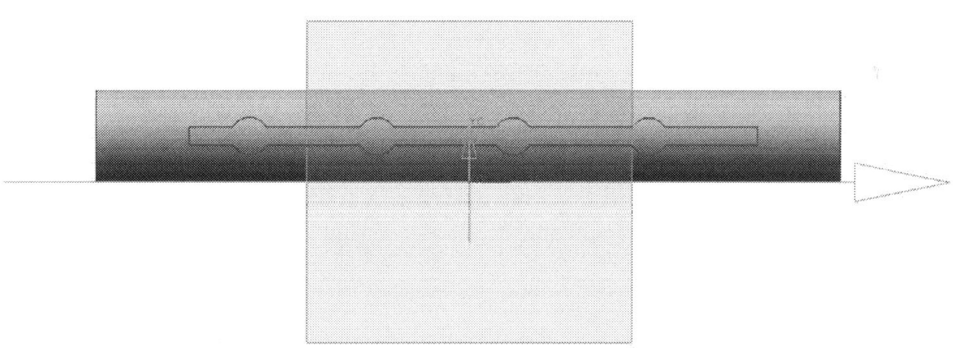

Select Insert / Curve from Curves / Project  .

***End of Project***

## 10.8 Practice: Creating Curve Geometry

The shape shown in the next figure can be built by using 'Basic Curves' combined with 'Offset Curve' operations.

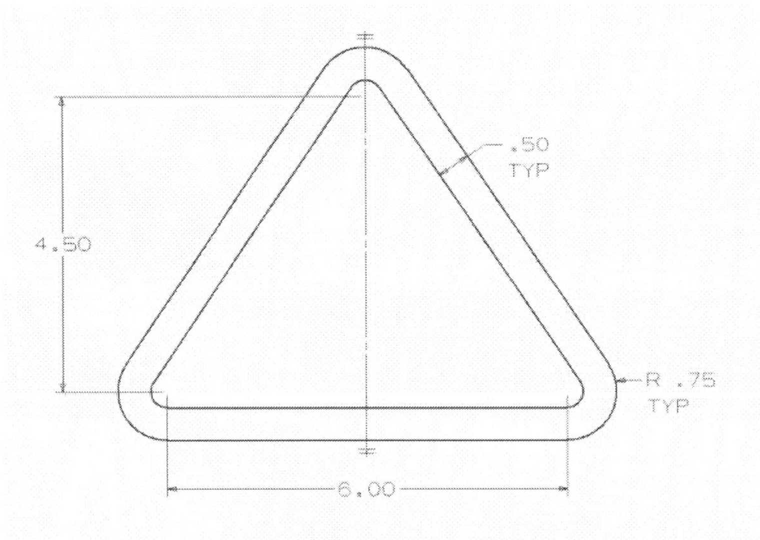


### *Practice Outline*

Position three circles of **0.75**R as shown.

Connect the three circles with tangent lines. Compare to the figure below.

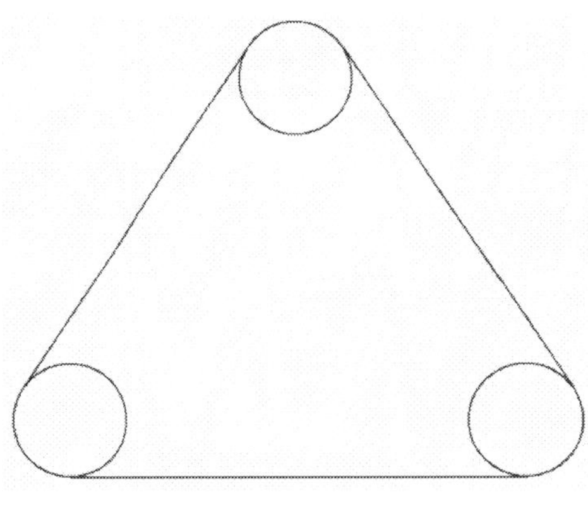

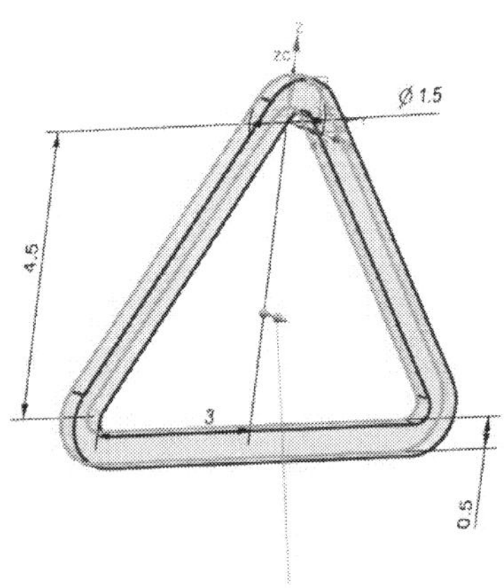


Trim the circles using the tangent lines as bounding objects.

Offset the curves inward by **0.5in.**

***End of Project***

## 10.9 Self-Test

(All questions are true/false. Answers are at the back of the book.)

1. The result of a 'Spline' function is a closed loop, parametric curve.

2. The result of the 'Offset' command is a string of coplanar curves. They may be parametric or non-parametric.

3. The 'Project' command results in non-parametric curves.

# 11. Combine Bodies

## 11.1 Introduction

Building complex solid geometry often requires combining multiple entities. In NX 8, a 'Boolean' operation is a feature operation that manipulates solids and sheets. There are three types of 'Boolean' operations – 'Unite,' 'Subtract,' and 'Intersect.' 'Unite' and 'Subtract' can only be used on solid bodies while 'Intersect' can be used on solids and sheets. We will also discuss the 'Sew' operation which works with sheets as well as solids. All of these can be found under the ***Insert / Combine*** menu as shown in the following figure.

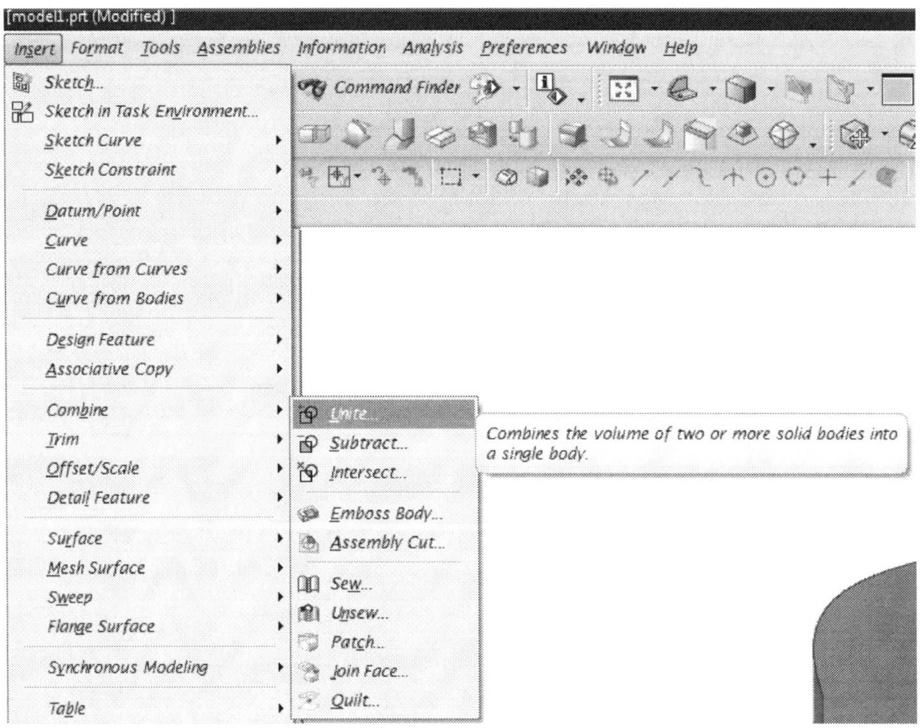

The next figure exemplifies some of the different Boolean operations on solids. You can add geometry to a model by 'uniting' solids together, or you could create a cavity in a *target* model by 'subtracting' a *tool* solid from it. The 'Intersect' Boolean operation is useful when you are modeling a solid based on different shapes in different views.

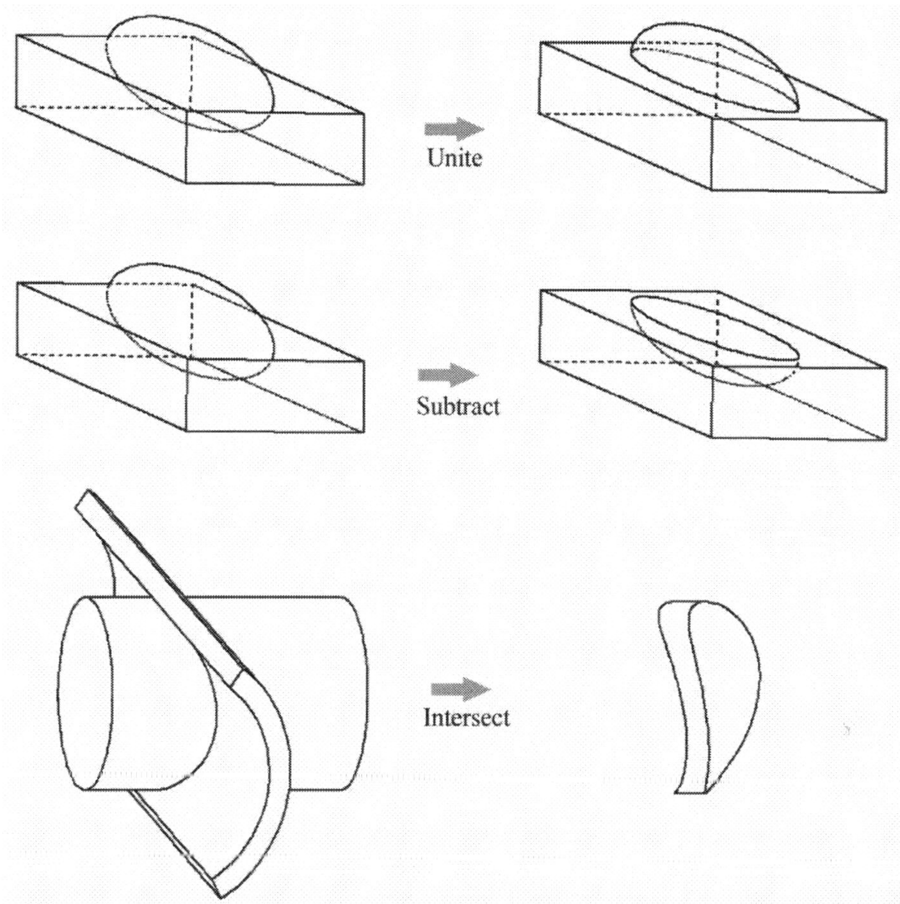


| **Note:** For 'Unite,' 'Subtract' and 'Intersect,' NX 8.5 allows you to retain the 'Target' and 'Tool' body designations. Even after the Boolean operation has been completed, it is possible to differentiate between the original two parts. |
|---|

## 11.2 Unite and Subtract

In this example, you will create three separate solids from a single sketch. You will then combine three bodies into a single solid body using 'Unite' and 'Subtract' operations.

### ***Detail: Using Unite and Subtract***

### *Project 11.2.1*

- Sketch the geometry shown in the figure below. Place all three strings in a single sketch. Don't worry about exact dimensions – just make it look roughly like the shape as shown, with a lower length of **6.5in.**
- Exit the Sketcher.

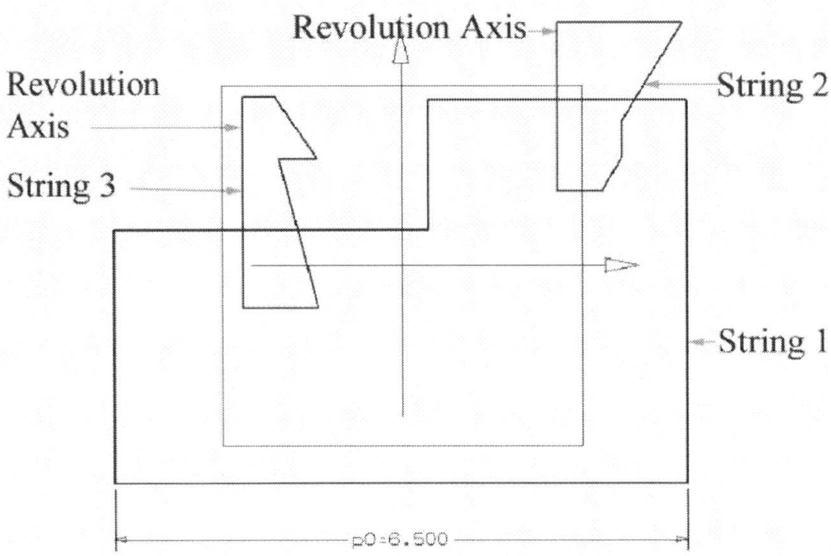


- Click the **Extrude** function.
- Since we want to extrude String 1 only, you will need to use the **Connected Curves** selection filter (as indicated below) to avoid selecting the entire sketch.

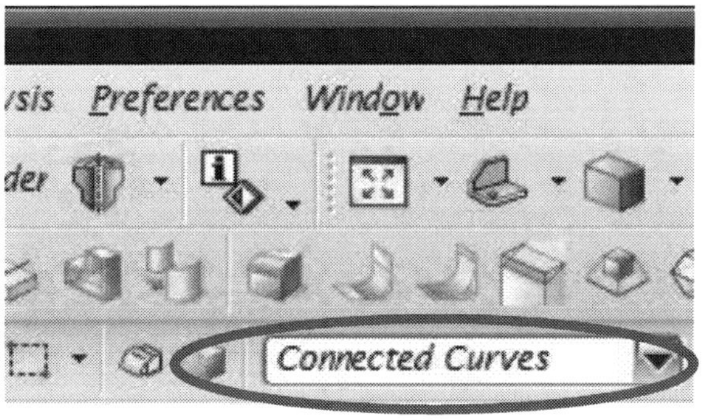


- Extrude String 1 from a start value of **–3** to an end value of **3**.

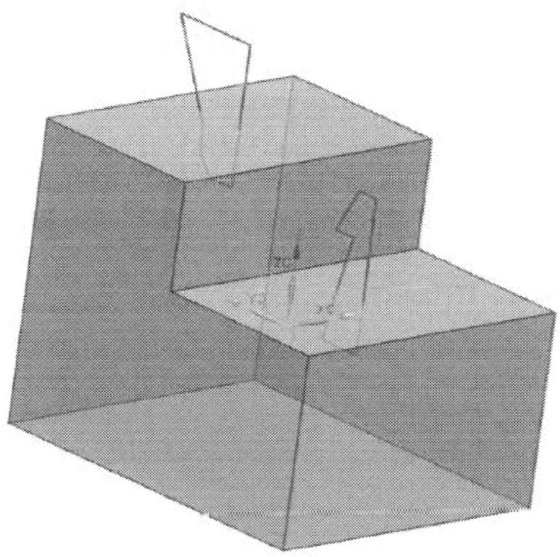

- Revolve String 2 using the ***Revolve*** command from 0 to 360 degrees. Make sure ***Connected Curves*** is selected in the selection intent menu.
- The 'Boolean Operation' drop down menu should be set to **None**. The 'Boolean Operation' drop down menu is selectable if solid geometry already exists in the part file. If no other solids were present, NX 8 would make this menu unusable.

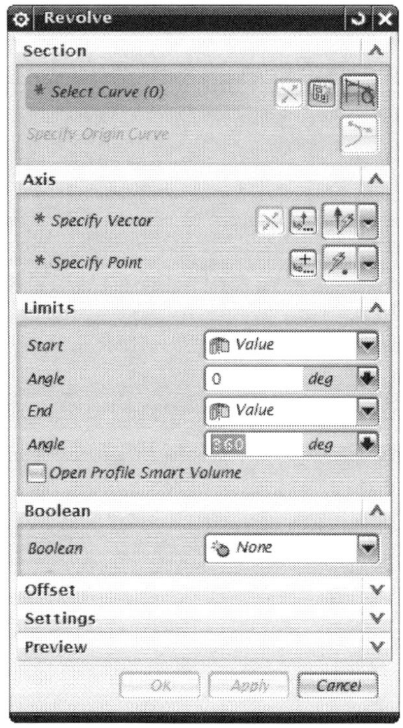


Revolve String 3 using the steps outlined above for String 2 from 0 to 360 degrees. Again, select None from the Boolean Operations options. Your model will consist of three solids, as seen in the figure below. Move your mouse over each of the solids and notice how NX 8.5 highlights them individually.

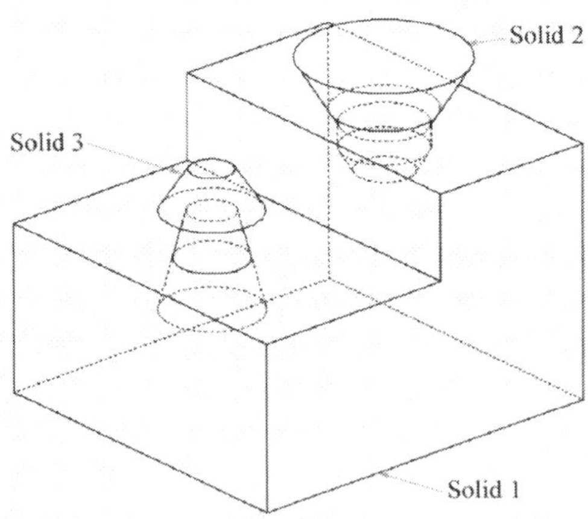


- Select **Insert / Combine / Subtract**  .
- Pick Solid 1.

The cue is prompting you for a target solid. The target solid is the entity from which the tool will be subtracted. There can only be one target body.

- Pick Solid 2.

This is the tool solid, which will be the entity that does the 'cutting.' You may select multiple tool bodies, if needed.

- Choose **OK.**
- Select **Insert / Combine / Unite**  .
- Pick Solid 1.
- Pick Solid 3.
- Choose **OK.**

You should now have one solid body, as shown below. The subtract operation created a hole in the target solid and consumed the tool solid, while the unite operation simply added the two solids together.

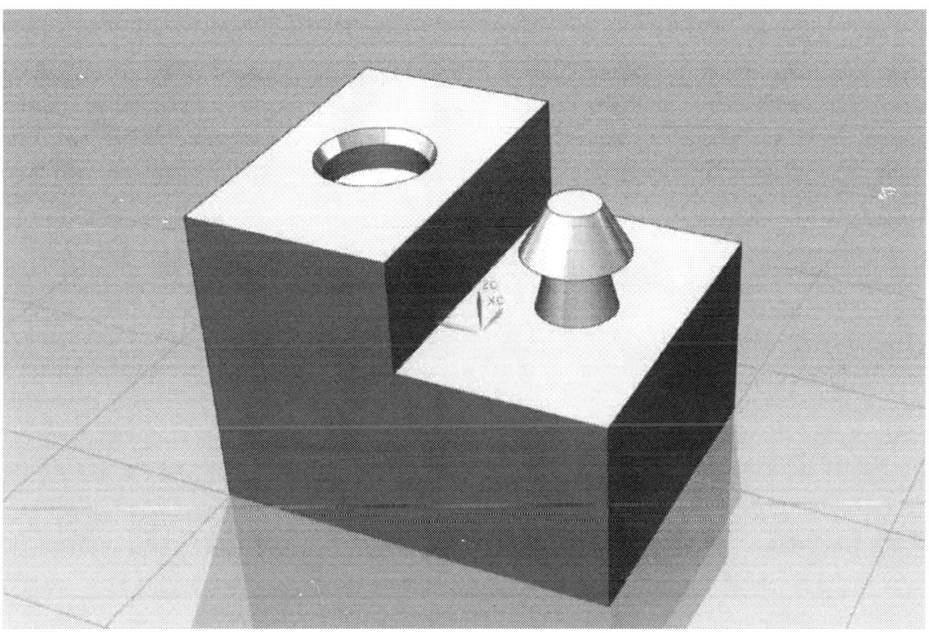

> **Note:** You could have selected Unite or Subtract while creating the revolved body and skipped the last few steps, but we want you to learn both options.

***End of Project***

> **Note:** Select Insert / Combine / Unite. Notice how there is only one solid body recognized by NX 8.5 by scrolling the mouse over the entities. NX 8.5 highlights the entire part as one entity, whereas previously there were three.

## 11.3 Intersection

Intersecting two solids is a great way to make complicated shapes. For example, the following figure details a tab that must be curved with a radius in one direction.

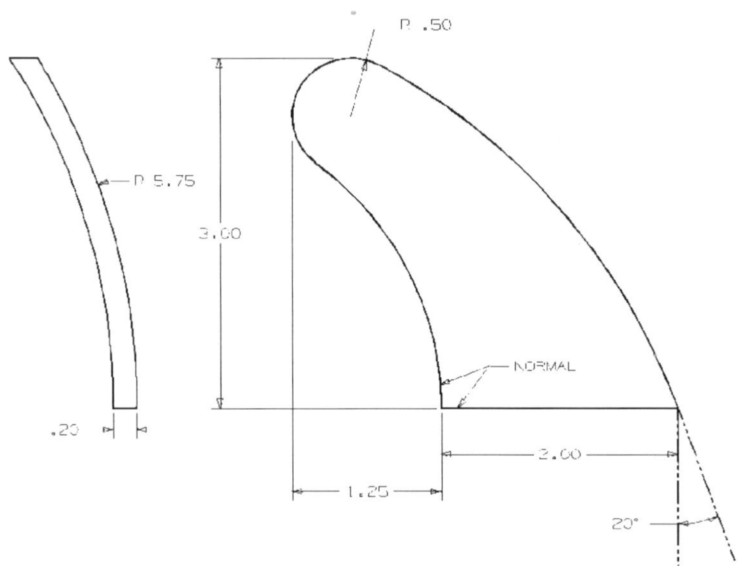


Since this geometry is curved in two planes, it is a good application for construction using the 'Intersect' operation. A very effective approach for this model is to create the two solids shown in the next figure and then perform an Intersect Boolean operation using them as tool solids.

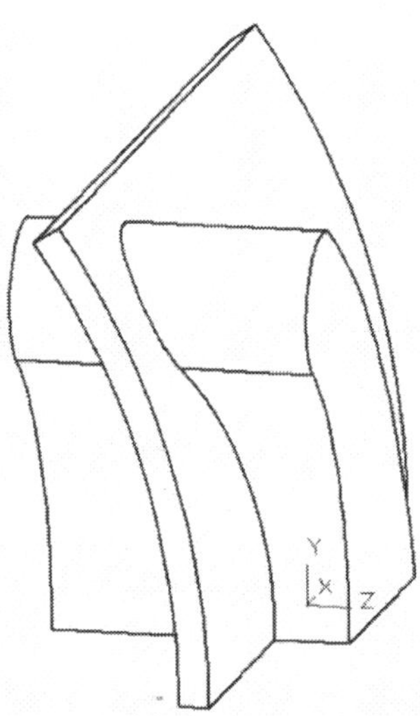

## *Detail: Using Intersection*

### Project 11.3.1

Create three fixed datum planes by selecting ***Insert / Datum/Point / Datum CSYS***. Select the Absolute CSYS option from the drop down menu.

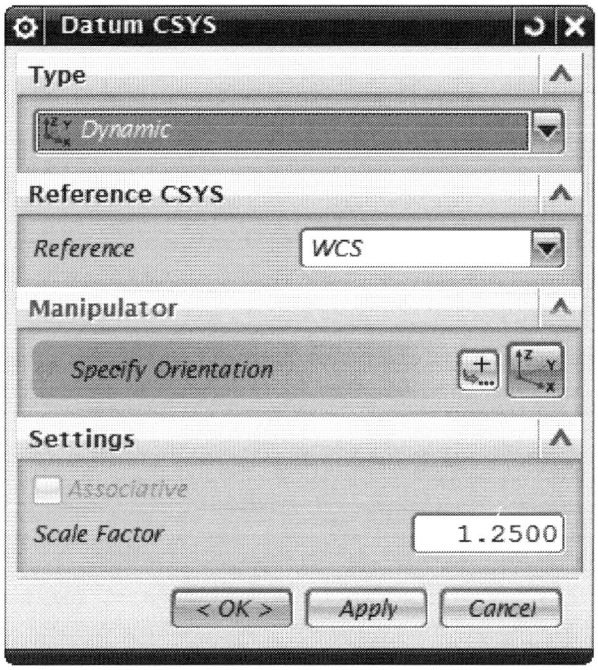


- Sketch the right view of the tab and position it as shown in the figure below. Arc 1 is normal to Line 1, so use a 'Point on Curve' constraint to align the center of the arc onto Line 1. Also, align Line 1 to the Y-Z datum axis with a collinear constraint. Create Lines 2 and 3 as reference lines and make Line 2 tangent with Arc 2.

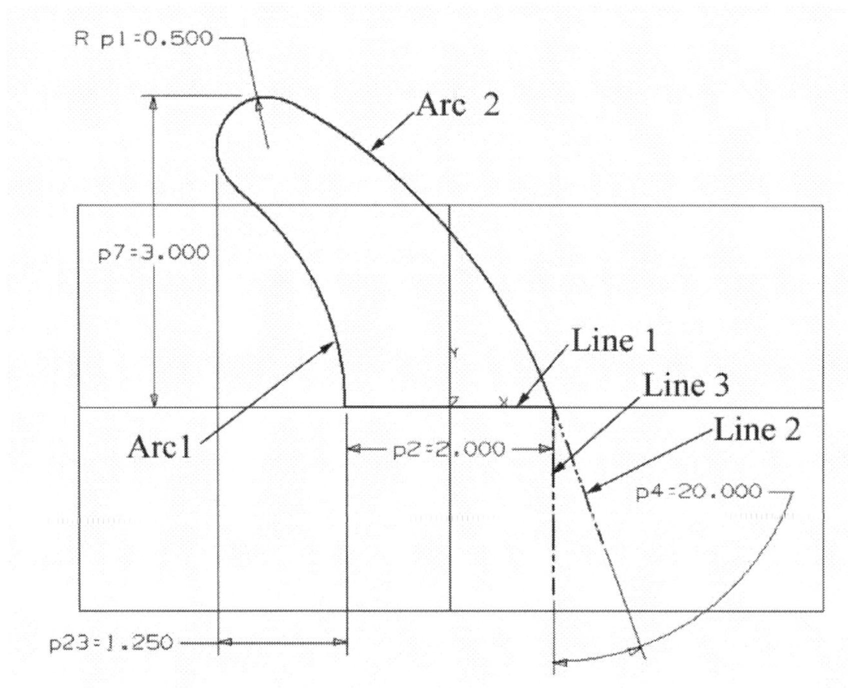


- Extrude your sketch from **-1.5** to **1.5**.
- Sketch the geometry shown in the next figure. Create this sketch on the datum plane that is vertically perpendicular to the plane of the first sketch. It only needs to consist of two entities: an arc and a vertical reference line. The reference line is tangent to the arc and is used to control the orientation of the arc. The height of the arc does not need to be a specific value as long as it is a bit taller than the height of the tab. For this example, the height of the arc is related to the first sketch by setting it equal to the height variable plus an arbitrary **0.25in**.
- Your sketch should be placed on the Z-Y plane.

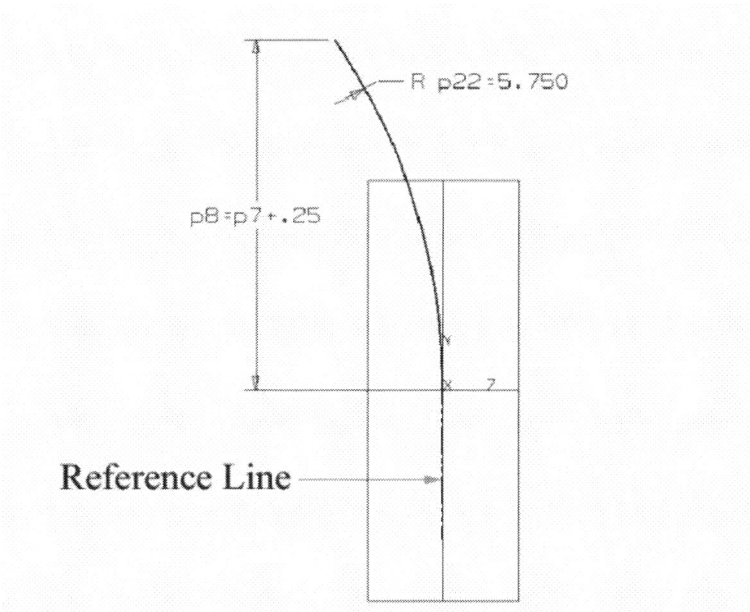


- Extrude the arc from –3 to 3 with an offset of 0.2 toward the arc's center. Observe the offset arrow to determine if you need to enter a positive or negative 0.2 offset dimension. Recall that this offset value can be entered as the first or the second offset.
- Select Insert / Combine / Intersect  .
- Pick the solid created by the first sketch.
- Pick the solid created by the second sketch.
- Choose **OK**. You should see the resultant solid shown below.

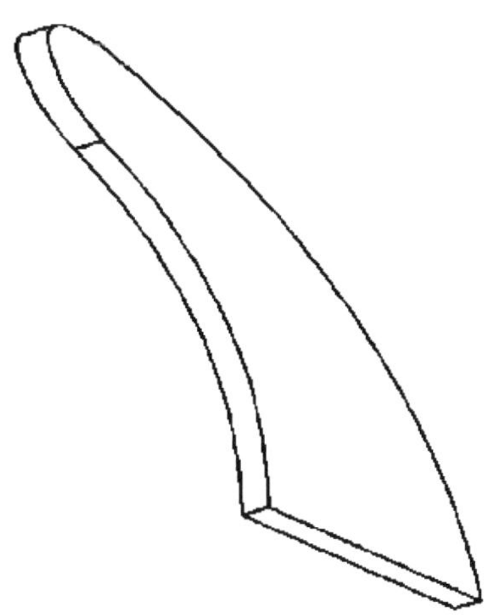

***End of Project***

Solid geometry that does not overlap is discarded when an 'Intersect' Boolean operation is performed. Thus, the outer extent of your tool solids is somewhat arbitrary – just make sure that they are large enough to create a complete intersection solid.

In these examples, you first created the solid geometry and then subsequently performed a Boolean operation using them as tool solids. You can also perform the Boolean operation at the same time that you create the new solid by choosing the appropriate selection from the Boolean Operation drop down menu (see third figure in 10.2).

## 11.4 Inferred Boolean

In this project we are going to look at the inferred Boolean option. If you set your Boolean option to 'Inferred', NX will predict the Boolean option that you want. For example, if you create an Extrude and there are no other solids in the model, the inferred option will simply create a solid. If you create an Extrude that intersects an existing solid, the 'Inferred' option will choose to subtract the Extrude from the existing solid. Finally, if you create an Extrude that touches face-to-face with an existing solid, the 'inferred' option will unite the Extrude to the solid.

### *Project 11.4.1*

Now let's try all of these techniques by creating some kind of pipe fixture.

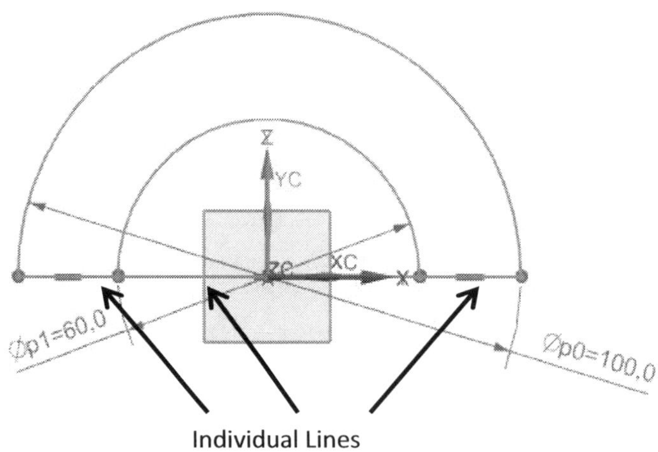


- To begin, click **Insert / Sketch in Task Environment** – select the XY plane and create the following sketch which consists of two concentric arcs and a line.

- Click **Insert / Design Feature / Extrude** to extrude the curves in the –Z direction, shown in the following figure, to a distance of 10 and set your
- Boolean option to 'Inferred,' also shown in the following figure.

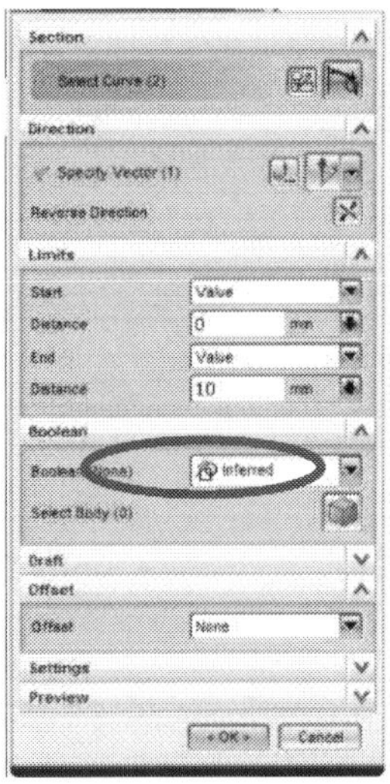

**Note:** It may be helpful to change your selection intent to 'Single Curve' before selection. You will notice that the 'Inferred' Boolean option has decided to create the solid.

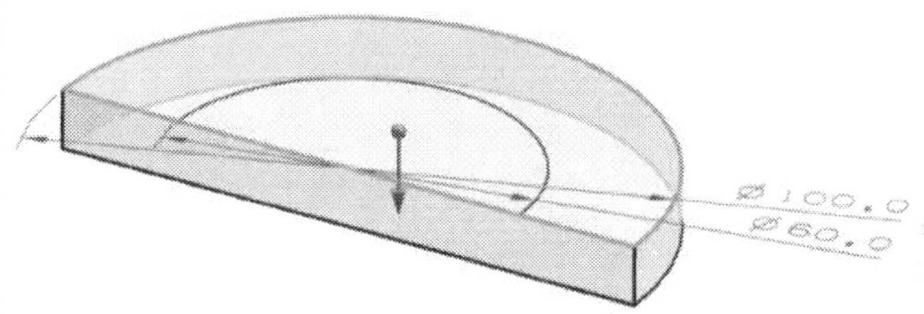

- Now create another extrude using the inside arc and part of the line. Note: Using 'Stop at Intersection,' highlighted in the following figure, will help your selection by only selecting a length of curve until it recognizes an intersection with another curve.

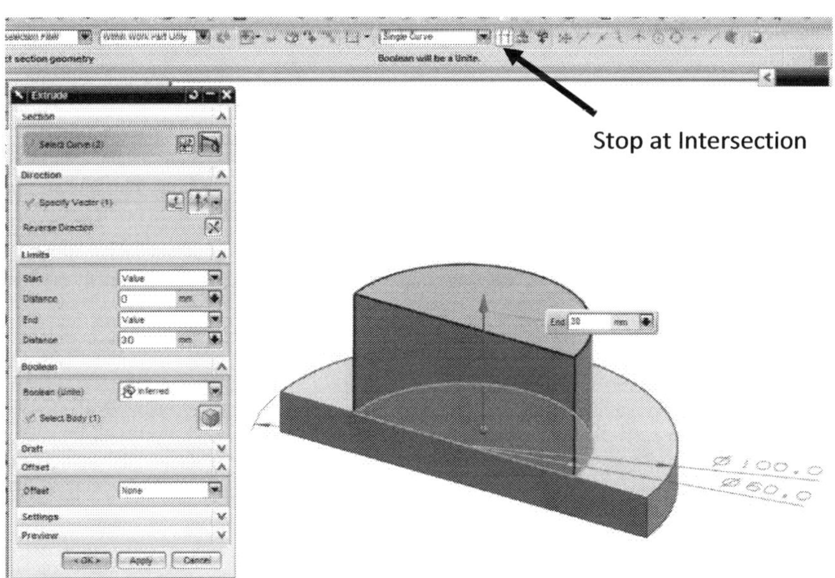

- Set the Extrude to a distance of 30 in the +Z direction, and use the 'Inferred' Boolean option as shown below. You will notice that NX inferred a Unite between the Extrude and the existing solid.
- Finally, we are going to use the 'Inferred' Boolean option to create a subtraction.
- Create the following sketch on the top face of the solid by selecting **Insert / Sketch in Task Environment**, and selecting the top face.

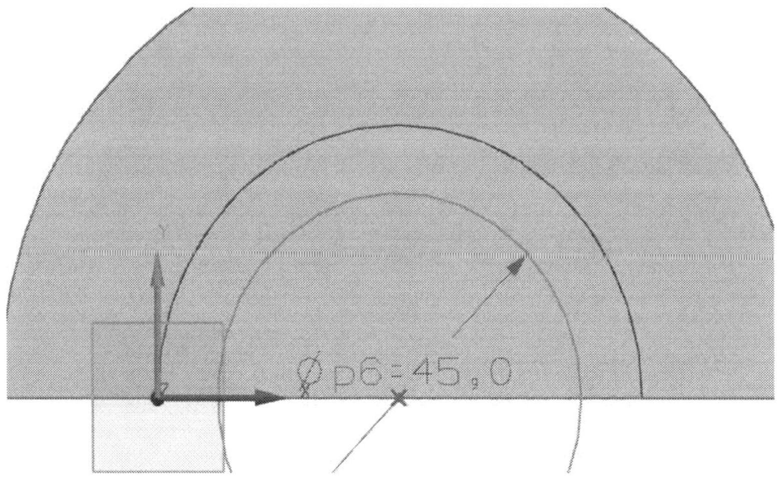


- Now use the extrude command with the Boolean option set to 'Inferred' and extrude a distance of 50 in the –Z direction as shown in the following figure.

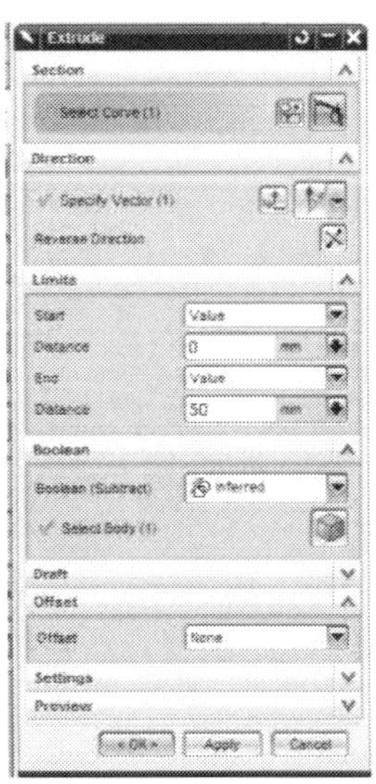

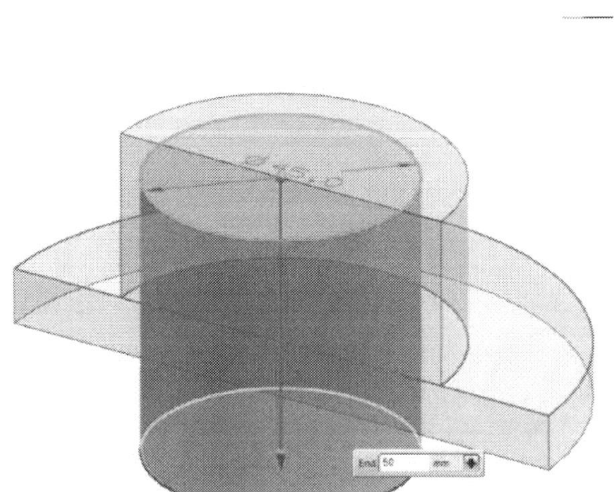

- Use NX 'Inferred' to subtract the Extrude from the solid, as shown in the following figure.

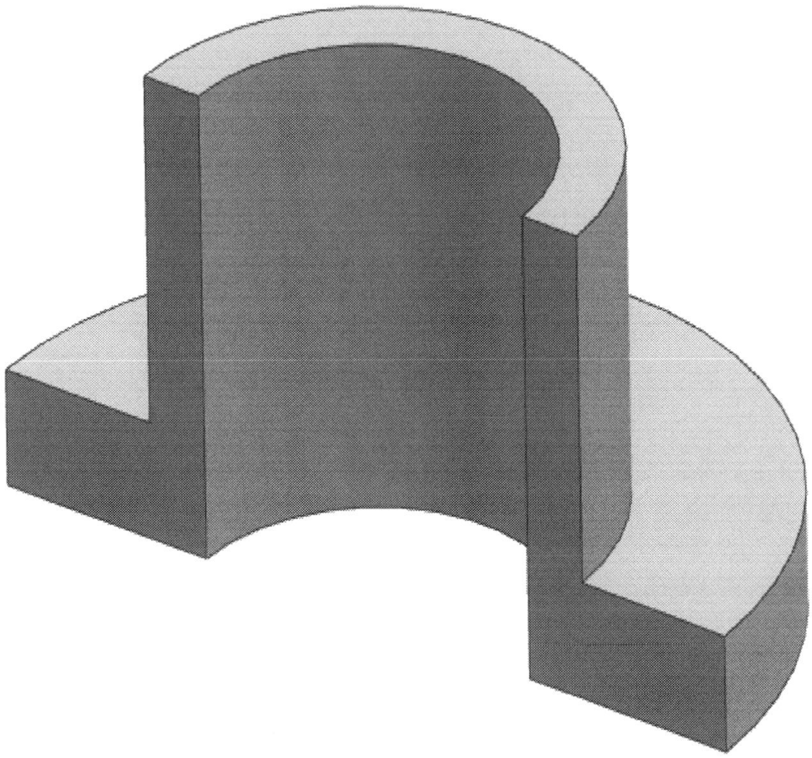

***End of Project***

## 11.5 Sew Operation

The 'Sew' option is similar to 'Unite,' but where 'Unite' can only be used to combine solid bodies together, 'Sew' allows you to combine sheet bodies as well as solid bodies. You can only sew solid bodies that have coincident faces.

### ***Detail: Sewing Sheets***

The most common use of the 'Sew' command is to sew sheets together. This is possible even if there are small gaps between the sheets. To sew any sheet bodies with adjacent edges simply use ***Insert / Combine / Sew***, select that target and sheet bodies, and hit **OK**.

### ***Project 11.5.1***

Sketch a rectangle like the one shown below.

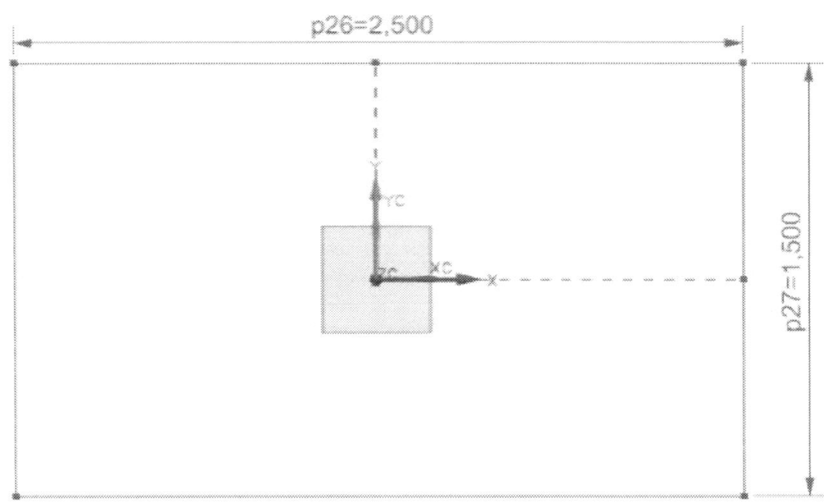


- Exit the Sketcher.
- Select Insert / Curve from Curves / Offset.
- Select any string from your sketch.
- Select **OK.**
- Input an Offset Distance of **0.05** inches and choose **None** for Trim. Be sure the 'Associative Output' radio button is NOT checked and 'Input Curves' is set to **Keep.**
- Select **OK**. Your model should look like the following figure.

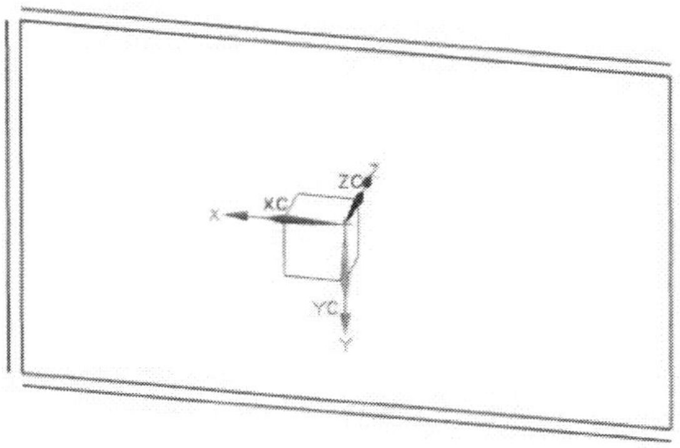


**Note:** You may need to 'Reverse Direction' in the Offset Menu to get a similar outcome.

- **Extrude** the four new curves **1in**. Compare your results to the following figure.

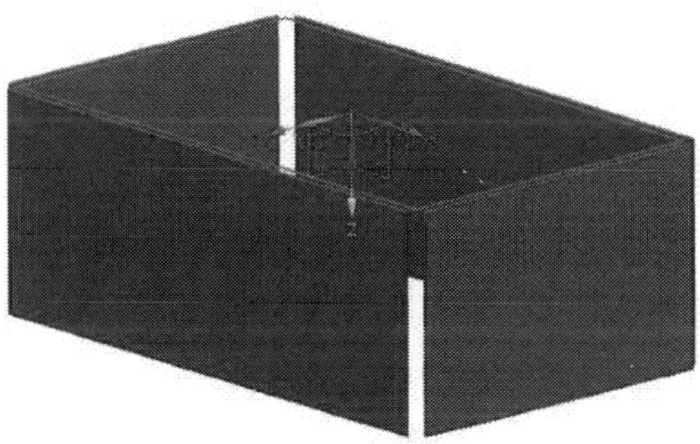

- Select **Insert / Combine / Sew**  . The 'Sew' window will appear, as shown next.

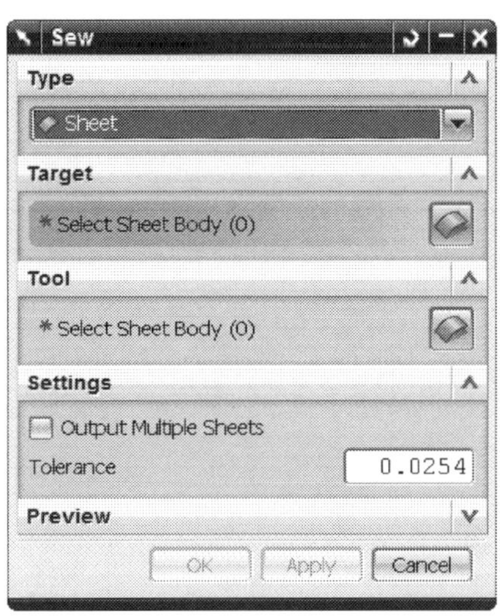

- Enter a 'Sew Tolerance' of **0.10.** This value is chosen because it is greater than the offset distance of 0.05 inches. The tolerance must be large enough to capture the seam.
- Read the cue line - NX 8.5 is asking you to choose the target sheet to sew. In this case, any of the four sheets can be selected as the target sheet.
- Select one of the extruded sheets and **Middle Click**.
- Read the cue line. NX 8.5 is asking you to choose the tool sheet(s) to sew. In this case, all three remaining non-target sheets should be selected as tool sheets.
- Select the three remaining sheets and select **OK**.
- Compare your results to the figure below.

> **Note:** Sewing sheets is best when sheets do not have gaps. Sewing sheets with gaps is possible by modifying tolerances as shown in example above, but it changes the geometry depending on the size of the gap. When gaps are bigger and surface geometry is critical, it is best to patch it up using a sheet from a new line or from the Through Curves feature.

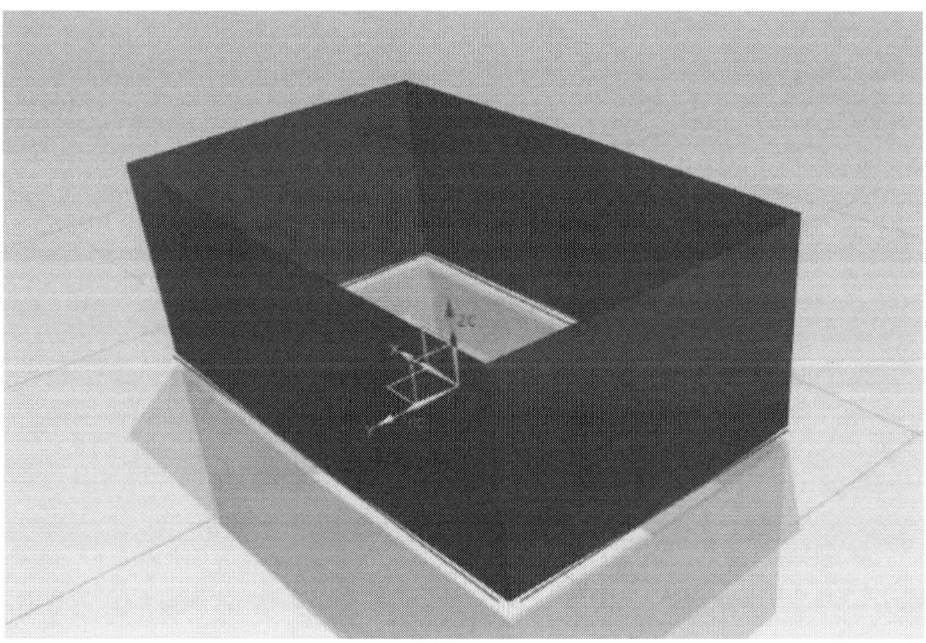

**End of Project**

## 11.6 Unsew 

The Unsew tool does the exact opposite of the Sew command. The basic principle of this tool is that when you select a face of a solid that you wish to 'Unsew,' the rest of your solids turn into sheet bodies as well. You can then separate the 'unsewn' sheet from the other sheets.

***Project 11.6.1***

- Start by selecting **Insert / Design Feature / Block** and creating a block 100x60x40mm, as shown in the following figure.

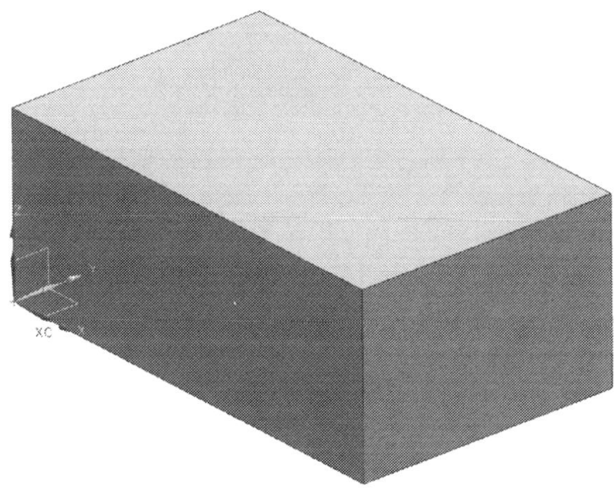


- We want to replace the top face of the block with something more exciting, so click on **Insert / Combine / Unsew**. Select the top face, as shown in the following figure, and Click **OK**.

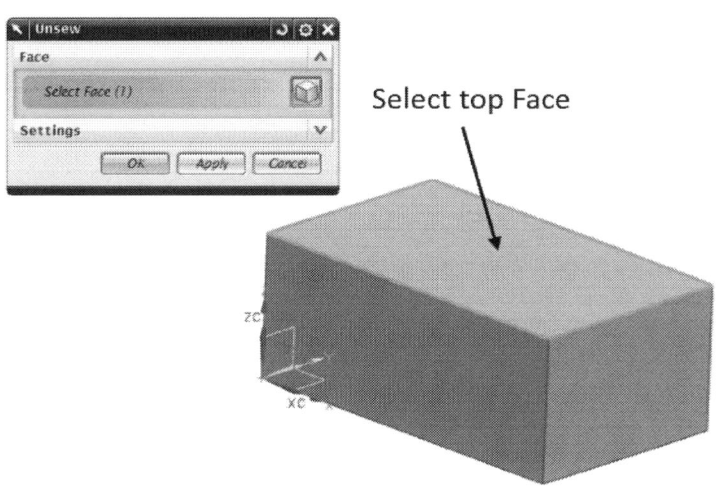

- It looks like nothing happened, but what we've actually turned this solid block into sheet bodies by 'un-sewing' the top face. Click on CTRL+B (which is the shortcut for the Hide tool), then select the unsewn face and hit enter. Your model should now look like the following model.

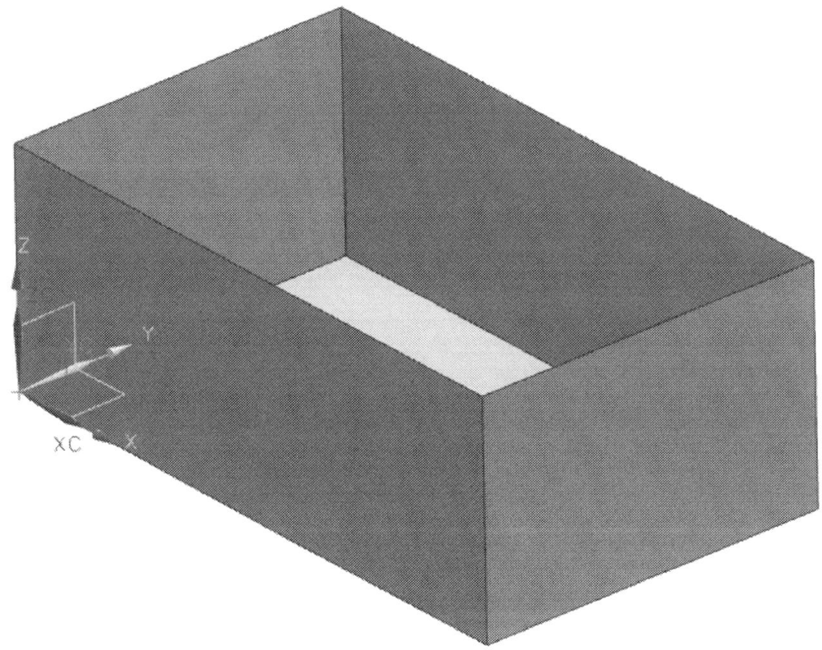


- To create a more exciting top face we need to sketch the profile we would like. To do this, we first need a datum plane to sketch on. To illustrate the new datum plane functionality, we are going to create a bisector datum plane between the two end faces with an offset of 10mm.
- To do this, select **Insert / Datum/Point / Datum Plane**. Pick **Inferred** as the type of datum plane and select the two small faces, as shown below. This will create a datum plane right in the middle of your model. Then, while you are still in the datum plane operation, check **Offset** in the datum plane window and type in 10mm. Click **OK** to complete.

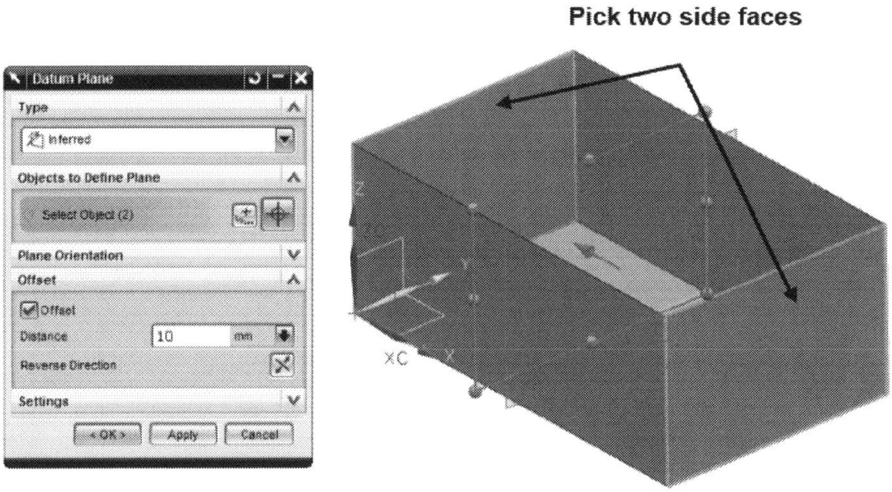


- Before we create the profile that will define our new face it would help to use some end points where we can connect our sketch too. Select **Insert / Curve From Bodies / Intersect** . Select the two side faces as **Set 1** of your intersect and the datum plane as **Set 2**. Click **OK** to complete.

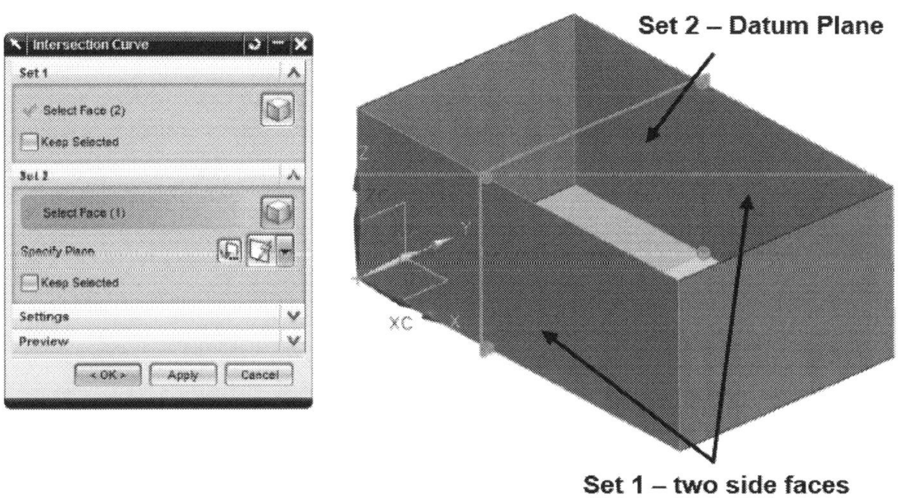


- Now create a sketch on the datum plane using the end points of the **Intersect** for the end points of your arc, as shown in the following figure.

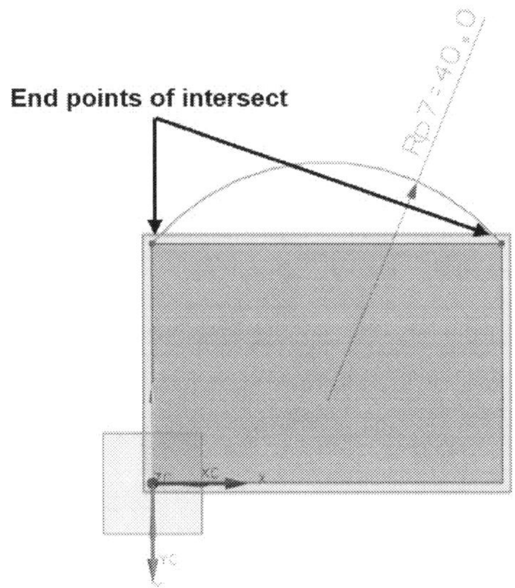


- Once you have finished the sketch we can begin creating a surface to replace the old one. Go to **Insert / Mesh Surface / Through Curve Mesh**. For this feature you have to pick **Primary Curves** and **Cross Curves**. First select one of the long edges as your **first Primary Curve**. Then click on the **Add New Set** button**.** Now you can select the opposite long edge as your second primary curve, as shown in the following figure.

| **Note:** Make sure the arrows are pointing in the same direction. |
|---|

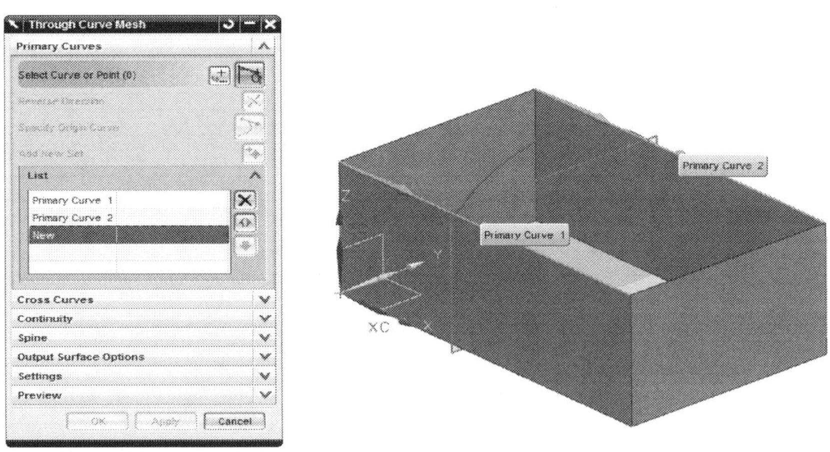

- Move onto the **Cross Curve tab** in the Through Curve Mesh Dialogue box. Select one of the short edges as the **first Cross Curve**. Click on **Add New Set** in the dialogue box, and then click on the Arc that you sketched as the **second Cross Curve**. Finally click on the **Add New Set** button, and then select the final short edge of the model as your **third Cross Curve**.

**Again:** make sure the arrows are pointing in the same direction. The three curves to select are shown in the following figure.

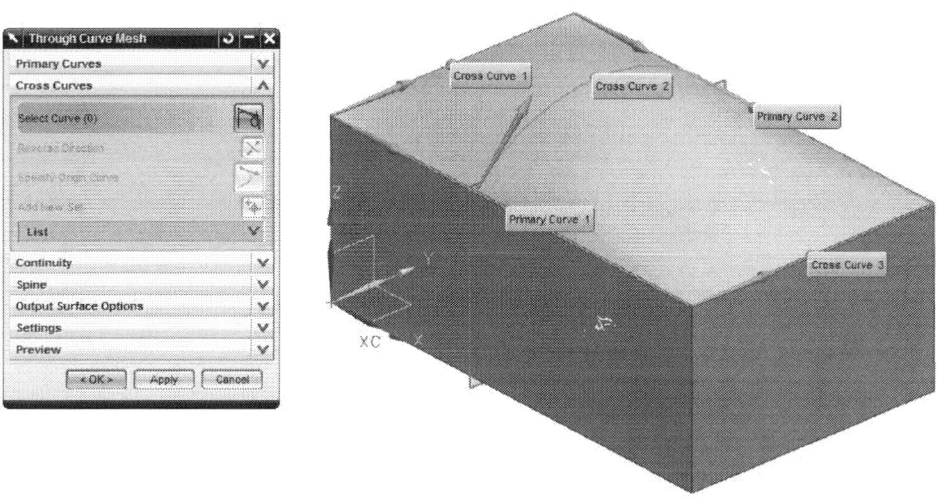


| **Note:** the surface is generated as you add Cross Curves. Click **OK** to complete. |
|---|

- Now we have to sew the separate surfaces together to make a solid body. Click on **Insert / Combine / Sew** 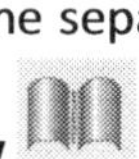. Select the new surface as your target and the existing surfaces as your tool. Click **OK** to complete.

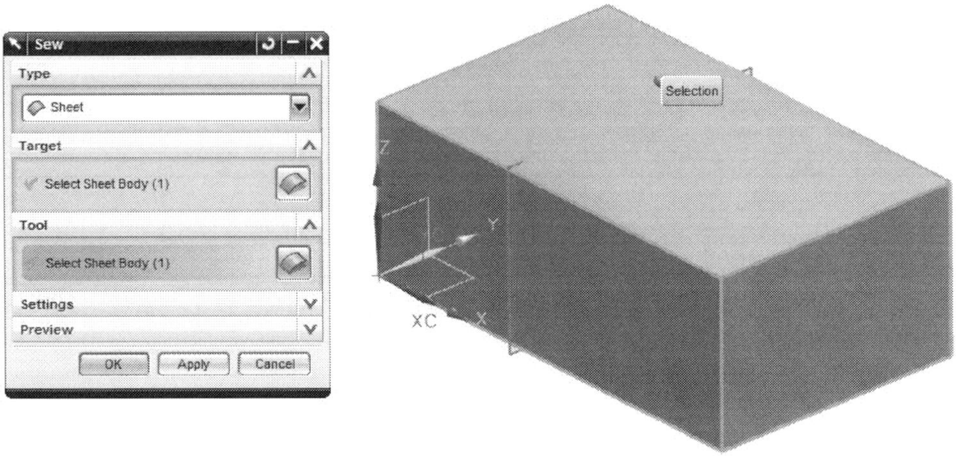


- Now you have a solid body with a beautiful new face.

***End of Project***

## 11.7 Emboss Body

The Emboss Body tool works similar to the subtract command. The basic principle of this tool is that when you select a target solid that you wish to 'Emboss,' the tool solid in a sense pushes it's shape through the solid to create a cut-out imprint into the target solid.

***Project 11.7.1***

- Start by selecting **Insert / Design Feature / Cylinder** and creating a cylinder 4in high by 2.5in diameter, as shown in the following figure.

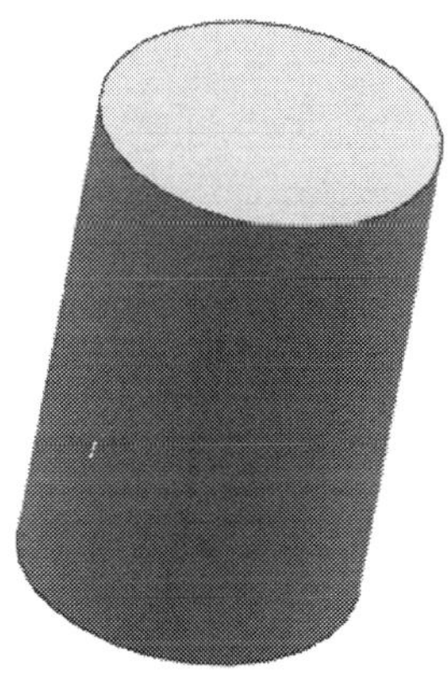

- Next sketch an arc and extrude it through the cylinder, thicken it by 0.5in.

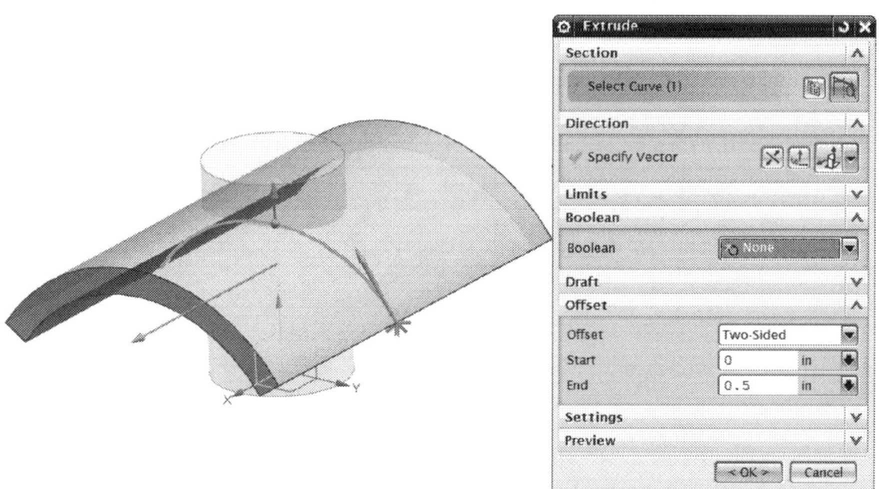

- Now that you have a cylinder and an extruded arc, the next step is to **select Insert / Combine / Emboss Body.** Select the extruded arc as the Target and the Cylinder as the tool. You may need to select the outer face of the cylinder as the Region.

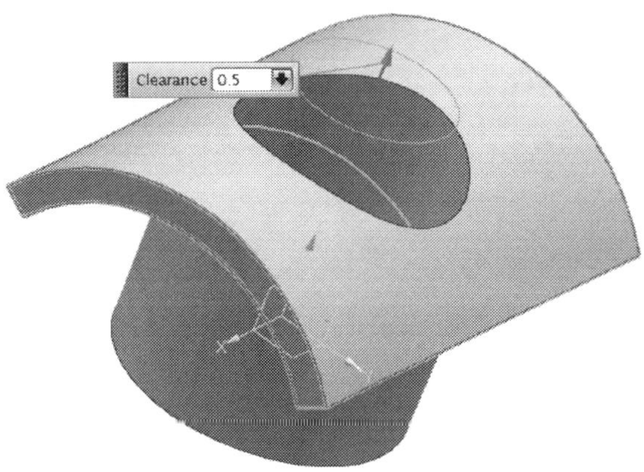


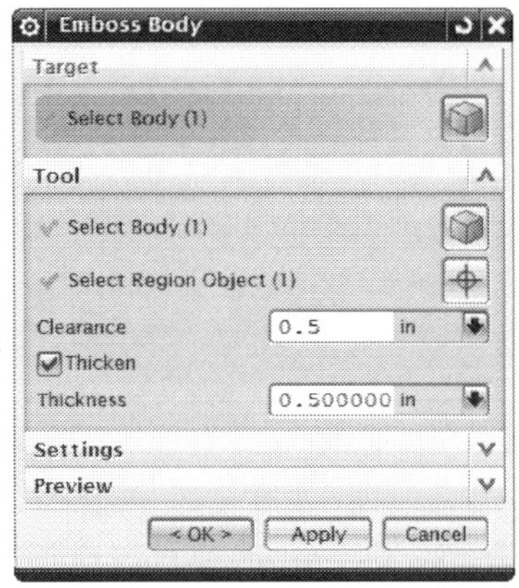


As you can see the cylinder not only subtracted itself from the extruded arc, it also embossed it's lower half as an imprint into the solid.

## 11.8 Practice: The Stepped Plug

The geometry shown below could be constructed by subtracting a number of blocks from a body of revolution. It is much quicker, however, to simply intersect two bodies.

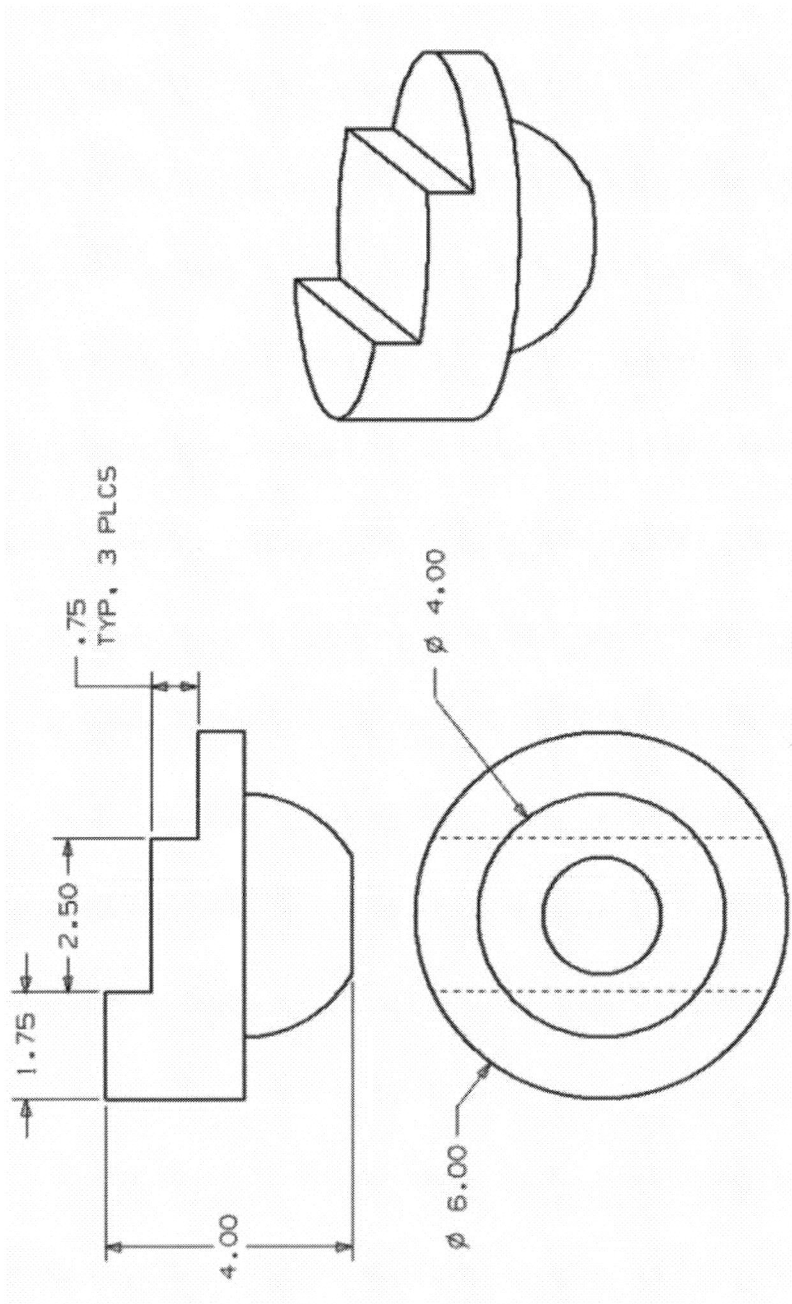

## Practice Outline

- First, create the sketch shown in the next figure.
- Notice how line 1 passes through the midpoint of line 2. You can enforce this constraint with the proper set of dimensions; however, there is a better use of geometric constraints.

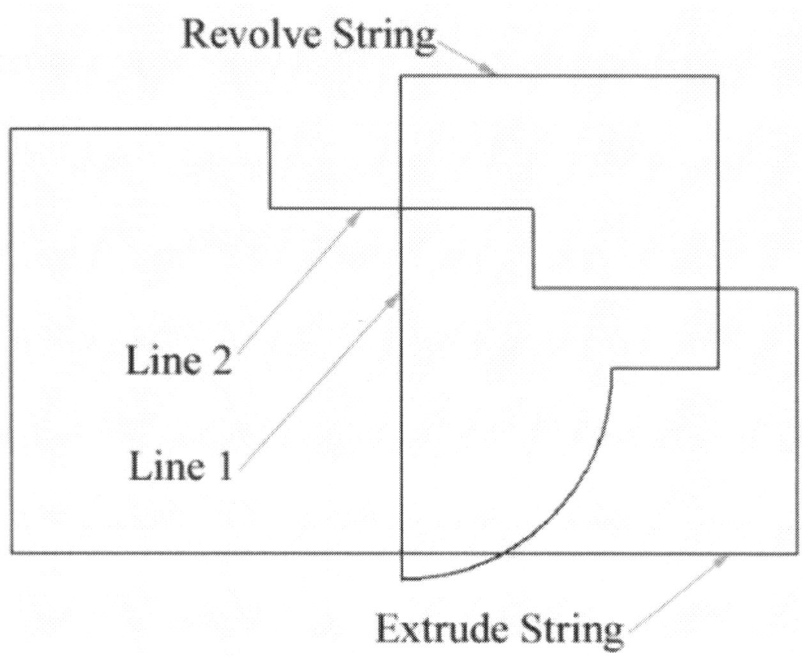


- Next, use the 'Revolve' and 'Extrude' functionality to create two tool solids.
- To complete the model use the 'Intersect' function, selecting the two bodies as target and tool solids (the order does not matter).

***End of Project***

## 11.9 Self-Test

(All questions are true/false. Answers are at the back of the book.)

1. 'Tool' and 'Target' designations may be retained when using the 'Unite' command.

2. The 'Sew' command can only be used for combining sheet bodies.

3. The 'Sew' command works even if there is a gap between the sheets and/or solid bodies you wish to unite.

# 12. Expressions

## 12.1 Introduction

In NX 8.5, an 'Expression' is an equation or value that defines the value or magnitude of a parameter. An 'Expression' can be written as a single numeric value (p24=4.125), a mathematical operation (p17=sin(30)), or a combination of other parameters (p49=p22+3*p7).

Each parametric feature created in NX 8 has a set of expressions that define its geometry. Changing the value of any of these expressions will result in changes to the parametric feature geometry. The feature expressions created by NX 8 are designated sequentially beginning with the name 'p0;' the second is 'p1,' then 'p2,' etc., and associates the expressions with the objects, solids, or sketches they are created on. It also automatically creates units based on how the expressions are used. You can edit expressions using the ***Tools / Expressions*** dialog box, shown in the following figure.

> **Note:** NX 8.5 provides you with several sorting functions in the 'Listed Expressions' pull down menu. Choose 'All' when unsure what method to use.

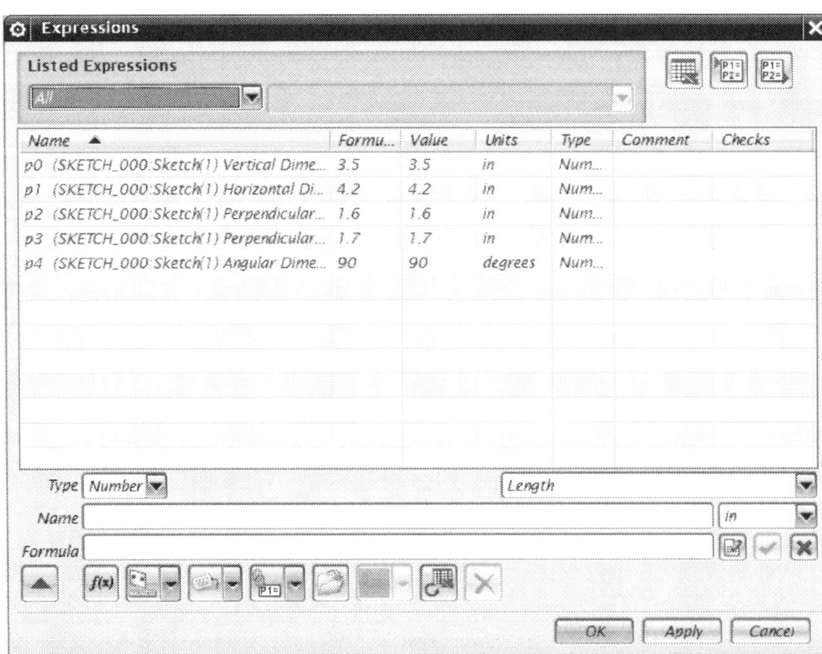

Using this function allows you to create relationships between feature dimensions. You can also create expressions manually by typing them into the 'Expressions Formula field.' Automatically created expression variable names may be changed here, enabling you to document your model with meaningful parameter names.

Clicking the 'Spreadsheet Edit' icon allows you to export the expressions into an Excel format file to edit. If several expressions need to be modified or viewed rapidly, use 'Listed Expressions.'

## 12.2 Pre-Determined Expressions

A model that has meaningful parameter names is easier to edit. For example, the parameter that controls the extrusion distance of a feature could be renamed "extrude_dist" instead of the standard "pXX" name, enabling immediate recognition of the feature to be edited.

Perhaps the simplest way to create meaningful parameter names is by entering equations into the Expressions menu prior to constructing a model. The parameter names can then be used during the definition of the solid feature geometry.

### *Detail: Basic Expressions*

The following figure illustrates a model constructed with a simple extrusion and a boss feature. All dimensions of the model are to be controlled by the equations that are listed. In this project, these expressions will be defined prior to creating the solid model.

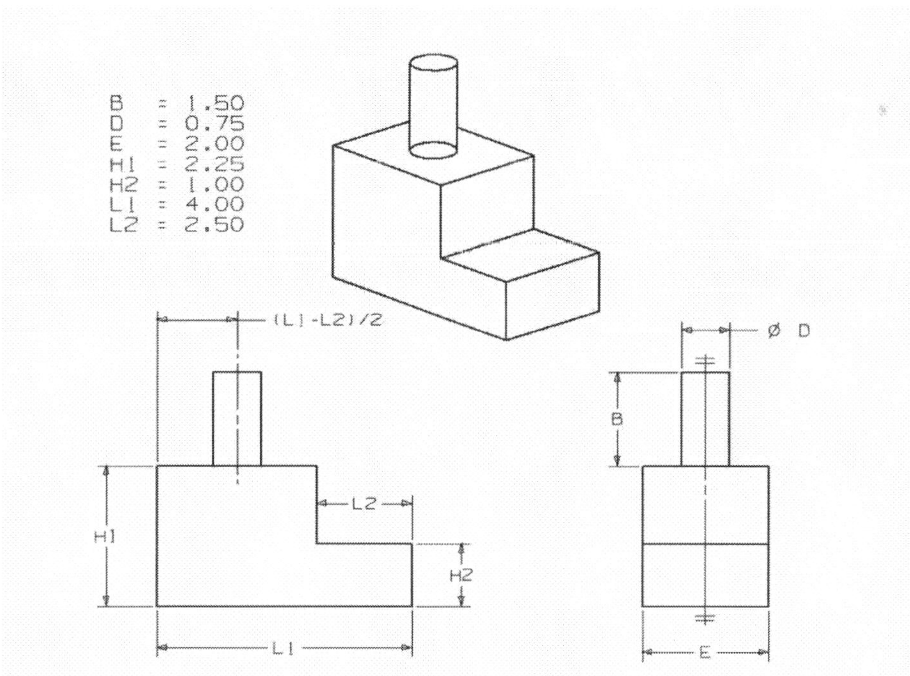

## *Project 12.2.1*

- Select Tools / Expression. NX 8.5 will display the 'Expressions' menu.

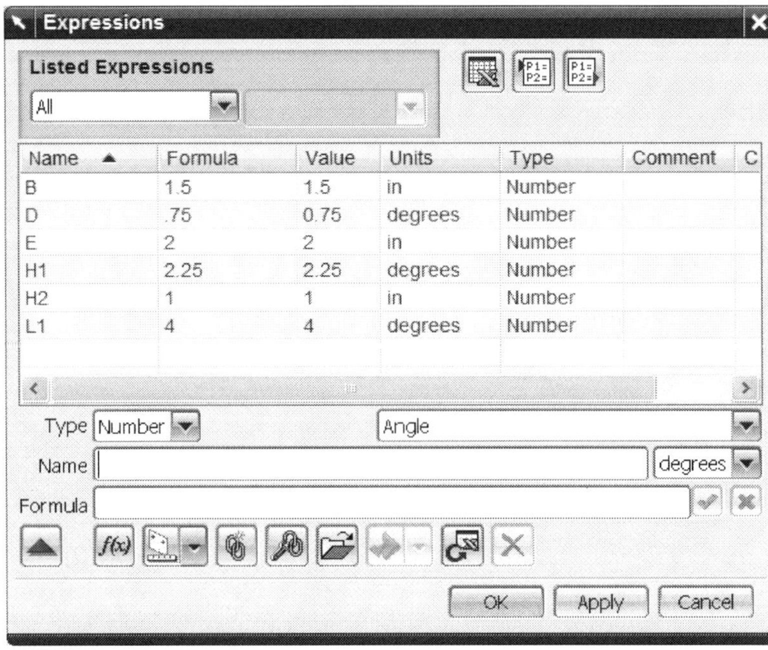


- Type **B** in the 'Name' input box, then type an equals sign "=" (which will automatically tab the cursor down into the 'Formula' box) and type **1.50**.
- Hit **Enter**. Your equation will be added to the list of expressions.
- Type in the other six expressions listed in the figure above. Hit **Enter** after typing each expression. You are finished when seven expressions are listed.
- Choose **OK**.
- Create the Sketch and apply dimensions and constraints according to the figure, but don't bother to adjust the dimension values yet.
- Modify the sketch dimensions, typing in variable names instead of dimensions (see below for guidance).

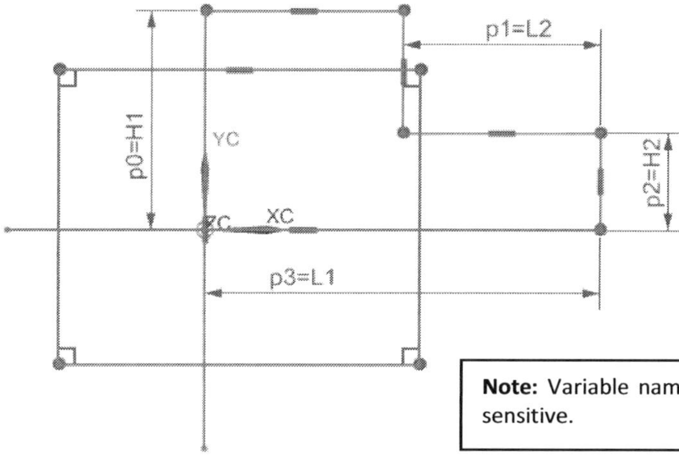


**Note:** Variable names are case sensitive.

- Extrude the sketch, entering Start Distance = **0** and End Distance = **E.**
- Create the boss, entering Diameter = **D** and Height = **B** (see below).

- Place dimensions on the boss using the variable names. It should be placed a distance of E/2 from Edge 1 and (L1-L2)/2 from Edge 2 (see next figure).

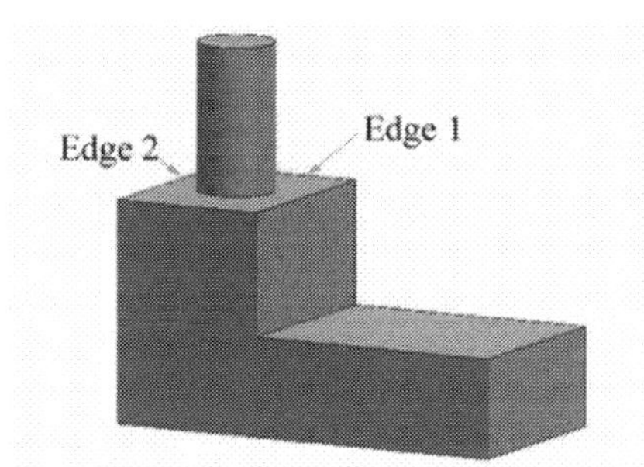


- Save the file and rename it as *'expression1_<your initials>.prt'* .

***End of Project***

## ***Detail: Changing Expressions***

The following figure shows a variation of the model that you created in the last detail section. To create it, simply change the variable values using the Expressions editor.

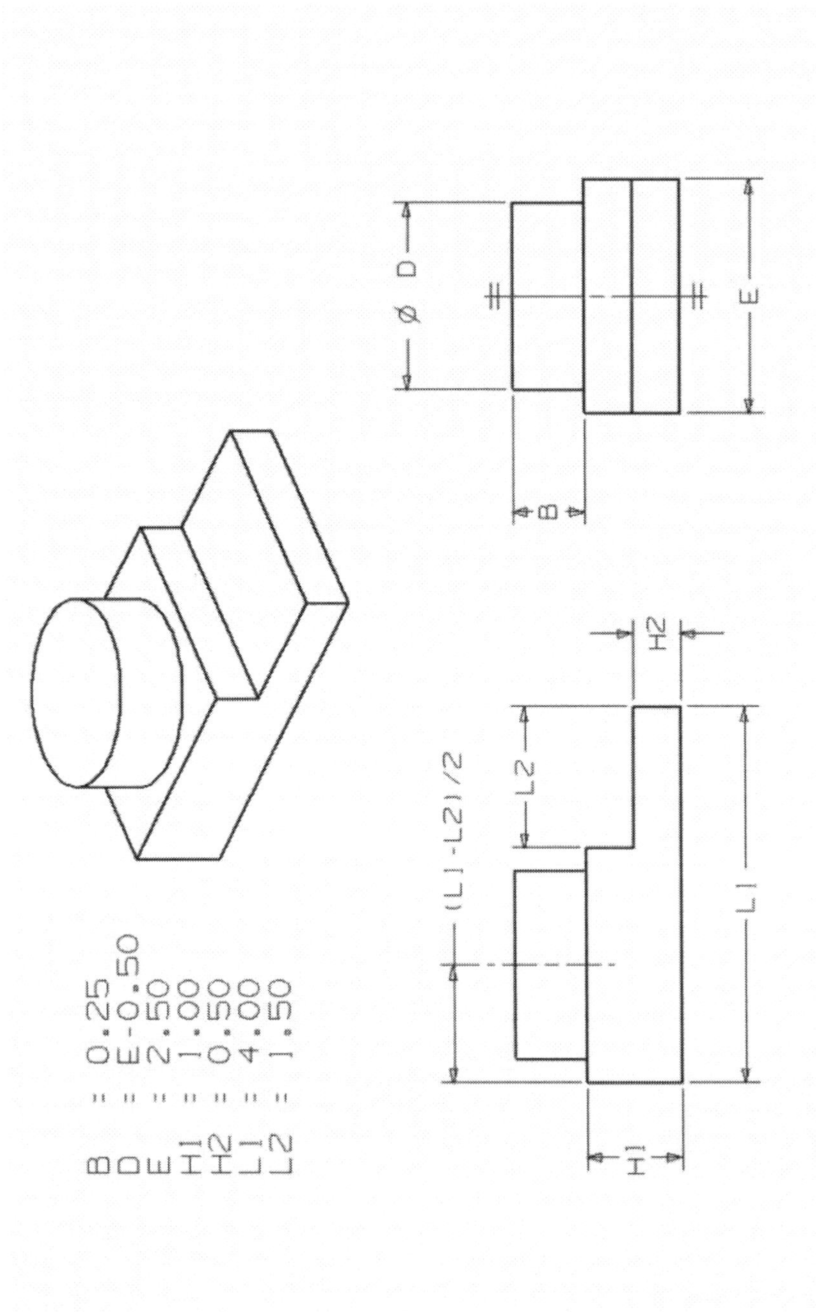

## *Project 12.2.2*

- Ensure that you are working in the file ***'expression1_<your initials>.prt'*** .
- Select Tools / Expression.
- Highlight the expression **B=1.5** in the 'Expressions' editor. *The expression will appear in the input box.*
- Click in the input box and change it to: **B= .25** and hit **Enter**.
- Edit the remaining expressions until they all match those listed in the previous figure. Choose **OK**. *The model configuration will change to that as shown above.*

> **Note:** The expression 'D=E-0.50' is an equation. Expressions can be used to set variables equal to a function of other variables and to use mathematical operators such as sin, cos, tan, etc.

***End of Project***

## 12.3 Creating Expressions during Modeling

You have seen that expressions can be directly input during feature definition. Most of the time, you may not know all the required variables at the beginning of the design process. In this case, you can create variables on the fly as features are defined.

### *Detail: Creating Expressions on the Fly*

You will create the model shown below using named variables. This time you will define the parameters as they are needed during the design process.

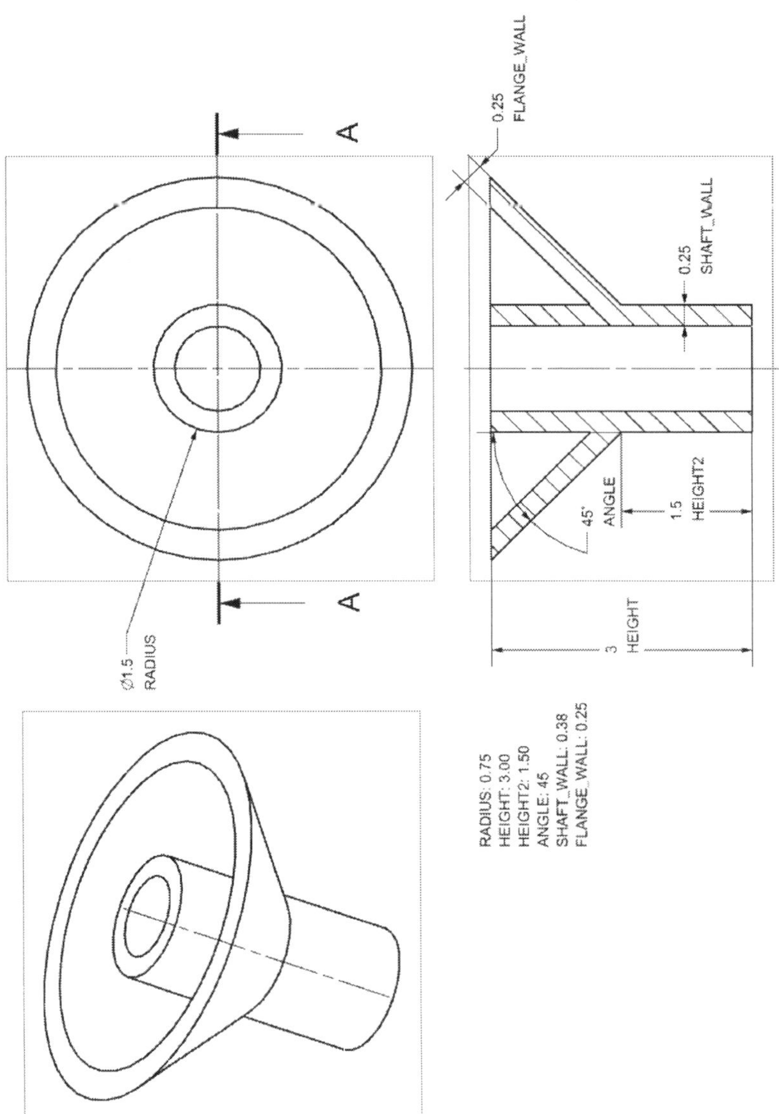

### *Project 12.3.1*

- Create the sketch shown in the next figure. The two horizontal reference lines are used to align the geometry in the vertical direction. Constrain the axis of revolution with a 'Constant Length' constraint or a 'Point on Curve' constraint to the top endpoint.

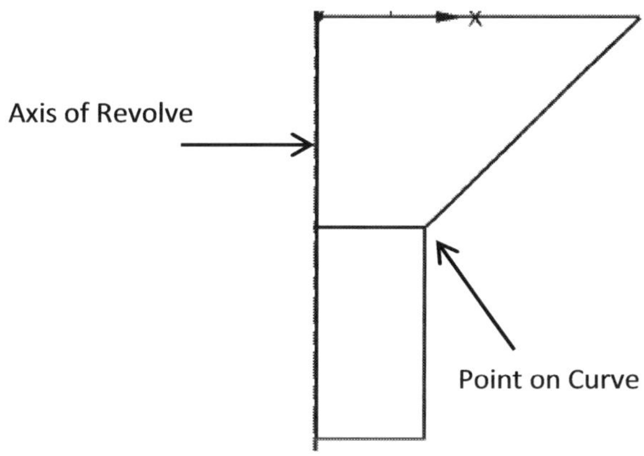


- Apply the vertical dimension as shown in the following figure. Do not choose Close so that the Dimensions dialog box remains displayed.

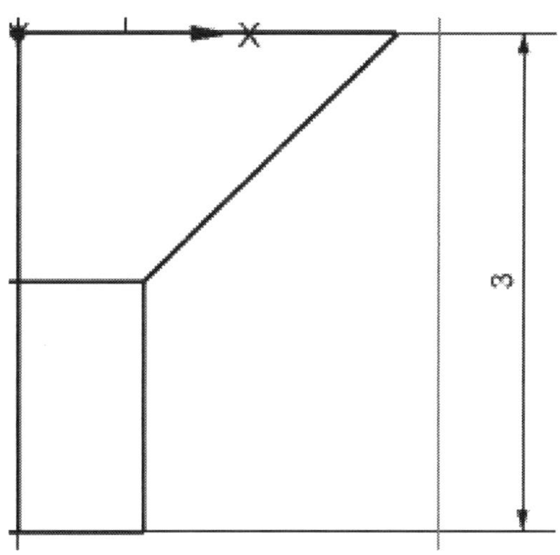


- Type HEIGHT in the left Current Expression box (see next page) and hit **Enter**. *Note that HEIGHT will replace the 'pXX' variable name in the sketch.*

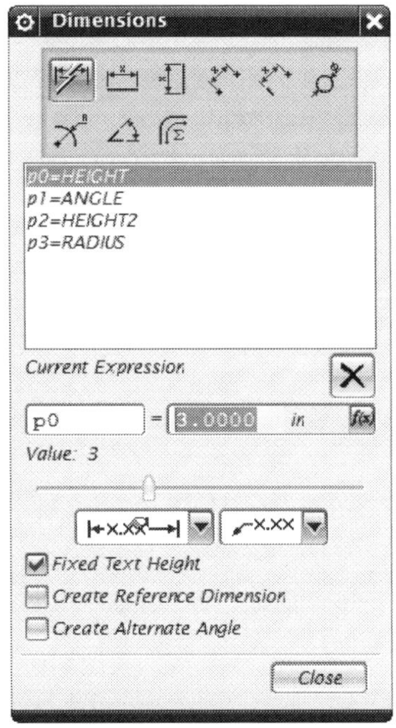


- Change the value of HEIGHT to **3** and hit **Enter.**
- Continue dimensioning the sketch and renaming the dimension variables until it appears as shown in the following figure.

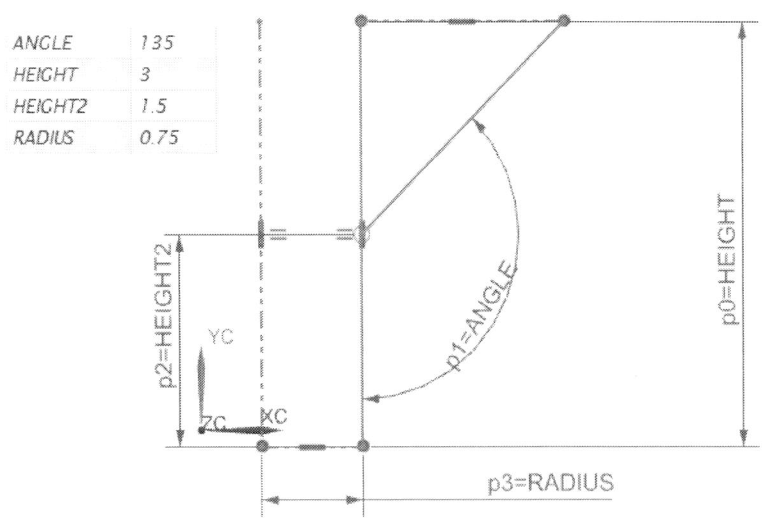


- Use constraints to position the sketch as seen below and exit the sketch command.

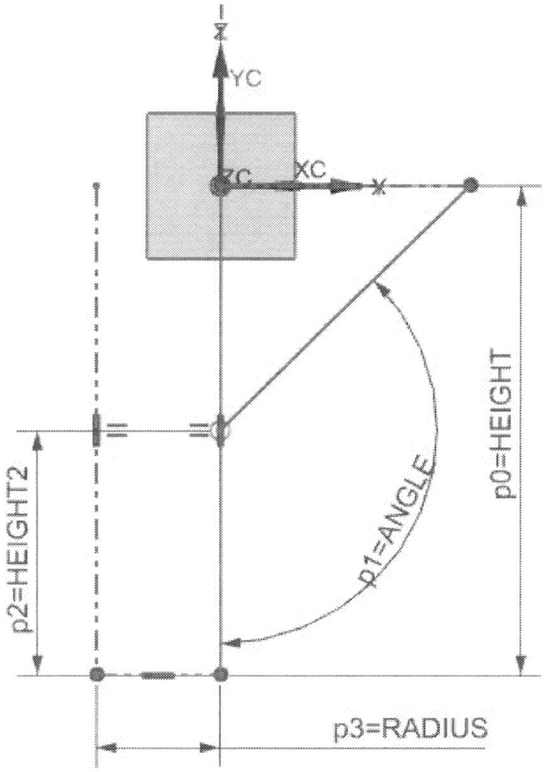


- Exit the Sketcher.
- Select Insert / Design Feature / Revolve.
- Use the **Single Curve** filter and pick the 3in vertical sketched line (not the reference line).
- Select the vertical reference line to define the object to infer a vector (axis of rotation).
- In the Revolved Body dialog box (shown below), enter Start Angle = **0**; End Angle = **360**; First Offset = **SHAFT_WALL = -0.25** and Second Offset = **0**.
- Choose **OK**.

**Note:** that you entered an expression rather than a value. This can be done anytime NX 8.5 requires you to enter feature parameters.

**Note:** You may need to enter **SHAFT_WALL = +0.25**, depending upon which direction the rotation axis arrow is pointing.

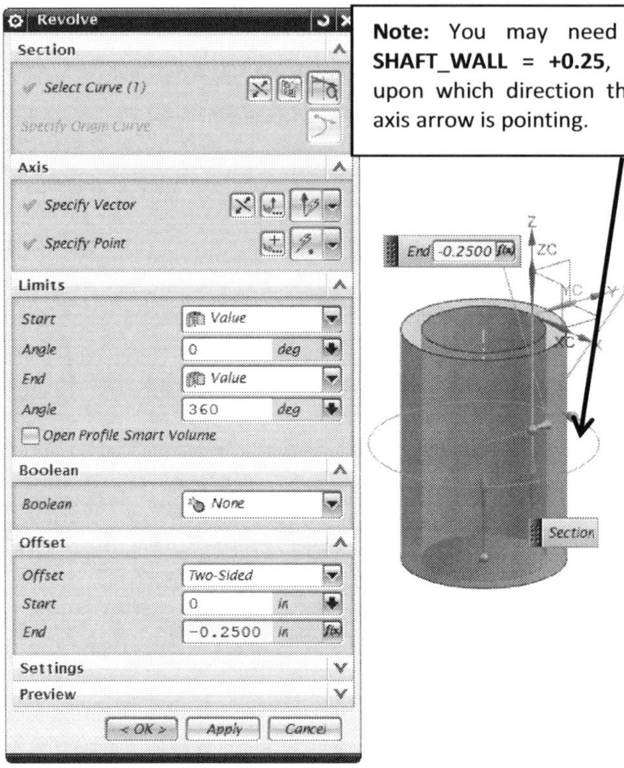


- Repeat the same revolve operation for the angled sketch line with a First Offset of **FLANGE_WALL= .25**.
- Unite the two bodies.

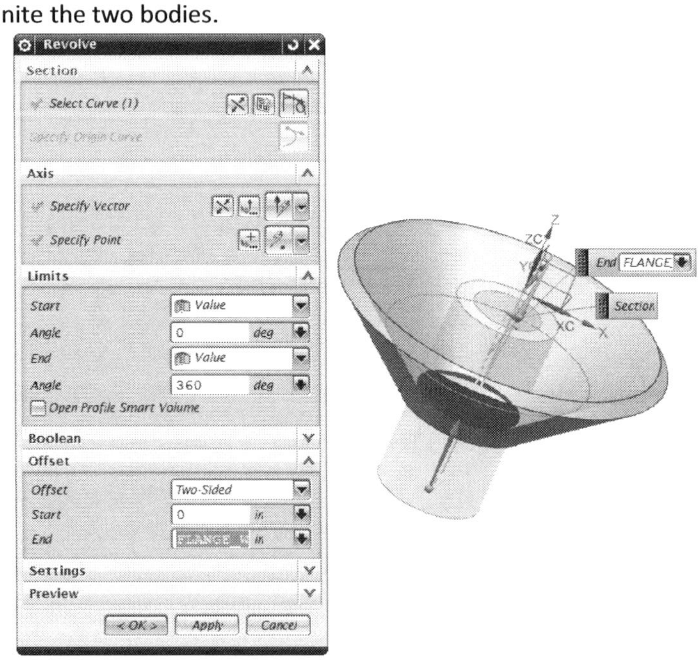

- Trim the solid to look like the following figure.

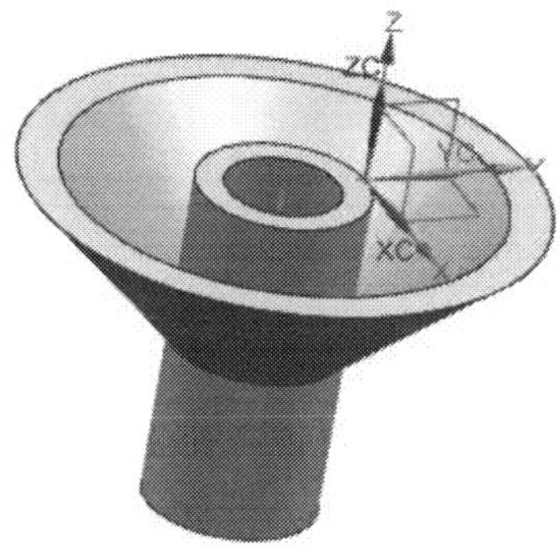


- To check your work, select ***Tools / Expressions***. The Expressions list should contain all the variables that you created, such as HEIGHT ANGLE, and SHAFT_WALL.

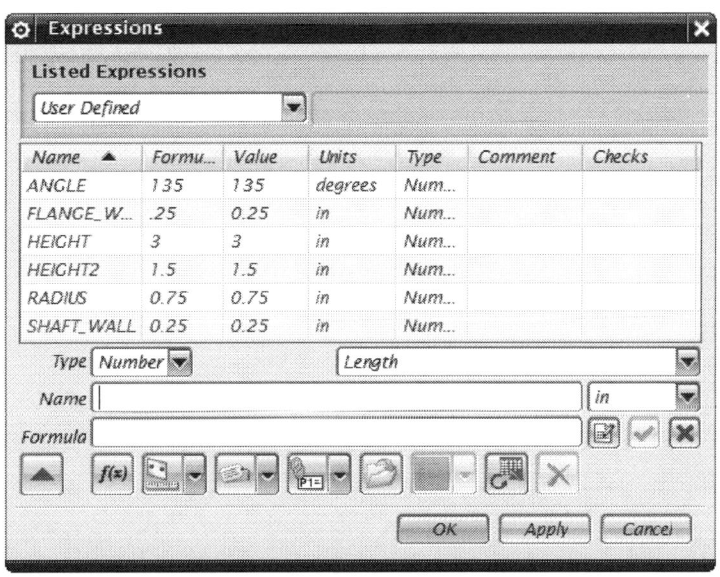


Save this model and rename it as ***expression2_<your initials>.prt***

> **Note:** Entering variables 'on the fly' is a great way to document important model parameters. As a practical example, consider the design of a machine element which contains a shaft. Quite often, the length of the shaft may change during the design process, so if you define a shaft length variable, modifications to this value are easy to make.

***End of Project***

## 12.4 Renaming Existing Variables

After building a model, you may find that renaming a few variables would help to document it. Renaming variables is easily accomplished with the Expressions editor.

### *Detail: Renaming Existing Variables*

Using the part you built in the previous project, you will rename two variables that control the rotation angles of the sketch lines to 'THETA1' and 'THETA2'. You will then change both of their values to 180 degrees.

### *Project 12.4.1*

- Open the file called 'expression2_<your initials>.prt'.
- Select Tools / Expression.
- Select the first expression that has a value of 360.
- Type **THETA1** into the 'Name' field of the dialog box as shown below.

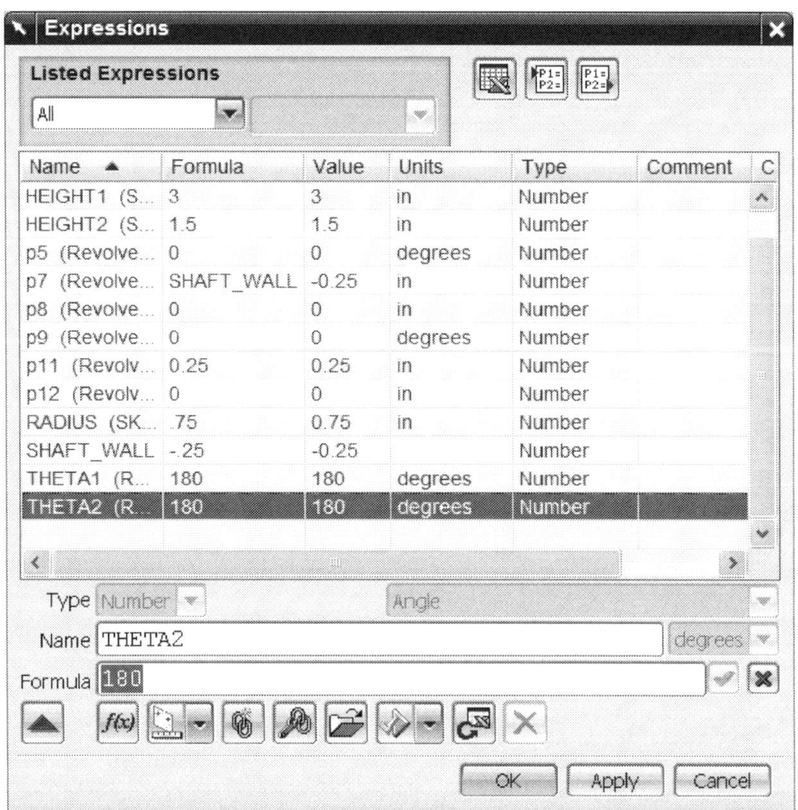


- Choose **OK**.

- Repeat this process to rename the second 'End Angle' parameter to **THETA2.**
- Change the values of THETA1 and THETA2 to equal **180** degrees. *(Use the method that you learned earlier in this chapter.)*
- Exit the 'Expressions' menu by choosing **OK**. Compare your results to the following figure.

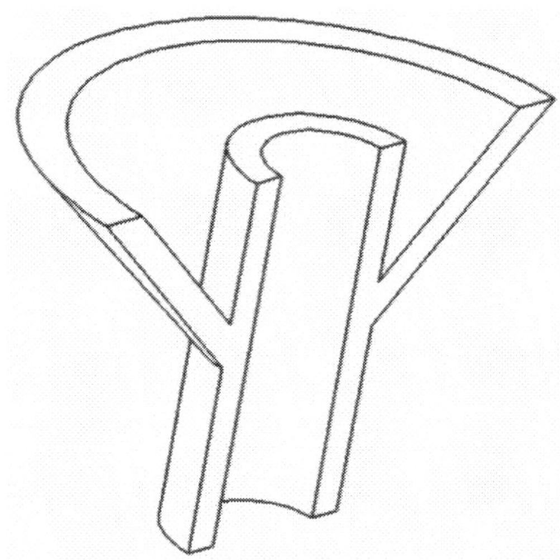

***End of Project***

Suppose that you wanted THETA1 and THETA2 to be equivalent. You can use the 'Expressions' editor to accomplish this by changing the expression **'THETA2=180'** to **'THETA2=THETA1**.'

Note: You do not need to change the variable names to edit their values. You can just as easily change parameters that retain their default 'pXX' name.

## ***Detail: Adding Comments to Variables***

NX 8.5 allows you to place comments next to the value of the variables. Comments should be placed in the 'Formula' entry box after two forward slashes "//". These comments are for the user's reference only as NX 8 does not recognize any text placed after the slashes.

## ***Project 12.4.2***

- Select Tools / Expression.

- Select the HEIGHT1 variable.
- In the formula line after the number 3, enter "**// This is the overall height.**"
- Select **Apply**. Notice NX 8 adds your comment to the 'Comment' column.
- Continue to add comments to other variables as shown next.

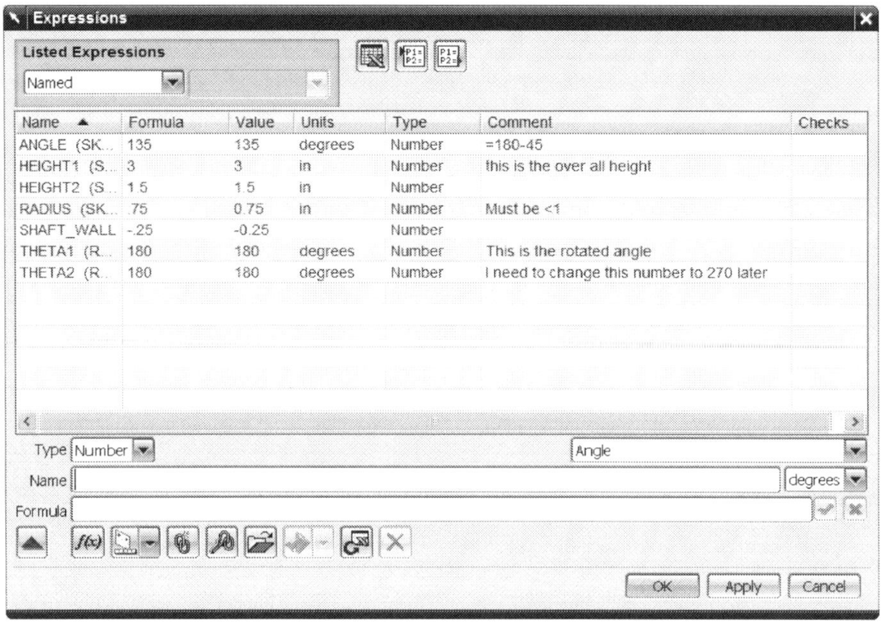


***End of Project***

> **Note:** Comments can also be added by **Right Clicking** over the name of the expression. Select Edit Comment and type in your comment in the provided text block and choose **OK**. The text is automatically converted to comment format with the necessary "//".

## 12.5 Practice: The Door Latch

The door latch model shown in the following figure is controlled by a single variable, 'E,' that was defined prior to modeling. Expression values are entered as a function of 'E' to control the thickness and other dimensions of the part.

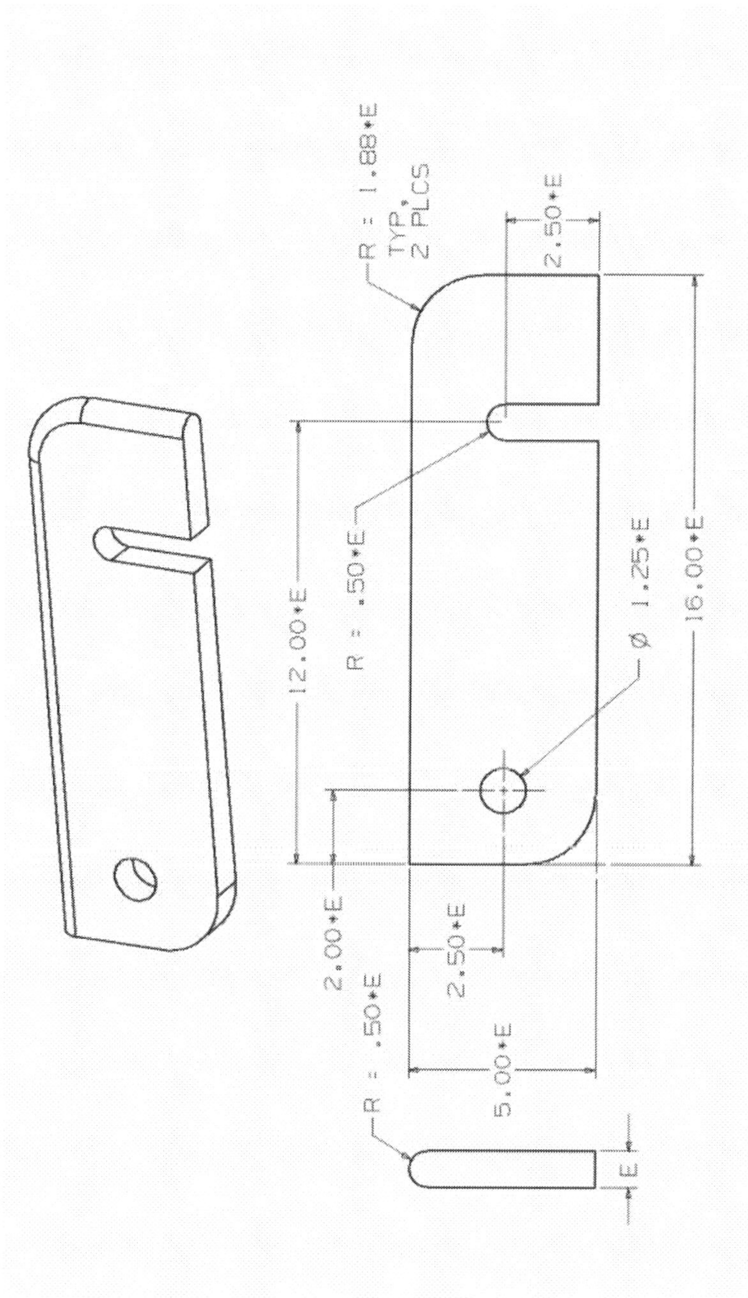

## ***Practice Outline***

- Using ***Tools / Expression***, create the expression E=1.
- Create a sketch similar to the figure below. Note that variable names, such as 'LENGTH' and 'HEIGHT,' use the Dimensions menu of the sketcher.

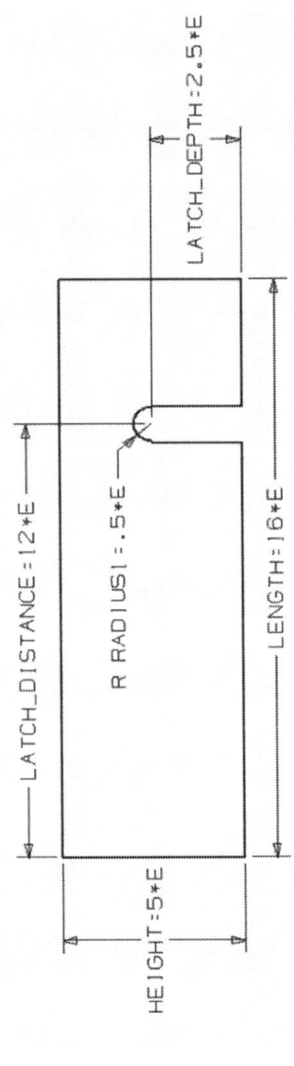


- Extrude and add features until you have the model shown in the previous figure. Enter feature parameters as a function of 'E.'
- Test your completed model by changing the value of E to **1.75**. Your model should change size.

***End of Project***

## 12.6 Self-Test

(All questions are true/false. Answers are at the back of the book.)

1. An expression is an equation that relates a variable and a value.
2. An expression value must be numeric.
3. Expressions must be created before modeling begins.
4. You cannot rename an expression variable once it has been used.
5. You cannot change an expression value if it still has the default 'p' name.

# 13. Edit Feature Operations

## 13.1 Introduction

The definition of any parametric feature can be modified using the ***Edit / Feature*** functionality or by accessing the necessary parameters through the 'Part Navigator.' Every parametric feature in NX 8.5 retains both the information required to define it as well as its history. This chapter discusses the commonly used functions to modify a feature's parameters, as well as to perform other types of feature operations such as reordering, suppressing or deleting.

## 13.2 The Part Navigator

The 'Part Navigator' is a powerful, easy-access organization tool for many common NX 8.5 commands. It is located in the Resource bar on the right side of the NX 8.5 interface. By default, the features are displayed in the order that they were created (**Right Mouse Button / Timestamp Order**).

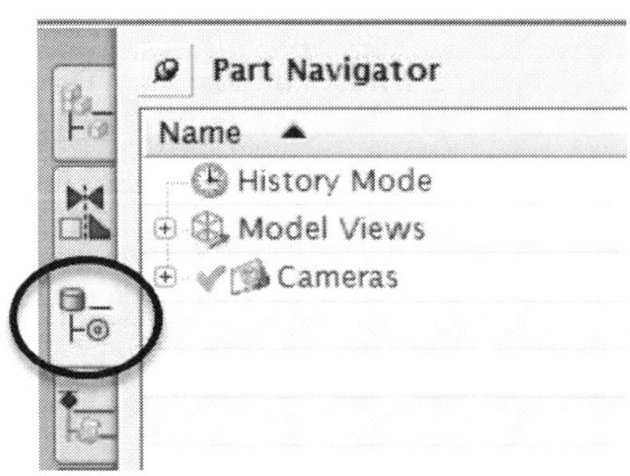


The most useful quality of the part navigator is the ability to access many commonly used feature edit operations. For example, to change a feature's parameters you can simply right click on a feature in the Part Navigator and select **Edit Parameters**.

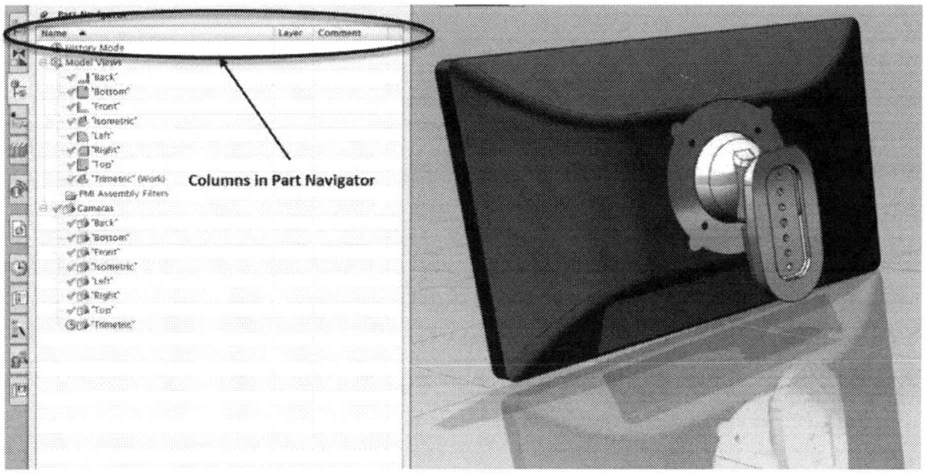


**Note:** All mouse commands referred to in this section assume you click in the Part Navigator title block or on the specified feature listed within the navigator itself.

Next to the name of each feature in the Part Navigator, there should be sections which provide information about the part (Timestamp, Layer, etc). You can add these columns and configure new user-defined attributes by **Right Clicking** the header of the **Name** column and selecting **Columns / Configure.**

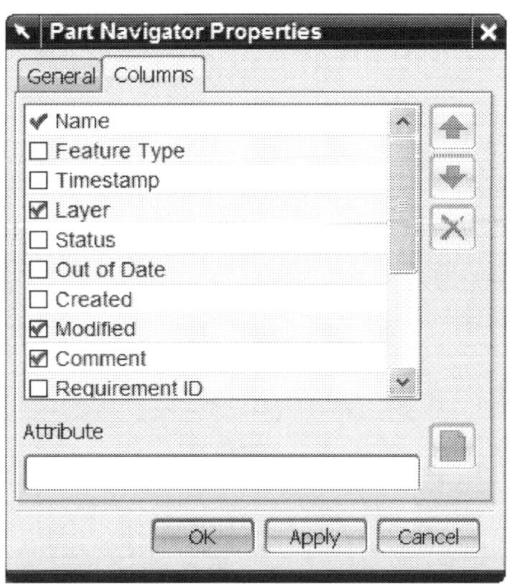

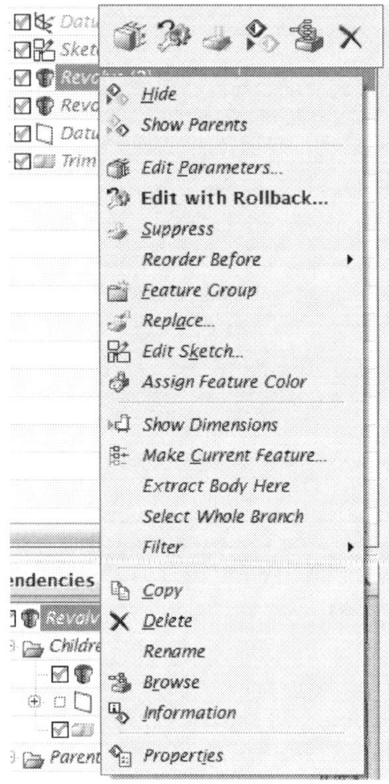


Several different actions are available by **right-clicking** on any given feature. These are illustrated in the figure on the right. Modifying sketches, parameters of the feature, highlighting, hiding, suppressing, deleting, editing, and object display can all be accomplished by use of this menu.

## ***Detail: The Part Navigator Dependencies Panel***

'Dependencies' help illustrate how changes to a component or steps in the modeling process impact other parts of the model. Although dependencies are shown in their own panel, NX 8.5 highlights parent and child dependencies in the main Part Navigator screen using different colors. These can be modified by **right-clicking** the 'Name' header bar in the part navigator and selecting ***Properties / Part Navigator Properties / General*** as shown below.

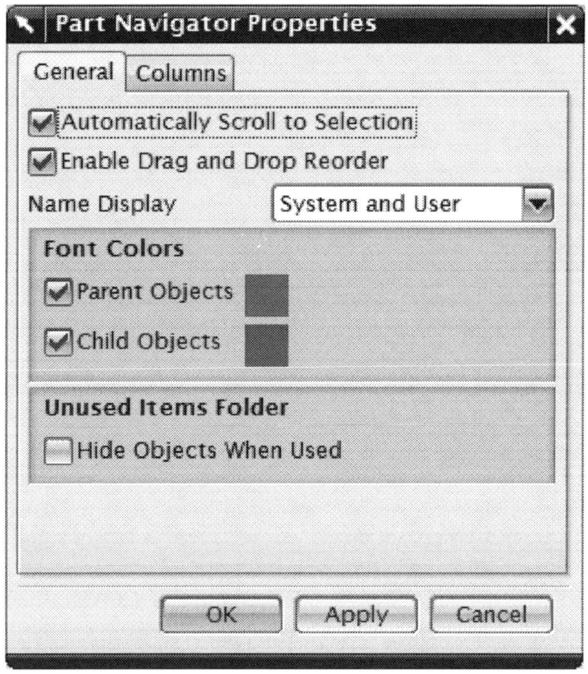


Attributes in the Dependencies panel may be modified by **right-clicking**, just like inside the Part Navigator.

## ***Detail: Model Views***

Another feature contained within the Part Navigator is the ability to change and create views of your model. **Left-clicking** on the "+" sign next to 'Model Views' expands a list of 8 standard views provided by NX 8.5 as shown below.

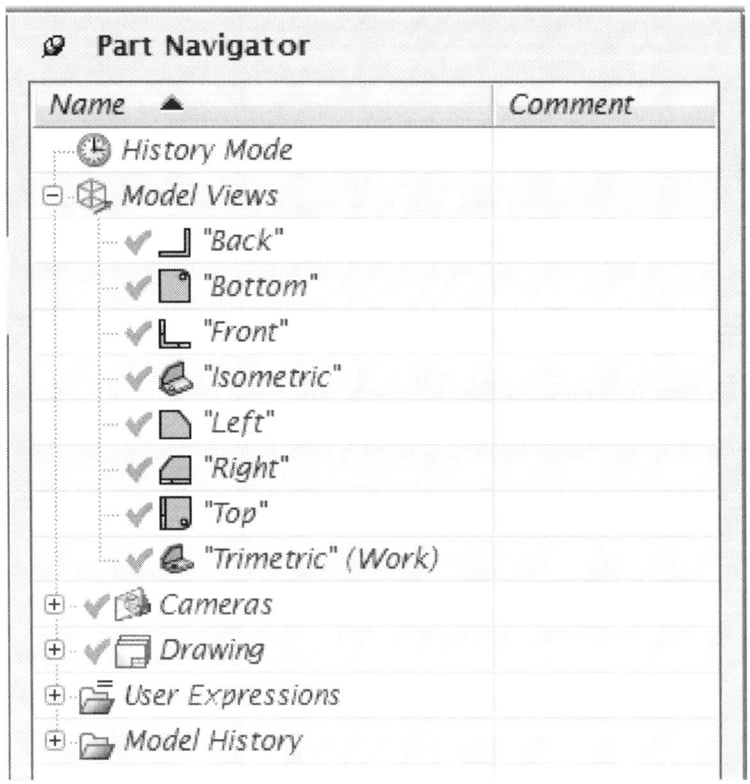


*Double-clicking (or **Right Click / Make Work Part**) makes the view selected your active work view.* NX 8.5 also allows you to create your own views. **Right-click** on the 'Model Views' expression and choose Add View. The orientation of the model will be saved and a new item list will come up to allow you to name your view. Anytime you want to recall it, simply double click on it.

## 13.3 Editing Feature Parameters

One of the most commonly used feature editing operations is 'Edit Parameters.'' This operation can be accessed either through the menus or by right-clicking on a feature.

### *Detail: Editing Feature Parameters*

The 'Edit Parameters' functionality allows you to "recreate" many of the defining attributes of a feature. For example, you can change the 'Pierced Face' for a Hollow feature or edit the defining curves for a Revolved feature.

You have already learned that one way to modify a feature's parameters is to use the Expressions Editor. Another useful way to accomplish this is to use the 'Edit Parameters' operations.

### *Project 13.3.1*

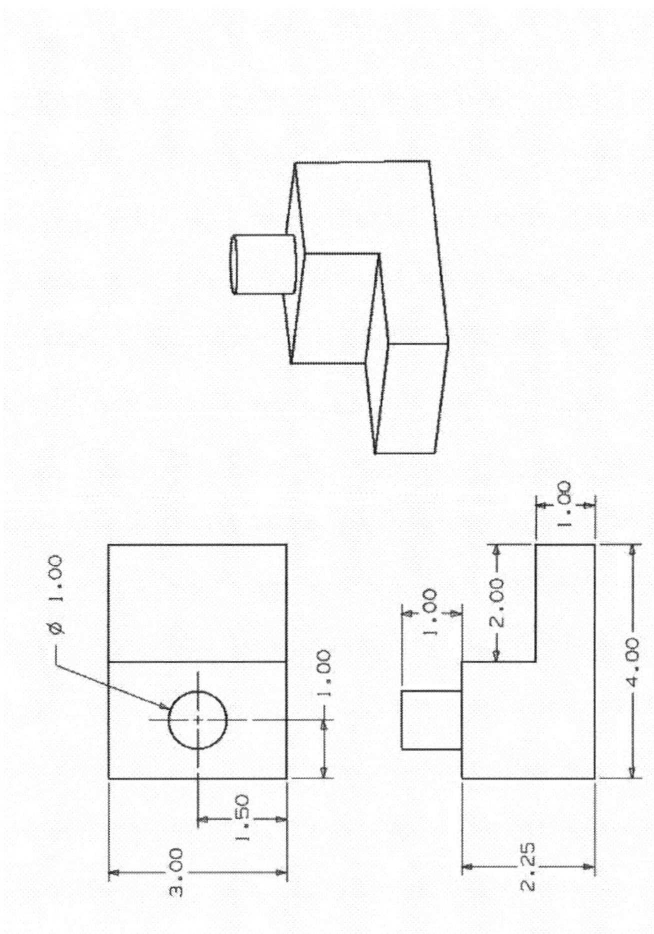

- Create the stepped block portion of the geometry shown above.
- Use the **Boss** feature to create the cylindrical boss geometry and parametrically position it at the center of the top face.
- Save your part file with the name ***modfeat_<your initials>.prt***.
- Select Edit / Feature / Edit Parameters.
- Select the boss feature. You can either select the boss in the Graphics Window or in the 'Edit Parameters' dialog box.
- Choose **OK.**
- Select **Feature Dialog**. A dialog box appears that allows you to modify the feature's defining parameter values.
- Enter Diameter = **0.5** and Height = **2**.
- Choose **OK.** (You should see the boss geometry update to reflect the new values.)

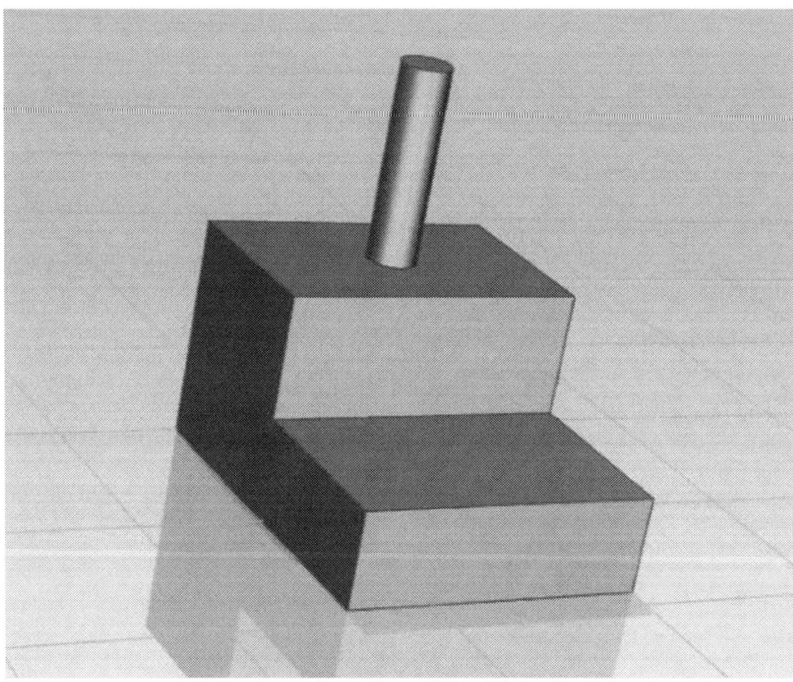

***End of Project***

## 13.4 Deleting, Suppressing, and Reordering Features

NX stores this information in the form of features and allows you to modify this "history" in many ways. You have already seen that you can modify numeric parameters. You will learn several techniques that allow you to change a model's geometry.

To understand the power of parametric models, it is useful to think of the final solid model as a linear history of operations (features). You can vary the size, shape, order, or even control whether or not an operation is present in the model's history.

## ***Detail: Deleting Features***

When you 'Delete' an individual feature in NX 8.5 it is *permanently* removed from the model's creation history. Consequently, the model will revert to a state it would have been in if the feature had never existed. The feature can never be retrieved unless an Undo operation is possible. You can delete a feature using ***Edit / Delete / Features*** or by right-clicking it in the Part Navigator and selecting 'Delete.'

## ***Detail: Suppressing and Unsuppressing Features***

When you **Suppress** an individual feature in NX 8.5, it is *temporarily* removed from the model's creation history. As with deleted features, the model will revert to the state it would have been in if the feature never existed. The difference is that a Suppressed feature can be retrieved using the **Unsuppress** operation. You can suppress individual features by using ***Edit / Feature / Suppress*** or by using the Part Navigator **right-click** pop-up menu.

### ***Project 13.4.1***

Here we will experiment with the Suppress and Unsuppress operations.

- Continue working with your part file ***modfeat_<your initials>.prt.***
- Select Edit / Feature / Suppress.

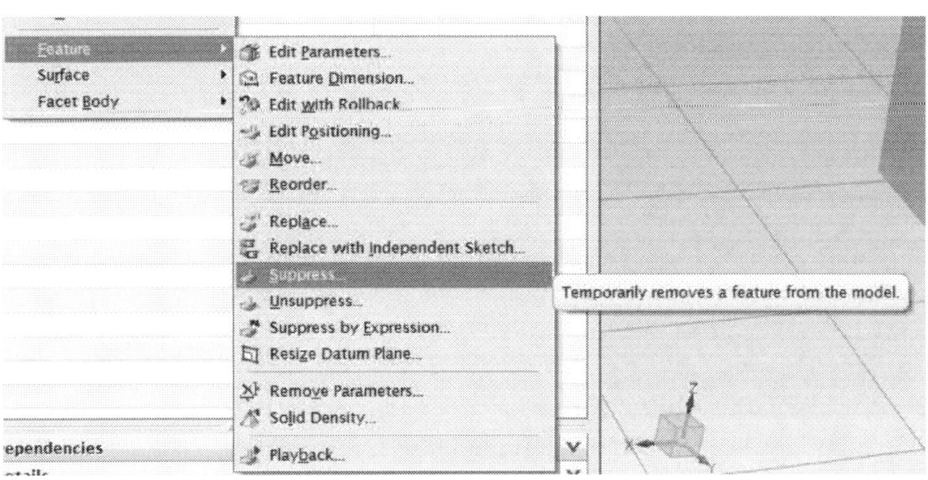

- Select the boss feature from the list displayed in the 'Suppress Feature' dialog box.
- Choose **OK** to suppress the feature.
- Choose **Undo.**
- Display the Part Navigator and find the boss feature.
- **Left-click** on the boss feature.
- **Right-click** and choose **Suppress.**
- Select Edit / Feature / Unsuppress.
- Select the boss feature from the list displayed in the 'Unsuppress Feature' dialog box.
- Choose **OK** to unsuppress the feature.

***End of Project***

## ***Detail: Parent/Child Relationships***

Parent/Child Relationships are the dependencies created when building up a CAD model. The skill of creating (and controlling) well-behaved parent/child relationships is what separates the expert-level user from the novice.

When you add a parametric feature to a model, NX 8.5 stores all of its information, including any pre-existing geometry that is referenced by the feature. If this geometry is modified, the feature must be updated. For example, consider a 2in x 3in x 4.5in Block feature with a 0.5in Blend feature on all its top edges. If the size of the block is changed, the Blend feature must update accordingly. Thus, the Blend feature is a child of the Block feature; conversely, the Block feature is a parent of the Blend feature.

In a geometrically complex model, it is often helpful to learn more about the Parent/Child relationship structure. This can be accomplished by selecting the feature in the Part Navigator and by expanding the Dependencies and Detail windows at the bottom of the Part Navigator. Once you understand the intricacies of a model's construction techniques, you can use the Edit Feature functionality to change the Parent/Child Relationship of features. This, in turn, allows you to perform parametric modifications that would otherwise be impossible.

## ***Detail: Reordering Features***

When you 'Reorder' an individual feature in NX 8.5, it is moved to a different place in the model's creation history. The geometry of the model is recalculated because of the Reorder operation, and the model's topology may or may not change. You can reorder individual features by using ***Edit / Feature / Reorder*** or by using the Part Navigator **right-click** pop-up menu.

## *Project 13.4.2*

In this project you will add a Shell feature to your model and reorder the boss to observe the change in the solid geometry.

- Continue working with your part file **modfeat_<*your initials*>.prt**.
- Insert a Shell feature with a thickness of **0.1**, as shown below.

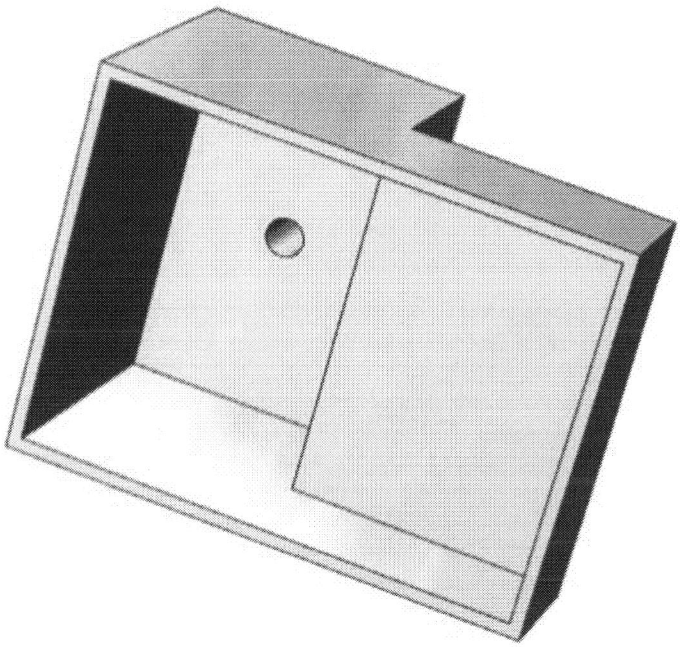

- Select Edit / Feature / Reorder.

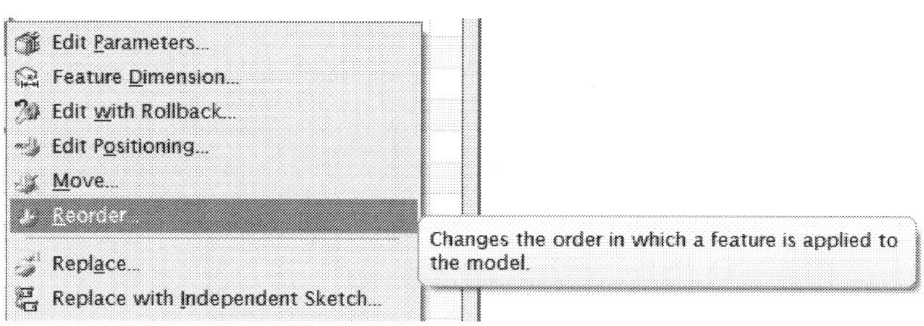


- Select the boss.
- Select the **Before** Choose Method.
- Select the Shell feature located in the 'Reposition Features' box.
- Choose **OK** to reorder the feature.

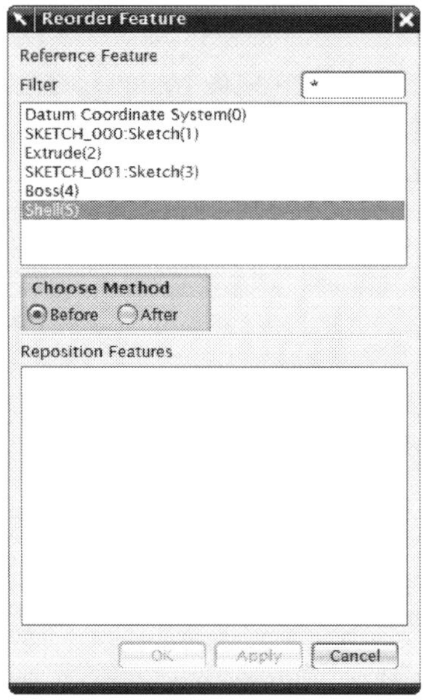


- Choose **Undo.**
- Display the Part Navigator and find the boss feature.
- Select the boss feature in the Part Navigator.
- Right-click and choose Reorder After / Shell.

- Compare your results to the figure above. Notice that the Boss feature is no longer hollowed out. This is a result of its being reordered after the Shell feature.
- Choose **Undo.**

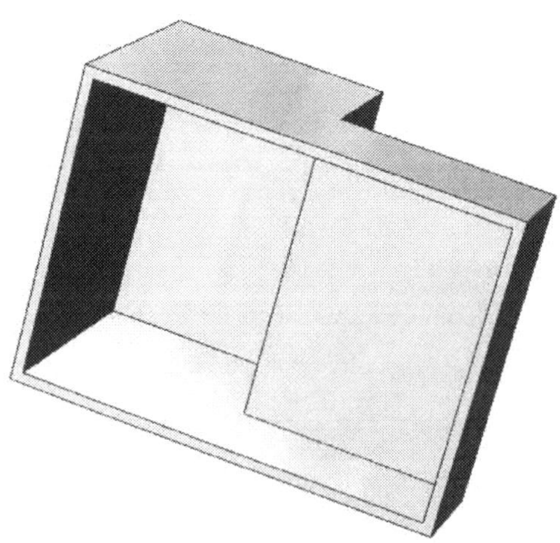

***End of Project***

## 13.5 Self-Test

(All questions are true/false. Answers are at the back of the book.)

1. Feature parameters can only be modified with the ***Edit / Feature*** menus.
2. You can only delete a feature with the ***Edit / Delete*** dialog box.

# 14. Synchronous Modeling

## 14.1 Introduction: A Double-Edged Sword

Synchronous modeling is a very useful and powerful tool. The principle of Synchronous Modeling is simple: taking surfaces and adding parametric information to them above and beyond their original definition – editing their location and shape to fit your needs. If there is one good argument against parametric modeling in general, it is that when used, models get too complicated. It becomes difficult to make edits because one must understand everything that was done before in that model. However, to perform Synchronous Modeling, it is not necessary to understand how a model was built. In fact, Synchronous Modeling can be performed on a model even if it's non parametric. The danger in using Synchronous Modeling is that all design rules that are captured in a parametric model get overridden and the model no longer works in the way that was originally intended to. The Synchronous Modeling dialog box can be seen in the figure below.

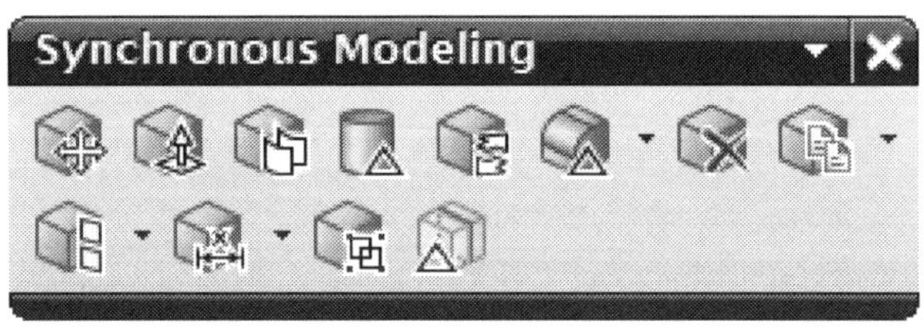

## 14.2 Move Face

There is a new face finder function in the move face tool which allows you to move additional faces to the one you selected. For example, you can select a face and then say that you want all tangent faces to move with it. The move face window is shown in the following figure with the face finder expanded. Just tick off what additional faces you want to move.

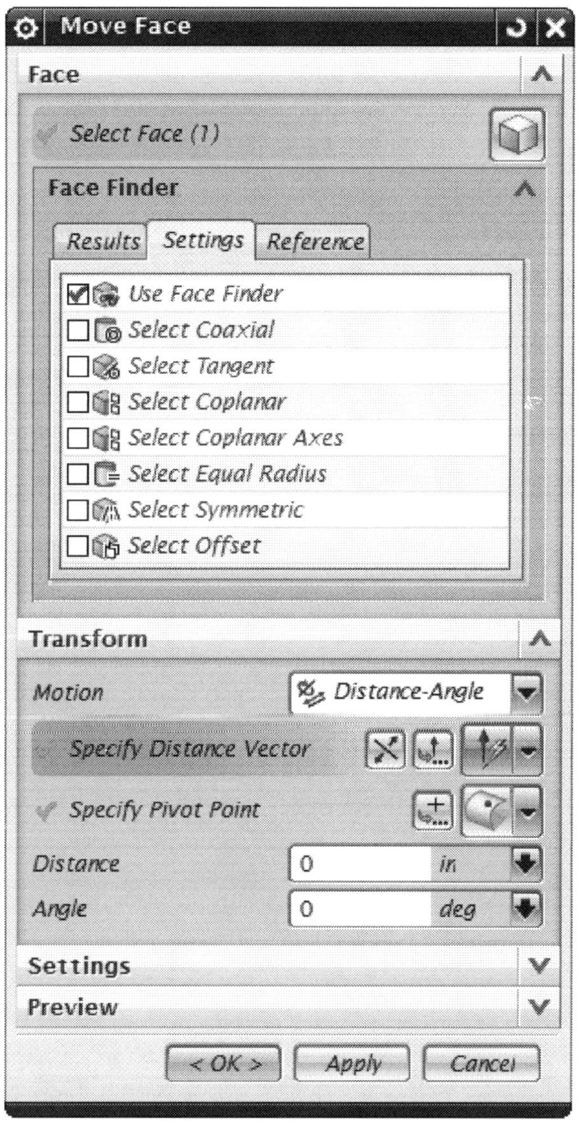

## *Project 14.2.1*

- Open the part 14.2.1 from the project files download or create the simple geometry shown in the following figure.

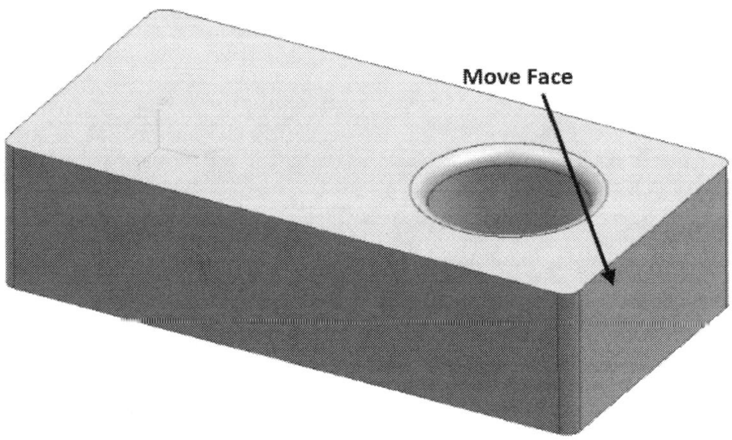


- Select to Insert / Synchronous Modeling / Move Face  .
- The first thing NX will ask you for is the face you wish to move. Select the face shown in the previous figure.
- Once you select the face your screen should look similar to the following figure.

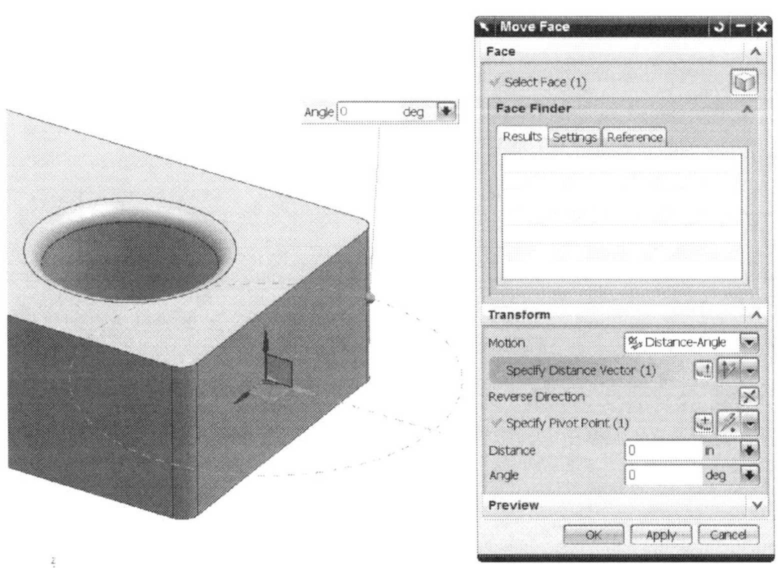

- Notice all the new options in **Move Face**. You can set the motion to Distance, Angle, Distance between Points, Radial Distance, etc.
- In this exercise, we are going to use **Distance – Angle** but feel free to experiment! So select **Distance – Angle** as the type of **motion** under **Transform**, as shown in the following figure.

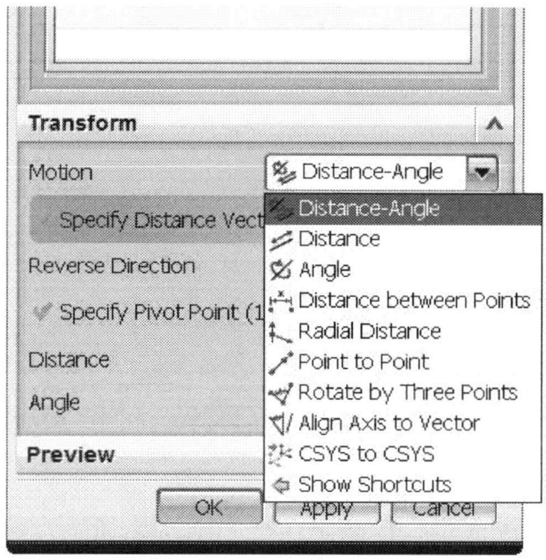


- Select the face as shown in a previous figure (the face you want to move) and you will view the following figure. You will see you have two options: the distance you want to move the face and the angle. You can either type them into the Move Face window or move the arrow and the ball around in the graphical display.

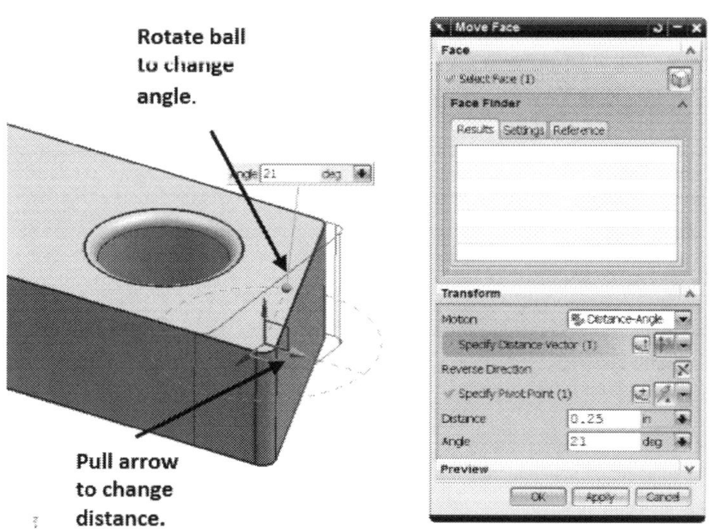

You can change the direction of rotation by changing the distance vector as shown in the following figure.

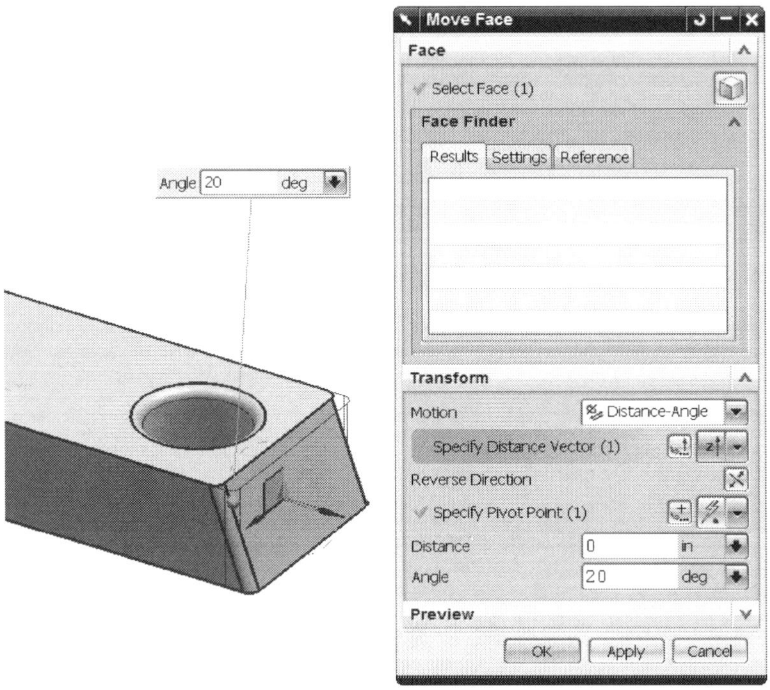


- As with the old Move Face, you can also change the position of more than one face. Select the same face and then the face of the hole as shown in the following figure.

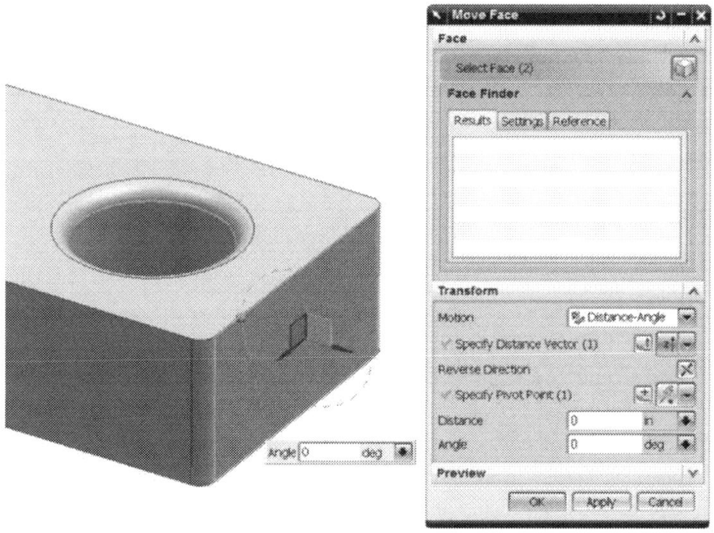


Making **Z** the distance vector, change the angle and see what magically happens.

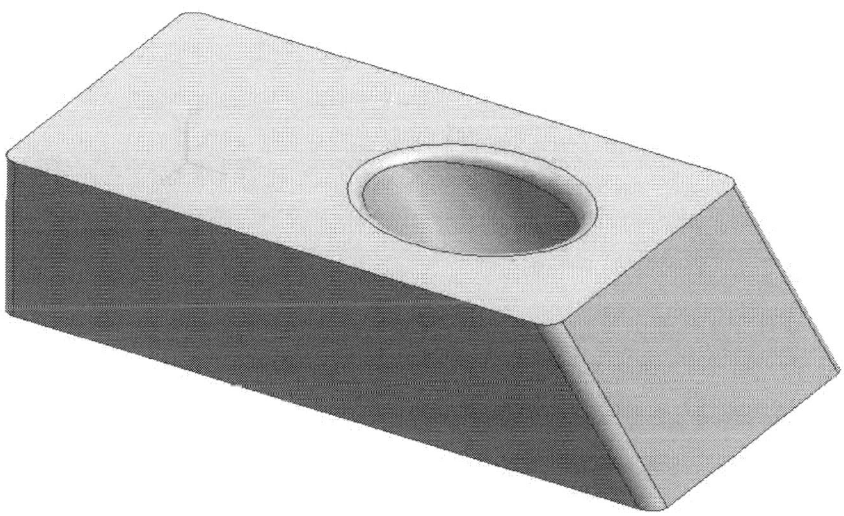

- Even the hole changed angle! That's the best thing since sliced bread!!! (It's hard to tell we're really excited about this.)

As with all synchronous modeling tools, it's best to experiment and see how endless the possibilities are.

***End of Project***

## 14.3 Resize Face

The 'Resize Face' Synchronous modeling feature may be used to change the diameter of cylindrical, conical or spherical faces. The position of the faces remains unaffected when the Resize Face feature is created.

### *Project 14.3.1*

- Open part 14.3.1 from the project files.
- Select Insert / Synchronous Modeling / Resize Face.

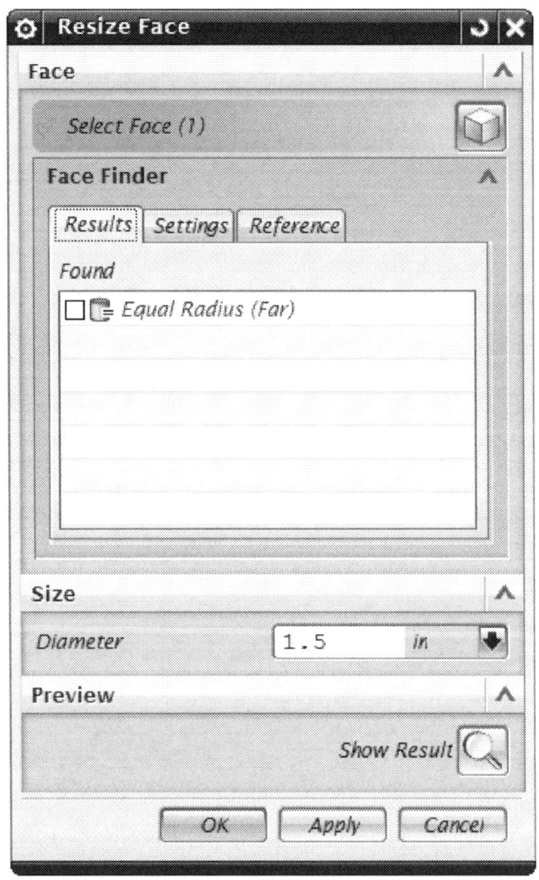


- Select the cylindrical face of the hole.

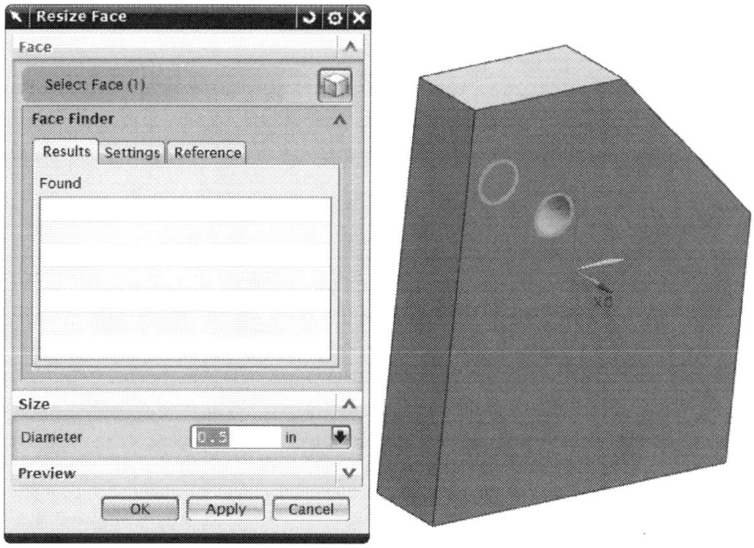


- Enter Diameter = **1** as shown to the right.
- Choose **OK.**

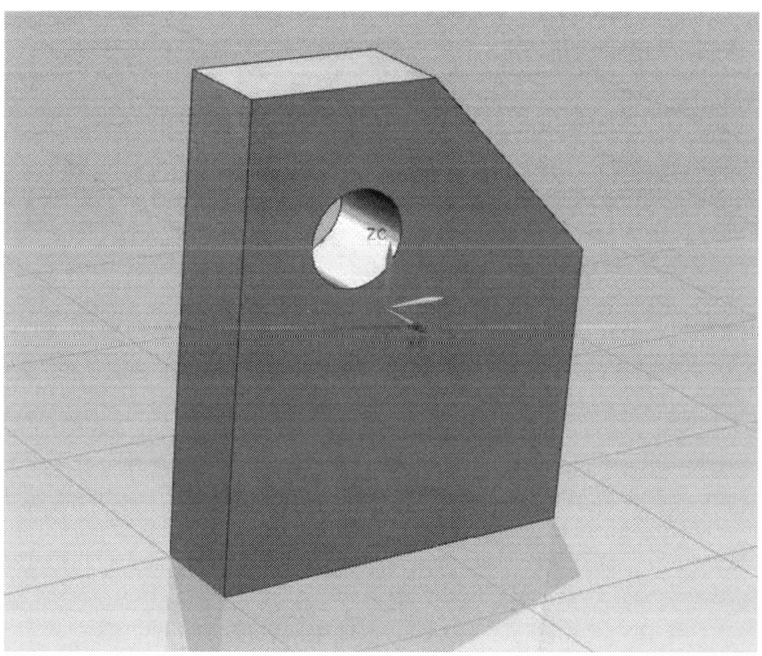


- Compare your results with the figure above.
- Choose **OK.**

***End of Project***

## 14.4 Offset Region

The 'Offset Region' Synchronous modeling feature allows offsetting one or more model faces. You can use it to increase or decrease the dimensions of a part.

### *Project 14.4.1*

- **Undo** the Synchronous modeling feature created in the last project.
- Select ***Insert / Synchronous modeling / Offset Region.***
- Enter Offset = **2in** as shown below.

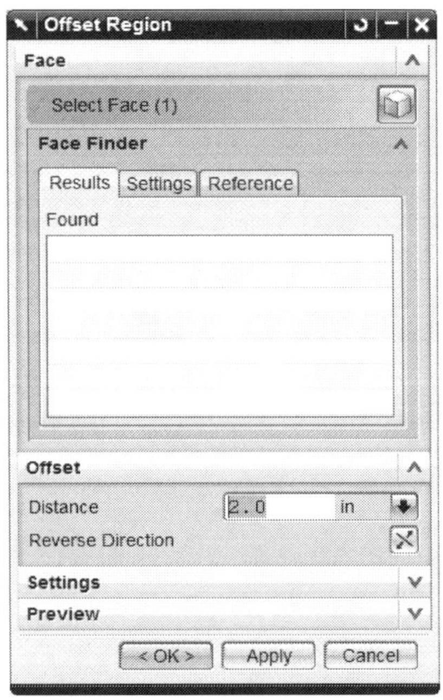


- Select Face 1 As shown on the following figure.

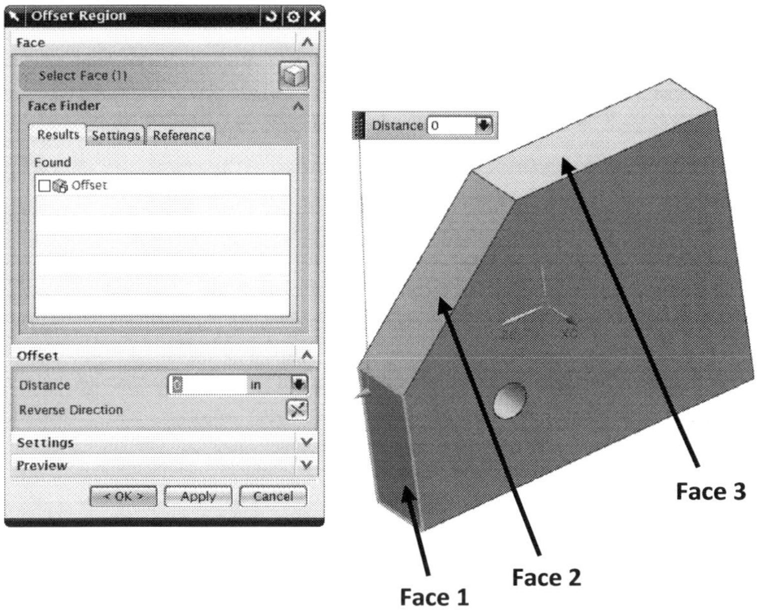


- Choose **OK.**

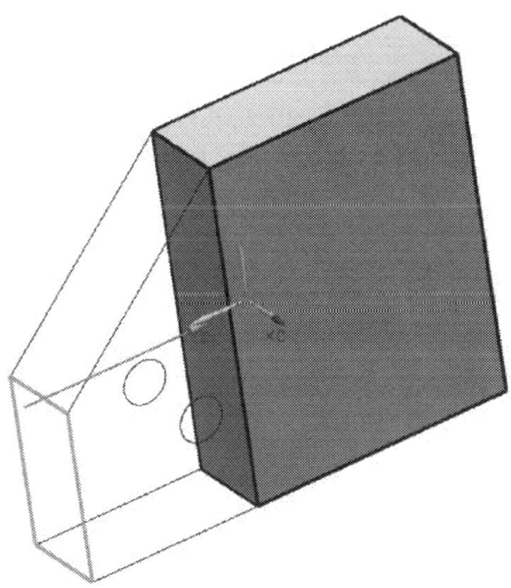

- Compare your results with the figure above.

***End of Project***

## 14.5 Replace Face

The 'Replace Face' Synchronous modeling feature enables the parametric replacement of a model face with that of a 'tool' face.

### *Project 14.5.1*

- **Undo** (or delete) the Synchronous modeling feature created in the last project.
- Create the following sketch

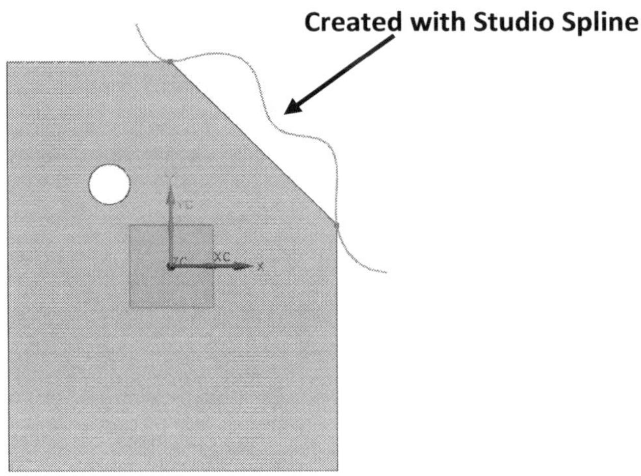


- Next, Extrude the Sketch by 1in.

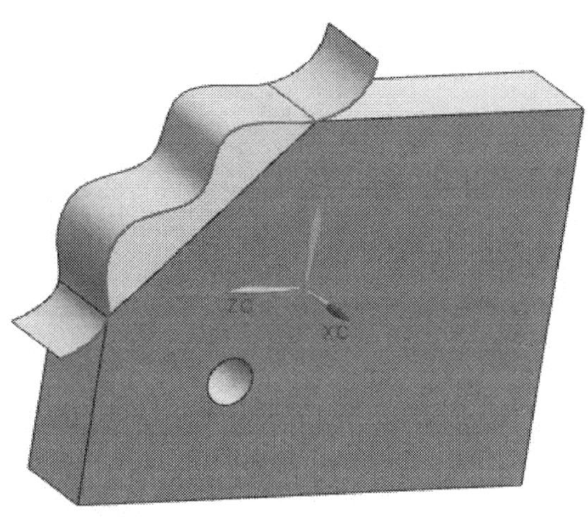

- Select Insert / Synchronous Modeling / Replace Face.
- Carefully observe the selection steps in the 'Replace Face menu,' shown below.

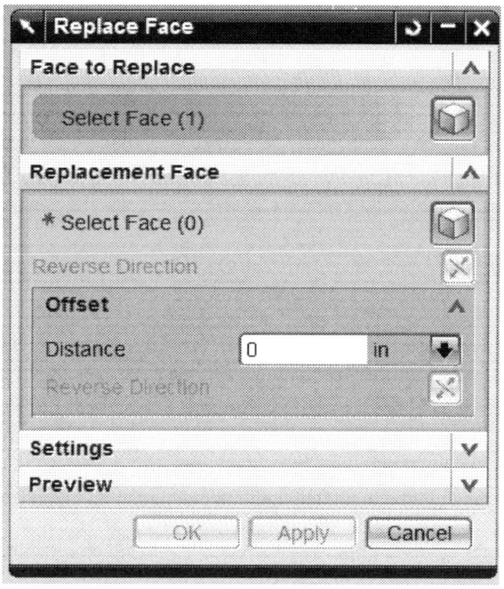


- Select Face 2 (as shown next) as the face to replace.
- Select Face 1 (as shown next) as the replacement face.
- Choose **OK.**

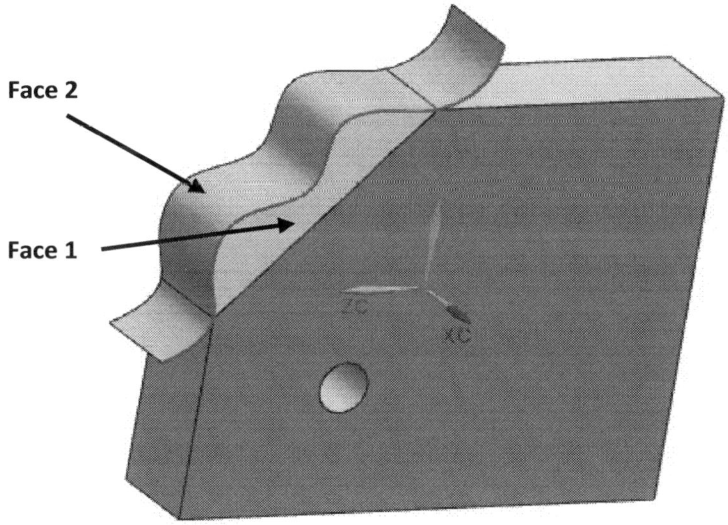

- Hide the Extruded curve.

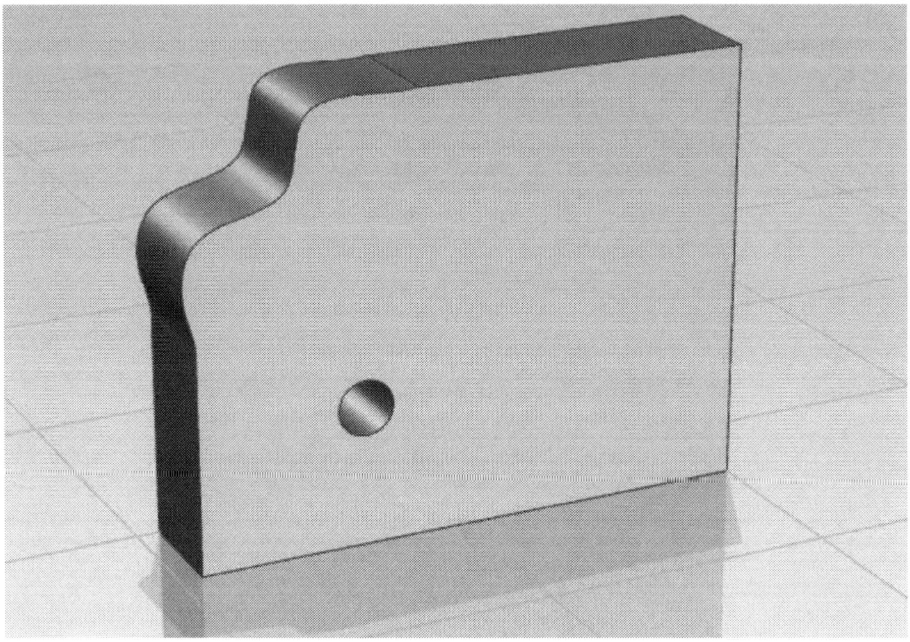

- Compare your results with the figure above.

***End of Project***

## 14.6 Pull Face

The **Pull face** command is similar to the old **Move face** command. It's very easy to use!

### *Project 14.6.1*

- Open part 14.6.1 from the project files.
- Select Insert / Synchronous Modeling / Pull Face.
- Select a face to move as shown in the following figure. Then specify a distance or drag the arrow to the position you want.

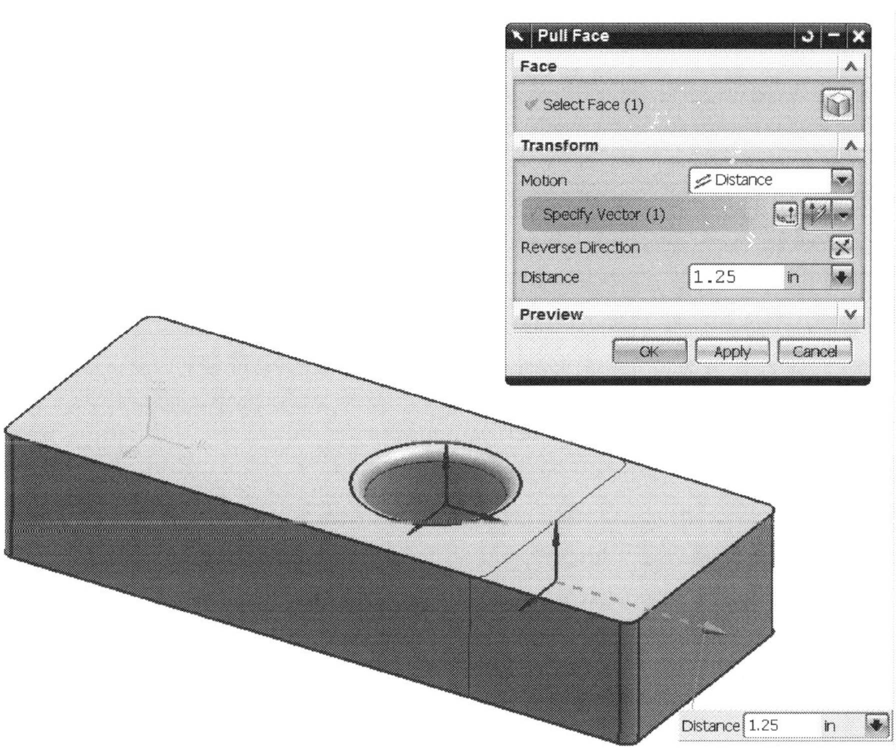


***End of Project***

## 14.7 Replace Face with Offset

The 'Replace Face' Synchronous modeling feature enables the parametric replacement of a model face with that of a 'tool' face. You can even replace a face with multiple connected faces. The following project shows how it works.

### *Project 14.7.1*

- Create this simple part (derived from only one sketch).

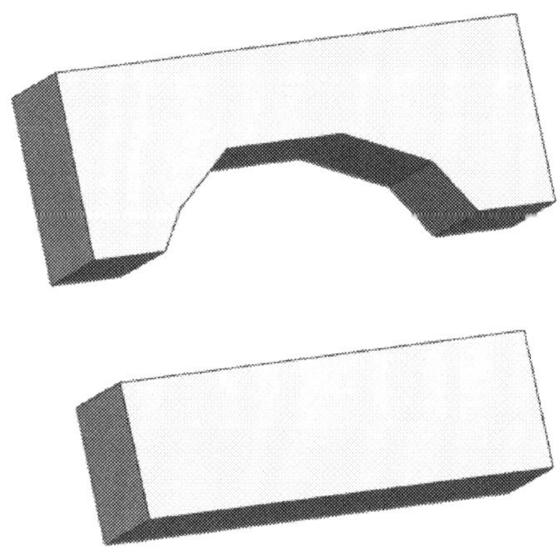

- Select Insert / Synchronous Modeling / Replace Face .
- First, select the face you want to replace as shown in the following figure.

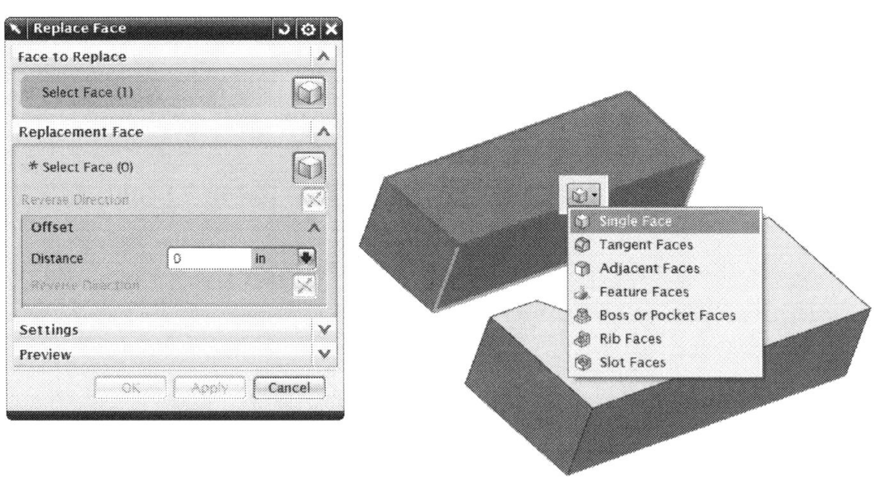


| **Note:** One relatively new feature to NX is the **selection filter** that shows up when you select geometry. You can pick between single face, tangent face, etc. |
|---|

- Select the replacement ***faces*** that you want. Note that we can now do faces. Also note that the faces have to be connected at their edges. Select the faces shown in the following figure. Change the offset distance, you can see that the faces move to the distance specified.

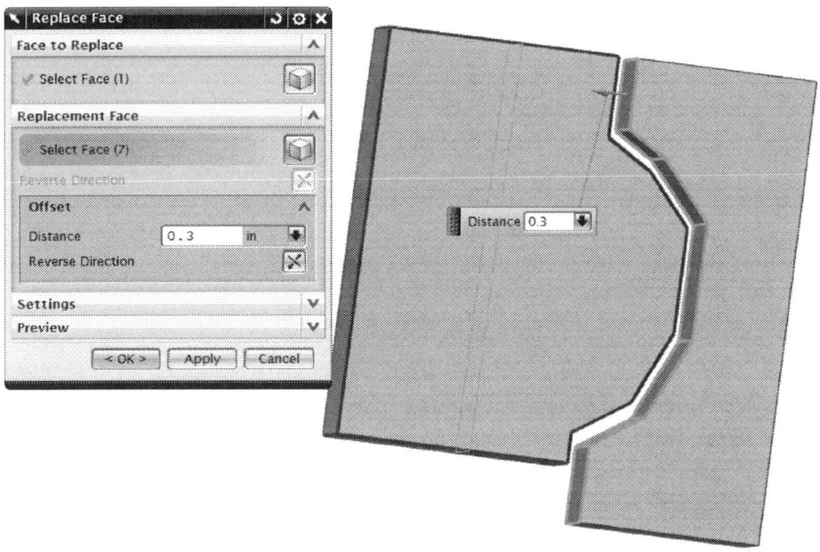


- Click on OK to complete the procedure and to view the following change (the other block has been made translucent in order to see the change more clearly).

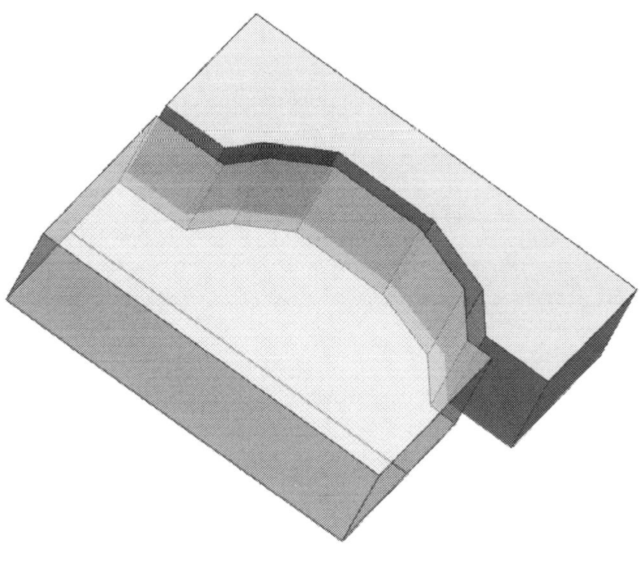

***End of Project***

## 14.8 Linear Dimension

The linear dimension tool allows you to change a dimension that you have placed between a point/edge/face and the face you wish to move. This function is not dependent on the history of your model and overrides the original placement of that face.

### *Project 14.8.1*

- Create the part shown in the figure below.

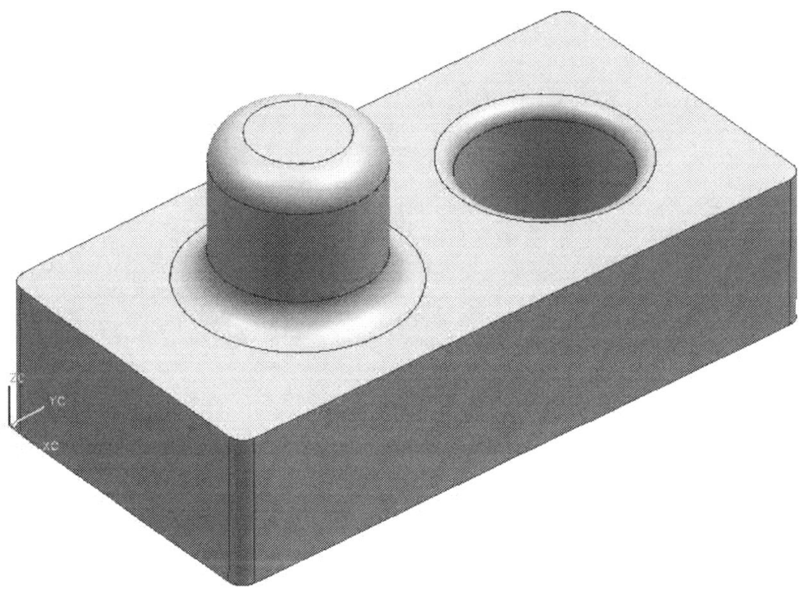

- Select **Insert / Synchronous Modeling / Dimension / Linear Dimension.**
- This will bring up the window shown in the following figure.
- It first asks you to select the origin of your dimension – this is where you're going to measure from. Select the edge as shown in the following figure.

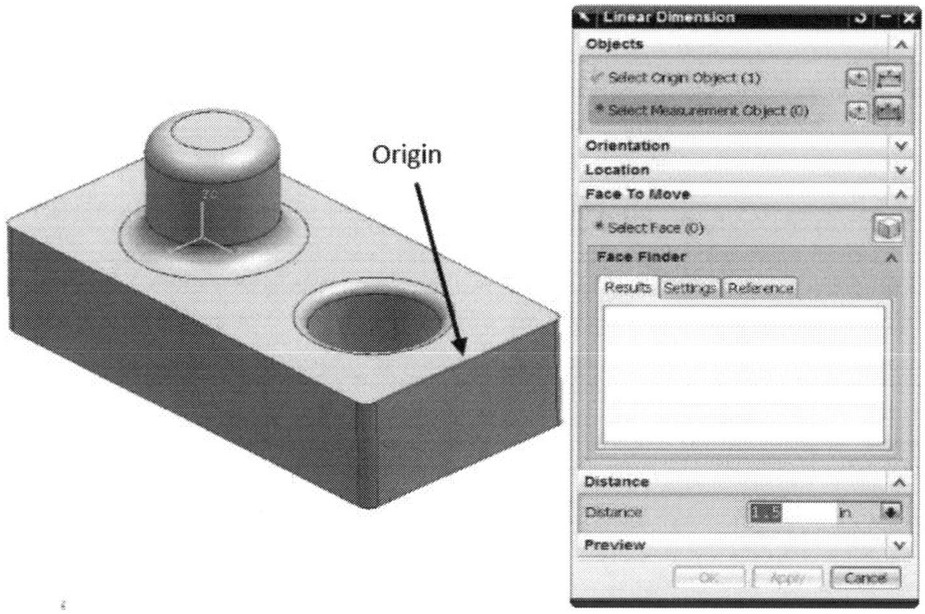


- Next, select the measurement object. This is what you're going to move.

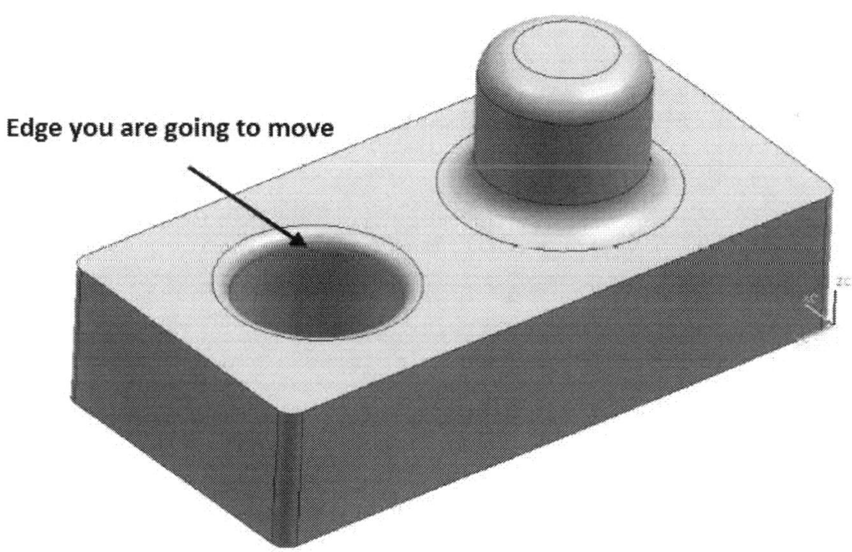


- As soon as you click on the circular edge you will see a dimension appear, as shown in the following figure. Select a place for that dimension by clicking on the screen.

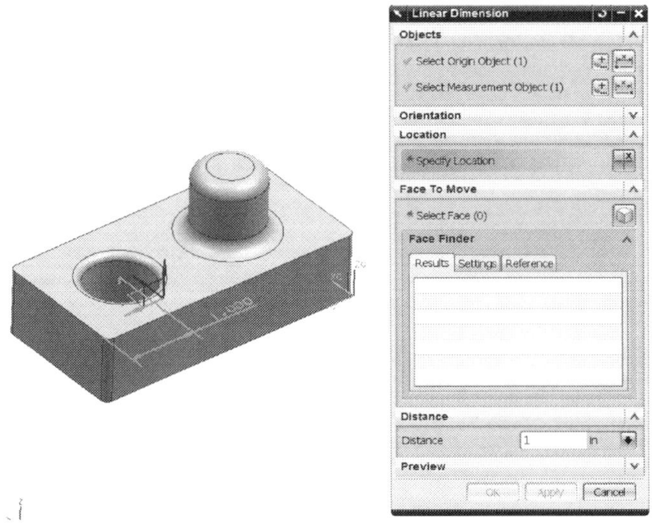


- Now you can change that dimension. The following figure was moved from **1in** to **0.75in** from the edge.
- Alternatively, you can just drag the arrow shown in the following figure. The automatic preview allows you to see what the changes are instantly.
- Your model should now look something like the following figure.

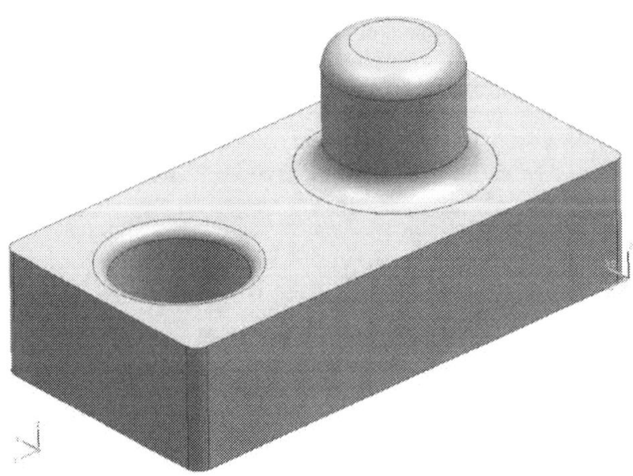

***End of Project***

## 14.9 Angular Dimension

The angular dimension changes the angle between two faces. You start by selecting the reference face, then selecting the face you want to move, and finally by defining the angle. Let's try it together!

### *Project 14.9.1*

- **Undo** (or delete) the Synchronous modeling feature that you created in the last project.
- Select Insert / Synchronous Modeling / Dimension / Angular Dimension.
- Select the **Origin face**. This is the face you want to reference from, i.e. it will not move. This is shown in the following figure.

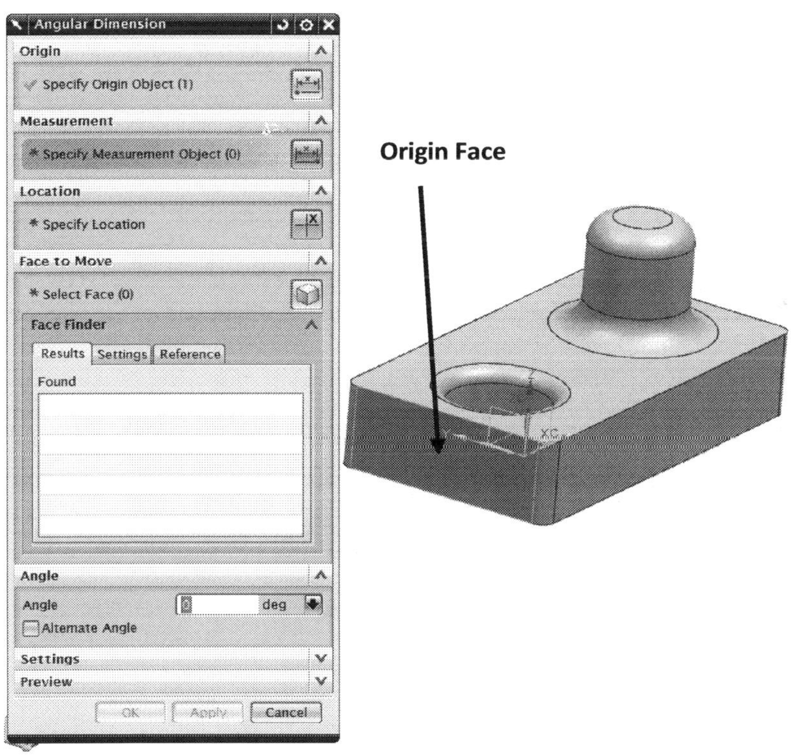


- The next step is to choose the **measurement object** (the face that will change angle).

- Select the face shown in the following figure.

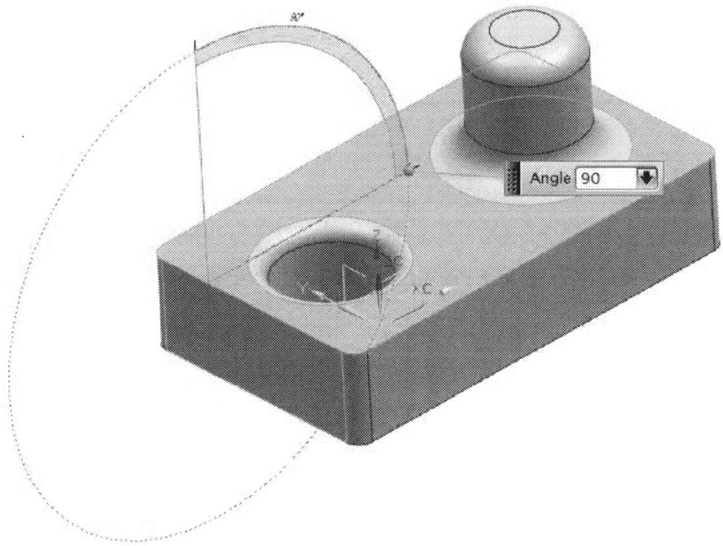


- Click on the screen somewhere to place the dimension you created (just like the last exercise).
- Then proceed to either specify the angle you wish or drag around the sphere that is circled below.

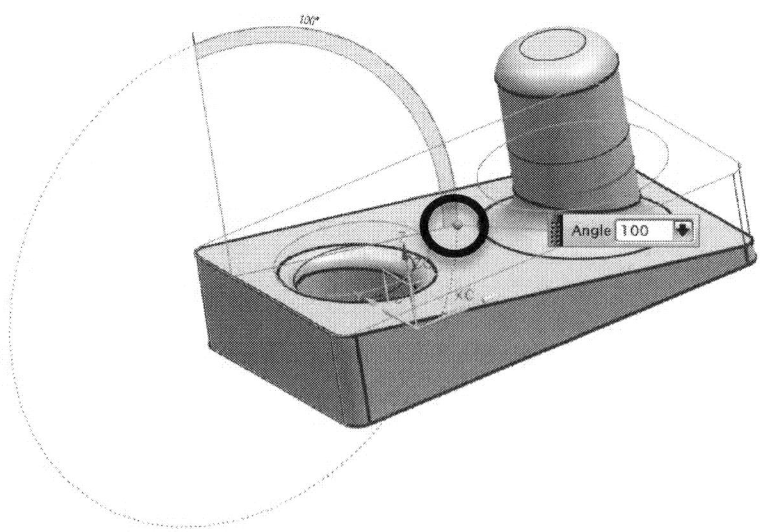

- Your model should now look something like the following figure.

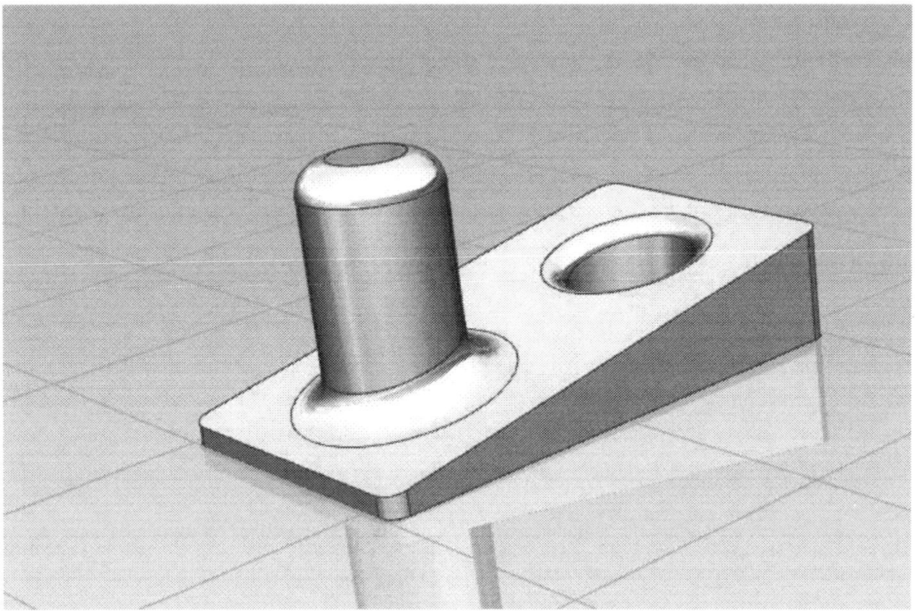

***End of Project***

## 14.10 Radial Dimension

The radial dimension tool allows you to change the size of cylindrical components and similar geometry.

### *Project 14.10.1*

- **Undo** (or delete) the Synchronous Modeling feature that you created in the last project.
- Select **Insert / Synchronous Dimensioning / Dimension / Radial Dimension**.
- Select the face you want to change the size of, as shown in the following figure.

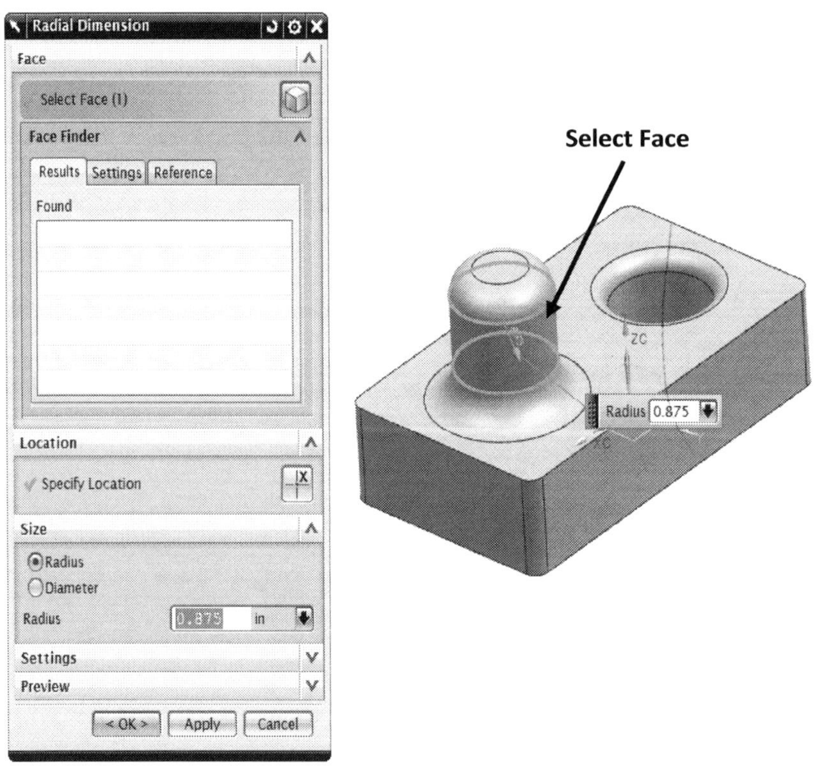


- Then simply change the size by either dragging the arrow or editing the value of the radius. Your model will now change, like the following figure.

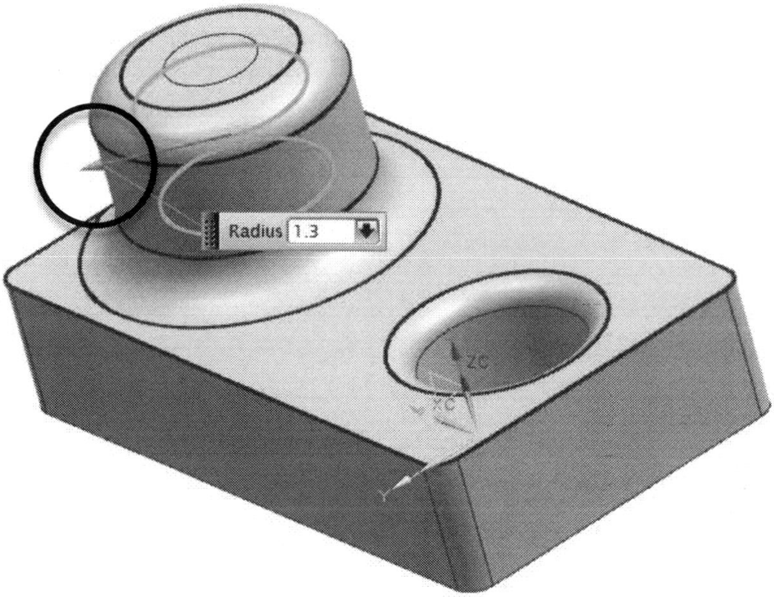


***End of Project***

## 14.11 Co-Axial Constraint

The constraint tools in synchronous modeling are very useful when making quick changes to your model. There is a plethora of easy to use tools under this command, from making faces parallel to making cylindrical faces coaxial. In this section, we are going to go through each of these tools. Because these tools are under the Synchronous Modeling menu, it means that it over-rides the parametric history of your model, allowing you to make changes easily.

### *Project 14.11.1*

This project is very simple. There are only two steps. First you select the face you want to move. Then you select the stationary face. This is like most of the constraint tools in synchronous modeling.

- Create a part shown in the figure below.
- Select Insert / Synchronous Modeling / Relate / Make Coaxial.
- Select the **face** you want to move (motion face), as shown in the following figure.

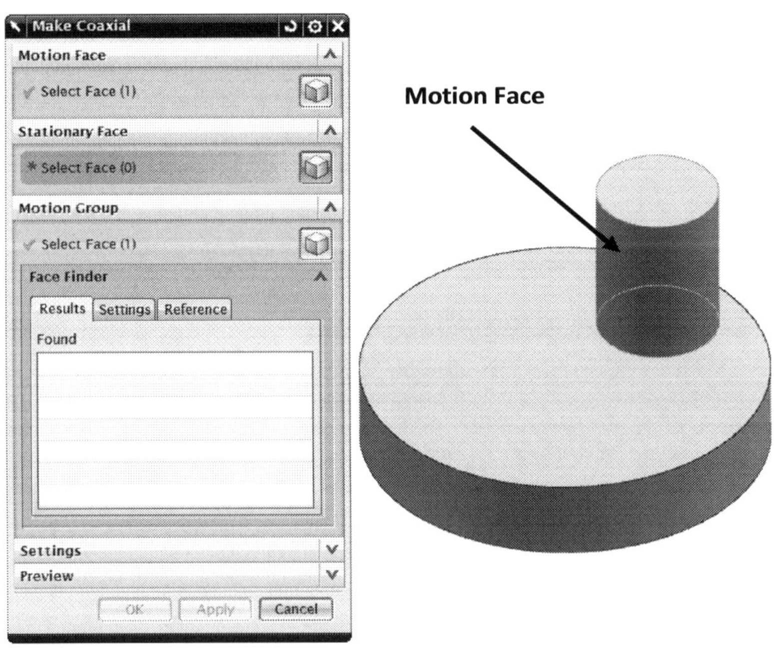

- Next, select the **stationary face**, the one you want to stay still, as shown in the following figure.

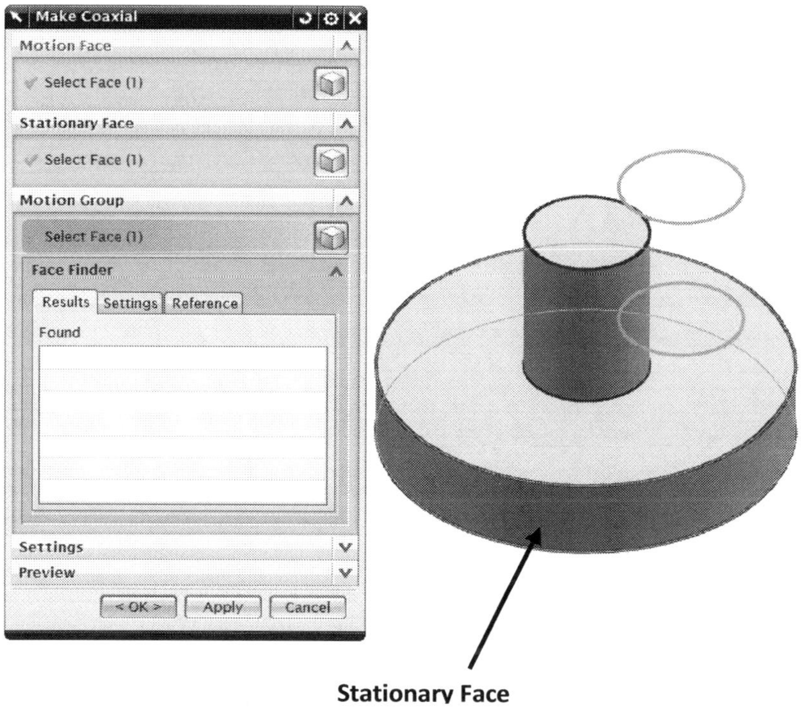


- As soon as you do so, the body will move to show a preview of what you are doing. Click **OK** to accept the operation.

***End of Project***

## 14.12 Make Tangent

A similar constraint tool in synchronous modeling is 'Make Tangent.' This does exactly as you would expect – it makes two cylindrical faces tangent.

### *Project 14.12.1*

- Create a part shown in the figure below.
- Select Insert / Synchronous Modeling / Relate / Make Tangent.
- First select the face you want to move, as shown in the following figure.

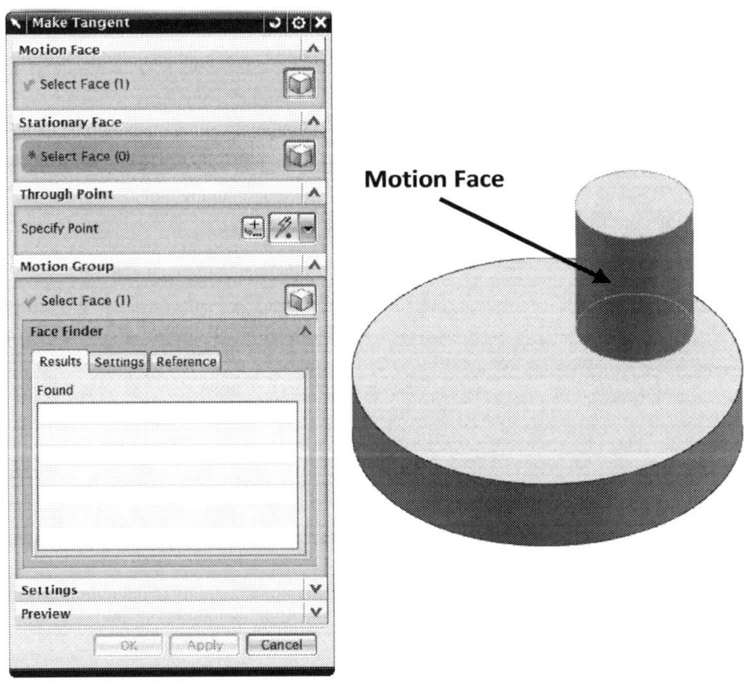

- Then select the face you want it to be tangent to, as shown in the following figure.

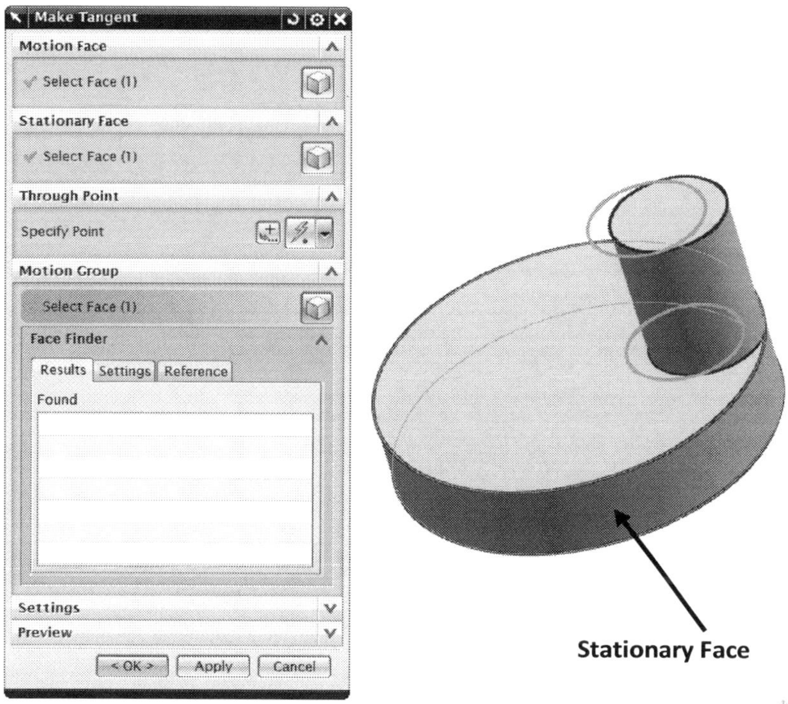


- Click **OK** to complete.

**Note:** There are several more constraint functions within synchronous modeling. They are as simple as the co-axial and the tangent constraints so go ahead and try them all!

***End of Project***

## 14.13 Pattern Face

The 'Pattern Face' Synchronous Modeling feature enables you to create patterned geometry using a region of faces. You can define three types of patterns: Rectangular, Circular, or Reflection. The advantage of the Pattern Face Synchronous Modeling feature over an Instance Pattern is that you can select surfaces from multiple features for the pattern.

### *Project 14.13.1*

- Create a part shown in the figure below.
- Create a CSYS in the corner of the model.

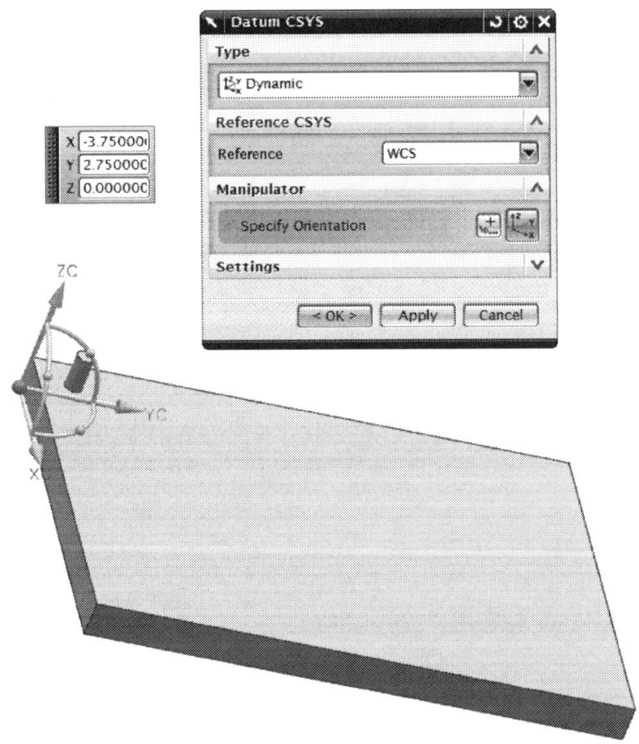


- First we have to create a Datum Plane so that we can create a Datum Axis. Select **Insert / Datum/Point / Datum Plane**.
- Create an angled datum plane by selecting the **Inferred** option as shown in the following figure.
- Create the plane by first selecting **Face 1**, then selecting the **Z datum axis** of the CSYS you created.
- Choose an angle of **120 degrees**.

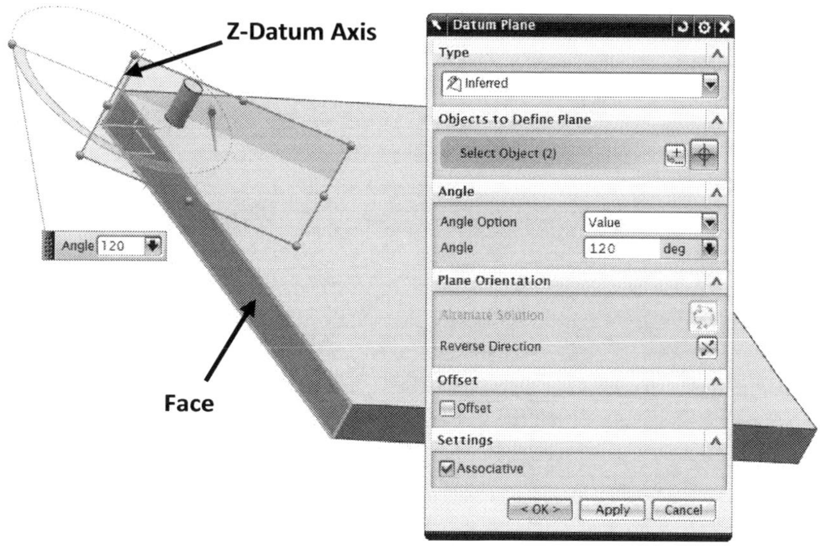


- Select Insert / Datum/Point / Datum Axis.
- Start by using the inferred axis as shown in the following figure.
- Select the **Face**, and then select the **Datum Plane**. Hit **OK**.

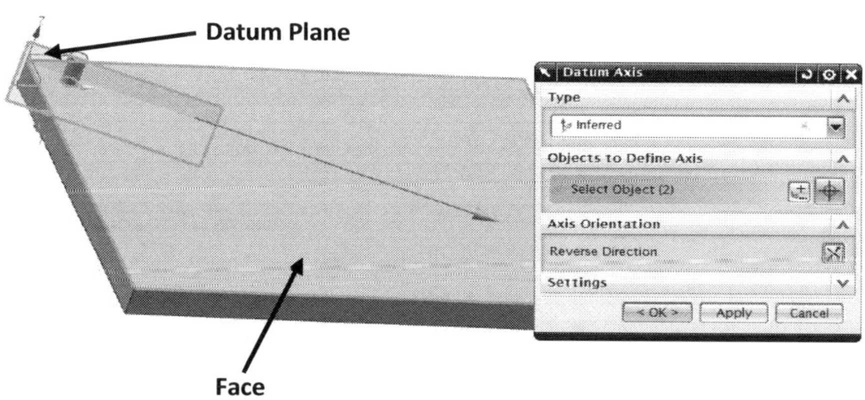


- We can now instance. Select Insert / Synchronous Modeling / Reuse / Pattern Face.
- Select the **Pattern Face**, as shown in the following figure.
- Select the **Rectangular Pattern** as the 'Type'

- Select the faces of the boss to create the instance, as shown in the following figure.
- Select the x datum axis as the **X direction**, also shown.
- And finally, choose the new datum axis you created as the **Y direction**.

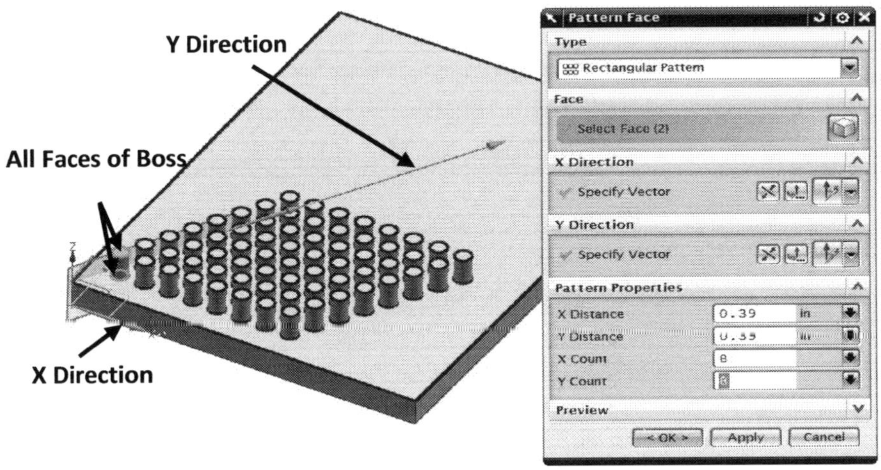


- Once you have selected everything, use **X and Y distances of .39** (If you get an error message, don't be afraid to change this value).
- Use an **X count of 10** and a **Y count of 8**, as shown in the following figure.

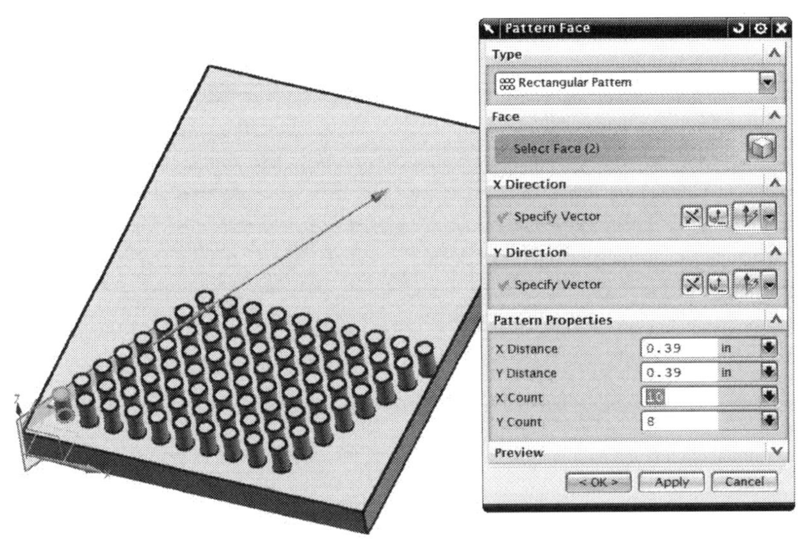

- And you're done!

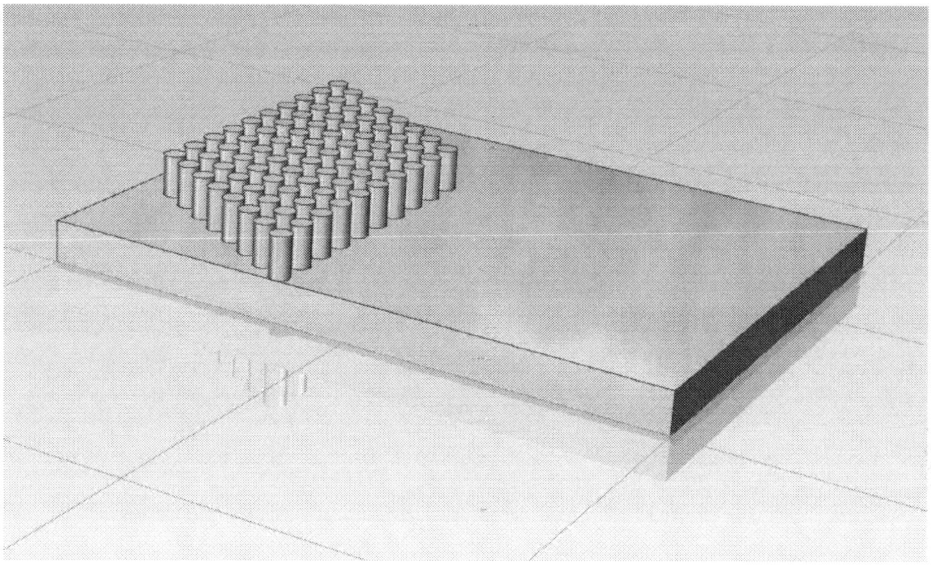

***End of Project***

## 14.14 - Resize & Label Chamfer

The **Resize Chamfer** tool allows the user to change the parameters of any angled face. If you make any other changes, such as moving a face that is adjacent to a chamfer, the **Label Chamfer** tool will make the whole chamfer move instead of elongating the chamfer.

### *Project 14.14.1*

- Begin by creating the following **sketch**.

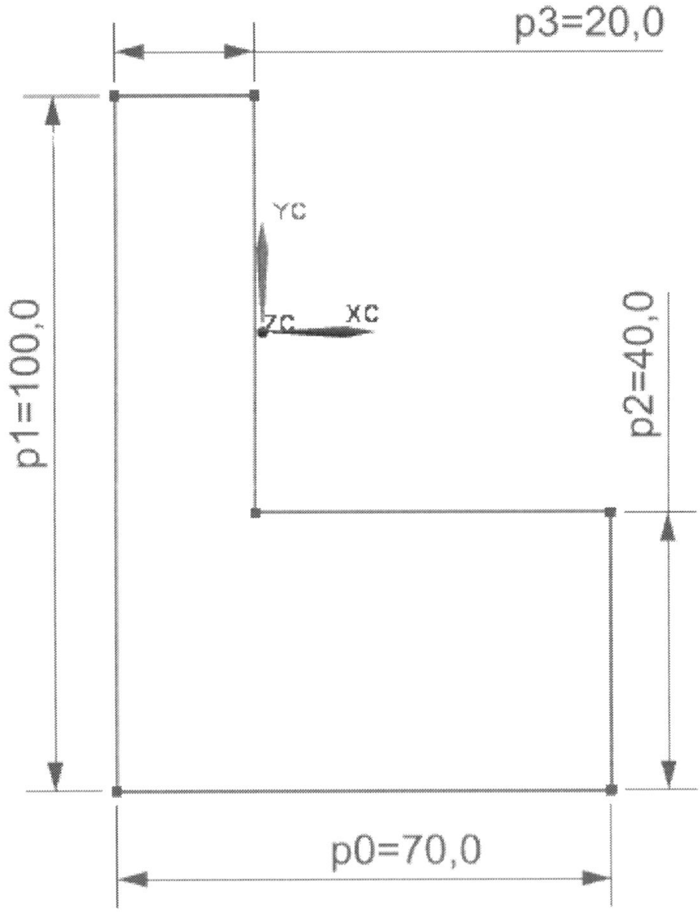

- **Revolve** the sketch **360 degrees** to create the following solid body.

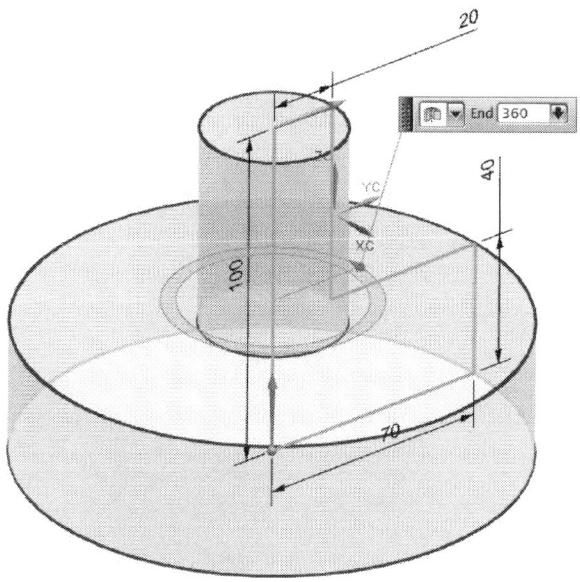


- Create the following **sketch** of a single line on the **XY plane**.

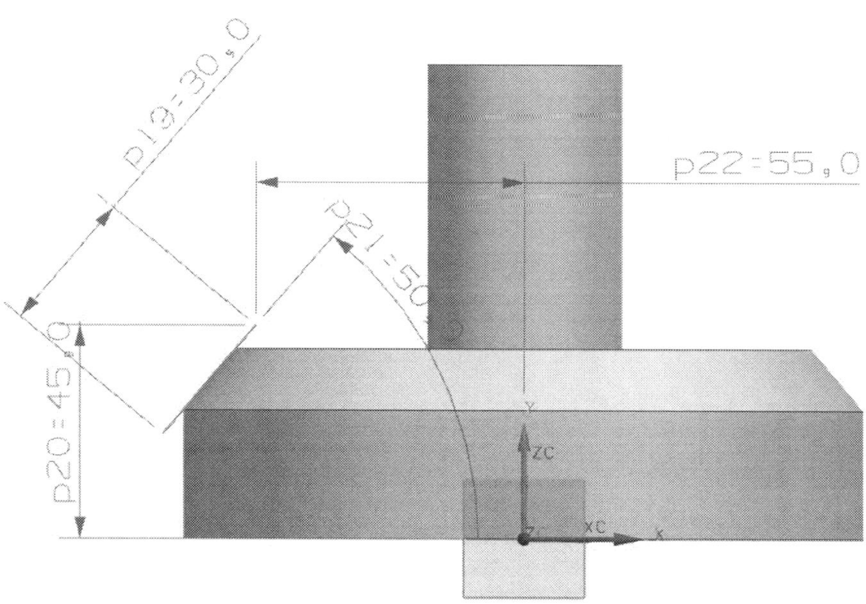

- Select **Insert / Sweep / Sweep along Guide**. Select the **sketch** as the section and the edge of the solid body as your guide. Make sure you change **Settings / Body Type** to **Sheet**.

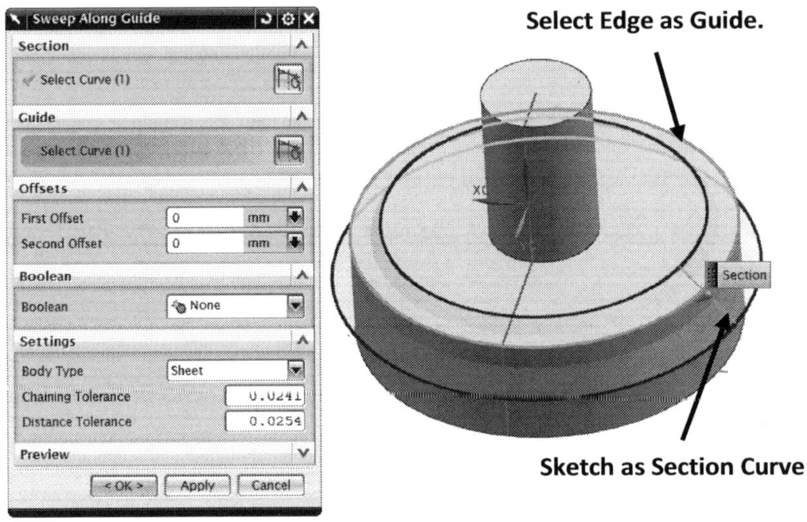


| **Note:** You could also use the revolve tool to create this sheet body. |
|---|

- Select **Insert / Trim / Trim Body**. Choose the solid body as the **Target** and the Sweep as the **tool**. Make sure you are choosing the right side to trim. Hide the sweep, and your result should look like the following figure.

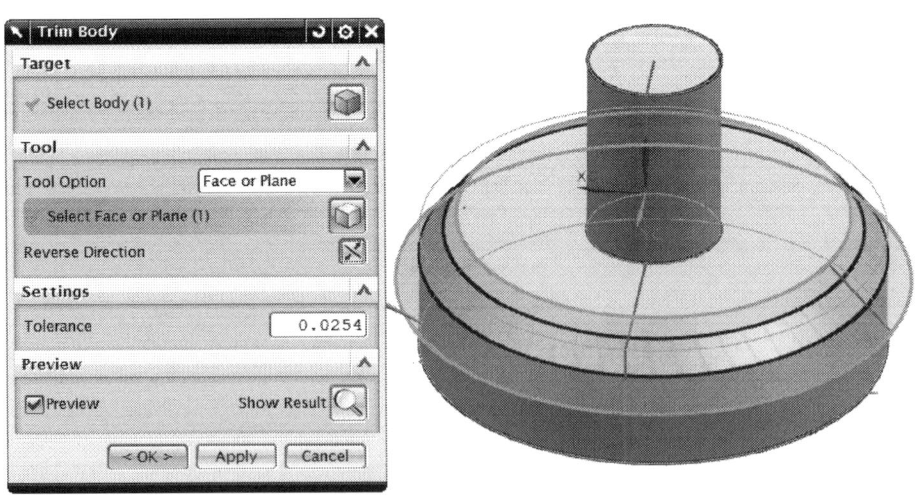


| **Note:** if the body trims in the wrong direction use the 'Reverse Direction' button. |
|---|

- Now go to **Insert / Synchronous Modeling / Detail Feature / Label Chamfer** . You accomplish this by selecting the **angled face** that you want NX to recognize as a chamfer. Most times that is all you will have to do as NX is smart enough to infer the construction faces. If that is not the case, just select the two faces that the desired chamfer is adjacent too. The selection is shown in the following figure.

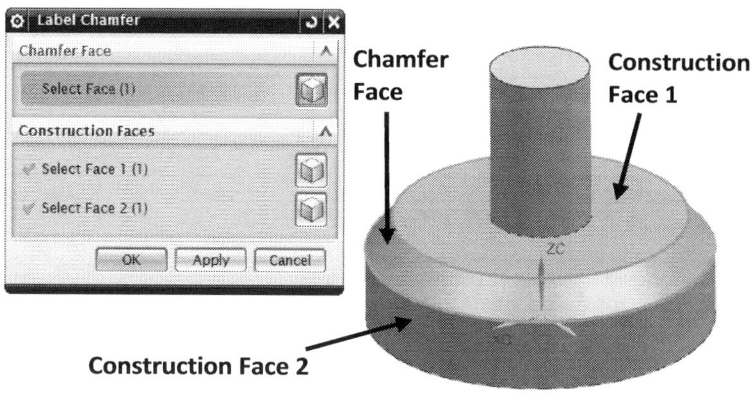


- Click **Insert / Synchronous Modeling / Move Face**. Select the face shown in the following figure and drag the arrow in the Z direction. Because we have labeled the chamfer, NX maintains the chamfer's parameters (2nd figure). The figure shown below (1st figure) illustrates the changes made when the chamfer has NOT been labeled as a chamfer.

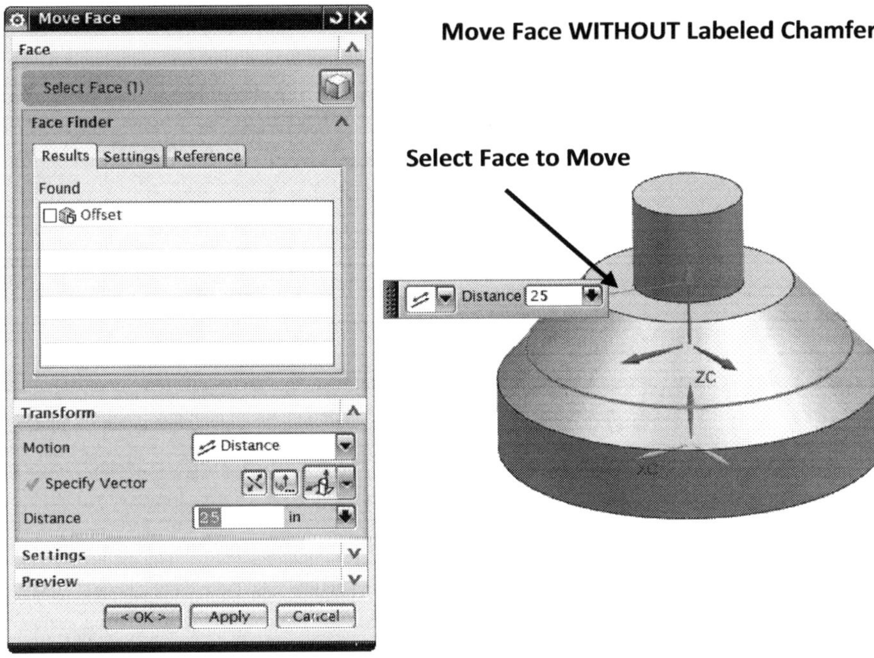
Move Face WITHOUT Labeled Chamfer
Select Face to Move
Move Face
Face
Select Face (1)
Face Finder
Results
Settings
Reference
Found
Offset
Transform
Motion
Distance
Specify Vector
Distance
25
in
Settings
Preview
< OK >
Apply
Cancel
Distance 25
ZC

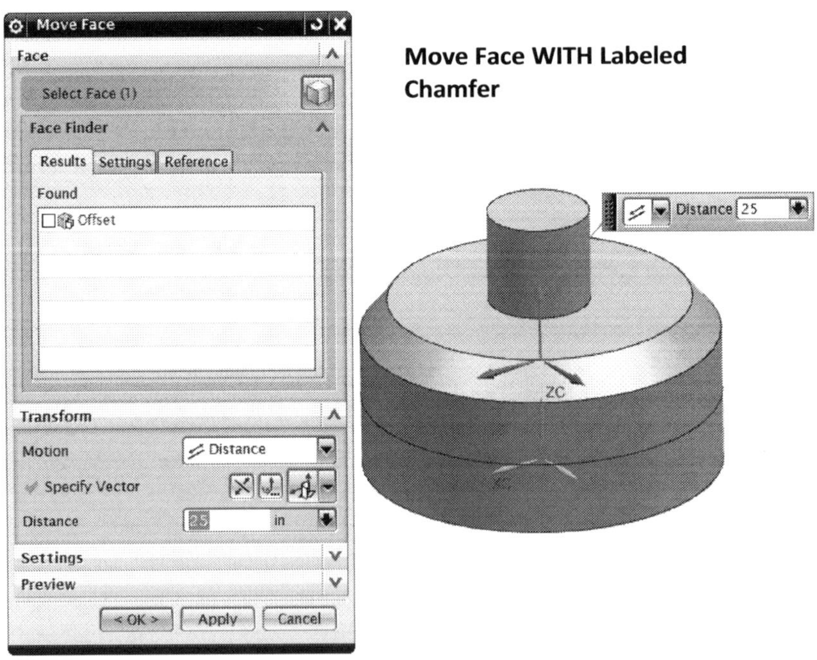
Move Face WITH Labeled Chamfer
Move Face
Face
Select Face (1)
Face Finder
Results
Settings
Reference
Found
Offset
Transform
Motion
Distance
Specify Vector
Distance
25
in
Settings
Preview
< OK >
Apply
Cancel
Distance 25
ZC

- Finally select **Insert / Synchronous Modeling / Detail Feature / Resize Chamfer**  . Click on the face of the '**Chamfer**,' choose the cross section as **Asymmetric** and fill out the following parameters.

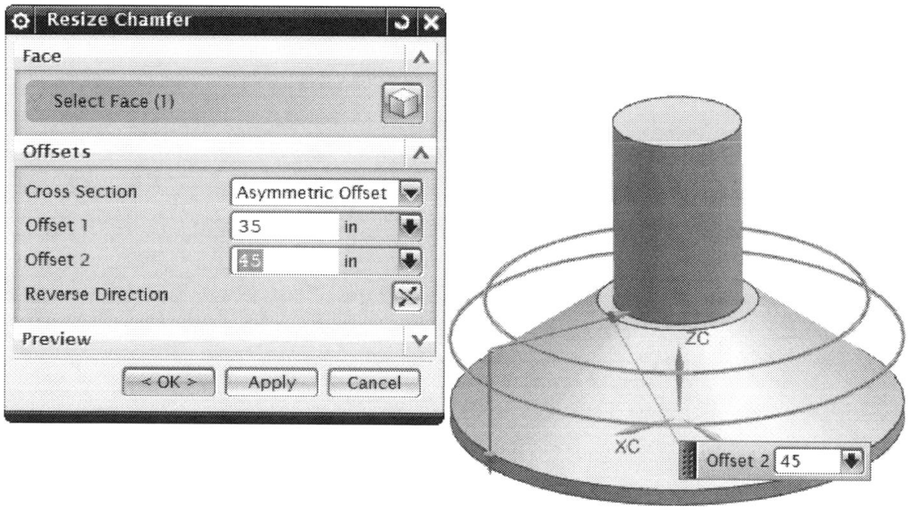


***End of Project***

## 14.15 - Make Offset

The **'Make Offset'** command is another fantastic synchronous modeling tool. One of the best things about the synchronous suite is that some of the tools are quite similar so that if one of the tools doesn't work, another one will. In this example we are going to use **Make Offset** in a case where **Replace Face** would not achieve the same effect. This type of tool is also excellent for working with solid exports from other CAD software or for manipulating industrial design surfaces.

### *Project 14.15.1*

- Create a **sketch** that looks something like the following figure. Feel free to be creative! This one includes **3 Lines** and a **Studio Spline**.

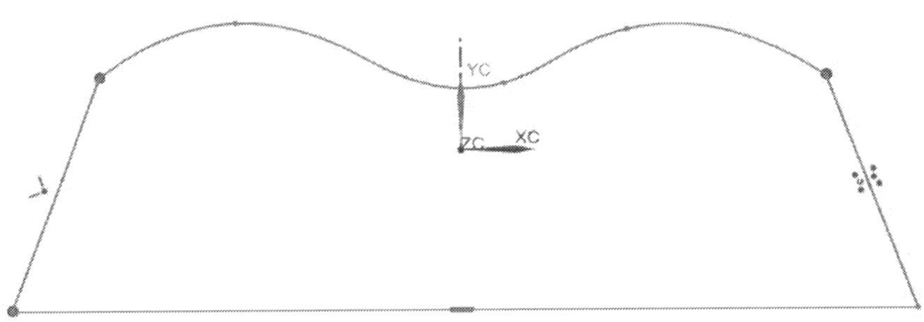


- **Extrude** the sketch so your model looks something like the figure below.

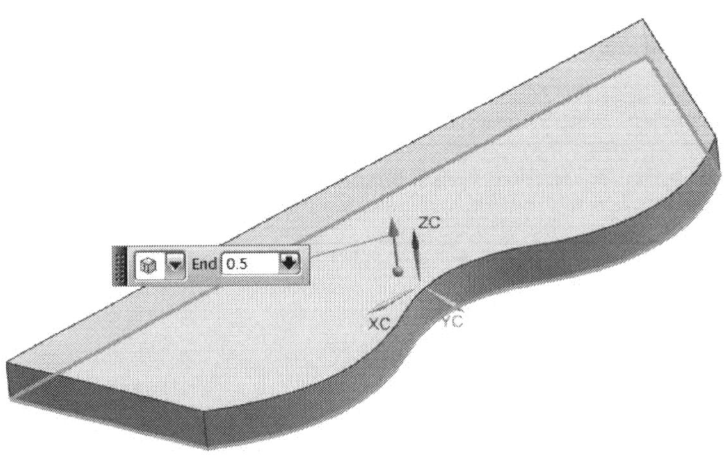

- Select Insert / Synchronous Modeling / Relate / Make Offset  .

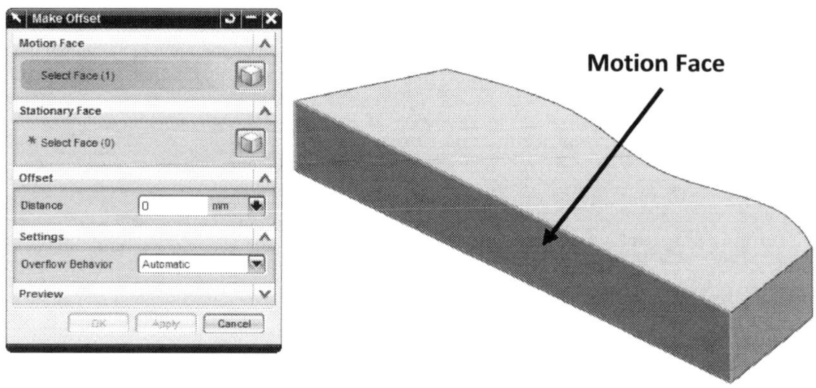


- Select the **flat face** as the **'Motion Face'** as shown in the following figure.

- For the next step select the **curved face** as the **'Stationary Face.'** You will be given the current distance between the faces. Change the value to roughly **1/3** of that value.

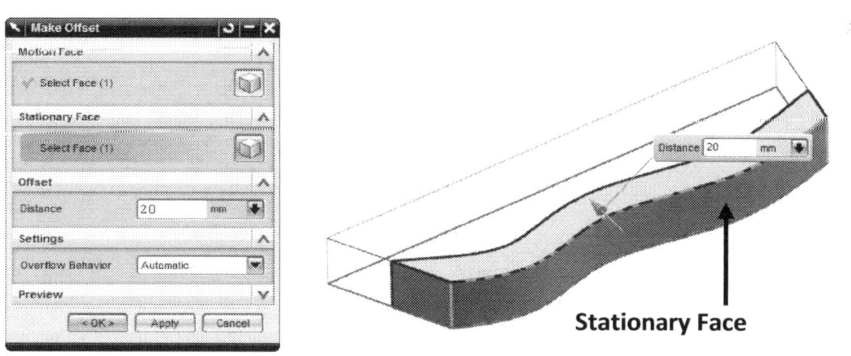


- Your model will look like the following figure.

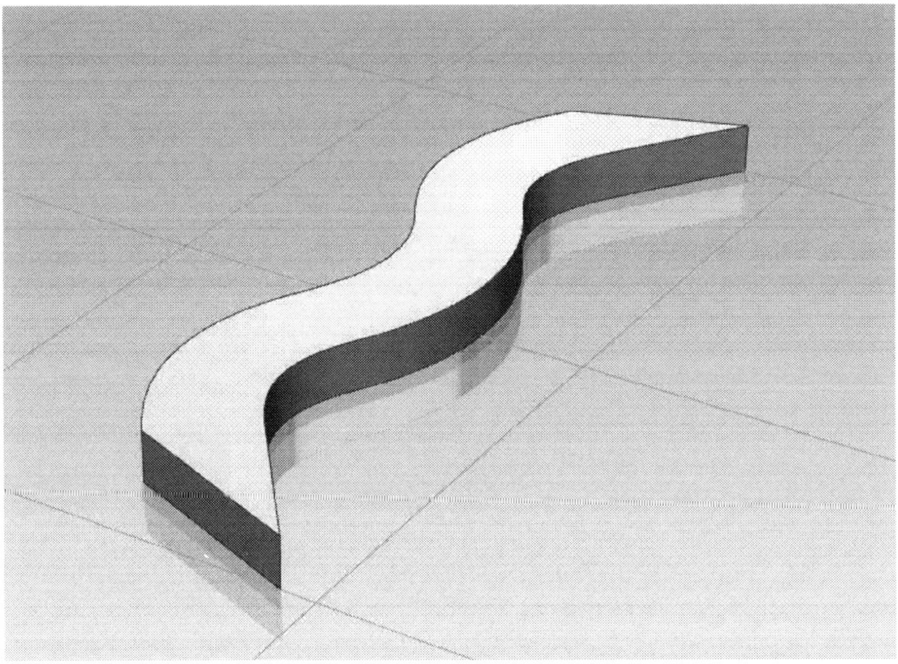

- Try and repeat this process using **Replace Face** and see if you get the same results.

***End of Project***

## 14.16 Self-Test

(All questions are true/false. Answers are at the back of the book.)

1. You can move multiple faces at the same time using move region.
2. Using synchronous modeling can convert a parametric model to non-parametric.
3. 'Pattern Face' and 'Instance' commands can be used interchangeably.

# 15. Surface Features

## 15.1 Introduction

'Surface Features' are powerful tools that are used to construct organically sculpted geometry. Surface Features include many types of geometry, including zero thickness, infinitely thin sheets (surfaces), closed profiles swept along a drive curve, certain kinds of blends, and offset faces. Surface Features are distributed under many sub-menus including 'Surface,' 'Mesh Surface,' 'Sweep' and 'Flange Surface.' The figure below illustrates two typical examples of Surface features.

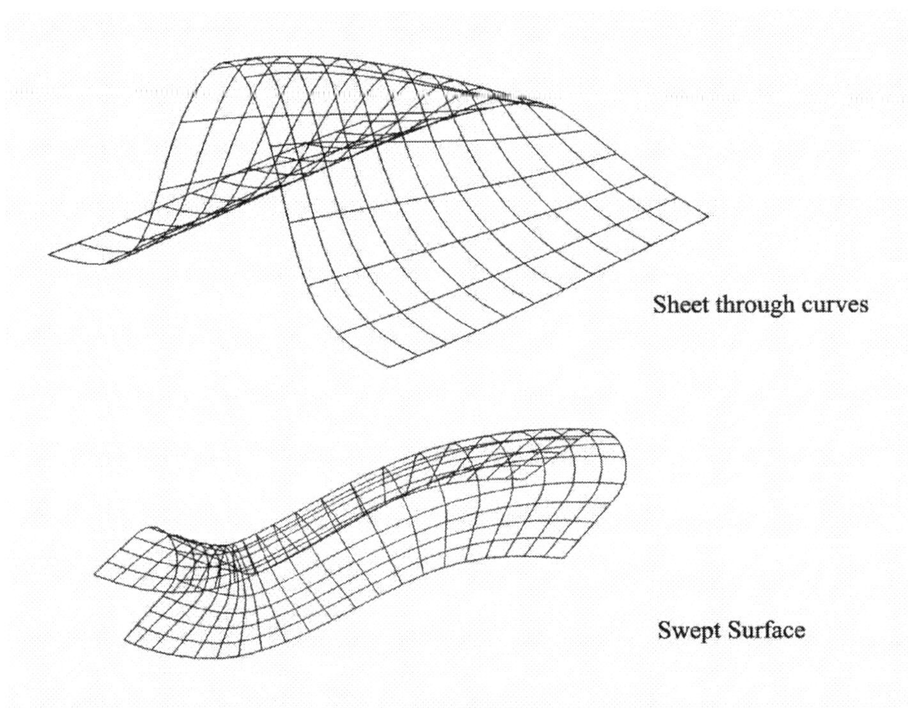


By their nature, Surface Features are highly dependent upon other geometry. In the most rigorous situations, Surface Features are created using fully parametric sketch curves, offset curves, or projected curves. By using these intelligent entities you can create the most difficult of geometries.

Surfaces are also known as sheet bodies in NX 8.5. Many other tool options that you have used so far to create solid bodies, such as extrude and revolve, can also be used to create simple surfaces or sheet bodies.

## 15.2 Surface Feature – Bounded Plane 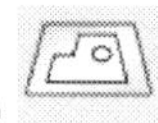

The 'Bounded Plane' option enables you to build a sheet surface by using connected curves. These section strings must be coplanar, closed and connected to form the sheet boundaries.

### ***Detail: Bounded Plane***

The Bounded Plane command may be used to create a glass block inside an existing solid as shown below.

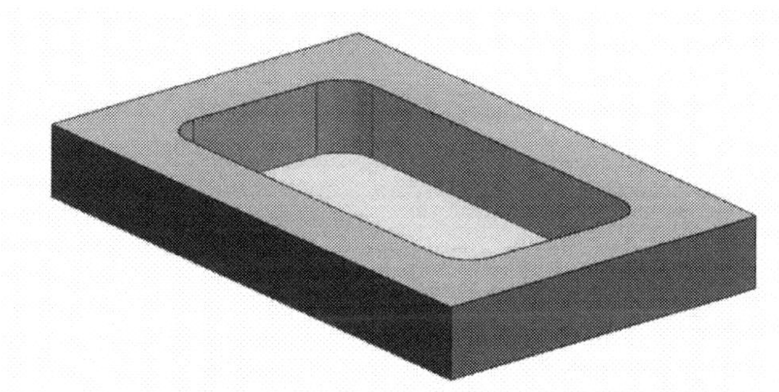

### ***Project 15.2.1***

- Sketch a rectangle with dimensions of **4in** x **2.5in**.
- **Finish** the sketch and **Offset** the curves by **0.5in** inwards.
- **Extrude** the curves by **0.5in**.
- Add **0.5in Edge Blends** on the inside of the model as shown below.

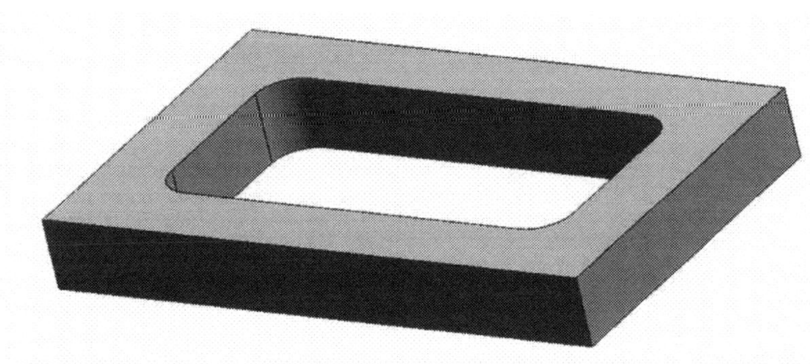

- Select Insert / Associative Copy / Extract Body.
- Select the face option from the **Extract** menu and select the internal faces and choose **OK**.
- Move the original solid to another layer and make it invisible. *The surfaces should look like the figure below.*

- Select Insert / Surface / Bounded Plane .
- Select all 8 (top) edges of the curved surface.
- Middle click to OK.
- Repeat the same to create a lower end cap.
- Select ***Insert / Combine / Sew*** and sew all the sheets together to create the solid.

Change the 'Object Display' to better illustrate the translucent property of glass.

- Select Edit / Object Display.
- Select the body you just created.
- Choose **OK**.
- Set the Translucency to **50** as shown on the next page.
- Make the original solid layer visible.

> **Note:** You might have realized that this block could have been created by simpler means, but the objective of this exercise was to get you familiar with the bounded plane command, which is used a lot with partially extracted or existing dumb geometry that is much more complex in nature.

***End of Project***

## 15.3 Surface Feature - Through Curves

The 'Through Curves' option enables you to build a surface that is "draped" over two or more rows of defining curves. Each row of curves must be continuous and are referred to as a 'Section String.' It is important to pay close attention to the feedback that NX 8.5 gives you during the process of creating a Through Curve Surface feature, because it is rather complex. In other words, *Read the Cue!*

To avoid defining a "twisted" surface, each Section String must be spaced progressively farther from the initial row. The direction of each Section String must also be defined in a consistent manner.

To obtain a fully parametric surface, each Section String should also be parametric. For example, each Section String may be a Sketch that is placed onto a parametrically-defined datum plane. When either the curves or the position of the datum planes is changed, the resultant Free Form surface will update accordingly.

### *Project 15.3.1*

Define the necessary construction geometry and create a Surface feature using the Through Curves option.

- Create a new file. Save it and rename it as **TC_<*your initials*>.prt**.
- Create at least four parallel datum planes, each spaced two inches apart.
- Create at least four sketches on the datum planes, as shown next. You may use spline geometry or lines and arcs; however, be certain that each sketch has contiguous strings (no gaps allowed).
- Select **Insert / Mesh Surface / Through Curves** . (Read the Cue – the system is prompting you to select a Section Curve.)

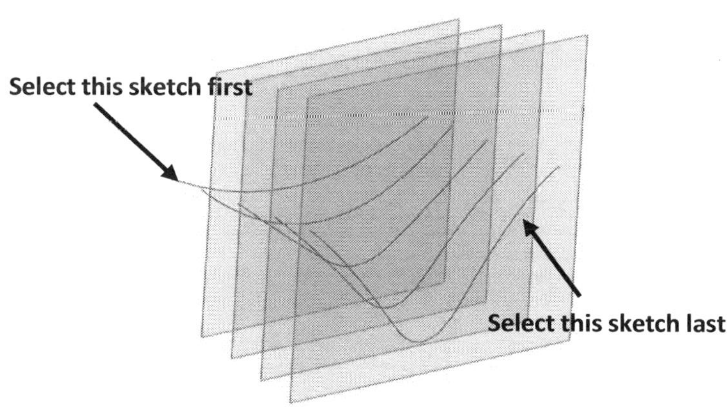


- Select the geometry at the left-hand side of the first sketch, as shown above. *(Read the Cue again – the system is giving you two options.)*
- Choose **OK** to complete the selection of Section Curve 1. (The system should display a direction vector at the end of the sketch you selected and you will see Section 1 displayed in the Through Curves menu.)

- Select at the left-hand side of the second sketch. Choose **OK** to complete the selection of Section String 2.
- Select at the corresponding location on the third sketch, and choose OK. (Check the direction vector – if it is ever incorrect, you should choose 'Back' and re-select the incorrect section string.)
- Select at the corresponding location on the fourth sketch, and choose **OK**.
- After selecting the fourth section String, choose **OK**. Note: The subsequent dialog boxes contain several options that are not within the scope of this exercise.

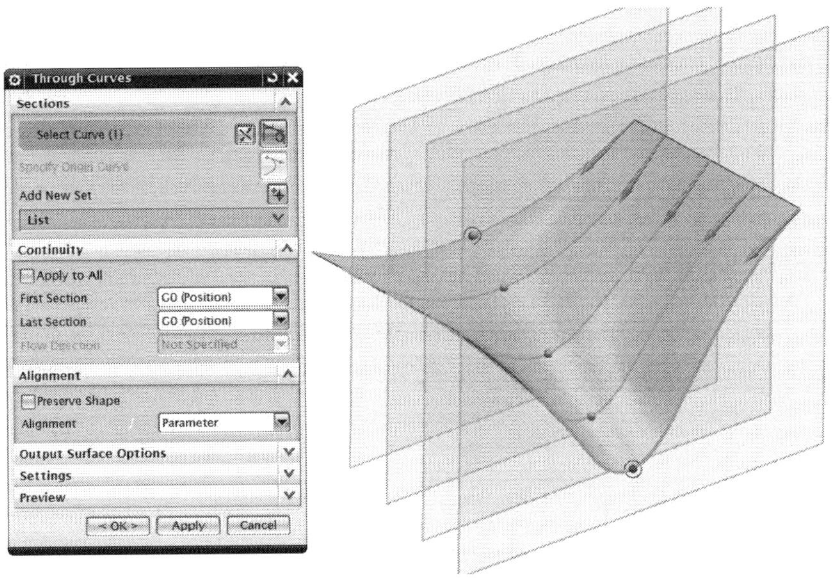


Compare your results to the figure below.

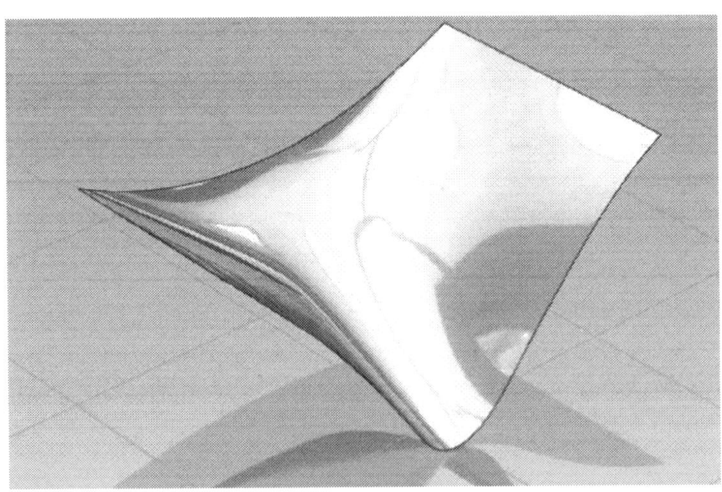

***End of Project***

## 15.4 Sweep Along Guide 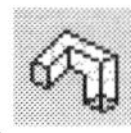

'Sweep Along Guide' can be used for creating any geometry that needs a uniform section to follow a curvilinear path. To use the Sweep Along Guide feature, you need two separate sketches in two different planes. One sketch will be the section string, which is the "C" shaped curve, and the second sketch will be the guide curve, which is the "S" shaped curve, shown below.

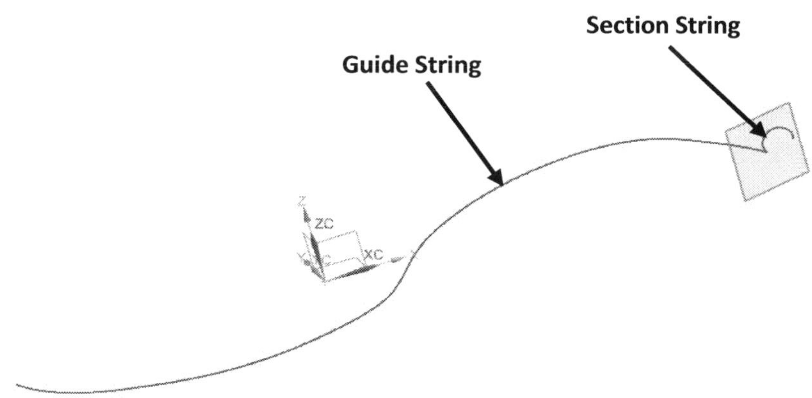


### *Project 15.4.1*

Using the Sketcher, create the **Guide String** on the default (X-Y) plane as shown below.

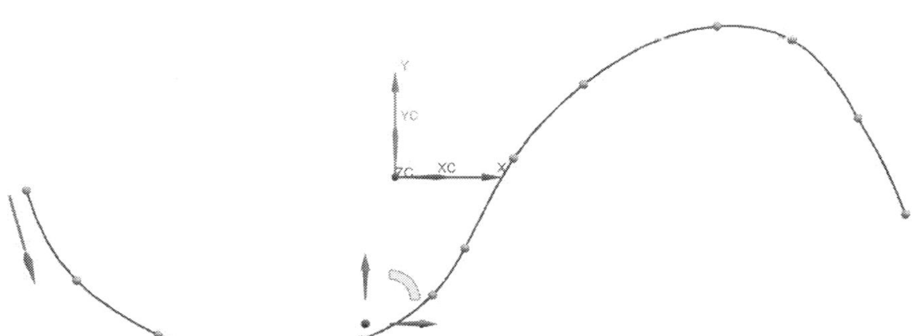


- **Finish** the guide string sketch.
- Create a datum plane at the end of the guide string by selecting the **Point and Direction** method from the 'Datum Plane' dialog box (see figure below).

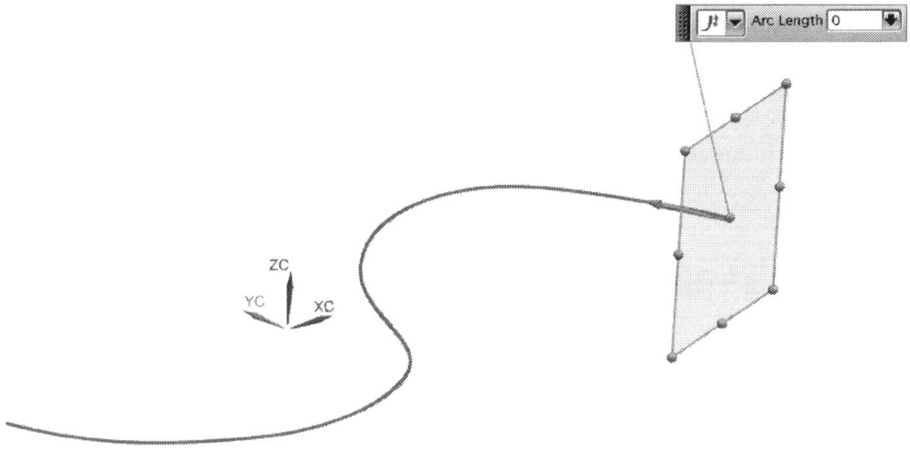


Create the "C" shaped section string sketch on the datum plane you just created, making sure that the end points of the curves are coincident.

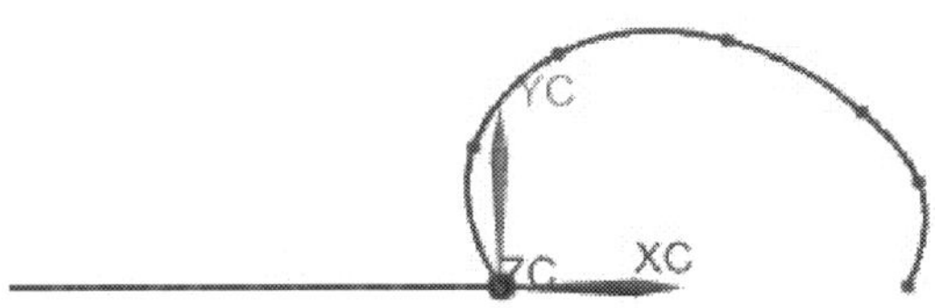


- Select ***Insert / Sweep / Sweep along Guide*** .
- Select the section string.
- Select the guide curve sketch.
- Ensure that both offsets are zero, and choose **OK.** If you enter an offset value you can create a solid body instead of a surface.

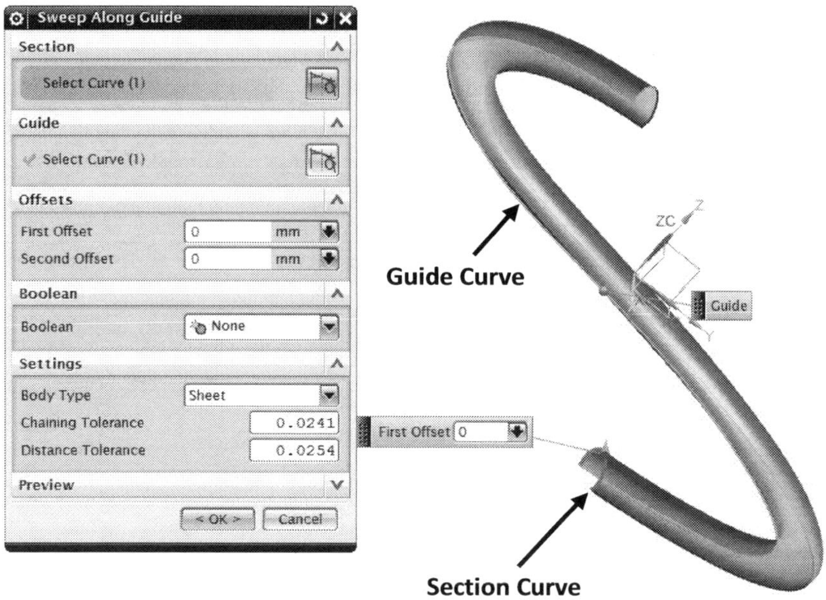


You should now have the following model.

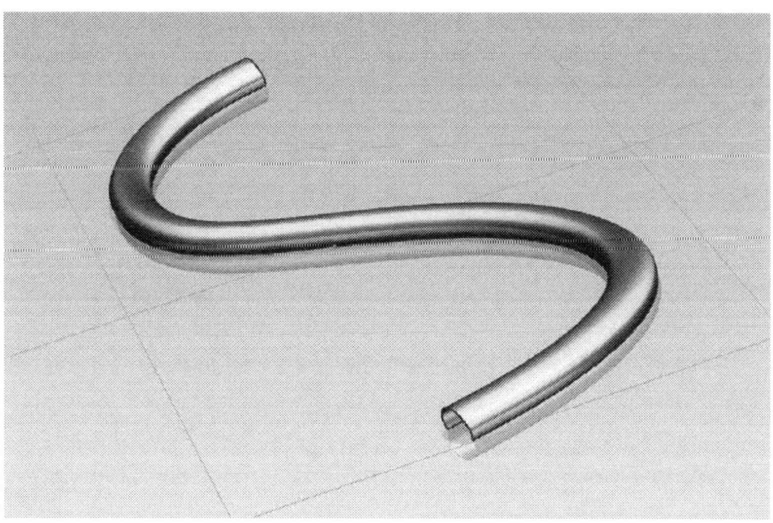

**Note:** Sweep Along Guide features require that the start point of the guide string must intersect the section string plane. This is often overlooked by novice users. Another common mistake is defining the radius on the guide string so that, when combined with the section string, it results in self-intersecting geometry.

***End of Project***

## 15.5 Swept Surface Features

The 'Swept' Surface feature allows you to specify not only multiple 'Section String' curves, but also 'Guide String' curves that control the shape in between them. The Swept Surface feature allows you to select one, two, or three Guide Strings and as many Section Strings as you want. The system first prompts you to define the Guide String(s), followed by the Section Strings.

It is worth mentioning a few points regarding the definition of Guide and Section Strings when defining Surface features:

- The direction of each Section String must be consistent; otherwise the resultant Surface feature will be "twisted".
- Section Strings may be composed of one or many curve segments.
- Section Strings are not required to have a corresponding number of curve segments, but if so, the resultant surface shape will often be smoother.
- Section Strings do not have to be planar, but if they are the resultant surface shape will often be smoother.
- If you use sketches to define the Guide Strings and Section Strings, the Surface feature will be fully parametric.

These comments are generally applicable to any type of Surface or Swept features created in NX 8.5. If the Section Strings are closed loops, the resultant geometry will be a solid feature. If the Section Strings are open loops, the resultant geometry will be a sheet body feature.

### *Project 15.5.1*

The first step is to build the feature construction geometry as shown below, which consists of the three default datum planes (CSYS) and one datum plane offset by 12 inches. Next, a square sketch is placed at each end with their centers aligned to the central datum planes. The final step is to create a sketch on the central datum plane that has a "hump" in the middle. Its endpoints must be aligned to the centroid of each end sketch.

- Create a new file. Save it and rename it as **swept_*<your initials>.prt***.
- Create the three datum planes and the **12in** offset datum plane.

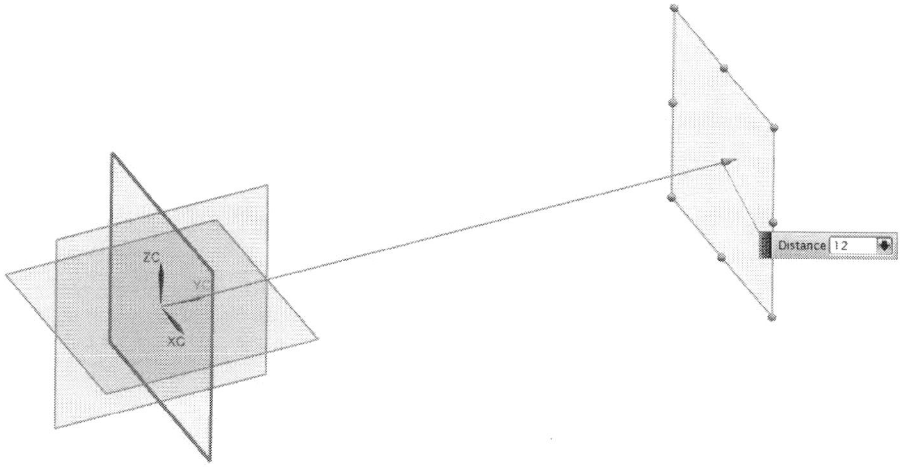


- Create the two Section String sketches as shown below. Be certain that the center of the rectangle is located at the center of the out-of-plane Datum Planes.

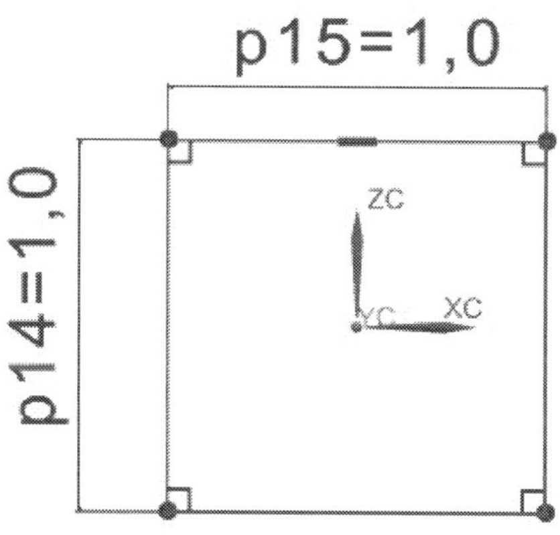


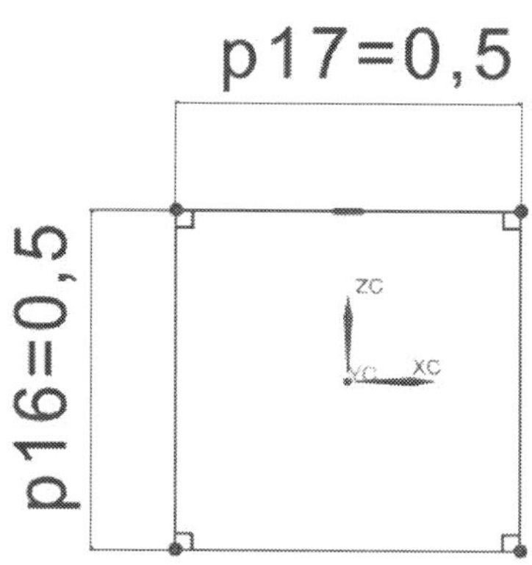

- Create the Guide String sketch, as shown below. Be certain that the ends of the sketch are aligned to the center of each Section String sketch.

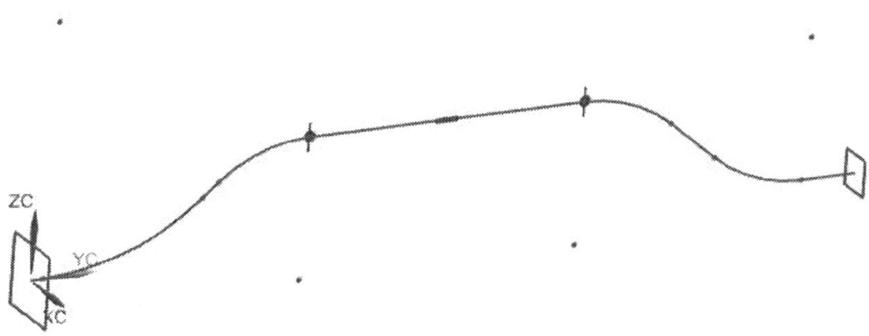


- Select Insert / Sweep / Swept .
- Select the Guide String.
- Select the Section String left side (pay attention to which entity and end you select – you must select the other Section String in a consistent manner).
- Select the Section String right side (select this section string in the same way you selected the left side Section String).
- Choose **Cubic.**

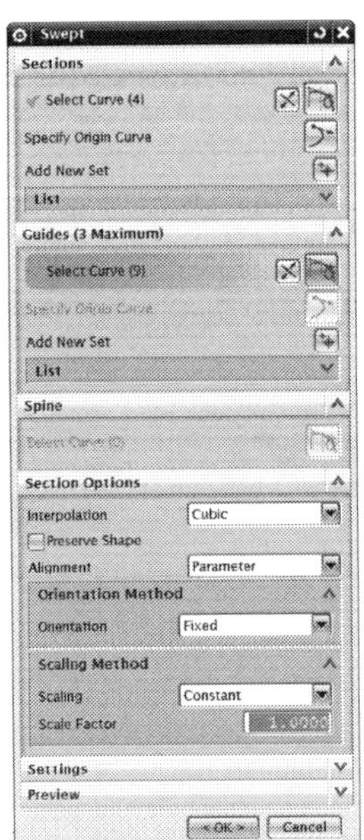


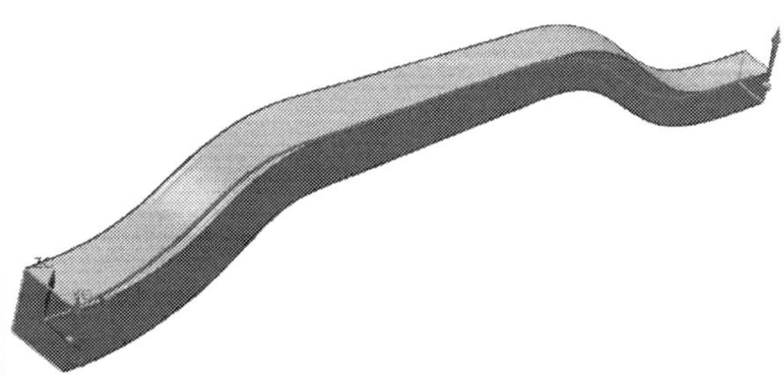

- Choose **OK** to accept other default options and the Swept solid will appear.

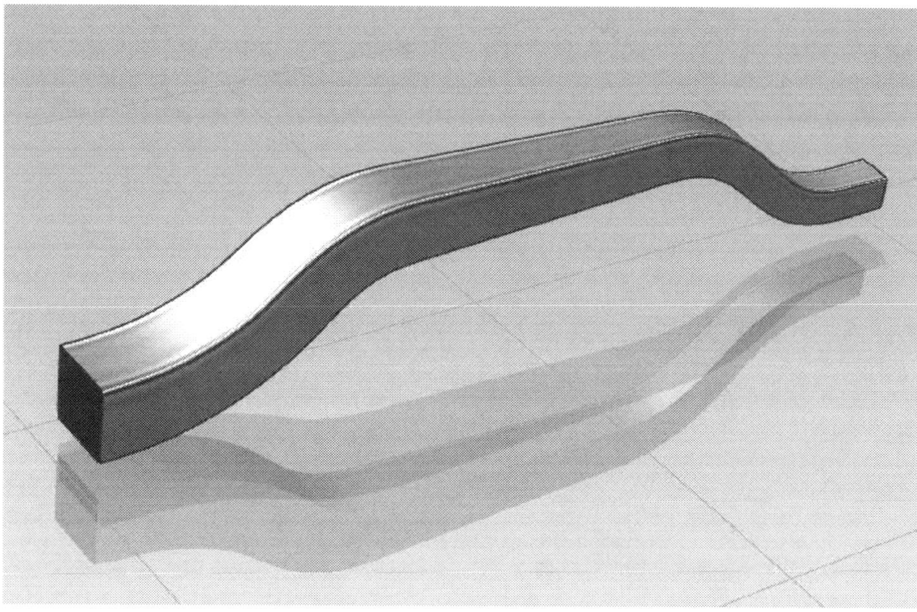

Compare your results to the figure above.

> **Note:** The Swept geometry defined by the sketches in this example resulted in the creation of a solid feature. If any of the Section Strings were not planar or closed, a sheet body would have been produced. Another important point is that if you do not select corresponding locations on each Section String, the Swept geometry will be twisted.

***End of Project***

## 15.6 Through Curve Mesh

The 'Through Curve Mesh' option enables you to create a solid or sheet body from existing strings running in two different directions. You can create complex surfaces and surface patches using this method. As per the usual, using closed section string entities allows you to create a solid body; using open section strings will result in a surface body.

The main difference between 'Through Curves' and 'Through Curve Mesh' is that you use the latter when you have guide curves for the surface to follow.

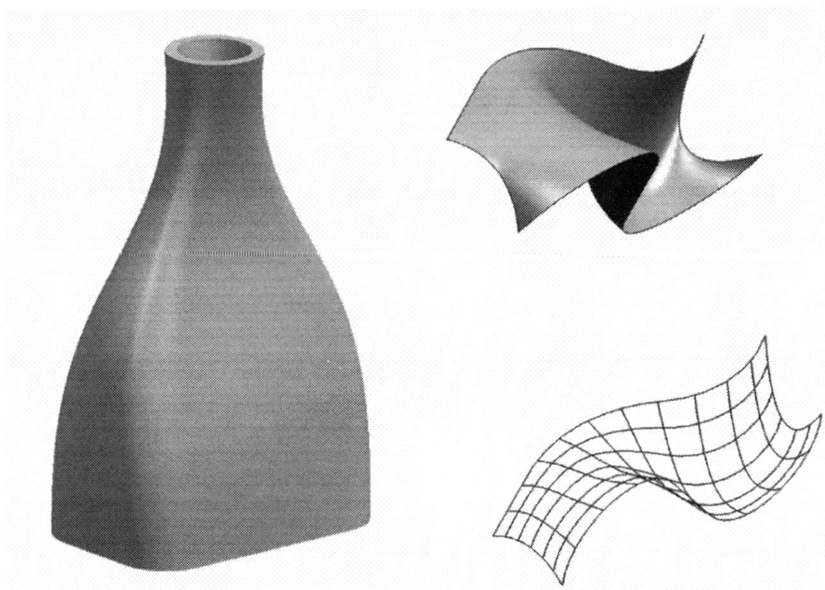

Solid Body Sheet Body

## *Project 15.6.1*

- Create the three standard XYZ datum planes using the ***Insert / Datum/Point / Datum CSYS*** command.
- Create a new datum plane offset by **5** inches in the Y-direction.
- Sketch a circle of Diameter = **1in** on the offset datum. Center it in the sketch plane.
- On the original datum plane, parallel to the offset plane, sketch a rectangle with dimensions **3in** x **2in** with **0.5in** fillets on all four corners as shown next. Center it in the sketch plane as well.

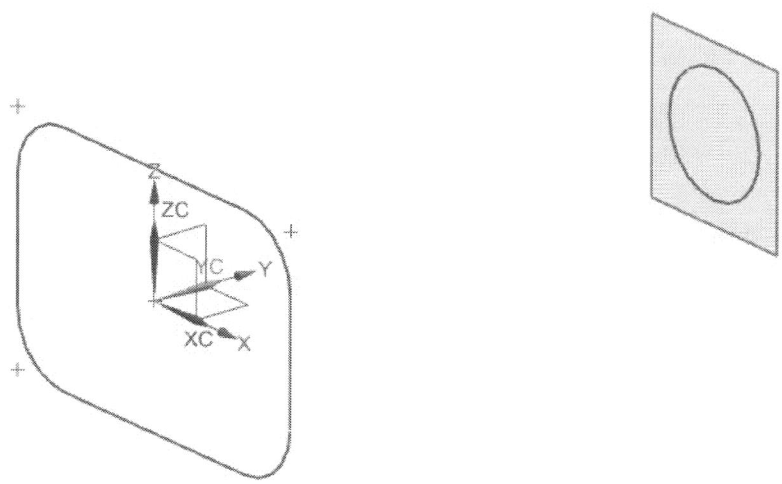


- Select ***Insert / Datum/Point / Point*** and place four datum points on the circle. Locate each datum point on a quadrant point of the circle.

> ***Note:*** *Be sure the* ***Quadrant Point*** *Snap Point option is selected.*

- Next, place four datum points on the rectangle. Locate each datum point on the midpoint of the four straight lines. *Note: Be sure the* ***Mid Point*** *Snap Point option is selected.*

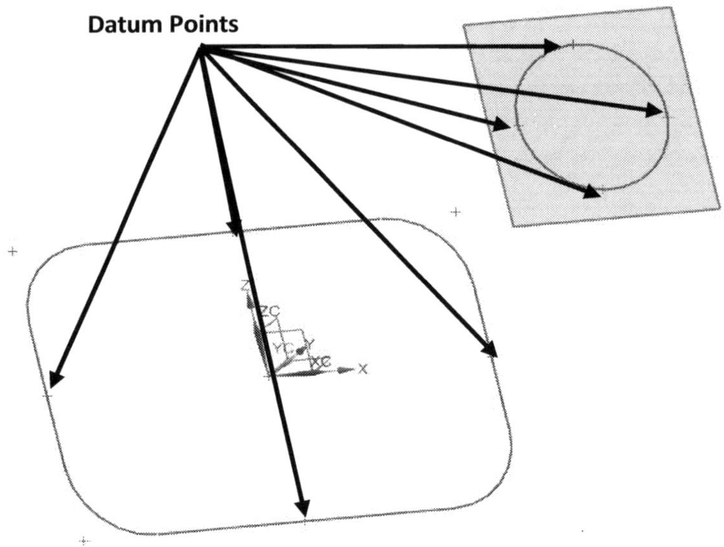


- Sketch two curves in the Y-Z datum plane to connect the datum points of the circle and rectangle as shown below. ***Hint:*** *You can mirror your sketch.*

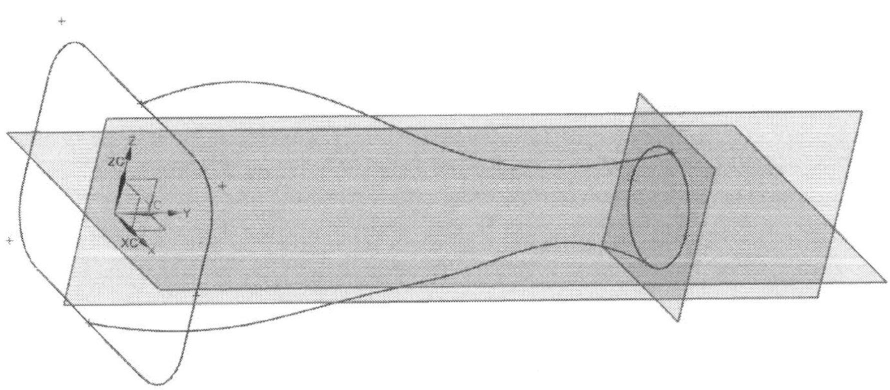

Sketch two curves in the X-Y datum plane to connect the datum points of the circle and rectangle shown below.

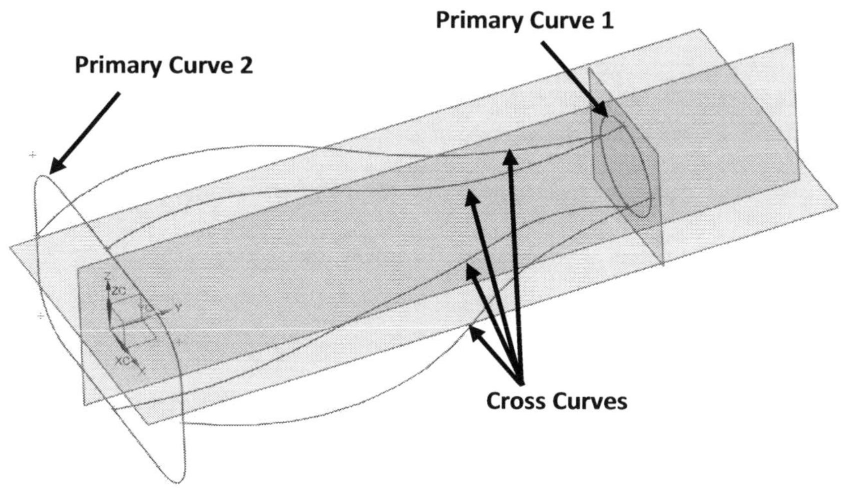


**Note:** Be sure the arrows associated with both strings are pointing in the same direction. If they are not, NX 8.5 will probably freak out. If they do not match, select **Back** and be careful to select in a similar location on both primary strings.

- Select Insert / Mesh Surface / Through Curve Mesh . Read the cue line - NX 8.5 is prompting you to select Primary String 1. Make sure the option in the selection intent menu is set to **connected curves**.
- Select the circle as primary string 1 and choose **Add New Set.** Select the rectangle as Primary String 2 and choose **Add New Set**.

Read the cue line – NX 8.5 is prompting you to select the first cross string.

- Select any of the cross strings as the first and hit **Add New Set**.
- Continue to select the remaining three cross strings in the same manner.
- When prompted for Cross String 5, re-select your starting string to let NX 8.5 know where to close the loop.

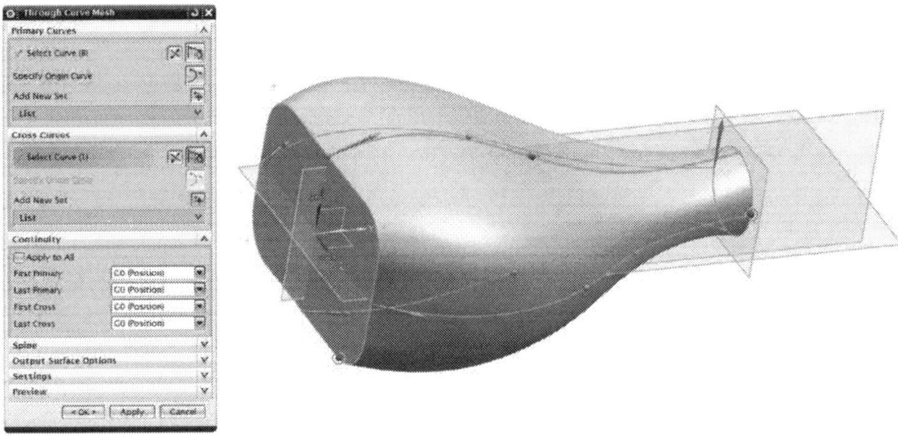

**Note:** Be sure to go clockwise or counterclockwise, according to the direction in which the primary stings are directed, when selecting the strings to ensure a smooth, continuous loop.

- Accept all the defaults in the 'Through Curve Mesh' window that comes up.
- Choose **OK**.
- Select Insert / Offset/Scale / Shell.
- Select the circular face as the face to pierce and select **Apply**.

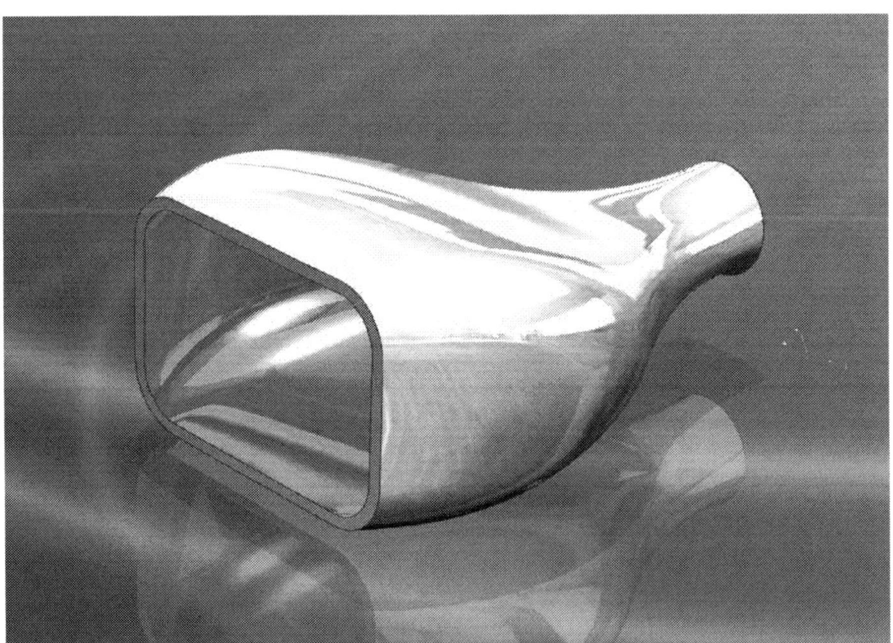

***End of Proje***

## *Project 15.6.2*

- Create the sketch shown below.

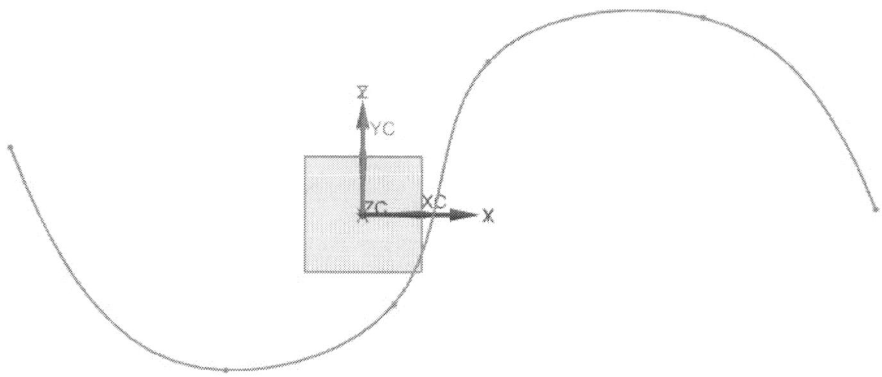


- Create another sketch on a datum plane perpendicular to the first one as shown below.

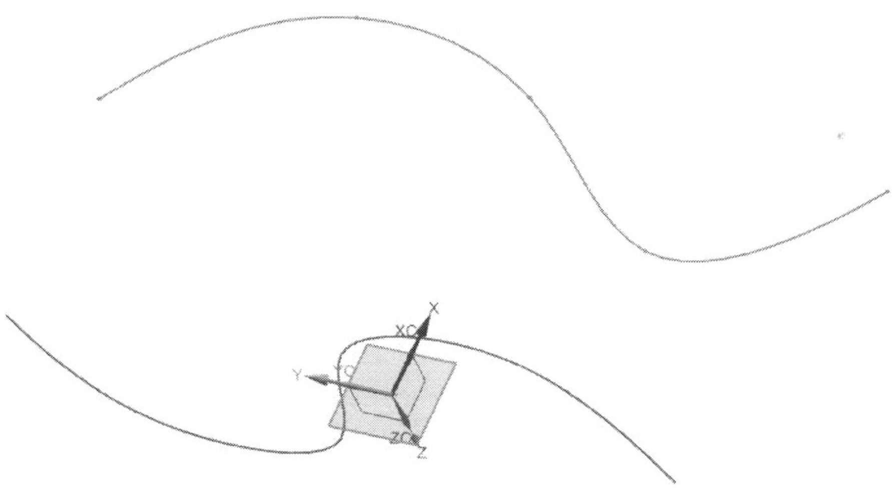

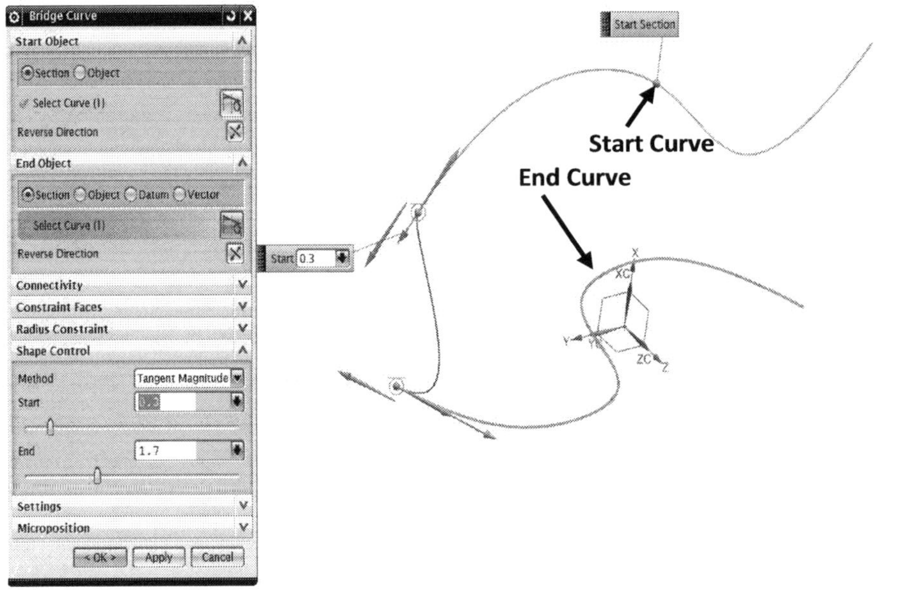


- Create the bridge curve to join the first two sketches using ***Insert / Curve from curves / Bridge Curve*** and select **OK**.
- Repeat the same process to join the other two ends.
- Select Insert / Mesh Surface / Through Curve Mesh.
- Select Primary Strings 1 and 2. Make sure that your arrows are pointing in the same direction.
- Select cross string 1 and cross string 2.
- Hit **OK**.

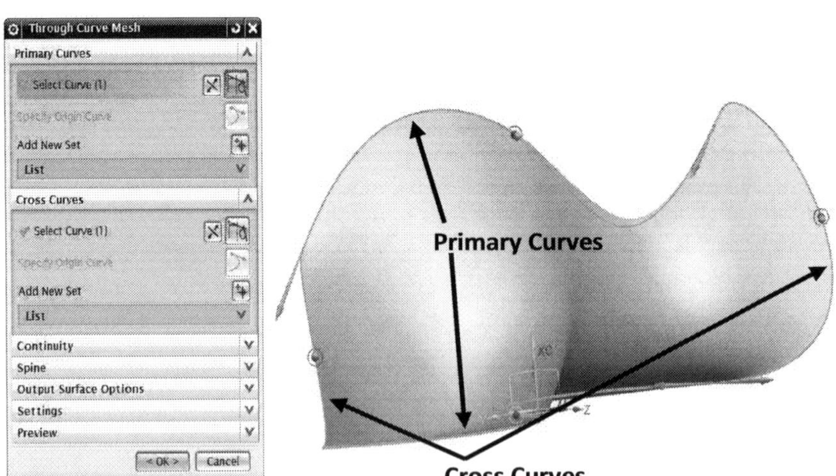

- Leave the options shown above. Notice the options for First, Last, Primary, and Cross Strings. When you create surface patches or are creating a surface adjoining another surface, you will need to define the constraint faces and other advanced options that are beyond the scope of this exercise.
- Hit **OK**.

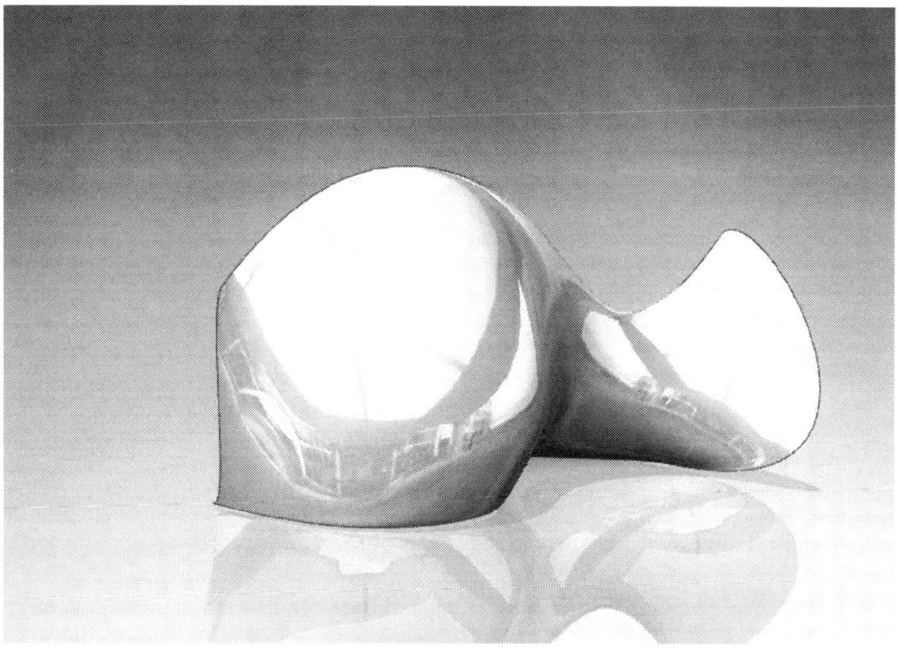

***End of Project***

## 15.7 N-Sided Surface Modeling Technique

The N-sided surface is a fantastic tool with many great techniques to control it. In the following exercise we are going to go through one of these techniques and discover how easy it can be. The following project will share a great technique to control an N-Sided Surface.

### *Project 15.7.1*

- Start by creating the following **sketch** of an **ellipse**. To get to the ellipse tool in the sketch environment simply click on **Insert / Curve / Ellipse**, then place the ellipse at the center of the datum CSYS as shown below.

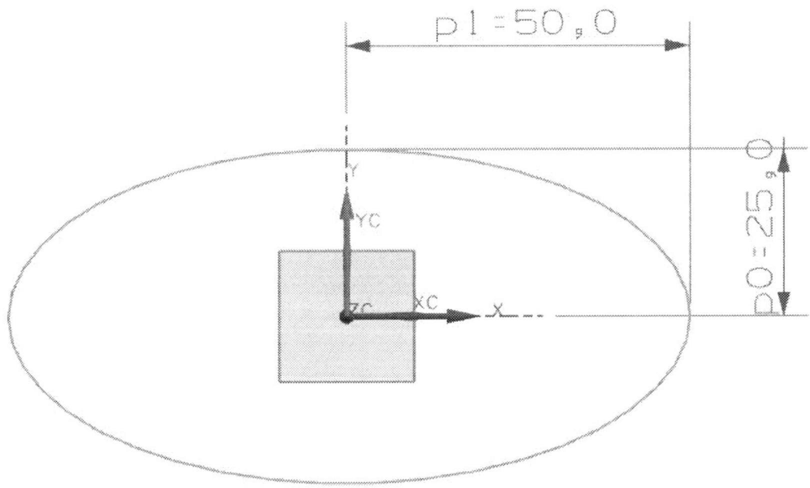


> **Note:** The trick to fully constraining an ellipse is to create a parallel constraint between an axis and the edge of the ellipse.

We want to end up with a domed surface that covers this ellipse. To help us control the shape of the dome, we are going to use an Extrude with a draft – where the draft angle will control the dome (N-sided Surface).

- Extrude the ellipse **10mm** in the –Z direction, include a **45 degree** draft, and choose **Sheet** as the body type, as shown in the following figure. This will create a sheet body instead of a solid.

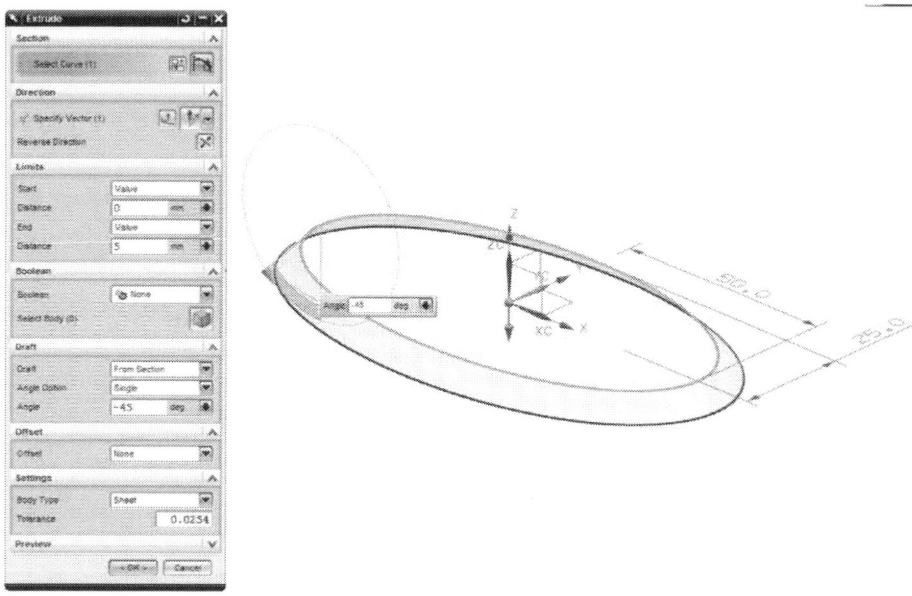

- You will now have a sheet body that looks like the following figure.

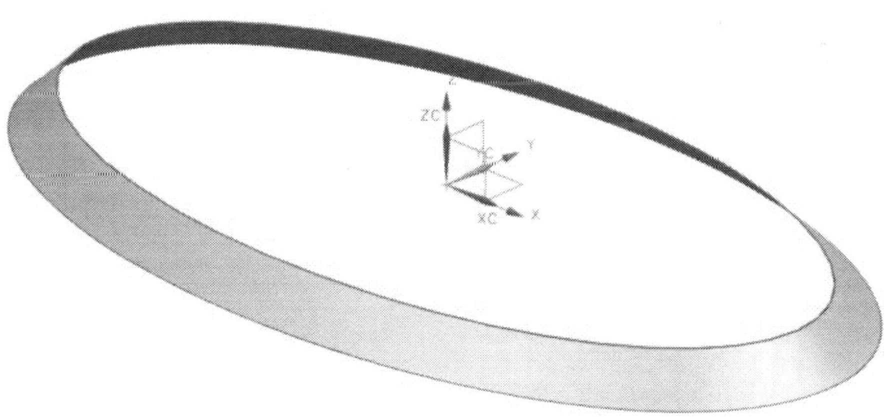

- Now select Insert / Mesh Surface / N-sided Surface 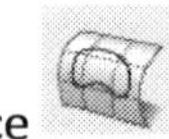 .
- Select the ellipse as the outer loop and the Extruded sheet as the constraint face, and finally check **Trim to Boundary** under settings, as shown below.

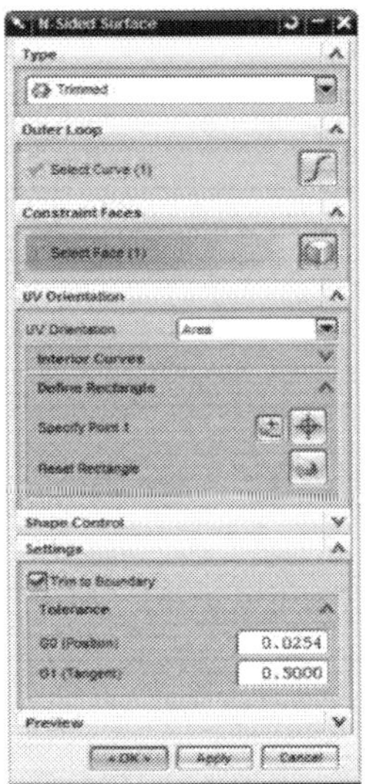


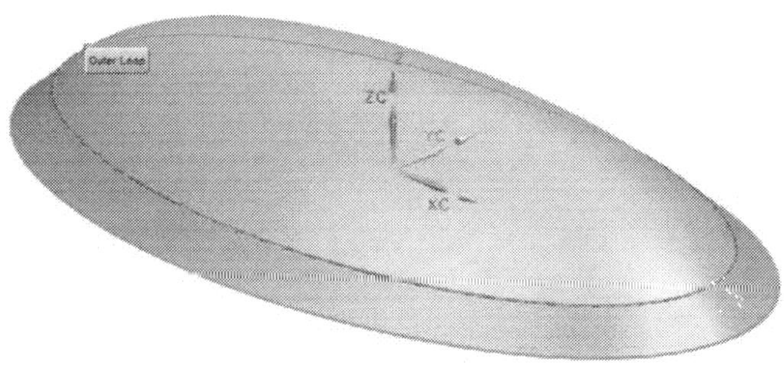


- Hide the Extrude and you will have your perfect dome shape – fit for any great consumer product. Now change the draft angle of the Extrude to 20 degrees and watch the change.

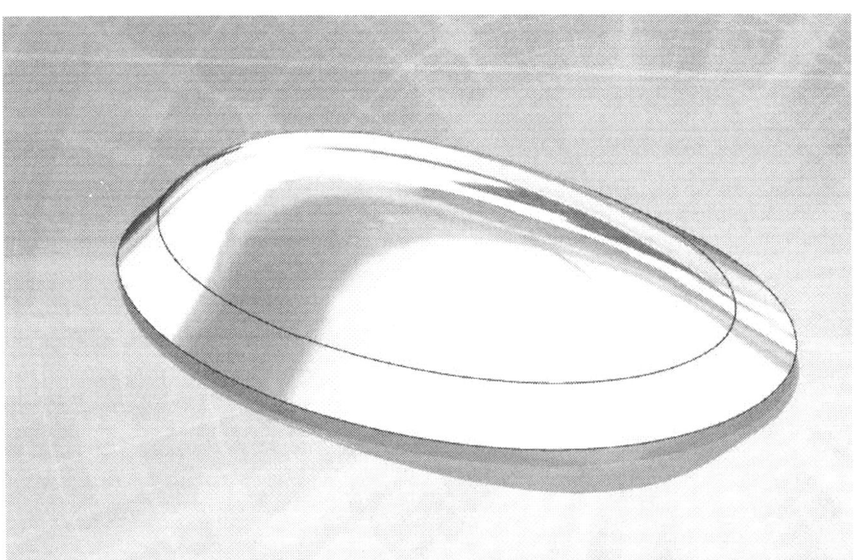

***End of Project***

## 15.8 More About Surface Techniques

It's amazing how good surfacing, coloring and texture can affect the look of a thing. As many industrial designers know, a product's form can attract and delight the end user. The form can capture the eye of a prospective customer, beacon them over to the shelf, or click on the link and input their credit card info then proceed to checkout. Product form can give a product owner a sense of pride as they are seen using the product by friends, family members or even total strangers. Form gives the user something that somehow appeals to an underlying sense of beauty and value. Although the function of a product is arguably more important than the form, and the function is somewhat inseparable from the form, it is instructive to consider the form as its own separate entity.

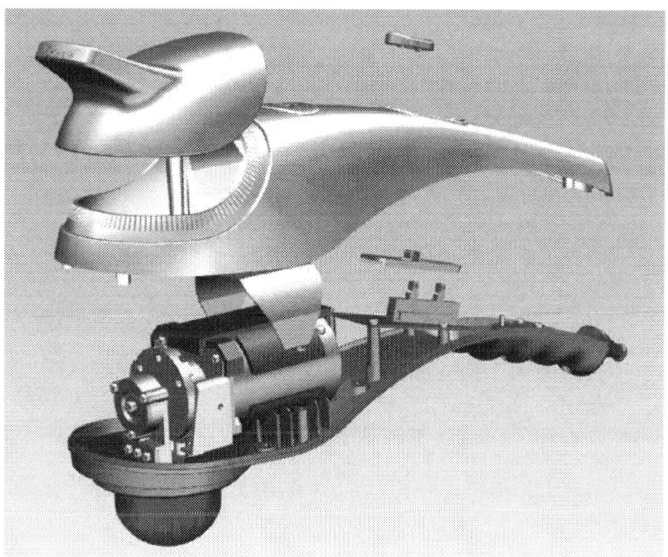

Exploded view of massager unit designed by Design Visionaries

Great Industrial design and the use of surfaces to achieve it is beauty and art. Unfortunately for the industrial designer trying to give a product a form that will help to sell it, art and beauty vary greatly from culture to culture, from age to age and from male to female. When defining a product that is supposed to sell worldwide, it is difficult to know what will be appealing to such a varied user community. That said, industrial designers who are trying to suit the tastes of a huge and varied audience benefit by considering what is universal. In NX8.5 there are many more ways of creating surfaces and many more options in those surface techniques than have already been covered. There are more than can be covered reasonably by this one book. However, here are a few more.

### *Project 15.8.1 Studio Surface*

The Studio Surface is an extremely powerful surface type. You create section curves and guide curves, assign continuity, adjust the output surface options and the settings and you will create a great surface. You get a preview as you select the various options which enable you to work very efficiently. It is the humble yet strong opinion of the author that regardless of the surface type employed, the design will only ever be as good as the imagination and skill of the creator. The imagination must come first. The CAD system can be a hindrance if the vision of what you are creating isn't clear before you start "cading". To that end it can be very useful to begin with a series of hand sketches. You can solve many problems with hand sketches way faster than with a CAD system.

In the example below, a device is required to extract thermometers out of acid bottles. The thermometers are longer than the diameter of the opening of the acid bottles. The thermometer has a bulb on it that is about .28 inches in diameter. The user must use the tool to grasp the bulb loosely but securely. As the thermometer is lifted it rotates in such a way that it will align with the tool and come out of the 2 inch hole. The tool has a button on top which allows the user to tighten and loosen the grip on the bulb. We can begin this task with a hand sketch.

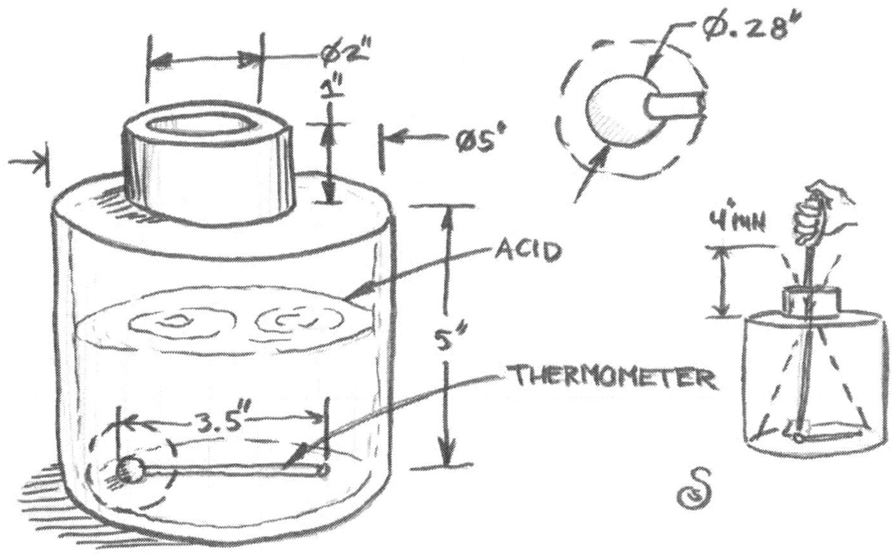

**Hand sketch of the problem**

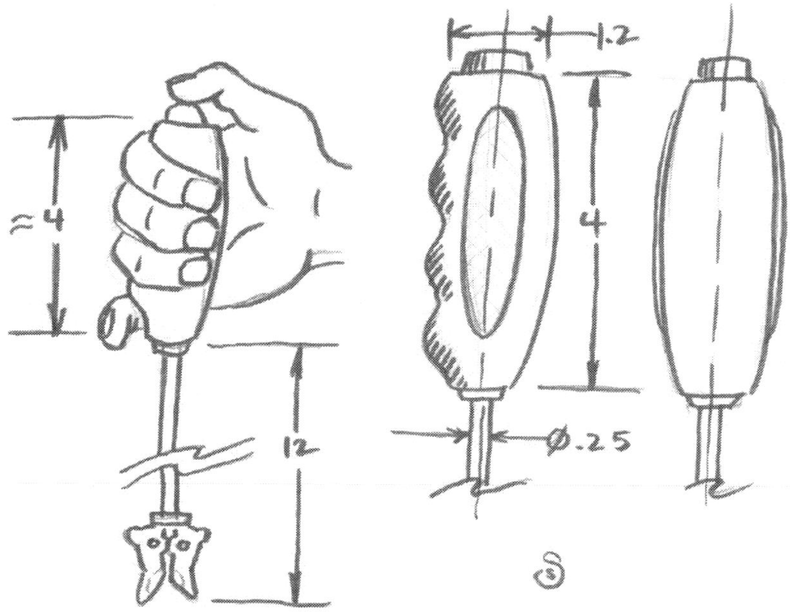


**Hand sketch the solution**

Once you've created the hand sketch, you can bring it into NX8.5 using the "**decal**" command. In this case and for this product a handle is needed. The shape of handles is extremely important because they provide both look and feel of a product. They are the most important interface that the user has with the product.

- To begin this exercise, create a hand sketch and scan it. You may use the scanned image in the example pack. It's called "hand sketches.JPG".
- Next create a bounded plane with the same aspect ratio as your scanned image. The aspect ratio can be found by bringing the image into a program like "Paint" and selecting on the lower left hand corner. The height and width will be displayed in the lower left hand corner. When you make your surface select dimensions that have the same ratio. Make sure the surface is large enough to fit a sketch of the entire product surface. For example. The product in this exercise is 4 inches long. The size of the jpeg is 483 by 561 or 1.1614. So make a surface that is 4 inches wide by 4 times 1.1614 (4.645) long.
- Next Make sure your view display is set to Studio  and choose **View/Visualizaton/Decal**
- Browse for the picture. It can be a JPEG or a Tiff. A preview will appear.
- Select the plane.
- Set the origin to "top left" and select the top left hand corner of the plane.

- You will have to play around with the **Scale** and the **Aspect Ratio** slider.

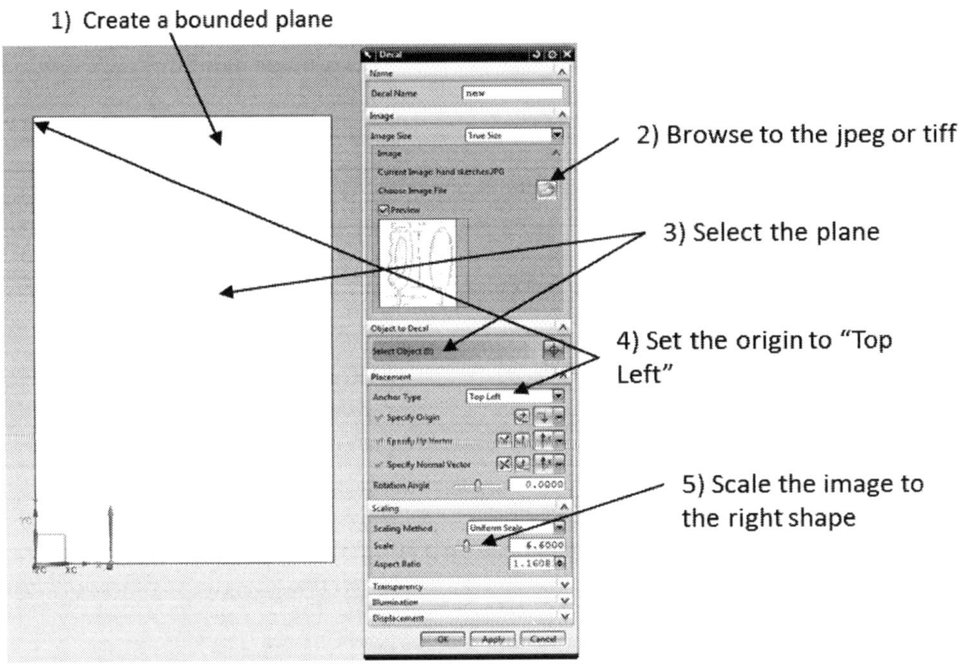


- Once you have placed the decal on the plane and scaled it using the scale button and the aspect ratio. You can start sketching over the image.

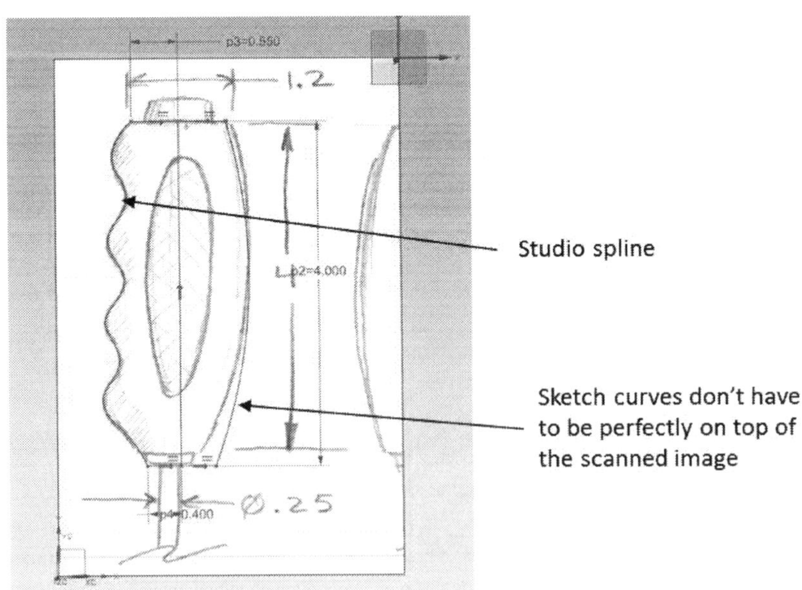

- Create a datum CSYS in the middle of the first sketch and create the sketch as shown on an orthogonal plane.

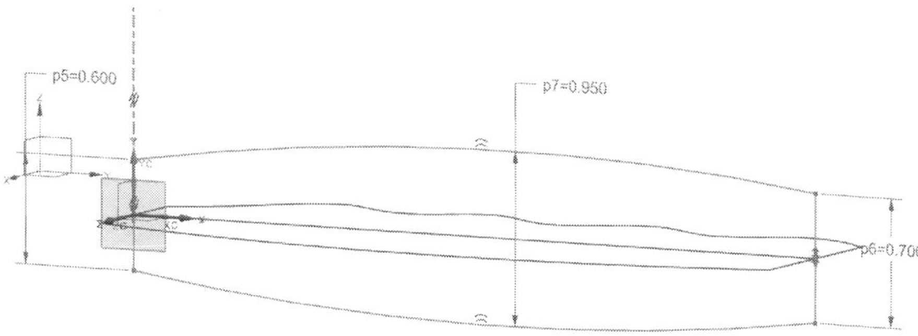


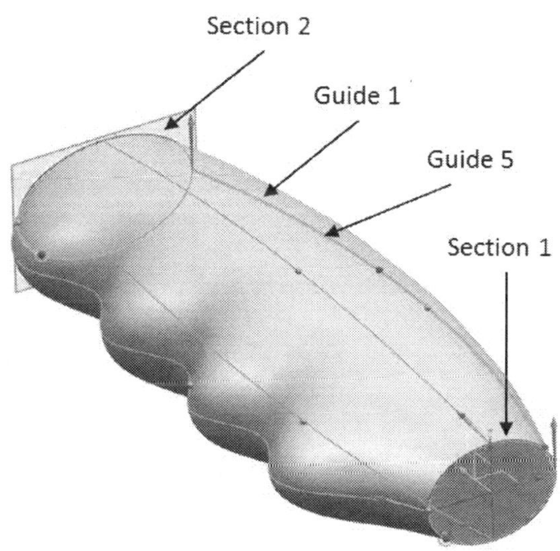


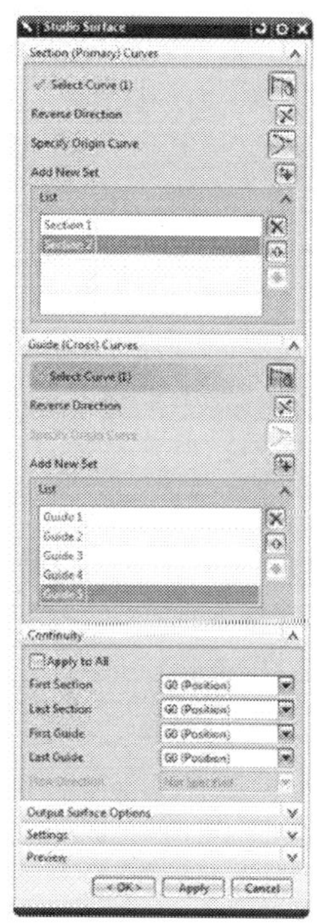

Next create an offset plane and sketch the raised area profile on a flat plane. The profile is projected on the face. Then divide the face at the projected profile location using **Insert / Trim / Divide Face**. Once the face is divided use Insert / Offset Scale / Thicken to raise the face by .1

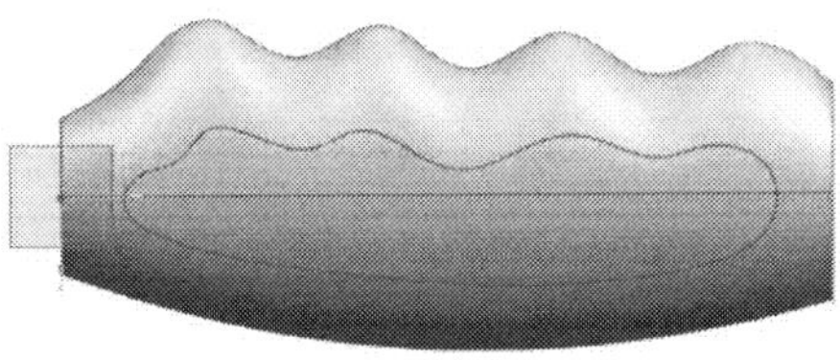
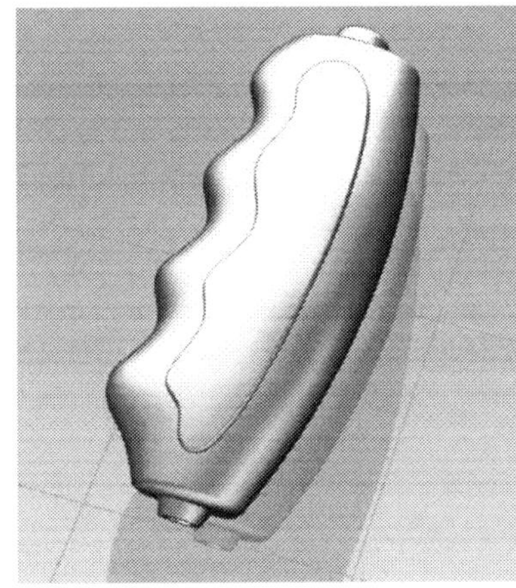

Assemble the rest of the pieces, get a prototype, impress the customer and collect your check.

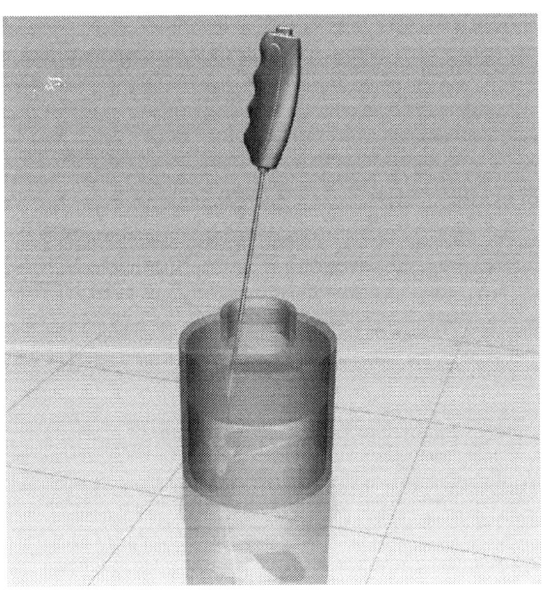

***End of Project***

## 15.9 Self-Test

(All questions are true/false. Answers are at the back of the book.)

1. You may create a 'Surface Feature' with the 'Through Curves' option by using sections containing different numbers of curves.
2. You may create a 'Swept Surface' feature using more than one drive curve.
3. You can create Solid bodies using sweep commands.

# 16. Basic Assemblies

## 16.1 Introduction

There are two major ways to assemble: by combining existing parts together ('Bottom-Up' Design) or by creating new parts on-the-fly in an assembly file ('Top-Down' Design).

In NX 8.5, there is no distinction between part databases and assembly databases, nor is there a separate drawing database. Thus, any part file can become an assembly, a drawing, or both! We suggest that to avoid confusion, you choose a good naming convention to help distinguish between parts, assemblies, and drawings.

## 16.2 Bottom-Up Assembly Design

To create Bottom-Up assemblies, first select ***Start / Assemblies*** to open the Assembly toolbar window. Then you can simply assemble pre-existing part databases (files) using ***Assemblies / Components / Add Component***

During the assembly process, NX 8.5 displays the component in a temporary Work View. You can either orient the part manually or constrain it to other components. To help visualize the assembly process, you may Rotate/Zoom/Pan in both the main Graphics Window and the component Work View at any time.

### *Detail: Load Options & Assembly File Management*

Since assembly components are stored in separate files, managing the data becomes a little more complex. You must consider modifications to each part as well as to the assembly. If you save the Top-Level assembly, NX 8.5 will save all of its modified components – which may lead to versioning problems. The three Load Options are 'From Folder,' 'As Saved,' and 'From Search Folders.' All three are accessed from the ***File / Options / Assembly Load Options*** dialog box. You can also define these options by clicking '***Options***' located in the bottom left corner of the 'open part file' while opening the assembly file.

In order to fully load an assembly, NX 8.5 must be able to load all of its component parts. An easy way to accomplish this is to store all files in the same directory and to set the Load Method to 'From Folder.'

> **Note:** NX 8.5 will not search in embedded folders when this option is chosen; therefore, the directory with the assembly file must also contain all of the components.

For the forward-thinking, there is the "As Saved' option. This requires that each component of an assembly remain in its original saved location.

If component part files are moved or stored in different directories, the 'From Search Folders' option provides the most flexibility. Select 'Add Folder to Search' to bring up an expanded menu that allows you to create one or more Search Directories. Any added search directories are only effective in the current session of NX 8.5. To make your changes permanent, choose the 'Save' option.

> **Note:** To search all subdirectories of a specified directory, use an ellipsis (three periods) "..." at the end of the chosen directory pathname. Otherwise, NX 8.5 only searches in the specified directory.

## ***Detail: Reference Sets***

In NX 8.5, a 'Reference Set' is a "named group" of geometric items that is used to control the display of assembly components. Reference Sets are defined in the part database but are not used until that part is added to an assembly. A common application is to filter out all of the construction geometry required to create a complicated part (datum planes, sketches, tool solids, etc.) and only display the resultant solid model. There are three pre-defined Reference Sets: 'Model,' 'Entire Part,' and 'Empty.' All other Reference Sets are user-defined.

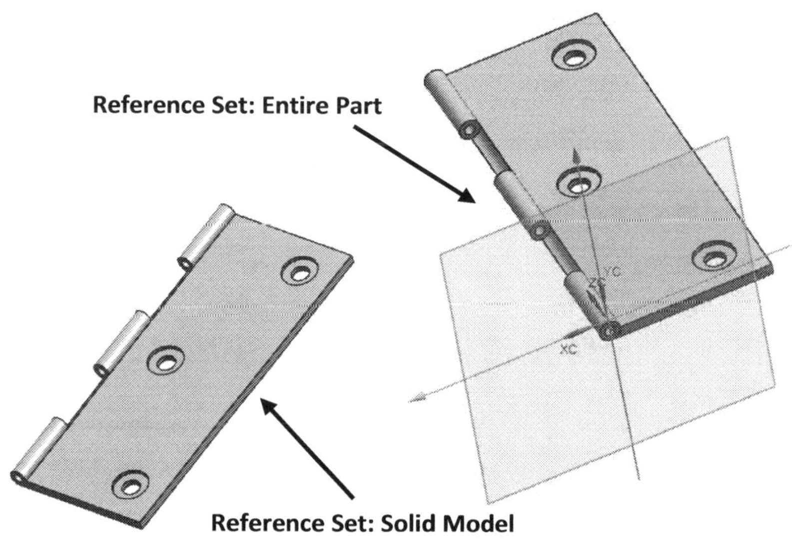


You may recall that in order to create the solid model of the hinge part, you created several datum planes and sketches. In many cases you can choose 'Model' reference set, which mainly consists of the solid or sheet body, but for different uses you can define different reference sets, especially in the case of parts with multiple solid bodies.

You can define a reference set to display only the desired solid geometry in an assembly. A component's displayed Reference Set may be easily changed (using ***Assemblies / Replace Reference Set***).

In the following project, you will first create Reference Sets in your hinge1 and hinge2 parts from Project 6.7 and assemble them together using Bottom-Up techniques.

> ***Note:*** *If you do not have the files available,* http://designvisionaries.com/goodies/workfiles/ *contains the completed parts for your reference.*

## *Project 16.2.1*

- Open the hinge1 part that you created in Chapter 6.
- Select Start / Assemblies.
- Select Format / Reference Sets.

The 'Reference Sets' dialog box appears as follows.

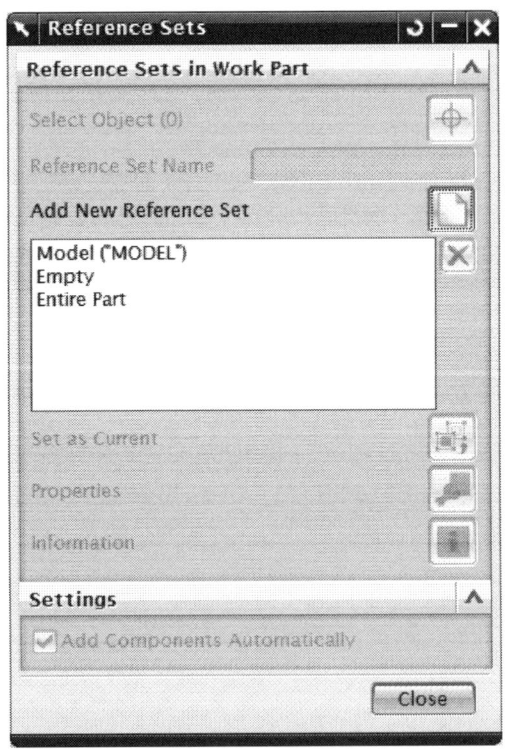


- Select the Add New Reference Set icon.

- Type SOLID and Hit Enter.
- Select the hinge solid body (be careful to not select any other geometry).
- Middle click to OK.

You have now successfully defined a reference set named SOLID. It should appear in the list displayed in the 'Reference Sets' dialog box.

- Choose **Close.**
- Repeat this procedure and create a **SOLID** Reference Set in your hinge2 part.
- **Save** both parts.

Now that you have successfully defined all of your reference sets, you are ready to build your hinge assembly.

- Select File / New. Name it hinge_assembly_<your initials>.prt.
- The ***Add Component*** window should open automatically.
- Select Open.
- Navigate to your hinge1 part, select it, and choose OK.
- Set the options as shown in the previous figure.
- Choose OK.

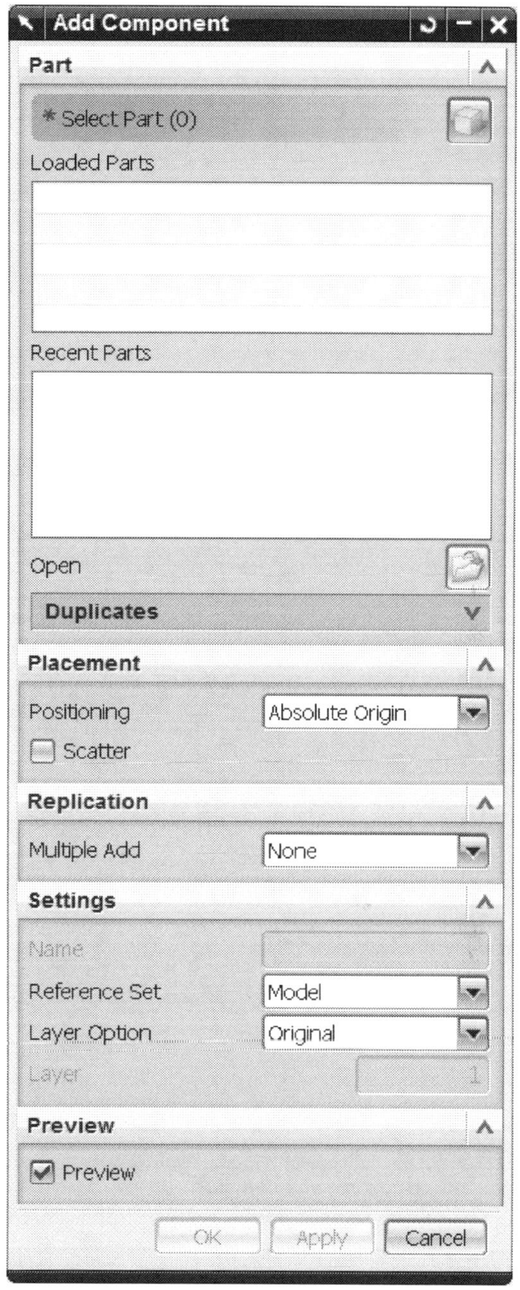


Your hinge1 part has been added to your hinge assembly. Rotate the view until it appears as shown in the following figure.

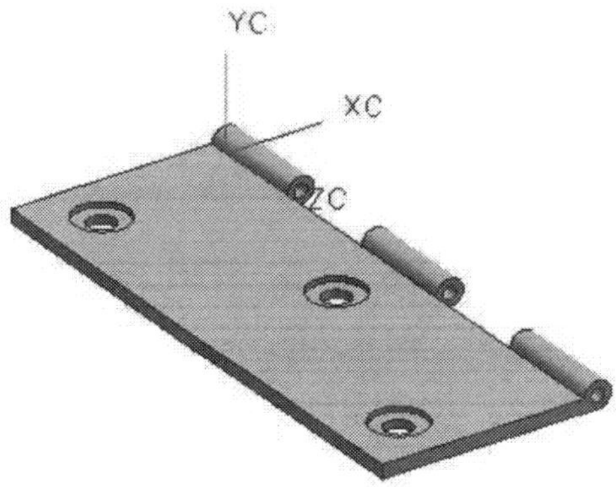


- Save your assembly file.

***End of Project***

## ***Detail: Positioning components***

Components may be repositioned in many degrees of freedom unless they are parametrically constrained by applying 'Assembly Constraints.' Normally, three Constraints are required to fully constrain a component, although you may apply fewer and allow the component to be under-constrained.

You are not required to apply any Assembly Constraints to the first part of an assembly but it sure helps. If all components are fully constrained, the positions of all the components will update when the first component is repositioned.

## ***Detail: Assembly Constraints***

To fully constrain a component, you must specify a set of Assembly Constraints. Each constraint removes one or more degrees of freedom from the component's movement. As an example, consider the 'Touch/Align' constraint, which attaches two components together. It removes the ability of one component to move independently of the other. Some of the most commonly used Assembly Constraints follow:

- **Touch** - Entities coincident; surface normal vectors in opposite directions
- **Align** - Entities coplanar; surface normal vectors
- **Infer Center Axis** - Aligns the center axis of cylindrical parts
- **Concentric** - Make circular edges coincident

- **Fix** - Hard fixes a component in the position it is in according to the WCS
- **Angle** - Entities oriented with an included angle
- **Bond** - Fixes components together to act as one
- **Parallel** - Entities constrained to be parallel
- **Perpendicular** - Entities constrained to be perpendicular
- **Center** - Entities centered
- **Distance** - Entities parallel with a specified offset
- **Fit** - Constrain two parts to fit together
- **Prefer Touch** - It is like Infer. NX guesses the direction it should mate/touch with the other assembly components

The order in which components are selected when applying Assembly Constraints is important. If you wish hinge2 to inherit its position from hinge1, the first face selected must be from the "follower" component, not the "leader" component.

| **Note:** You can mate faces, edges, datum planes, datum axes, points, lines and curves. |
|---|

## *Project 16.2.2*

Continue working in your hinge assembly, and add another part to it using Assembly Constraints to constrain it parametrically.

- Open your hinge_assembly part (if required) and select ***Start / Assemblies.***
- Select Assemblies / Components / Add Component or select the  icon.
- Select Choose Part File.
- Navigate to your hinge2 part, select it, and then choose **OK.**

You will see your hinge2 part displayed in the 'Component Preview' area in the Graphics Window. Move your mouse into this window and Rotate, Pan, and Zoom until your model appears as shown below.

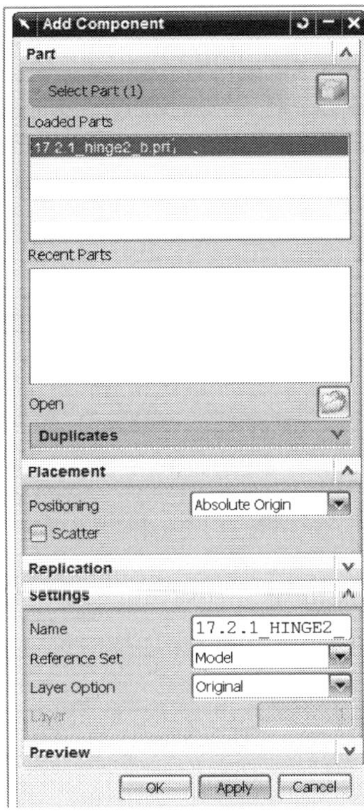


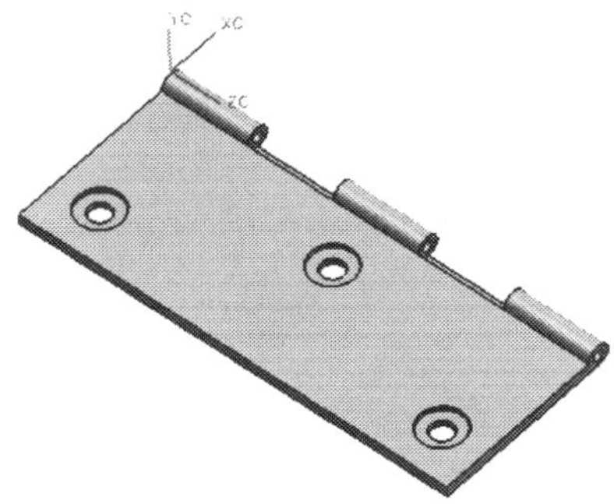

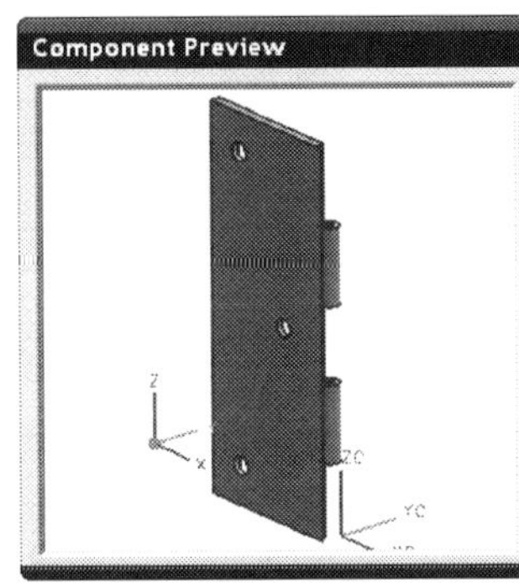


- Select Reference Set: **SOLID**, Positioning: **By Constraints**, and Layer options: **Work** from the 'Add Existing Part' dialog box, as shown.
- Choose **OK.**

The **Assembly Constraints** dialog box appears, as shown below.

Read the Cue – NX 8.5 is prompting you to select your first object.

- Select the Type as: **Touch**  , shown above.
- Zoom up on hinge2 and select Face 1 of hinge2 as shown below.
- Now, select Face 2 of hinge1 as shown.

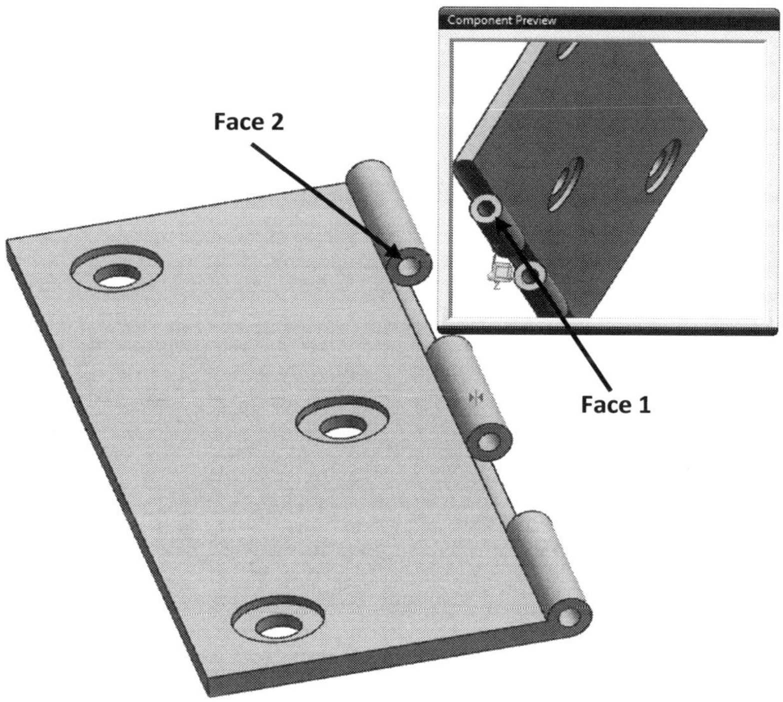


The two parts should now be touching in the graphics window. Click **Cancel** to exit constraint mode.

- Go to ***Assemblies / Component Position / Move Component***  , click on hinge2, and select "Specify Orientation." A dynamic CSYS should appear. Clicking and dragging the arrows of the CSYS tells you exactly what degrees of freedom hinge2 still has. *You will learn a bit more about moving components later on.*
- Add a new constraint. In the Assembly Constraint window, the **Type** should be 'Touch/Align' and the **Orientation** should be "Infer Center/Axis.'
- Hover over the edge of the leftmost cylinder of hinge1, shown next with an arrow. An axis should appear. Select this axis, called the "Centerline in hinge1.'

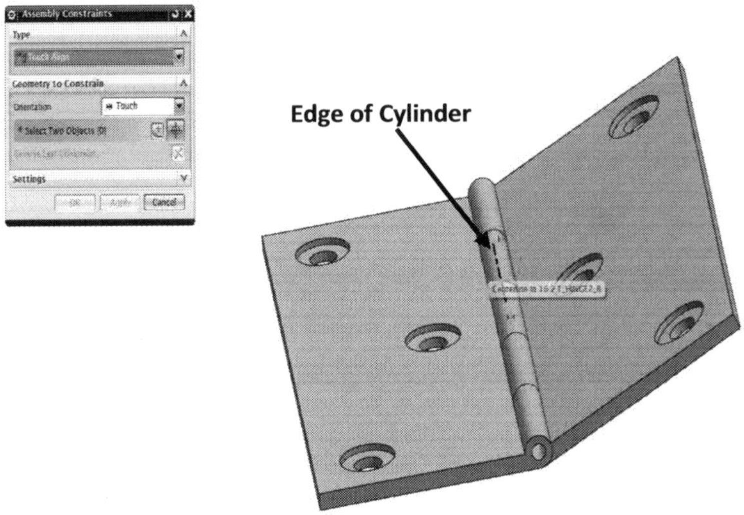


- Use the same method to select the 'Centerline in hinge2.'
- Choose Cancel to finish assembling new components.
- Compare your results to the following figure. If you **Move Components** again, you should see the hinge properly constrained.

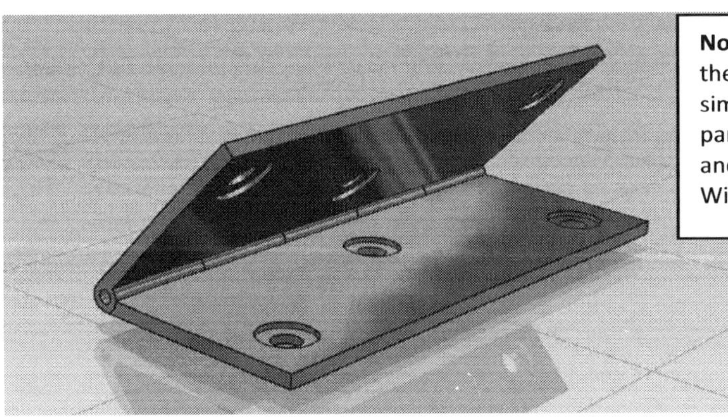

> **Note:** You can also initiate the assembly process by simply dragging an existing part straight from its folder and onto the Graphics Window of an open part.

- Save your assembly file.

There are many different constraints that can be used to position parts in assemblies. This project only covers a few of them. Play around with the assembly constraints to see what other cool tricks you can do in assemblies!

***End of Project***

## 16.3 Top-Down Assembly Design

To perform Top-Down assembly design in NX 8.5, first enter Assembly Mode by selecting ***Start / Assemblies.*** Create new parts using ***Assemblies / Components / Create New Component***. Once you have a new empty component, you can proceed to add geometric features to it. Note that when you create a component in this manner, it inherits its units from the parent assembly. If you wish to use different units, you must create it outside the assembly and add it using Bottom-Up assembly design techniques.

The ability to create components directly within an assembly is a very powerful tool. To grasp this concept fully, you must first understand the concept of 'Work Part' and 'Displayed Part.'

### ***Detail: Work Part vs. Displayed Part***

The 'Displayed Part' is the part that is currently displayed in the Graphics Window. The 'Work Part' is the part to which any command is applied (i.e. ***Insert / Design Feature***).

> **Note:** The Displayed Part is always the work part, but the work part is not always the only displayed part.

### *Project 16.3.1*

In this project, you will create a new hinge pin component in your hinge_assembly.prt and define its geometry using Top-Down design techniques.

- Open your hinge_assembly part (if required) and select ***Start / Assemblies.***
- Select Assemblies / Components / Create New Component.
- Type in the new model file name: ***hinge_pin_<your initials>.prt***.
- Choose **OK.**

The 'Create New Component' dialog box appears.

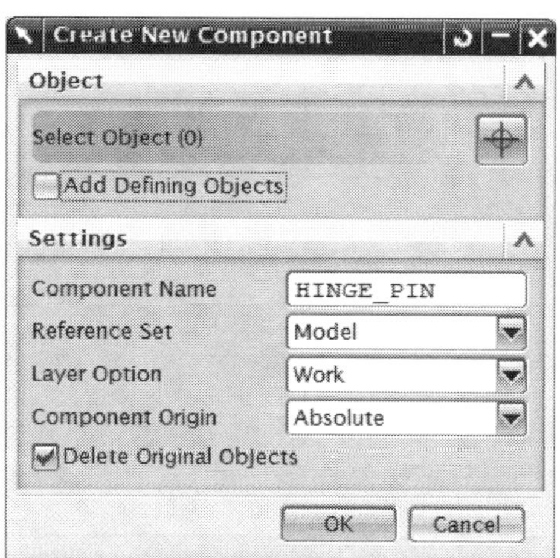

- Set up your new component as shown above.
- **Right-click** on hinge_pin in the Assembly Navigator and select ***Make Work Part***.
- The hinge parts will change their color to indicate that they are not the Work Part.
- Insert a sketch on the YC-ZC Datum Plane.
- Create the sketch shown next.

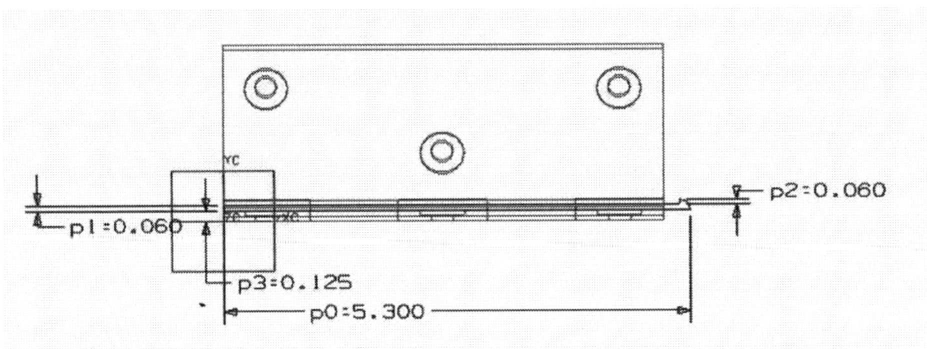


- Create the hinge pin solid by revolving the sketch 360° about the hinge center axis.

Your pin may not be in exactly the right spot. We'll get to the positioning a few sections from here.

- Compare your results to the figure below. Note that your hinge_pin part is displayed in the normal system colors since it is currently the Work Part.

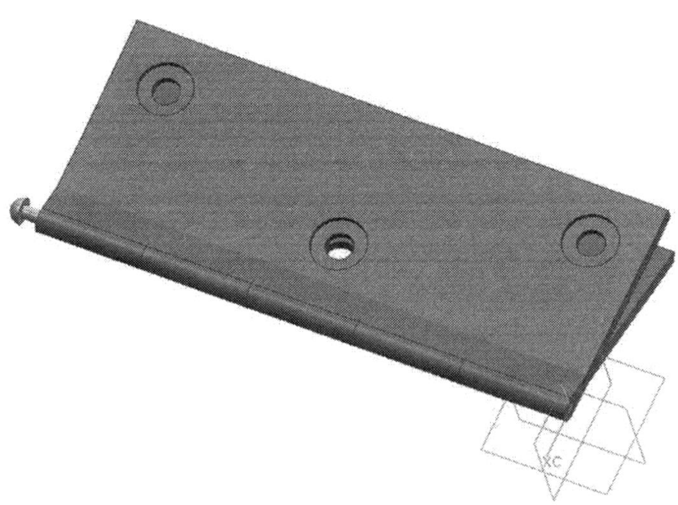

- Make the hinge_assembly the work part. Compare your results to the figure below.

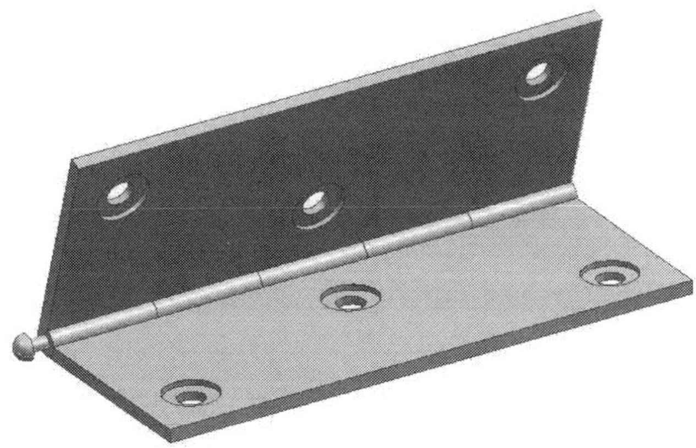

- Save your assembly file.

***End of Project***

## 16.4 Editing Assembly Components

### *Detail: Changing the Color of a Component*

To change the color of a component, select ***Edit / Object Display*** or ***Ctrl+J***, just as in modeling mode. There is one difference – when you set the color of an assembly component, it does not change the color of the original part (in other words, the component color is context-sensitive).

### *Detail: Moving Components*

You can reposition any assembly components that have not been fully constrained. When you move a component, NX will constrain the component's motion according to any existing constraints.

### *Project 16.4.1*

Change the color of your components and reposition the hinge2 part.

- Open your hinge_assembly part (if required) and select ***Start / Assemblies.***
- Select ***Edit / Object Display.***
- Select your hinge_pin part and choose **OK.**
- Change the color of hinge_pin and choose **OK.**
- Change the color of your hinge2 part.
- Select Assemblies / Components / Move Component.
- Select your hinge_pin part and click on 'Specify Orientation.'

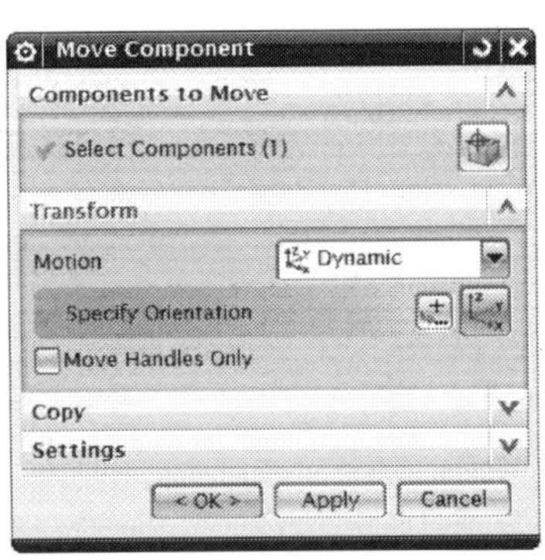


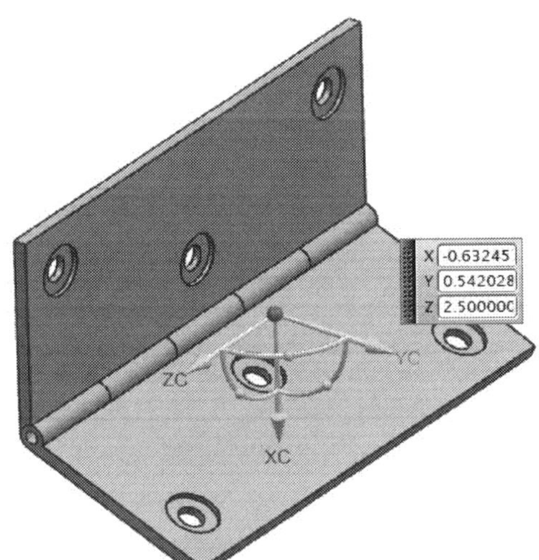


*If you select one of the arrows, the component will be constrained to move along that axis only.*

- **Left click** and drag hinge2. Notice that it will move around only the cylindrical axis. Drag the hinge into the position shown below and choose **OK**.

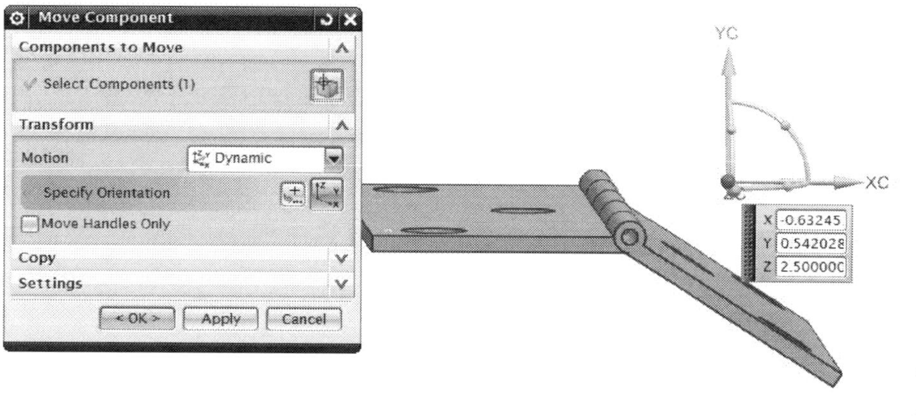


- Save your assembly file.

***End of Project***

## ***Detail: Adding Assembly Constraints to a Component***

You can edit the constraints of components that are not yet fully constrained. This allows you to have components that are partially constrained and then fully constrain them when you have proceeded further into the design process.

### *Project 16.4.2*

In this project you will fully constrain your hinge_pin component by adding one 'Touch' and one 'Infer Center/Axis' constraint between the hinge pin and the hinges.

- Open your hinge_assembly part and select ***Start / Assemblies.***
- Select Assemblies / Component Position / Assembly Constraints .
- Select the Type icon: **Touch Align** .
- Select the faces from your hinge_pin and hinge2 parts as shown below.

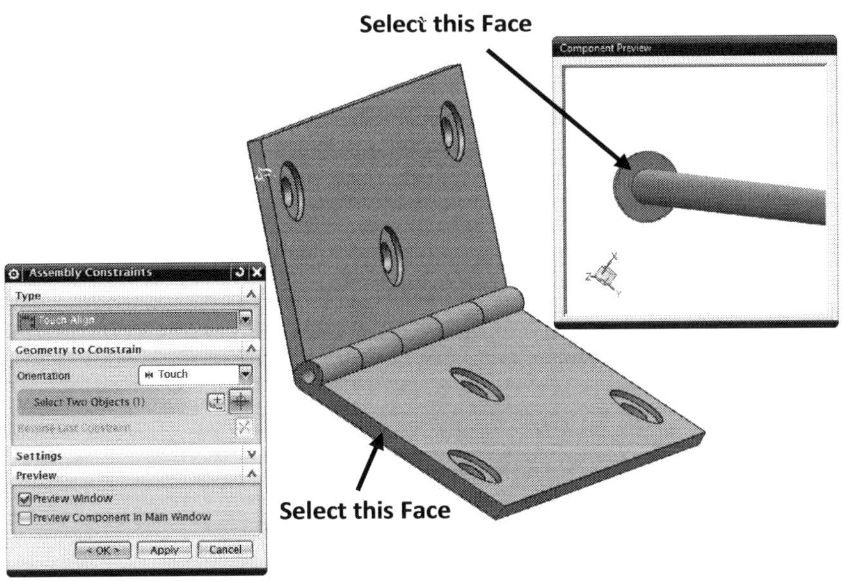


- Select the Type Orientation icon: **Infer Center/Axis** .
- Select the centerlines in your hinge_pin and hinge2 parts.

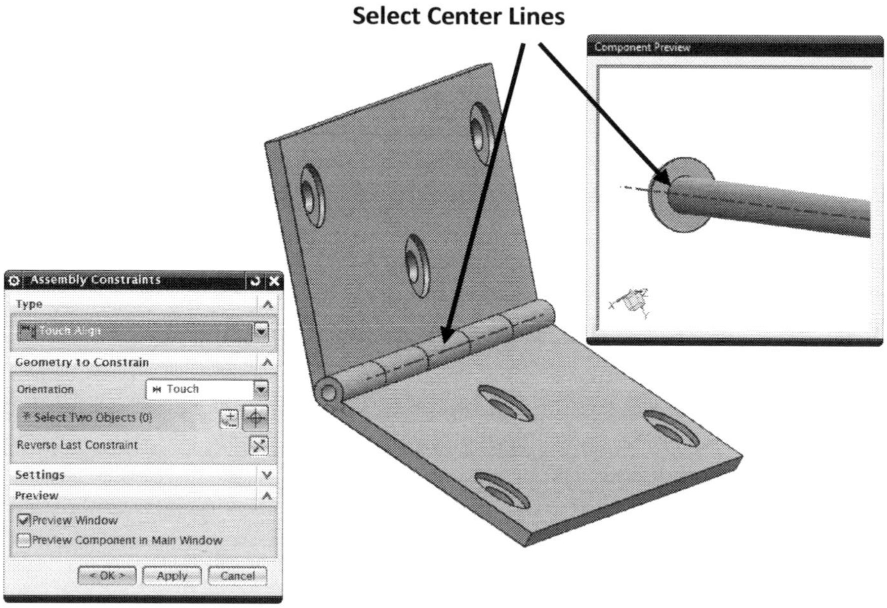


- Choose **OK** to finish specifying constraints.
- Compare your results to the figure below.

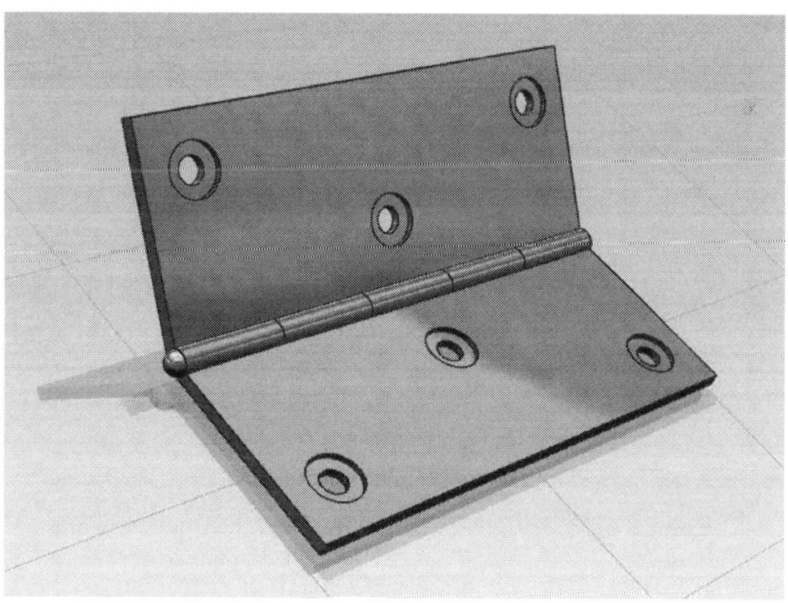

- Save your assembly file.

***End of Project***

## ***Detail: Creating Variable Component Dimensions***

You have learned that you can move a component in any degree-of-freedom in which it is not constrained. In some cases, you may wish to further constrain this movement and vary it using a parameter. This can be accomplished by adding Assembly Constraints that have offset parameters.

## *Project 16.4.3*

In this project you will add an angular dimensional constraint between the two hinge parts and vary its value.

- Open your hinge_assembly part (if required) and select ***Start / Assemblies.***
- Select Assemblies / Component Position / Assembly Constraints.
- Select the **Type** as 'Angle.'
- Select the faces from your hinge1 and hinge2 parts as shown below.

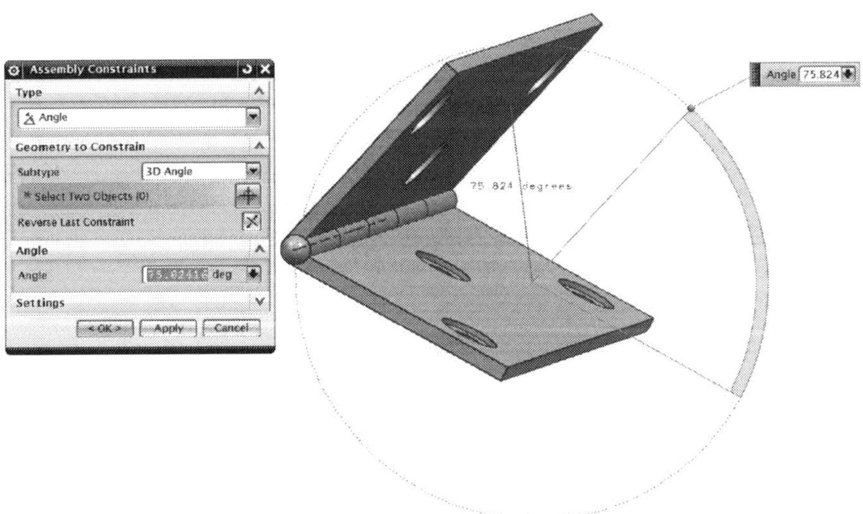


- Enter an angle of **150**.
- Choose **OK**.

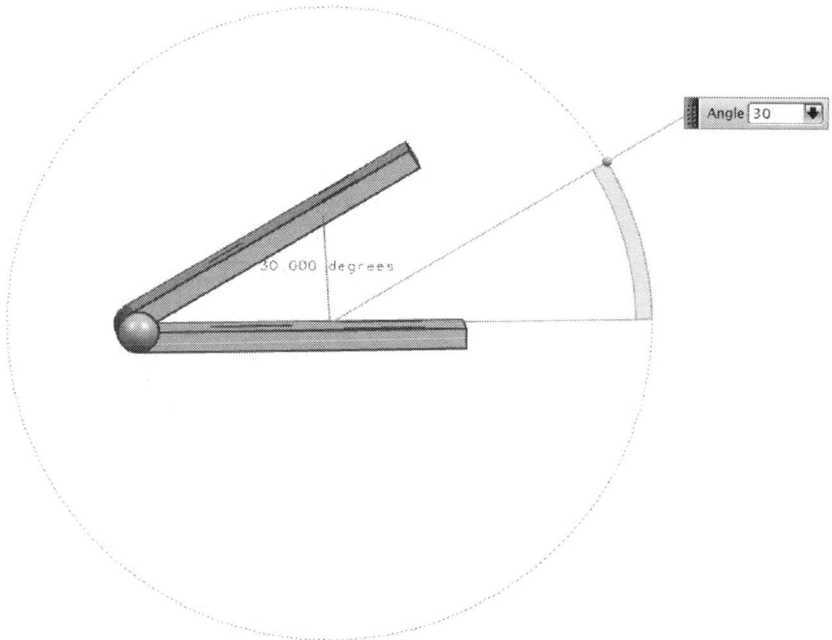


- Compare your results to figure above. In some cases, you may need to change the angle value from 150° to 30° to achieve this result.
- Save your assembly file.

> **Note:** You may need to use negative values for the angles, depending on your configuration.

***End of Project***

## 16.5 Modifying Assembly Components

NX allows you to modify assembly components in many ways. You have already learned how to Move components and to add Assembly Constraints to existing components. You may also modify existing Assembly Constraints by deleting them, suppressing them, or changing their type. Components may themselves be deleted, suppressed, or modified. Some of the commonly-used techniques are covered in the next few paragraphs.

### *Detail: Deleting and Suppressing Components*

Assembly components may be deleted or suppressed. Just as in modeling mode, a 'Suppressed' component may be 'Unsuppressed,' but a deleted component is gone forever. A suppressed component will retain its Assembly Constraints, so you may choose to first suppress a component until you are certain that it no longer will be used in the assembly. It is important to note that deleted components themselves are not removed from the hard drive – they are simply deleted from the assembly file.

To suppress a component, select ***Assemblies / Component / Suppress Component***. Similarly, to Unsuppress a component, select ***Assemblies / Component / Unsuppress Component***. You can delete a component by selecting ***Edit / Delete*** and then choosing a component. The Assembly Navigator functionality also allows you to delete a component by **right-clicking**.

### *Detail: Suppress and Deleting Assembly Constraints*

You can suppress an Assembly Constraint by unchecking the box next to the constraint, as shown below. The system remembers the suppressed constraint, so you can restore it by simply checking the box again.

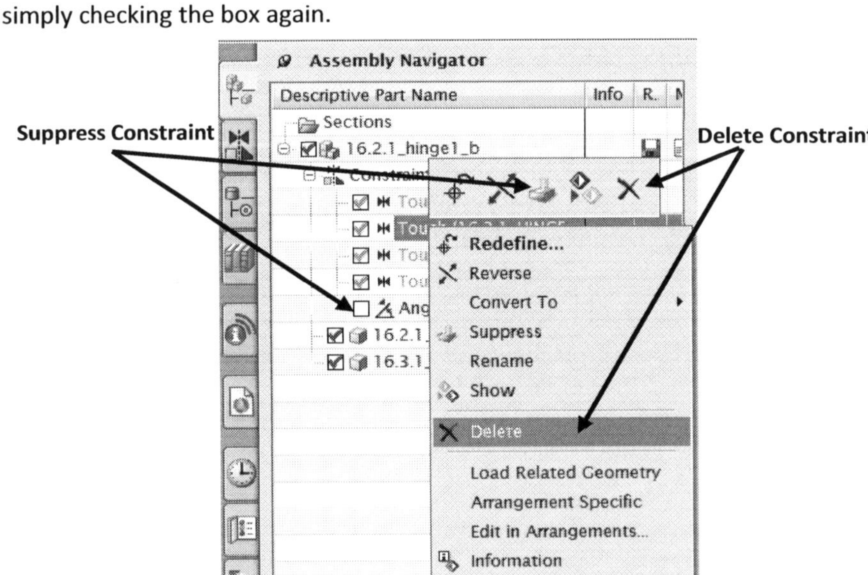


If you are certain that you want to completely remove an Assembly Constraint, you may delete it by **right-clicking** on the constraint and selecting 'Delete,' as shown above.

## ***Detail: Changing Properties of a Constraint***

You can change the property of a constraint by selecting the ***Right Click / Convert To*** option as shown below. For example, you can convert a 'Touch' constraint into a 'Distance' constraint, thus enabling you to specify an offset value for the constraint.

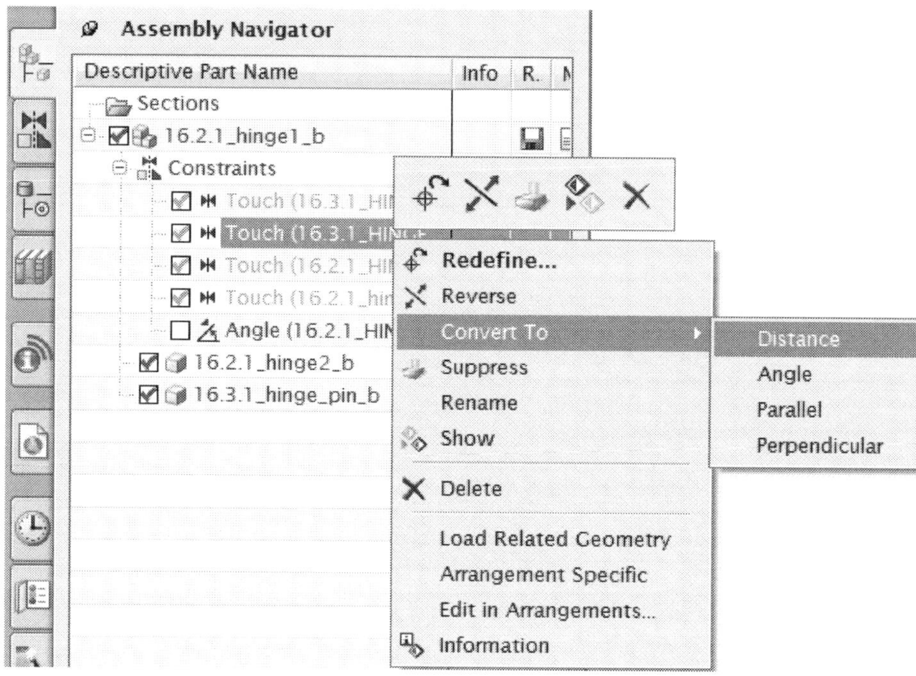


## ***Detail: Editing Part Features in an Assembly***

You can edit any part's feature in the assembly by making it the Work Part. When a component is set as the Work Part, you have access to most of the NX 8 modeling functionality, with the added benefit of seeing the part within the context of its parent assembly. This process is known as Design-In-Context.

## *Project 16.5.1*

In this project you will adjust the length of the hinge pin in the assembly, using Design-In-Context.

- Open your hinge_assembly part and select ***Start / Assemblies*** (if required).
- Set your hinge_pin part as the Work Part.
- Select ***Start / Modeling.***
- Select Edit / Sketch.

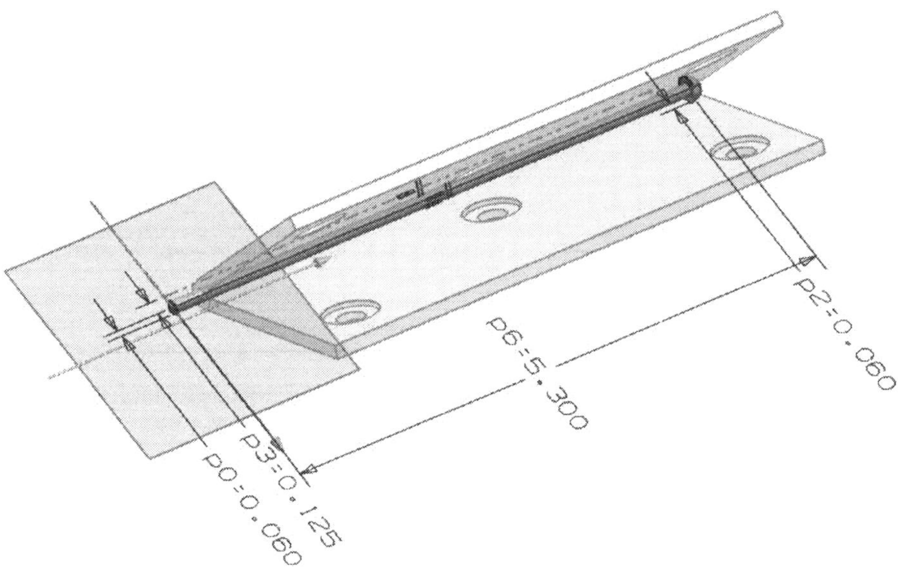


- The figure above shows the original sketch. Edit the sketch to make the pin **5.00in** long from the end of the pin to the shoulder of the pin head.

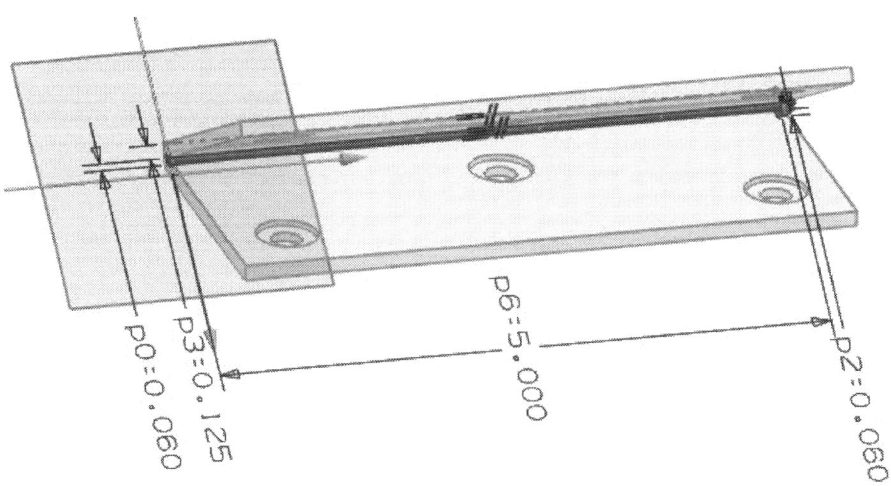

- Compare your results with the figure below.

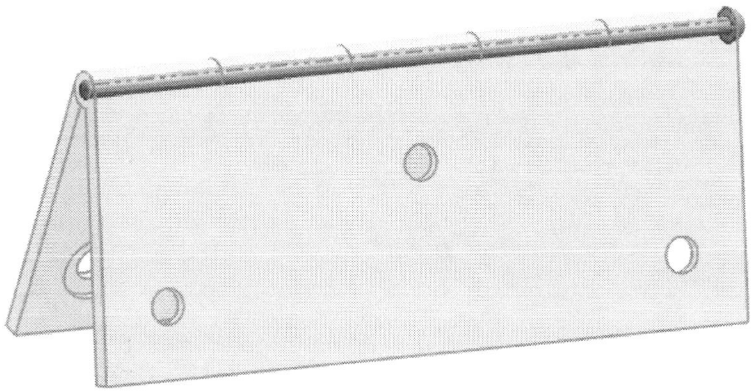

- Make the assembly file the work part.
- Save your assembly file.

***End of Project***

## ***Detail: Changing the Reference Set***

By assigning a Reference Set to a part, you can load only the features of a part that you need into the assembly instead of the entire part. NX 8.5 gives you an option to use Reference Sets when you add an existing component to the assembly. Most of the time, using the Model Reference Set is sufficient unless you have solid / surface bodies that you don't want to load in the assembly. You can also create and replace Reference Sets after a part is assembled.

## *Project 16.5.2*

- Open your hinge_assembly part and select ***Start / Assemblies*** (if required).
- Set your hinge_pin part as the Work Part.
- Select Format / Reference Sets.
- Select the **Create** icon from the 'Reference Sets' dialog box.
- Enter the name: **SOLID,** and choose only the revolved solid of the pin.
- Press **Enter.**

The Reference Set named **SOLID** will be added to the 'Reference Sets' dialog box, as shown below.

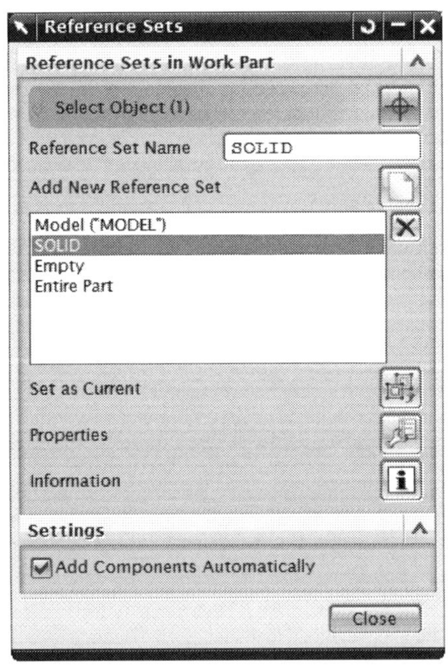


- Choose **Close**.
- **Right-click** on the hinge_pin part in the Assembly Navigator and select ***Replace Reference Set/ SOLID.***
- **Right-click** on the hinge_assembly part in the Assembly Navigator; and select **Make Work Part**. Compare your results with the figure below.

- Save your assembly file.

***End of Project***

## ***Detail: Adding Translucency to a Component***

Adding translucency can be a great help in visualizing complex assemblies. Before you add translucency to a component, make sure that the 'Translucency' option under 'Session Settings' is checked in the ***Preferences / Visualization / Visual*** dialog box, as shown in the figure below.

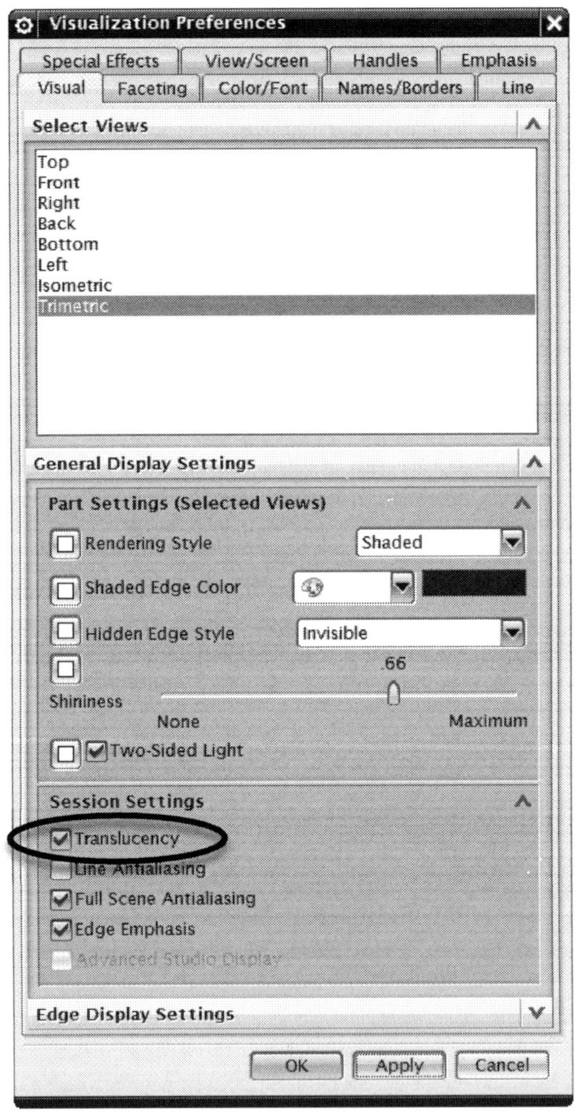

## *Project 16.5.3*

- Open your hinge_assembly part (if required) and select ***Start / Assemblies.***
- Set your hinge2 part as the Work Part.
- Select *Preferences / Visualization / Visual* and check the Translucency box.
- Select ***Edit / Object Display.***
- Select the hinge2 solid and choose **OK.**
- Move the 'Translucency' slider bar to 50 in the 'Edit Object Display' dialog box, as shown below.

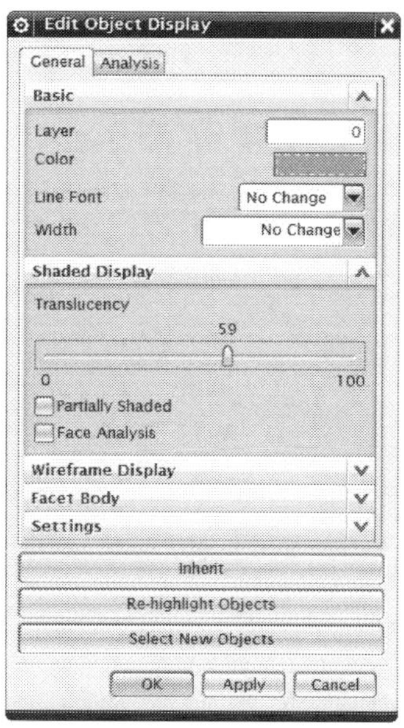

- Choose **OK.**
- Set your assembly part as the work part.
- Compare your results to the following figure.

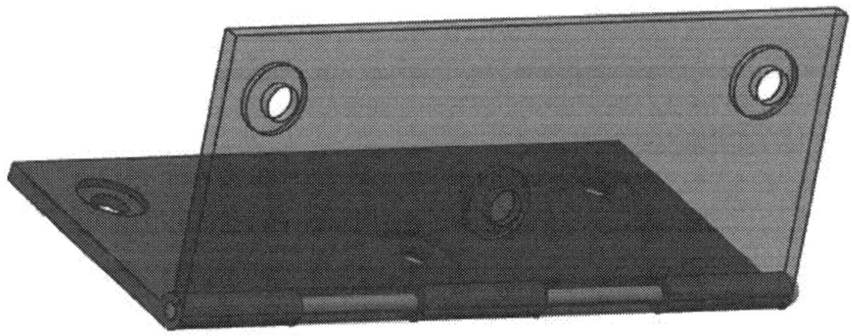

- Save your assembly file.

> **Note:** You can **hide** a part in the assembly to make it invisible and **show** it to make it visible again.

***End of Project***

## 16.6 Exploded Views in an Assembly

'Exploded' views allow you to "explode" parts away from their assembly location to better show how parts are assembled and/or used in assemblies. These are helpful for assembly schematics and instructions.

Detail: Creating an Exploded View of an Assembly

You can create an exploded view of the assembly components. They are stored as a "snapshot" of the assembly that includes the exploded positions of the components. Exploded views can even be used for display in assembly drawings.

### *Project 16.6.1*

In this project you will create an exploded view of your hinge assembly components.

- Open your hinge_assembly part (if required) and select ***Start / Assemblies.***
- Select Assemblies / Exploded Views / New Explosion.
- Choose **OK** to accept the default name 'Explosion 1.'

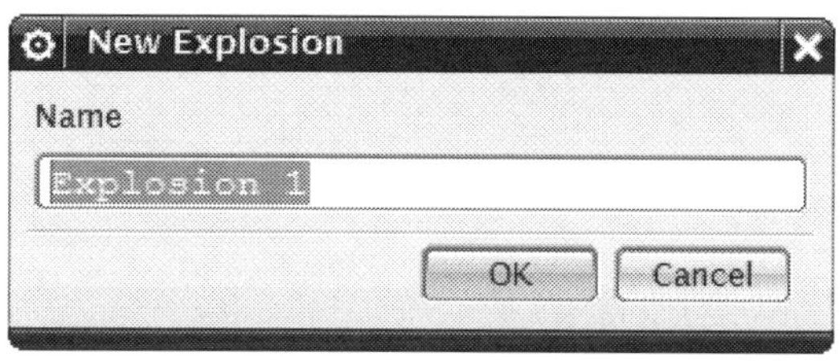


- Select Assemblies / Exploded Views / Auto-explode Components.
- Select your hinge_pin component.
- Choose **OK.**
- Enter Distance = **6** as shown next.

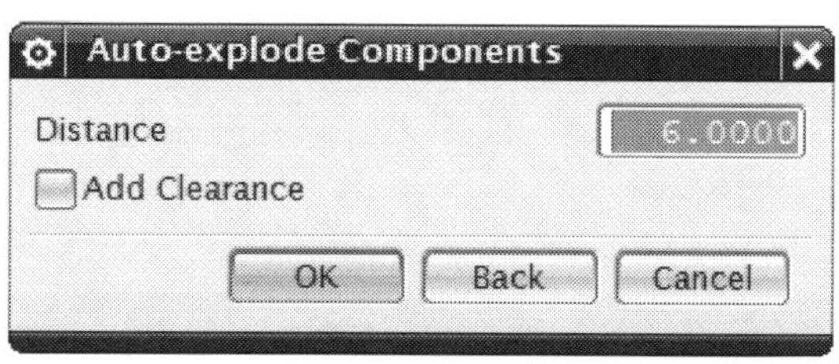


- Choose **OK.**
- Compare your results with the following figure.

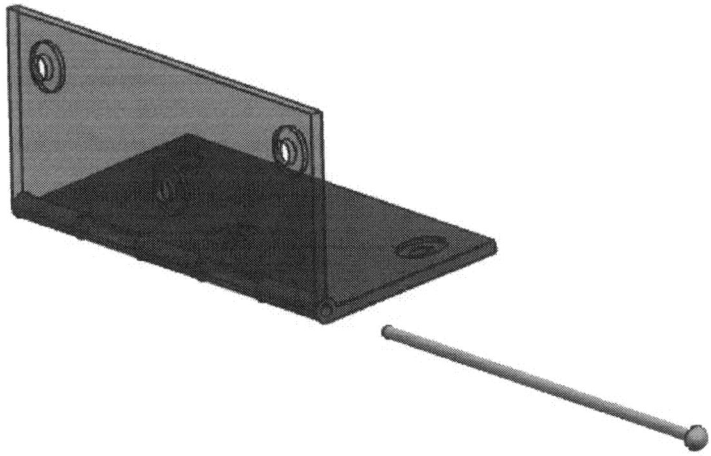

- Select Assemblies / Exploded Views / Edit Explosion.

The 'Edit Explosion' dialog box appears, as shown below.

- Toggle the radio button to **Select Objects**, and select your hinge2 component.
- Toggle the radio button to **Move Objects.**

The Dynamic WCS appears, as shown below.

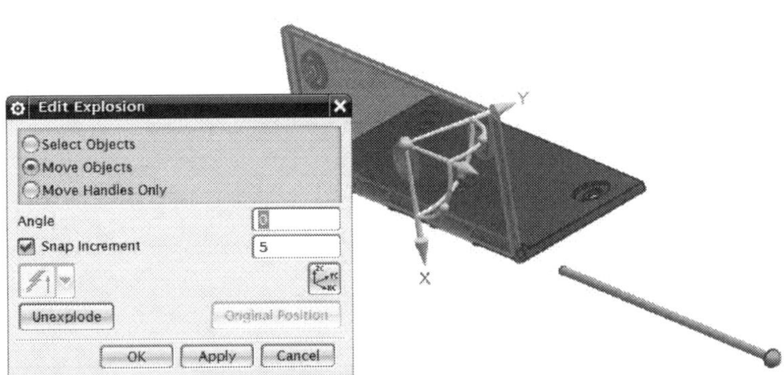

- Drag your hinge2 part up a short distance.
- Choose **OK** to finish editing the Exploded View.
- Compare your results with the figure below.

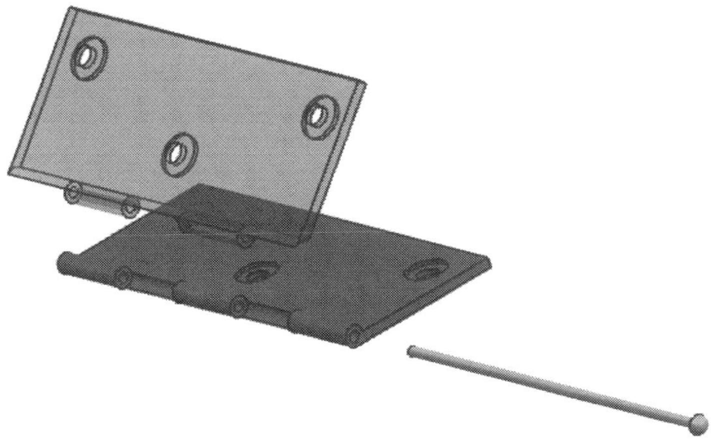

- Select Assemblies / Exploded Views / Unexplode Component.
- Select your hinge_pin part, and choose **OK** to unexplode the selected component.

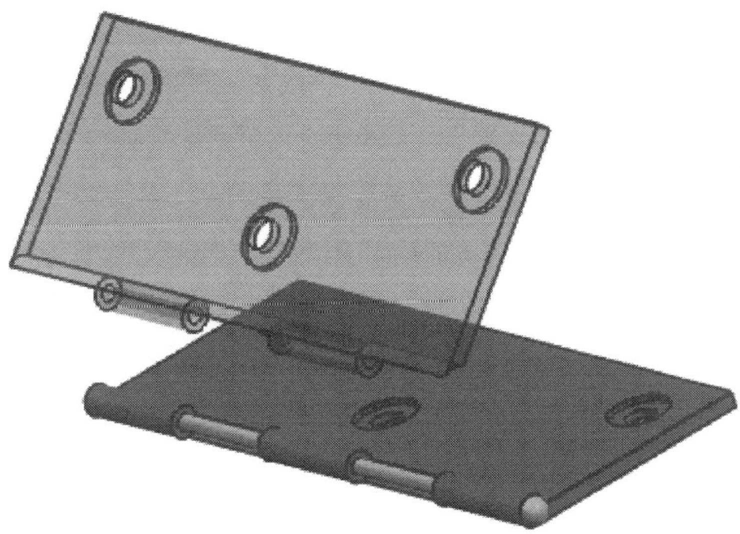

- Select ***Assemblies / Exploded Views / Hide Explosion*** to return the assembly to the unexploded state.
- Save your assembly file.

***End of Project***

## 16.7 Assembly Constraints

Now you have the skills to do nothing! No really; this next toy is called the "Do Nothing."

### *Project 16.7.1*

In this project we are going to use Assembly Constraints to create the following assembly. This is a small mechanism, often used as a children's toy. If we use the assembly constraints properly, we will be able to show the motion of the mechanism without having to do a complicated motion simulation.

Luckily for you, the parts are already created. All you have to do is create the assembly.

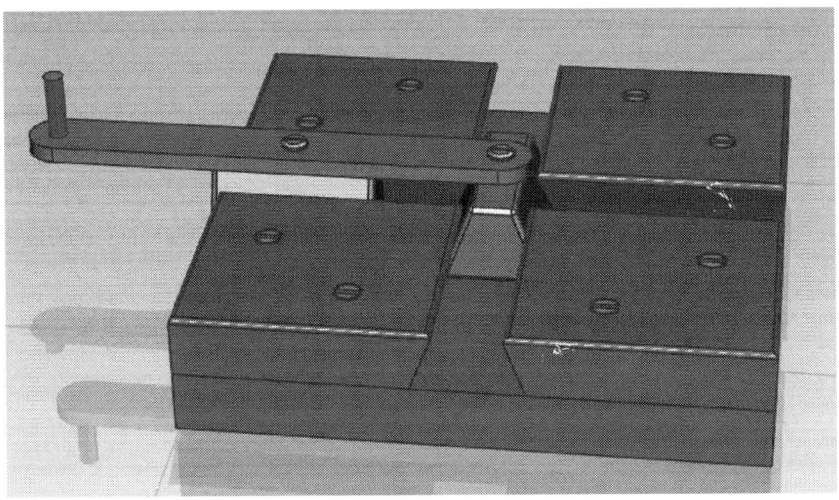

- Start with File / New to create a new assembly file.
- Name the new assembly *cool toy.prt* as shown in the following figure.

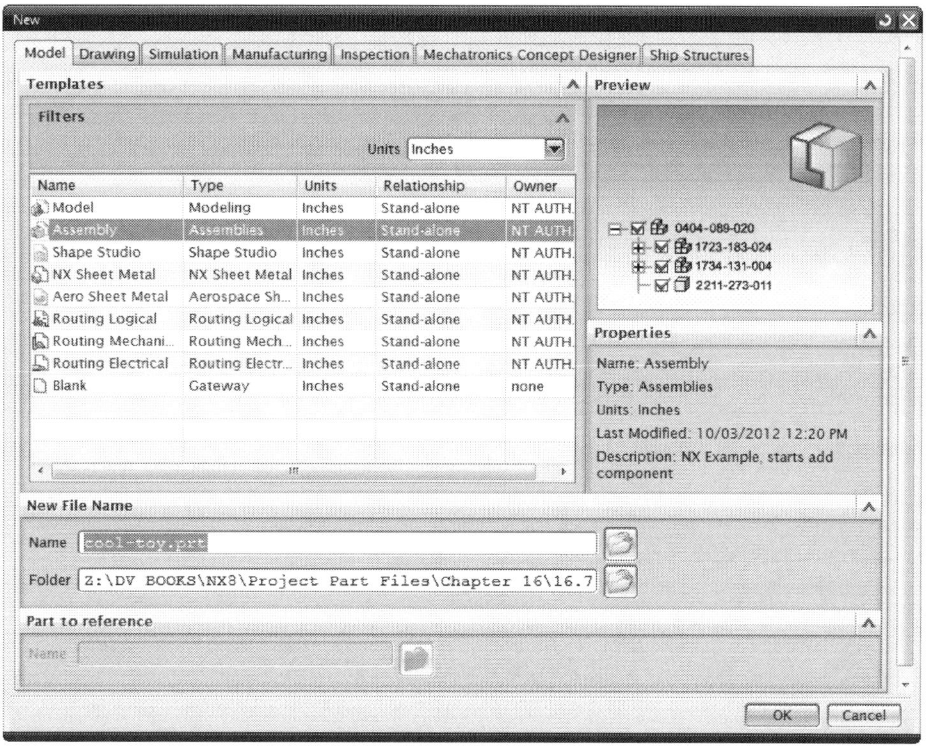

The first window that shows up is asking if you would like to add a component. Click on the folder button next to Open, navigate to the directory where your parts are (downloaded from the website) and then select *Base.prt.* Once the part is loaded, it will show up in the Component Preview window, as shown in the following figure.

- Select Absolute Origin in the Add Component window for its placement, as shown in the previous window.
- Then Select OK to fully import the part into your assembly.

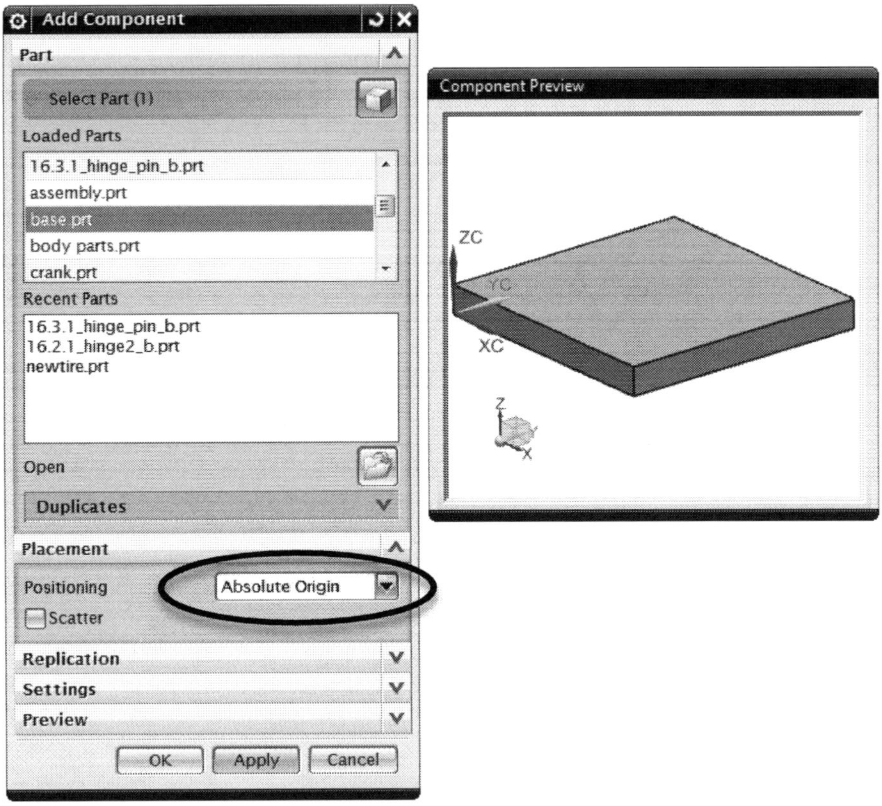


Before we start adding new components to this assembly we have to Fix the first component. If this is not done, the base part is likely to move around.

- To Fix the component, go to Assemblies / Component Position / Assembly Constraints. Or click on the assembly constraints button in the Assemblies toolbar, as shown in the following figure.

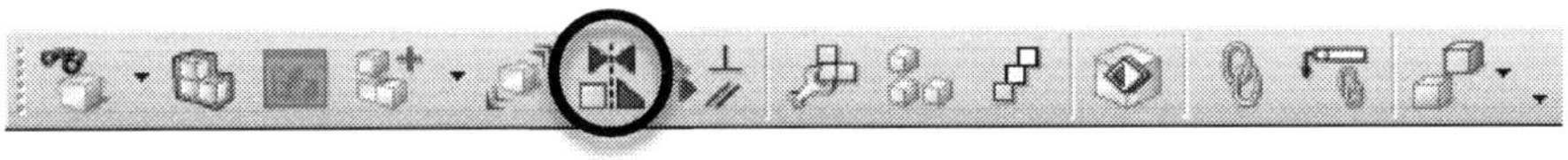

This will show you the Assembly constraints window as shown in the following figure. The drop down arrow shows all the constraints available.

- Select the Fix constraint, and then click on the base in the graphics window. You will now see the 'Fix' symbol attached to the base in the following figure.

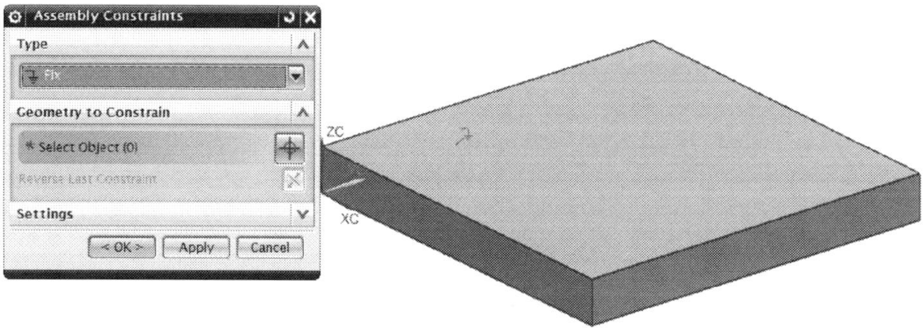


You can add or remove these little constraint indicators by right clicking on the constraints tab in the assembly navigator and selecting or deselecting 'Display Constraints in Graphics Window.' We prefer to turn them off, as when the assembly gets more complicated it gets messy with them there!

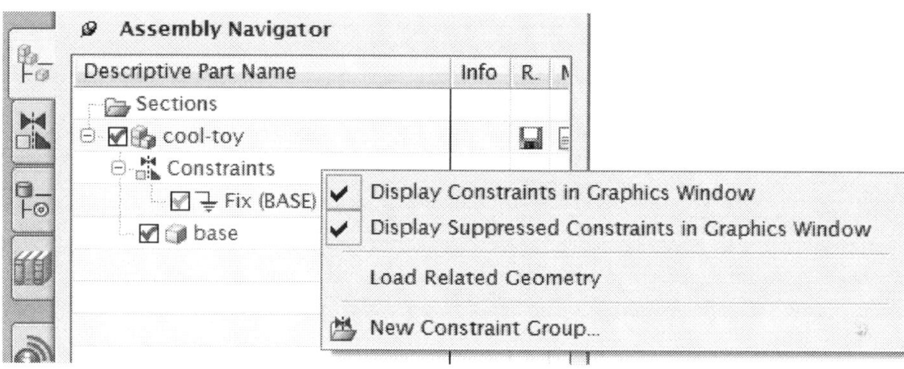


- Now we will add the next component. Go to **Assemblies / Components / Add Component**. Click the same button as before to navigate to another part file. This time choose body parts.prt.
- In the placement section of the Add Component window, select By Constraints, as shown in the following figure.

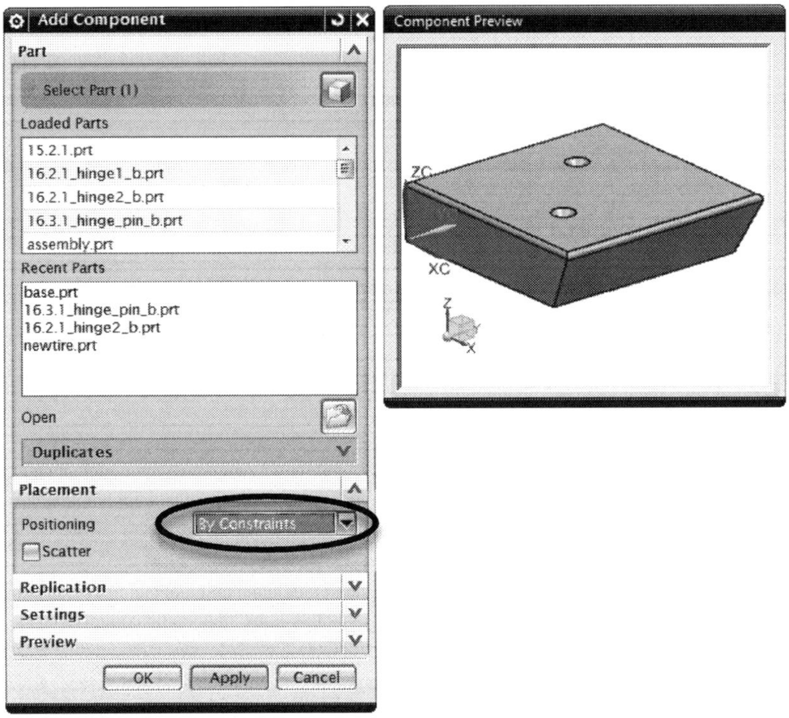


- Once you hit Apply, the Assembly Constraints window will appear with the preview window.
- Select the Touch Align constraint and set the orientation to Prefer Touch, as shown in the following figure.

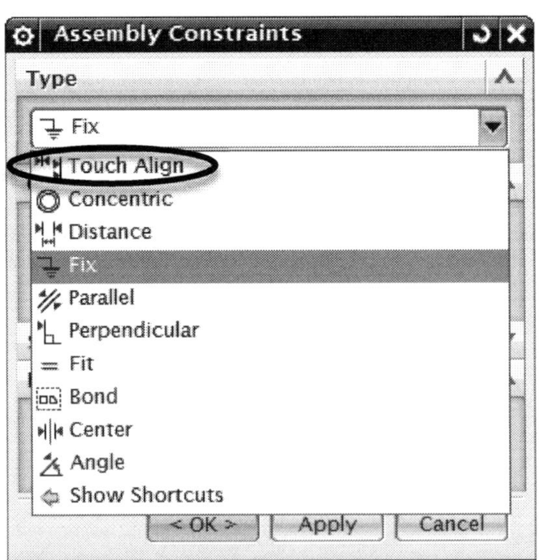

- First, select the bottom face of the body part, as shown in the following figure.

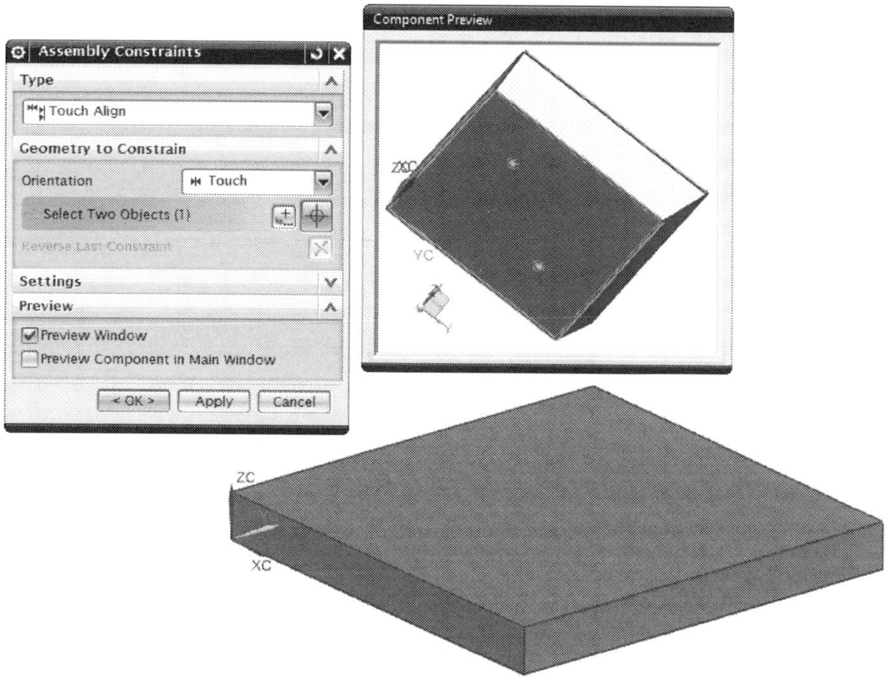


- Then select the top face of the base. You do not have to click apply or OK yet; only do that once your model is fully constrained.

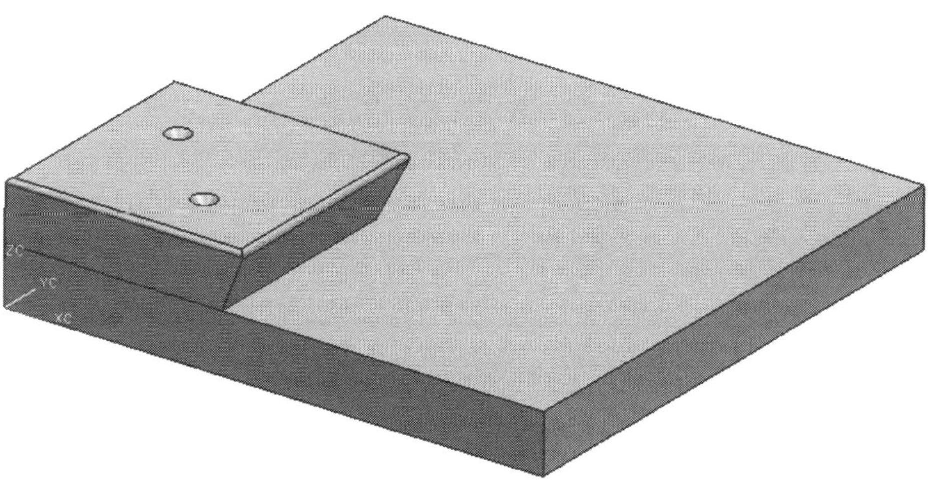

- The model will now jump into position. If your model flips the wrong way, you can click the Reverse Last Constraint button on the assembly constraints window, as shown in the following figure.

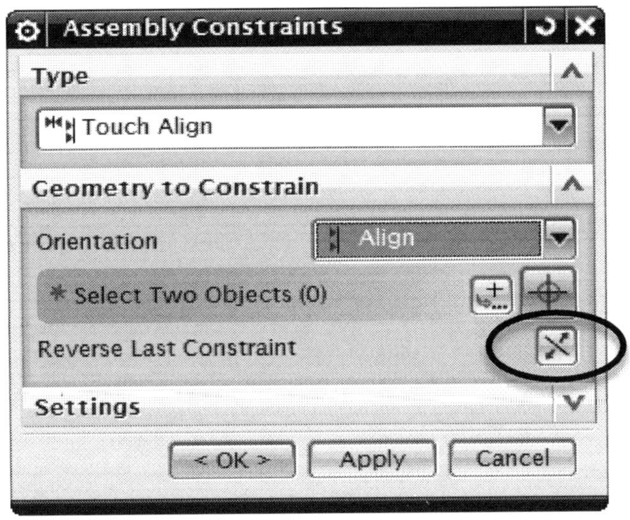


The model is not fully constrained yet – we have two left to do. The next two are also Prefer Touch constraints. Make sure the orientation is set to Align. Select face 1 shown in the following figure, then select face 2.

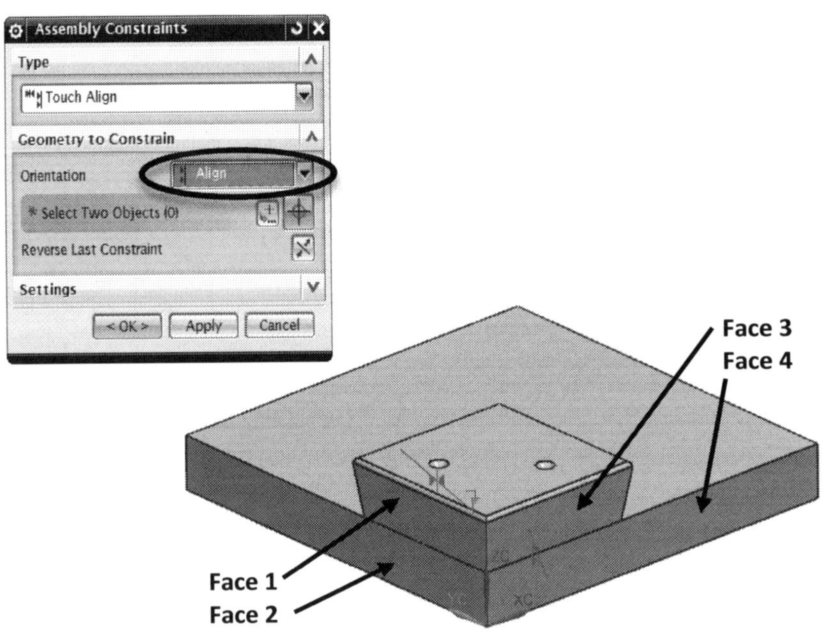

> **Note:** You don't have to choose the model in the preview window, though it is useful when you can't see the model clearly in the normal graphics window.

- Then do the same with Faces 3 and 4, shown in the previous figure. Click OK to complete.
- Repeat this process, adding 3 more body parts.prt to the base as shown in the following figure.

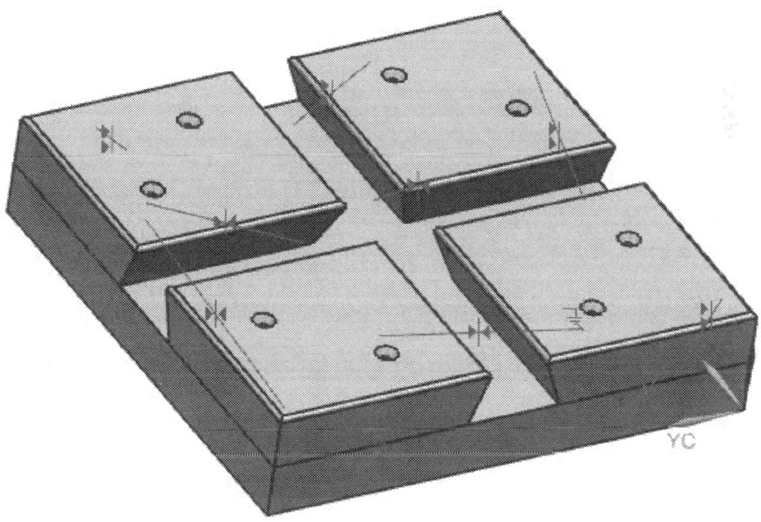


- The next step is to add a new component. This time add the slider.prts using placement By Constraints.
- The first constraint will be a Touch Align between the large bottom surface of the slider and the top surface of the base, as shown in the following figure.

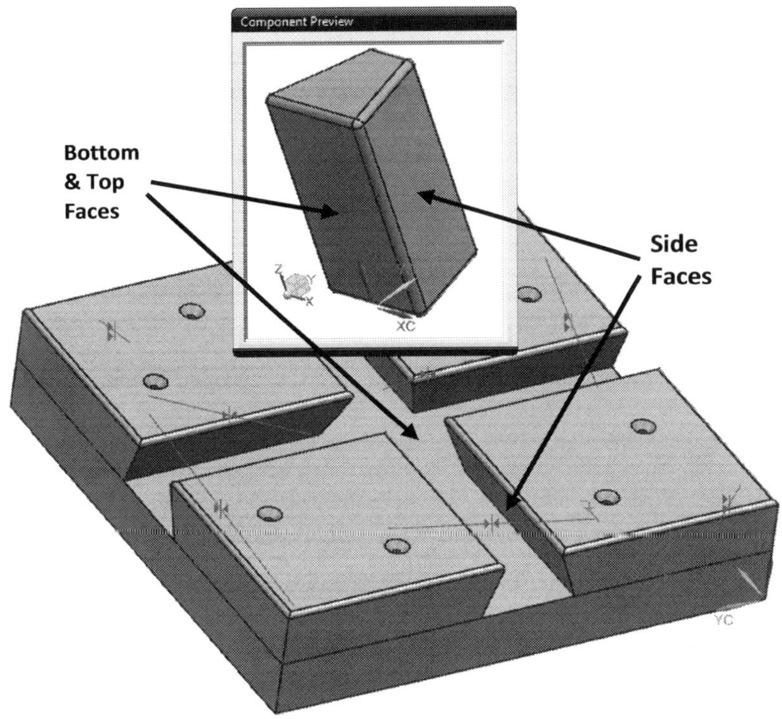


- Then Touch Align a Side Face on the Slider with the Side Face on Body Part.
- Repeat this process with another slider.prt so that you get the following model.

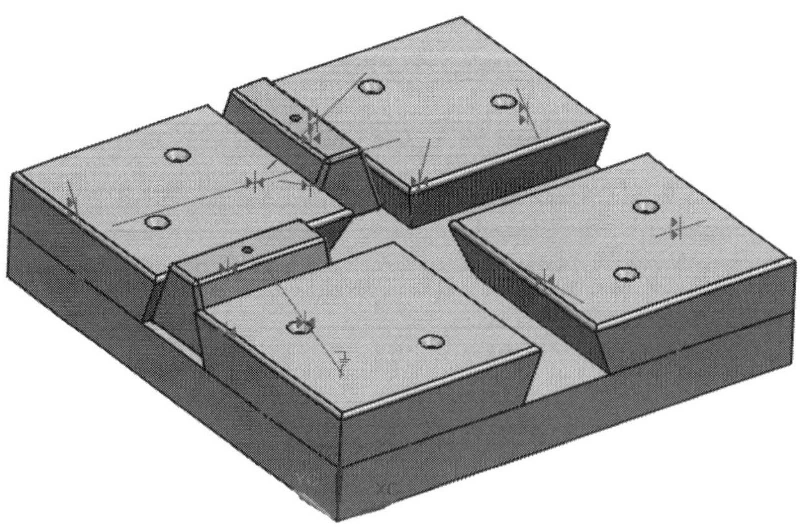

- Notice all the constraints you have been defining in the Assembly Navigator. From here, you can redefine them or convert them to a different type of constraint, as shown in the following figure.

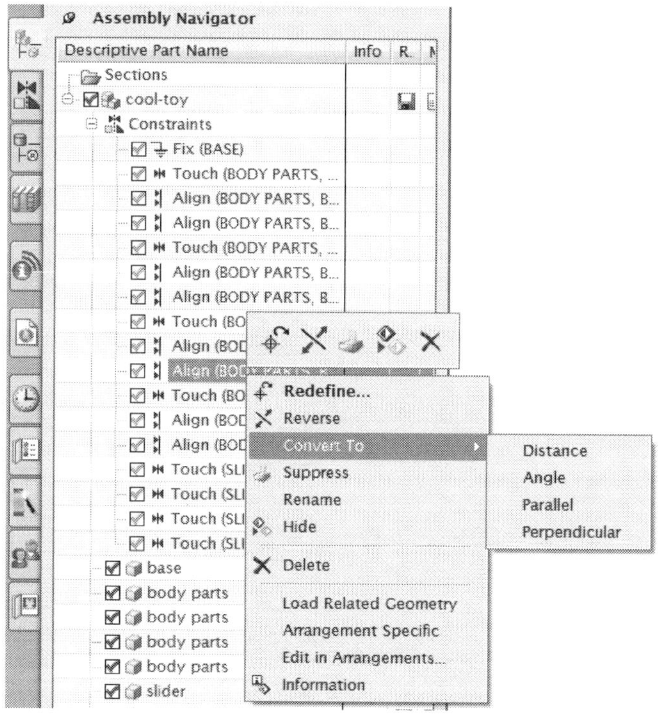


- Next, add the *crank.prt*, using the Touch Align placement. However, this time, choose the 'infer center/axis' command.

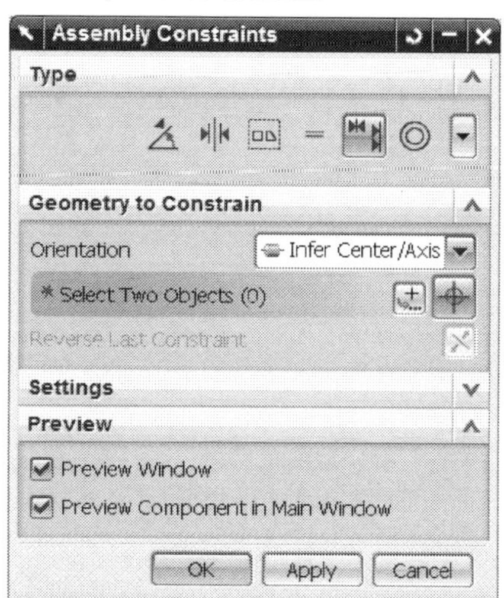


- Connect the pairs or faces shown in the following figure by clicking on a face of pair 1 then selecting the corresponding face of pair 1. For the next constraint, choose a face of pair 2 with the corresponding face of pair 2.

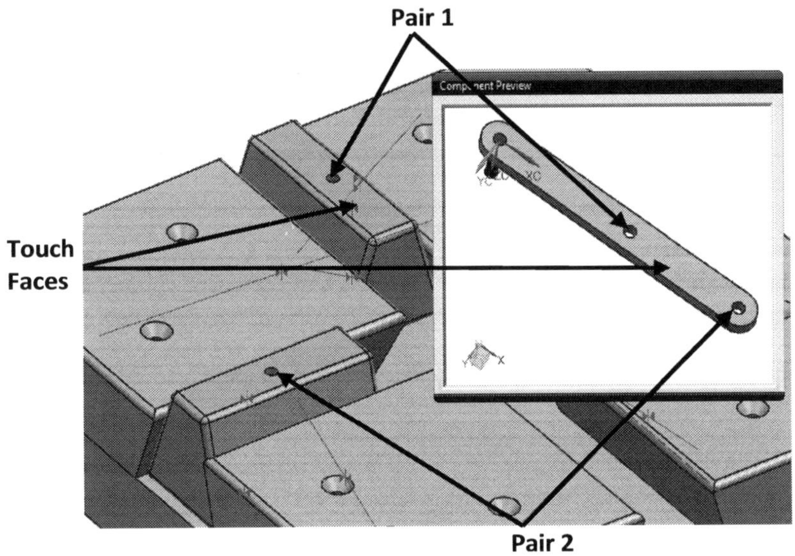


- Then use the Touch Align constraint, specifying the touch command, shown in the following figure.

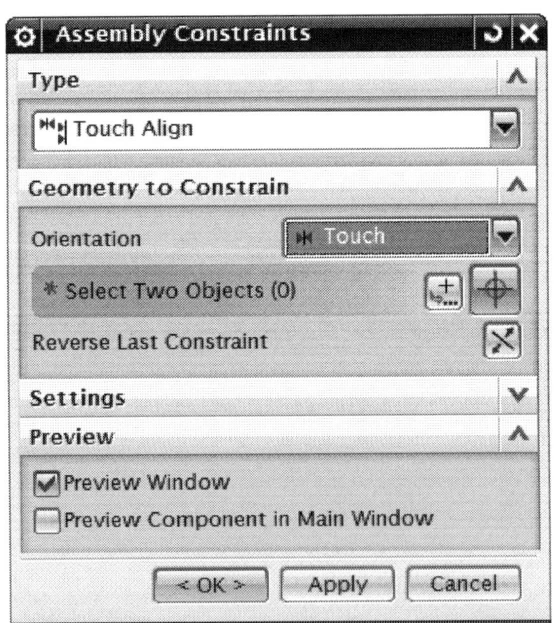


- Now your model should look something like the following figure.

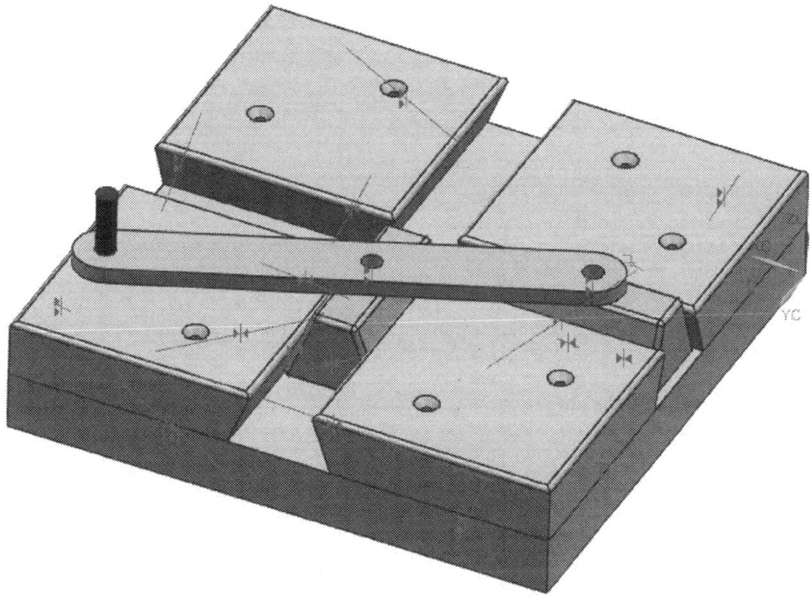


- Let's add screws to this model, since there are screw holes! We will screw in the body parts first.
- Go to **Assemblies / Components / Add Component** and open *wood screw.prt* from the folder.

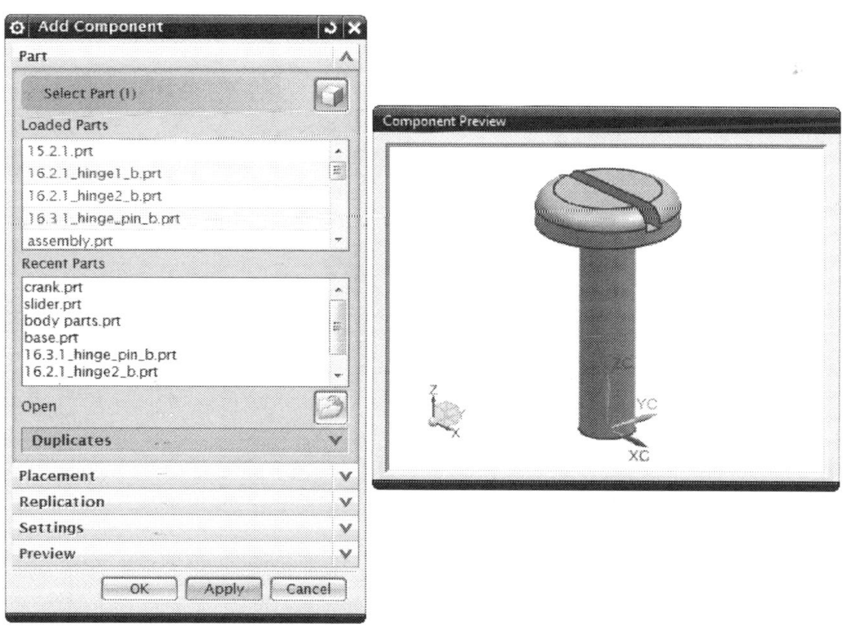


- For Positioning, choose By Constraints. Click Apply, as shown in the following figure.

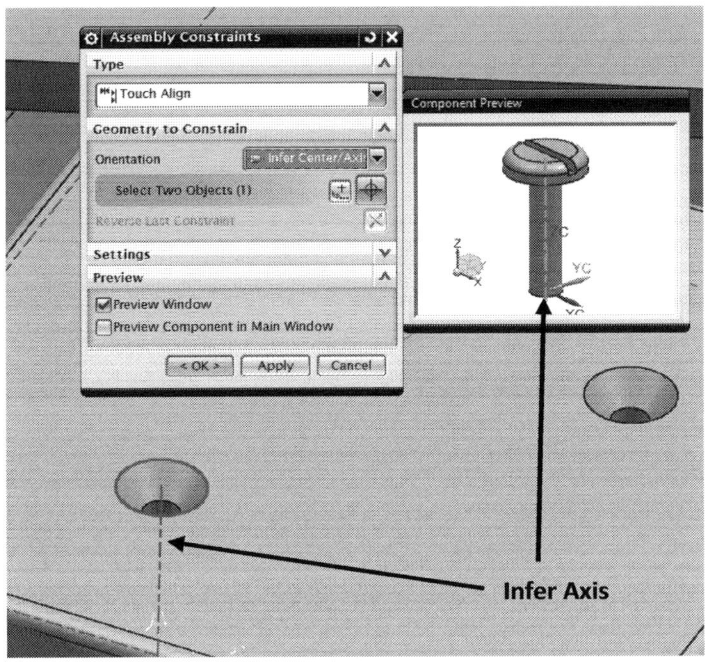


- Choose the Touch Align assembly constraint along with the Infer Center/Axis option.
- Then choose the Prefer Touch assembly constraint to align the Top Face of the screw with the Top Face of body part.prt.
- You may have to flip the direction to get what you want; if you have done it correctly you should have something similar to the following figure.

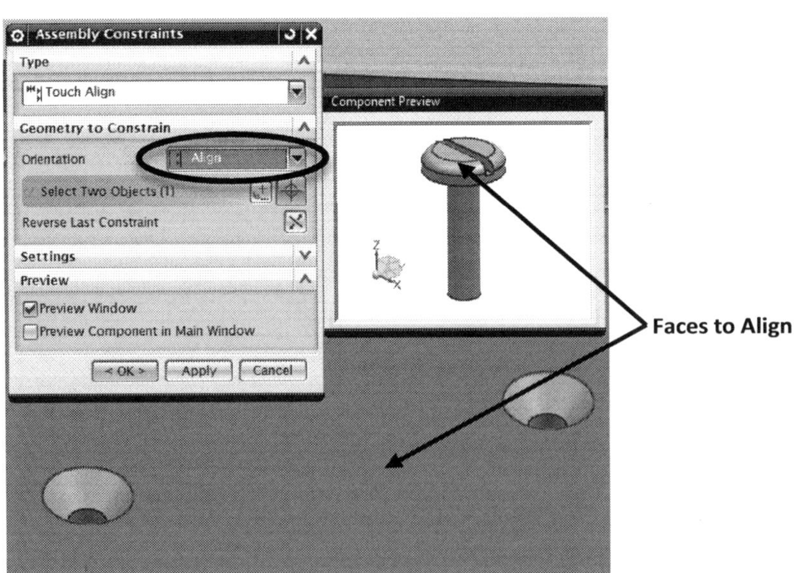

- Last but not least, assemble *screw.prts* in the holes of the crank exactly the same way as you did the other screws. Your model should look like the following figure. When you choose Move Component to rotate the crank, you'll see the beauty of the Do Nothing.

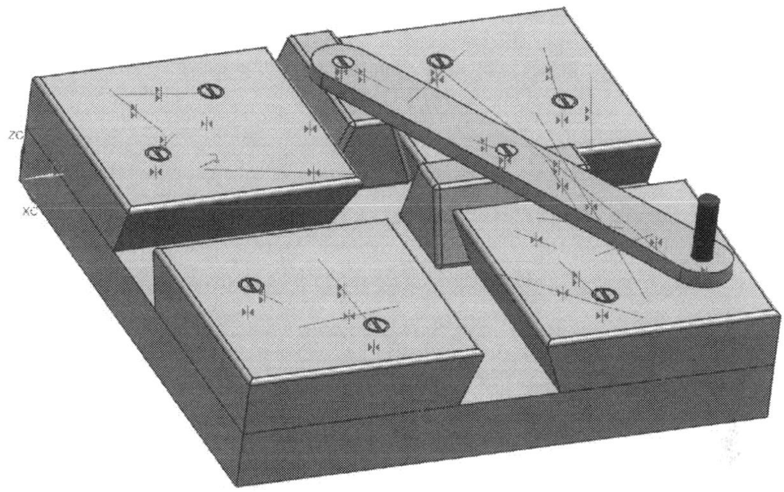

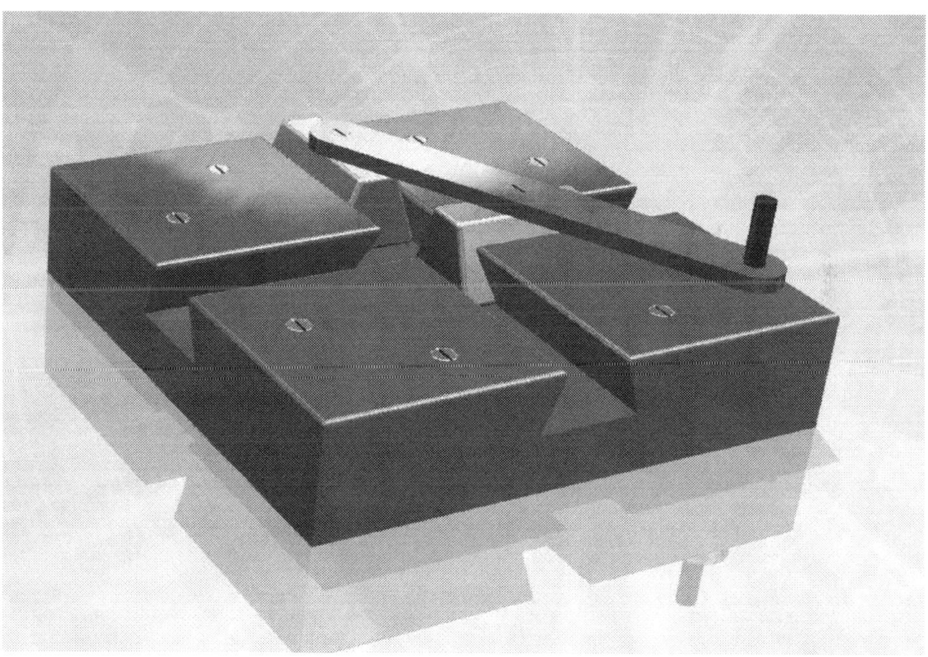

***End of Project***

## 16.8 Make Unique

The "Make Unique" command is an extremely useful new ability that is now available in NX8.5. A good way to describe it is a way of dealing with the situation that arises when you have an assembly with a number of components that you first thought would be all identical, then you subsequently found it necessary to make some of them different. For example, if you have a table that has four identical legs, and then you decide that two of the legs should have casters on them so you could just tilt the table and roll it away.

The BOM for the typical table would have all four of the legs be identical as shown below. Notice that the table leg has as quantity of 4.

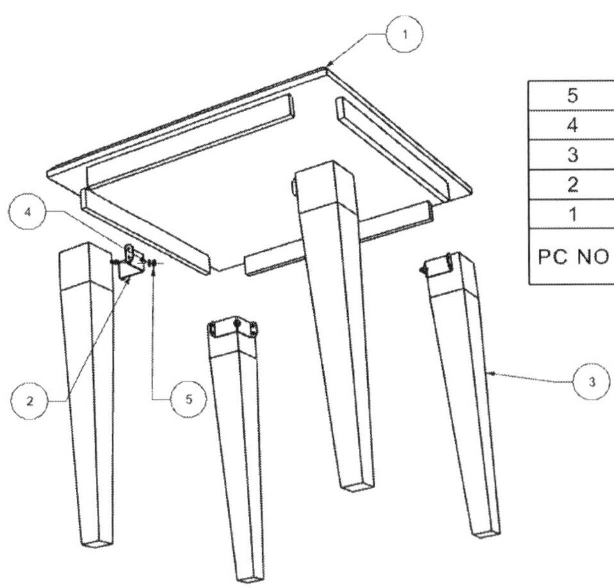

| 5 | BOLT | 4 |
|---|---|---|
| 4 | WASHER | 4 |
| 3 | TABLE LEG | 4 |
| 2 | BRACKET | 4 |
| 1 | TABLE TOP | 1 |
| PC NO | PART NAME | QTY |

Using the Make Unique command all you need do is select the two legs that you want shortened and select the command. Assemblies / Components / make Unique. The following menu appears.

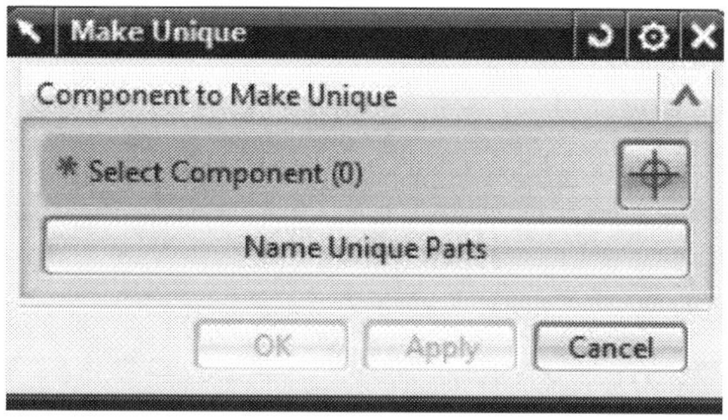


With the assembly the work part, you are prompted to select the components that you want to alter. Once selected, you click on the "Name Unique Parts" button. You are prompted to enter a new component name. In this case "Table leg Shortened" was chosen. You must type it into the "Name" box and select Enter.

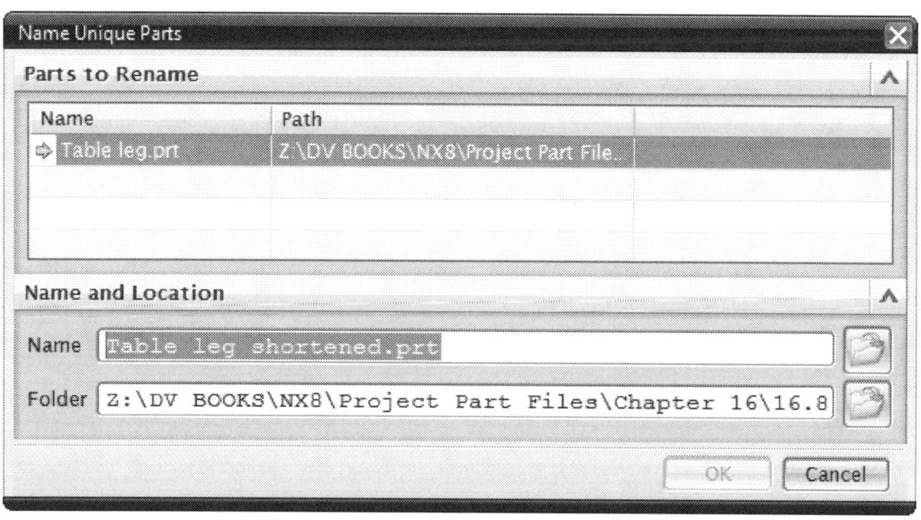


Then you make the shortened leg the work part and chop the end of it off where the caster will go. Finally you assemble the casters onto the end of the shortened legs and you're good to go. The new assembly is complete.

The bill of materials updates to reflect the new structure.

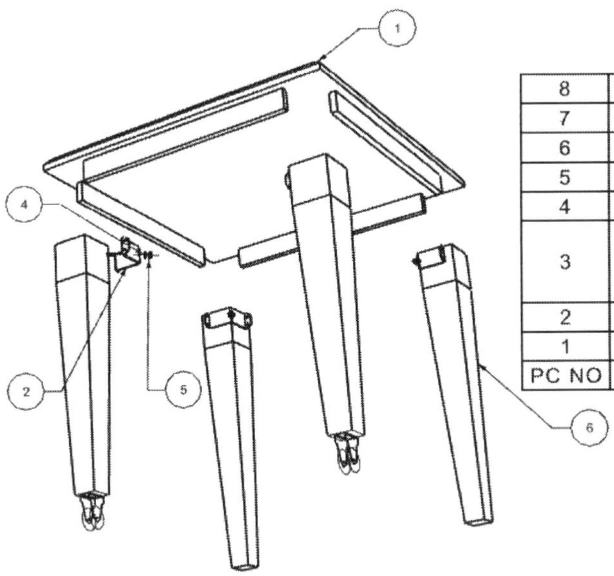


| PC NO | PART NAME | QTY |
|---|---|---|
| 8 | WHEEL | 2 |
| 7 | CASTER | 2 |
| 6 | TABLE LEG | 2 |
| 5 | BOLT | 4 |
| 4 | WASHER | 4 |
| 3 | TABLE LEG SHORTENE D | 2 |
| 2 | BRACKET | 4 |
| 1 | TABLE TOP | 1 |

The Make Unique command can be found in the Assemblies / Components menu. To give it a try download the Make Unique Table Assembly. It is a zip file that contains the original table assembly plus the caster. You will have to shorten the two legs by 3.25 inches.

## 16.9 Self-Test

(All questions are true/false. Answers are at the back of the book.)

1. You can select units while creating a component in the assembly.
2. You can create mating conditions only while adding a component in the assembly.
3. You can have a number of same subassemblies with different reference sets.

# 17. Basic Drafting

## 17.1 Introduction

Once you have a solid model of a part or assembly, you can use it to create an engineering drawing. Drawings are defined using the 3D data, and are updated whenever you update the associated solid model. Many types of drawings can be created in NX 8.5; however, in this manual we will discuss only detail part drawings.

## 17.2 Creating Drawings

To enter the drawing mode, select ***Start / Drafting***. If no drawing exists, NX 8.5 will prompt you to create one. You may have as many drawing sheets as you require, however you may view or edit only one at a time.

NX 8.5 enables you to define standard drawing templates that can be imported into any drawing. Typically, a template may contain a drawing border, title and revision blocks, and other standard information. Setting up templates is a useful technique to standardize engineering drawings across a company, but this is an advanced topic not covered in this manual. You can ask your CAD System Administrator for information about your company's standard drawing templates.

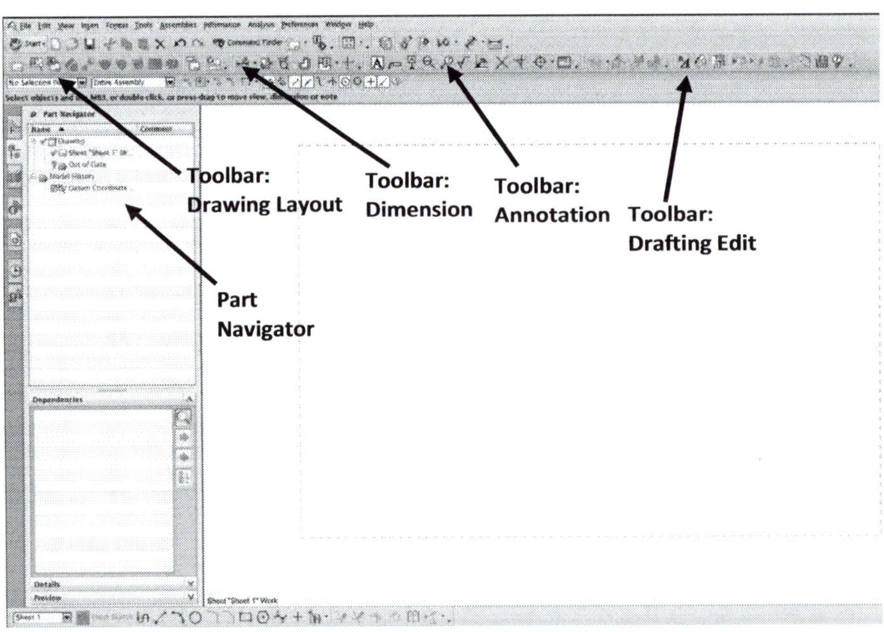

## ***Detail: The NX 8.5 Drawing Environment***

The NX 8.5 Drafting Application is shown above. As always, drawing functions are invoked by clicking on an icon or selecting from the top menus. Several different toolbars will appear that are specifically tailored for the drafting mode. You can right click in the toolbar area and customize it. You can also drag and drop toolbars at your desired locations.

## ***Detail: The Drawing Process***

Every drawing is created with a process much like this:

1. Create a solid model
2. Create drawing sheet 1
3. Import the company-standard drawing template
4. Add views to drawing sheet 1
5. Add dimensions
6. Add GD&T annotations (if required)
7. Add drawing annotations (notes, symbols, etc.)
8. Add drawing sheet 2, and repeat the process

> **Note:** You may begin creating a drawing at any time during the design process. Because the drawing views and dimensions are associative, they will update as the model evolves.

## ***Detail: Defining the Drawing Parameters***

Before you start adding views to a drawing sheet, you should define the drawing name, units, scale, size, and projection angle. You may change these parameters at any time; however, you may need to re-arrange the views and annotations quite a bit as a result.

## *Project 17.2.1*

Create the model shown below, and start an engineering drawing of it.

- Create a new part. Save it and rename it as: **drw_block_<your initials>.prt**.
- Change all dimensions to match those shown in the figure below. The size is **3in** x **2in** x **1in** with a **0.75in** diameter hole at the center and a **1in** x **1in** corner chamfer.

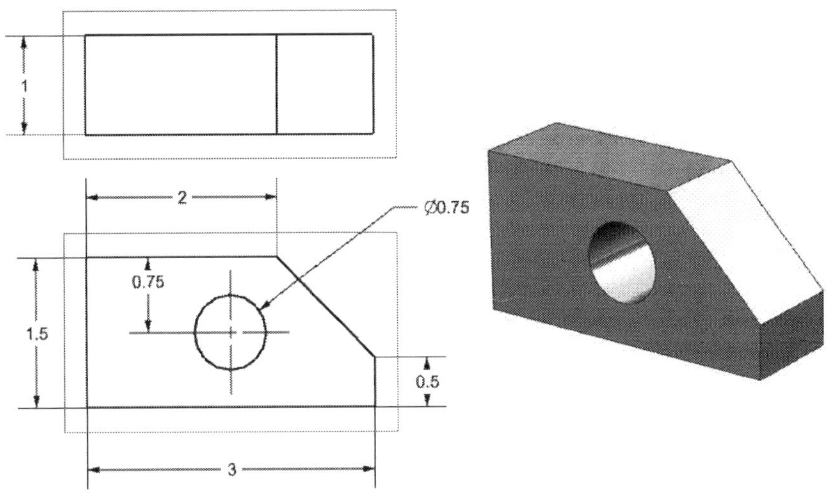


- Save your part model.
- Select ***Start / Drafting*** 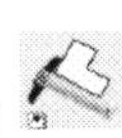 . A new drawing sheet dialog box appears as shown below.

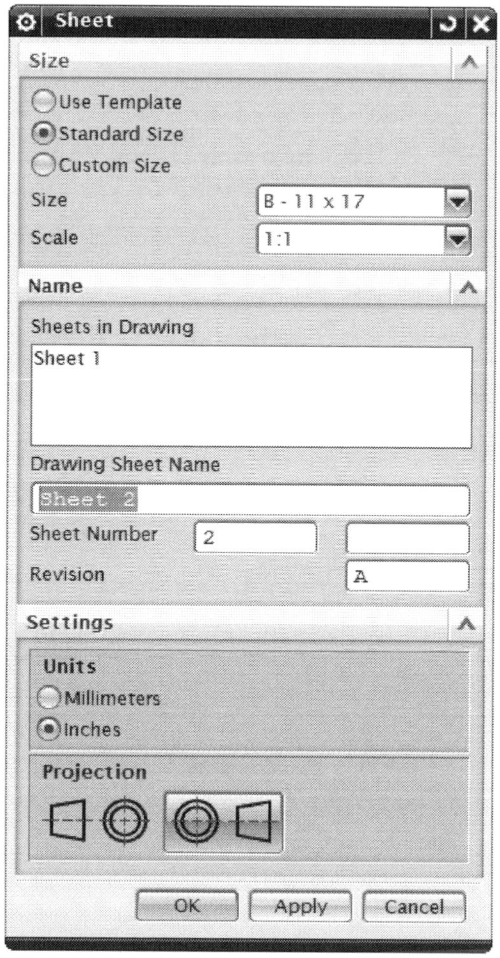


| **Note:** If no dialog box appears, select **Insert / Sheet.** |
|---|

- Enter the drawing sheet name if you don't want to use the default name.
- Select the drawing size, units, and projection angle options according to the figure above.
- Enter Drawing Scale = **2:1**
- Choose **OK.**
- Save your drawing file.

***End of Project***

## ***Detail: Drawing and Modeling Layer Standards***

Most companies have standard procedures for modeling and drafting. An important component of these standards is Layer conventions. Usually, a range of layers is reserved for modeling entities and another range for drafting entities. It is a good practice to organize your work in this fashion. Please contact your CAD System Administrator to obtain more information on how this should be done at your site.

## 17.3 Creating Drawing Borders

If your company has standard drawing borders, you can import them into a drawing or you can use the border template depending on your company policy. The procedures for this vary widely between companies, so no further discussion of this would be appropriate in this manual. Please contact your CAD System Administrator to obtain more information on how this should be done at your site.

## ***Detail: Creating Drawing Borders***

In the following project you will learn how to create a border and save it in order to use it as a template or import it into the current drawing. *If you already have drawing borders from your company you can skip this project.*

### *Project 17.3.1*

- Create a new part "**B_size_in_SH1.prt**" part and select **Start / Drafting.**
- To create a border select Tools/ Drawing Format / Borders and Zones, as shown in the following figure.

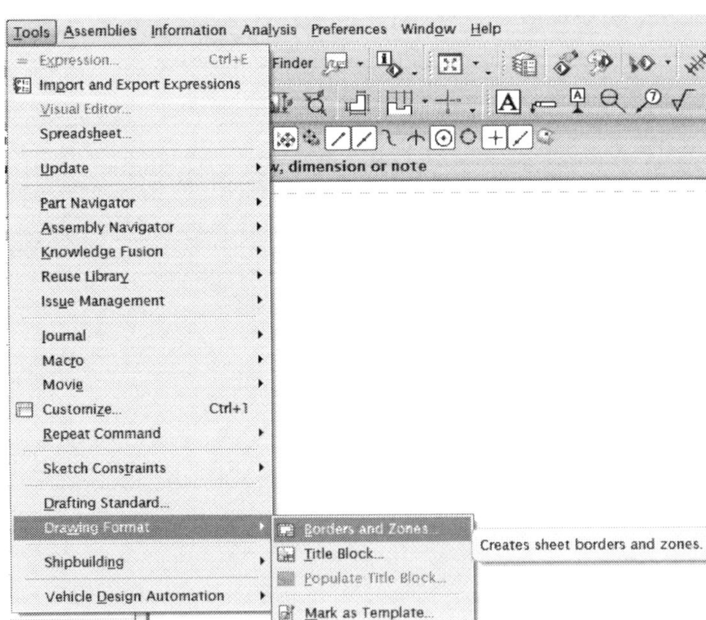

> **Note:** For a B-sized drawing, select the **Inches** radio button and select **B – 11 x 17** from the pull-down menu.

- A dialog window will be presented containing many different options how to style the border as shown in the following figure.

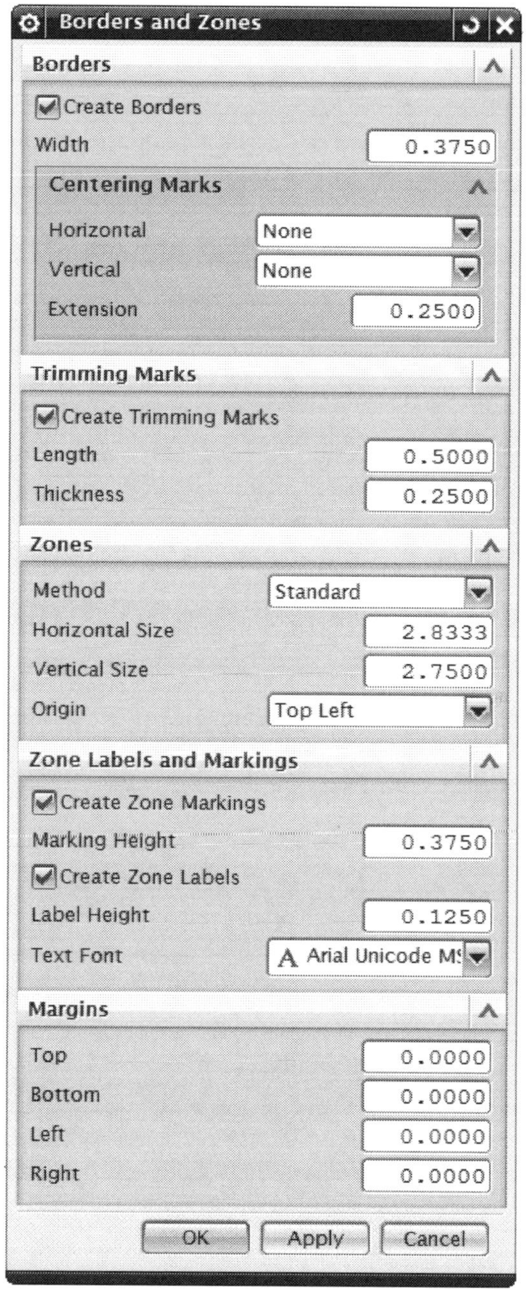

- By pressing OK a default border will be created in NX.

- In order to create a Title, a Table must be created. Select Insert / Table / Tabular Note. When selected the following dialog window will be presented.

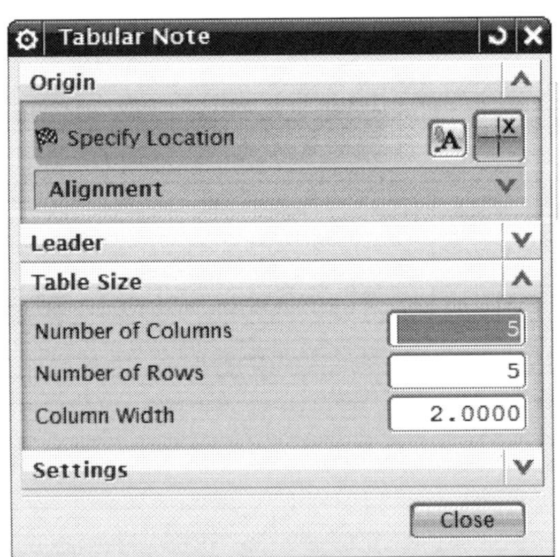


- Read the que: NX is asking for the Location to place the table. Select a point inside the border that aligns the table on the left. Set the Number of Columns to 3 and the Number of Rows to 4.

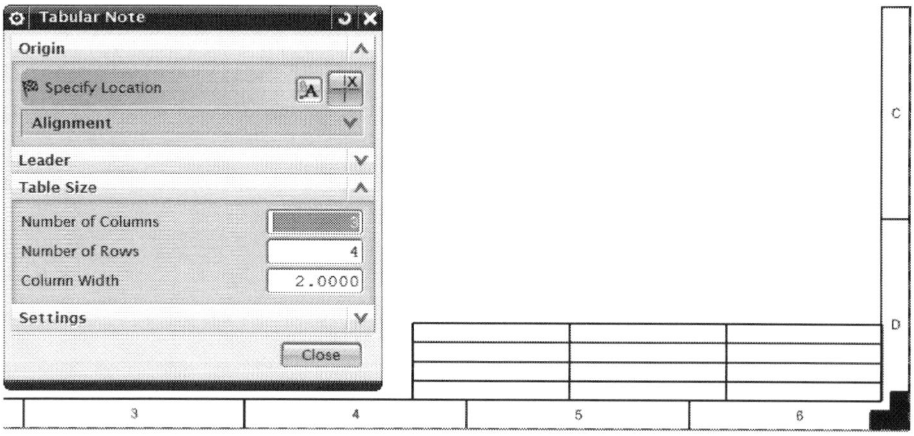

**Note:** By left clicking and dragging the mouse on the dividers between cells and rows, the table can be resized to the styles of data the must be inserted.

- Each Cell of the Table can be clicked and information can be added. Now that there is a table in the drawing a title can be added. Select Tools / Drawing Format / Title Block.

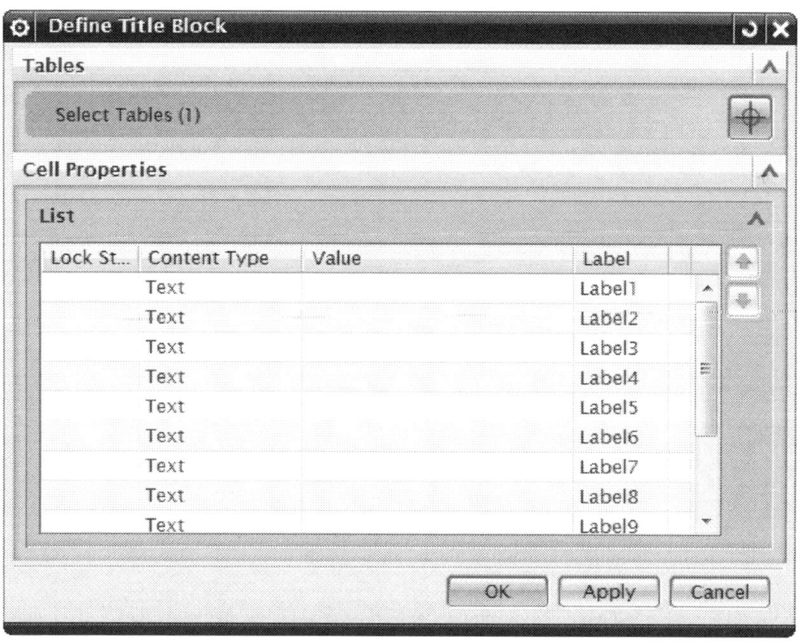

- Select the table that was previously created. Below Labels can be entered into the List. Create a Label for Title, Revision and Designed by as shown in the following figure.

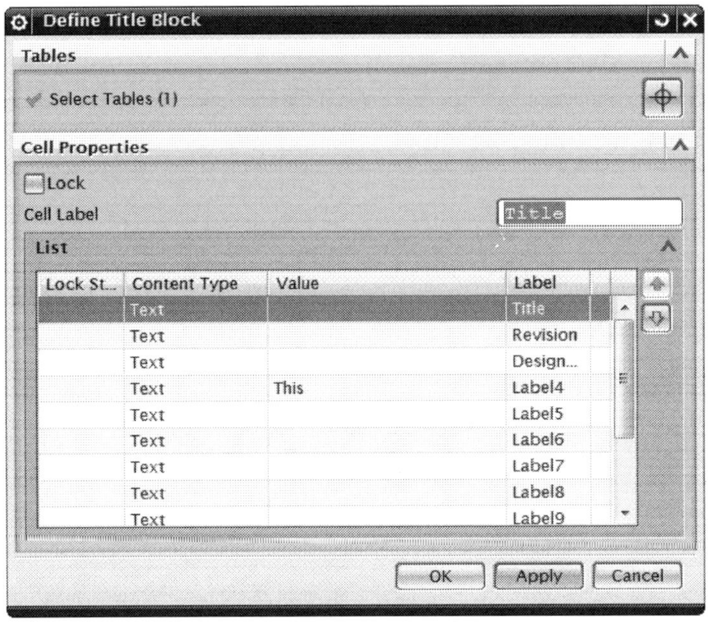

- Press OK
- Double click on the Table, a new dialog window appears allowing you to populate the fields of the table. Populate the table as shown below.

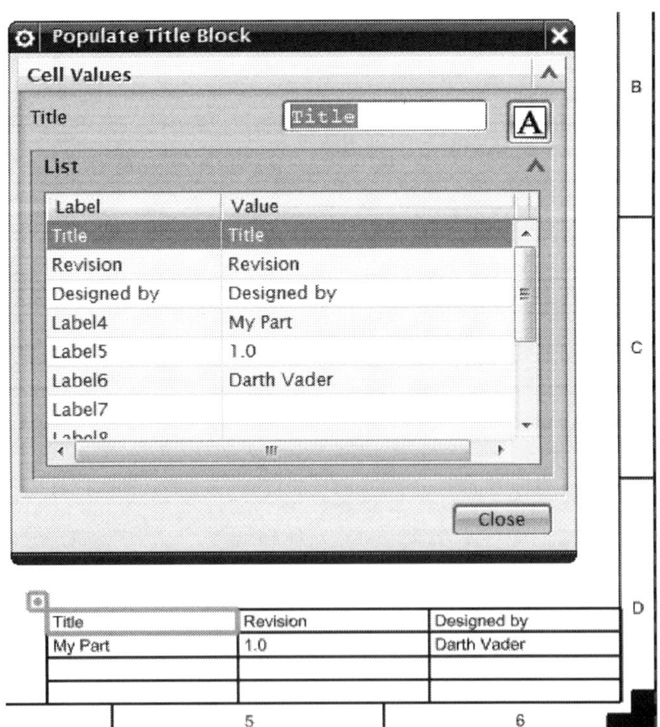

- Create a new folder "**Drawing_Borders_Import**" and save your file.
- You just created a file that can be used to import into your current drawing. A better way to create drawing templates is to create a file with all the sheets and basic views that you usually need. You can create several of these files and save them into a folder that you can define as a template folder, but using this method you cannot add a drawing border to the part file you are already working in.

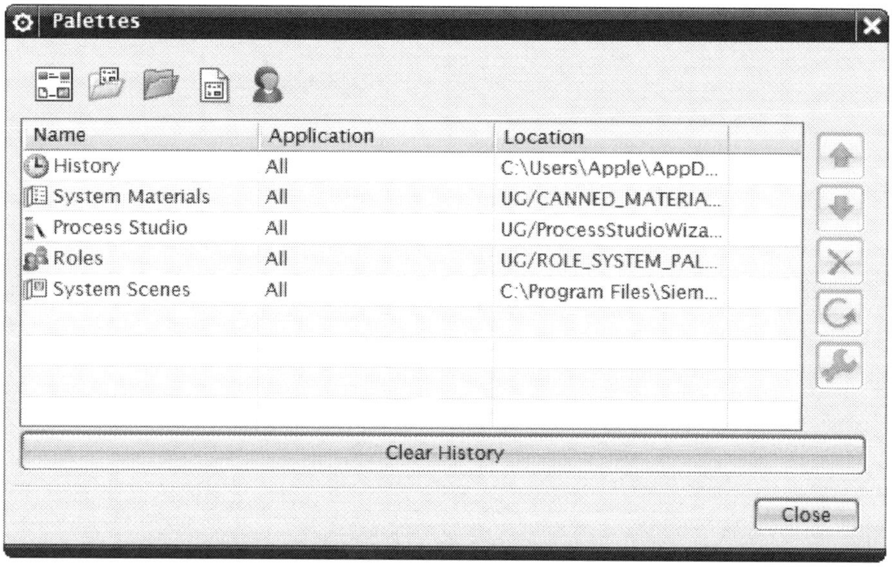


- Create another file with two sheets. Add borders and Titles Save it. Re-name it "***B_size_in_<your initials>.prt***", and place it in another folder "**Drawing_Border_Templates.**" *Later you can add some views to it.*
- Select *Preferences / Palettes* and click Open Directory as Palette as shown in the figure above.
- Browse to folder "**Drawing_Border_Templates**" and select it. You will notice that it is added to the list as well as to the resource bar.

***End of Project***

## ***Detail: Importing Drawing Border***

In the following project you will learn how to import the drawing border into your current drawing.

## *Project 17.3.2*

- Open your drawing from **Project 17.2.1** and select ***Start / Drafting.***
- Select File / Import / Part.

- Leave the default options as shown above and hit **OK**
- Select the template file you just saved and hit **OK**
- A Point Constructor menu pops up; make sure that XC, YC and ZC are set to zero. *You can use the **Reset** button to set it back to zero if they are not.*
- Hit **OK.**
- You should see your drawing border in this part file.
- Hit **Cancel** to get out of the Point Constructor menu.

***End of Project***

## Detail: Importing Sample Palettes

In the following project you will learn how to import the sample palettes provided by NX8.5. Adding your own files to palette files requires editing or creating an xml file with the ".pax" extension, which is a more advanced topic to be covered in our future books.

## Project 17.3.3

- Select ***Preferences / Palettes*** and click the ***Open Palette File*** icon as shown in the figure below.

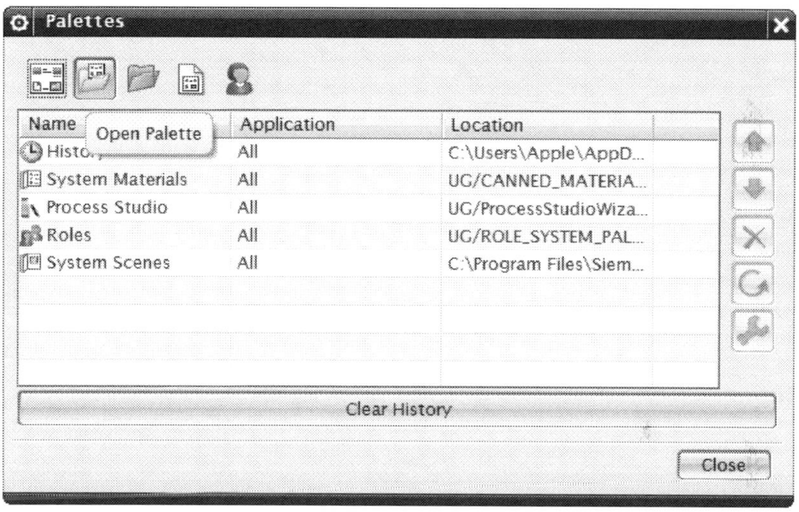


- Browse to the location where NX 8.5 is installed (usually "/Program Files/ Siemens") and select the file "english_drawing_templates.pax" from folder "\NX 8.5\UGII\html_files".
- Hit **OK.**

You will notice that it is added to the palette list as well as to the resource bar.

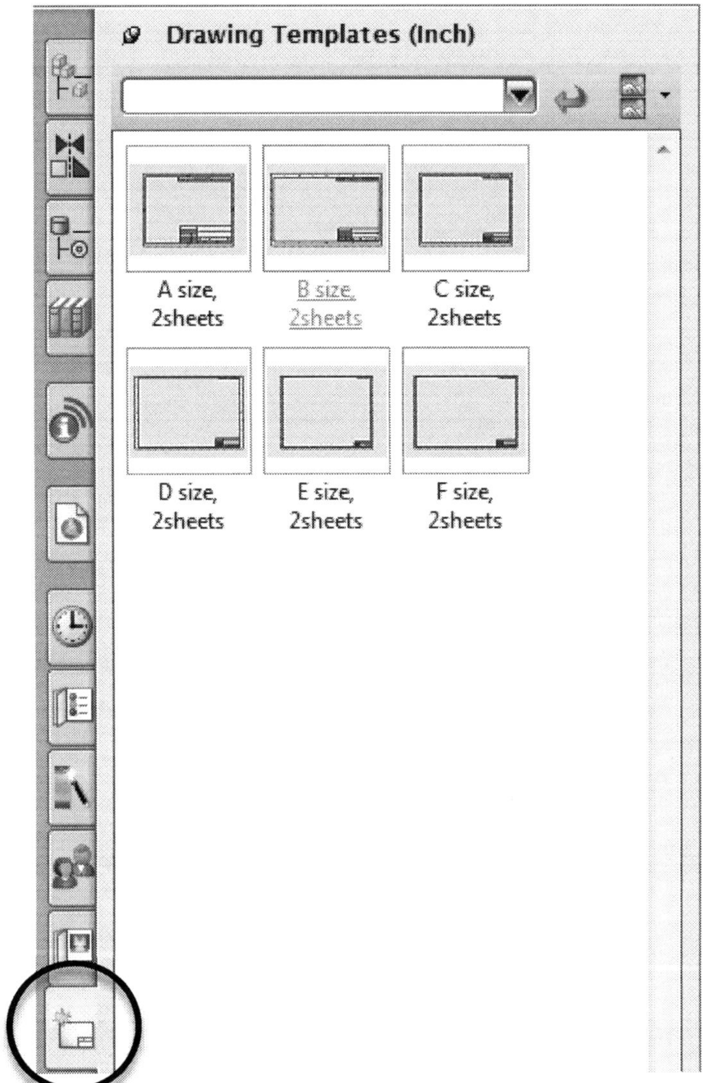


You can add other sample palette files available in the same folder in the same way.

- Open any part file containing some geometry that you have created in one of the previous projects.
- Click the icon ***Drawing Template (inch)*** as shown in the following figure to open the template and click on the file "B size, 2 sheets".

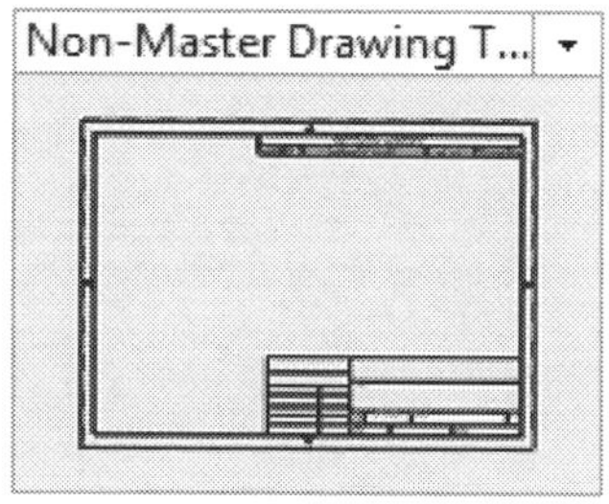


A size, 2sheets

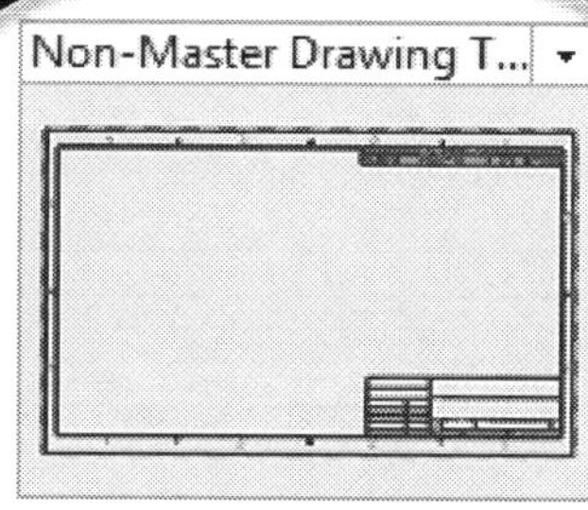


B size, 2sheets

C size, 2sheets

- This will automatically create a new file with your base part assembled in it. Notice the Assembly Navigator and the Title Bar. This is the concept of master modeling that we will be discussing in Chapter 19.
- Open the Part Navigator and notice how the drawing sheets with the drawing border and the standard views defined in the template file have been automatically added to this new part file. *This is a very fast way to produce drawings and lets you work independently in a separate file without making any changes to the original file.*

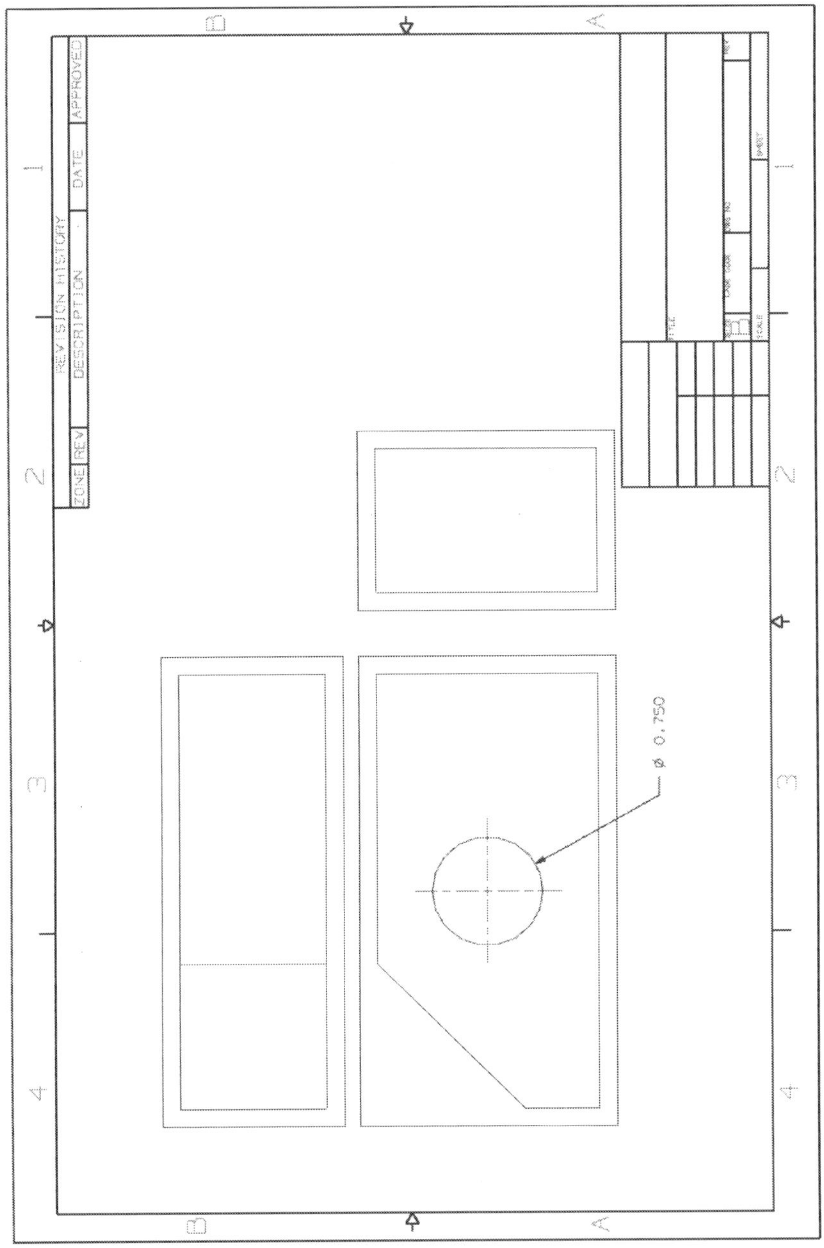


- Save the file.

***End of Project***

## 17.4 Adding Basic Drawing Views

In NX 8.5, views of the solid model may be added to drawings using several methods, including:

- Importing pre-defined or saved model views
- Adding projections of an existing drawing view
- Adding a section view based upon an existing drawing view
- Creating an enlarged detailed view of an existing drawing view

Drawing views may be moved, edited, and deleted using functionality that will be discussed in a later part of this chapter.

### ***Detail: Layer Masks and Drawings***

The Layer Masking functionality is used in **Drafting** mode to control the visibility of items in drawing views. Layer Masks are accessed through the ***Format / Layer Visible in View*** menu (see the figure on the next page). This allows you to control which layers are visible in every view and drawing sheet contained in a part file. Manipulating these settings allows you to control the display of your drawing views.

By default, when you create a new drawing, NX 8.5 sets the 'Visible in View' status to 'Global' for the drawing sheet according to the following rule: all layers that are marked 'Selectable' become 'Visible,' while all others are 'Invisible.' The same thing occurs when you import a view. What this means is that any entity on a 'Visible' layer can be seen in the drawing. Often, you will need to modify these settings to get the proper geometry displayed on the drawing.

Follow these rules to reduce errors and problems related to the 'Visible in View' functionality:

- Place the resultant solid geometry on a specific layer, and don't change it
- Move any construction geometry to layers that are Invisible before adding views
- Create all drawing entities on a specific layer or layers
- When possible, all drawing views should have the same layer mask
- Assign a layer mask to all finished drawings to isolate them from layer settings changes

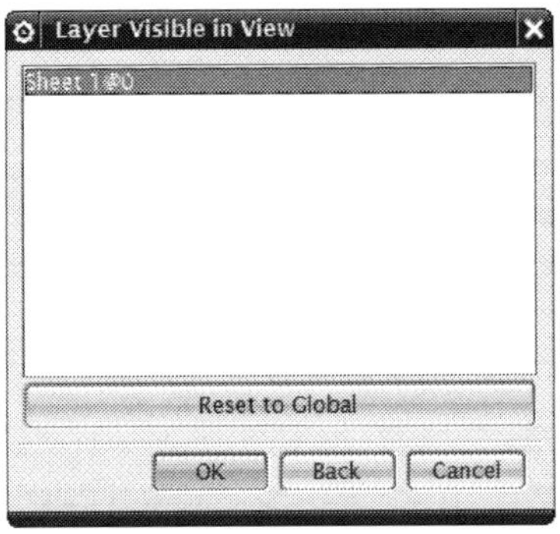


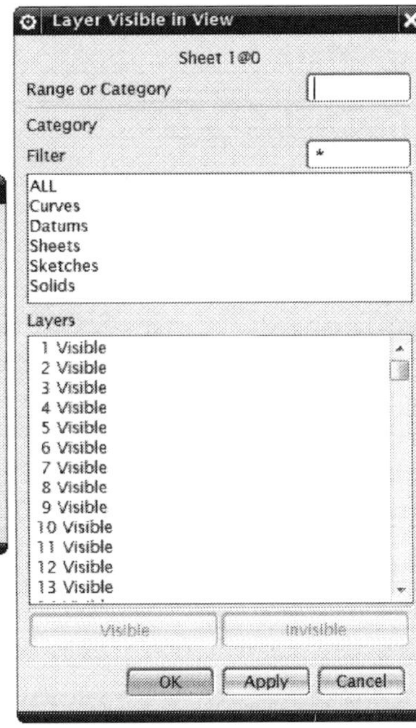


## ***Detail: Updating the Drawing Views***

The NX 8.5 drawing is created using views of the 3D model. When you add, change, or delete a feature from the model, the drawing views become "Out of Sync" with respect to the model. If this happens, you will see an 'OUT-OF-DATE' message appended to the drawing sheet name when you enter the drafting application. It is important to keep the drawing views up to date. To update the drawing views at any time, select ***Edit / View / Update*** and select the views you need to update or select one of the options shown in the figure below. You can also select a view and **Right Click** to open a pop-up that has a list of commands including update views.

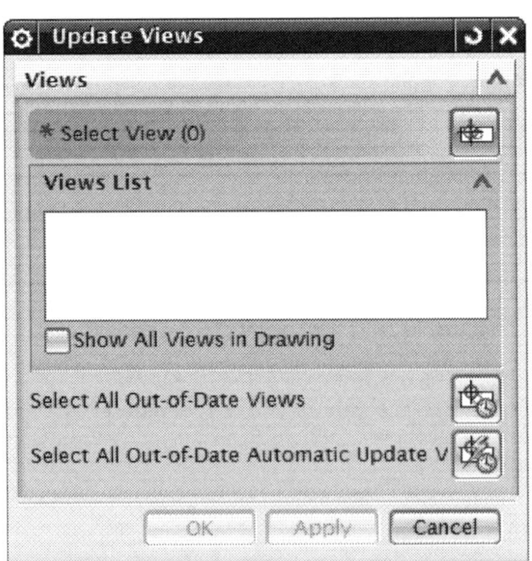

## ***Detail: Importing the Base View and Orthographic Views***

The first view added to a drawing sheet must always be imported from the model. You can use one of the pre-defined views (TOP, FRONT, RIGHT, etc.), or you may use a view that you have saved. The first view is commonly referred to as the 'Base View,' since it becomes the basis for the remaining drawing views.

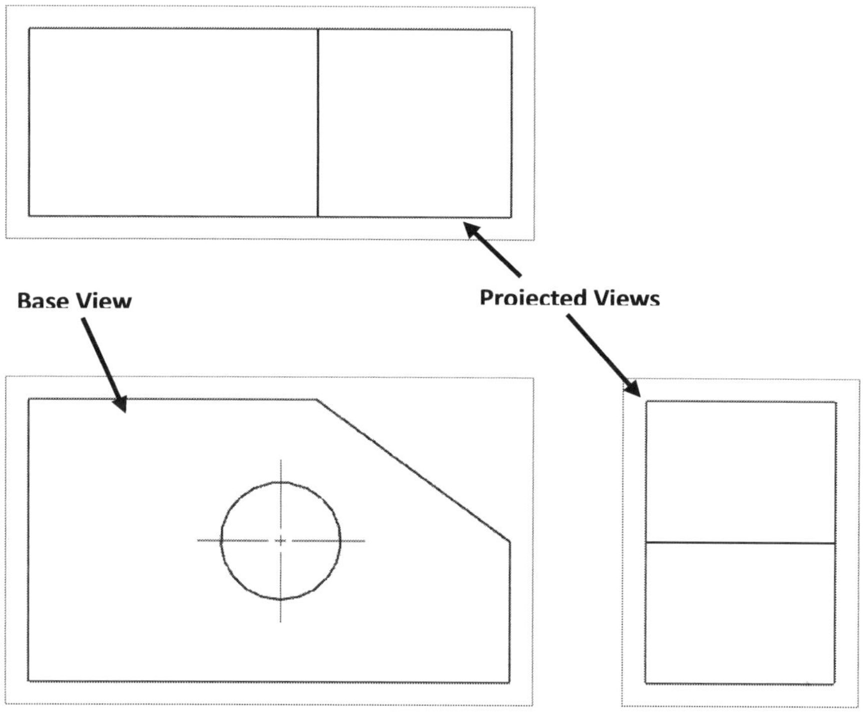


Once you have added the Base View, you can create projected views using the 'Orthographic View' option. Orthographic Views are created by selecting the Base View as the 'Parent View,' and using either the 'First-Angle' or 'Third-Angle' projection method, according to the drawing settings. As such, the Base View essentially becomes the 'Front' view on the drawing sheet, and all of the projected views become the 'Top,' 'Right,' 'Bottom,' etc. views.

You may define Orthographic views using any view type as the Parent View (although in some cases this would be a fruitless task!). All Orthographic projected views are aligned to the parent view. By default, they also inherit its Layer Mask.

## *Project 17.4.1*

Here you will create the Base View and several projected views. Move all your construction curves, sketches, and datum planes to invisible layers, and display only the solid body on the drawing.

- Open any part you have with a solid body and datum planes (In this example we have used a solid from a previous exercise and added a datum plane to trim the solid) and select **Start / Drafting.**
- Select **Insert / View / Base**.
- Select the 'TOP' view from the list.
- Select scale 1:1.
- Move your cursor onto the drawing sheet, and place the view by Left Clicking. Notice that when you move your cursor, NX 8.5 automatically generates another view based on the one you just placed.
- Choose OK or Middle Click to let NX 8.5 know you are done placing views.

You should see that a view has been added to your drawing sheet, as shown in the figure below. If your TOP view does not match this, delete it and add the Base View again, this time importing the FRONT, RIGHT, or any saved view that gives you this result. If the orientation doesn't match you can click on the **Orient View** icon and reorient your view.

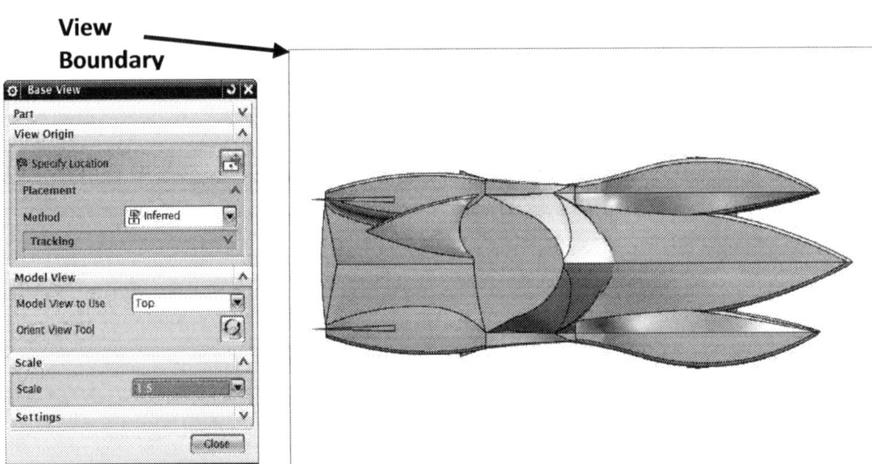


Notice the rectangle around the view – this is the 'View Boundary'. It is updated automatically, based upon the size of the model displayed in the view, and does not show up in the final printed drawing. You can hide/unhide the view boundaries from **Preferences / Drafting / View / Display Borders.** The entities displayed in your Base View may be different, depending upon the Layer Mask. If the solid body is on a layer by itself, and the construction entities are on hidden layers, you will not see them in the drawing.

- Select ***Start / Modeling.***
- Select ***Format / Move to Layer.***
- Select the solid body and choose **OK.**
- Enter **10** for the Destination Layer and choose **OK.**

The solid body should disappear, since it was moved to a layer that is hidden. Follow the next steps to make Layer 5 the Work Layer, mark Layer 1 'Invisible,' and make Layer 10 'Selectable.'

- Select ***Format / Layer Settings.***
- Enter **5** in the 'Work Layer' field and choose **Apply.**
- Double-Click on Layer 1 to mark it as **'Invisible.'**
- Double-Click on Layer 10 to mark it as **'Selectable,'** and choose **OK.**
- Select ***Start / Drafting.***

Notice that there are some construction entities visible in your base view; however, the solid body is no longer visible. The next steps will correct this problem.

- Select Format / Layer Visible in View.
- Select your base view and choose **OK**.

The 'Visible Layers in View' dialog box appears. Use it to mark all layers except layer 10 'Invisible.'

- Select the **ALL** category.
- Select **Invisible** to mark all layers as 'Invisible.'
- Double-click on layer 10 to mark it as 'Visible,' and choose **OK.**
- Choose **Cancel** to exit the 'Layer Visible in View' dialog box.

Notice that an (OUT-OF-DATE) message appears at the lower left corner of the Graphics Window. This indicates that one or more drawing views has been modified, and requires an update.

- Select ***Edit / View / Update.***
- Choose **OK.**

The solid body should now be displayed, and all construction entities should be invisible. Notice that the View Boundary is updated to reflect the entities shown in the Base View. Now, proceed by adding several projected views.

- Select Insert / View / Projected.
- Notice that an arrow appears from the parent view. This arrow can be rotated 360 degrees around the base view to create projected, secondary views on the drawing. Move the cursor to the right of the Base View, and click to create the projected Orthographic view shown in the figure below. NX 8.5 will "snap" the view horizontally in place.
- Create the remaining three projected views using this technique, as shown below.
- Save your drawing file.

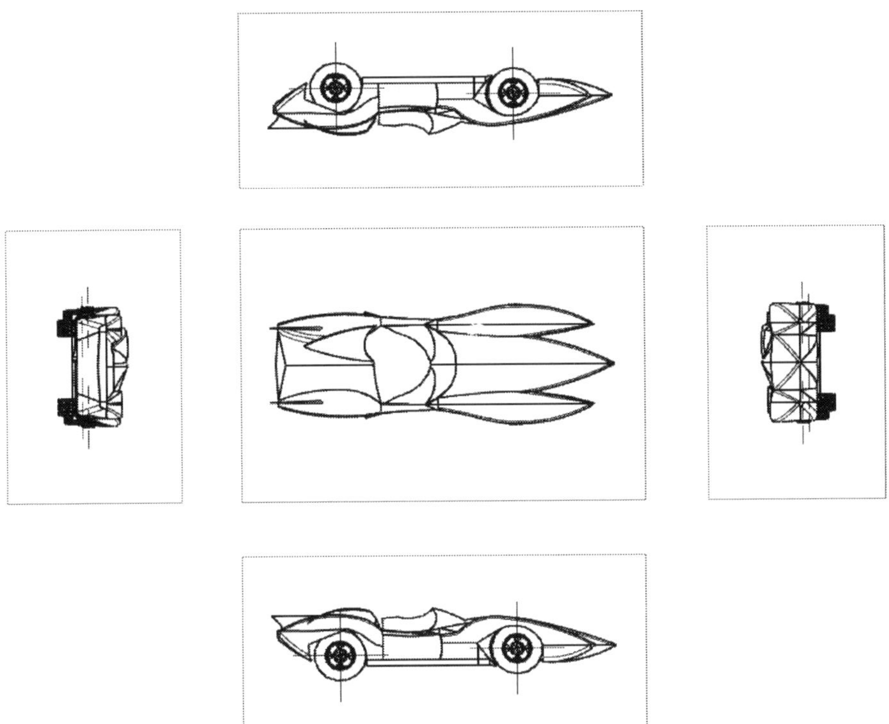

***End of Project***

## 17.5 Editing Drawings and Views

Creating engineering drawings is really an art form. The intent is to clearly communicate the function and/or manufacturing dimensions of a product. To accomplish this usually requires some iterations of the drawing, especially in the early stages of documentation.

### *Detail: Editing Drawing Sheets*

You can change options of existing drawings using ***Edit / Sheet***. A common operation is to increase or decrease the size and/or the scale of a drawing. Note that if the size is decreased, any existing drawing views must fit within the new boundaries, or the operation will fail.

### *Detail: Edit View Style*

You can edit drawing views by selecting ***Edit / Style*** and then selecting an object to modify. You can change its angle, scale and other parameters from the menu shown below.

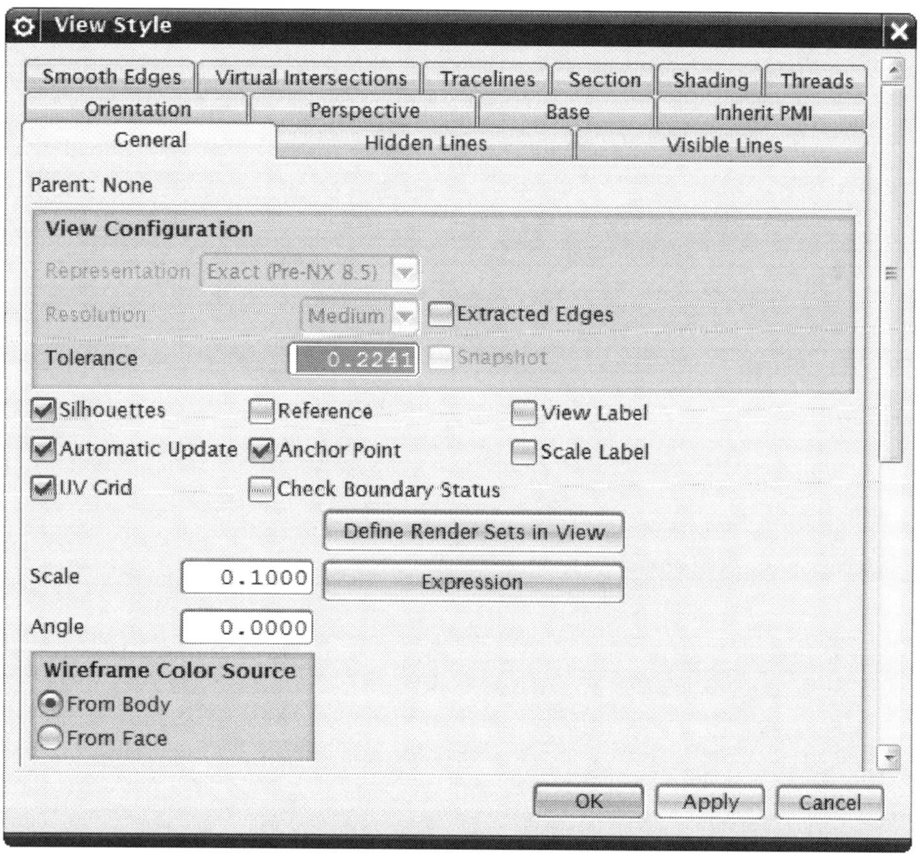

You can also add or remove the View or Scale Label of a drawing view, change parameters such as Hidden Lines, Visible Lines, Smooth Edges, and Threads, and Virtual Intersections using this dialog box.

## ***Detail: Move View***

One or more Drawing Views may be moved using ***Edit / View / Move/Copy View*** (see the figure below). The movement of the drawing views may be free or restricted to certain directions, such as either horizontal or vertical. You can also move a view just by clicking and dragging it.

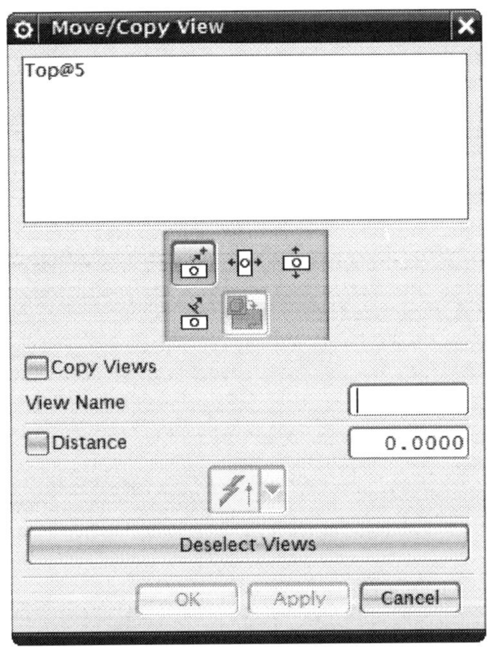


You may also duplicate drawing views or move them to another drawing sheet using the 'Move/Copy View' dialog box. Whenever you copy or move a drawing view, all dimensions, notes, and symbols associated with the view move along with it. You can also move a view from one sheet to another by simply cutting and pasting it.

## ***Detail: Remove View***

To delete a drawing view, simply select the view(s) that you wish to delete and hit Delete on your keyboard or right click the view and choose the delete option. You can also select the ***Edit/Delete*** option or delete the view from the Drawing Navigator. When you remove a drawing view, all dimensions, notes, and symbols associated with the view are also deleted.

## *Detail: Align View*

Drawing Views can be aligned using ***Edit / View / Alignment***. Views may be aligned horizontally, vertically, or perpendicular to a line. There are three alignment methods available when the Overlay Method is selected: Model Point, View Center, and Point-to-Point. NX 8.5 also shows dotted alignment lines if you click and drag a view to align it.

The most frequently used alignment method is 'Model Point', which allows you to select an entity from the 3D model that serves as the alignment point for all selected views. The 'Point-to-Point' method is similar, except that you must select the corresponding point from each view to be aligned. NX 8.5 also lets you align the views by simply dragging them to approximate locations with respect to other views. A dotted line with arrows appears and the view snaps in to align with the other views.

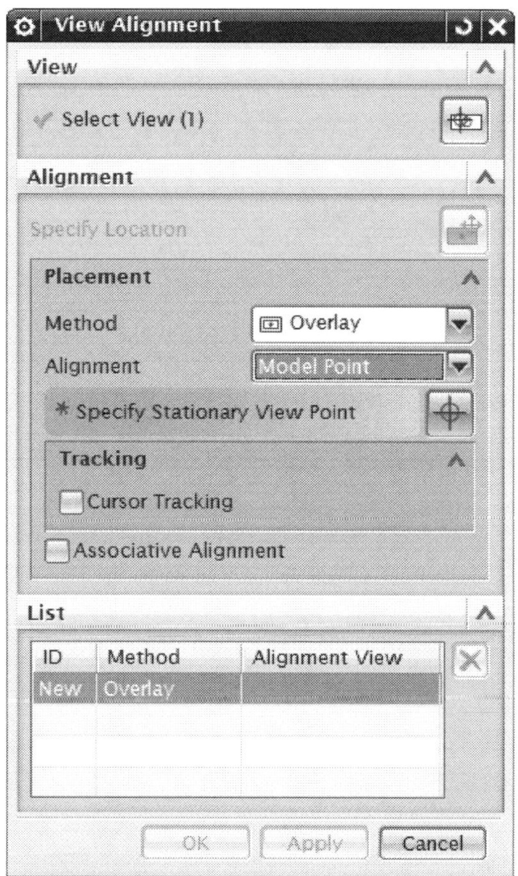

## *Project 17.5.1*

Remove several extra views from your drawing, change the drawing scale, and re-arrange the remaining views.

- Open the part from previous project (if required) and select ***Start / Drafting.***
- Select the Top and Left projected views and delete them.
- Select ***Edit / View / Move/Copy*** .
- Select the three remaining views, and select the 'To A Point' icon .
- Move the views to the lower-left corner of the drawing and click to drop them.
- Choose Deselect Views.

Select the 'Bottom' projected view, and select the 'Vertically' icon .
Move the view up and click to drop (see figure below).

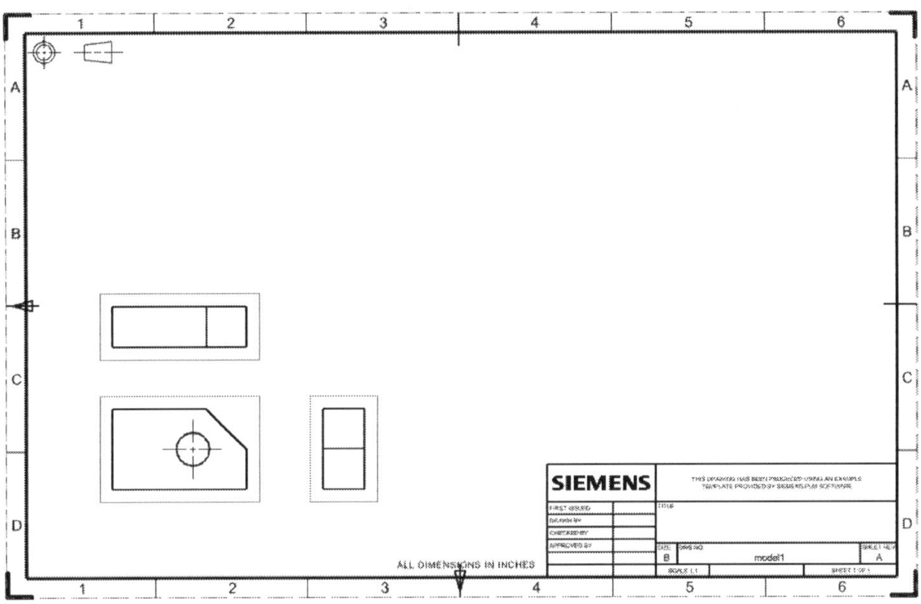


- Choose Deselect Views.

Now intentionally misalign the Right projected view, and then re-align it...

- Select the 'Right' projected view, and select the 'To A Point' icon.
- Move the view so that it is no longer aligned to the Base View.
- Choose **Cancel.**
- Select ***Edit / View / Alignment*** .
- Select the Stationary Model Point as shown in the figure below.

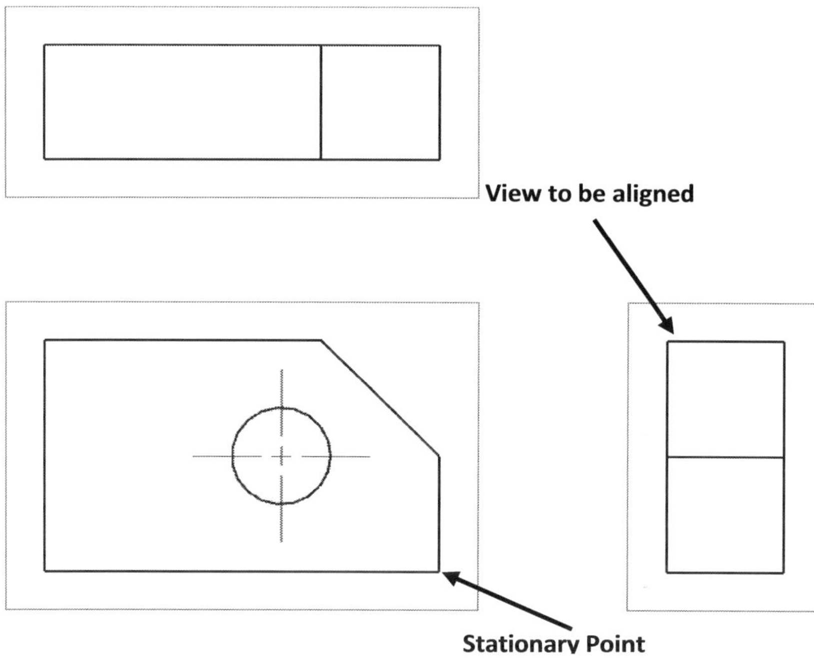


- Select the View to be aligned, as shown in the figure above.
- Select the 'Inferred' alignment icon  .

The misaligned view should become re-aligned to the Base View. You may need to use 'Move View' to change its horizontal location...

- Save your drawing file.

***End of Project***

## 17.6 Advanced Drawing Views

You have learned how to create and manipulate basic drawing views. To complete most drawings, you will be required to create more advanced view types, such as Section and Detail Views. Each of these view types requires a Parent View, just as for Projected Views.

### *Detail: Projected Views at an angle*

Auxiliary Views are projected views along non-orthographic directions, as shown in the figure below. They are very useful to show the form of a model that has features located on angled surfaces.

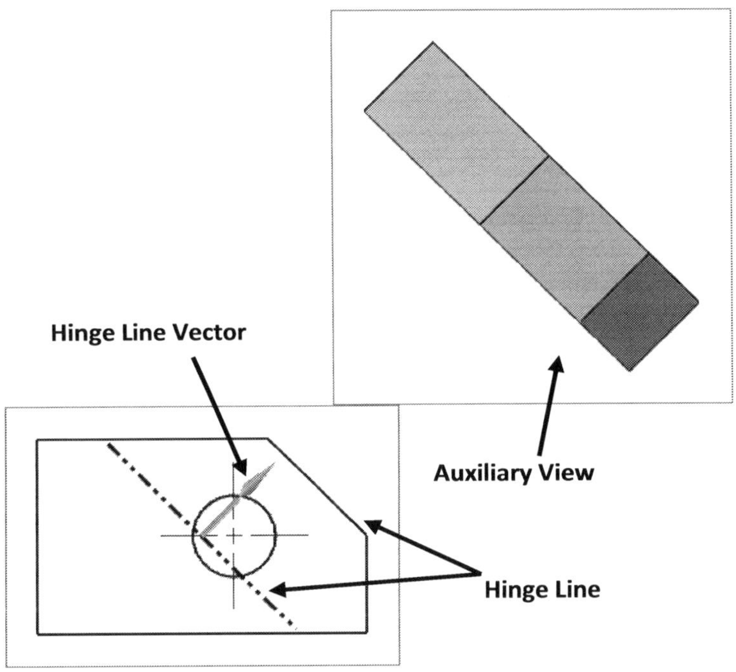


In NX 8.5, the projection direction of Auxiliary Views is controlled by a 'Hinge Line.' The Hinge Line is usually a linear model edge that is normal to the desired projection direction ('Hinge Line Vector'), although it may be defined using other techniques. If the Hinge Line is defined using model geometry, the view angle is parametric; thus, the view orientation will automatically update if the angle in the model is modified.

The direction of the Hinge Line Vector determines which side of the model is visible once the auxiliary projection is done. If it is in the wrong direction, you can select the 'Reverse direction' option.

## ***Detail: Simple Section Views***

NX 8.5 has the ability to create Section Views of solid models. The section cuts are automatically generated, and updated along with the solid geometry. Section Views are a very effective technique for illustrating internal geometry of complicated parts. There are several types of Section Views available in NX 8.5; here we will discuss only Simple Section Views.

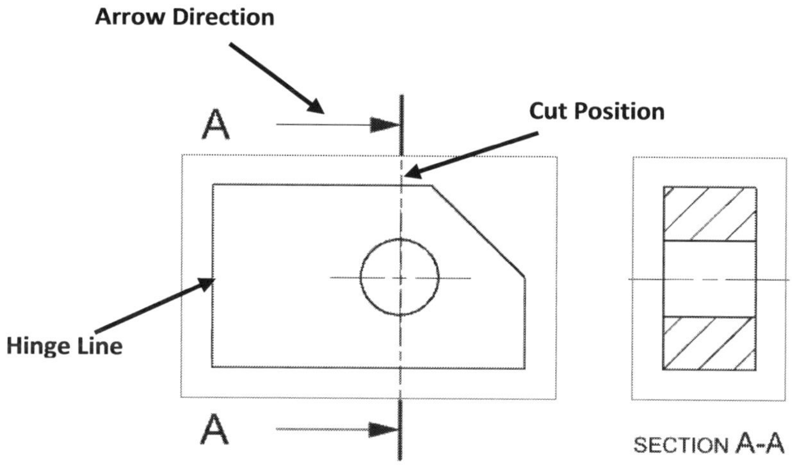


Simple Section Views are similar to Projected Views in that they require a Hinge Line. In addition, you must specify a Cut Position and the Section Arrow Direction, as shown in the figure above. NX 8.5 creates the section cut by constructing a plane that is parallel to the Hinge Line and passes through the Cut Position.

The location of the Section View Arrows is automatically determined by NX 8.5 if you don't select it. You can modify these locations, as well as the Arrow Direction and Cut Positions of existing Section views using ***Edit / View / Section Line***. The Section View Label display, View Letter, and crosshatch parameters can be changed via various Preference settings.

## Detail: Detail Views

Detail Views are essentially enlarged regions of existing drawing views. You are able to specify the boundary and scale, and NX 8.5 creates a set of parametric view entities as shown in the figure below.

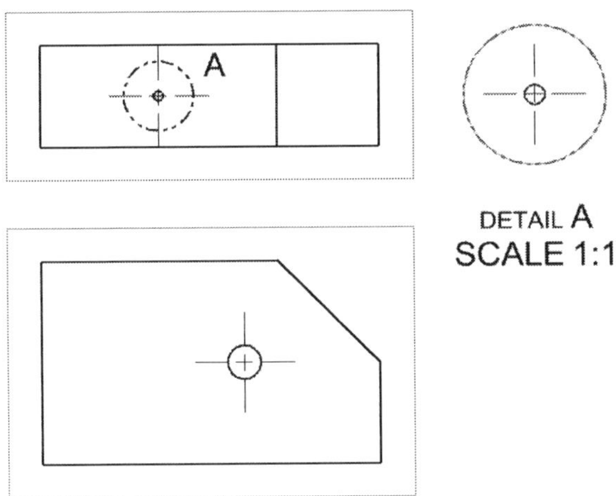


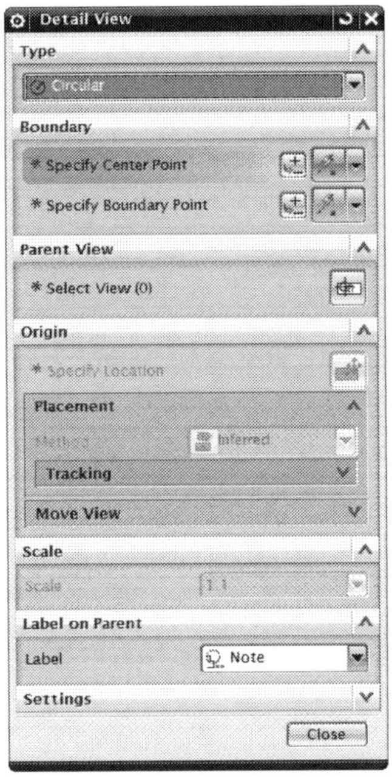


Detail Views are useful for illustrating small features of a model. Any changes to the solid model are automatically cascaded through the Parent View and into the Detail View. The Detail View Label display, View Letter, and View Boundary options can be changed via various Preference settings.

## Detail: Adding 3D Views

You can add 3D views of a model to a drawing by using the 'Insert View' option. You can insert one of the standard views (i.e. Isometric or Trimetric), or you may import a customized orientation. If you save a view in the modeling application, you can then use it in the drawing. In NX 8.5 you can also add an image (i.e. jpeg, png and tif) in to a drawing. To add an image use **Insert / Image.**

## Detail: View label

When you add a view, if the 'View Label' and 'Scale Label' boxes on the **View Style** dialog box (see the figure below) are checked, the View Name and Scale will be shown on the drawing.

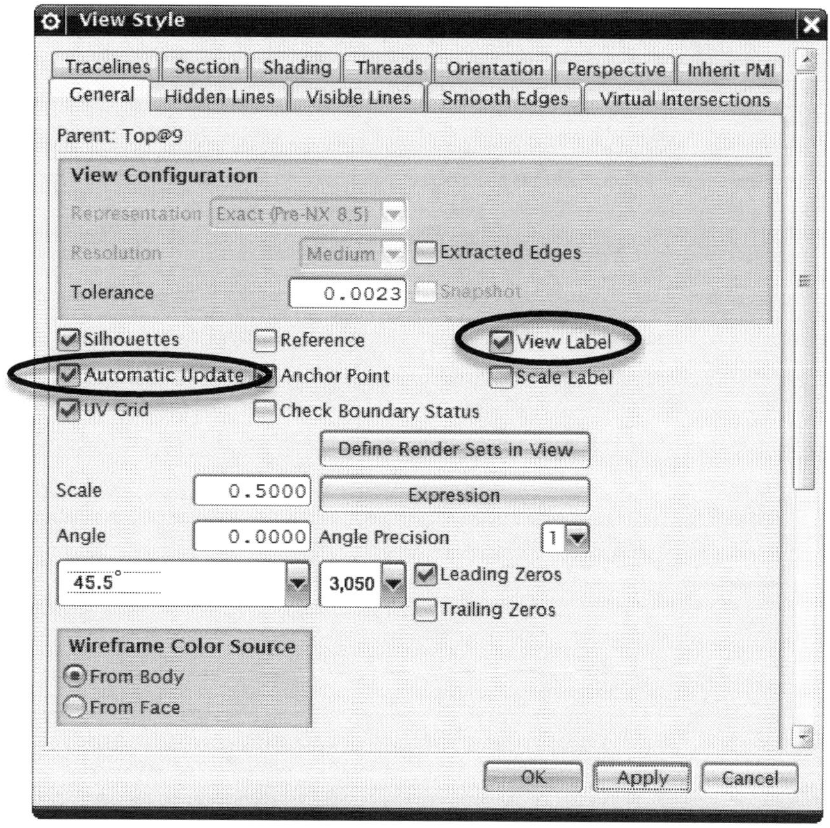


You can change these options for views that have already been created using ***Edit / Style***. If you display the View Name, you may also wish to assign a meaningful name to it.

You can modify the View Labels with ***Preferences / View Label***, as shown in the figure below. There are different options available for Section, Detail, and Other views.

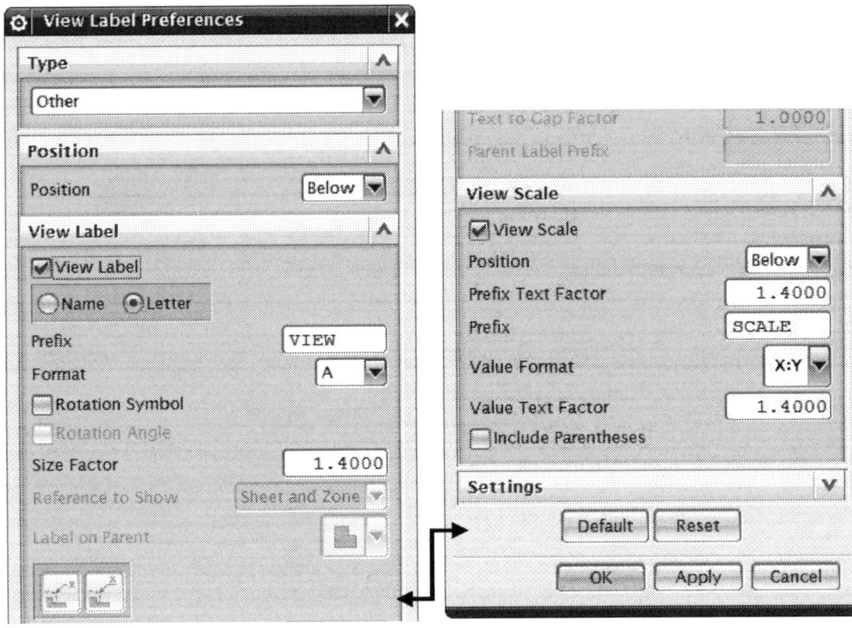


Some commonly used options are to change the existing View Letter and modify the Scale display format. Labels on existing views may be turned on or off using ***Edit / Style.***

## *Project 17.6.1*

Add several features to your solid model and update the drawing by adding three new views. Then modify an existing view to add a section cut.

- Open the part from the previous project (if required) and select ***Start / Modeling.***
- Rotate your model into a good 3D orientation.
- Select View / Operation / Save As.
- Enter '**NICE**' into the 'Save Work View' dialog box, and choose **OK.**

You have now saved a view that can be used later on for a drawing view.

- Add a **0.20in** chamfer to the existing hole, and add a **0.125** x **1.0** hole in the top, as shown in the following figure.

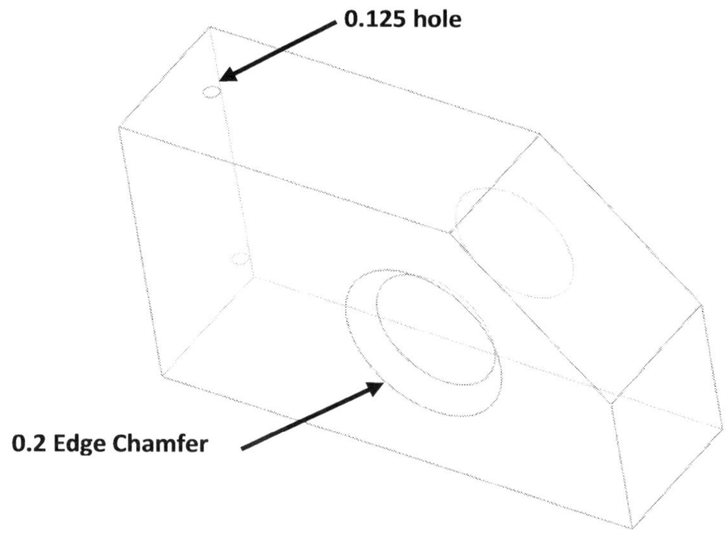


- Select ***Start / Drafting.***
- Select Edit / View / Update.
- Select All Views and choose **OK** to update all of the drawing views with the modeling changes. *You can also select All Out-Of Date Views option.*
- Select **Insert / View / Projected.** A preview of the projected view appears.
- Move the cursor to the right of the *Base* View and **click** to create the Auxiliary View, as shown in the following figure.

**Note:** NX 8.5 will align the projected view to the angled edge by "locking" the view in place. Move the cursor around near the angled edge until it locks. The orthogonal face will highlight when locked.

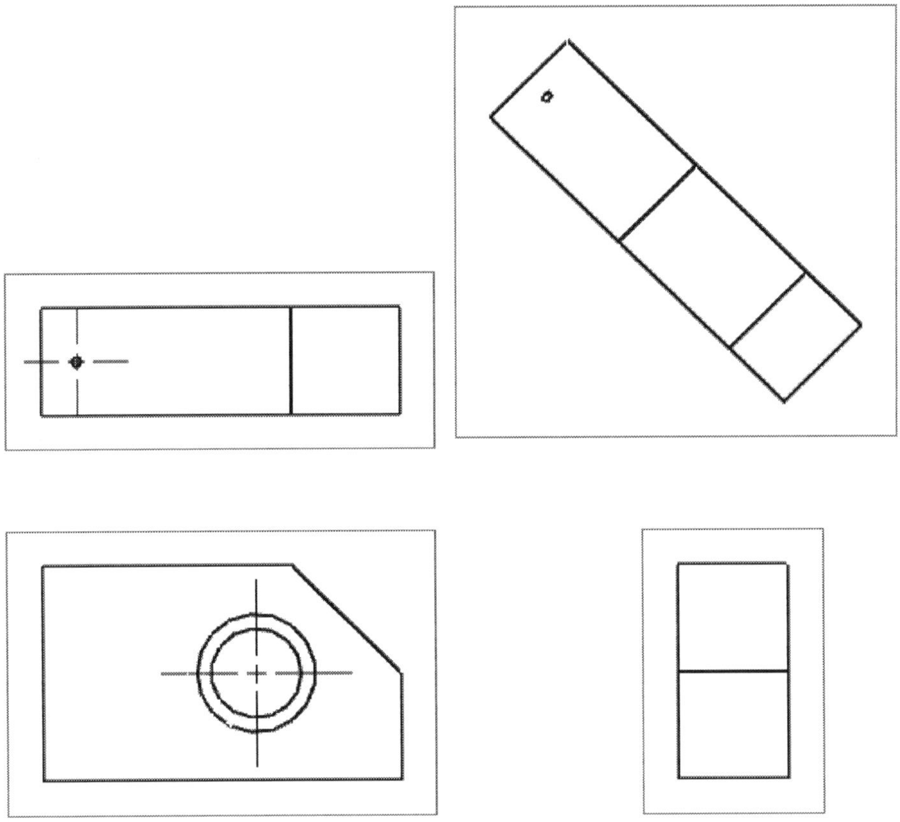

- Hit Escape.

Now, add a Detail View of the small hole.

- Select the ***Insert / View / Detail***  from the 'Drawing Layout' menu bar.

Read the Cue. NX 8.5 is prompting you to select the detail center on the Parent View.

- Select the center of the small hole as the detail center and move your cursor slightly above the top edge of the block to define the size of the Detail View Boundary.
- Select the scale 2:1 from the menu displayed in the top left corner.

Read the Cue. NX 8.5 is prompting you to select the location of the Detail View on the Drawing.

- Place the center of the Detail View as shown below.

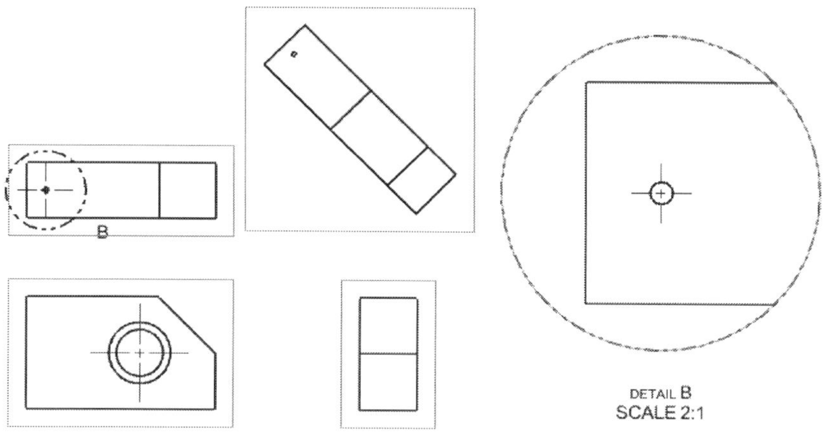


- Click near the edge of the circular block surrounding the Detail View Boundary.
- Choose **OK.**

Compare your results to the figure above. Now, add a Section View through the middle of the block.

- Select ***Insert / View / Section***  .
- Select the top projected view as the Parent View by clicking.

Read the Cue. NX 8.5 is prompting you to define the Cut Position.

- Click the center of the small hole to define the Cut Position.
- Move the cursor above the 'Top' projected view and click to create the Section View, as shown below.

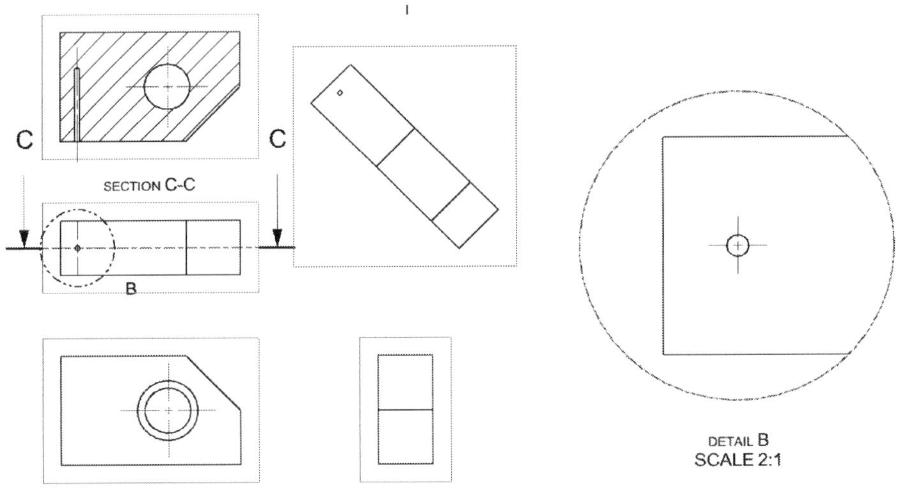


- Hit Escape.

Compare your results to the figure above. Now add a 1:1 scaled 3D view of the model.

- Select Insert / View / Base.
- Select Scale 1:1.
- Select the 'NICE' view from the drop down list of available views.
- Click on the style button and check the 'Scale Label' box to add a Scale Label.
- Move your cursor into the Graphics Window and place the view as shown below.

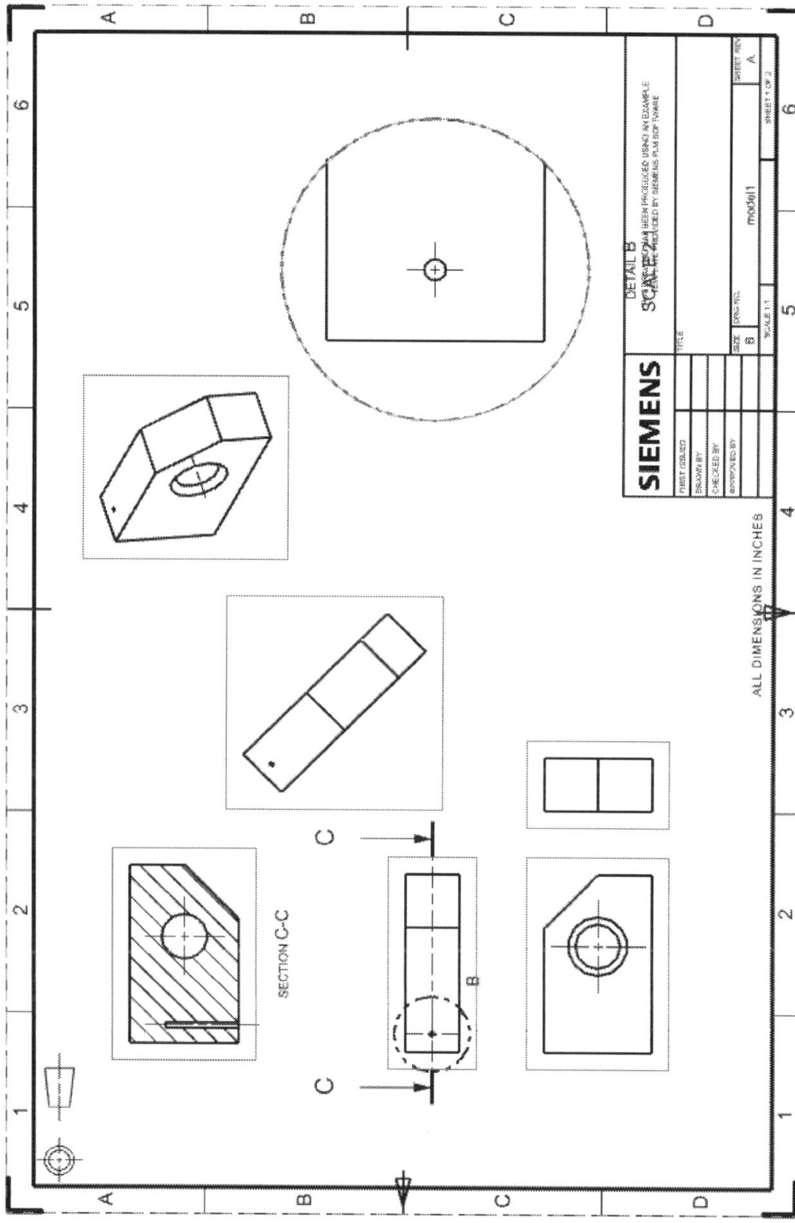


- Save your drawing file.

***End of Project***

## 17.7 Creating Dimensions

Placing dimensions on a drawing is a very similar process to dimensioning a sketch. In drawing mode, you can also include information related to the part's manufacturing and inspection processes, such as the number of decimal places and tolerance specification.

There are a number of Dimension icons available in NX 8, including Horizontal, Vertical, Diameter, Radius, Perpendicular, Parallel, etc. In this manual, we will discuss the 'Inferred' option, as its function encompasses most of the others.

### *Detail: Inferred Dimensions*

The 'Inferred Dimension' dialog box is shown in the figure below. You can find it in ***Insert / Dimension*** menu.

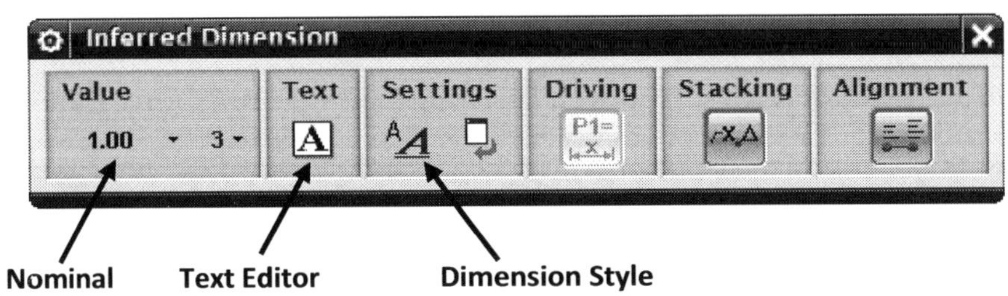


A general outline of the dimension creation procedure is as follows:

1) Select the **Dimensions** tab, and set Dimension Placement to 'Automatic Placement' 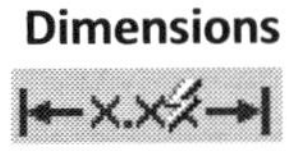 from the dimension style menu.
2) Set the Tolerance Type and Dimension Precision.
3) Add Appended Text (if required).
4) Select the entity (or entities) to be dimensioned.
5) Place the dimension in the Graphics Window (by left-clicking).
6) If the dimension requires modification, select it and change the desired settings.

You will find that the units, text justification and extension line/arrow settings are not frequently modified once you have started a drawing.

## *Detail: Moving / Modifying Existing Dimensions*

Existing dimensions can be moved at any time by simply clicking and dragging them to a new location. You can also modify the attributes of existing dimensions using the Dimension dialog box. Simply select one or more dimensions, change the desired attributes, and choose Apply.

## *Detail: Appended Text*

You can append text to dimensions with the 'Appended Text' functionality of the **Annotation Editor**. The figure below shows several icons that allow you to place text in various orientations on a dimension. You may add appended text to one, or all of these locations.

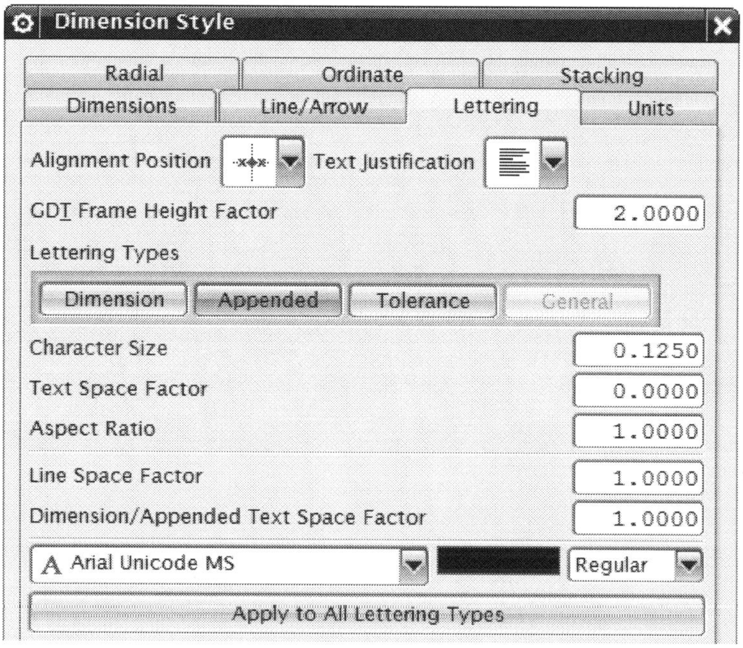


This option is useful for adding notes, GD&T callouts, and special drafting symbols. The functionality of the Annotation Editor will be covered in more detail later in this chapter.

## *Detail: Dimension Placement and Leader Direction*

In general, the 'Automatic Placement' option yields good results; however, some dimensions require manual placement. There are two types of manual placement options: 'Arrows In' and 'Arrows Out.' If you set the Dimension Placement to one of these options, you may also specify either the 'From Left' or the 'From Right' leader placement method. You can change these options from the ***Dimension Style*** menu.

## ***Detail: Drawing-Related Preferences***

Many properties of NX 8.5 drawings are set using 'Preferences.' Select ***Preferences / Drafting*** to set the Drafting Preferences. See the figure below.

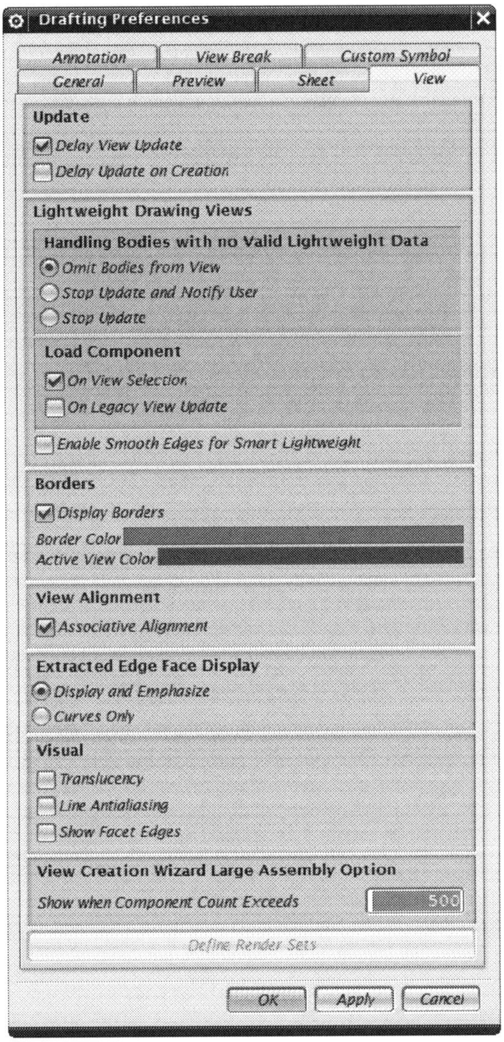


The default values for most Preferences settings are controlled by the settings in the NX 8.5 'Customer Defaults' file. Most 'Preference' settings are stored in the part database, so your changes will remain after you have saved the part. It is good practice to verify all preference settings before creating any drafting entities, thus avoiding a lot of tedious re-work.

Other commonly used preferences are 'Origin,' 'Section Line Display,' 'View' and 'View Label.' You can find it in Preferences menu. To change the preferences of a drafting object follow this procedure:

- Choose ***Preferences / Annotation***
- Select the object(s).
- Change the desired Preference Settings.
- Choose **OK.**

The figure below shows the 'Annotation Preferences' dialog box. Here you can control many settings for Dimensions, such as Line / Arrow Style, Lettering Sizes, Units, etc.

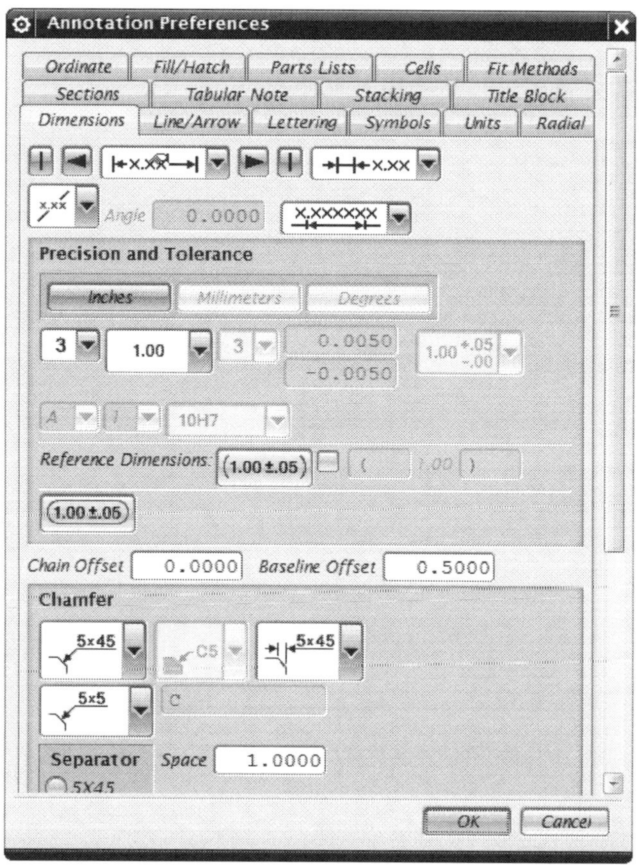


You will also notice that certain Preference settings are duplicated in other dialog boxes. This provides for convenient access to commonly used functionality, such as changing the dimension tolerance mode or text height.

## ***Detail: Preferences Action Buttons***

Most drafting-related preference dialog boxes have a set of buttons on the bottom that can be used to synchronize the settings of the current Preferences dialog box.

The 'Inherit' button allows you to select an existing entity to set up the Preferences dialog options on the current tab. The 'Inherit All' inherits or resets ***all*** properties on ***all*** tabs based on the selected entity. The 'Load Defaults' option restores the Preferences dialog options on the current tab according to the customer defaults file. The 'Load All Defaults' does the same for ***all*** properties on ***all*** tabs. The 'Reset' and 'Reset All' options will restore the dialog box settings to those most recently set for the part.

## *Project 17.7.1*

Use the 'Inferred' tool to add several dimensions to your drawing, as shown in the figure below. Use the View Label Preferences to re-name the drawing views.

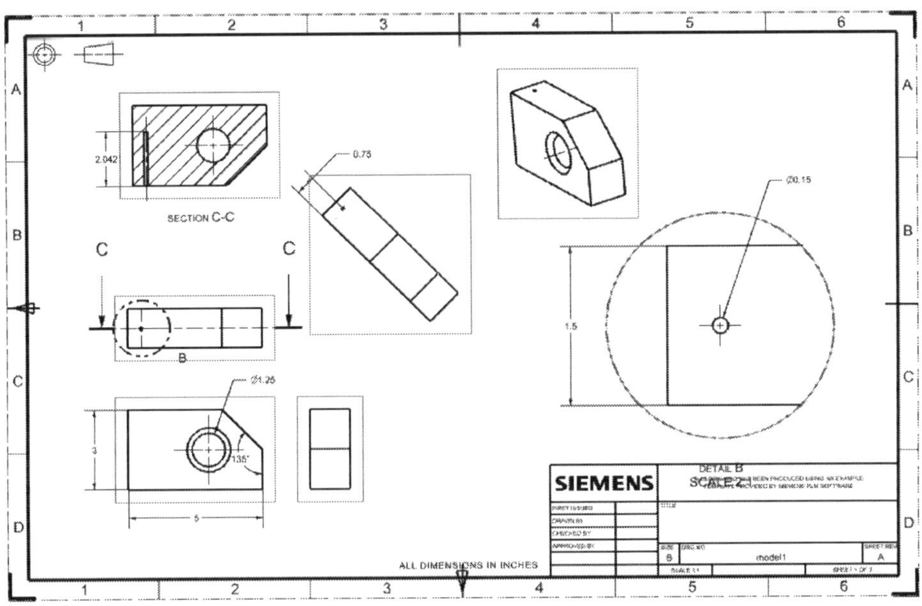


- Open the part from last project (if required) and select ***Start / Drafting.***
- Delete the right projected view and add a section view as shown in the figure above.

Turn off the View Boundary Display.

Select Preferences / Drafting.

- On the View tab, uncheck **Display Borders** to turn off the view boundary display.

Set up the default dimension style.

- Select ***Preferences / Annotation.***
- Select the **Inches** from Units menu.

- Select the 'No Tolerance'  Tolerance Type setting from the Dimension menu.
- Select the three places (3) dimension precision setting from Dimension menu.
- Select the 'Automatic Placement' 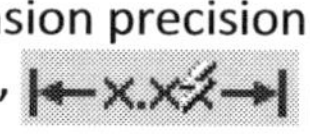 Dimension Placement method.
- Choose **OK.**

| **Note:** You may also change these Settings from the 'Inferred" Dimension dialog box. |
|---|

- Select ***Insert / Dimension / Inferred*** or select the  icon.

Create the width and height dimensions of the block, as shown below.

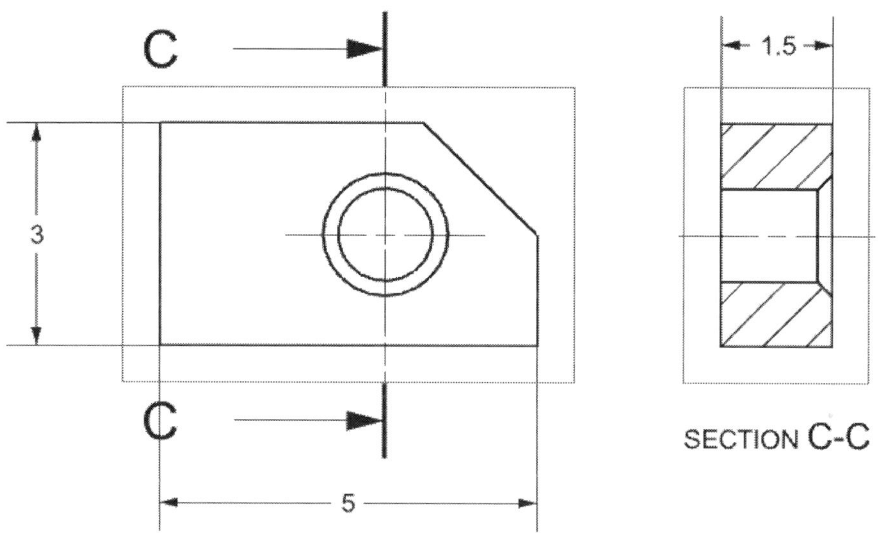


- Click on the bottom edge of the block (selecting it once is enough).

Move the cursor into the Graphics Window to preview the dimension placement.

- Click to place the width dimension.
- Select the left edge of the block and create the height dimension.

Create the thickness dimension of the block.

- Select the top edge of the block in Section C-C and click to place the thickness dimension.

Now, create the diameter dimension on the small hole.

- Click on the **Hole** icon in the dimension toolbar.
- Select the edge of the small hole in the detail view.

Notice that as you preview the diameter dimension its leader length is constrained.

- Double Click the dimension leader line, Dimension Style window will open.

- Before you place the dimension, select **Manual Placement – Arrows In**  Dimension Placement method.
- Move the cursor into the Graphics Window and place the dimension as shown below.

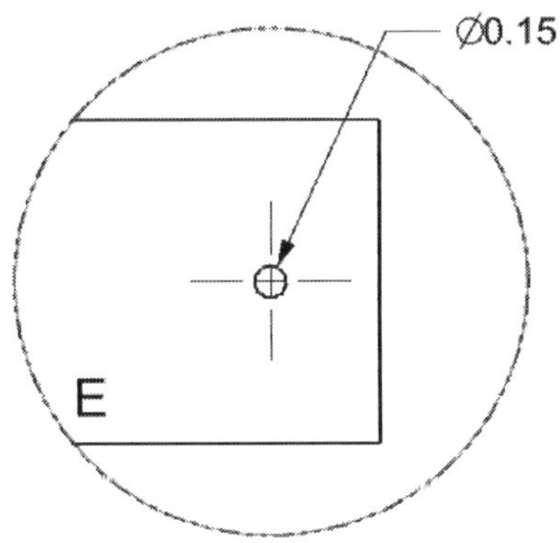


DETAIL E
SCALE 1:1

Now we will create a two-place depth reference dimension on the small hole.

- Set the 'Tolerance Mode' to Reference and the 'Precision' to Nominal – Two Decimal Places, as shown in the Figure below or in Dimension Style window.

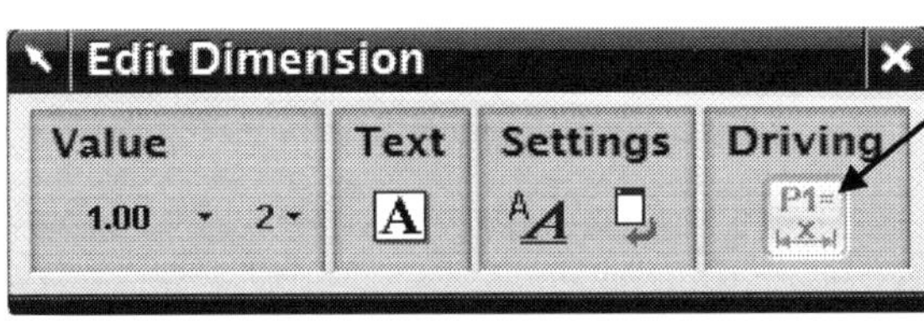


Change Driving to Reference

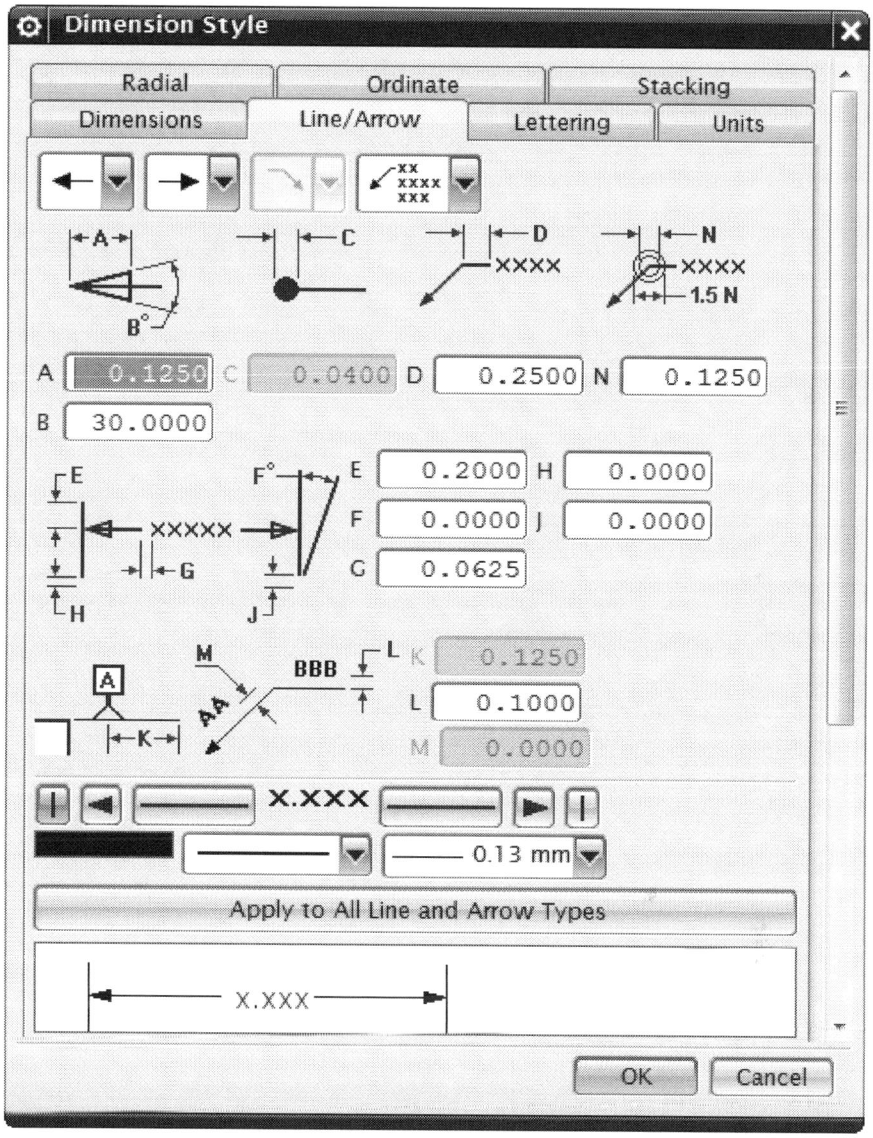

- Select the 'From middle' Leader Direction from the Dimension Style Menu found under the ***Line/Arrow*** tab. Hit **OK**.
- Move the cursor into the Graphics Window and place the dimension as shown below.

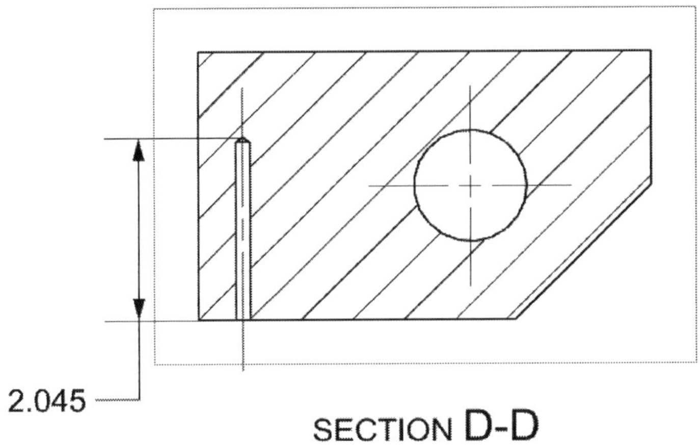


- Choose **Reset** (from the toolbar) to restore the dimension preference settings to their default values.

Create an angular dimension on the side chamfer. Use a counter-clockwise selection order to get the dimension orientation correct.

- Select the edges in the order shown below, and place the angular dimension.

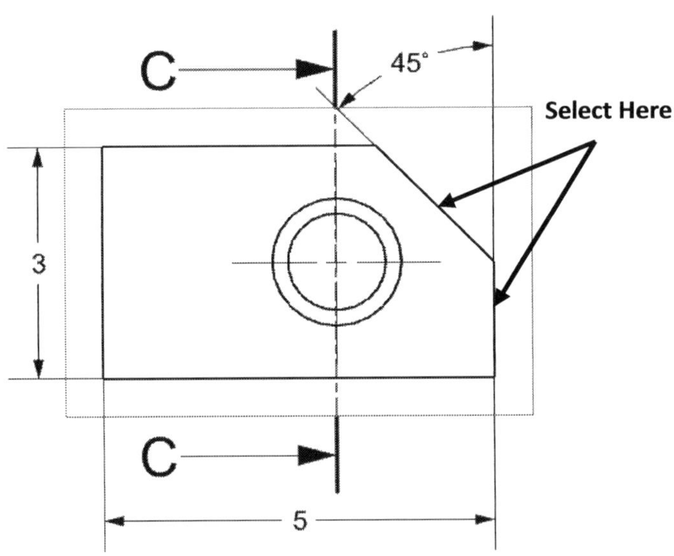


- Choose Escape.

Create a diameter dimension on the large hole using the 'Cylindrical' dimension tool.

- Select ***Insert / Dimension / Cylindrical*** or select the 'Cylindrical' dimension icon .
- Select the edges as shown in the figure below and place the diameter dimension.

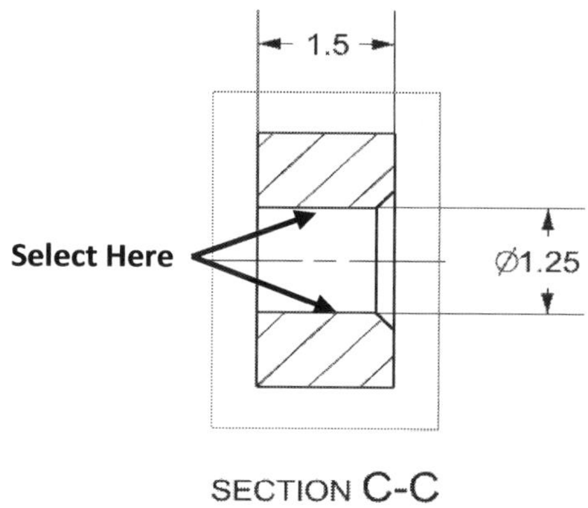


- Hit Escape.

Create a dimension to the center of the small hole using the 'Perpendicular' dimension tool.

- Select ***Insert / Dimension / Perpendicular*** or select the icon.
- Select the edges as shown in the figure below.

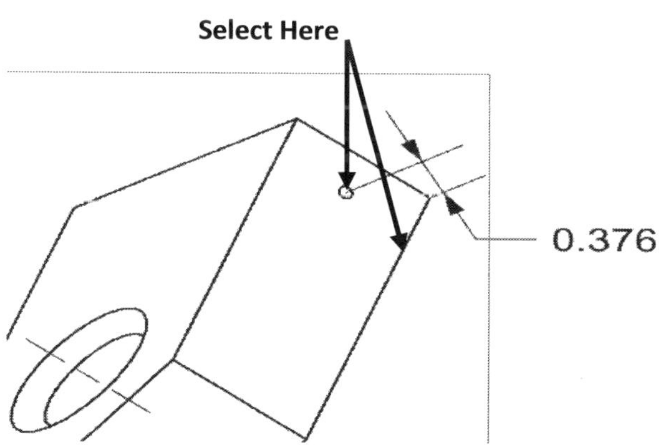


- Before you place the dimension, select **Manual Placement – Arrows Out** from the Annotation Preferences.
- Place the perpendicular dimension as shown in the above figure.
- Choose **Escape.**

Now we will modify the View Labels for Section C-C and Detail A.

- Double-click the 'Detail A' View Label Style window will pop up.
- Set the 'Value Format' pop-up menu to **Ratio** .
- Set the 'Label on Parent' pop-up to **Label** .
- Change the 'Letter' to **D** .
- Choose **OK.**
- **Double-click** the 'Section C-C' view label.
- Change the 'Letter' to **A** as shown in the figure below and choose **OK.**

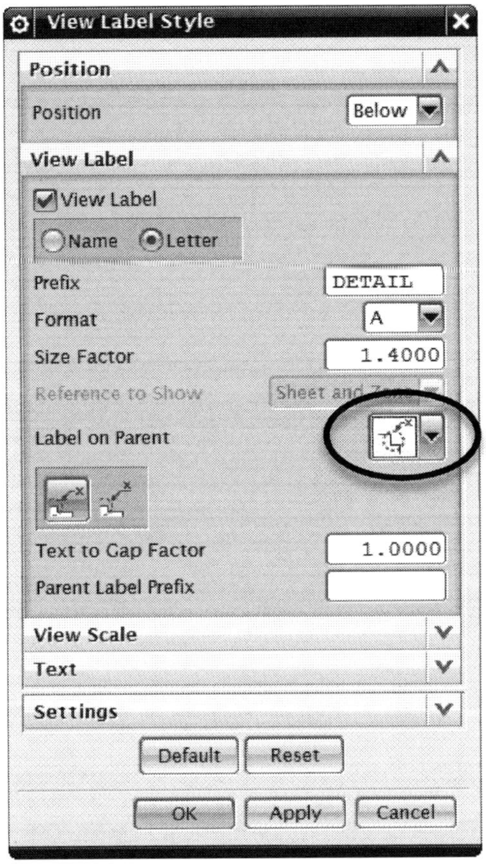


- Now we will add a View Label to the Auxiliary view.
- Double click the auxiliary view.
- Check the **View Label** box as shown in the figure below.
- Choose **Apply** and then choose **Cancel.**

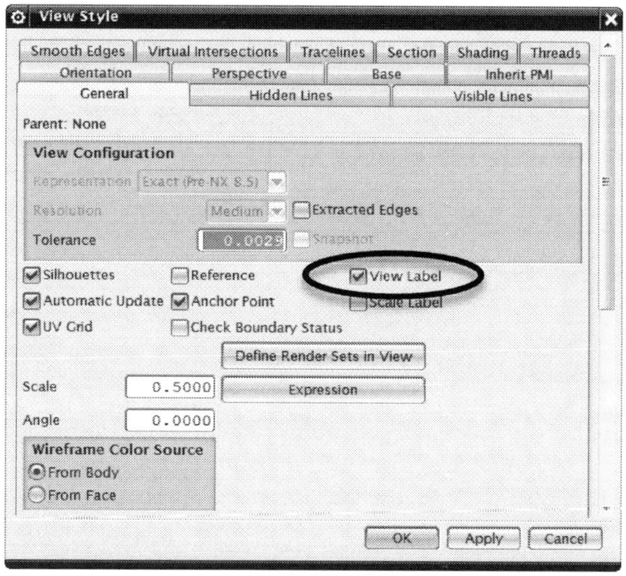


- Compare your drawing to the figure below.

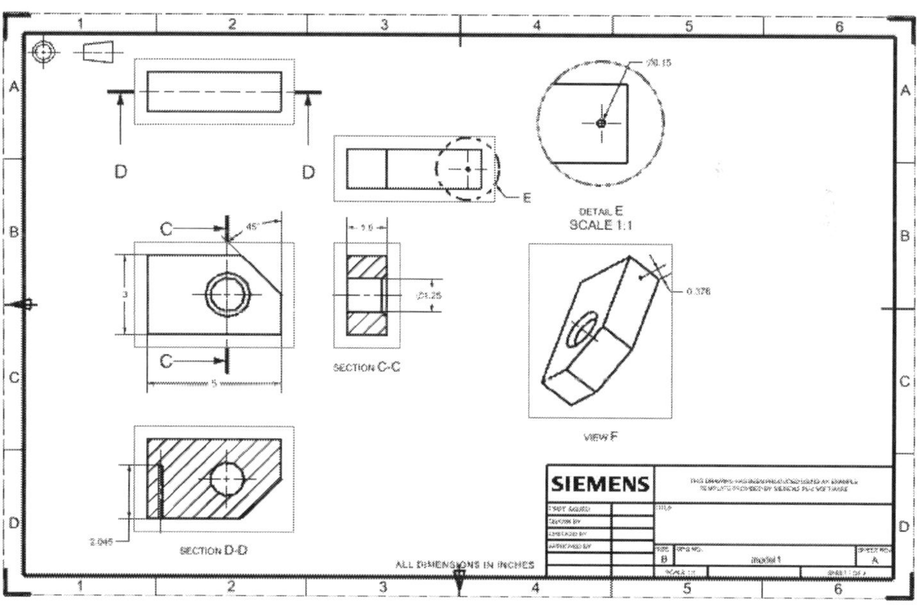


- Practice creating dimensions by adding more.
- Save your drawing file.

***End of Project***

## 17.8 Creating and Editing Notes

Adding various types of notes is an important part of creating an engineering drawing. With them, you can add explanatory text to dimensions, features, and drawing views. The note text may contain a variety of standard drafting symbols as well as alphanumeric text. In NX 8.5, you may define notes with or without leaders, depending upon your requirements.

### *Detail: The Annotation Editor*

In NX 8.5, you can create or edit notes using the 'Annotation Editor,' shown below. All note text is entered in the center area and can be previewed in the area immediately below. You can enter special characters by using one of the various symbol palettes, such as the Drafting Symbols, GD&T Symbols, or User Defined Symbols. If you have a note that is stored as a text file, you can import it into the Annotation Editor, and then place it on the drawing.

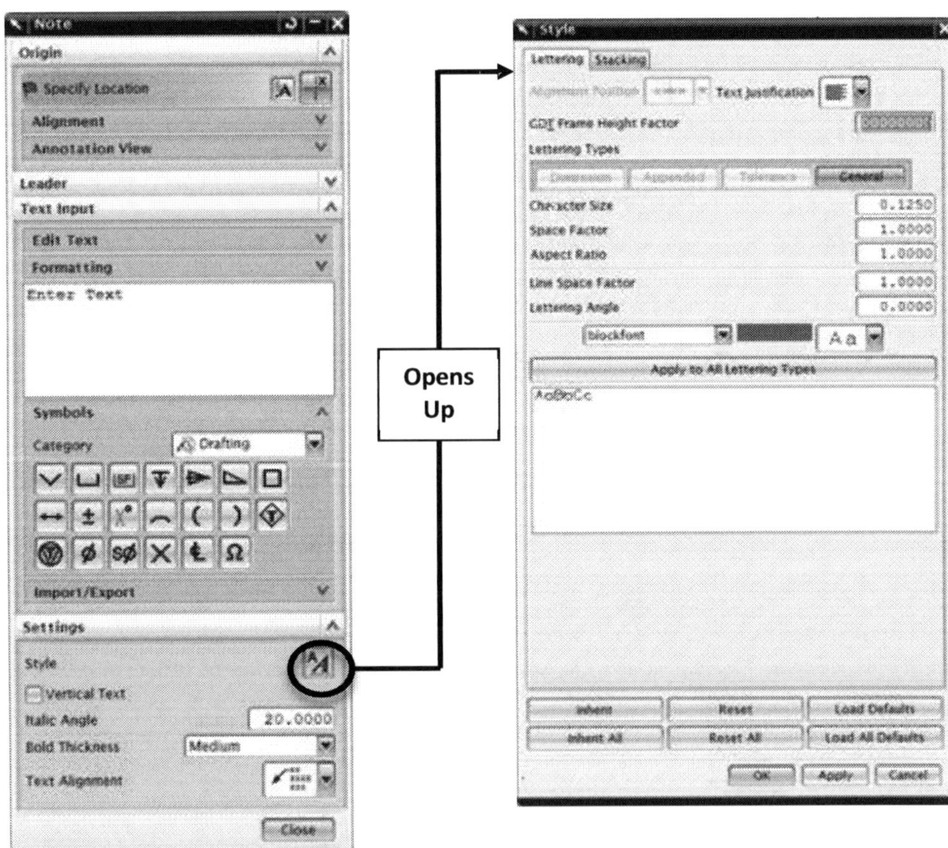

To insert a note, follow this procedure:

- Select **Insert / Annotation / Note.** The left menu above appears.
- Type in the note text, including any special symbols.
- Click on the symbol shown in the figure to open the Text Editor if you need more options.
- Change any desired Text Preference or Justification Settings in the Annotation Style dialog box by clicking the second symbol from the Annotation menu.
- Place the note in the drawing at your desired location. If you want to create a leader, check **Create with Jogs** under 'Leader' in the Note menu.

The Text Editor may also be used to edit existing notes. You can either double click the note or select the note and select ***Insert / Note***.

## ***Detail: Free Note Placement Options***

Free notes (with no leader) are placed using the 'Origin Tool' dialog box shown in the figure below. The 'Alignment' pop-up menu allows you to specify the alignment of the note relative to its origin. Three of the most commonly used Origin options are:

- **Drag** – place the note anywhere on the drawing.
- **Offset View** – place the note relative to a drawing view.
- **Offset Character** – place the note relative to a dimension or symbol.

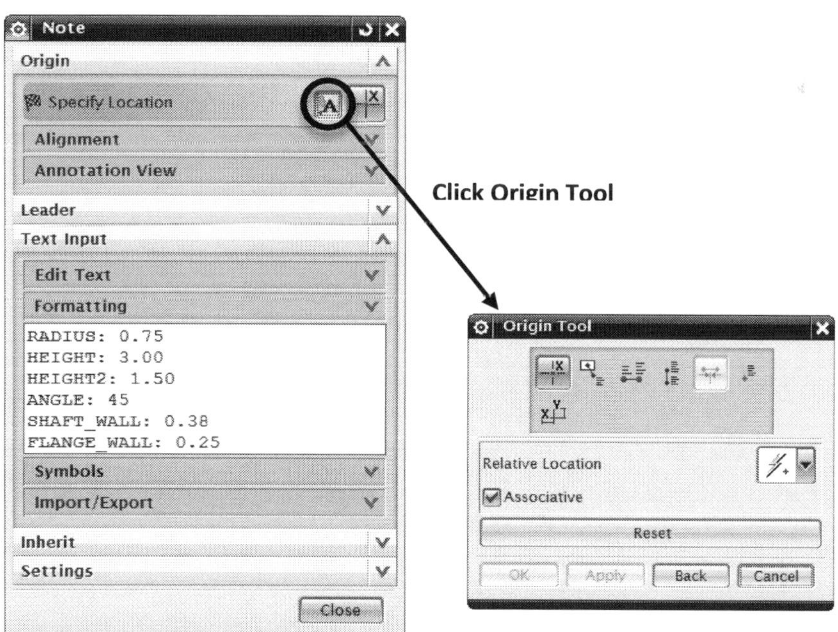

If you select the 'Associative' drop down option, the note will be tied to the selected entity. Using this option is a recommended practice, since dimensions and drawing views are rearranged frequently.

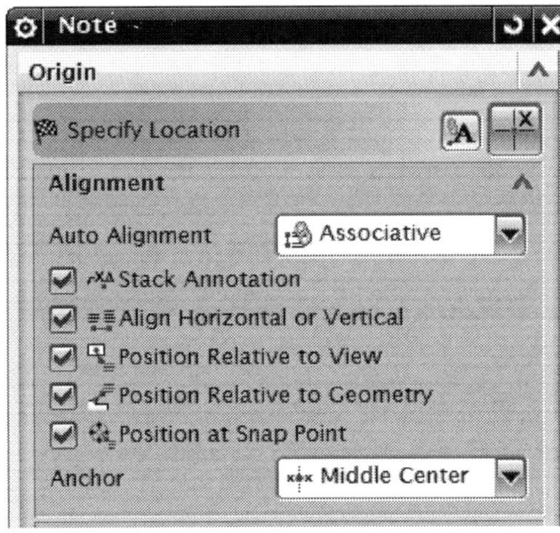


## ***Detail: Leader Note Placement Options***

Notes with a leader are placed using the Leader section of the Note toolbar shown in the figure below. You can select one of five leader types:

- **Plain** – Left- or Right-Hand leader
- **All Around** – same as **Plain**, but with an 'All-Around' symbol
- **Flag** – leader which extends parallel from a model edge
- **Datum** – leader that extends perpendicular to an edge
- **Dot Terminated** – leader that terminated to any point you select

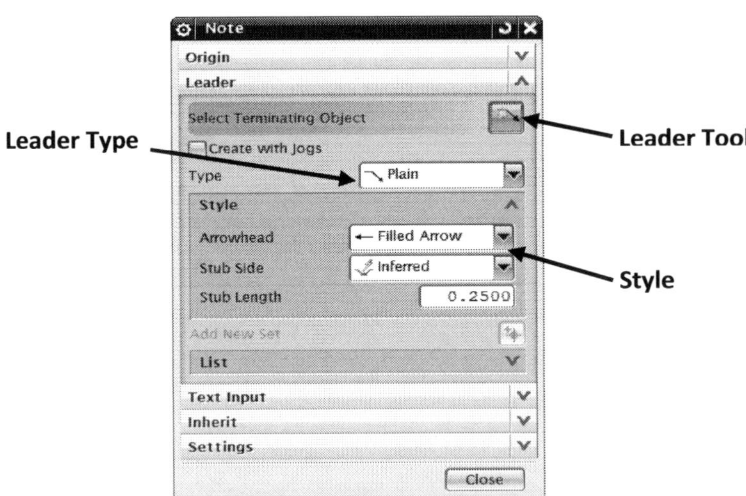

## Project 17.8.1

In this project we will use the Text Editor to create several notes on your drawing.

- Open the part from the previous project (if required) and select ***Start / Drafting.***
- Select ***Insert / Annotation / Note.***

*The Text Editor will appear.*

- Under Settings, select the **Style** icon  .
- The Annotation Style menu allows you to change many attributes of a note, including the lettering type, color, character size, and lettering angle.
- Enter Character Size = **0.375** and hit **OK.**
- Click in the white text box and type "**I Can Draw!**"
- Move your cursor into the Graphics Window and place the note as shown below.

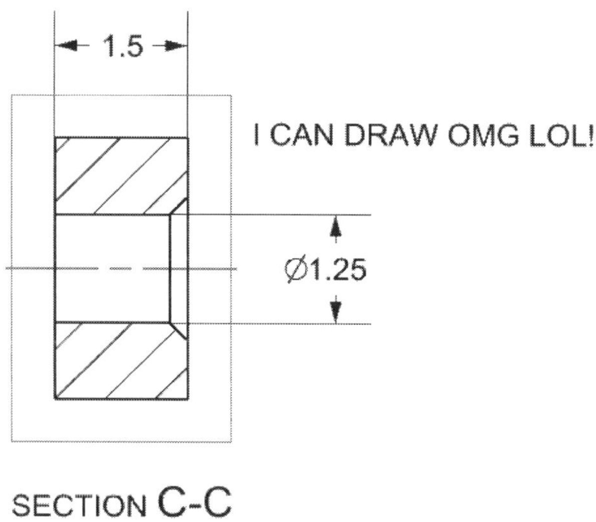


Now we will add two notes that are associated to the 3D view.

- Select the Reset button to restore all of the text settings (located in the Annotation Style menu). Click in the white text box and type "BREAK EDGE".
- Under Leader, hit the 'Select Terminating Object' button.
- Select the 'Plain' Leader type

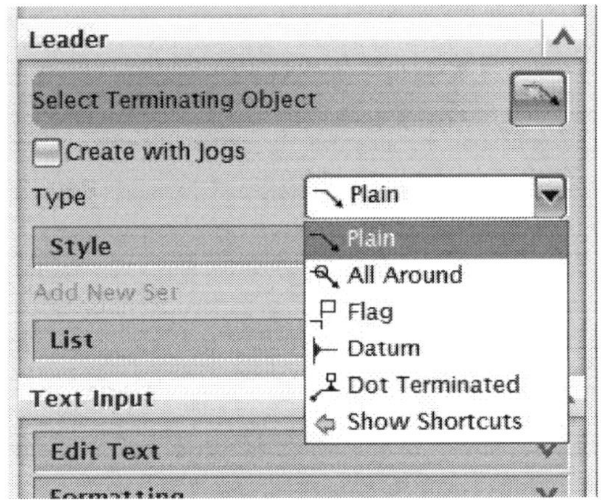


- Move the cursor into the Graphics Window and select the edge as shown below.

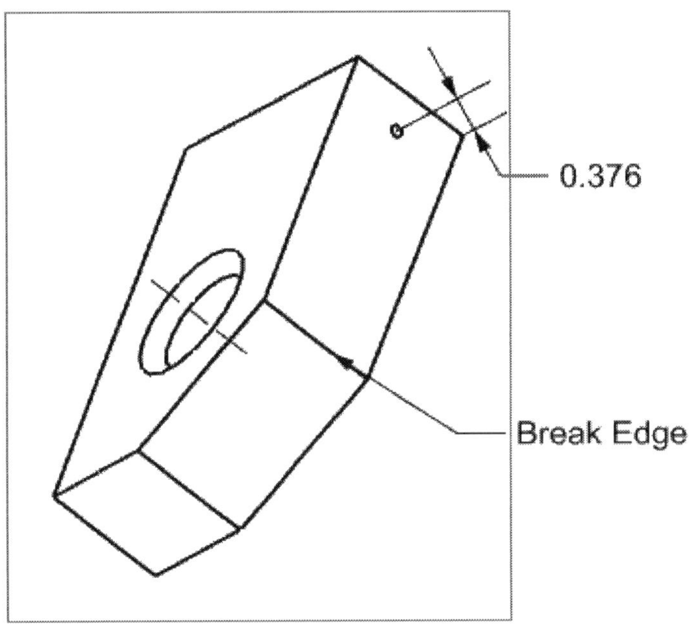


VIEW F

- Choose OK and drag the note to the location shown above.
- Click in the white text box and type "**REFERENCE VIEW**".
- Select the **Origin Tool** from the Annotation Placement Toolbar and select the **Relative to View** icon  .
- Select the 3D view, and place the note as shown in the following figure.

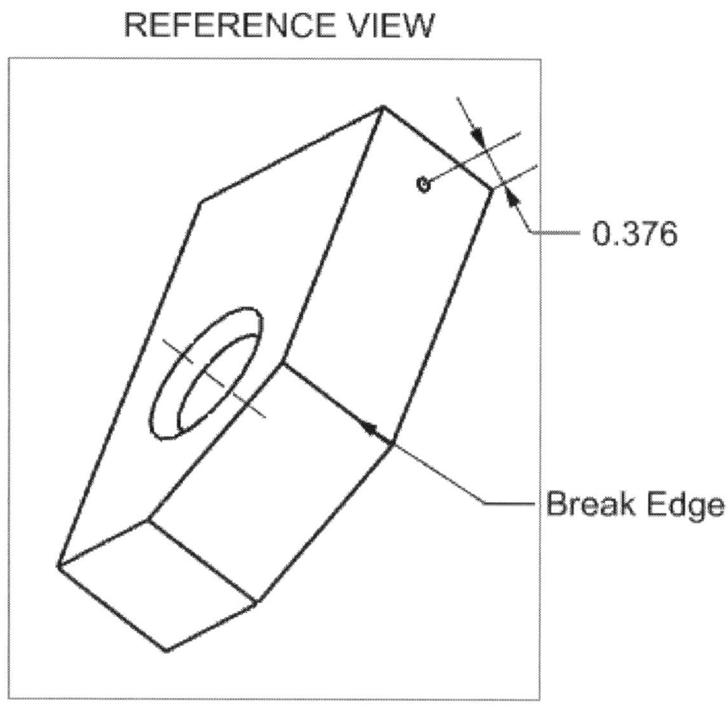


- Choose **Cancel.**
- Move the 3D view to a new location using ***Edit / View / Move / Copy***. Both notes will move along with the drawing view. (Select **Cancel** when done).

*Now we will use the Text Editor to add Appended Text to a dimension.*

- Select ***Insert / Dimension / Inferred*** .
- Select the 0.11.5
- 5 dimension on the small hole. Select the 'Text Editor' icon .
- Select the **After** icon .
- Click in the white text box and type "**x 1.0**," then click the **Depth Drafting Symbol** as shown in the following figure

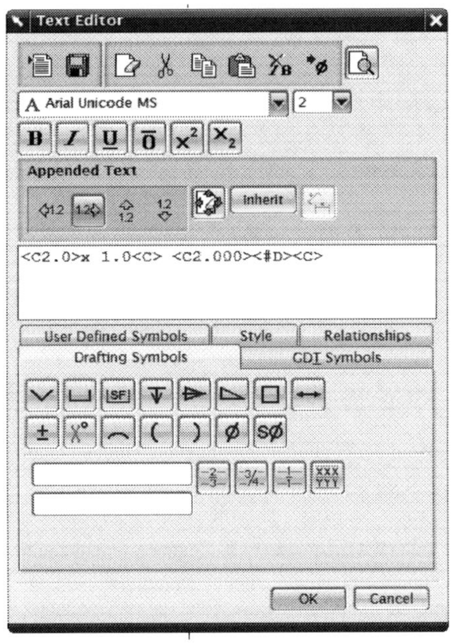


- Choose **OK** to close the Text Editor.
- Click and drag the dimension to the lower right corner of the view, as shown below.

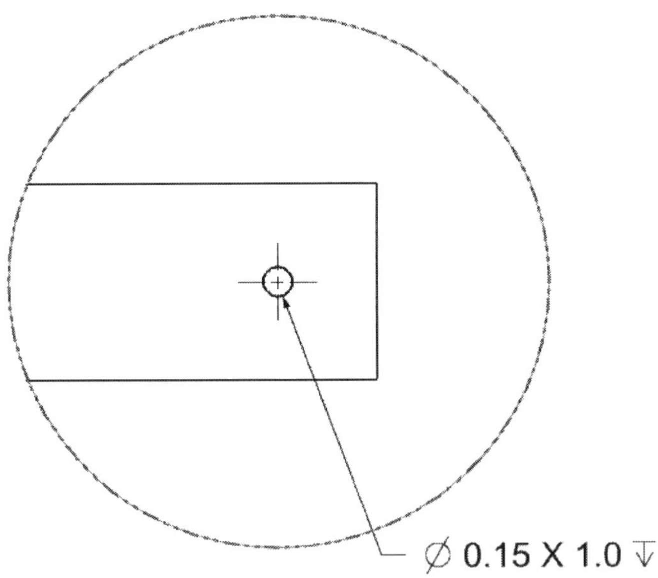


DETAIL A
SCALE 2:1

- Save your drawing file.

***End of Project***

## 17.9 Drawing Symbols

In NX 8.5, you can add several types of symbols to drawings that help you to document the engineering intent. Two of the most important are **Utility Symbols** and **ID Symbols**. The same dialog boxes are used to create or modify existing drawing symbols. In previous versions of NX the Utility symbols could be found under **Insert/ Centerline**.

### *Detail: Utility Symbols*

You can add certain pre-defined symbols such as 'Center Mark,' 'Bolt Circles,' and 'Target Points' found under Insert/Symbols or Insert/Centerline.

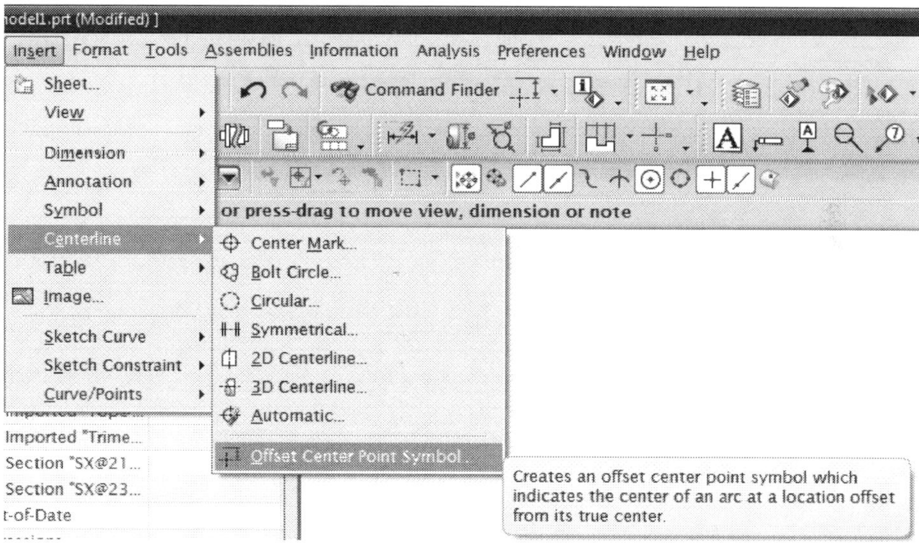


Each Utility Symbol has a set of parameters that control its appearance. The default values are set at the time the part is created (inherited from the NX 8 customer defaults file) and are stored in the part database. The defaults may be modified in the same manner as Drawing Preferences.

To add a Utility Symbol to a drawing, simply select ***Insert / Symbol***, select the desired symbol, then place it using the appropriate drawing entities. It is important to read the cue when adding Utility Symbols so that you understand what NX 8.5 expects you to select.

## ***Detail: ID Symbol***

In NX 8.5, 'ID Symbols' are geometric shapes (such as circles or hexagons) that are often used in conjunction with drawing notes to annotate features. ID Symbols may be created either with or without leaders, just like drawing notes. The menu in the figure below appears by selecting ***Insert / Annotation / Balloon*** .

## ***Detail: User-Defined Symbols***

NX 8.5 allows you to define customized symbols. 'User-Defined Symbols' are useful for automating the creation of commonly used items, such as company logos or standard drawing items. They can be used in NX 8.5 by selecting ***Insert / Symbol / User-Defined.*** Check with your company's CAD System Administrator to see if you have a symbol library.

## ***Detail: Weld Symbols***

You can define different weld types by adding weld symbols in your drawing. To create a weld symbol select ***Insert / Annotation / Weld Symbol*** 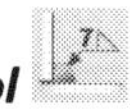 and specify weld type, size, finish and any other parameters you need to define a weld.

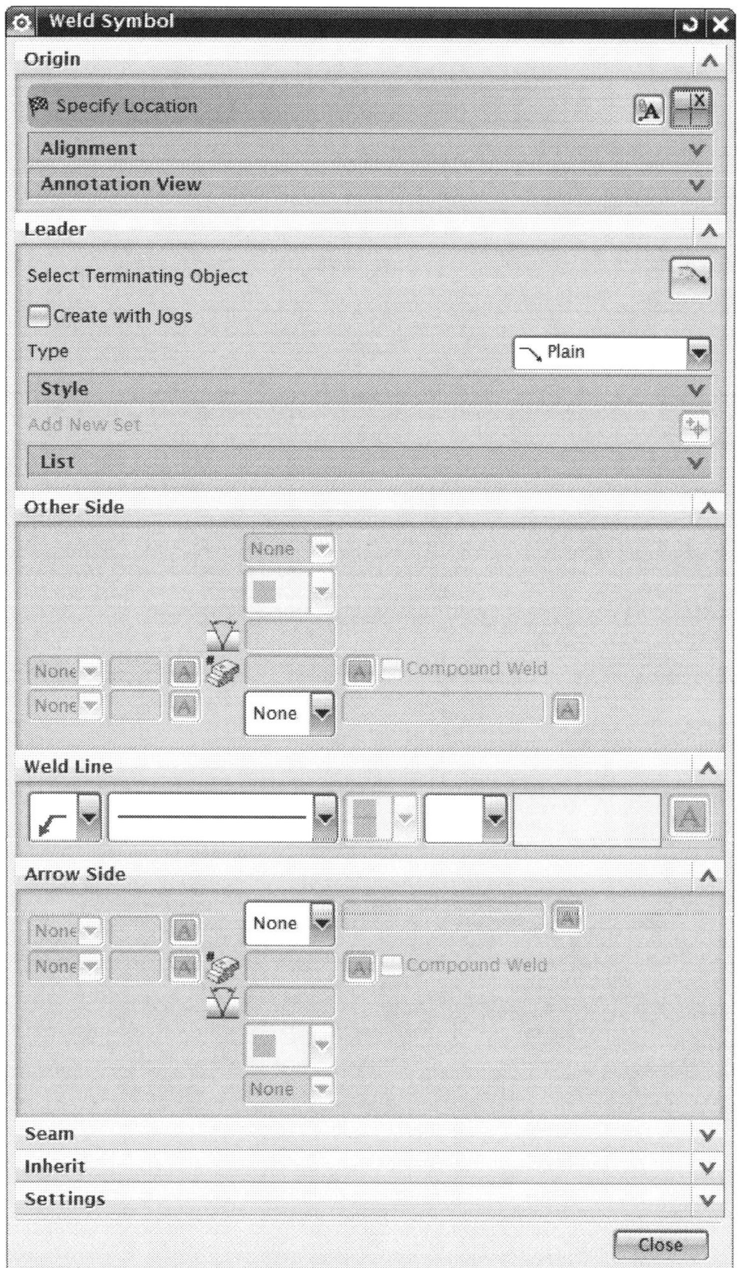

## *Project 17.9.1*

Create several Utility Symbols and ID Symbols on your drawing.

- Open the part from the previous project (if required) and select ***Start / Modeling.***
- Add an R0.75 Blend feature to the model, as shown below.

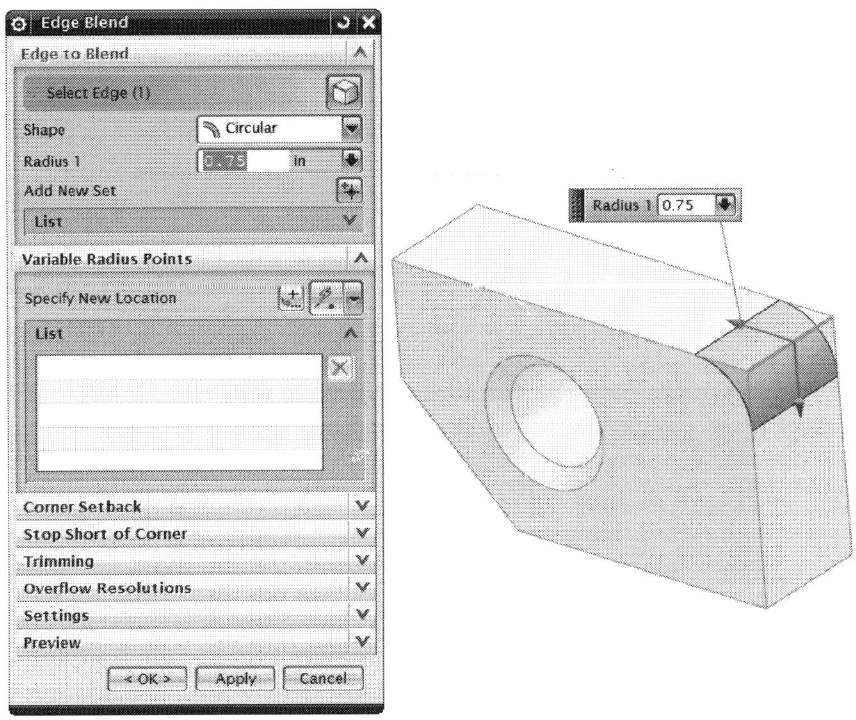


- Select ***Start / Drafting.***
- Select **Edit / View / Update** to update the drawing sheet. Notice that any dimensions that referenced the edge removed by the Blend are now incorrect, as shown in the following figure.

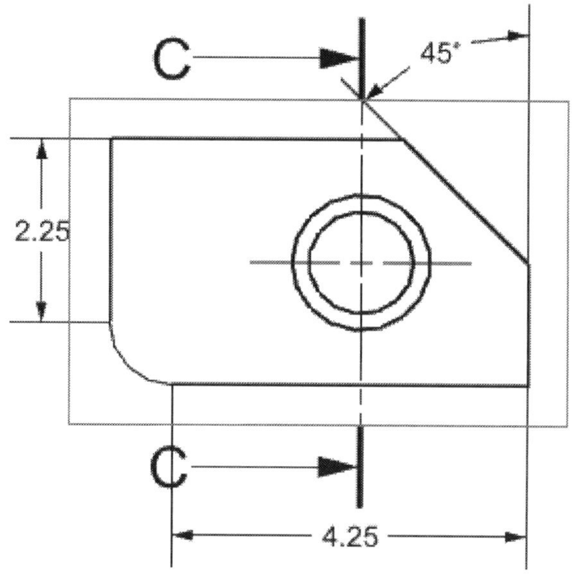


- Delete any incorrect dimensions.
- Select Insert / Annotation / Intersection Symbol  .
- Select the two edges as shown below to create the intersection symbol.

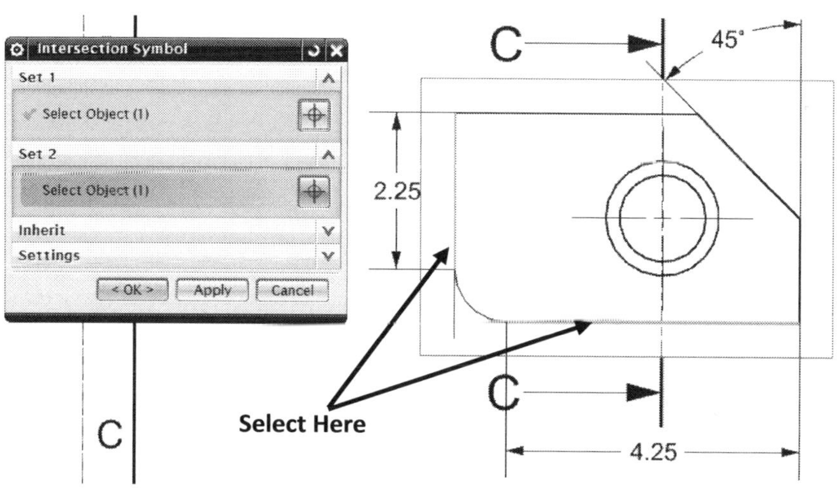


- Re-create any dimensions that you deleted, and add a dimension to the new Blend feature.
- Select the ***Automatic Centerline*** icon 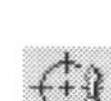 or by choosing **Insert / Centerline/ Automatic** .

- Select the view shown below and choose **Apply.**

Notice that a centerline symbol appears on the large hole

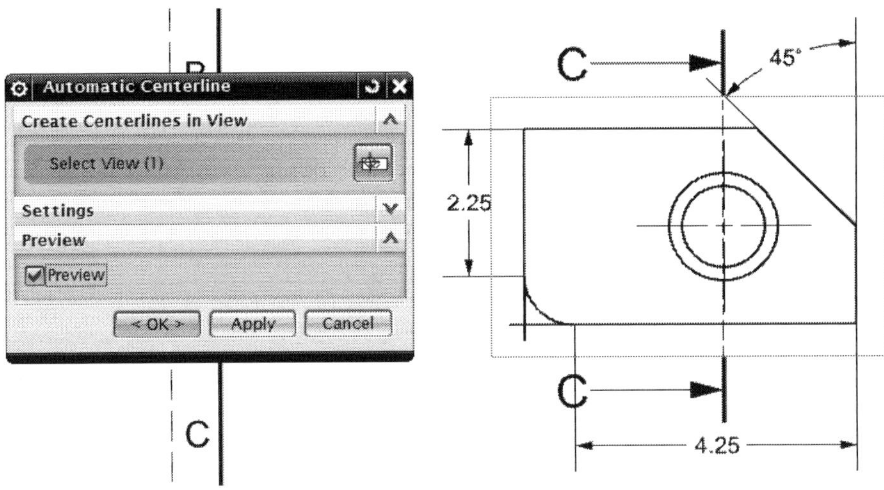


- Select the 'Top' and 'Right' projected views and choose **Apply.**
- Select the Insert / Centerline / ***2D Centerline*** icon .

Read the Cue throughout the symbol definition process. NX 8.5 is prompting you for the first side edge object of the block centerline.

- Select the Auxiliary View's two defining edge sides, as shown below.

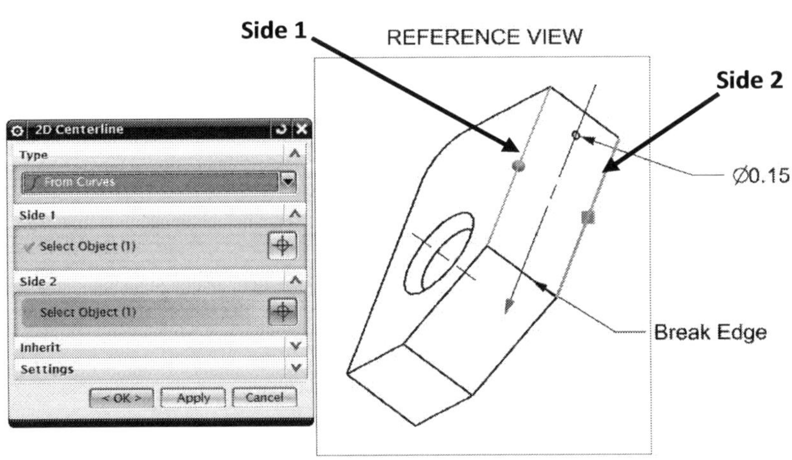


Read the Cue throughout the symbol definition process. NX 8.5 is prompting you for the first two defining edges of the block centerline.

- Select the two edges, as shown above.
- Choose **Apply.**
- Choose **Cancel.**

Create ID symbols that annotate part features. First, create one with a leader.

- Select ***Insert / Annotation / Balloon*** .

- Select the **Hexagon** icon .
- Enter 'Upper Text' = **3** and 'Symbol Size' = **0.75** in the 'ID Symbols' dialog box.
- Select the small hole in the "Top" projected view.
- Click and drag the cursor into the Graphics Window and place the ID Symbol below the view.

Now we will create an ID symbol without a leader.

- Enter 'Text' = **2** in the 'Identification Symbol' dialog box.
- Move your cursor into the Graphics Window and place the ID Symbol near the 'BREAK EDGE' note.
- Compare your results to the following figure.

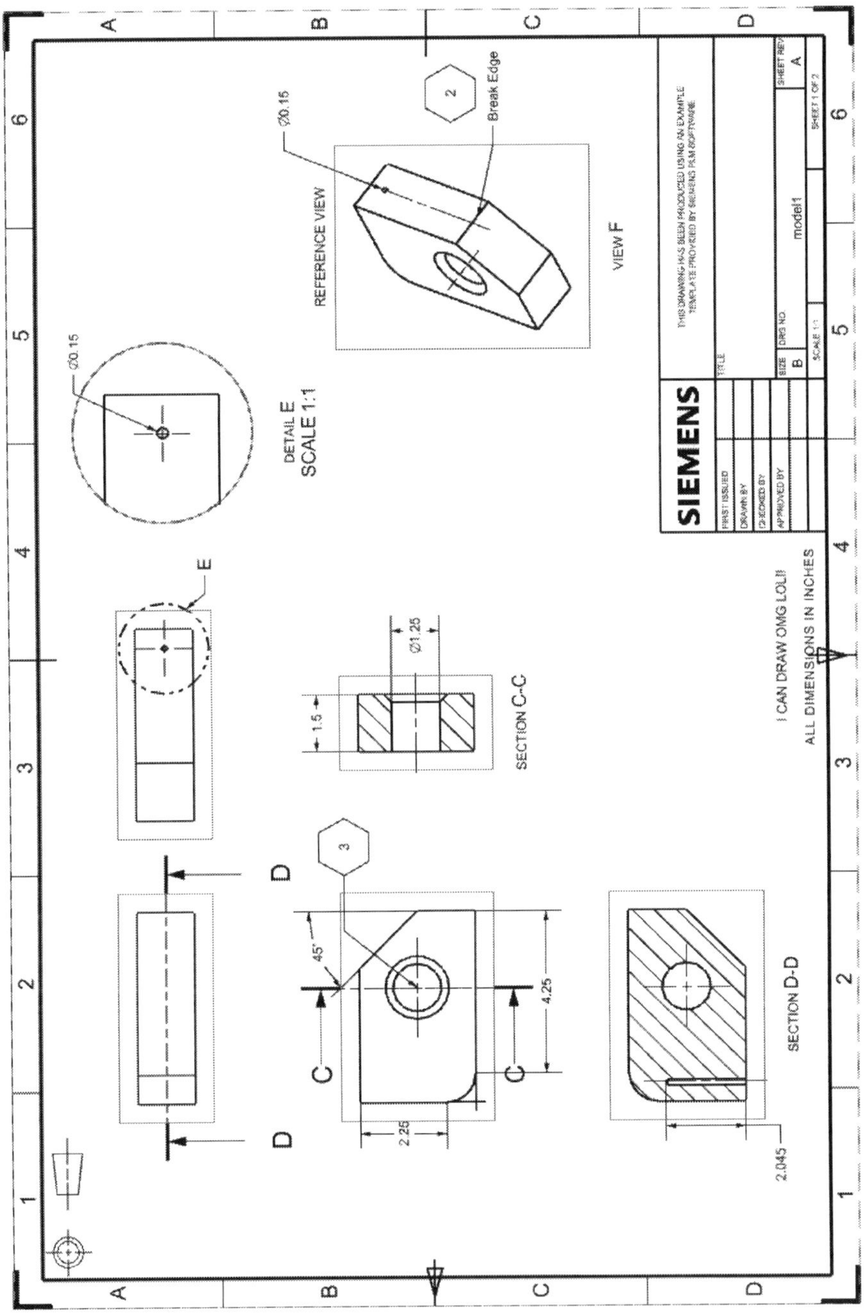


Save your drawing file.

***End of Project***

## 17.10 Self-Test

(All questions are true/false. Answers are at the back of the book.)

1. You can update a drawing if your model is changed.
2. You can add text to a dimension.
3. You can link your text to the part attributes.

# 18. Master Modeling

## 18.1 What is a "Master Model?"

Master modeling is a simple technique that enables concurrent engineering. It is best understood by an example. Suppose you have a model of a modem unit, just the outside surfaces or the ID, and in other files you had all the other components that were related to the outside shape of the master model file. Every time you change the master model file, everything related changes – that illustrates master modeling working for you.

The following figure shows the master model along with each part created directly from the master model. The last figure shows what happens to all the components when the master model is changed.

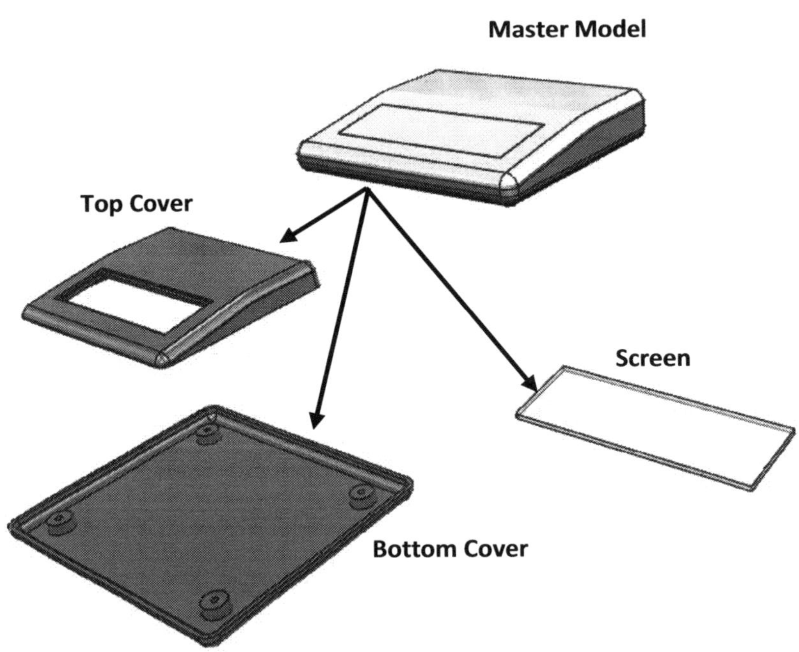

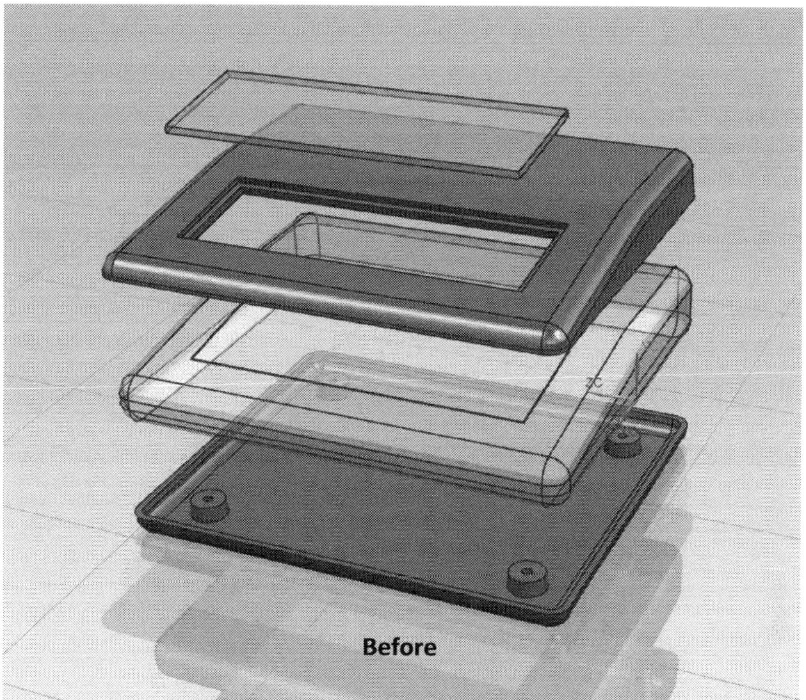

**Before**

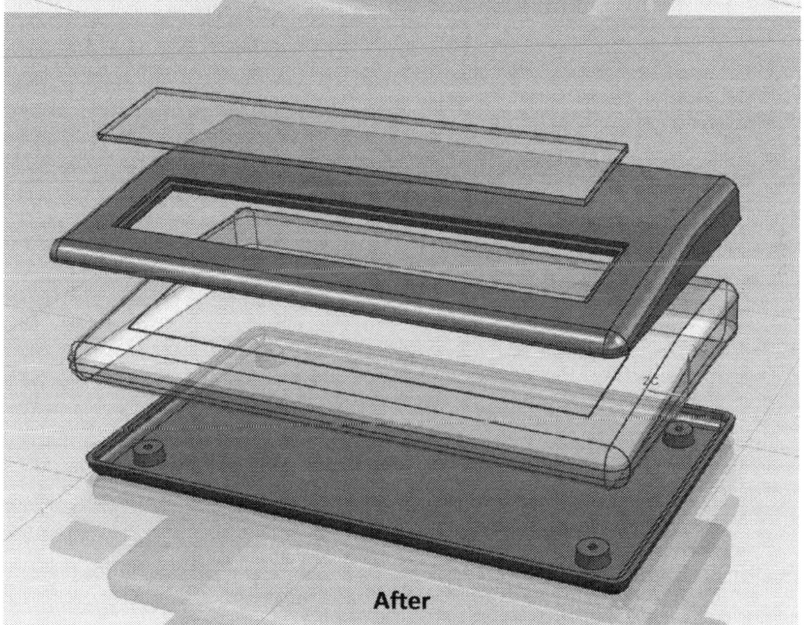

**After**

## 18.2 How to make an Associative Assembly using a Master Model

As depicted in the previous figures, Master Modeling can begin with a single part file that captures the main form of the geometry. Then you can add as many non-master part files that use that geometry. All you have to do is create your master part, create parts that reference the geometry of your master, and then your master modeling will be complete.

### *Project 18.2.1*

- Create a new part file and call it **Master**.
- Create a simple sketch that has the geometry and dimensions shown in the figure below.

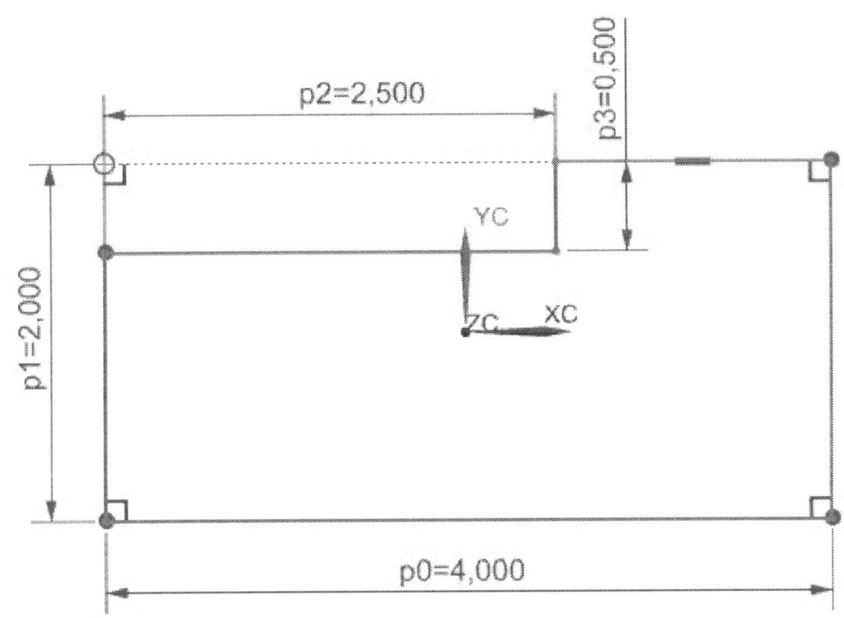


- **Extrude** it **1in** in the **Z direction** with a **10** degree draft.
- **Extrude** it **2in** in the **–Z direction** with a **10** degree draft.
- **Unite** them to create a model similar to the one shown in the following figure.

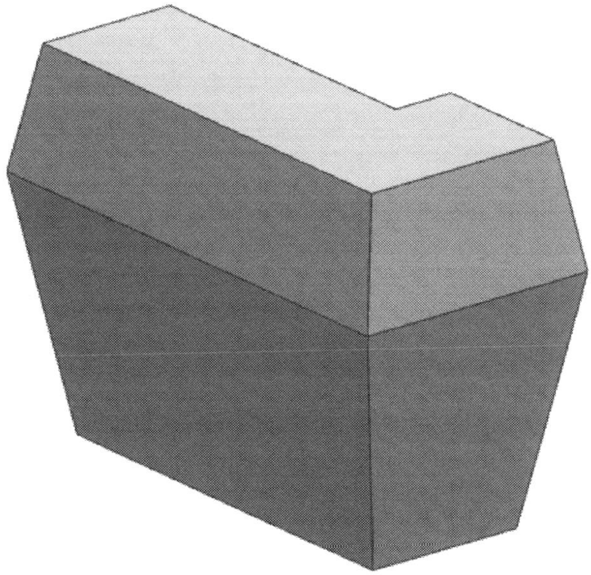

- **Save** the model.
- Make sure the Assemblies application is turned on (**Start / Assemblies**).
- Now go to **Assemblies / Components / Create New Parent.**
- Call the new component **Assembly**.
- Next Create a new Blank Component called **Top Cover**

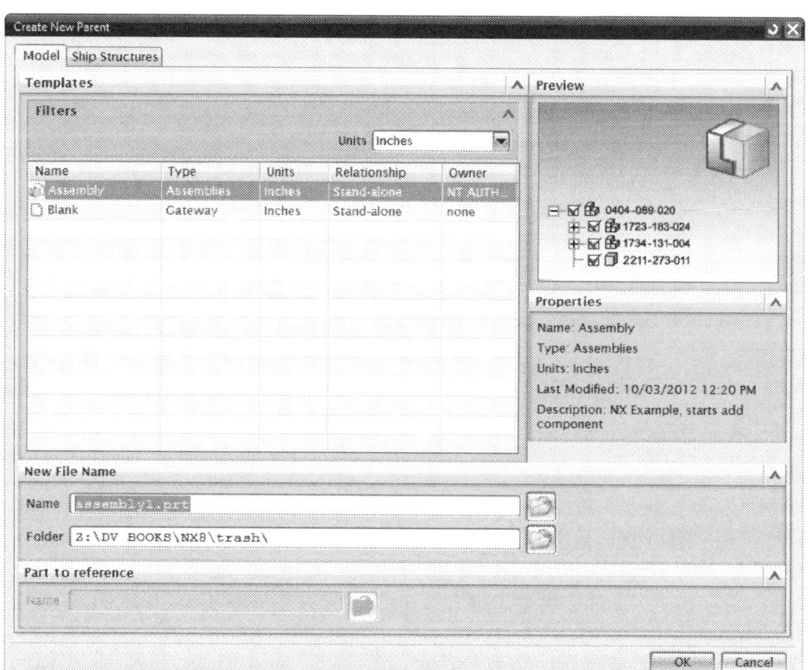

- Make the top cover the work part and you're ready to steal the geometry you need to create your new component.
- Select Insert / Associative Copy / WAVE Geometry Linker. Select Body as your type.
- Select the entire body (as shown in the following figure) and click **OK**. Now that body will be linked to your **Top Cover** part file. Make sure that the **Associative** box is checked, as shown.

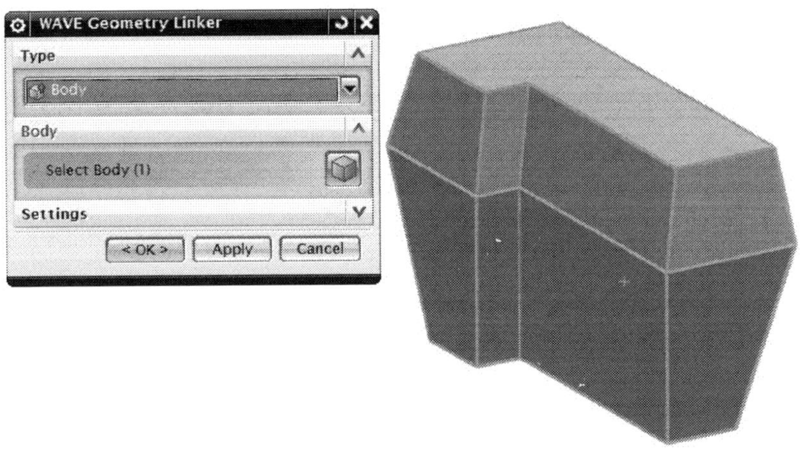


- When you hide the master part file you will see that you have a stolen body to work with, as shown in the following figure.
- Create a datum plane at the parting line of the body, as shown in the following figure. Use **Insert / Trim / Trim Body**, select the **body to trim**, and then select the **datum plane** to trim with. Discard the bottom section, as shown below.

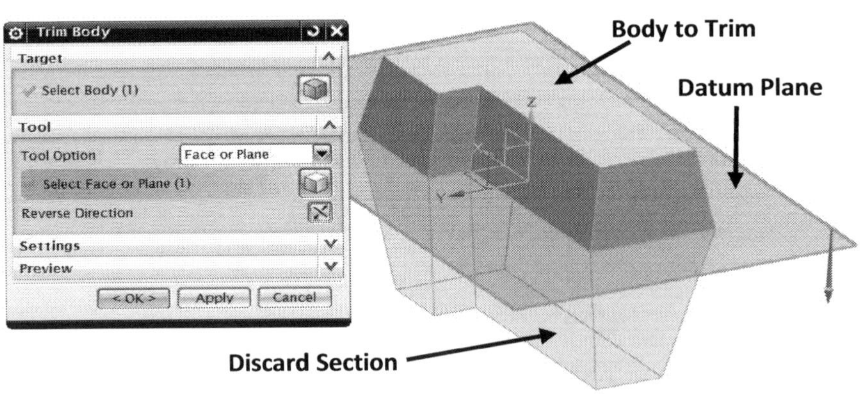

- Now blend all the **edges** with a **0.1in radius** and **shell** the bottom face with **0.05 in** wall thickness.

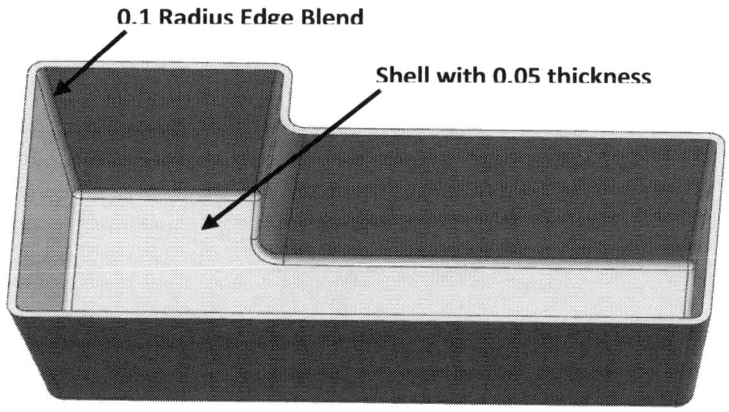


- Next repeat this process. **Create a new part** called **bottom case**. Steal the bottom body using the **WAVE geometry linker** from the master model. Hide the master body first. Create a **datum plane** at the same position and then **trim** the body, discarding the **top** section this time. Finally, **blend the edges** and **shell** it using the same values as the top cover. The bottom cover should look like the following figure.

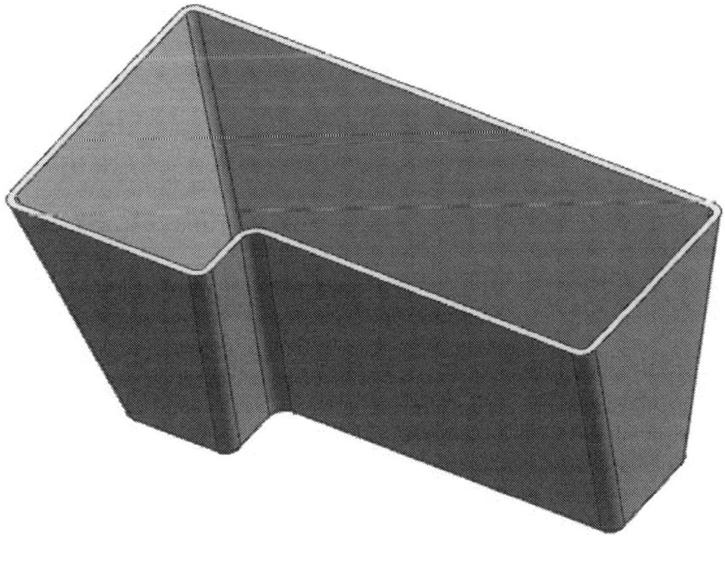

- At this point you should have three different part files: Master, Top Cover and Bottom Cover. The top and bottom covers rely on the master for its overall dimensions.

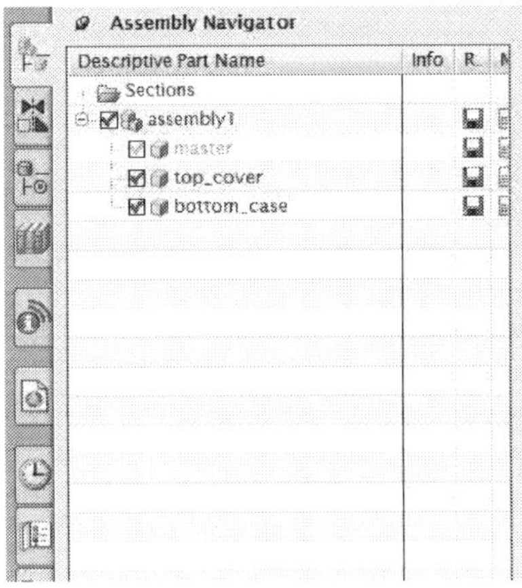

- Now change the **length** of the original sketch in the Master part file to **7**. The corresponding model will appear as shown in the figure below.

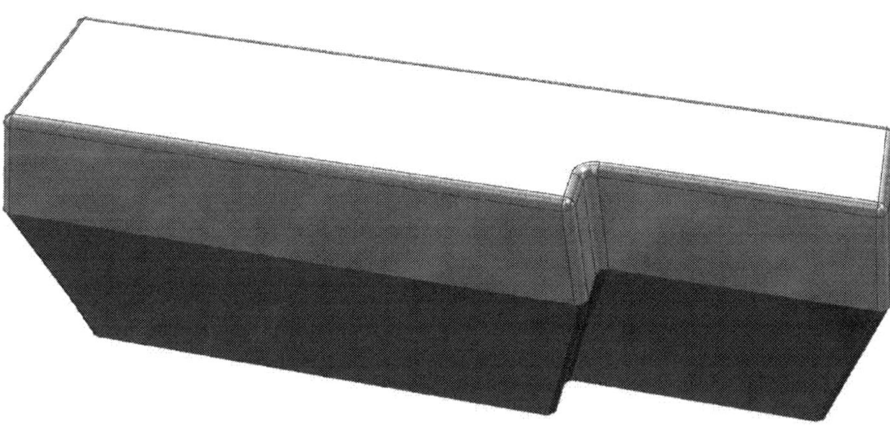

***End of Project***

## *Detail: Master Modeling for Various Applications*

The most common application of master modeling is when it is used for drafting. NX 8.5 is one of the only CAD packages that gives you the choice to perform your drafting in a separate non-master part file or right there in the actual component part file. Imagine having a component part file, such as the top cover of the modem housing mentioned above as your master part file, and a separate non-master part file for the drawing data. The top cover geometry is assembled into the drawing part file and all the drafting is performed in it. This way, the drafter and the modeler can work at the same time without overwriting each other's changes.

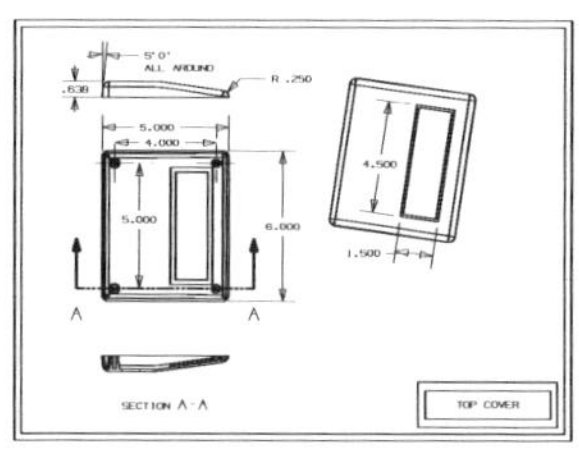

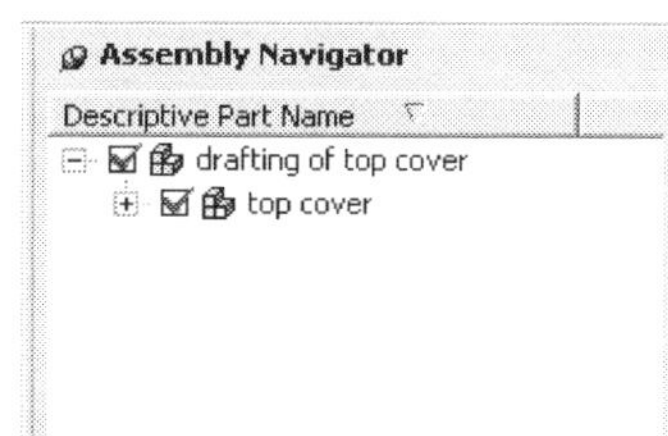


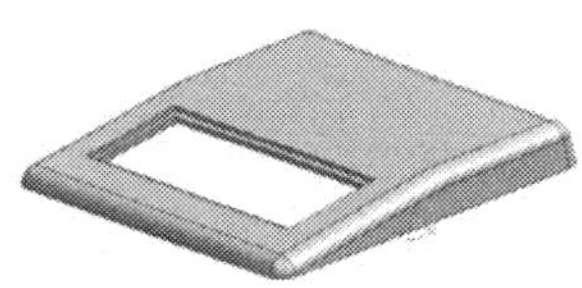

| Drafting of top cover | Assembly Navigation tool file | Top cover master part |
|---|---|---|

## *Detail: Master Modeling for Drafting*

Companies that have a separate drafting department from their design engineering department love master modeling because they can gang up on a design job and realize the vastly under-utilized functionality of concurrent engineering.

- Create the simple geometry shown below in a new part file that you call *"drafting master,"* or use one of the part files from a previous exercise.

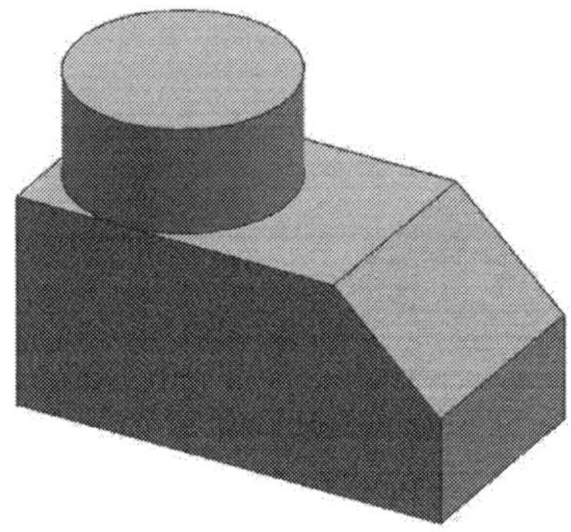

- Now create a non-master part file as explained in the previous project.
- Create a new part file that is a *drawing*, not a model: Click on the drawing tab at the top, and at the bottom, in the "Part to create a drawing of' box, select the non-master part file you just made. Click **OK**.
- Create and save your drawing. It should look something like the picture below.

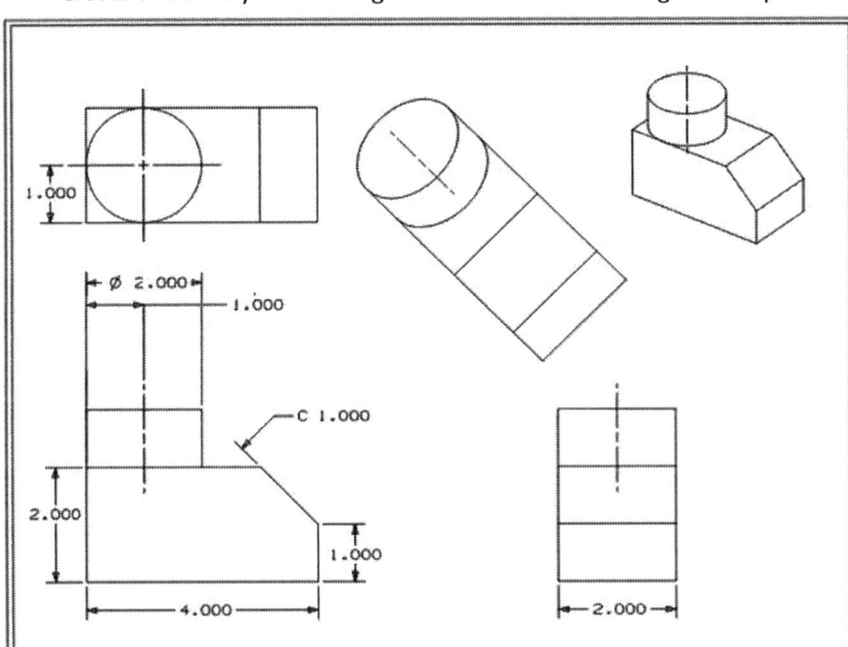


- If you change your master model, the drawing should update too!

## 18.3 Wave Geometry Linker

### *Project 18.3.1*

Here we are going to use the **Extract Virtual Curve** tool to create a centerline on a part in an assembly. Then we will use the **parametric centerline** to create a wire that will run between two identical parts in the assembly. During this exercise we will use the **WAVE Geometry Linker** to create links between parts and create a fully parametric assembly.

- Begin by creating the following component named '**flange**.'

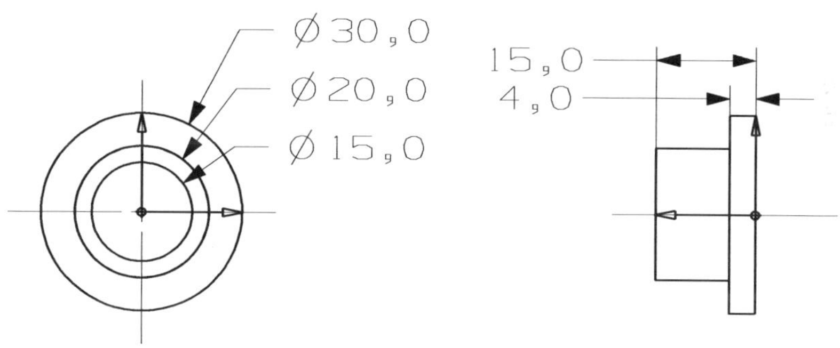


- Now create this component named '**container**.'

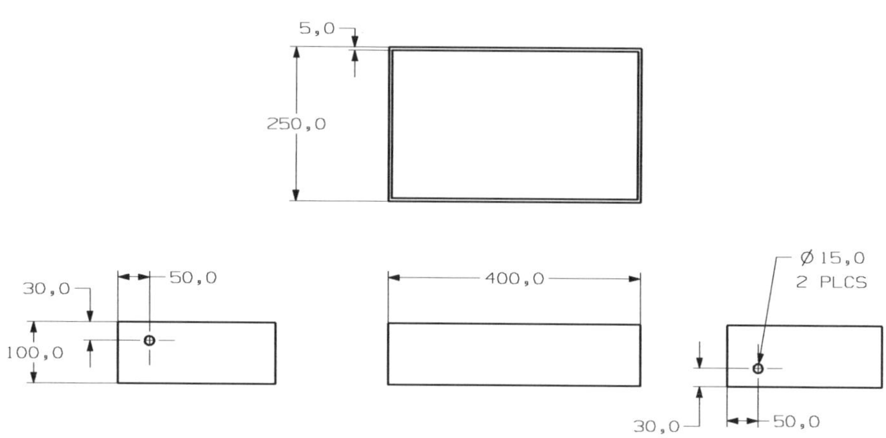

Now that you have created both of the parts, put them into an assembly. Use Assembly constraints to **Fix** the container. Then constrain the two flange components inside the container at the hole locations as shown in the following figure (*One is hidden in the following view*).

- Make the **flange** the **displayed part**. Select **Insert / Curve from Bodies / Extract Virtual Curve**. Select **Rotation Axis** as the type, then select the cylindrical face shown below and click **OK**. You will now view your Extracted Curve, shown in the following figure.

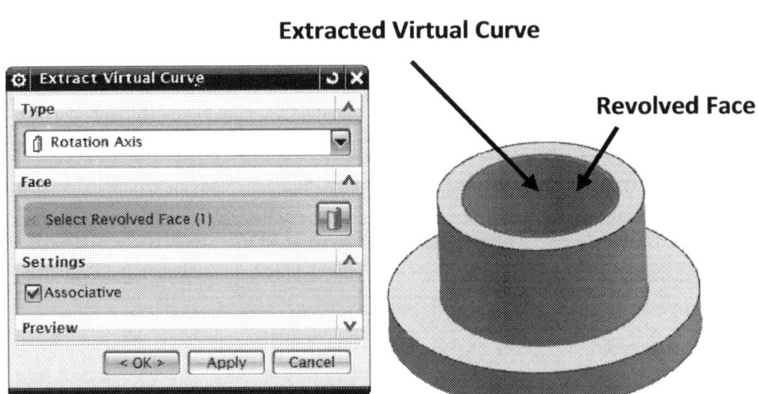

- Now select **Format / Reference Sets** from the menus. Click the **Add New Reference Set** button. Type in **Extract Curve** for the name, select the solid body and the extracted curve to include in your new reference set, and then click **Enter** on your keypad to complete.

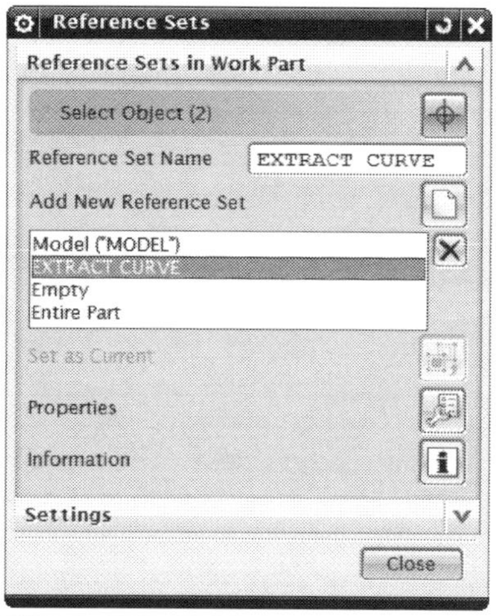


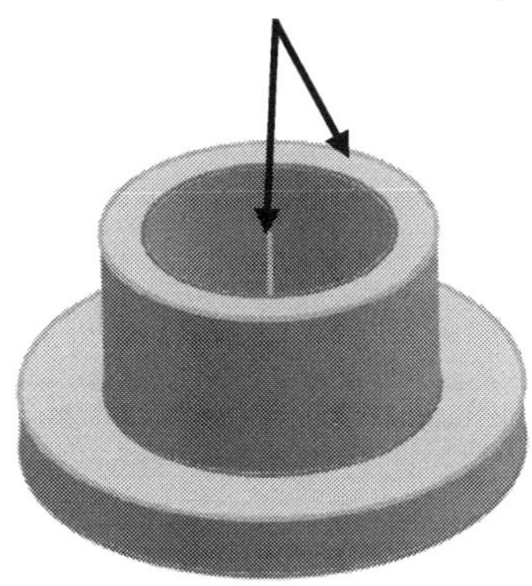


**Note:** Automatic centerlines may already have been created upon view insertion in your drawing depending on NX 8.5 defaults. If you already have centerlines, skip ahead to the insertion of ID symbols on the next page.

- Go back to the assembly. Right click on the **flange component / Replace Reference Set / Extract Curve**. Now you will see the extracted curve as well as the flange solid in the assembly, as shown below. Repeat this for the other flange in the assembly.

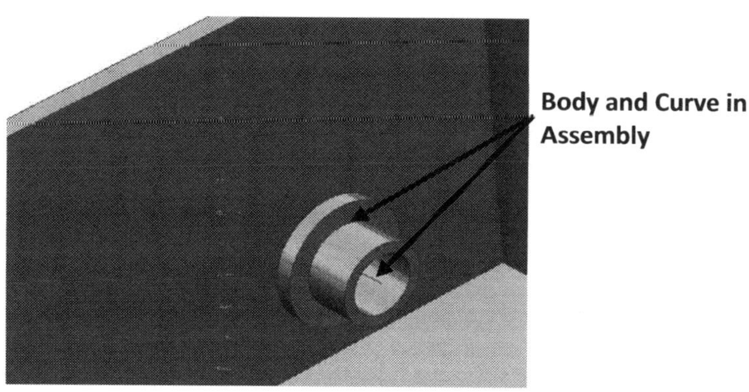


- While in the assembly, select **Assemblies / Components / Create New Component**. Name the new component **Wire** and click **OK**.

- Make the new component the **Work Part** and select **Insert / Associative Copy / Wave Geometry linker**. Choose **Composite Curve** as the type and then select the extracted curve from the flange component. Under settings make sure **Associative** and **Fix at Current Timestamp** is checked. Click **Apply,** then repeat the process for the other flange part in the assembly.

**Note:** Reference sets are one way to control what you see of a part file at the assembly level.

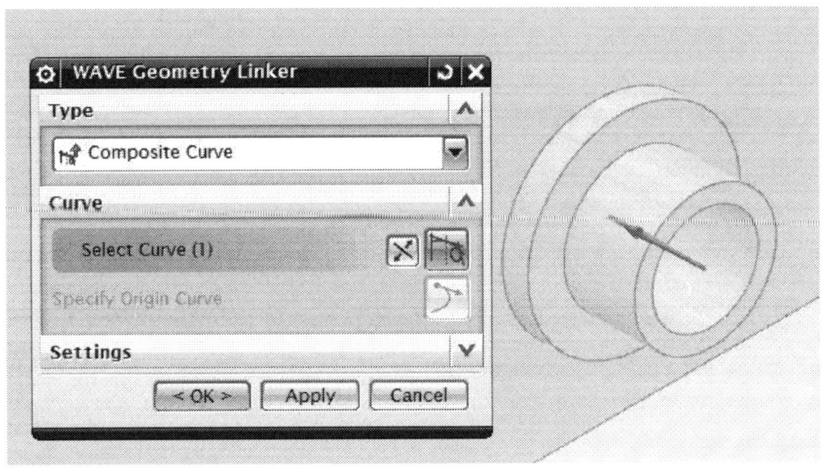


- Now select **Insert / Curve from Curves / Bridge**. We are going to use the bridge curve to connect the WAVE'd lines from each **flange**. Zoom up on one of the flanges and select the line.
- Now select the line from the other flange in the assembly and you will see the curve that connects the two flanges, as shown in the following figure. Click **OK**.

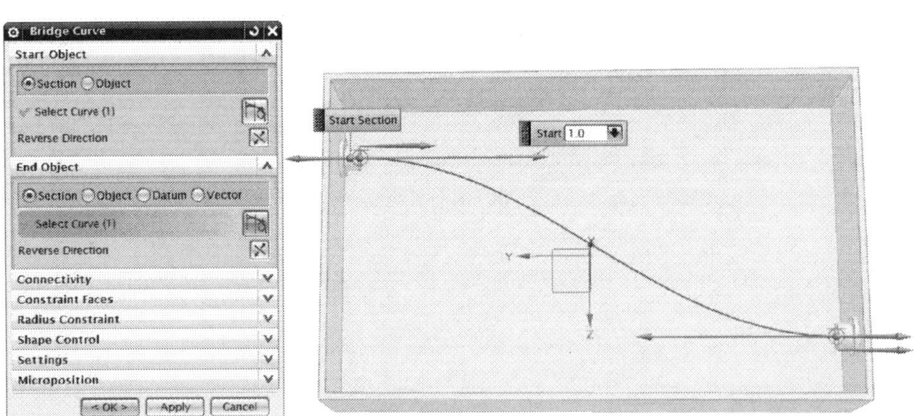

- Finally select **Insert / Sweep / Tube**. Select the **bridge curve** with an outer diameter of **15**.
- Your assembly will now look like the following figure.

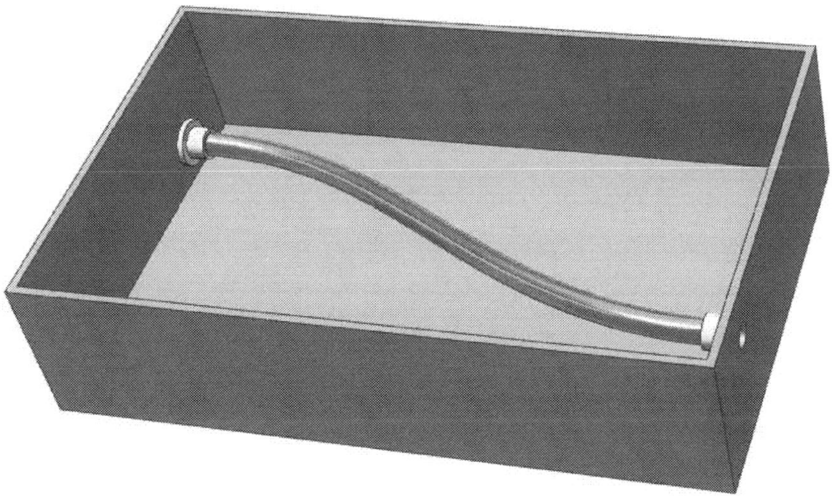

***End of Project***

## 18.4 Self-Test

(All questions are true/false. Answers are at the back of the book.)

1. Master Modeling is synonymous with 'concurrent engineering' in NX 8.5.
2. Selecting the 'Non-Master Part' option allows you to use NX 8.5's Master Modeling abilities.
3. Master Modeling enables two or more individual users to be accessing the same part files or assemblies concurrently.

# 19. Sheet Metal Features

## 19.1 Introduction

To create sheet metal parts you have to use the NX Sheet Metal module found under the START menu.

Sheet Metal features let you add details that are parametric, can be added to any solid model, and can show bending and distortion due to forming. Using Sheet Metal Features, you can create flanges, beads, bends, brackets and other geometry that are commonly used in sheet metal operations.

## 19.2 Sheet Metal Flange Features

Start all the projects in this chapter by going to the Start Menu / NX Sheet Metal. To create a flange, select ***Insert / Bend / Flange***. You can create a simple flange by selecting the edge of a solid block, then choosing the direction and attachment face as shown in the figure below.

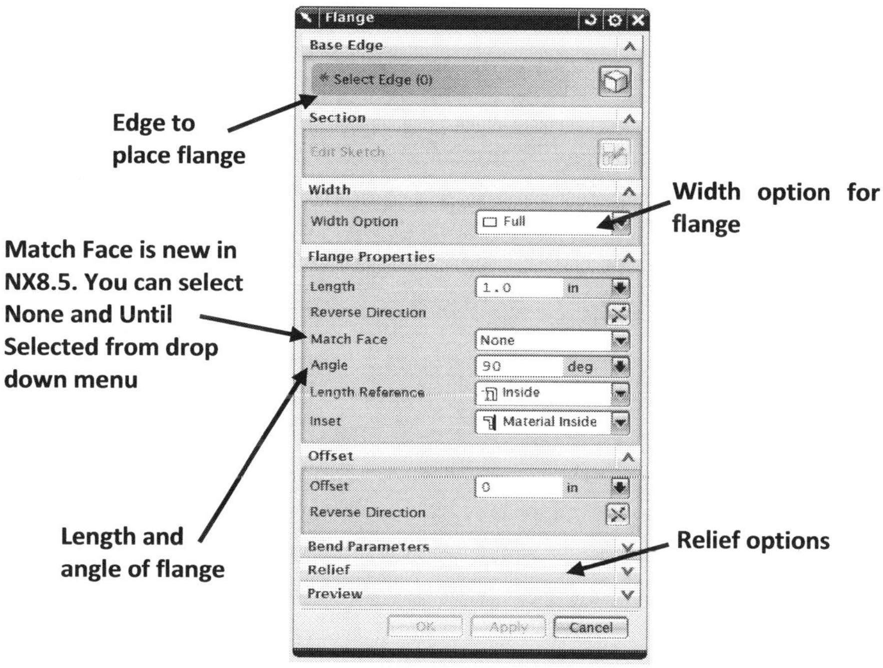

## *Project 19.2.1*

In this project you will create a flange and define the width using options.

- Click on **Insert / Tab**. The tab tool asks for either curves to create a tab or a plane on which to sketch the profile of your tab.
- Click on the XY axis and sketch the following sheet metal Tab using a rectangle that is **5in** x **2in**.
- Edit the thickness to **0.05in** in the Tab window after you have finished the sketch.

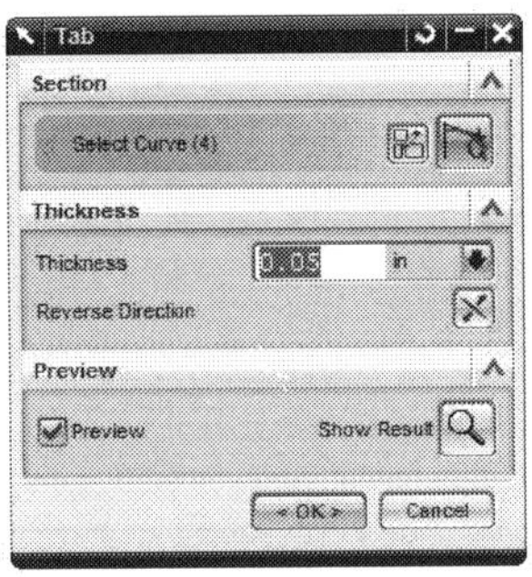


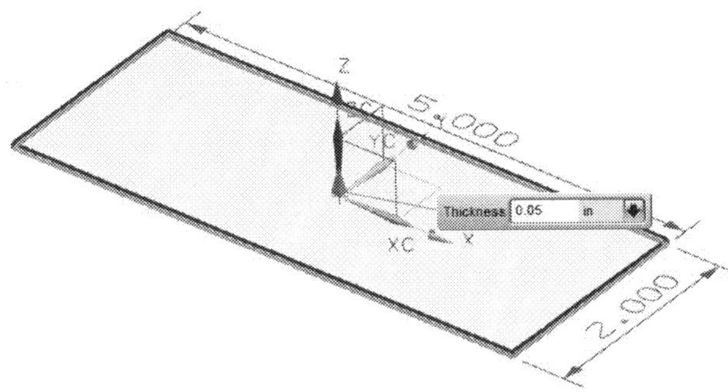


- Select Insert / Bend / Flange.
- Select the top edge of one of the long sides as the **Bend Edge**.
- Type in the **Width Option, Length, and Angle**, as shown in the previous figure.

- Set the **Width Option** to '**At Center**.' This will center the flange on the edge that you selected.
- Also set the Flange properties inset to **Bend Outside**.

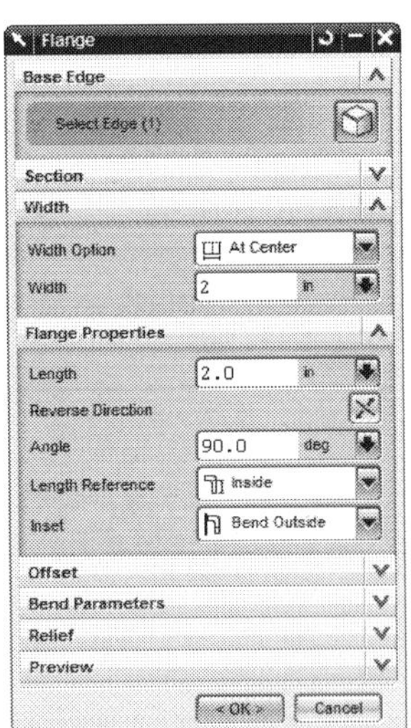


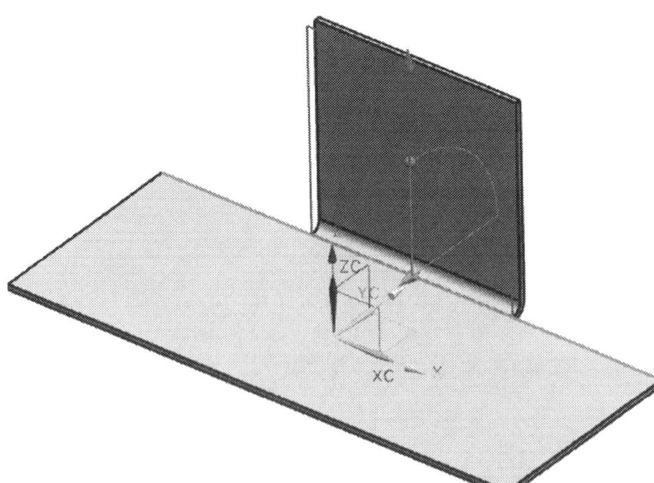

- You can also add reliefs to the sides by choosing an option from the Relief tab in the flange menu. Try creating a **0.1in** round relief.

***End of Project***

## 19.3 Inset Flange

Using this option, you can insert a flange on any planar face. This is different from the 'Flange' option because **Flange** adds the geometry on a planar face, while the **Inset options** in the **Flange** insets it. See the figure below.

**Inset Flange**

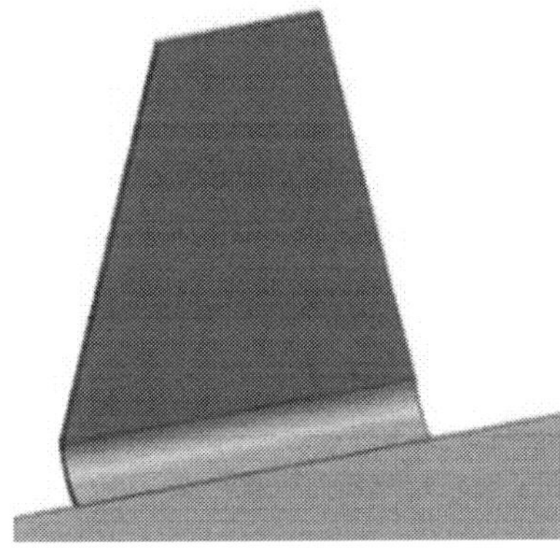

**Flange**

### *Project 19.3.1*

In this project you will create a flange using inset flange with relief on the edges.

- Create a Tab with dimensions **5in** x **2in** x **0.05in**.
- Insert / Bend / Flange.

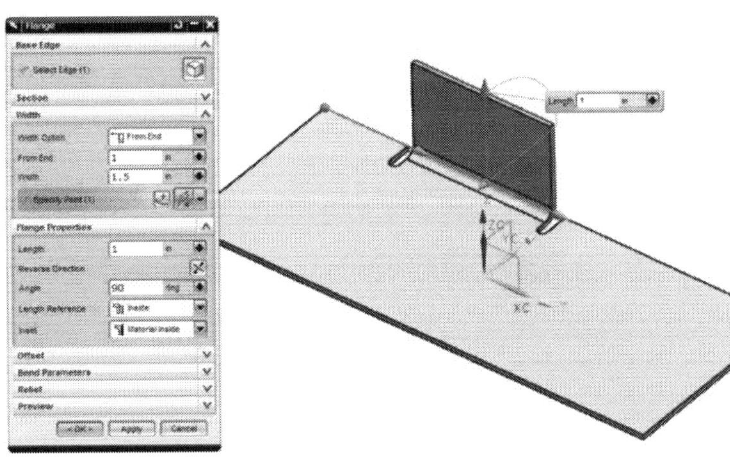

- Create a flange with a length of 1.5in. Change the Width Option to From End. Choose the left edge and give it a distance of 1in. This is the distance away from the edge your flange will be placed.
- To make it an Inset Flange, change the Inset options under Flange Properties to Material Inside as shown in the previous figure.
- Finally, change the **Relief** option to have a **Round** of 0.1in in width and depth.

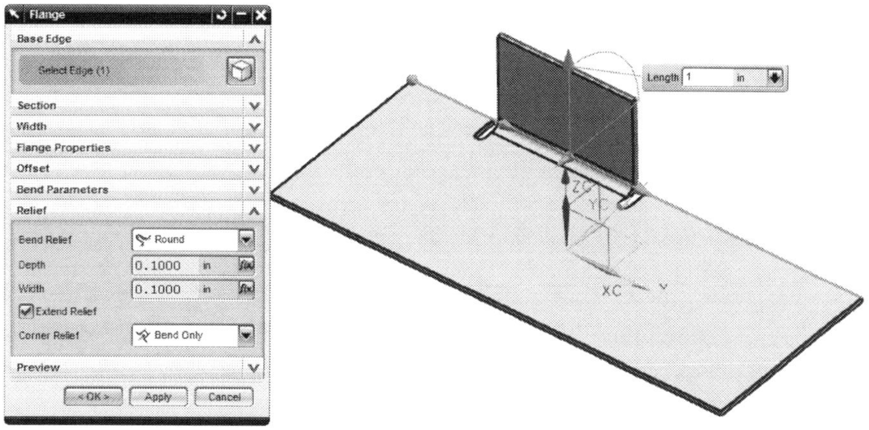


- Once you click OK, your model should look like the following figure.

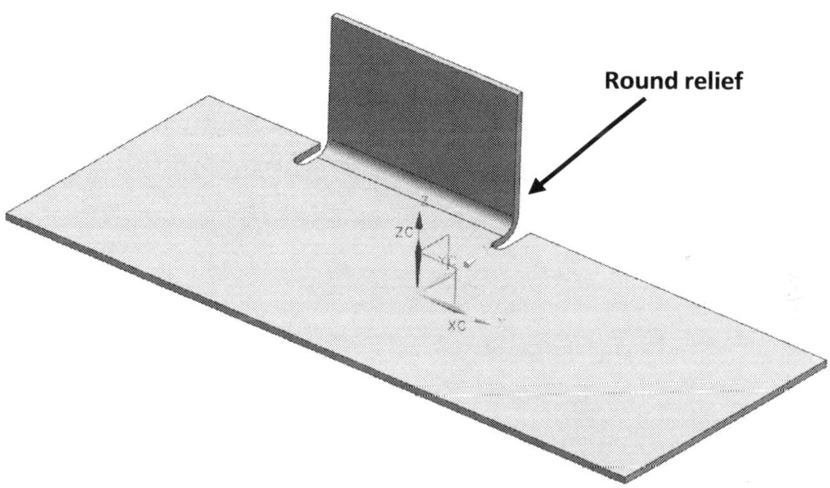


***End of Project***

## 19.4 Sheet Metal from Solid

**Sheet Metal from Solid** is used to create a sheet metal feature using reference geometry. This is a really good technique especially when you have a part that needs a sheet metal inlay or something to that effect. You simply select the faces that are to be included and sheet metal is generated.

### *Project 19.4.1*

In this project you will create a sheet metal part using the reference geometry.

- Create geometry by extruding the sketch shown in the figure below by **2** inches. Note: You can create this in the NX Sheet Metal environment.

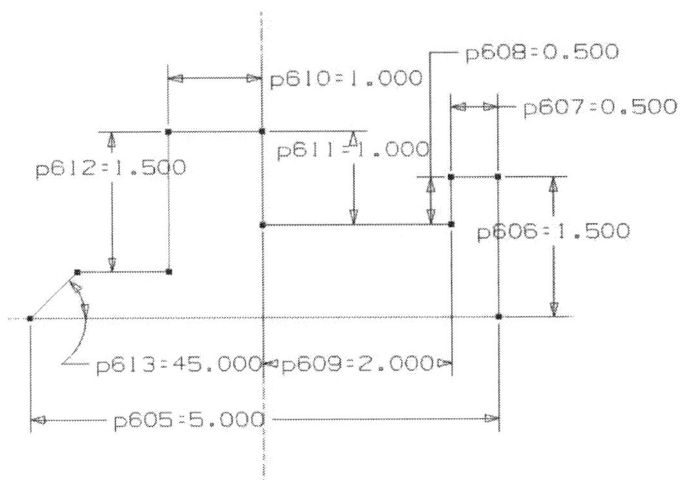


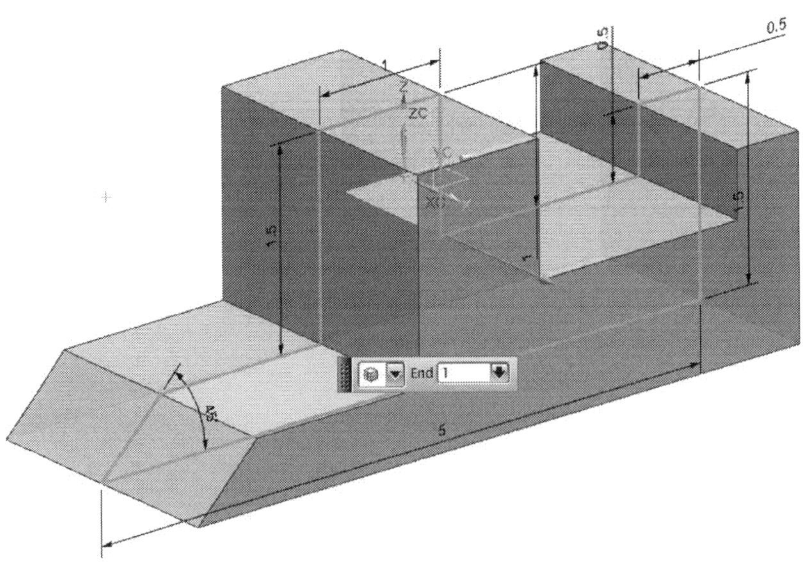

- Insert / Bend / Sheet Metal from Solid.
- Sequentially select all the faces on the outside of the solid, as shown below, from one side of the part to the other. After you click the 2nd adjacent face, the sheet metal will begin creating.

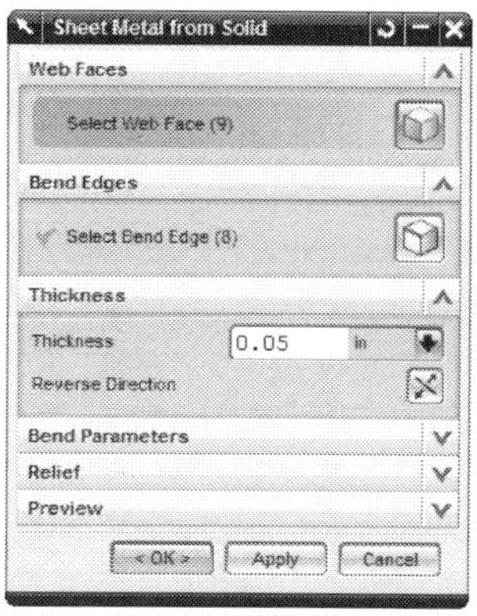


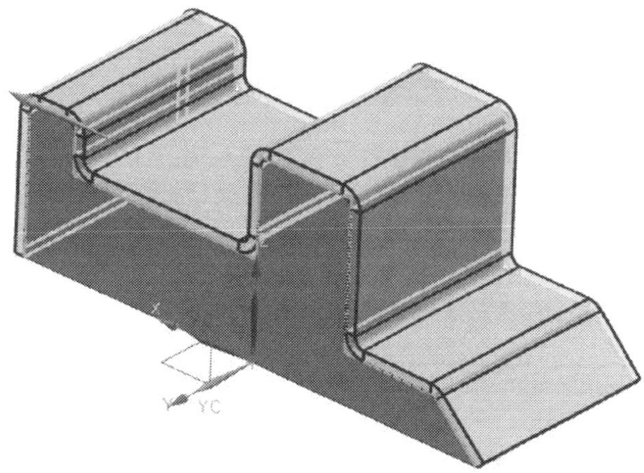

- Change the thickness of the sheet metal to **0.05in**.
- Click OK to complete. Your sheet metal will now look like the following figure.

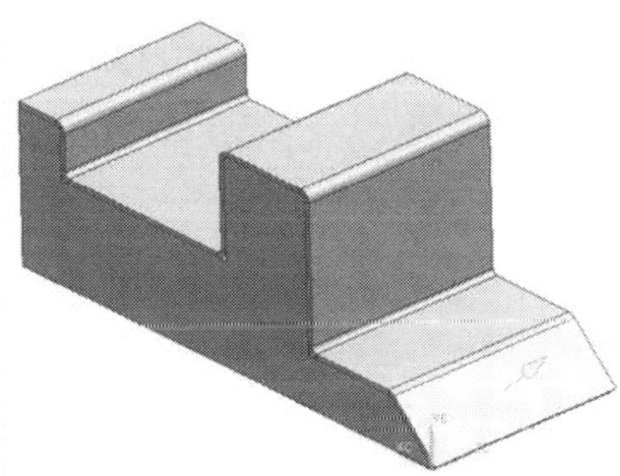

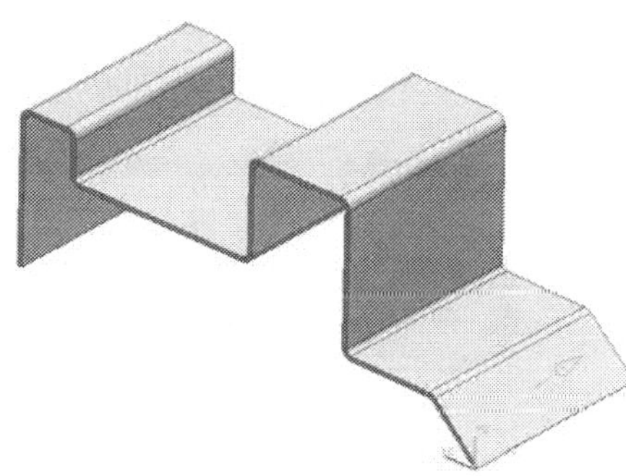

***End of Project***

## 19.5 Contour Flange

The **Contour Flange** is a great tool that can create a custom flange around the entire perimeter of a tab.

### *Project 19.5.1*

In this project you will create a **Contour Flange** with a custom profile.

- Create a **Tab** that is 10in x 5in x 0.1in.
- Insert / Bend / Contour flange.
- Choose **Secondary** as the **Type**. The next stage is prompting you to select an edge to sketch your custom flange profile on. Select the edge shown, define the postion as 50% along the edge in the following figure, and you will be thrown into the sketch environment.

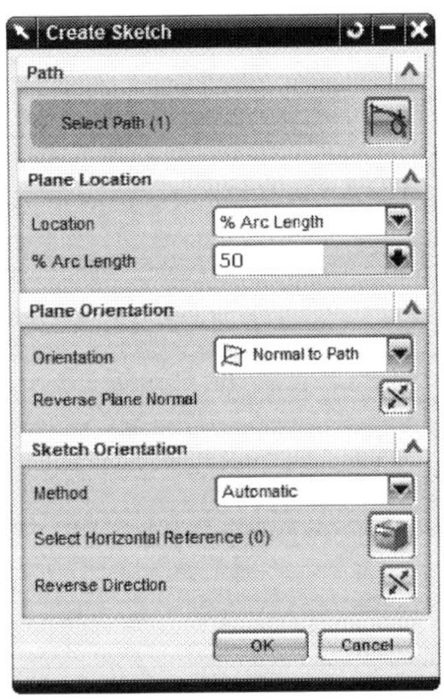


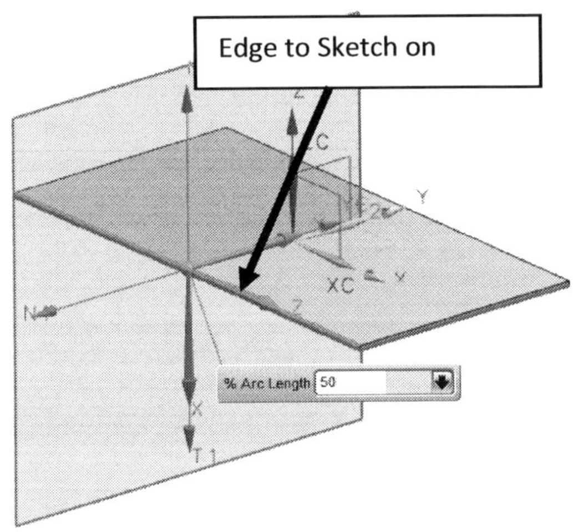

- **Draw** the profile shown in the following figure. Note: Start the profile from the edge that you selected for the flange in the previous step.

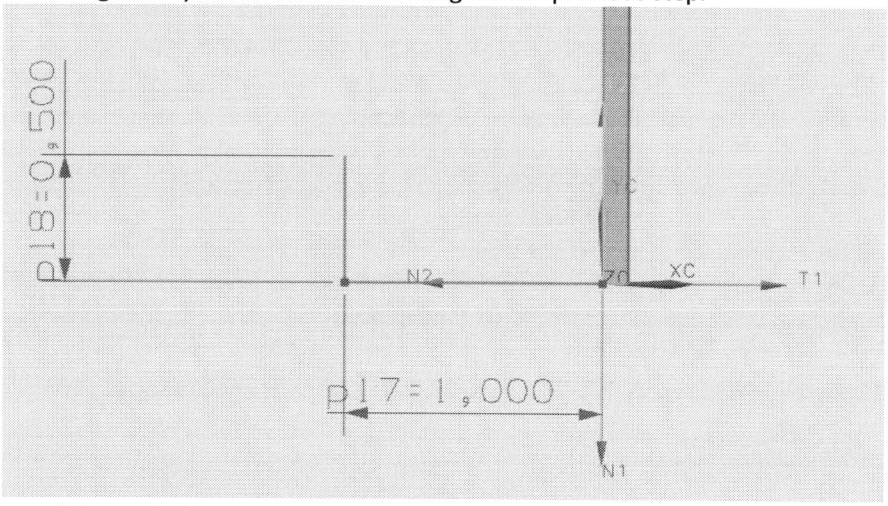


- Click Finish Sketch to move onto the next step.
- At the **Width** step, select **Chain** as the option.

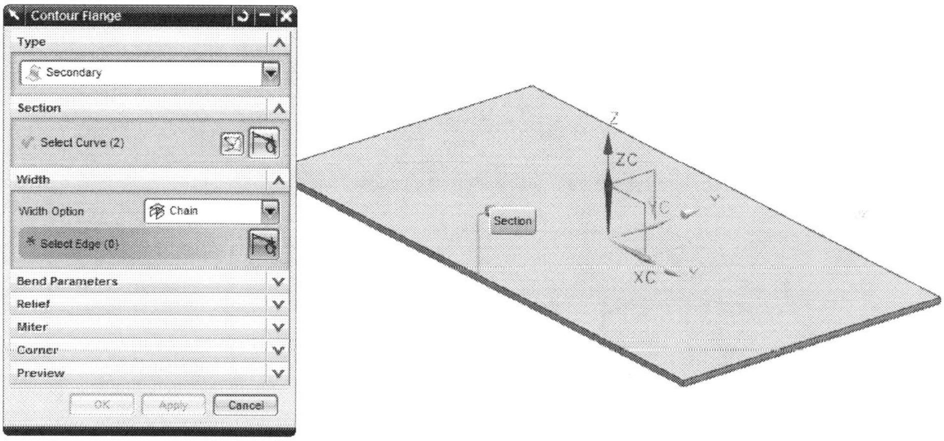

- In the Width option step select the top edges of the Tab. The flange will be created around the perimeter, as shown in the following figure.

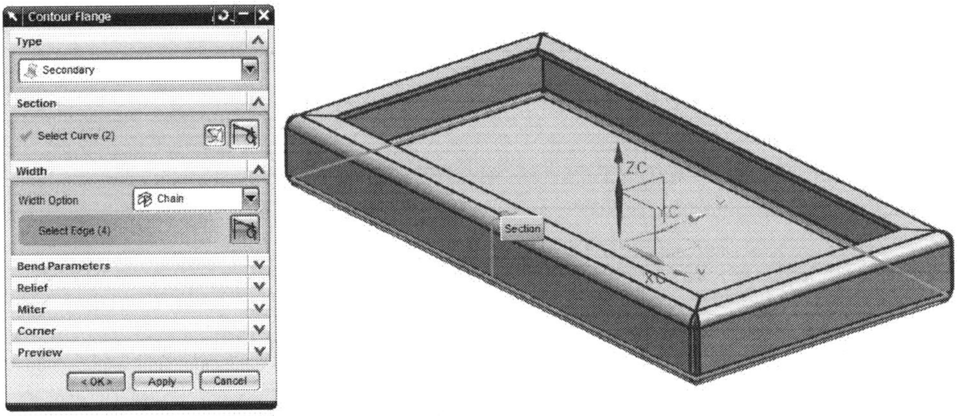


- Your model will now look like the following figure.

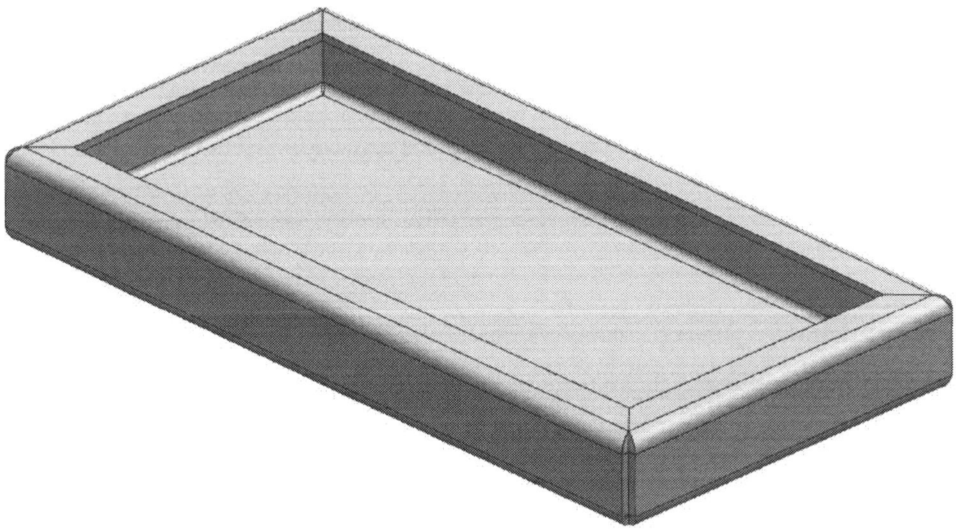

***End of Project***

## 19.6 Bridge Bend

The Bridge Bend feature is used to connect two bodies. You can define constraints at base and target profiles to create a bridge between two sheet metal components.

### *Project 19.6.1*

In this project you will create a bridge using Z-bend.

- Create two separate tabs that are 6in apart in the X direction and 2in apart in the Z direction.

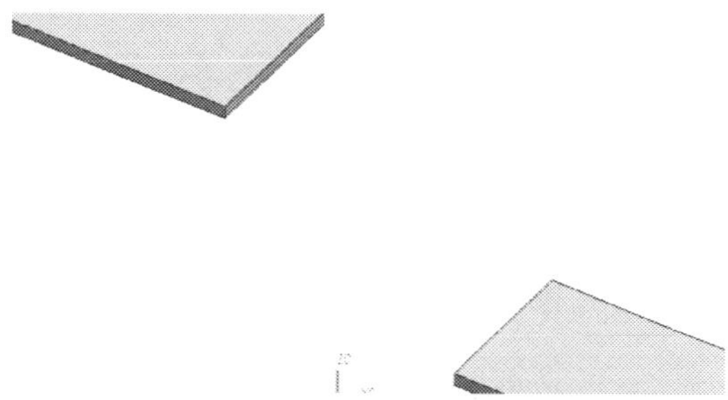

- Insert / NX Advanced Sheet Metal / Bridge Bend.
- Choose **Z or U Transition** as the type. Then select the lower edges from each of the separate tabs for the **Transition Edges,** as shown in the following figure.
- Select the **Reference Point** on the lower corner of the edge shown in the following figure.
- Make the **width** of the **Z and U Transition** the same as the tab.

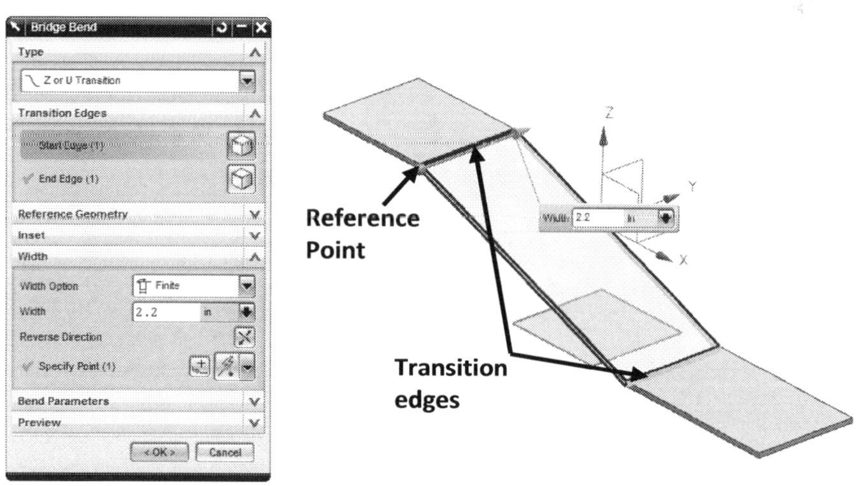

Select **OK** and compare your results with the figure below.

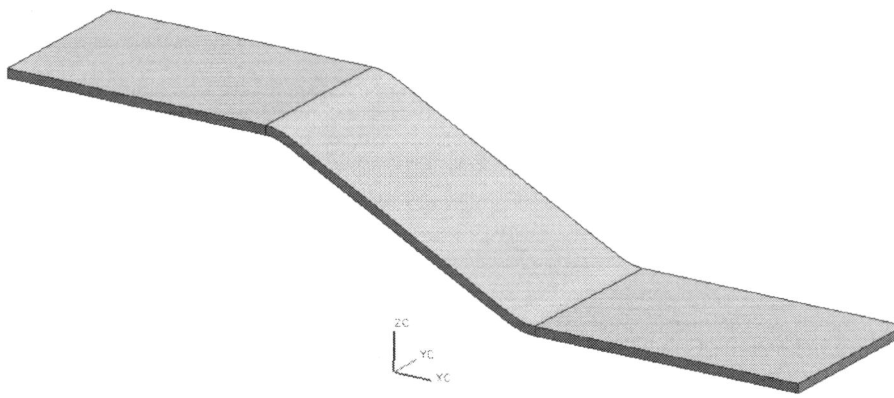


***End of Project***

## 19.7 Bead

'Beads' are created using open-ended curves as centerlines. These are fully associative with the body. You can create 'U-shape,' 'V-shape' and circular beads.

### *Project 19.7.1*

In this project you will create a U-shape hollow bead and then pattern it.

- Start with the model you created in the previous project.
- **Insert / Punch / Bead**. Select the top face shown below and you will be thrown into the sketch environment.
- Sketch a centerline to create a bead as shown in the figure below.

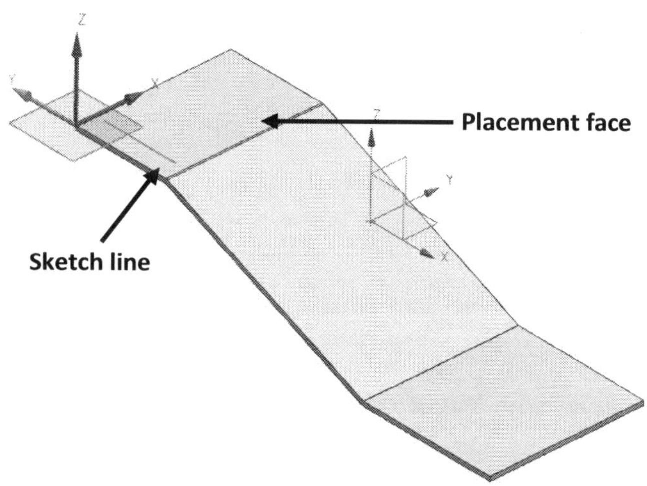


- The figure on the following page shows the Bead window.
- Choose U-shape as the Cross Section and fill out the parameters shown in the following figure.

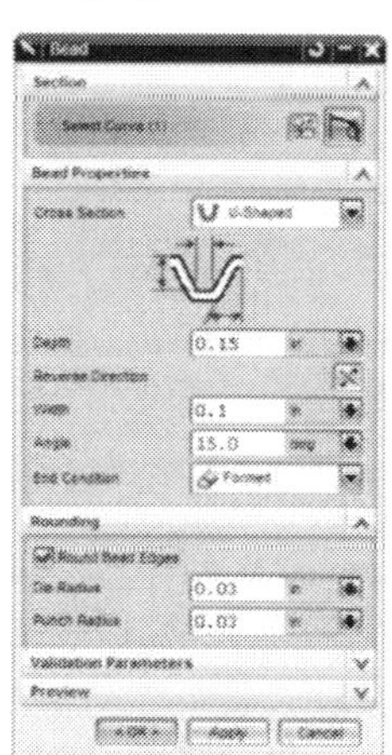

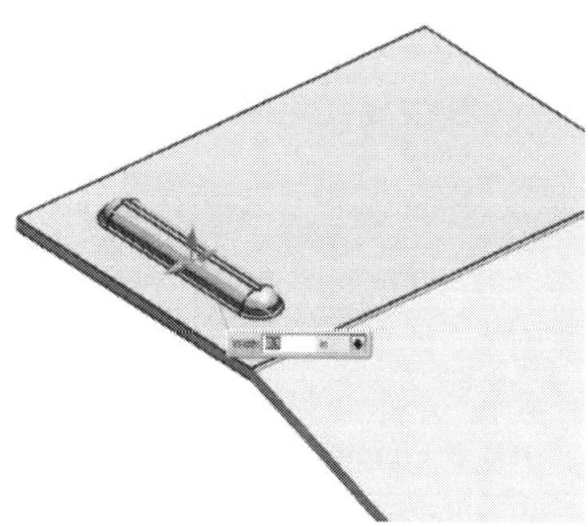

- **Click OK.** Your model will now look like the following figure.

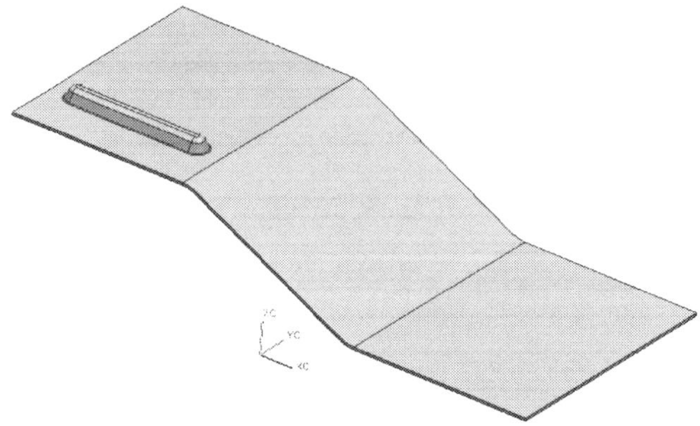

- Use **pattern-face** to repeat the beads as shown in the following figure. Select all the faces that make up the bead feature. Choose the long edge of the top tab as the X and the short edge as the Y. Make the number along X = 3 and Y=1. Specify an appropriate distance for X and click OK. Your model should now look like the following figure.

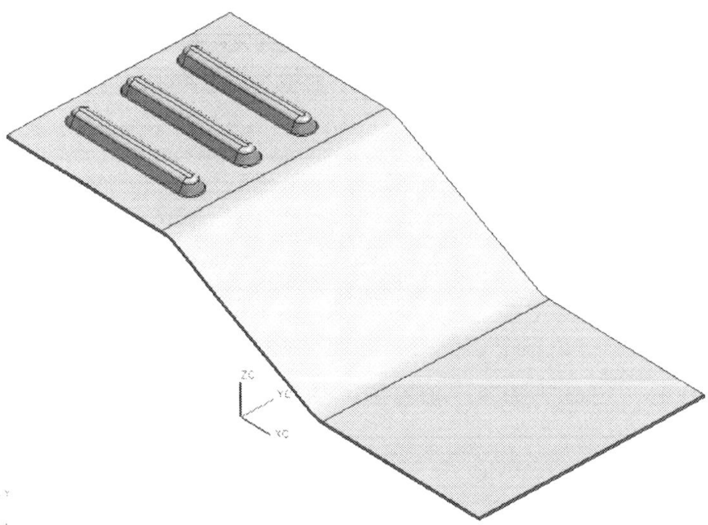

***End of Project***

## 19.8 Drawn Cutout

'Drawn Cutout' is one of the many Punch tools available in NX. You can also use the solid punch tool that will be shown in a later exercise.

### *Project 19.8.1*

Use the embossed punch to create a sunken area with a flanged cutout.

- Create the sketches shown in the following figure using the model from the last exercise.

- Insert / Punch / Drawn Cutout.
- Select the edges of the square profile as the **section**. Change the **Depth** to 0.2in and the **Rounding** to 0.05in.

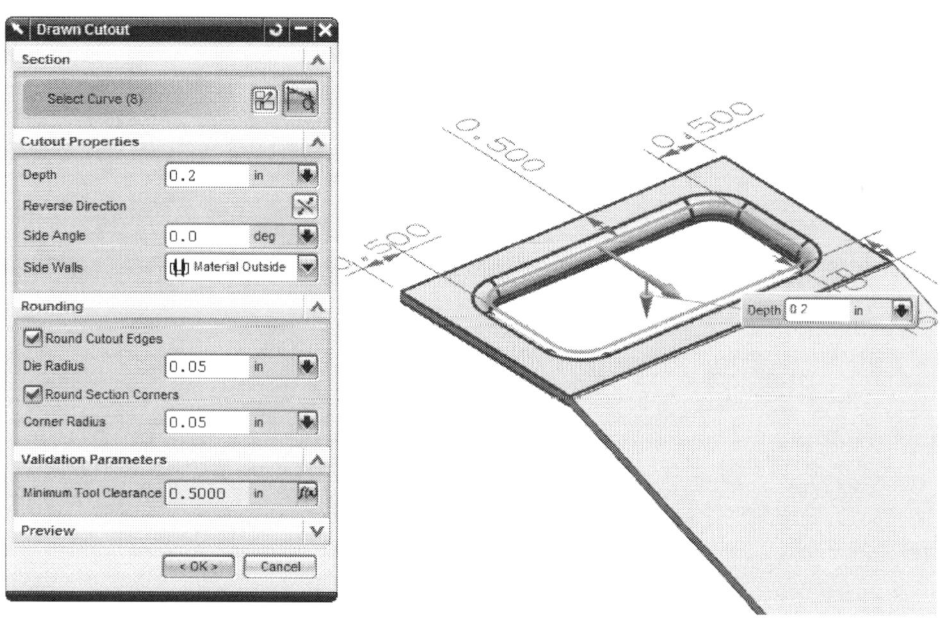

- Your model should now look like the following figure.

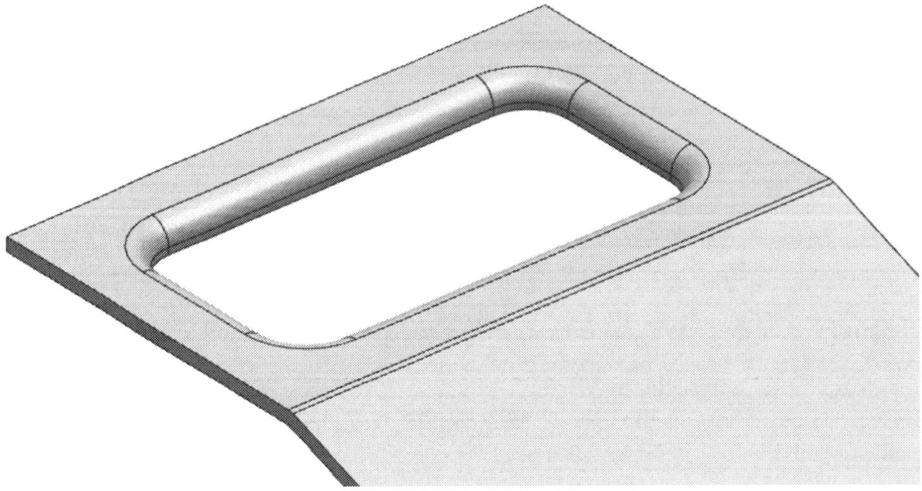

***End of Project***

## 19.9 Corner

The **Corner** feature can be used to create *Butt, Machinery, Simple Miter* and *Full Miter* corners between two flanges.

### *Project 19.9.1*

In this project you will create a corner between two flanges.

- Create a Tab of **6in** x **4in** x **0.05in**.
- Create a **1in** high flange on all four sides as shown in the following figure.

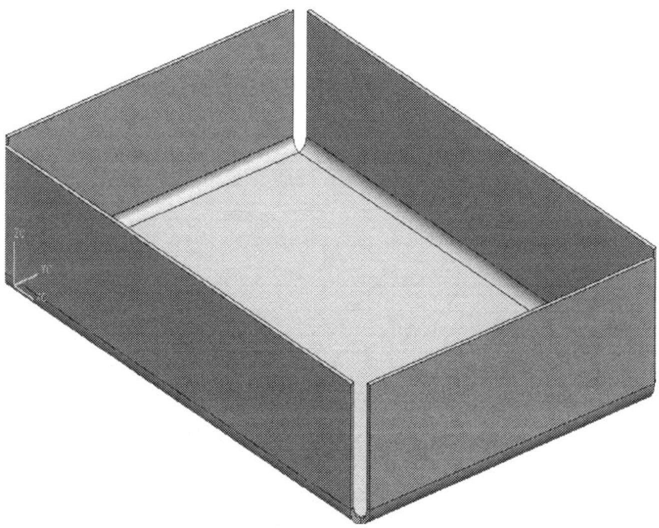

- **Insert / Corner / Closed Corner.**

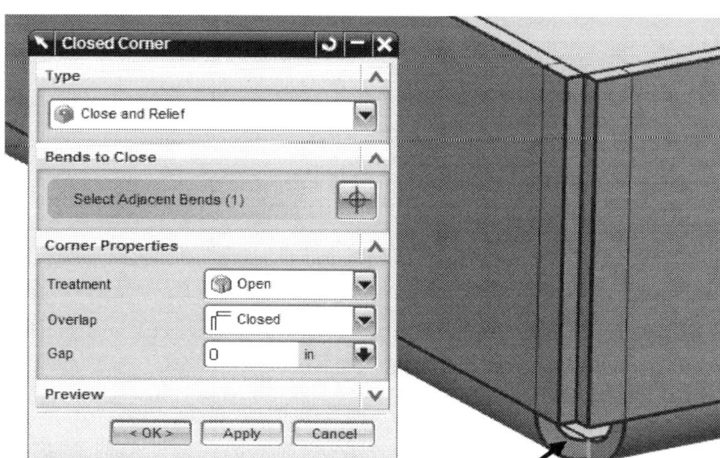


**Select both bends**

- Choose **Close and Relief** as the **Type** and the **Treatment** step as **Open**.
- Make two corners **Open** (shown below on right) and two corners with **Closed** (shown below on left).

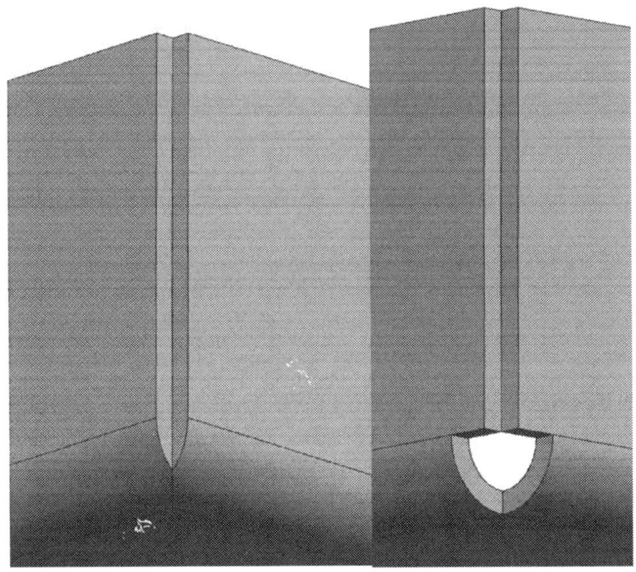

Compare your results with the figure above.

***End of Project***

## 19.10 Unbend/Rebend

**Unbend/Rebend** can be used to form or unform self-forming features. This is a time-stamped operation so you can Unbend a flange, go through an operation where you need a flat surface, then Rebend the flange after.

### Project 19.10.1

- Open the previous project.
- Insert / Form / Unbend.
- Select a **Stationary face** (as shown below).

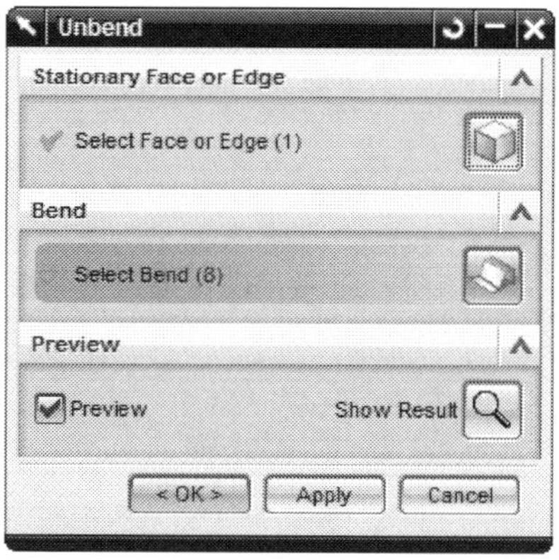


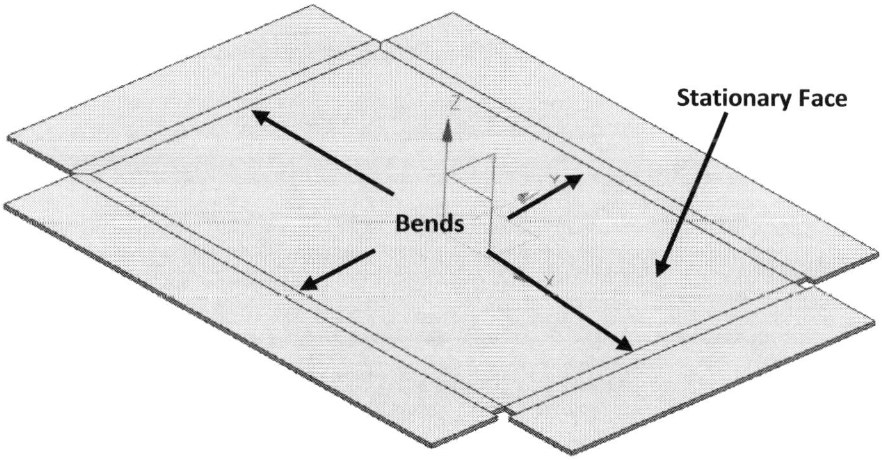


- Select the bends that you want to Unbend (see figure above).
- Compare your results with the figure above.

***End of Project***

## 19.11 Relief

In NX 8.5 you can create reliefs within the Closed Corner tool. You can create a Circular, U, V, and Rectangular Relief features between bend areas that are fully associative with the part.

### *Project 19.11.1*

- Use the part file from the last exercise and delete the **Unbend** operation and the **Closed Corner** operations.
- Now we are going to create new Closed Corners, this time with reliefs. ***Insert / Corner / Closed Corner.***
- Select the two bends around one of the corners like you did in Project 19.9.1.
- For **Treatment,** choose **Circular Cutout** and copy the dimensions shown in the following figure.
- Click OK to complete. Your corner should now look like the following figure.

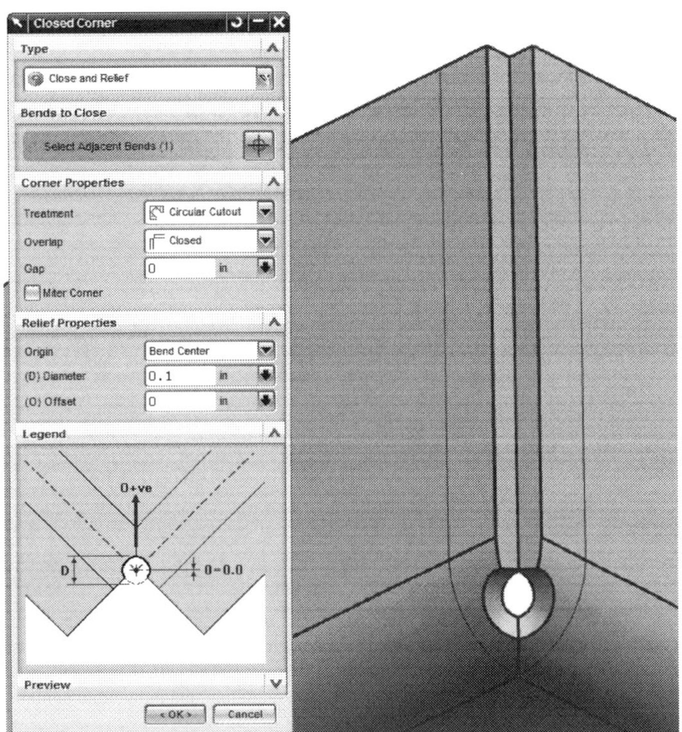

- Use the other 3 Relief styles on the other corners in your model.

**U-shaped Relief**

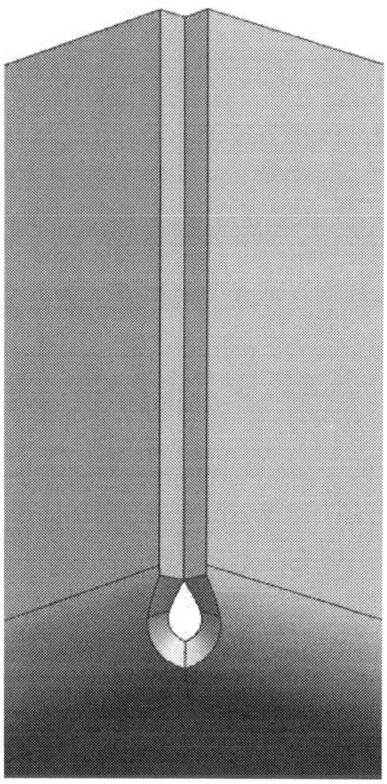

**V-shaped Relief**

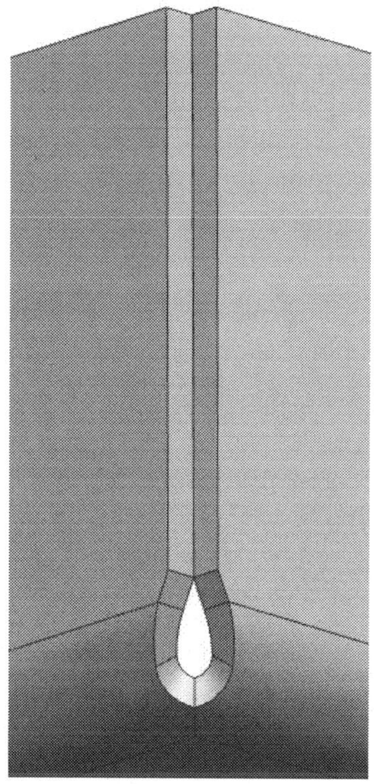

**Rectangle-shaped Relief**

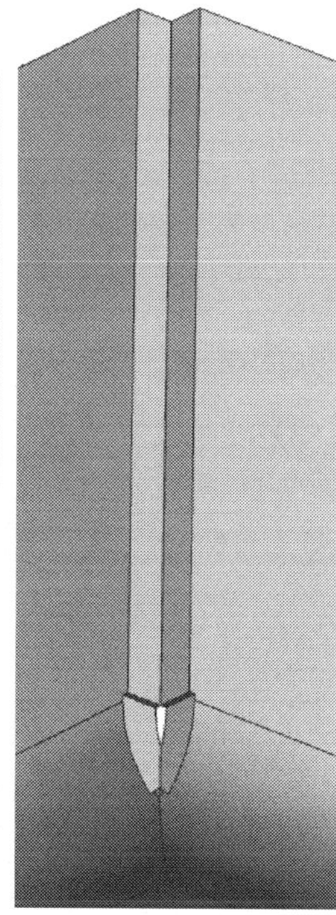

***End of Project***

## 19.12 Solid Punch

A **Solid Punch** can be used when you want to inherit a shape from a tool body.

### *Project 19.12.1*

- Open the part file from the previous exercise.
- Create a tool solid, as shown in the figure below.

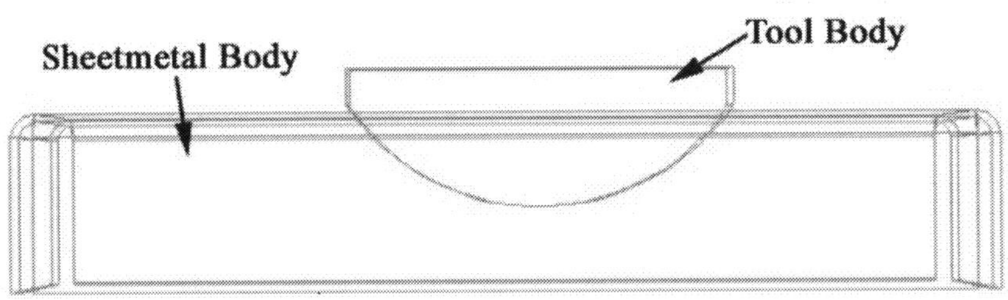


- Select Insert / Punch / Solid Punch.
- Select the Target Face and Tool Body as shown in the figure above.
- Check the Hide the tool body option, along with Infer Thickness and Auto Centroid.

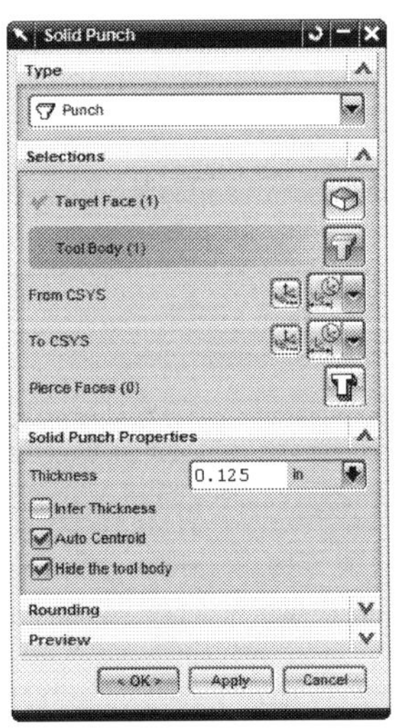


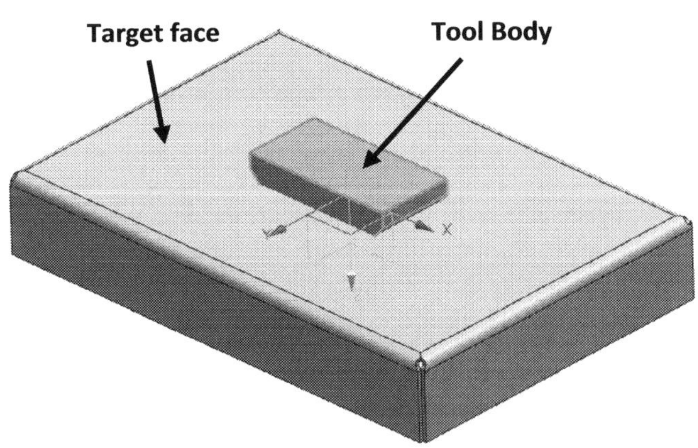


- Click **OK** and compare your results with the following figure.

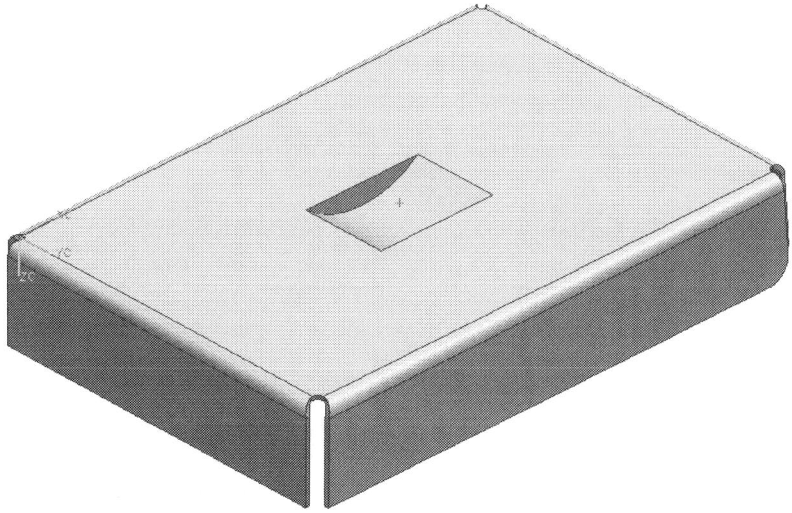

***End of Project***

## 19.13 Bend

You can bend a solid body of uniform thickness along any straight curve using the **Bend** option.

### *Project 19.13.1*

- Create the following sketch, that is roughly **5in x 5in** and extrude the square **0.05in** so your model looks like the following figure. The lines are going to be used to create the bends.

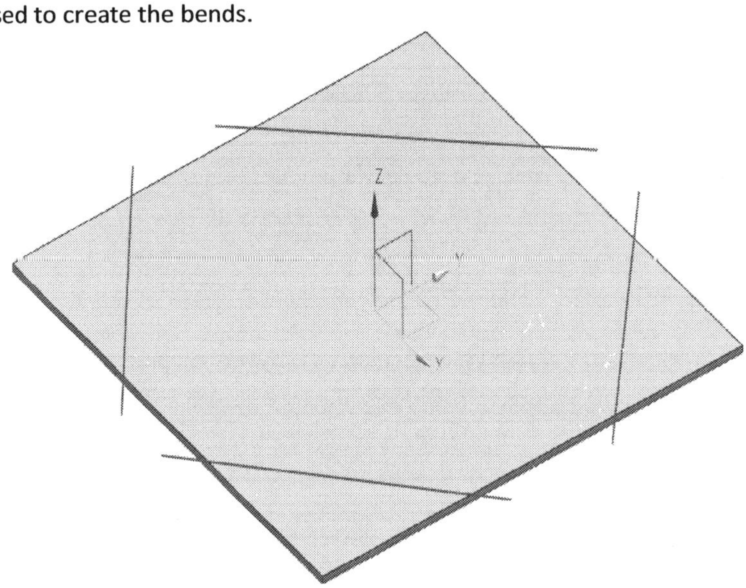

- Select Insert / Bend / Bend.
- Select one of the bend lines and a bend will be automatically created.
- Make your model look like the following figure.

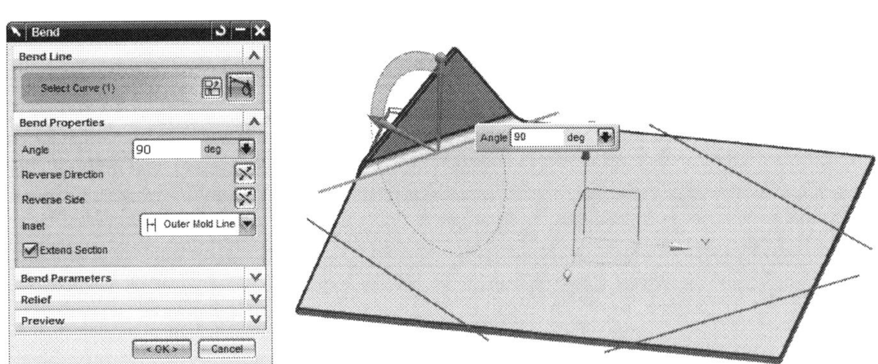

- Click OK to complete.
- Repeat the process for the other lines, changing the bend direction for each bend.
- Your model should look like the following figure.

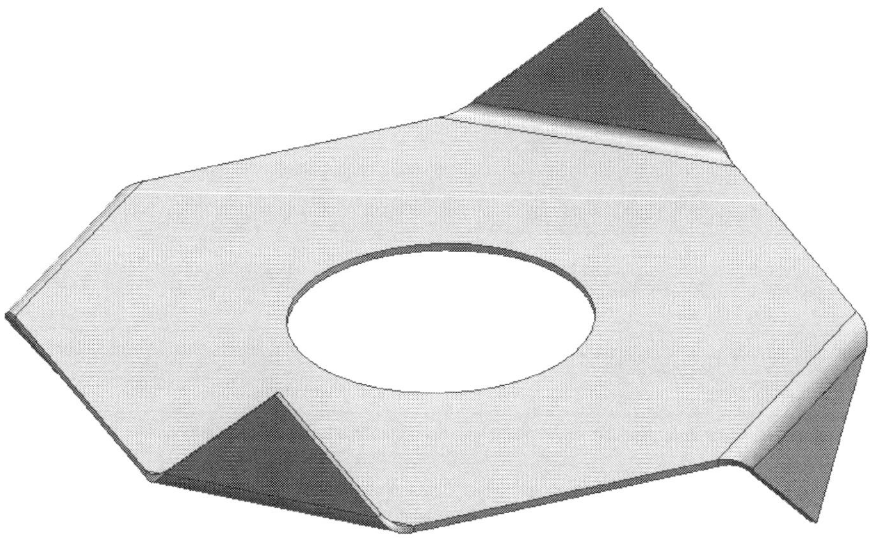

> **Note:** The Reverse Direction button will change the direction of the Bend. The Reverse Side button will change what side of the line is to be bent.

***End of Project***

## 19.14 Self-Test

(All questions are true/false. Answers are at the back of the book.)

1. You can create an Inset Flange in the Normal Flange tool.

2. There is no difference between a normal hole feature and a sheet metal hole feature.

3. You can create a relief while inserting a flange.

4. You need to sketch the profile of a square to create Bead Features.

# 20. Injection Molded Parts and Castings

## 20.1 Introduction

Injection molded and die cast components hold a special challenge for modelers and designers. Proper designing and location of ribs and parting lines can save untold hours and money. The features that you design over and above those methods for the part to actually function are critical. This chapter is a compilation of some of the modeling techniques that are necessary to create such features.

## 20.2 Draft from Edges

The 'Draft from Edges' option allows you to Draft the faces from selected edges. This option is of great value when the edges (from which you want to draft) are not coplanar.

### *Project 20.2.1*

- Sketch the geometry shown in the figure below.

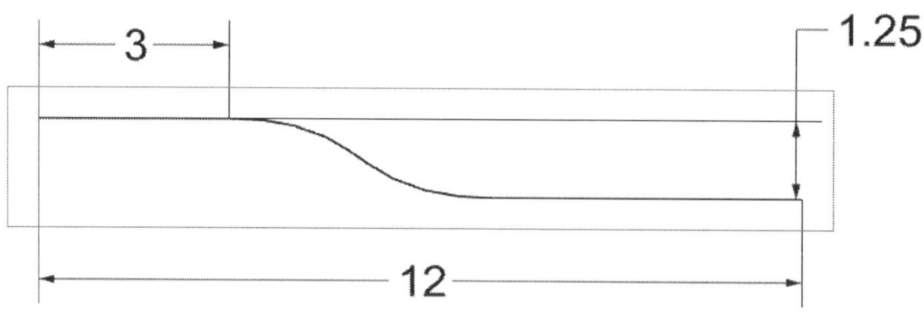

- Extrude the sketch from **0** to **2** inches with a **1** inch downward offset.
- Select Insert / Detail Feature / Draft.
- Select the **From Edges** type and the edges shown in the figure below.
- Enter an angle of **10** and specify a draw direction as shown in the figure below

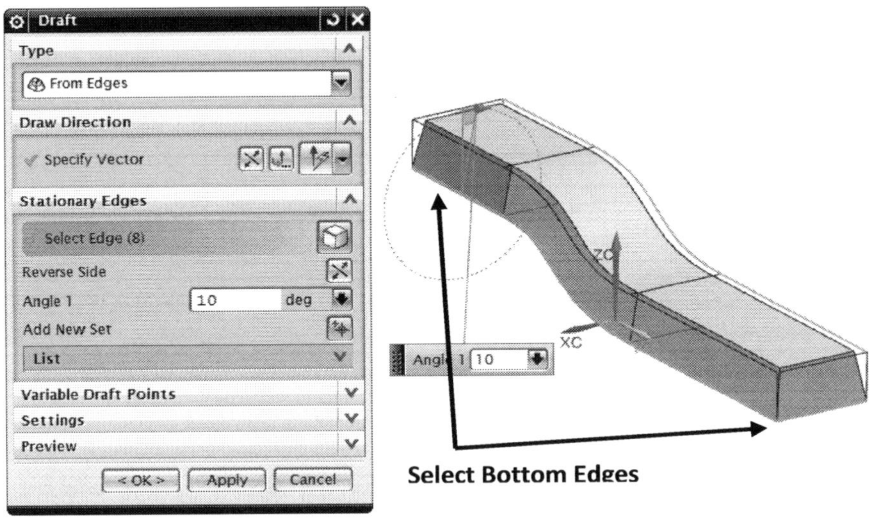


**Note:** You can also change the angle if you're feeling adventurous.

- Shell the resulting block with a thickness of **0.1In** as shown in the following figure.

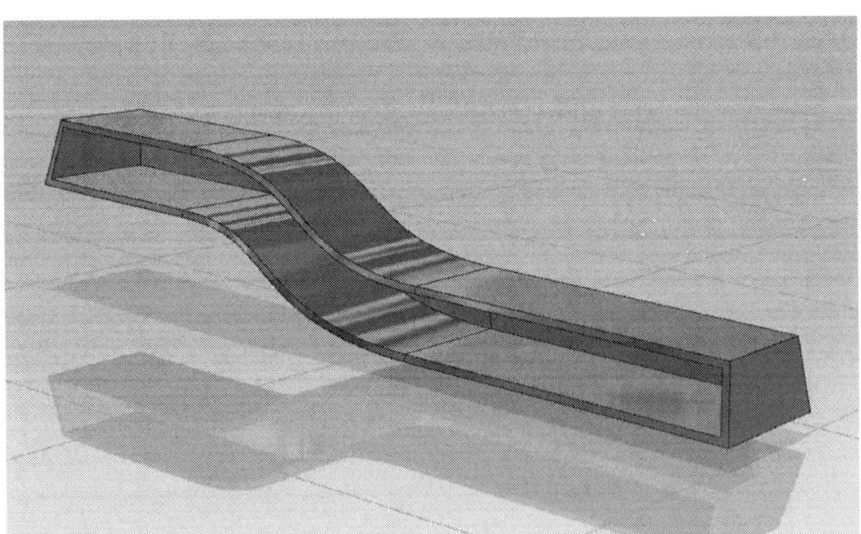

***End of Project***

## 20.3 Body Draft

The 'Body Draft' function is very useful when you are developing a complete form that needs molding or casting. Once you have a datum plane or surface to define the parting plane, you can create drafts on both sides. You can match the drafts and even add material to any undercut areas.

### *Project 20.3.1*

- Create a solid as shown in the figure below.

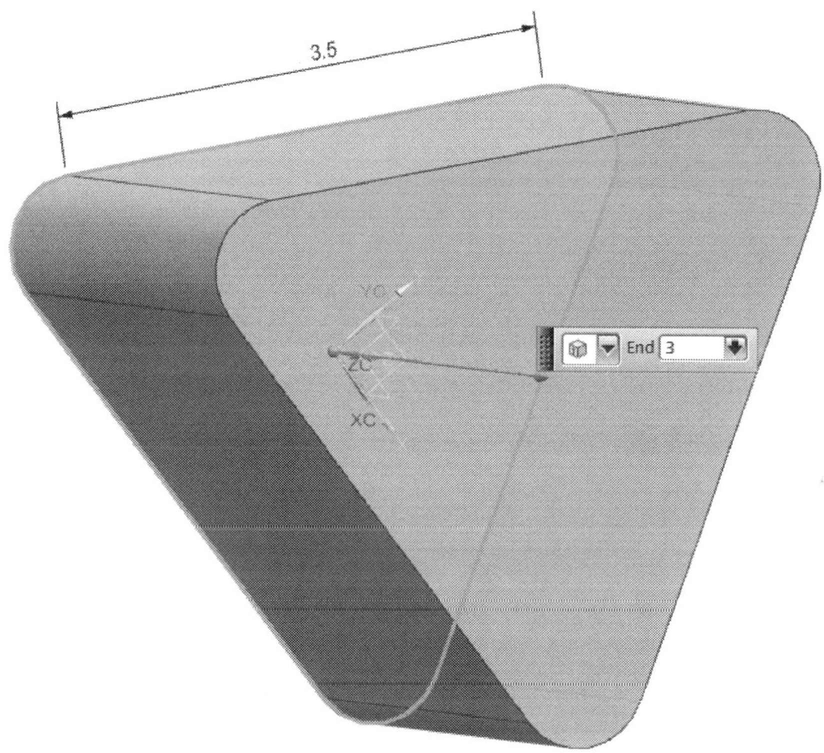


- Create a parting surface as shown in the following figure.

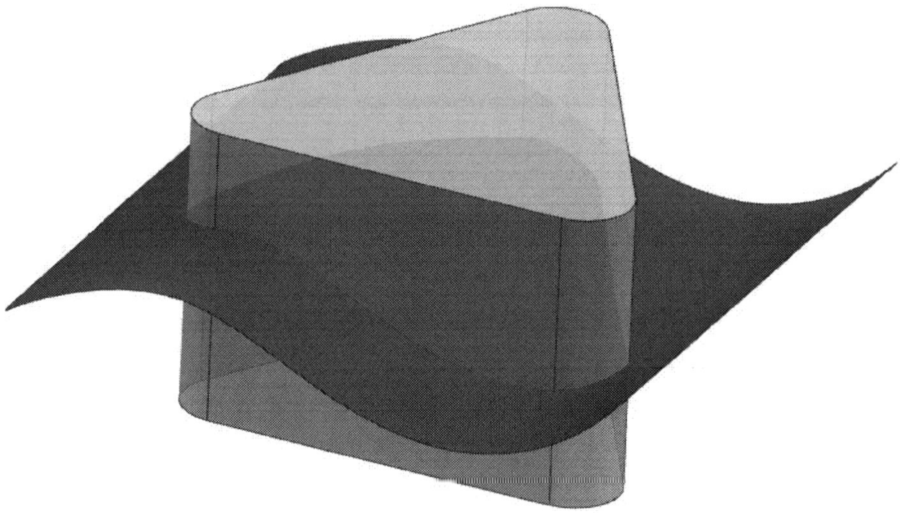

- Select Insert / Detail Feature / Draft Body.
- Select the parting surface as shown in the following figure.
- Select the upward direction, and select the faces to be drafted.
- Select ***Match All*** and ***True Draft*** and select **OK**.

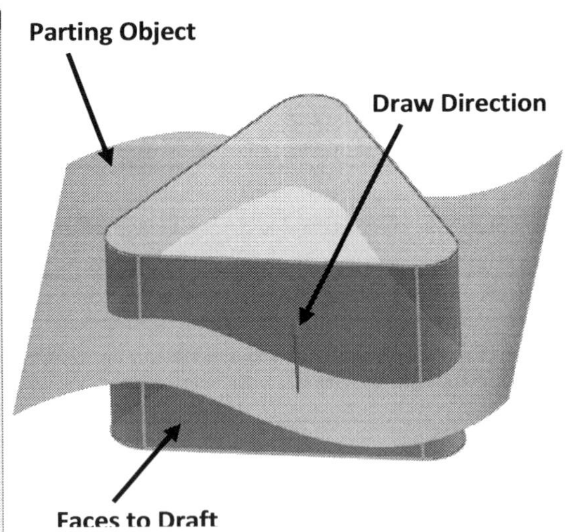

Notice the split draft in the part as shown in the following figure.

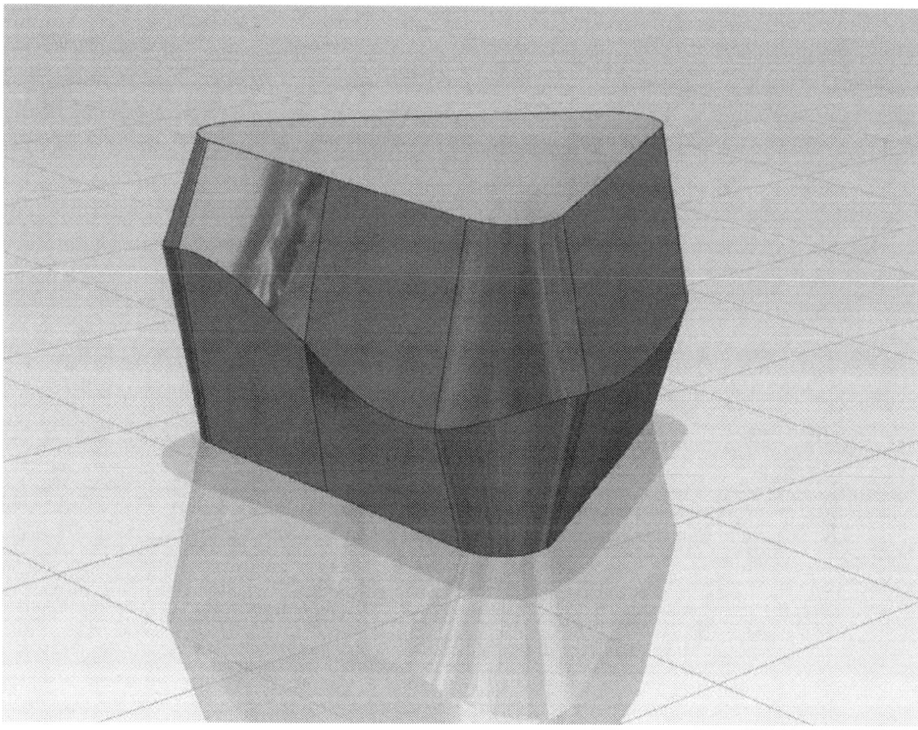

***End of Project***

## 20.4 Isocline Curves

'Isocline Curves' are best used to check undercuts in a part on either side of a parting plane because the draft angle on a set of faces is constant along an isocline curve. 'Isocline Curves' can also be used to create associative extracted curves that update with any modifications in the geometry (creating associative curves is optional). You can also use them to split a surface or to create a "draft angle map" of a face. Overall, they can be used for very detailed analysis of a face.

### *Project 20.4.1*

Create two sketches as shown in the figure below.

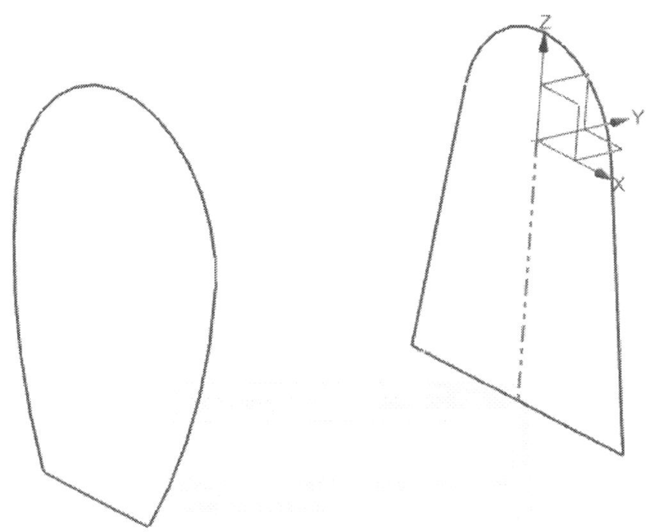


- Use ***Insert / Mesh Surface / Through Curves*** to create a form shown in the following figure.

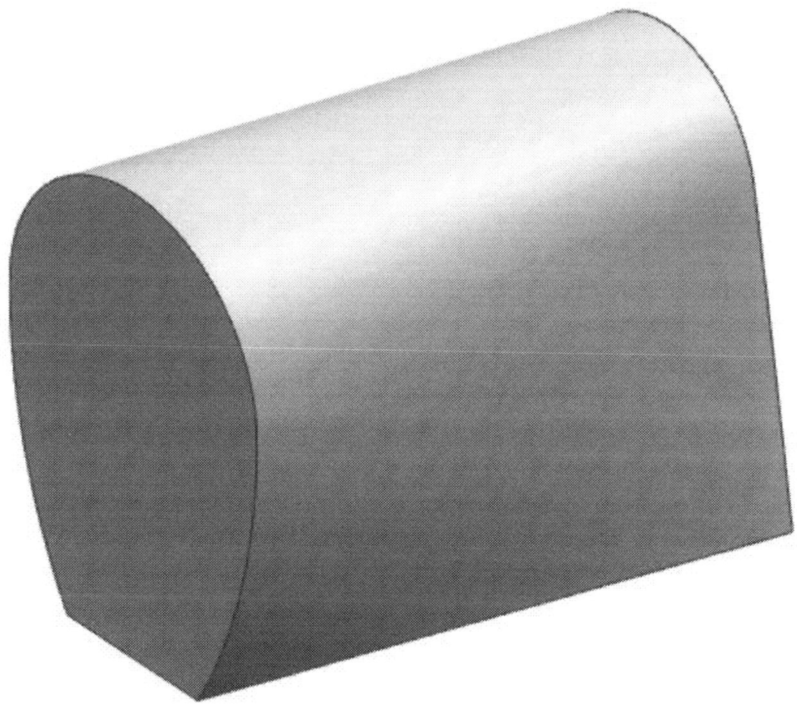

- Select **Insert / Curve from Bodies / Extract**.
- Select ***Isocline Curves*** from the options as shown in the figure below.

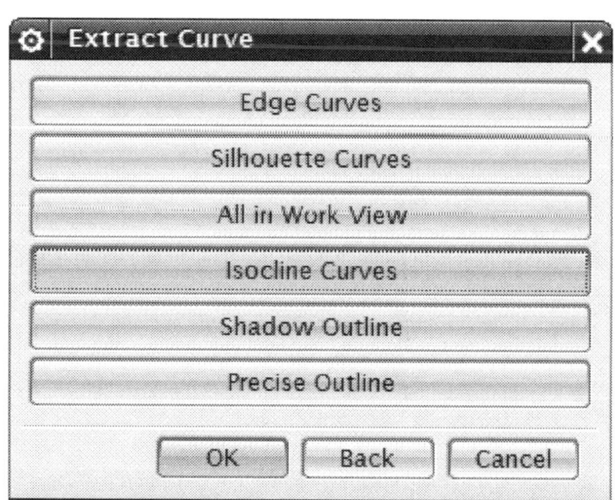


- Select ***YC*** from the ***Vector Constructor*** for the isocline direction as shown in the figure below.

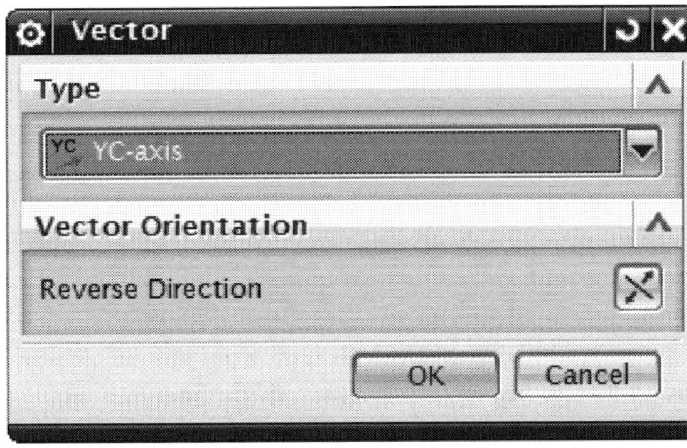


- Select **OK** and define the isocline angle as shown in the figure below.
- Check the **Associate** box.

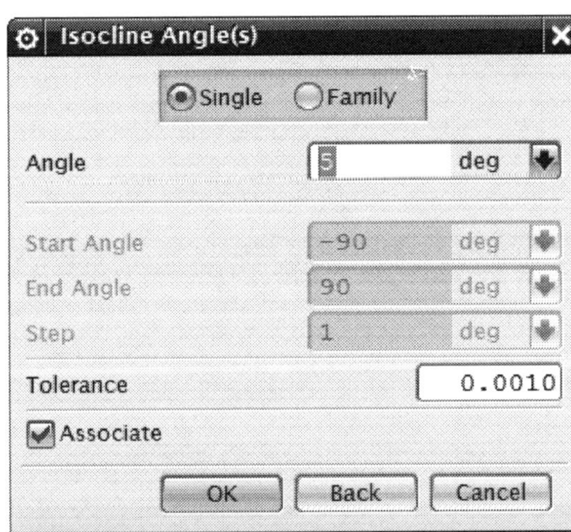


- Select **OK** and select the faces that you want to include to create the curve, shown in the following figure.

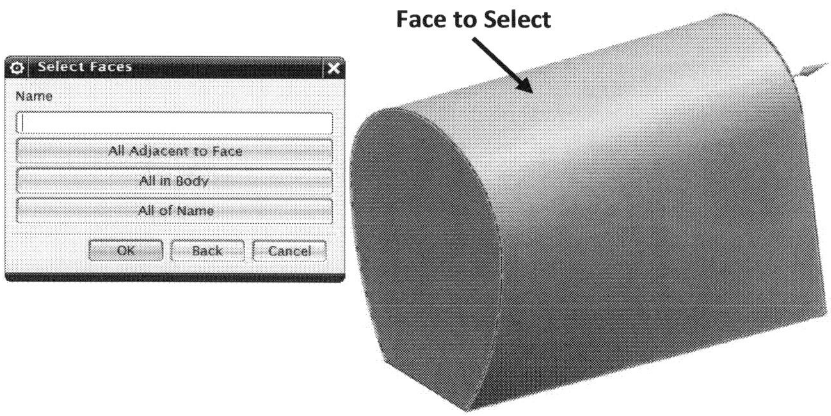


- When you select OK the curves will be created, as shown in the figure below.

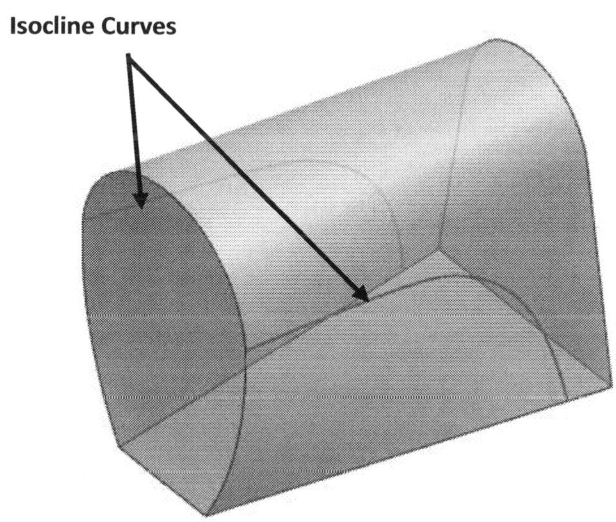


- You can further use these curves to split the surface or create a parting line.

If you make any changes in the geometry, the isocline curves will update, since we created associative curves. This is shown in the following figure.

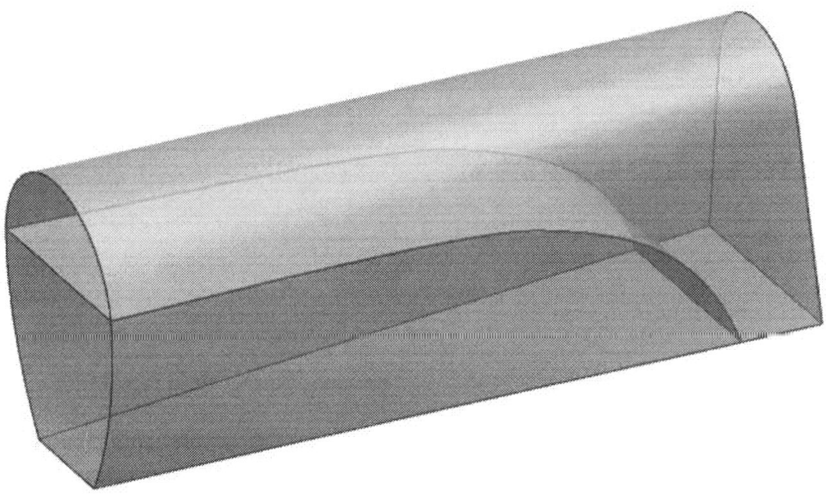

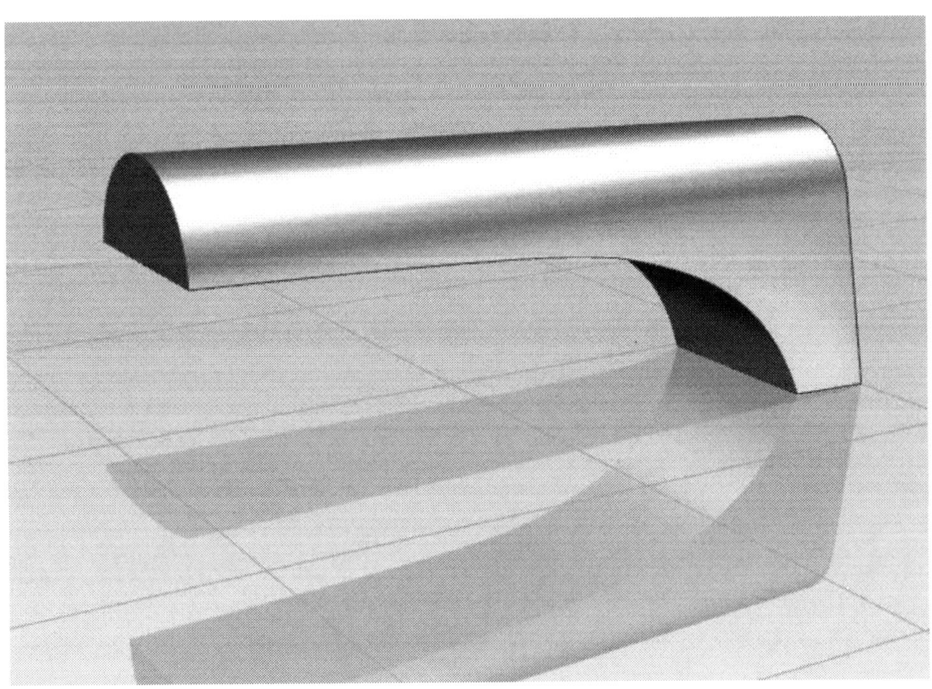

**End of Project**

## 20.5 Divide Face

'Divide Face' is a very important face editing feature that lets you divide a face using existing curves. This is used when you don't want to work with a whole face but only a part of it. 'Divide Face' can also be used to limit problem areas to a small, single face.

### *Project 20.5.1*

Create a surface as shown in the figure below.

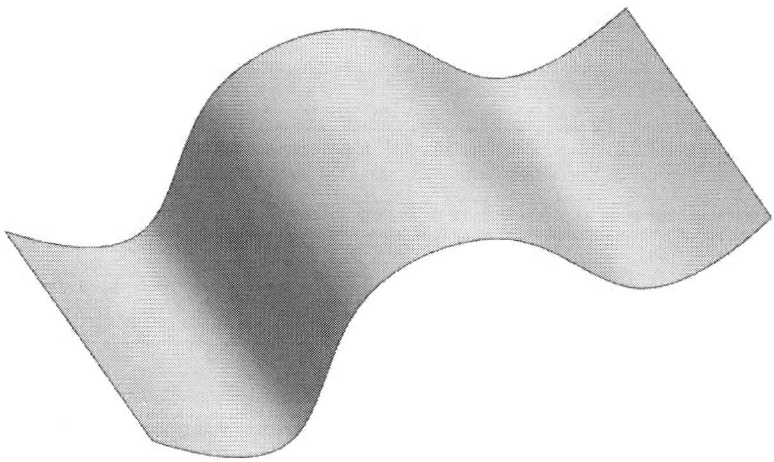

- Create curves to divide the face as shown in the figure below.

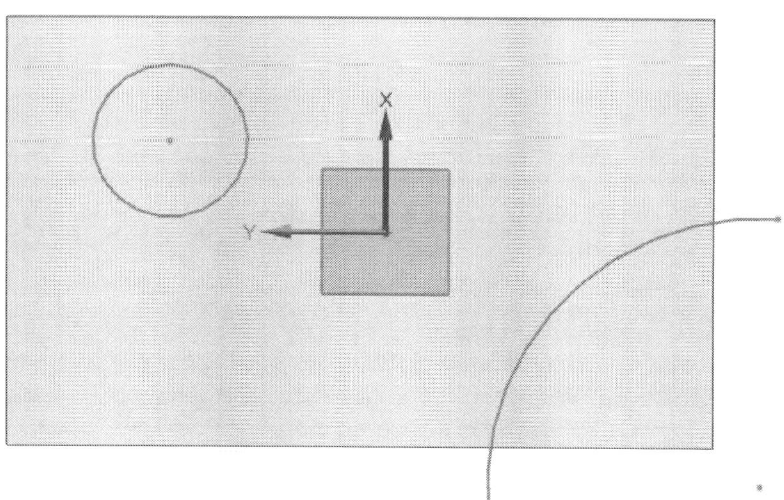

- Select Insert / Trim / Divide Face.
- Select the **face** you want to divide, as shown in the following figure.
- Then select the **dividing objects**, also shown on the following figure. And select **OK**.

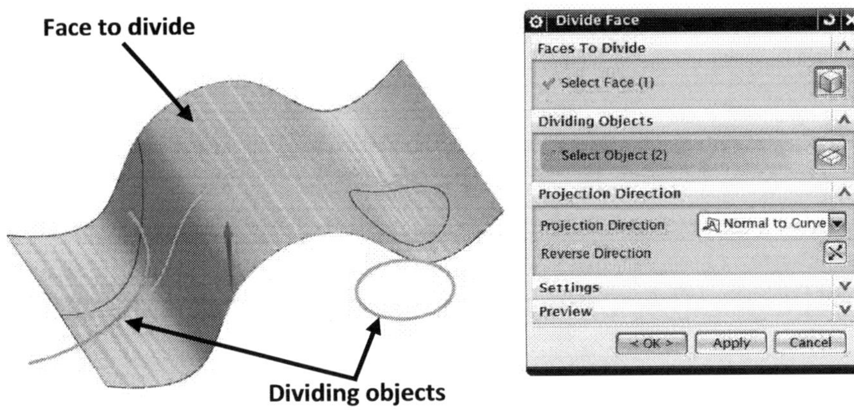


- Once you divide the face you can use it in different ways. For example, select ***Insert / Associative Copy / Extract Body*** and extract the circular face.
- Now, you can use the ***Insert / Offset/Scale / Thicken*** command to create a solid from the extracted face as shown in the figure below.

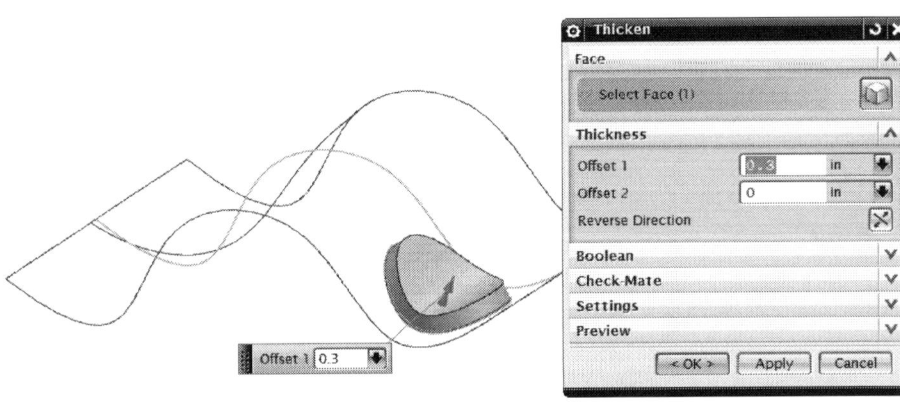


- Also, you can use these faces to create a sheet with an offset. Select the face shown in the following figure and select ***Insert / Offset/Scale / Offset.*** Enter a distance to create a sheet as shown in the following figure.

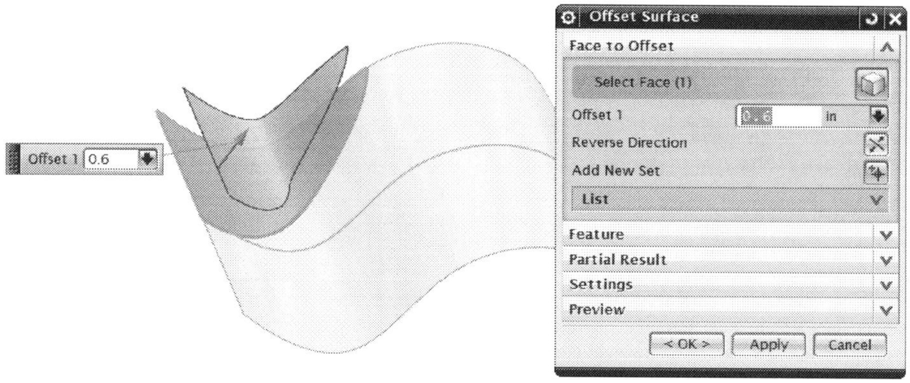


- You can also extend any iso-parametric edges of these faces. For example, select ***Insert / Trim / Trim and Extend.***
- Select By Distance *and* Select Natural Tangent.
- Select the edge as shown in the figure below.
- Enter the length value and hit **OK**.

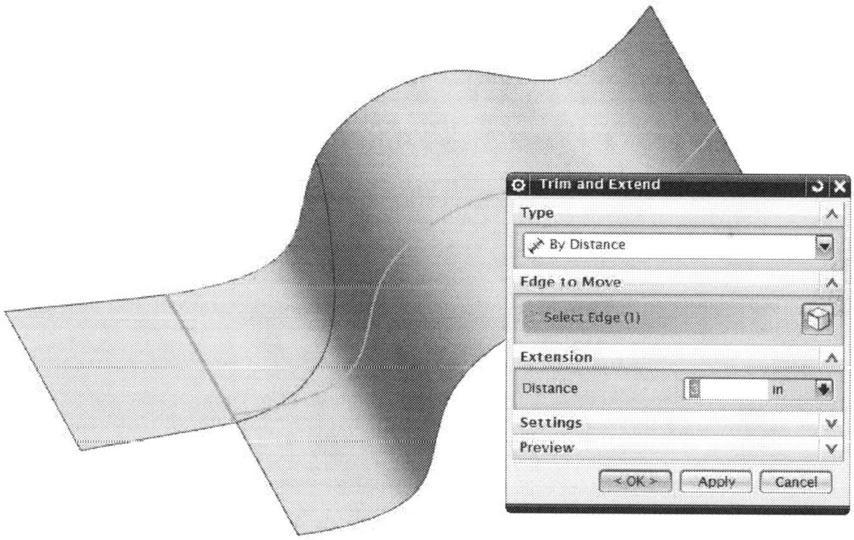


***End of Project***

## 20.6 Sewing Surfaces

You can join multiple sheets together to create a single sheet or solid. If the sheets you are sewing together create a closed volume, 'Sew' creates a solid. In all other circumstances, the resultant is a surface. Sewing is a vital command when importing surfaces into NX 8.5, but be careful of modeling tolerances.

### *Project 20.6.1*

- Return to **Modeling** mode. Create a sketch as shown in the figure below and **Extrude** it from **0** to **1in**.

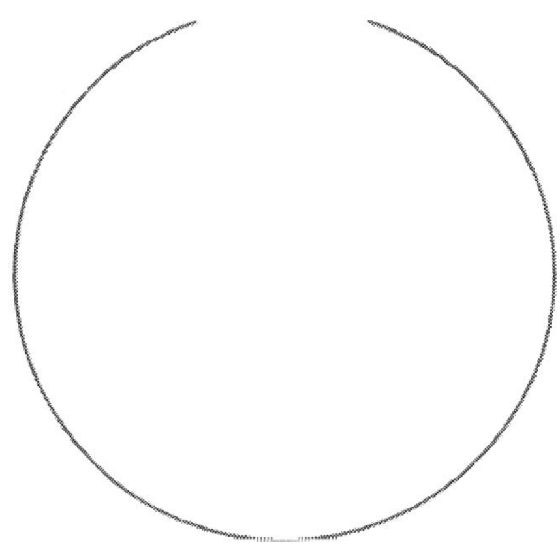

- Create an ***Insert / Detail Feature / Bridge***.

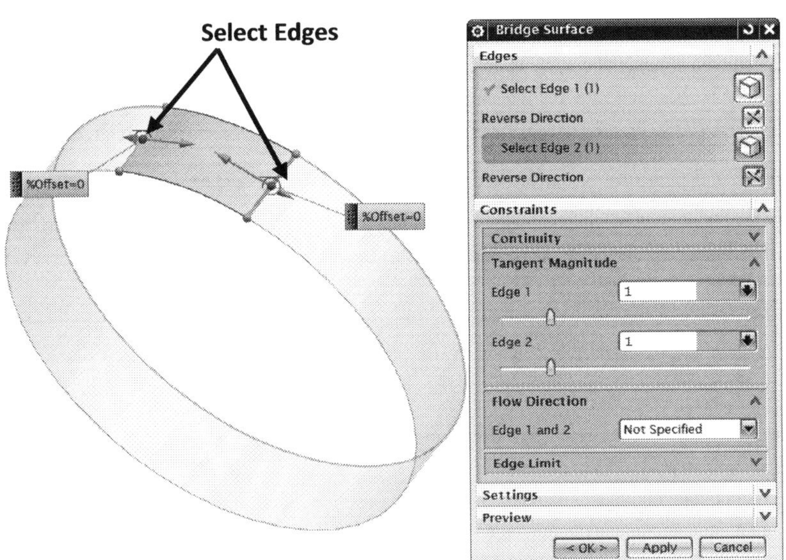

- Select ***Insert / Combine / Sew*** and select the **Target** and the **Tool** sheets as shown in the figure below.
- Select **OK** to **Sew**.

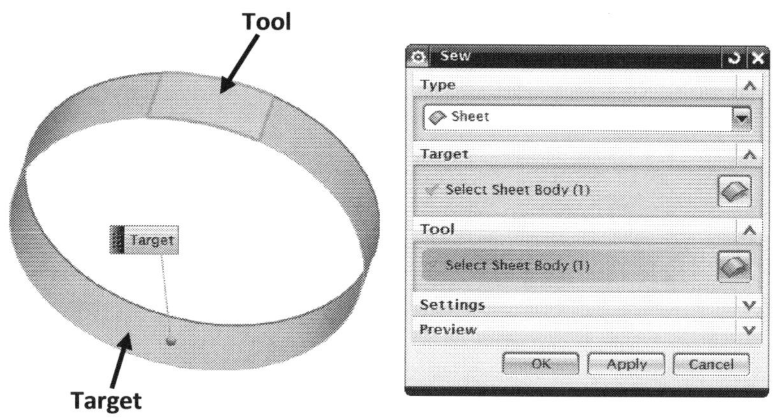


Notice that the result is a sheet body because the sewed sheets don't enclose a volume. Now we will create two more sheets to create a closed volume and sew them together to create a solid.

- Select ***Insert / Mesh Surface / N-sided Surface*** and select the boundary curves.
- Use the **Triangular** type and check the **Merge Faces if Possible** box.
- Select **OK** and leave this surface flat at the center.
- Repeat the process with the other side. Select the boundary curves as shown in the figure below and hit **OK**.

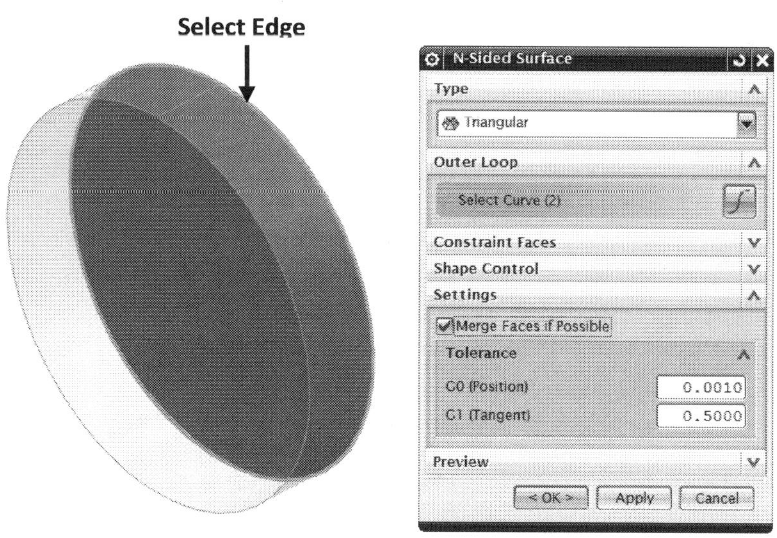

- Use the shape control slider to elevate the center of the sheet. See the figure below.

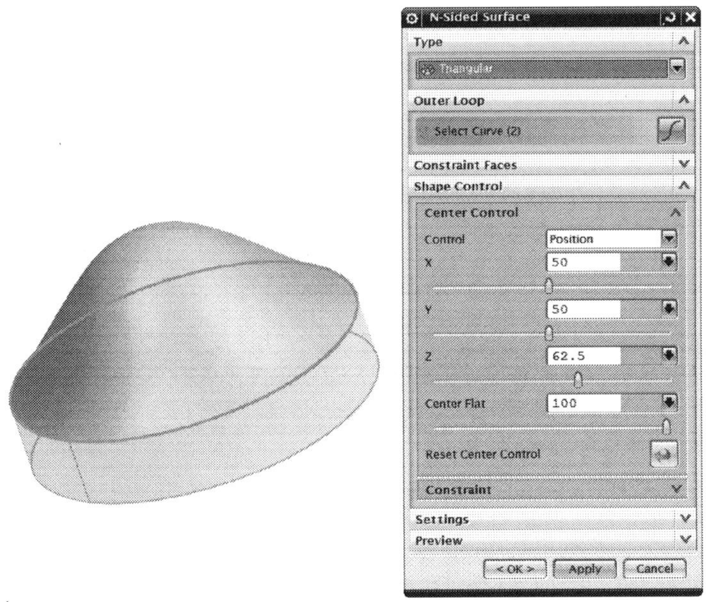


- Select ***Insert / Combine / Sew*** and select one of the surfaces as the target and all other as tool sheets.
- Select **OK** to sew them together to create a solid.

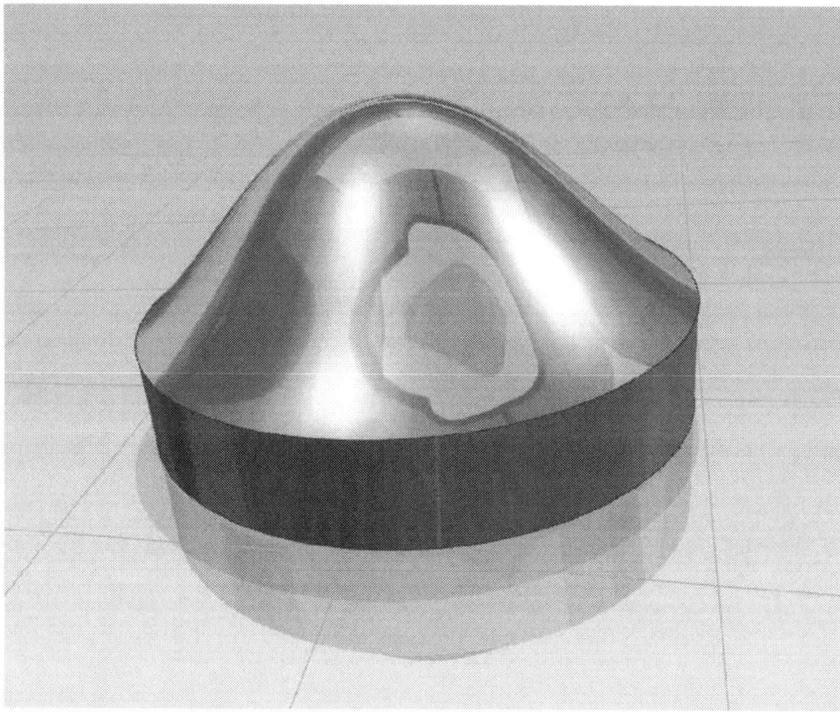

**Note:** Sometimes changes in tolerance can help sewing sheets that have some gaps.

***End of Project***

## 20.7 Patching Surfaces and Solids

You can use a surface to replace a part of a solid or sheet body by using the 'Patch' command. Patching can be very useful when you are working with problem areas of surfaces. It is also a good option to add or subtract solid geometry with a single command.

### *Project 20.7.1*

- Create a **4in**x **4in**x **4in** Block.
- Create two sketches as shown below and create a surface using ***Insert / Mesh Surface / Through Curves.***

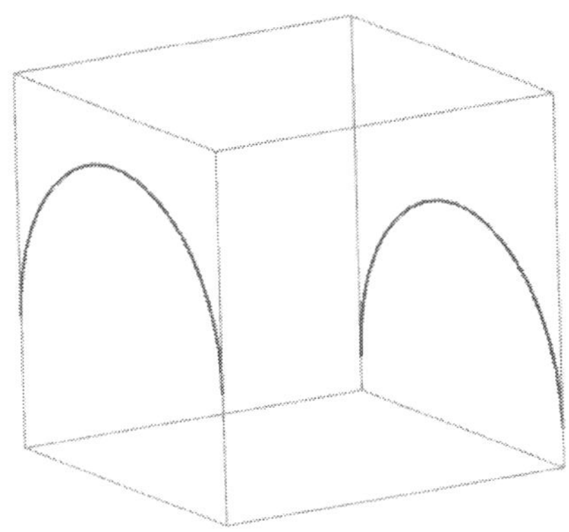

- **Move** the block to an invisible layer or make it translucent.
- Create two more sketches as shown in the following figure.

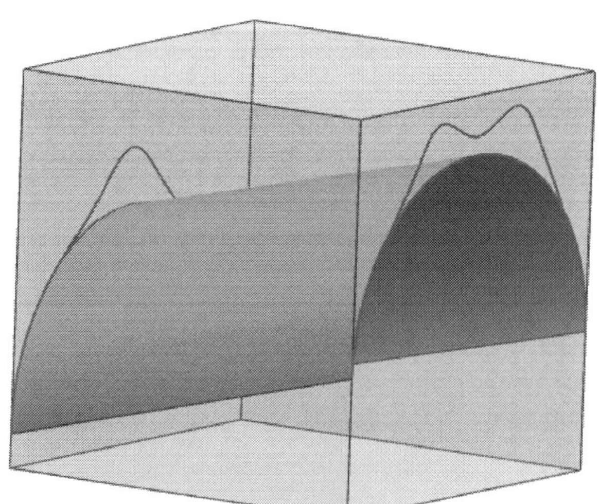

- Create a surface using ***Insert / Mesh Surface / Through Curves*** using these new sketches.
- Select ***Insert / Combine / Patch*** and select the **Target Body** and the **Tool Sheet** as shown in the figure below.

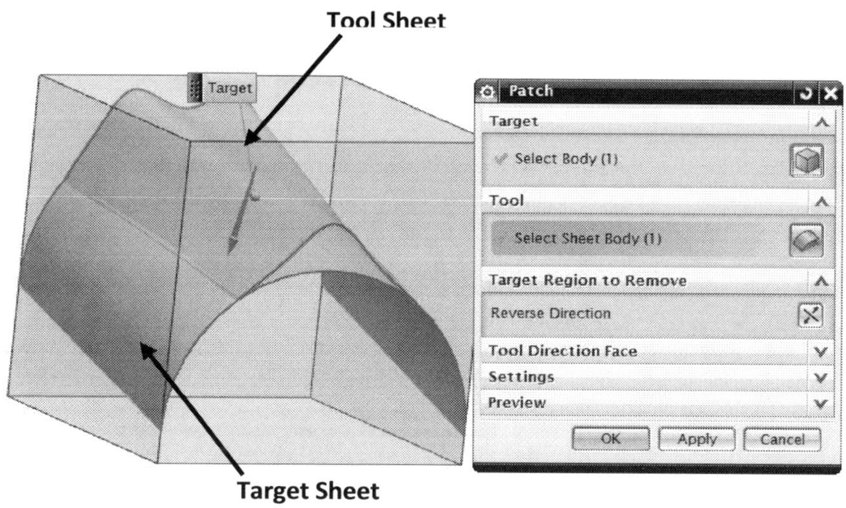


- Compare your results with the figure below.

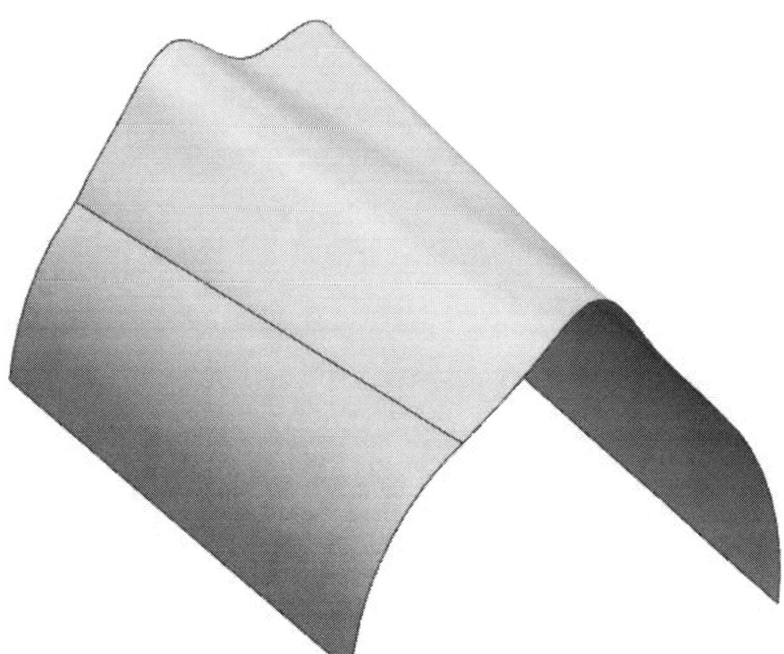

Now, let's try using the patch option with the solid block you created.

- Make the solid block visible.
- Select ***Insert / Combine / Patch*** and select the **Target Body** and the **Tool Sheet** as shown in the figure below.

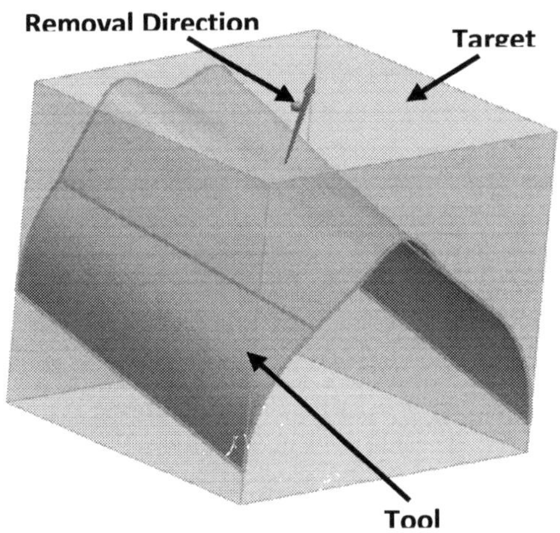

- Make sure that the removal direction arrow is pointing outwards. If it is pointing inward, click the **Reverse Removal Direction** button.
- Select **OK** and compare your results with the figure below.

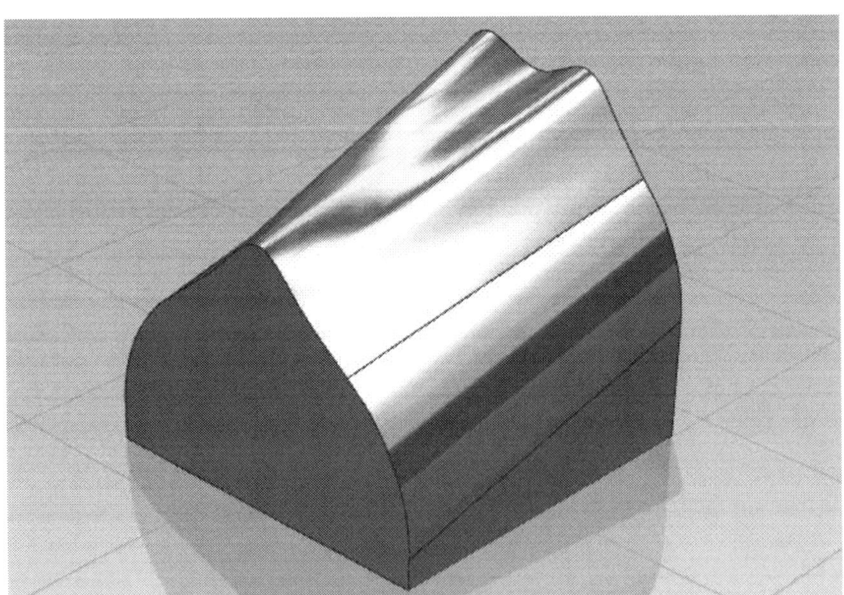

***End of Project***

## 20.8 Ribs

Ribs are one of the most important structural elements of an injection molded part. Creating correct rib geometry can be critical for manufacturability. As a rule of thumb, the base thickness of a rib is 2/3 (two-thirds) the overall thickness of the part to avoid sink marks. There is a lot to know about designing ribs that you can find in design books; the point of this topic is only to make the reader familiar with the general techniques of creating ribs in NX 8.5.

### *Project 20.8.1.*

Create two curves as shown in the figure below.

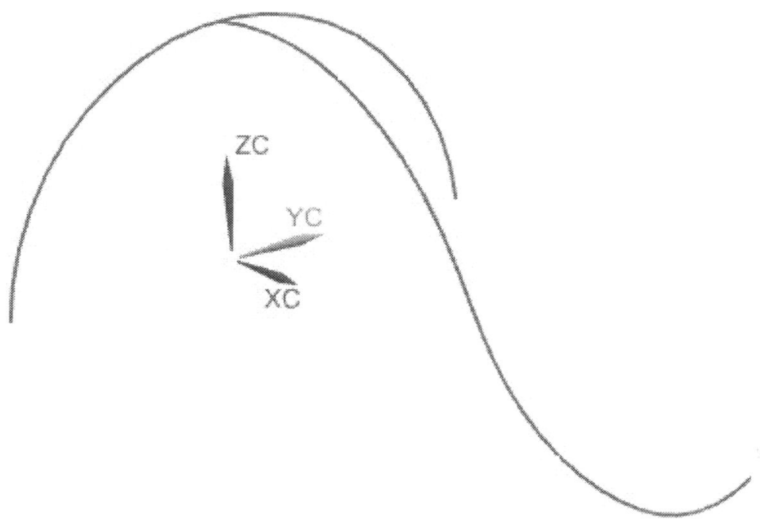


- Select Insert / Sweep / Swept.
- Select the **guide string** and **section string** as shown in the following figure. Don't forget to read the cue – simply select with the middle mouse button when it asks for the second guide or section string.
- For Section Location, select **Anywhere along Guides,** and for Alignment Method select **Parameter**.
- Select **Forced Direction** from the Orientation Method menu shown in the following figure.

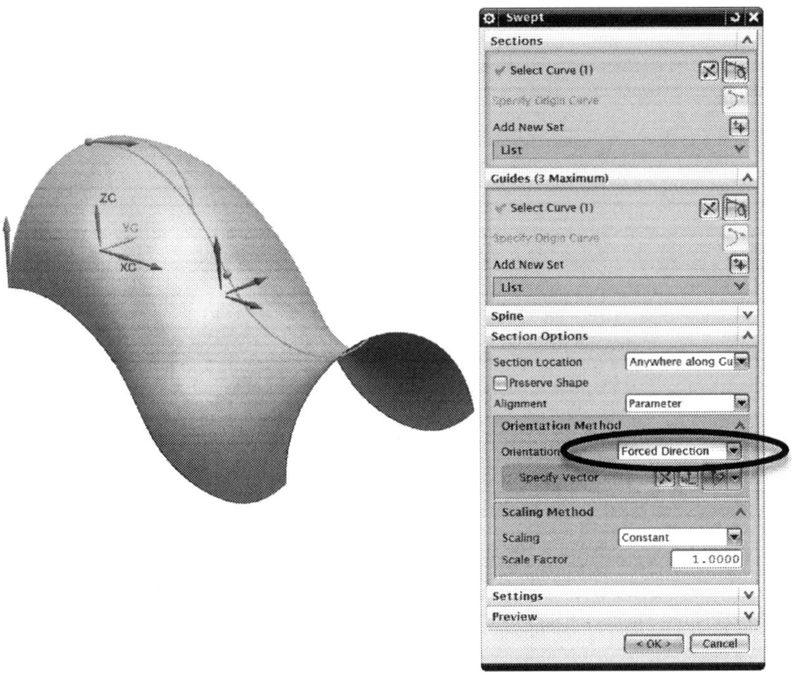


- Select **YC** and select **OK**. We are assuming that YC is the pull direction for this part.
- For the Scaling Method, enter Scale Factor **1**.
- Specify a thickness "**t**" in the expressions.
- Use ***Insert / Offset/Scale / Thicken*** to add thickness "t" to the sheet body.

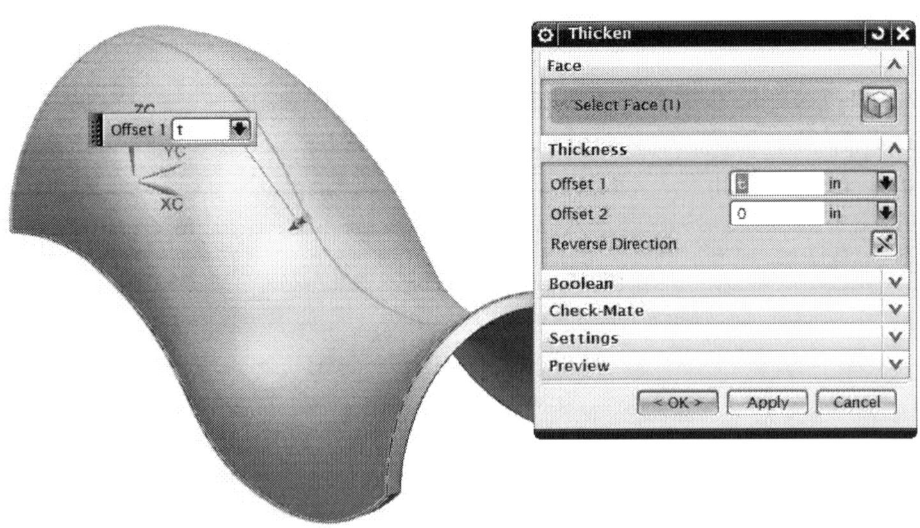

- Create a sketch in the XC-ZC plane as shown in the following figure.
- Select Insert / Curve from Curves / Project.

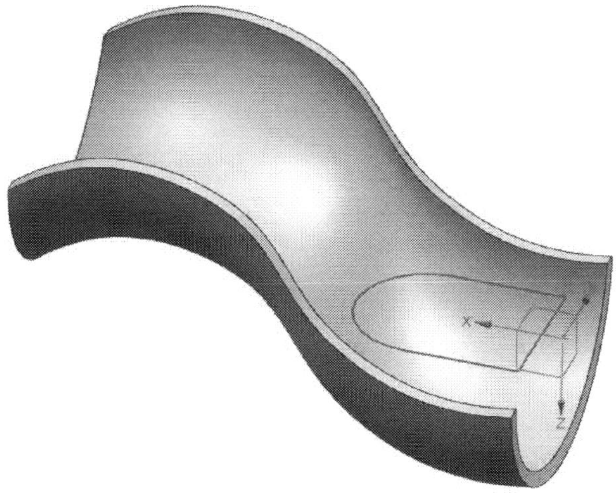

- Select the curve to project.
- Select the Direction Method **Along Vector**, as shown in the following figure.
- Select the **ZC** direction for projection.
- Select the face on which to project the curve and select **OK** to create the curve.

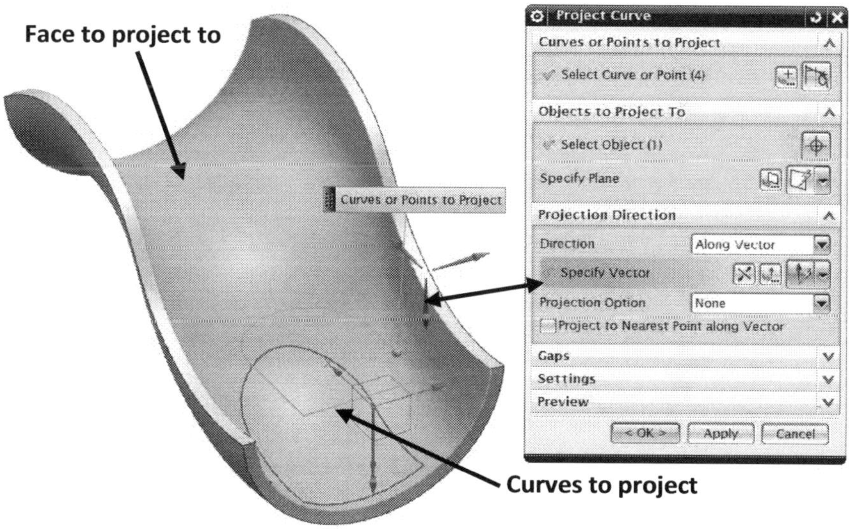


- Create a sketch representing the rib section using the **Sketch On Path** functionality. Sketch on the projected curve you just created at a **% Arc Length** of **0**.

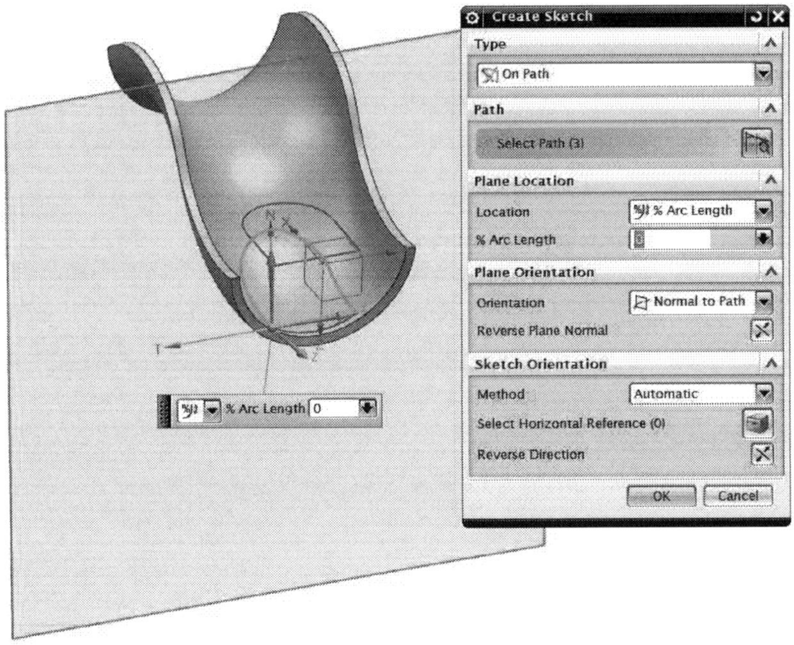


- Then create the following sketch on that path.

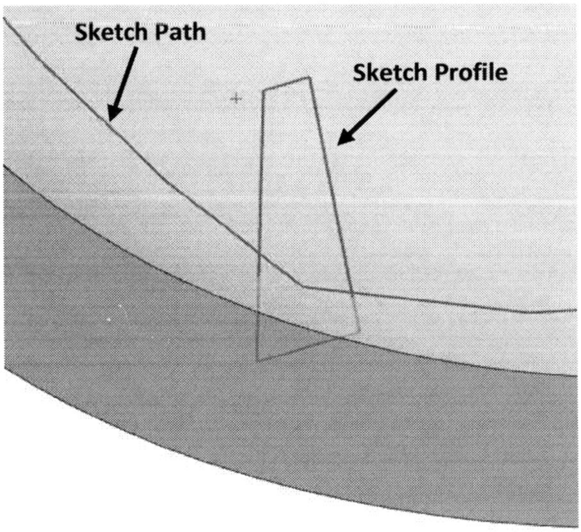


- Select Insert / Sweep / Swept.
- Select the guide string and section string as shown in the figure below.
- Select the 'Alignment Method' **Parameter** and the 'Section Location' **Anywhere Along the Guides**.
- Select **Forced Direction** from the orientation method menu.

- Select the Z arrow as shown in the following figure for the vector.

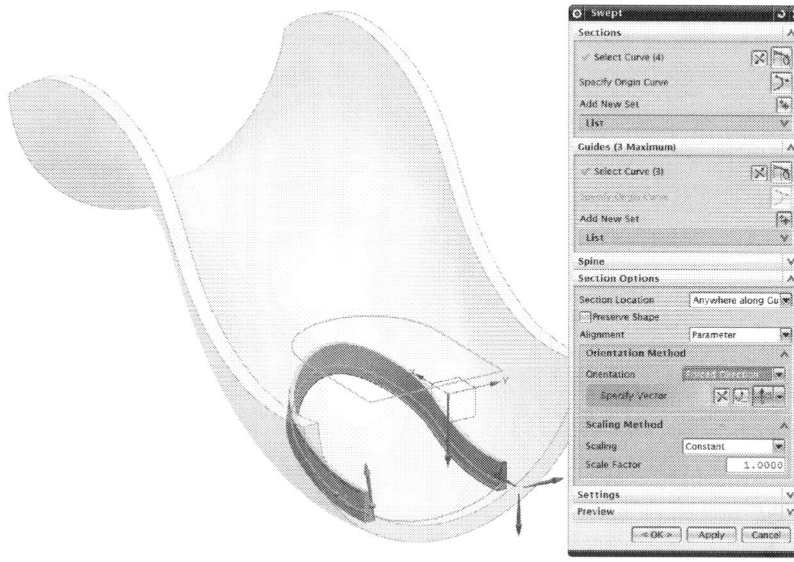


- Select **OK** for the scaling method and enter scale value **1**.
- After it is created, **Unite** the two bodies together.

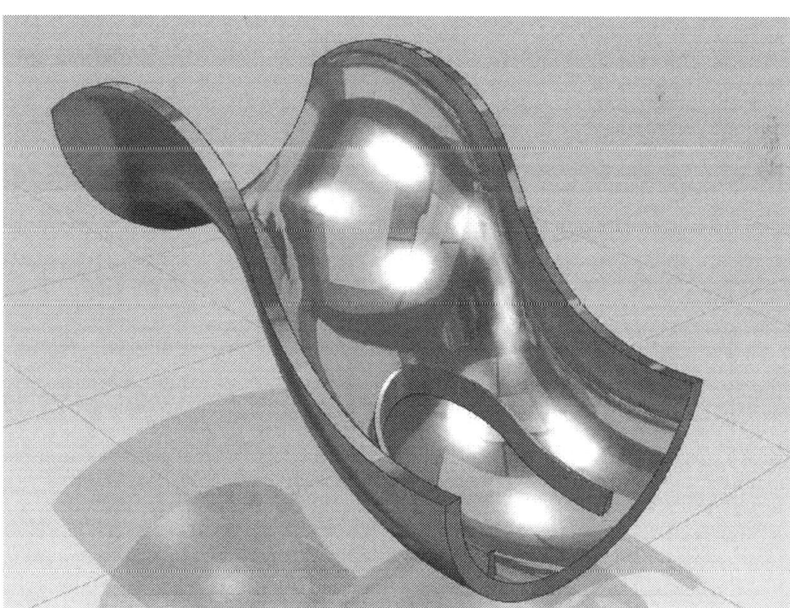

***End of Project***

## *Project 20.8.2.*

Create the following sketch:

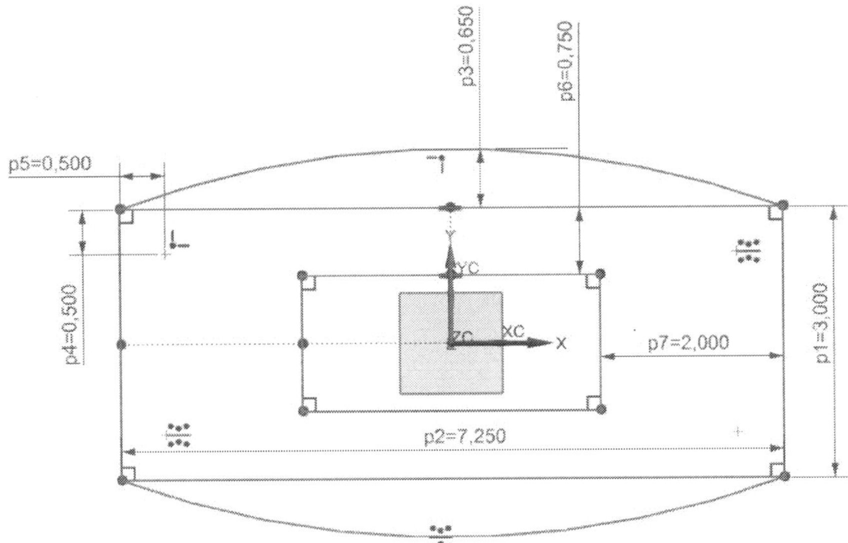


- Extrude the outer and inner curves by 1in, make sure to draft the extrude by 5 degrees.

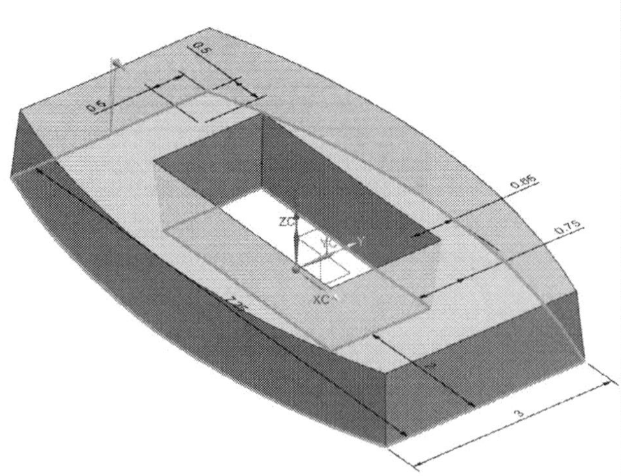


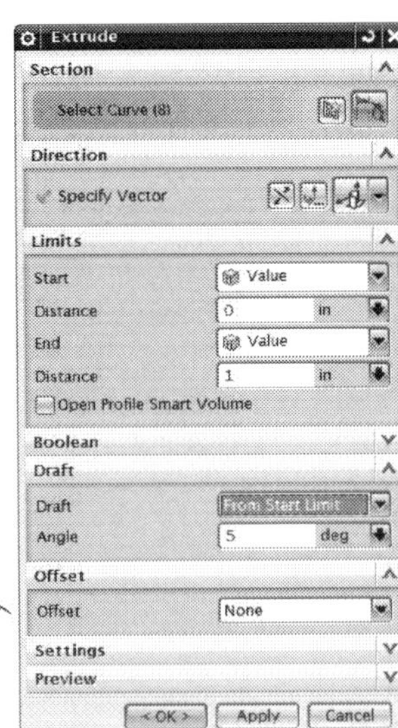

- Next shell the bottom face by 0.125in.
- Create a boss on each datum point that is 0.5in diameter, 1in tall.
- Create a 0.25in and 5 degree tapered hole on each boss.

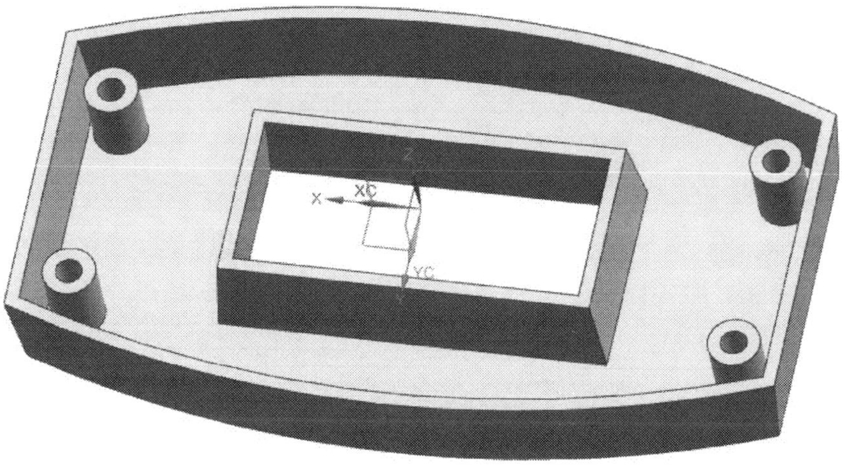

- Next sketch 4 lines for the ribs that will extend from the center outward through each boss as shown below:

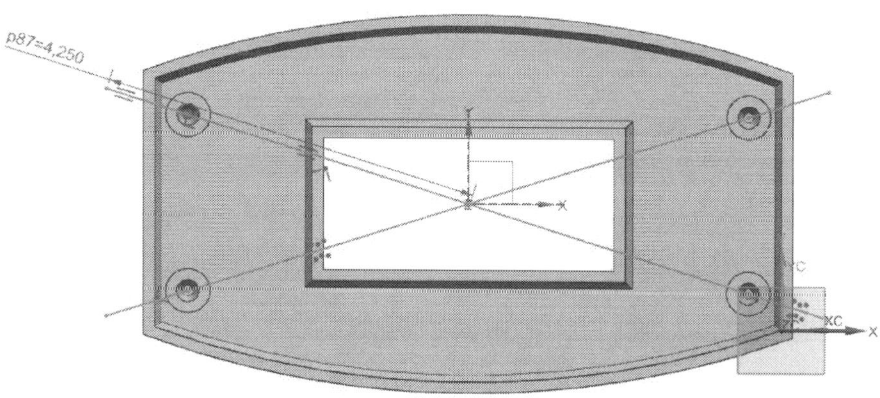

- Extrude each line by 1in individually with an offset of a symmetrical 0.125in.
- Draft each rib and boss by 2.5 degrees.

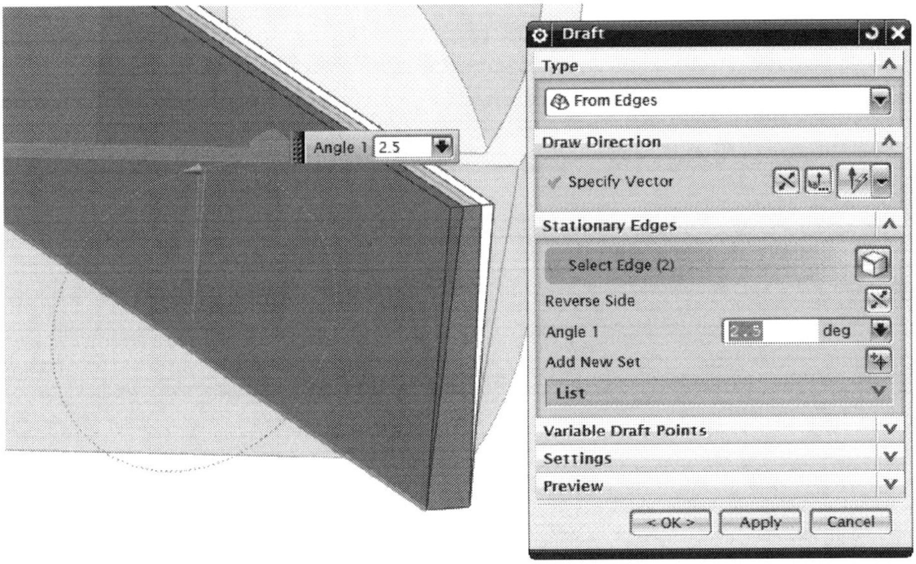


- Next the Select Insert / Combine / Unite.
- For this Combine Select Define Regions checkbox. Select the Shelled body as the Target, Select each rib as the tool.
- For the Region, select only the interior sections of the ribs as shown below.

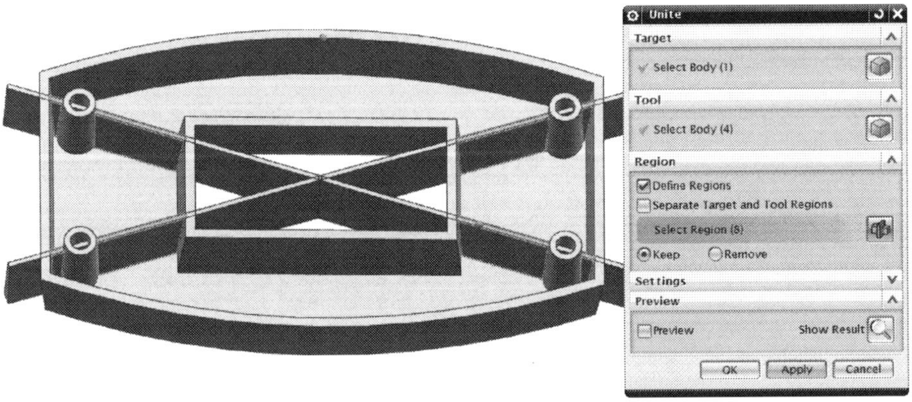


- Press **OK**.

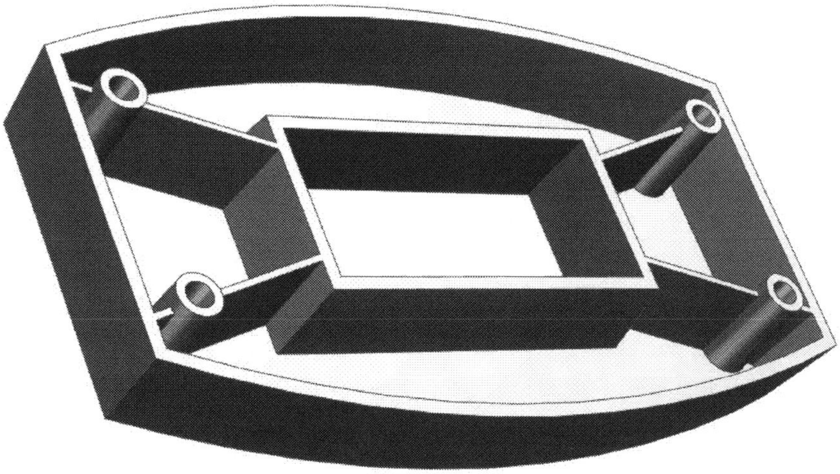

***End of Project***

## 20.9 Reducing Surface Patch Count using Join Curves

When you create a surface using curves that are either created or extracted from other objects, you can just use the curves as if they are tangent to each other. Using such curves results in multiple surface patches instead of one single, smooth surface. This makes the geometry unnecessarily heavier. By joining the curves before creating a surface, you can produce much cleaner and more desirable geometry.

### *Project 20.9.1*

- Create two curves.
- Move the WCS and create two more curves on a different plane as shown in the figure below.

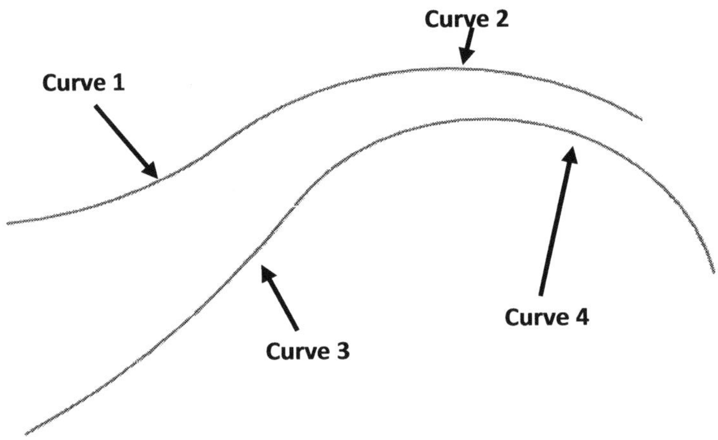


- Create a surface using ***Insert / Mesh Surface / Through Curves***. See the following figure.

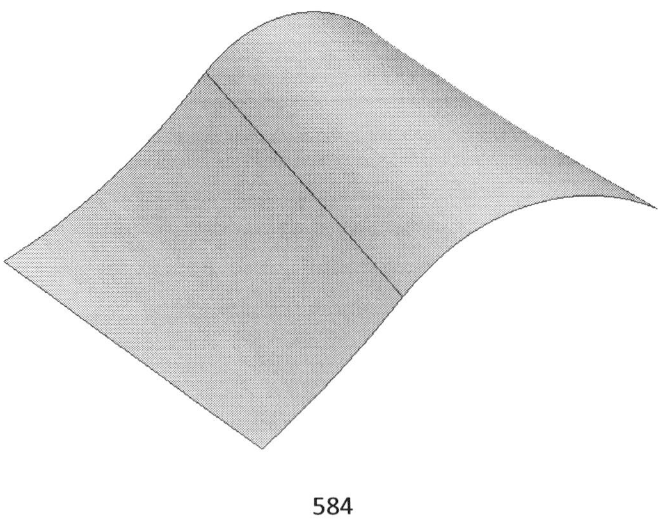

- Prior to NX 8.5 two different faces are created using the curves as it is.
- In the event such a situation arises in NX 8.5, select Insert / Curve from Curves / Join.
- Pick Curve 1 and Curve 2 to join and select **OK**.
- Select **Hide** from the **Input Curves** pull down menu and check the **Associative Output** checkbox. See the figure below.
- Select **OK** and repeat the same process to join Curve 3 and Curve 4.

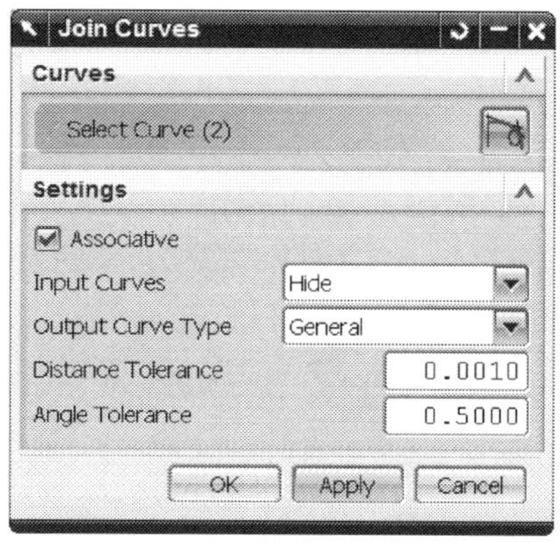


- Create another surface using ***Insert / Mesh Surface / Through Curves*** using the joined curves. See the figure below.

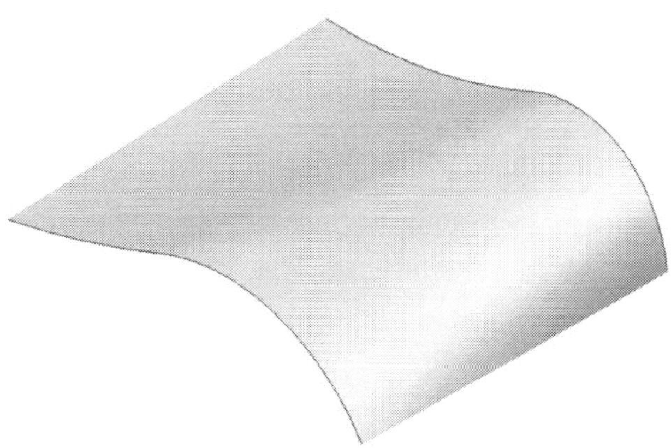

Notice the new surface with a single face.

***End of Project***

## 20.10 Lips

Lips are most commonly used along the edges of a part that mate with other parts. You can create a lip by simply extruding the edges if they are in the same plane. Creating lips can be a difficult task if the edges don't lie in one plane.

### *Project 20.10.1*

- Create a solid and **Shell** it to a **0.1in** thickness.
- Insert / Design Feature / Extrude
- Select the edges, as shown in the figure below.

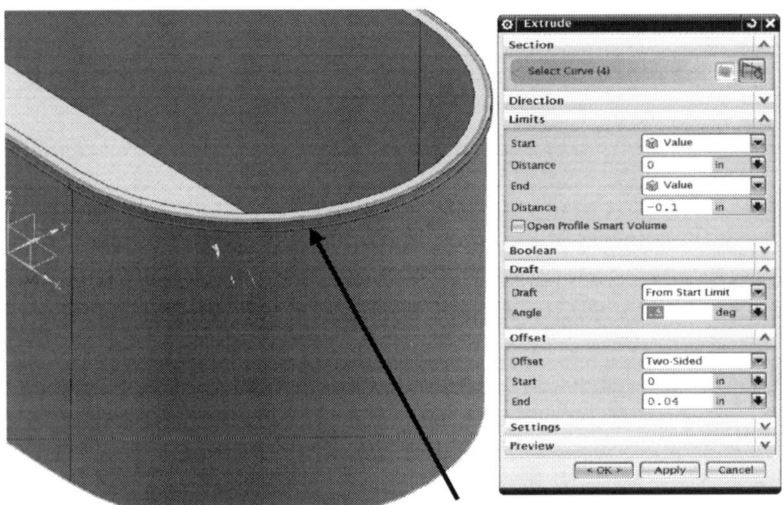


**Edges to Extrude**

- **Extrude** the edge by **0.1in** with ± **0.04in Offse**t and a **0.5º** draft angle.
- For Boolean, select **Subtract**.
- See the figure below.

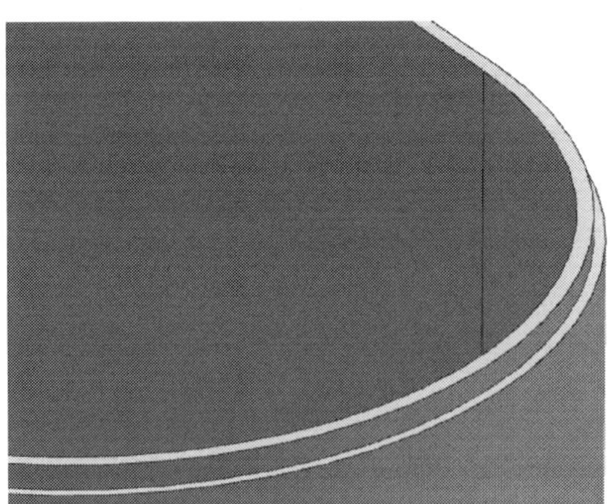

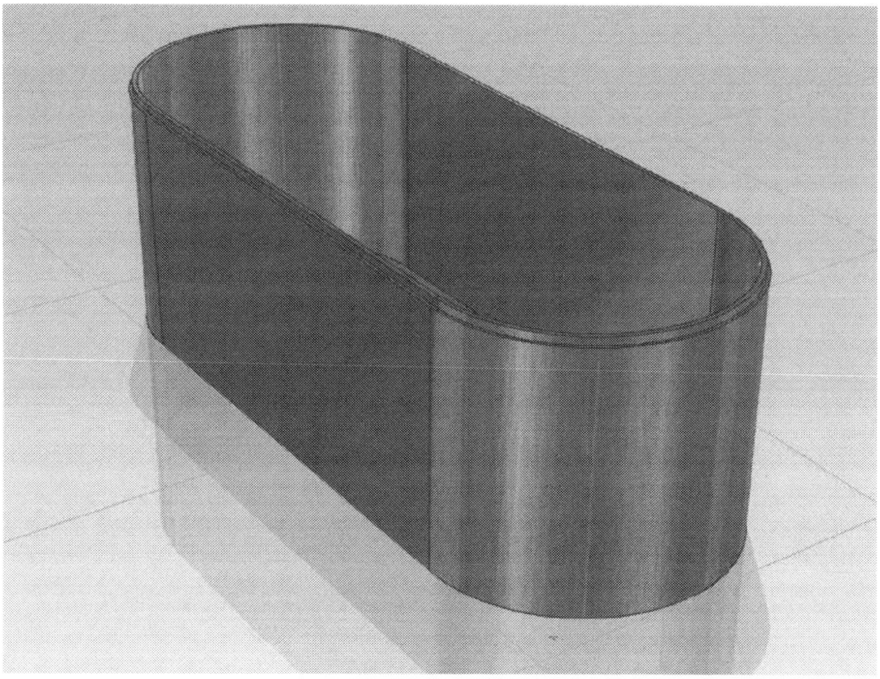

***End of Project***

## *Project 20.10.2*

Use the original solid from the previous project and **Trim** it using a surface. See the figure below.

- Select ***Insert / Sketch.*** Then use the **Sketch On Path** function to place your sketch on the edge of the body at **0 % Arc Length**, as shown in the following figure.

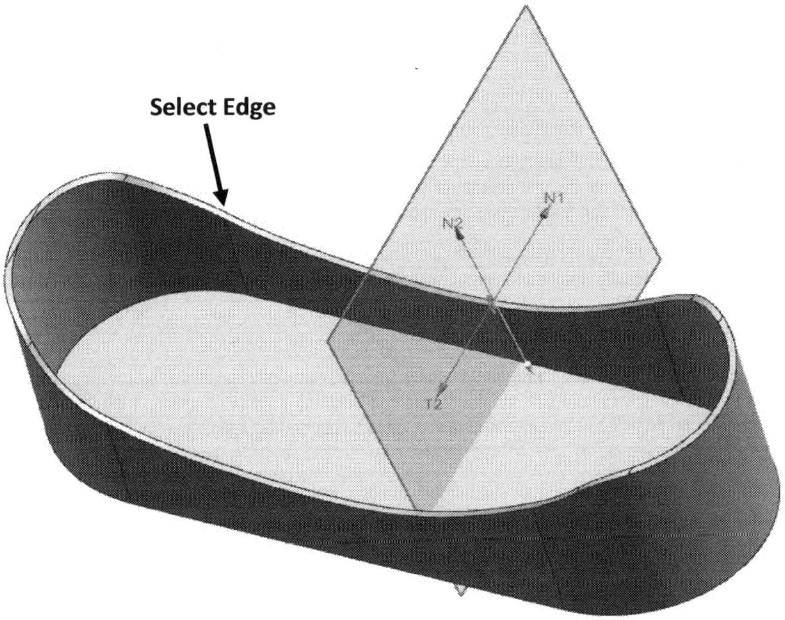


- **Create the following sketch.**

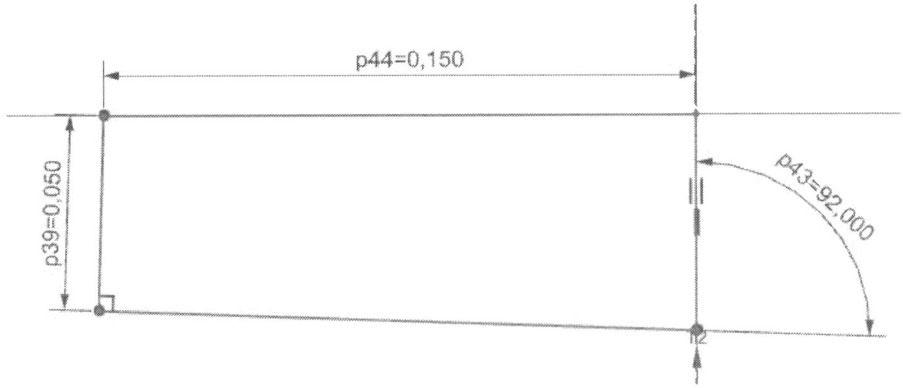


- Insert / Sweep / Swept.
- Select the **outer edge** of the solid as the **Guide String** and the **sketch** as the **Section String**.

- Specify parameters as shown in the figure below.
- Choose **Vector Direction** as the orientation method, choosing the **Z-axis**. This will ensure that all of the section sweeps along the guide so that it is always perpendicular to the pull direction.

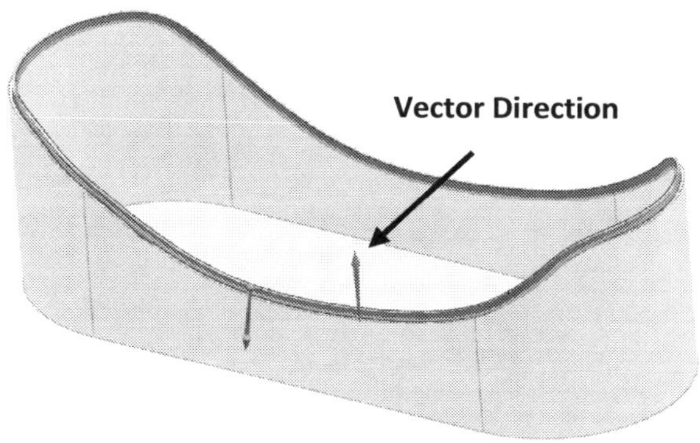


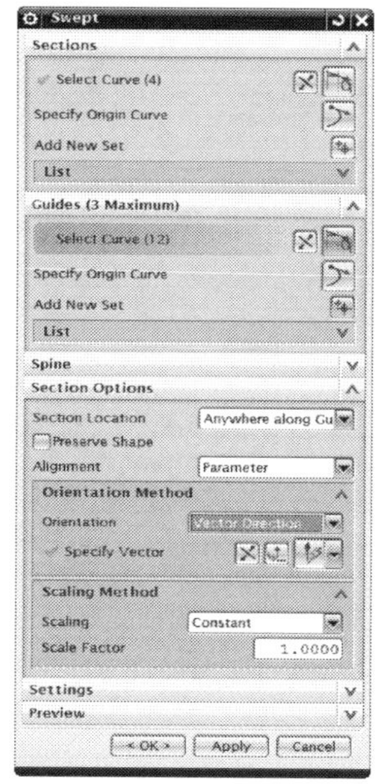


- Select the Scaling Method to be **Constant** and select **OK**.
- Then, simply subtract the swept body you created from the existing body.

Compare your results with the figure below.

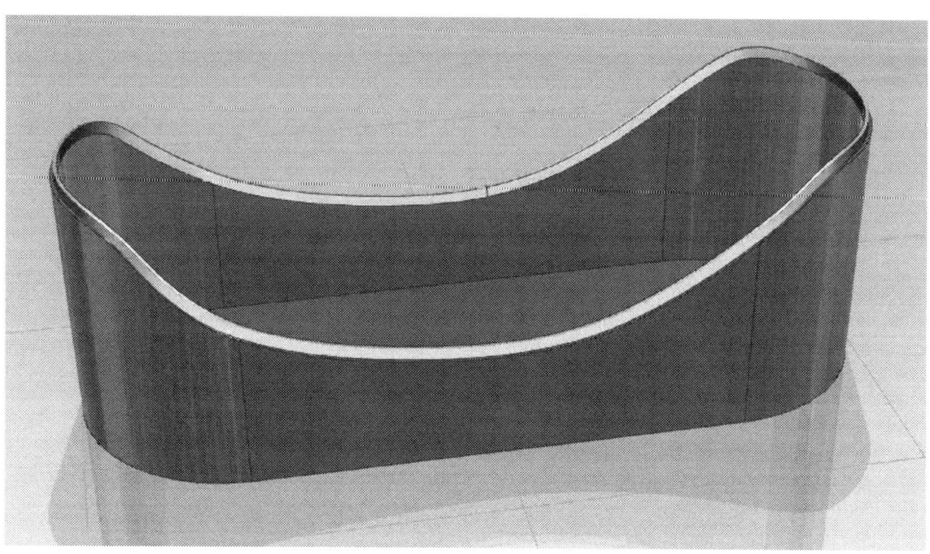

***End of Project***

## 20.11 Heat Stakes

'Heat Staking' is the process of applying heat and pressure to stake or seal small posts used in assembling many types of components. Not only does this reduce the cost of the part as there are less components, but it also avoids degradation of the components.

The design for creating the posts for the heat staking process is very simple. The height of the portion to be staked should be 1-2.5 times the diameter of the post (1.5 times should work in most cases). Posts can be of any shape, but round and square posts are generally used. Hollow posts can be used for high retention and better finish. See the figure below.

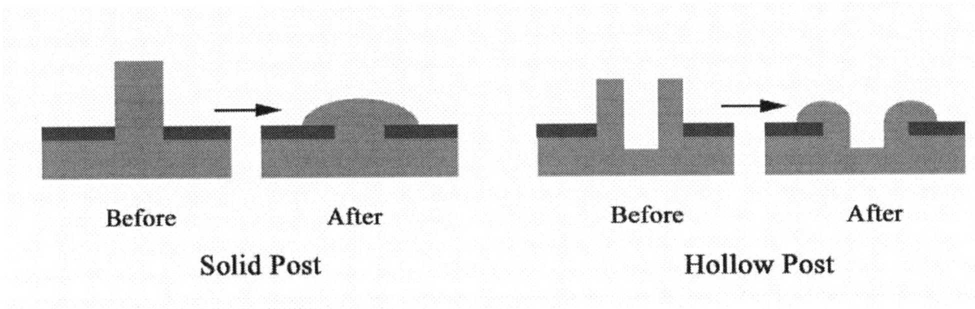


For high volume operations where aesthetics and accuracy may not be high priorities, a knurled profile is often used. For better aesthetics and high break strength, a dome shaped profile works best.

In the following subtopics you will learn how to model a heat stake and easily control the shape of the post in assembly and part files.

In this next project, you will also learn a cool new method of feature control called **'Suppress by Expression.'** This creates a Boolean expression that suppresses a feature when the value is **0** and unsuppresses it when the value is **1**.

For that matter, when using Booleans in NX, it is important to note that **0** is **false,** and **1** is **true**.

## *Project 20.11.1*

- Create a **4in** x **4in** x **1in** block in a new part file to create a basic geometry. **Blend** it and **Shell** it with **0.1** thickness.

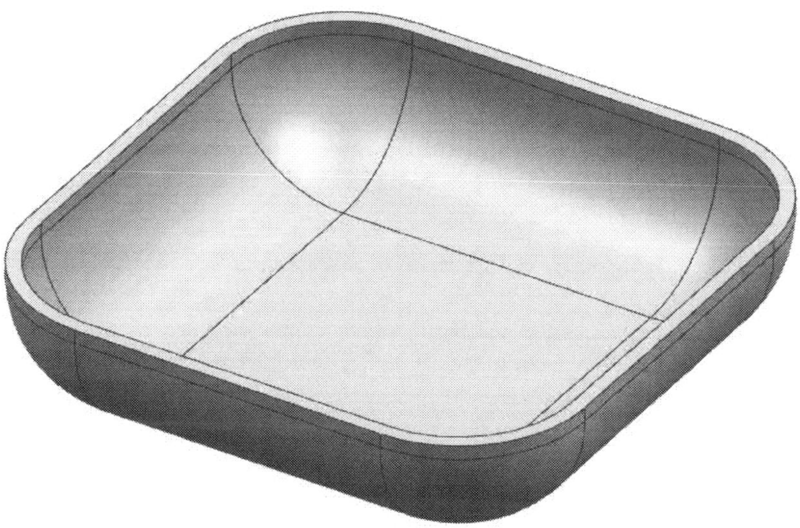

- **Create Expressions**:
  - Stake = 1
  - y = If ( stake=1 ) Then ( .1 ) Else ( .3 )

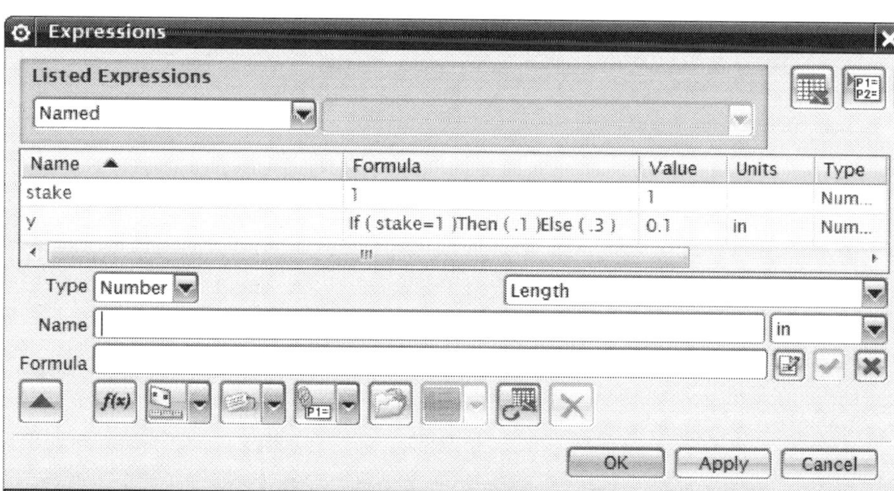

- Create a boss of Diameter = **0.06**, Height = **y** and taper angle **0.5** and position it **1.5in** and **2in** from the outer edges.
- Create another **Boss** on top of the first boss with Diameter = **0.15**, Height = **0.06** and Draft angle =**10**. Then **Blend** the edges.
- Select Edit / Feature / Suppress by Expression.
- Select the **expression option** button and choose **Create Shared**.
- Select the last **Boss** and **Blend**. **Apply.**
- Open the 'Expressions' menu. Notice the new expression, in this case: p38 = 1. Change this to **p38 = Stake**. See the previous figure. *It will have (boss(5) Suppression Status as the name.*

- Create a Datum plane in the center and mirror the two bosses and the blend feature.
- Change the expression value to **Stake = 0**. Notice the post shape when you change the value from 1 to 0 or 0 to 1. See the figure below.

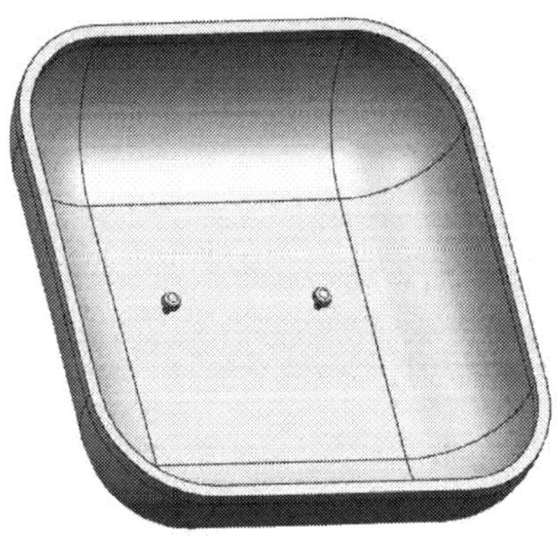

**Stake = 1**

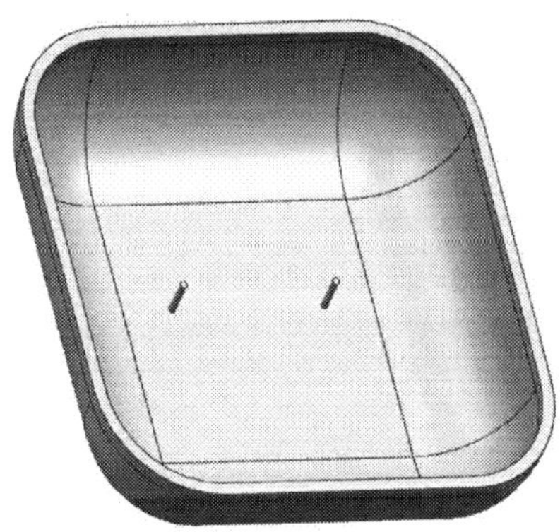

**Stake = 0**

***End of Project***

## *Project 20.11.2*

- Use the part from previous project. Select ***Tools / Define Deformable Part.***
- Define a name for the deformable part. Click **Next**.
- Check the **Add Children Features** box. Select **Block** and click the **Add Feature** arrow. All other features will be added automatically. See the figure below. Click **Next**.

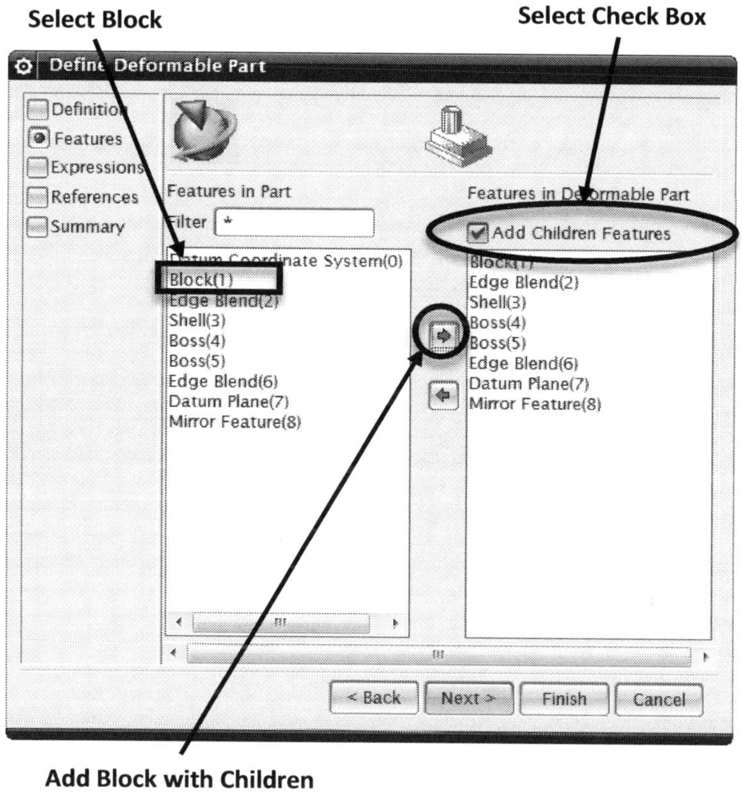


- Add the Expression **Stake** from the available Expressions. See the following figure.

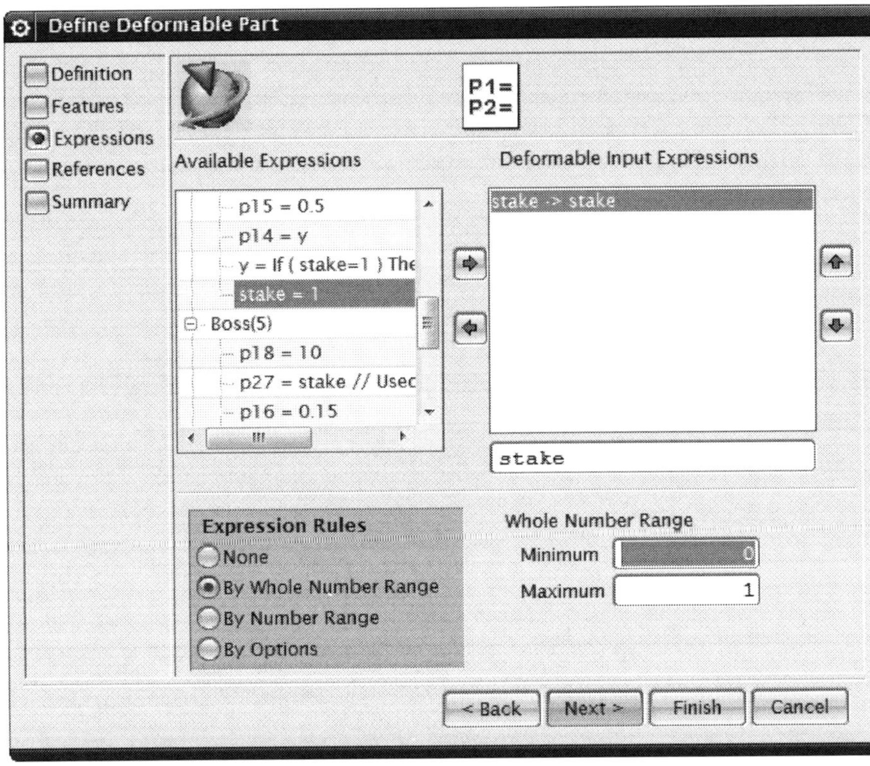


- Select the expression rule **By Whole Number Range**.
- Define the 'Whole Number Range:' Minimum = 0, Maximum = 1.
- Finish defining Deformable part. Notice the 'Deformable Feature' added to the 'Part Navigator.'

## *Project 20.11.3*

- Create a new assembly. Create a new part in the assembly. Create a bracket with two holes **1in** apart. See the figure below.

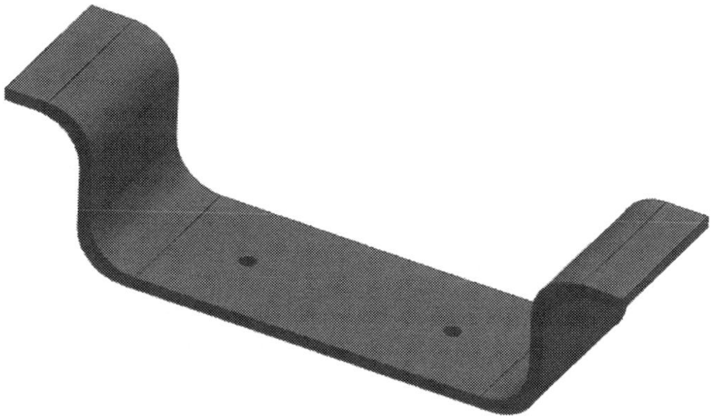

- Add the **Deformable part** that you created in the previous exercise.
- **Mate** it to the bracket.
- **Right click** on the **deformable part** in the assembly navigator and select **Deform**. It will ask you again what part you want to deform, so just double click on the part. Then drag the cursor to change the deformations and click OK. Your model will look like the following figure.

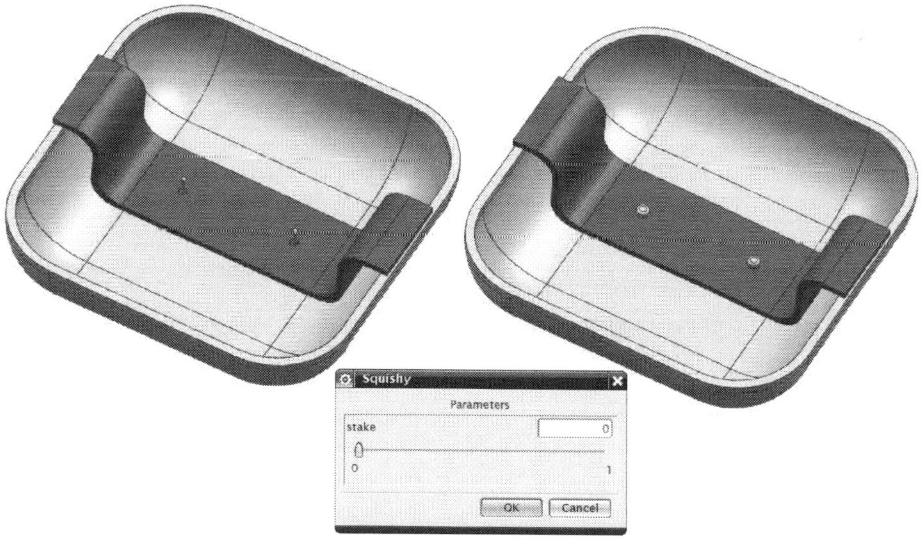


Select **OK**.

***End of Project***

## 20.12 Snap-Fit

The snap-fit is a simple, quick and cost effective method of assembling plastic parts (thermoplastics are ideal). Snap-fits use the inherent flexibility of plastic to eliminate fasteners needed for assembly. They can be used to mount, locate and join components. Snap-fits usually produce an undercut in the part geometry that makes the mold more expensive to manufacture. With careful design, the undercut can sometimes be eliminated, as shown in the figure below.

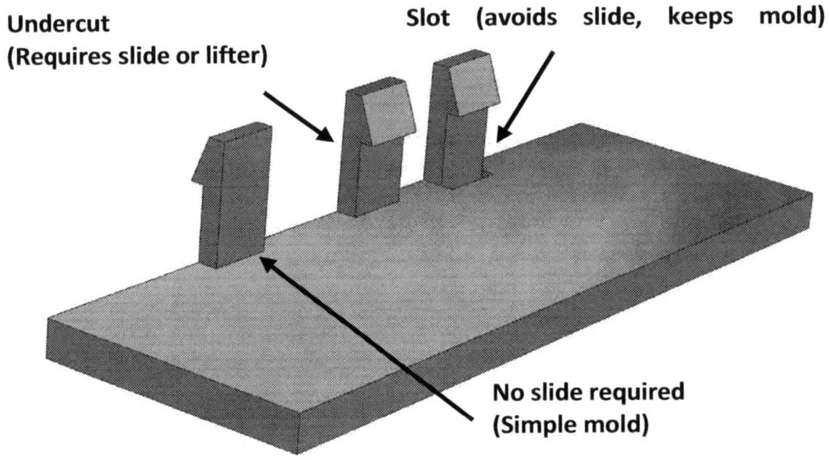


To calculate the bending force and required stiffness, use the following equations for stress and deflection. For a cantilever beam with constant cross section:

Stress $\sigma = \frac{F_b L d}{I}$

Deflection $\gamma = \frac{F_b L^3}{3I}$

For an actual snap-fit member, use the strain formula:

Strain $\varepsilon = 3\frac{\gamma d}{L^2}$

Where: $\sigma$ = Maximum Stress

$\gamma$ = Maximum deflection of a constant section beam

$\varepsilon$ = Maximum fiber strain

E = Modulus (Flexural)

$F_b$ = Load causing deflection

I = Moment of inertia of cross section

L = Length of beam

d = Distance from neutral axis to extreme fiber (d=t/2)

See the figure below.

It is important to determine the insertion force for a snap-fit design. If you are designing for multiple insertions and removal, you need to determine the pullout force as well. The next figure shows some simple formulae for assembly forces.

**Blending Force: $F_b - yEbt^3/4L^3$**

**Pullout Force: $F_p - F_b/\tan(90\text{-}C)$**

**Insertion Force: $F_i - F_b/\tan(90\text{-}C)$**

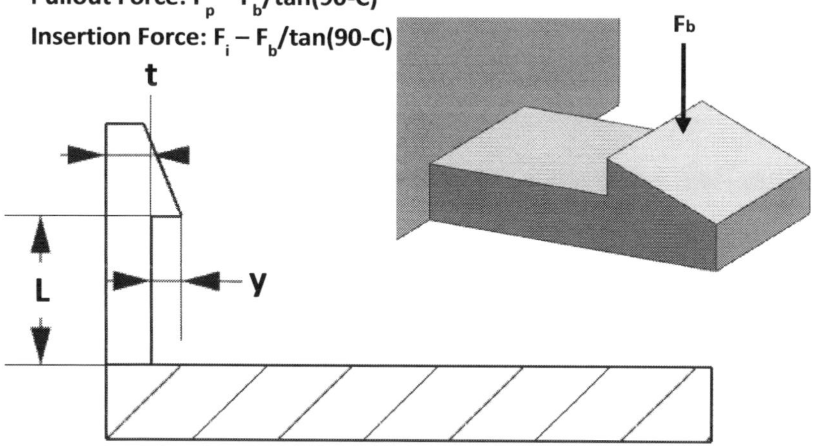


**Where E = Modulus (Flexural), C and D = incident angles**

It may appear very easy to design snap-fit using linear beam equations, but there are many more factors involved. Nonlinear stress-strain effects often exist, especially with complex geometry with varied base radius, angles and thickness. You can also create U or L shaped snap-fit features based on your design needs. Adding more details about snap-fits is outside of the scope of this book, but there is ample literature available in the market for your further study.

## *Project 20.12.1*

In this project you will create a snap-fit on which you may perform your calculations. You will also learn how to use the copy paste command.

- Create a new part and make a simple **Block**.
- Create a snap-fit on top of the block, as shown below. Use any method you choose.

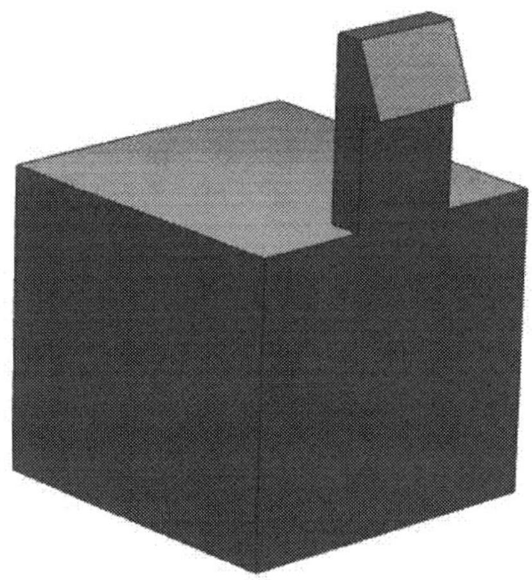

- Copy the snap feature you just made by using ***Edit / Copy Feature***.

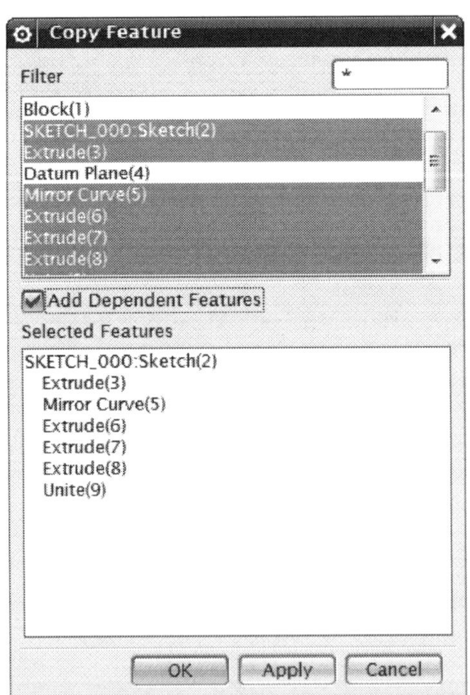

- Now create a new part file. Create another simple block.
- **Select *Edit / Paste*.** You will see the window below. When prompted, select a face and a curve on which to place your snap. This command can be a little tricky. If it doesn't work the first time, try playing around with your selections.

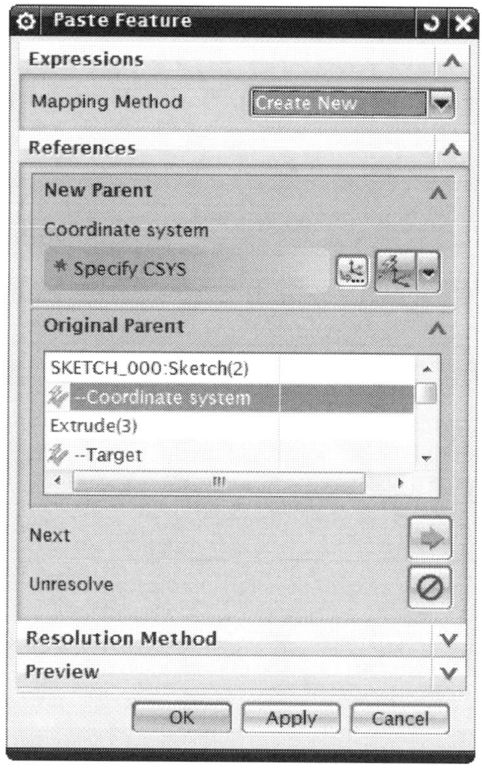


- You should now have a great snap created using the very useful copy/paste feature. You didn't have to use the copy paste feature in order to create the snap, but now you know that you can copy and paste features between parts. Now that's cool!

***End of Project***

## 20.13 Plastic Rules of Thumb

Designing for injection molded parts and casting can be fairly complex and detail oriented. Design of the different features can vary depending on the context, but there are a few rules of thumb that can be useful in general.

### 20.13.1 *Effect of Surface Texture on Draft Angle*

A draft is required on almost all the vertical surfaces that need to be injection molded. Amount of draft needed for a part can vary significantly depending on the depth, location of the walls, parting line and surface texture.

A minimum draft angle of 0.5° should be provided, even for extremely smooth surfaces. In certain applications, the use of draw polish on the mold surface may allow a smaller angle. Parts with a molded-in deep texture, such as leather-graining, require additional draft as part of their design. As a rule of thumb, an additional 1° of draft should be provided for every 0.001 inch (.025 mm) depth of texture. It is wise to discuss these parameters with the mold texturing supplier as each part has unique needs.

### 20.13.2 *Nominal Wall Thickness*

A uniform wall thickness should be maintained throughout the part. If you need to change the wall thickness in certain places, use a transition zone with a very liberal angle as shown in the figure below. Abrupt changes in wall thickness can create stress concentrations which may cause the part to warp, twist, or even crack.

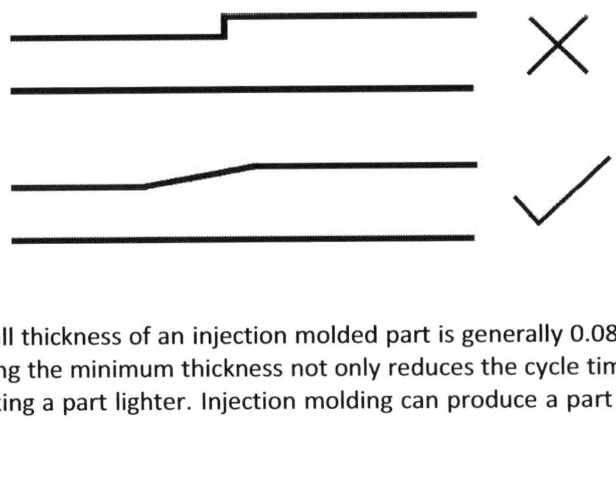

The minimum wall thickness of an injection molded part is generally 0.08 to 0.16 inches (2 to 4 mm). Using the minimum thickness not only reduces the cycle time and cost but also helps in making a part lighter. Injection molding can produce a part as thin as 0.02 inch (.5 mm).

## 20.13.3 Radius

It is very important to use a generous radius to avoid stress concentrations at corners. Sharp corners can also affect the flow of plastic in the mold. Generally, an inside radius should be at least half of the material thickness around the corners and at least 0.015 inches, or a quarter of the material thickness around the bosses and ribs.

## 20.13.4 Ribs and Gussets

Ribs and gussets are important structural features that add strength to the part. Usually a given ribs' thickness is 60% of the part's nominal wall thickness. See the figure below. It is better to increase the number of ribs instead of increasing the height to increase the stiffness.

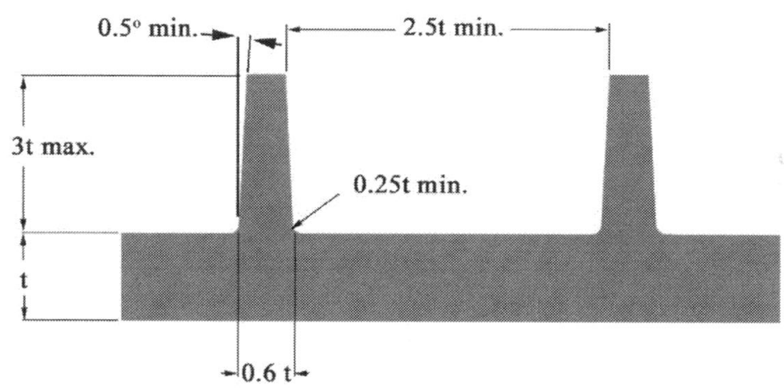


## 20.13.5 Bosses

Wall thickness for bosses should follow the same rule as ribs. It should generally be 60% of the nominal wall thickness. Boss height may vary according to design, but try to keep the boss height less than 2.5 times the boss diameter. Avoid stand-alone bosses; try to connect them to the nearest wall or rib or use gussets if the boss is too far away from the wall. When connecting a boss to the side wall of a part, be sure you are not increasing the wall thickness. See the following figure.

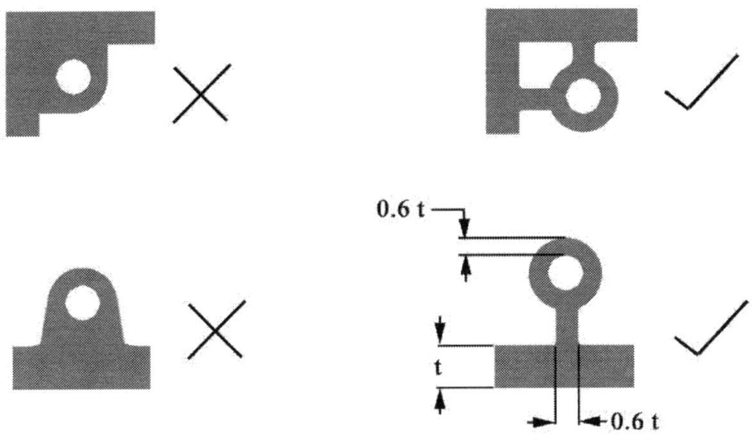
0.6 t
t
0.6 t

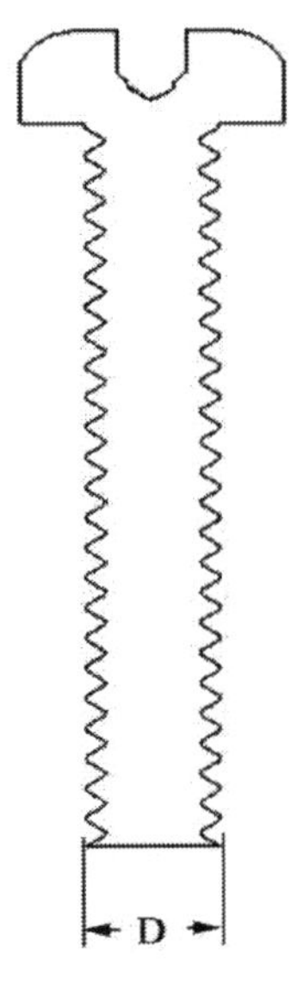
D

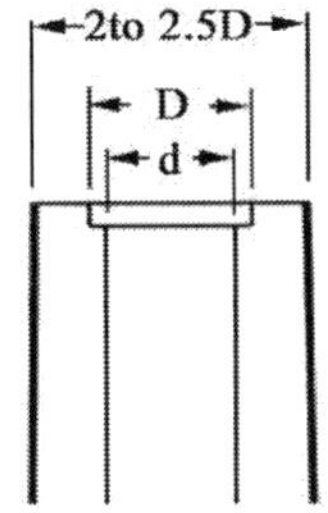
2to 2.5D
D
d

A counterbore hole can help reduce hoop stress around the hole area, especially for bosses that are designed for self-threading screws. (Avoid using countersunk head screws because the wedging action they create when connected to the part causes high hoop stress that can cause splitting around the edge.) The counterbore diameter should be equal to the major diameter (D) of the screw thread and the depth should be equal to one pitch. The outside diameter of the boss should be 2D to 2.5D. The hole diameter should be equal to the pitch diameter (d) of the screw and the minimum thread engagement should be 2.5d.

### ***20.13.6 Sink Marks***

Sink marks normally occur because of the local increase of thickness that suffers a higher than average shrinkage. To avoid sink marks, follow the rules for bosses and rib design, and keep them thin. The amount of shrinkage also depends on gate design, material type and process conditions. The easiest way to avoid sink marks is to add a groove or step on the opposite side of a rib and to keep ribs near the gating points.

## 20.14 Final Project 'PCB Component Assembly'

Use the following design prints and NX 8.5 to make the PCB Component Assembly. All components are made using English units. The Top and Bottom covers are plastic injection molded parts, which is why they have a minimum of 1 degree drafts to simplify extraction from the mold.

The PCB Main Board has many features that can be sketched, dimensioned and extruded all from a single sketch. The LED Card can also be created using one sketch. The Top and Bottom covers have many features that are drafted, so make sure that your draft angles are in the correct direction.

After creating each part you should be ready to assemble them. It is recommended when mating the Top and Bottom covers to use the top faces of the bosses as your mating surfaces. This will ensure the proper 0.01in outer surface gap between the Top and Bottom pieces. After completing the PCB Component Assembly, compare your results to the assembly drawings on the figures below. Good luck!

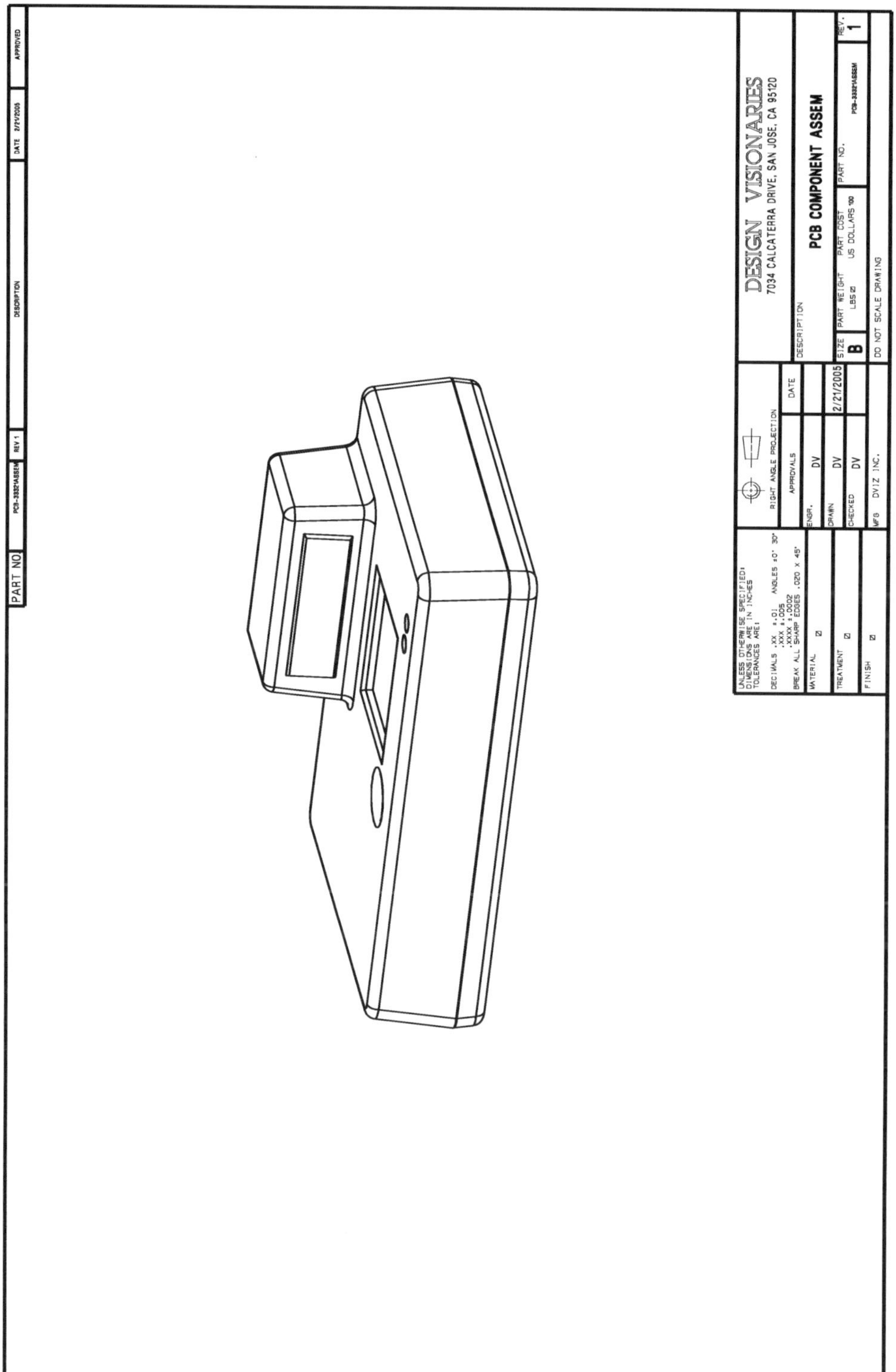
PART NO.
DESIGN VISIONARIES
7034 CALCATERRA DRIVE, SAN JOSE, CA 95120
PCB COMPONENT ASSEM
RIGHT ANGLE PROJECTION
APPROVALS
DATE
DESCRIPTION
ENGR. DV
DRAWN DV 2/21/2005
CHECKED DV
MFG DVIZ INC.
SIZE B
REV. 1
DO NOT SCALE DRAWING
MATERIAL
TREATMENT
FINISH

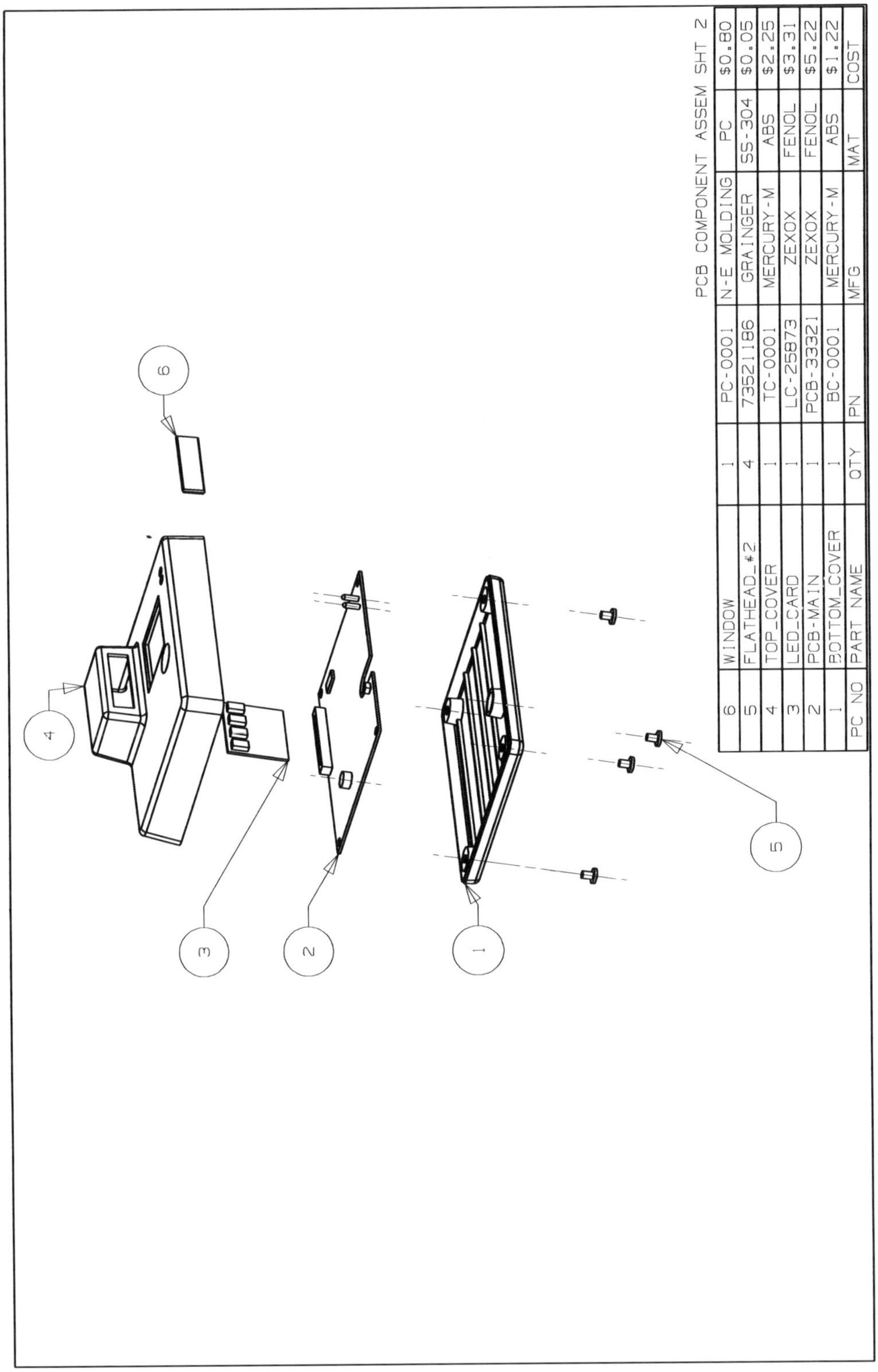
PCB COMPONENT ASSEM SHT 2
6 WINDOW 1 PC-0001 N-E MOLDING PC $0.80
5 FLATHEAD_#2 4 73521186 GRAINGER SS-304 $0.05
4 TOP_COVER 1 TC-0001 MERCURY-M ABS $2.25
3 LED_CARD 1 LC-25873 ZEXOX FENOL $3.31
2 PCB-MAIN 1 PCB-33321 ZEXOX FENOL $5.22
1 BOTTOM_COVER 1 BC-0001 MERCURY-M ABS $1.22
PC NO PART NAME QTY PN MFG MAT COST
1
2
3
4
5
6

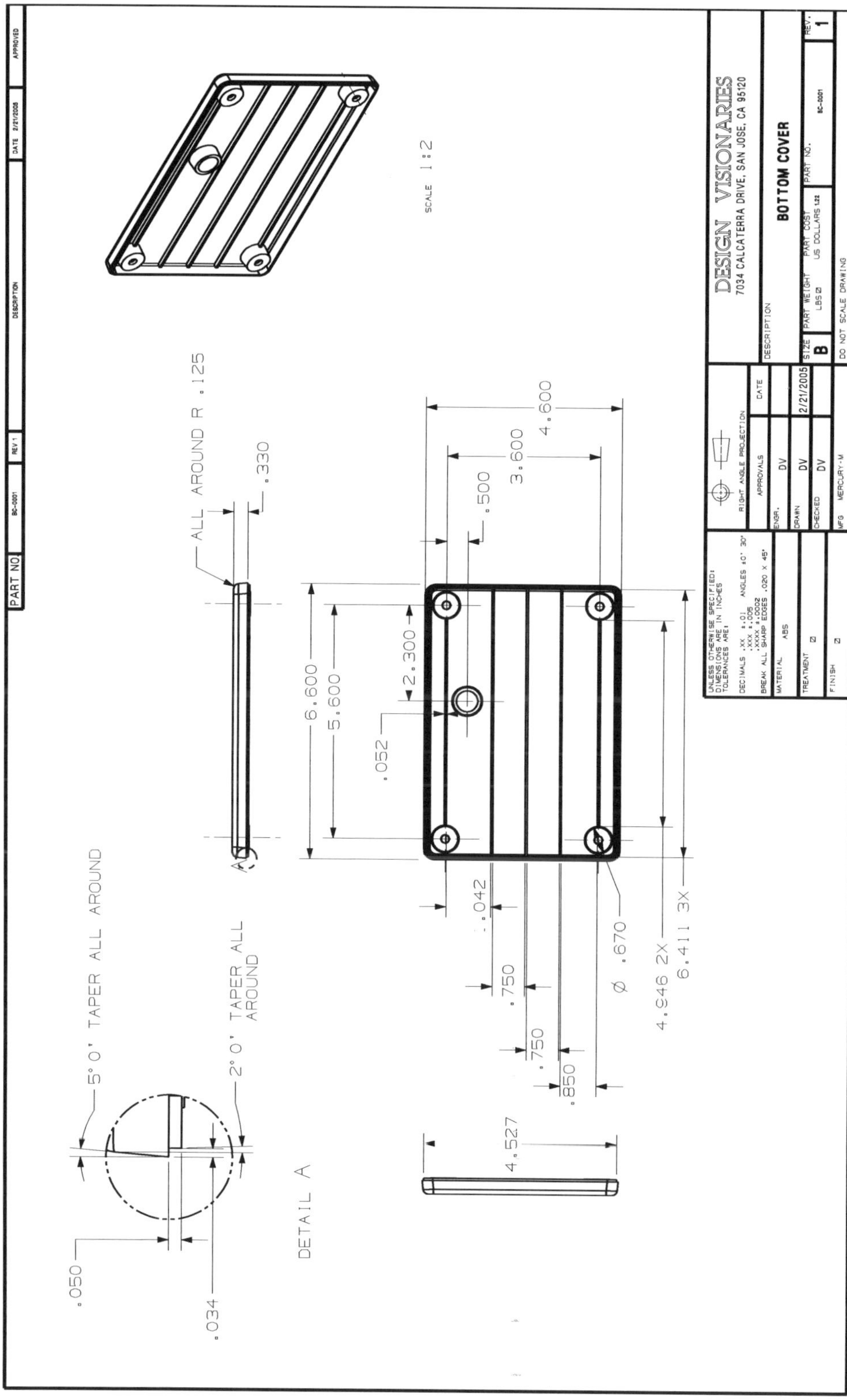

PART NO
BC-0001
REV 1
DESCRIPTION
DATE 2/21/2005
APPROVED
SCALE 1:2
ALL AROUND R .125
.330
.500
3.600
4.600
6.600
5.600
2.300
.052
.042
.750
.750
.850
Ø .670
4.946 2X
6.411 3X
4.527
5° 0' TAPER ALL AROUND
2° 0' TAPER ALL AROUND
.050
.034
DETAIL A
DESIGN VISIONARIES
7034 CALCATERRA DRIVE, SAN JOSE, CA 95120
BOTTOM COVER
RIGHT ANGLE PROJECTION
MATERIAL ABS
DV
2/21/2005
MERCURY-M
SIZE B
DO NOT SCALE DRAWING

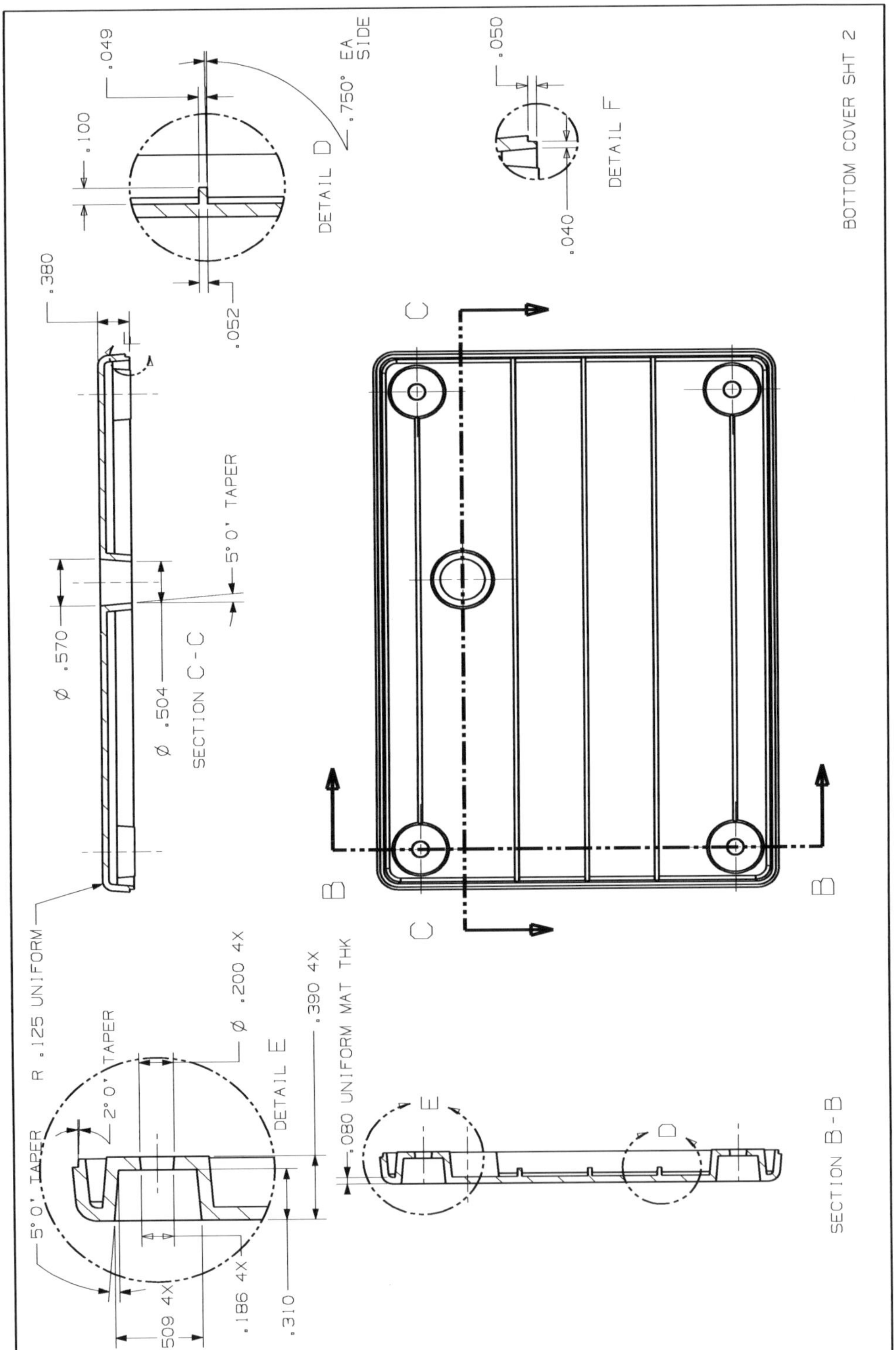
.049
.100
.750° EA SIDE
DETAIL D
.052
.050
DETAIL F
.040
BOTTOM COVER SHT 2
.380
5°0' TAPER
Ø .570
Ø .504
SECTION C-C
C
B
R .125 UNIFORM
Ø .200 4X
.390 4X
.080 UNIFORM MAT THK
2°0' TAPER
5°0' TAPER
DETAIL E
.509 4X
.186 4X
.310
E
D
SECTION B-B

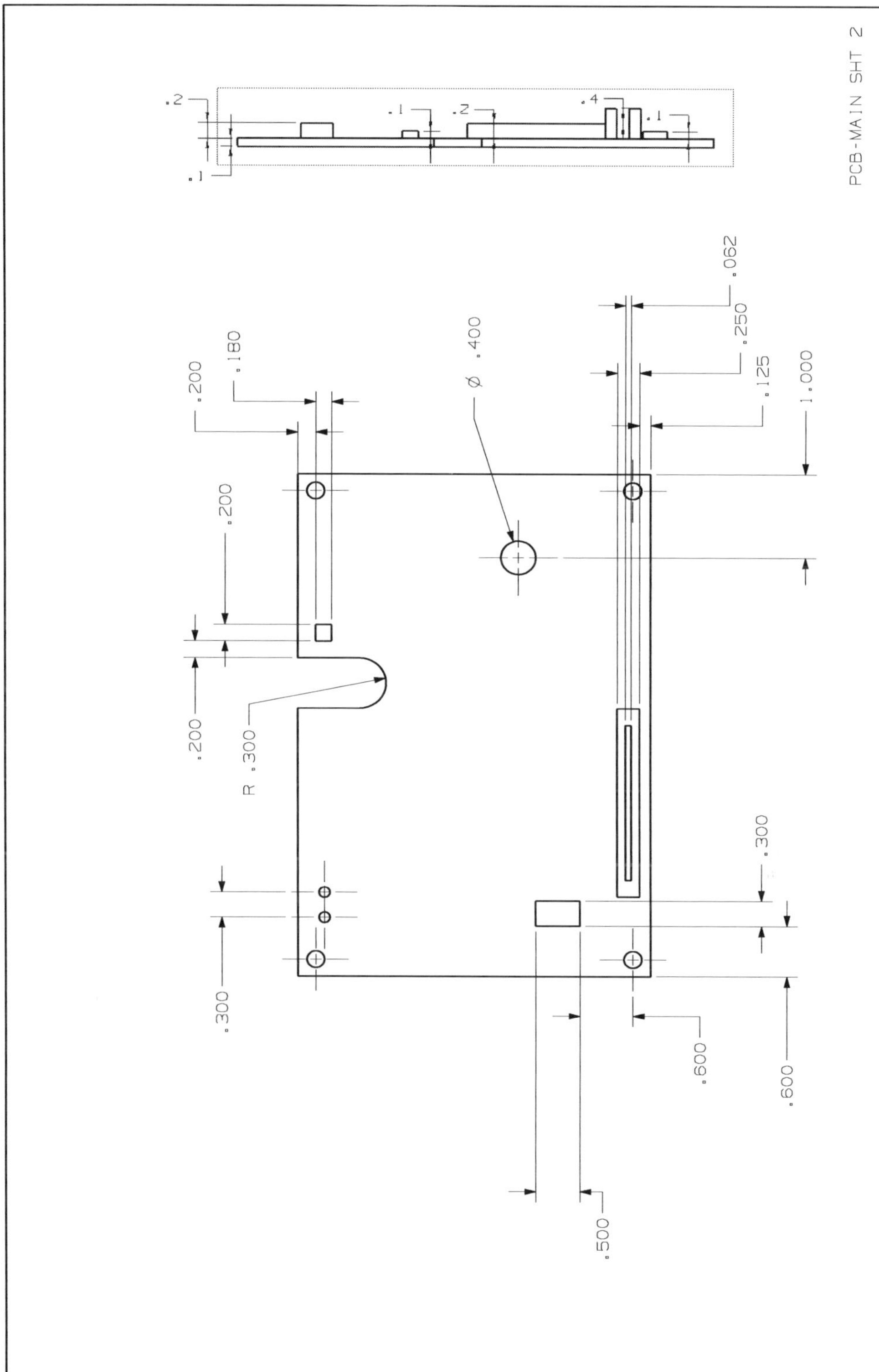
PCB-MAIN SHT 2
.062
.250
.125
1.000
Ø .400
.180
.200
.200
.200
R .300
.300
.300
.600
.600
.500
.2
.1
.4

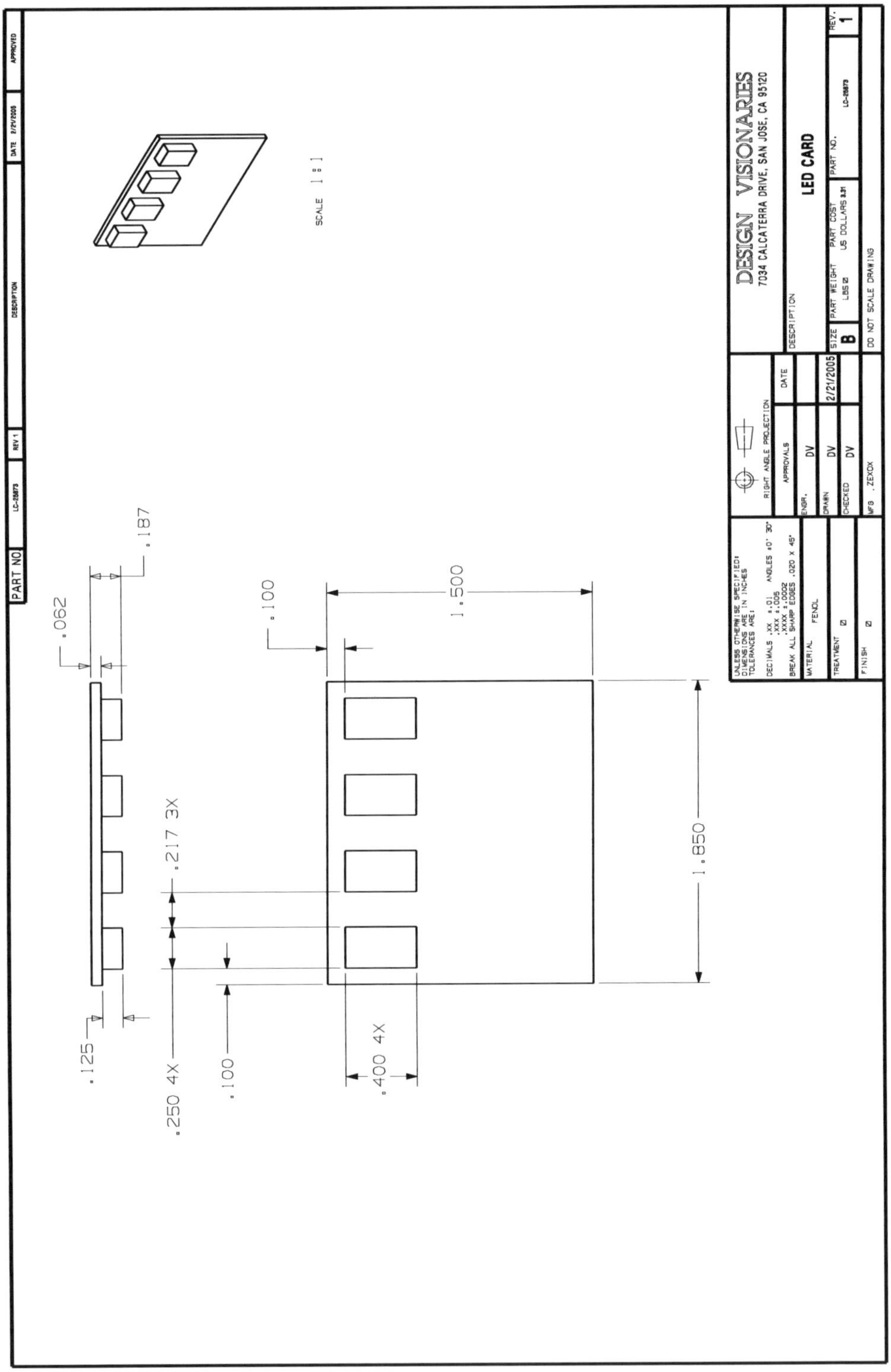
DESIGN VISIONARIES
7034 CALCATERRA DRIVE, SAN JOSE, CA 95120
LED CARD
SCALE 1:1
PART NO
REV 1
DESCRIPTION
APPROVED
RIGHT ANGLE PROJECTION
APPROVALS
DATE
DV
2/21/2005
SIZE
B
PART WEIGHT
PART COST
PART NO.
REV.
1
DO NOT SCALE DRAWING
MATERIAL
FENOL
TREATMENT
FINISH
ENGR.
DRAWN
CHECKED
MFG
.ZEXOX
1.500
1.850
.100
.062
.187
.125
.250 4X
.217 3X
.100
.400 4X

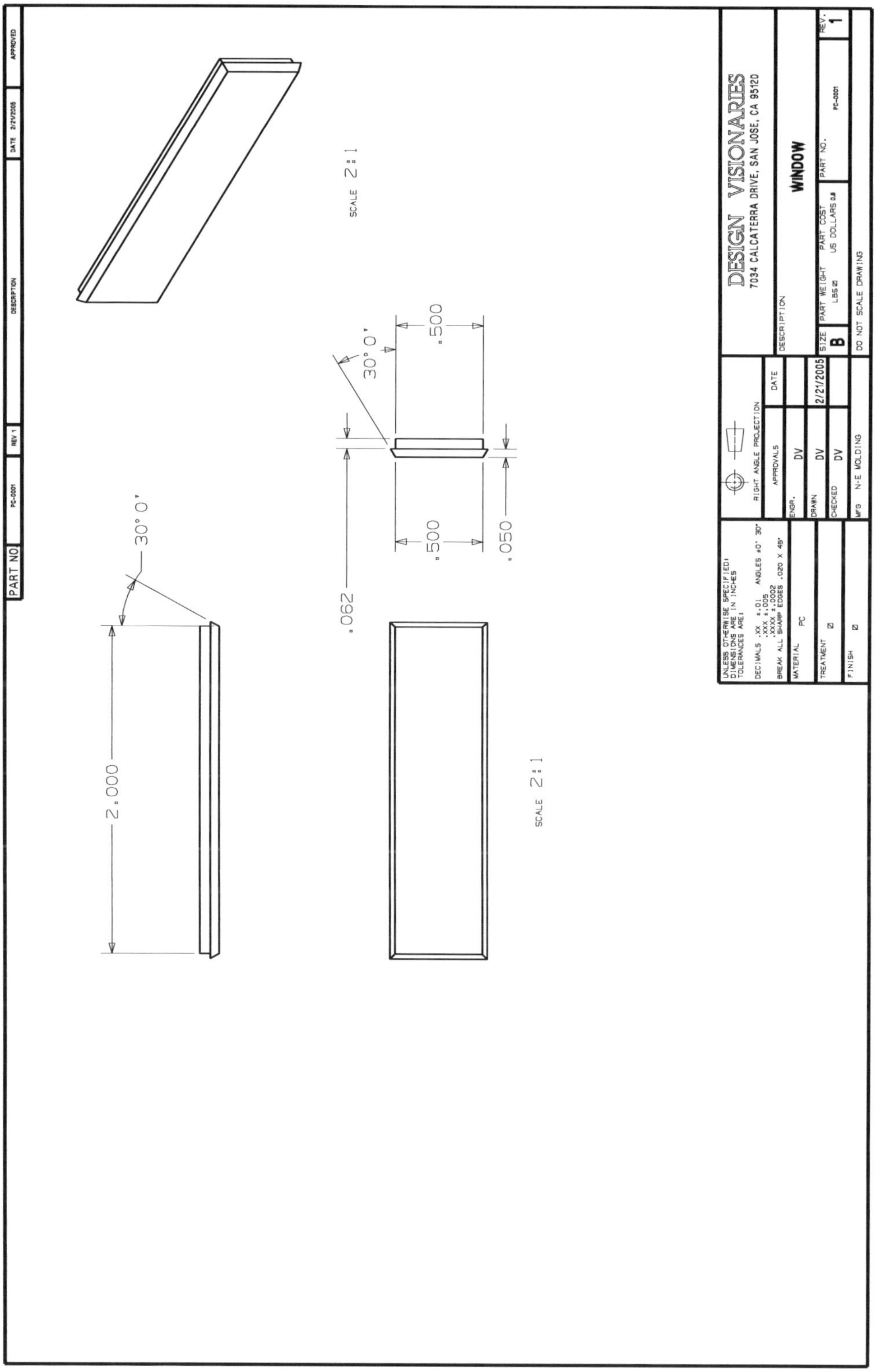
DESIGN VISIONARIES
7034 CALCATERRA DRIVE, SAN JOSE, CA 95120
WINDOW
SCALE 2:1
SCALE 2:1
30° 0'
30° 0'
2.000
.500
.500
.062
.050
PART NO
PC-0001
REV 1
DESCRIPTION
DATE 2/21/2005
APPROVED
UNLESS OTHERWISE SPECIFIED:
DIMENSIONS ARE IN INCHES
TOLERANCES ARE:
DECIMALS .XX ±.01 ANGLES ±0° 30'
.XXX ±.005
.XXXX ±.0002
BREAK ALL SHARP EDGES .020 X 45°
MATERIAL PC
TREATMENT
FINISH
RIGHT ANGLE PROJECTION
APPROVALS
DATE
ENGR. DV
DRAWN DV 2/21/2005
CHECKED DV
MFG N-E MOLDING
DESCRIPTION
SIZE B
PART WEIGHT LBS
PART COST US DOLLARS
PART NO. PC-0001
REV. 1
DO NOT SCALE DRAWING

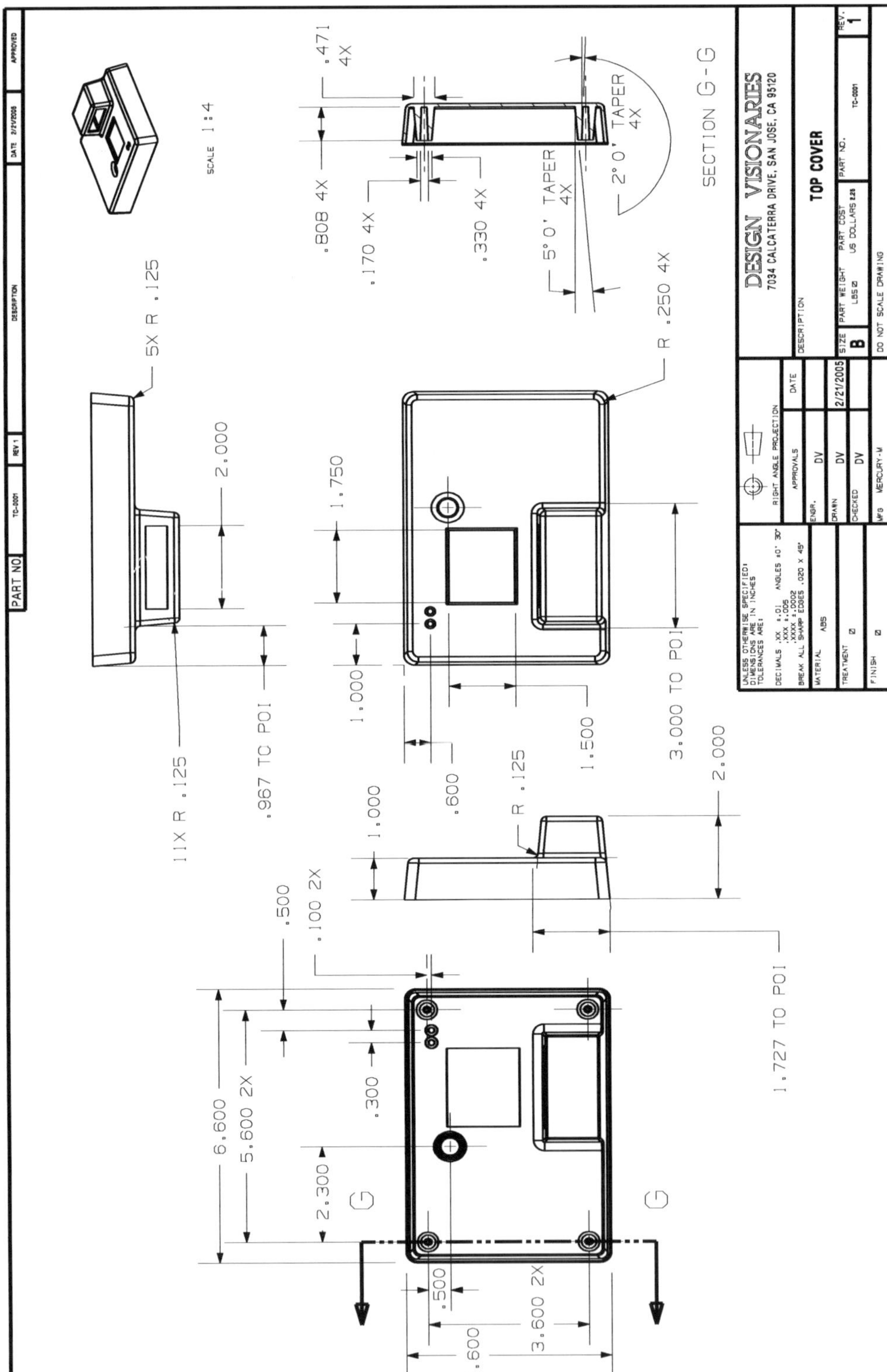
DESIGN VISIONARIES
7034 CALCATERRA DRIVE, SAN JOSE, CA 95120
TOP COVER
SECTION G-G
SCALE 1:4
PART NO TC-0001
REV 1
DO NOT SCALE DRAWING

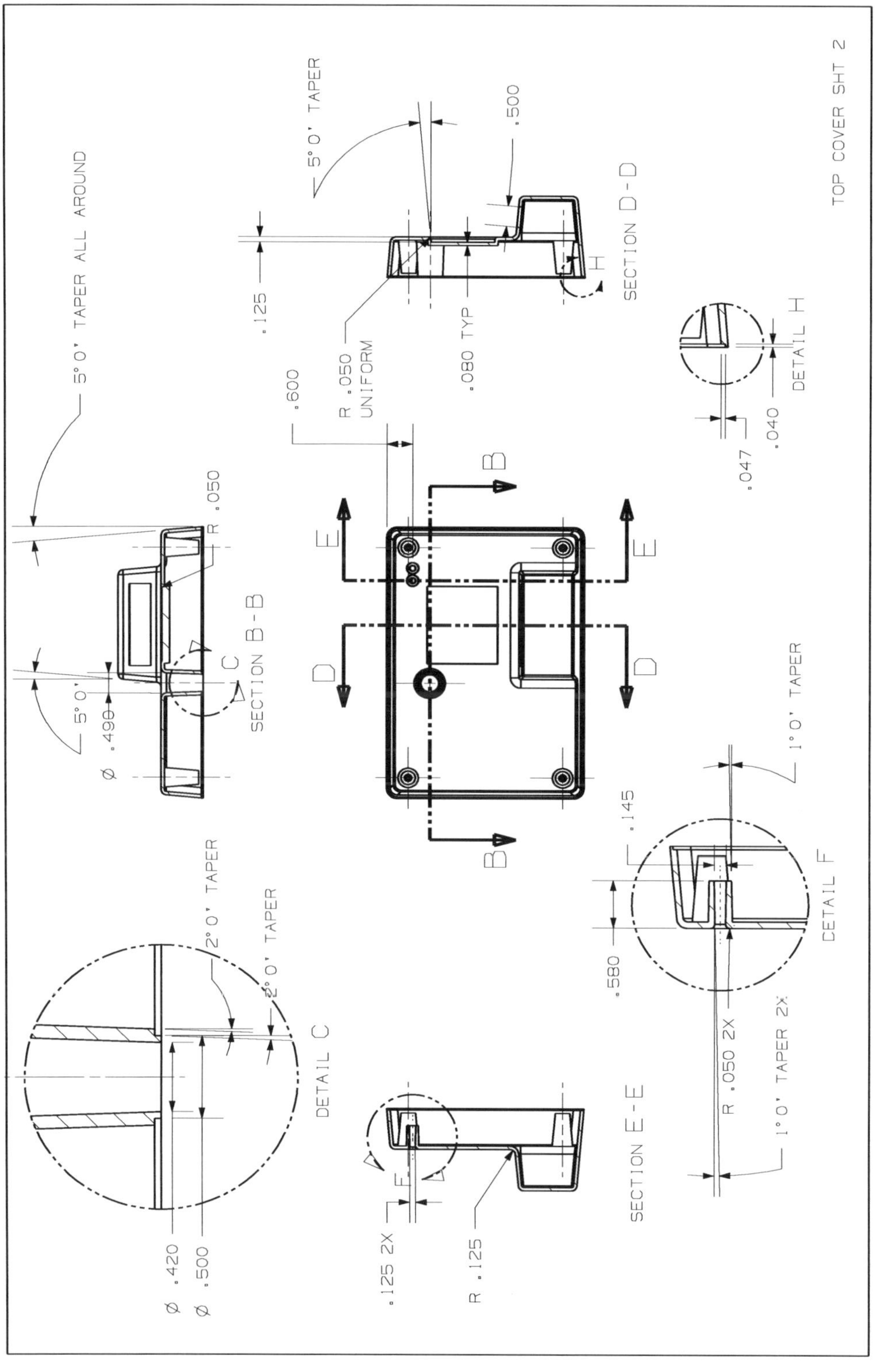
TOP COVER SHT 2
SECTION D-D
5°0' TAPER
.500
.125
R .050
UNIFORM
.080 TYP
.600
DETAIL H
.040
.047
5°0' TAPER ALL AROUND
R .050
SECTION B-B
5°0'
Ø .490
C
B
B
D
D
E
E
1°0' TAPER
.145
DETAIL F
.580
R .050 2X
1°0' TAPER 2X
2°0' TAPER
2°0' TAPER
DETAIL C
Ø .420
Ø .500
SECTION E-E
.125 2X
R .125
F
H

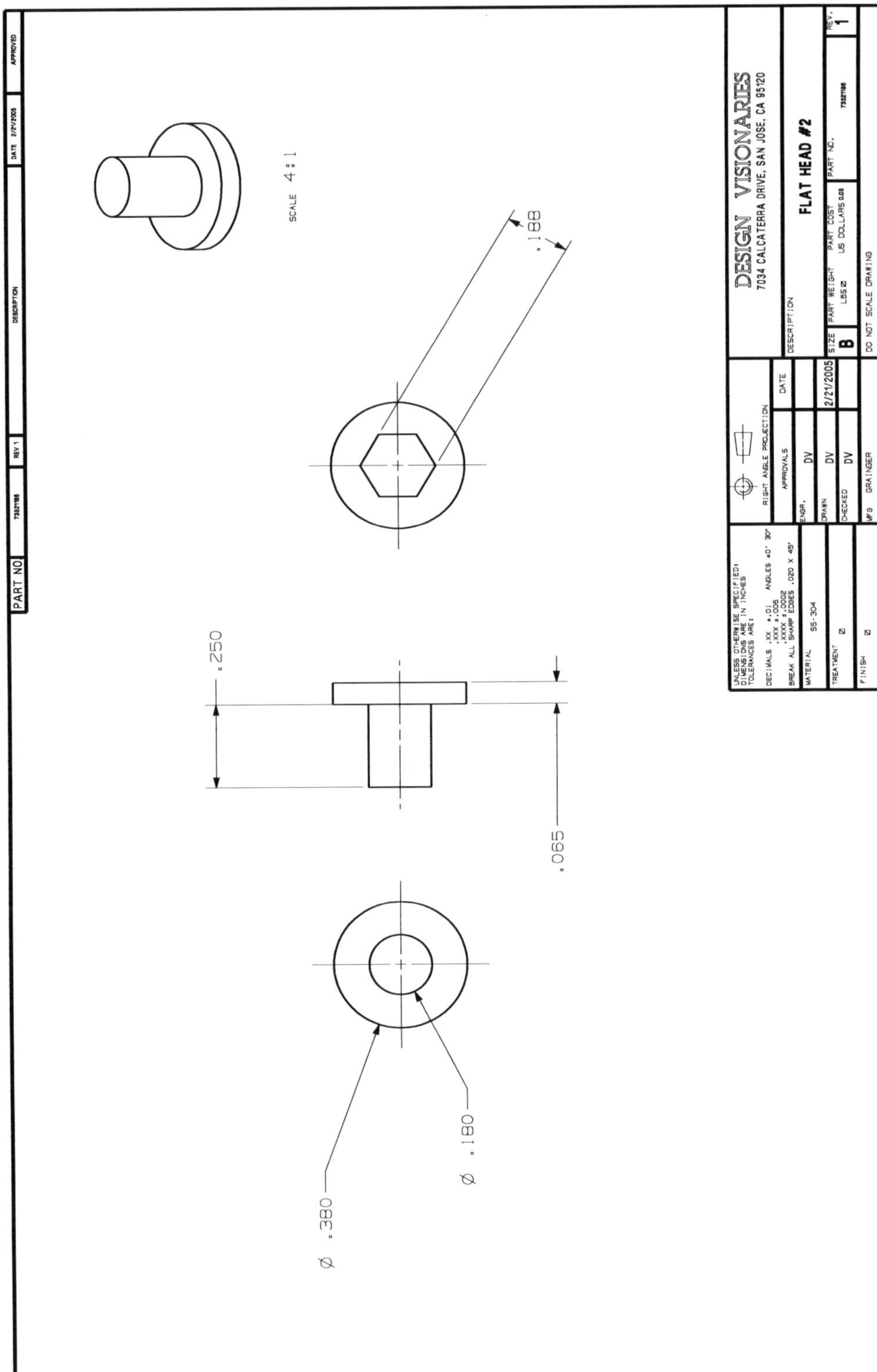
DESIGN VISIONARIES
7034 CALCATERRA DRIVE, SAN JOSE, CA 95120
FLAT HEAD #2
SCALE 4:1
.188
.250
.065
Ø .380
Ø .180
PART NO
DO NOT SCALE DRAWING

## 20.15 Self-Test

(All questions are true/false. Answers are at the back of the book.)

1. All the edges should be in the same plane to use Draft from edges.
2. Isocline curves always update with the geometry change.
3. You can define a forced direction while creating swept features.

# 21. How to Make NX 8.5 Run Great!

## 21.1 Introduction

Regular users of any software learn "tricks of the trade" over time that can help increase their productivity greatly. Throughout this book, we have tried to introduce shortcuts to give you more options. In NX 8.5 you can customize and create your own keyboard shortcuts. Along these same lines, we would like to show you a few more tricks to speed up your NX 8.5 efficiency in no time!

Better yet, blog about it on nxtutorials.com and get the word out to everyone!

## 21.2 Customized Keyboard shortcuts

Often, keyboard shortcuts can eliminate mouse clicks or unnecessary scrolling and time consumption. To customize them, select ***Tools / Customize*** and click the ***Keyboard*** button at the bottom of the ***Options*** tab at the bottom. The menu shown below will appear.

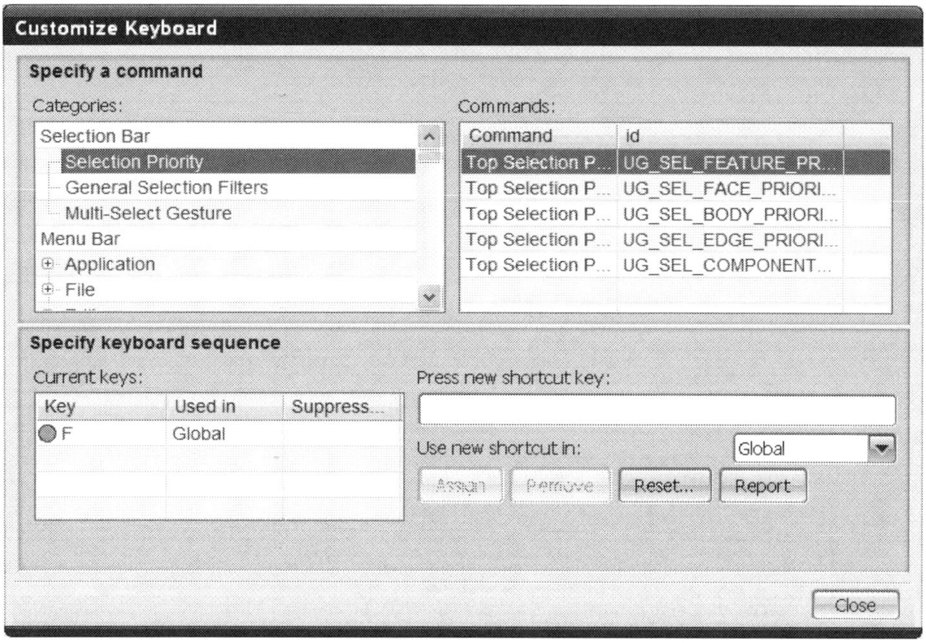


To create your own shortcuts, select the category from the left and command from the right and type in new shortcut key. Once you press the keys it will appear in the 'Press new shortcut key' field and the '**Assign**' button will become active. If that key is already assigned to some other command it will appear in the bottom left area. Press the '**Assign**' button to create the shortcut and close the window.

## 21.3 Full menus

Similar to many Microsoft Windows applications, NX 8.5 offers short menus designed to provide rapid access to commonly used items. However, waiting for the full menus to appear can be frustrating, especially when exploring new commands or using less frequented selections in NX 8.5. To customize menu and toolbar preferences, select ***Tools / Customize***. Click on the 'Options' tab and click on 'Always Show Full Menus,' as shown below.

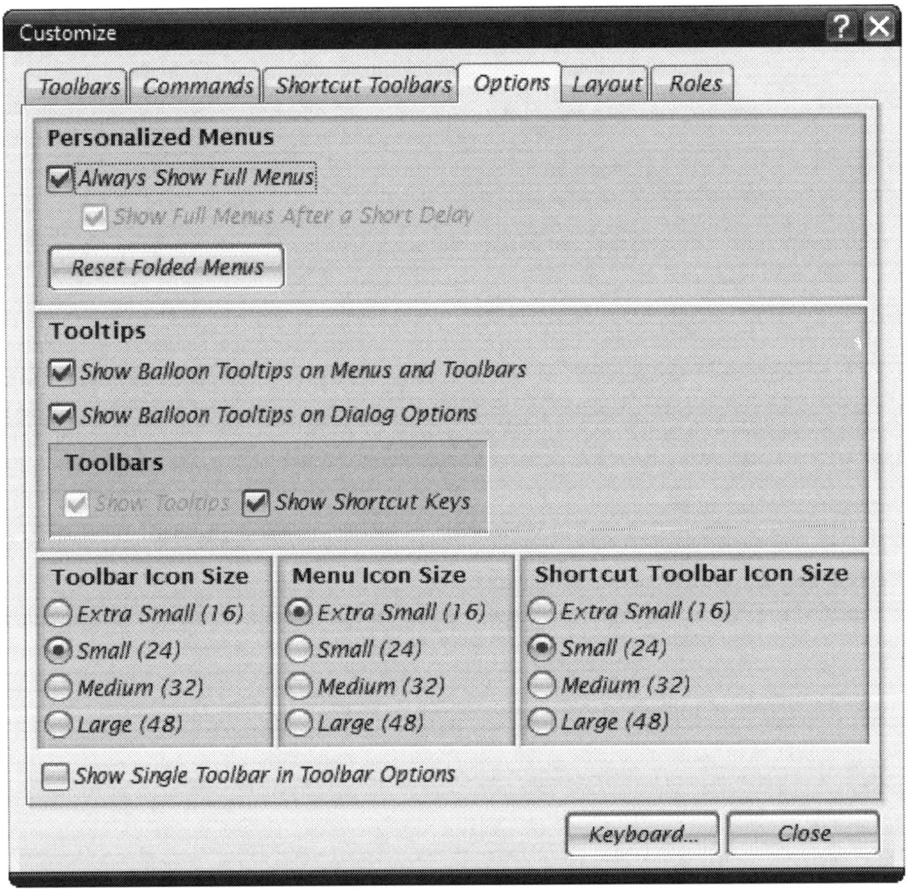

## 21.4 Save Layout

Another useful feature under the Customize window is the ability to save a specific layout. This allows you to set up your "ideal" layout, save it to a reference file, and recall that layout at any time. This is quite useful when creating several parts or drawings or for simply saving your personal preferences each time you use NX 8.

NX 8.5 allows different layouts to be created and saved for different functions. Commonly modified layouts include modeling, drafting and sketching. To access these layouts, first select the application (i.e. ***Start / Modeling***) and then follow the directions below.

Select ***Tools / Customize***, and choose the 'Layout' tab in the 'Customize' window as seen in the figure below. From here, you can save specific layouts to the following directory on your hard drive: **C:\...\ NX 4.0\UGII\menus**. Within this directory, NX 8.5 creates default layouts for all NX 8.5 operations.

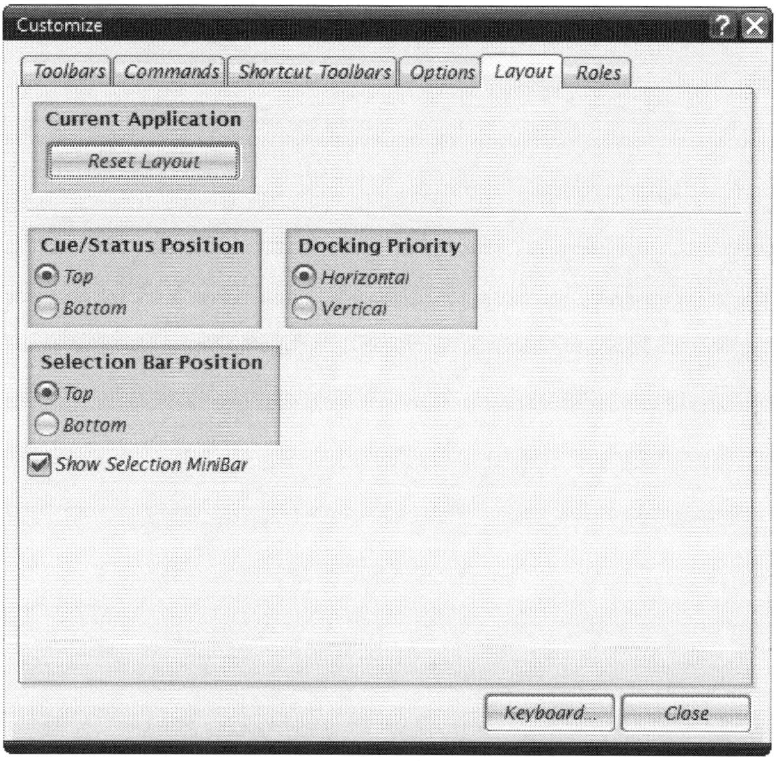


**Note:** We encourage you to create your own shortcuts and insights into how to make NX 8.5 run smoother and faster for you. If you think you've found a useful tip, feel free to drop us a line at info@designviz.com.

Selecting 'Reset Layout' resets all the built-in toolbars and menu-bars to their original NX 8.5 setting, regardless of whether a layout has been previously saved or not (it erases and sets to default any previous layouts).

An easier way is to save the layouts that you have customized for any of your applications is to make sure that the 'Save Layout at Exit' check box is checked (see figure below). Select ***Preference / User Interface*** to open the menu shown below, click on the 'Layout' tab, and check the box to 'Save Layout at Exit' located at the bottom. This way your customized layout will be saved automatically when you exit NX 8.5 and the next time you open or create a part file, you will have your customized toolbars displayed.

**Note:** Saving files anywhere else will not allow you to access or recall them.

## 21.5 Selection Delay

NX 8.5 has built-in delays on specific selection commands. Select ***Preferences / Selection*** to activate the menu shown below. Setting the Rollover Delay to 0 will ensure that when you roll your mouse cursor over any object/line/selection, there is no delay in highlighting your intended target. This ensures rapid selection of your desired entity.

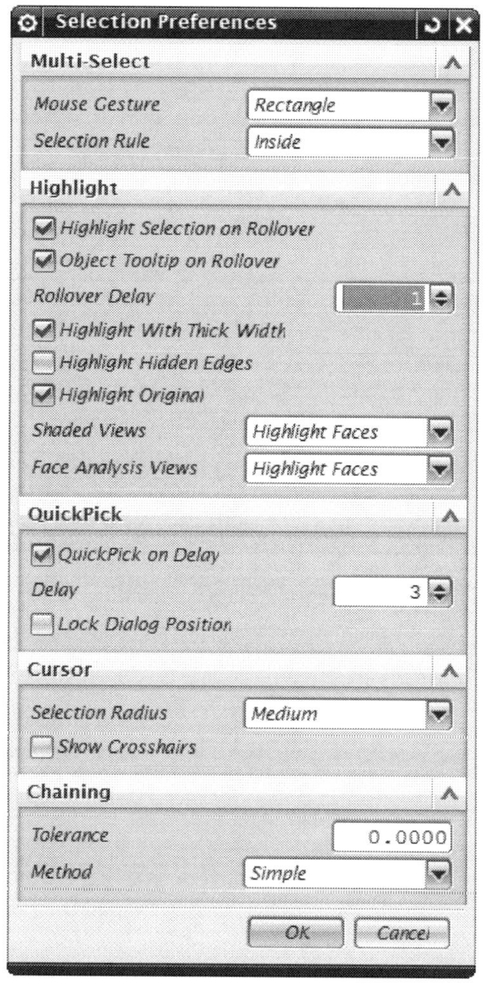


The 'Quick Pick Delay' bar allows you to specify how long NX 8.5 should wait before providing the **Quick Pick window**, like the one shown in the figure below.

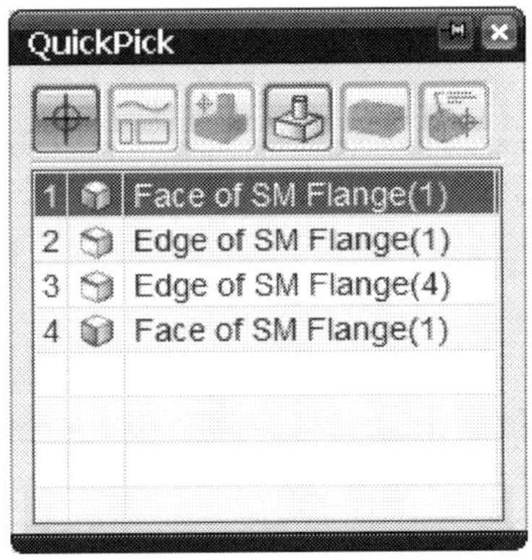


For fastest working speed, set all delays to 0.

## 21.6 View Animation Speed

You may have noticed when changing views in NX 8.5 that the part is slowly rotated around as the view changes. For simple parts, this can be done quickly and may be more convenient if you are unsure of each view point. However, as parts and assemblies become complex and consume more CPU time, it is advisable to disable the smooth view change function so NX 8.5 does not need to calculate and graphically update the views as it is rotated or zoomed into or out of. Most importantly, you will be able to move a lot faster with low view animation speed.

To access the Visualization Performance Preferences, select ***Preferences / Visualization Performance / General Graphics.*** For best Visualization Performance, move the slider all the way to the right.

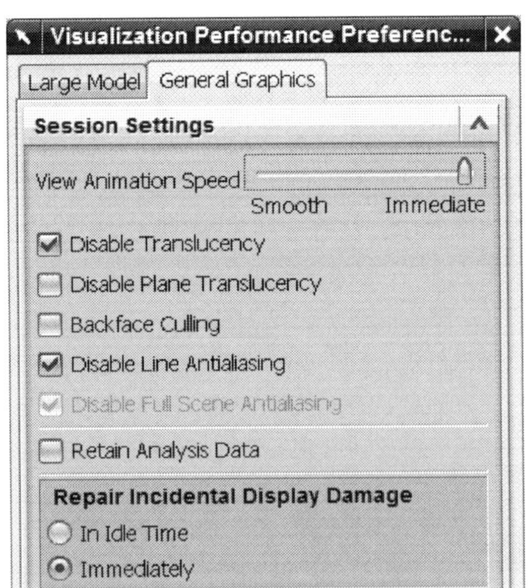

## 21.7 Shaded Face Edges

When viewing complex 3D models in NX 8.5, selecting to turn 'Shaded With Edges' off will not only improve the speed of NX 8.5 but will also allow you to view the model in a more realistic format. Select and turn Shaded Face Edges Off from the View toolbar. You can also select Shaded Edge Color 'None' from the pull down menu shown below. To access this menu select ***Preferences / Visualization***.

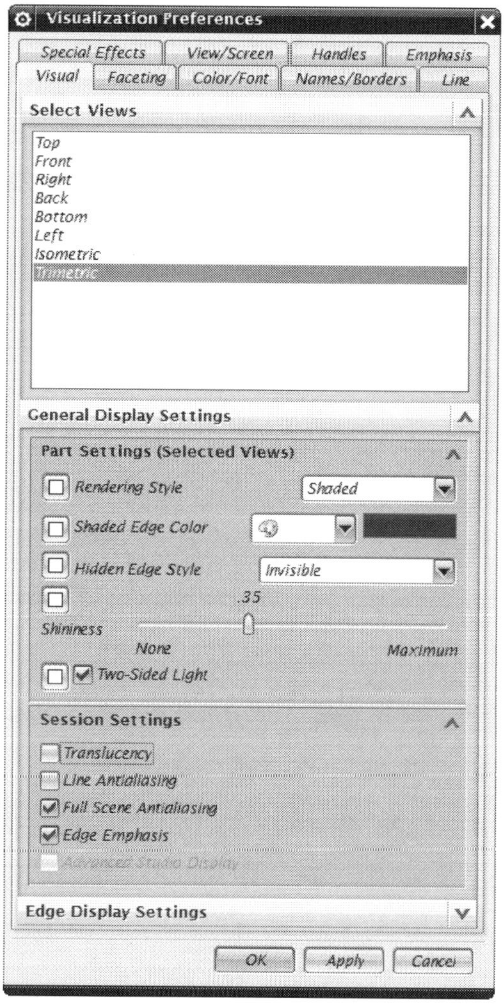

## 21.8 Assigning Feature Names in a Part

It is very easy to work in a complex part file if you assign names to different features. It works best if you need to select certain features while working on the part for operations like ***Show and Hide***.

- To assign a name to a feature select ***Edit / Properties***.
- Select the desired feature on the part in the graphics window or in the part navigator and hit **OK**.
- Click on the 'General' tab and type any name you want as shown below.

- Hit **Apply** and repeat the same process for any other features you want.

Now that you have assigned different names to different features, it is very easy to select a particular feature from the 'Class Selection' menu by typing the name of the feature you want.

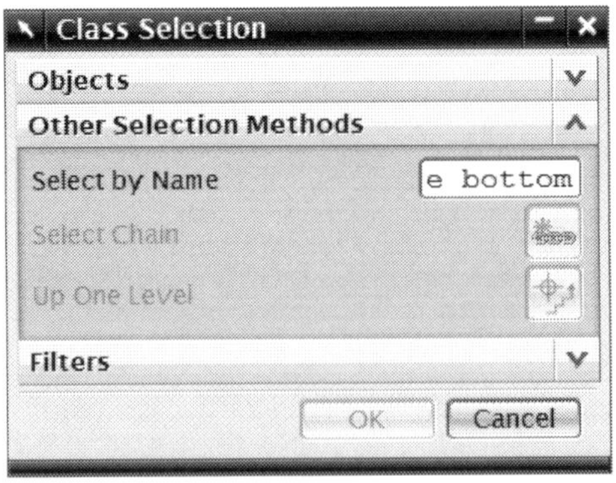

## 21.9 Using Grids

For some of the projects you can work much faster if you work on a grid. To define the grid, select ***Preferences / Grid***. A Grid menu will pop up as shown below.

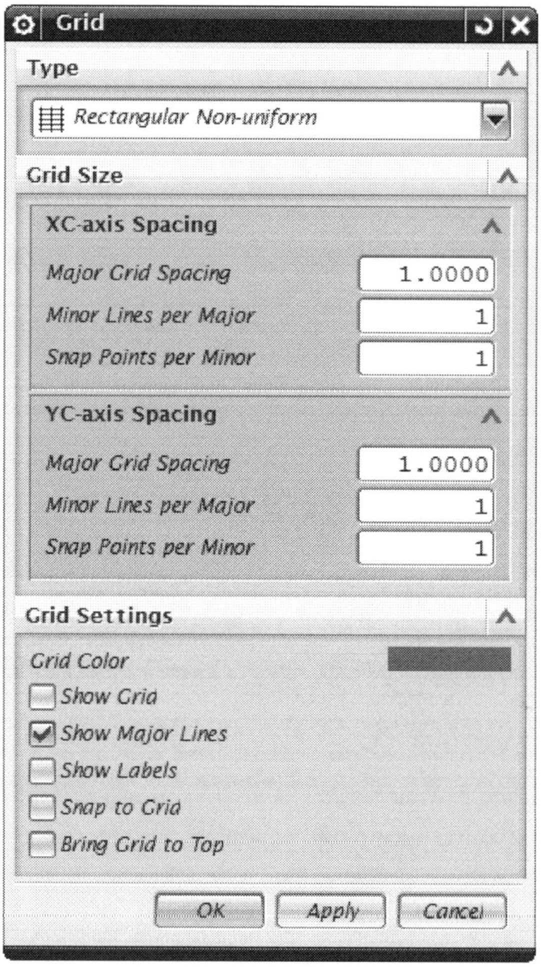


You can define a uniform or non-uniform grid based on your needs. This Work Plane Preferences menu offers the options for Display Grid, Snap to Grid, Show Emphasis Line, and more as shown above. The Grid option is also available for the drafting application.

## 21.10 Section Tool

Whether it's checking the clearances in assemblies, selecting a face that can't be seen, or using it in advanced simulation to constrain an element, the section tool can make your job easier every day!

### *Project 21.10.1*

- Start by opening any part. This one is a Thumb Piano.

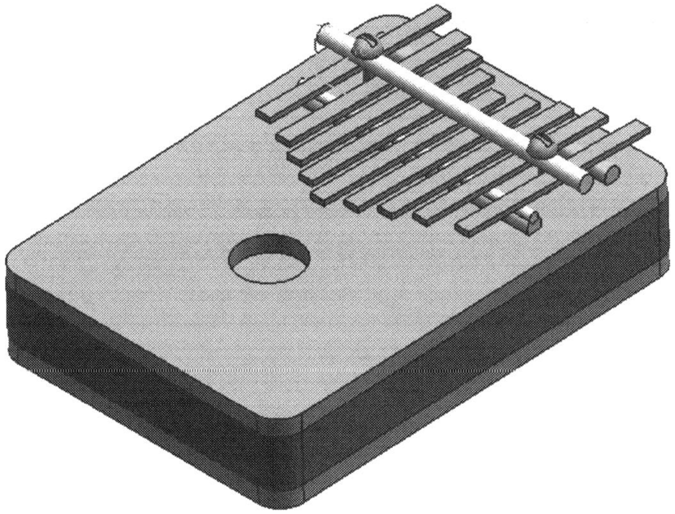

- Now, click on View / Section / New section.

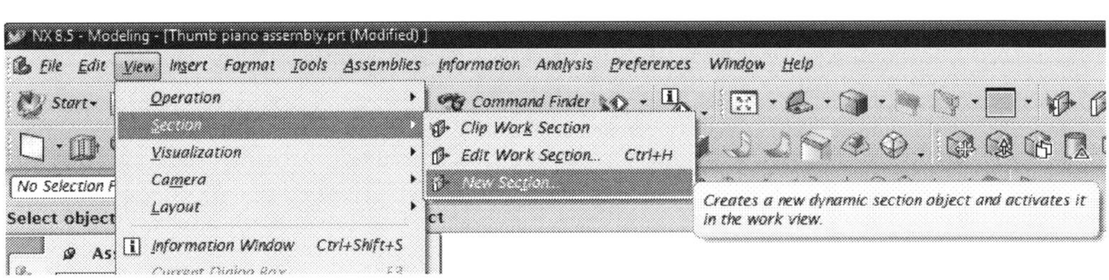


- The default plane section will appear. This is usually the XY plane, as shown in the following figure.

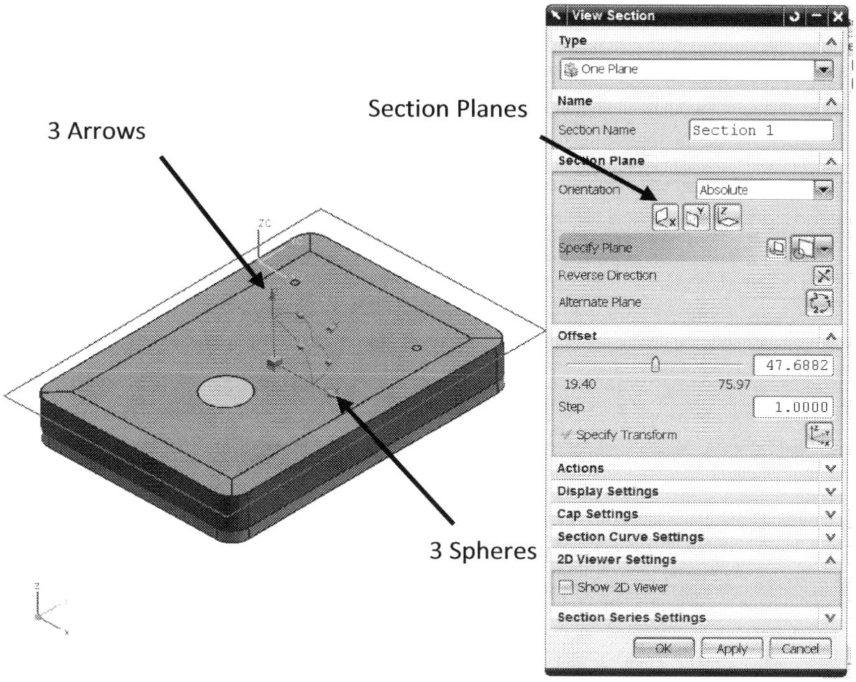


- You can change the plane by choosing a different one under Section Plane.
- The easier way to control it is to use the arrows and spheres. Like extruding curves, you can simply **drag the arrows** around to move the cutting plane of the section tool.
- Likewise, you can **rotate the spheres** around like angled datum planes to change the rotation of the cutting plane.

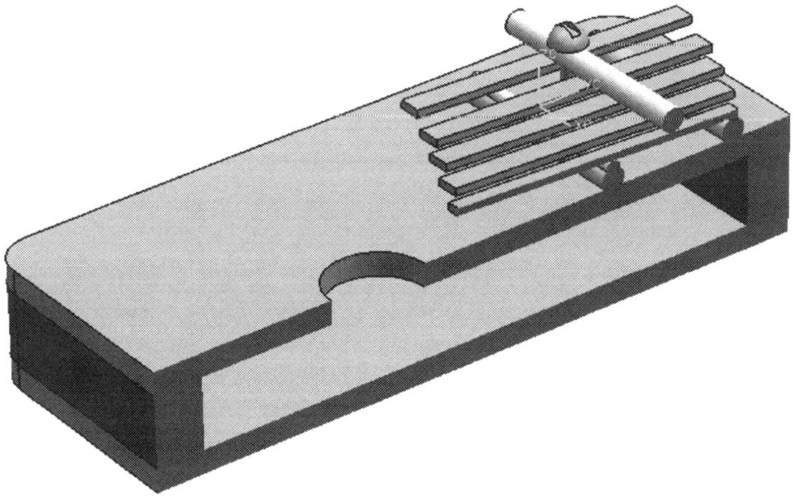

- If you press apply and the section disappears, you may have to go to **View / Section / Clip Work Section** to turn on the section view permanently.
- Another amazing part of the section tool is the 2D Section Viewer. If you check the **Show 2D Viewer** button you will see a drawing of your section.

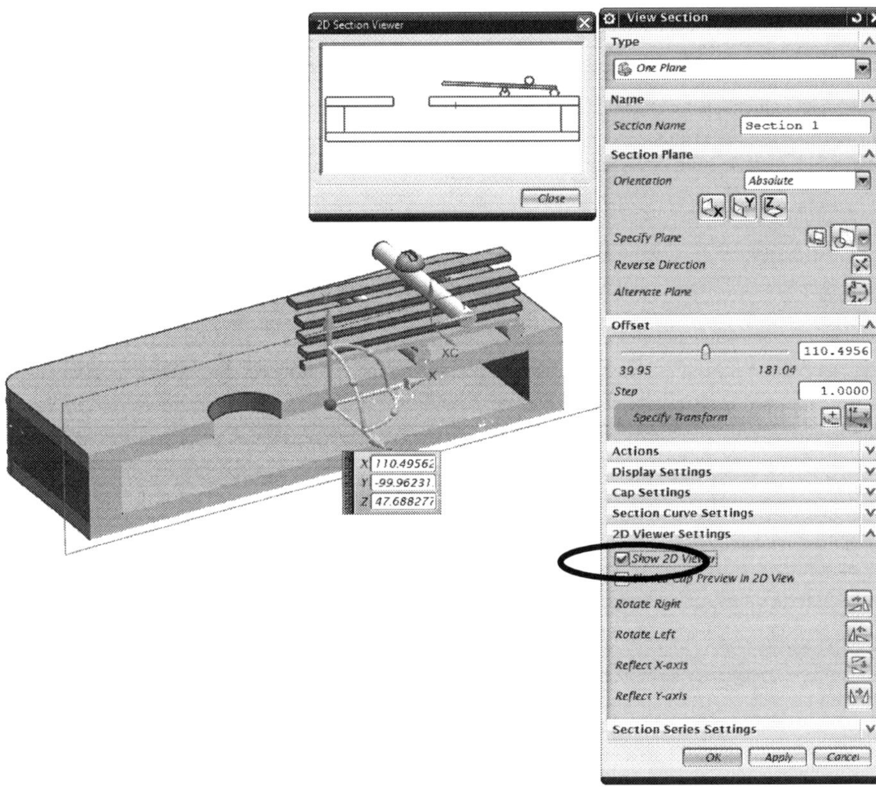


There are quite a few cool little settings in the Section tool. Mess around with it and see what else you can do!

***End of Project***

## 21.11 True Shading Editor

There is a fantastic viewer available in NX 8.5 that takes your model and makes it look like a real life model with very little effort. It's easy to use and is especially great for showing your customers how amazing your design is.

### *Project 21.11.1*

Using the Thumb Piano assembly, this quick project will show you the basics of how to use the True Shading Editor.

- Open the Thumb assembly in NX.

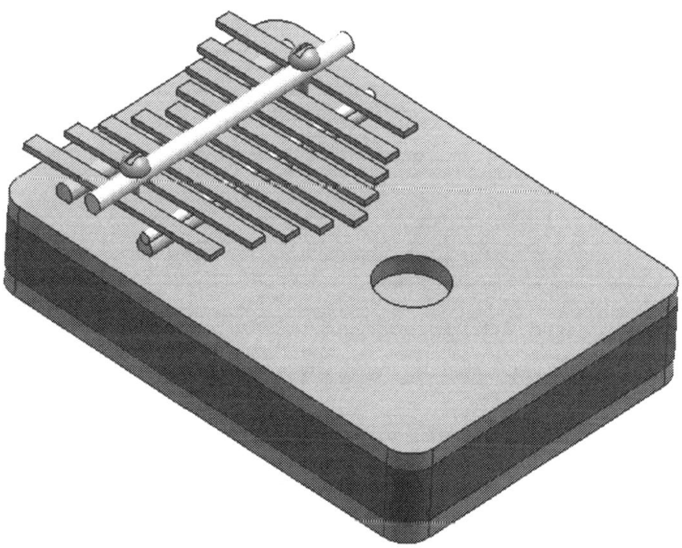

- Click on **View / Visualization / True shading Editor**.

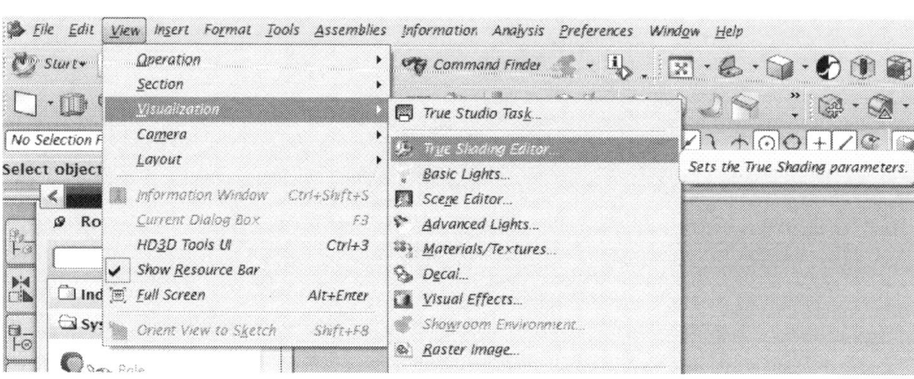


- You will now see the joy of the True Shading Editor. Your model will look amazingly realistic.

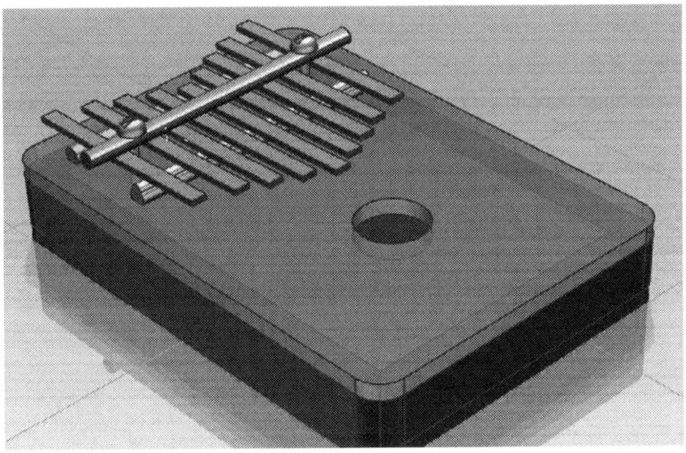

- Let's look at some of the options you have to edit the look of not only the model but the NX environment it's in.
- The following figure is the True Shading Editor box. It appears as soon as you click on **View / Visualization / True Shading Editor**. If you close it for some reason and want to use it again, just re-click the aforementioned menu clicks.

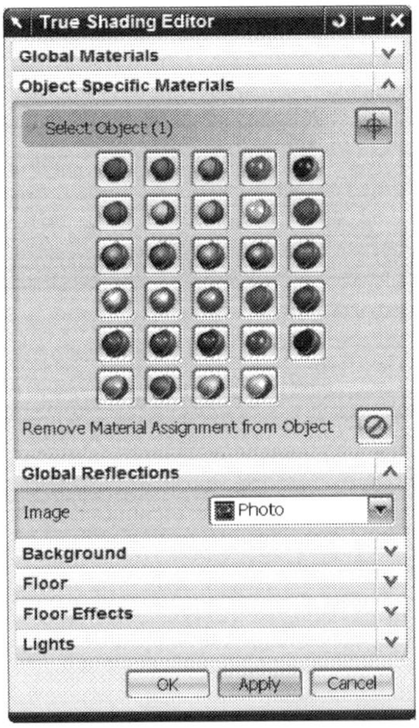


- The Global Materials tab is to change the look of the entire model. What is more useful is the **Object Specific Materials** tab. **Expand this tab** if it's not already open.
- **Highlight** one of the **legs** as shown in the following figure. *Not just a face of the legs – the whole body.*

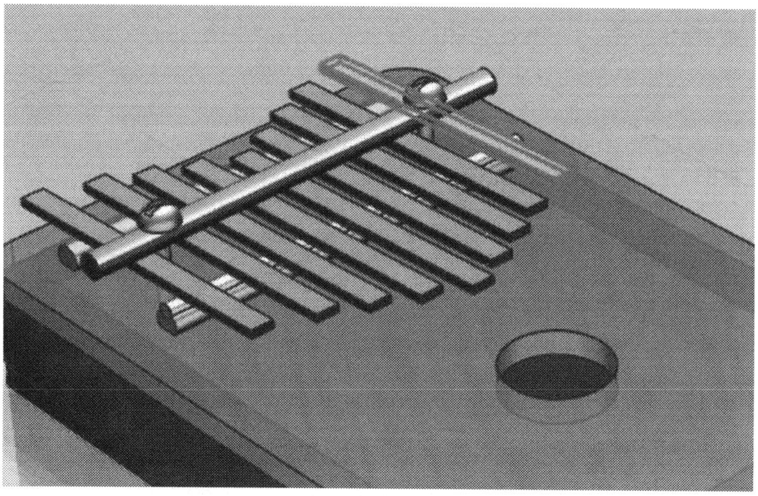

- Whilst the leg is highlighted, select **copper** as the material in the TSE box.

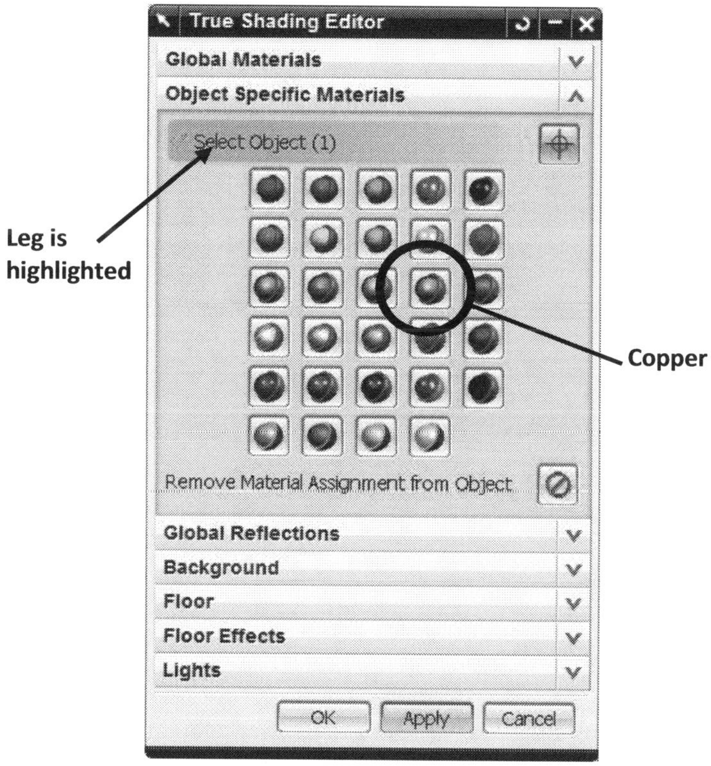


- As soon as you click on the **copper** button, the body will change color and texture.

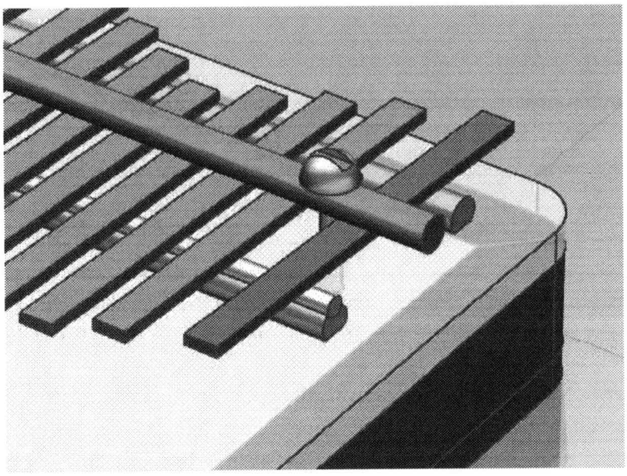

- Feel free to change the color of all the bodies in your model. There is a huge list of options.
- There are also different reflections that you can get from your model, as shown in the following figure. Photo tends to be the most glamorous, but change them around to see what looks good for your model.

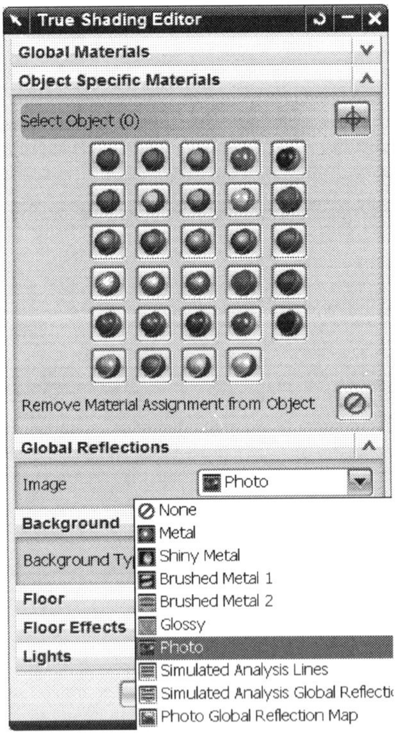


The next great feature is the ability to control the background. There is a big selection to choose from. I suggest the **Graduated Dark** or the **Light** options. It really depends on what mood you are looking for.

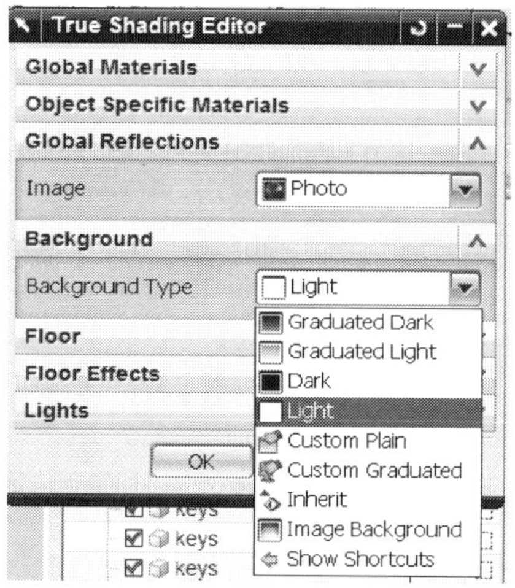


- The floor options and effects are quite self-explanatory. You can define the floor and include shadows and reflections, which are usually turned on as the default.

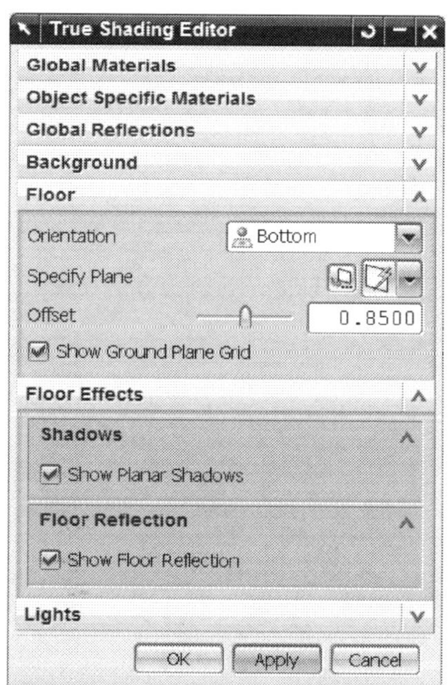


- Finally, the lighting options are available for you to play around with. You can change the lighting position and the brightness. Enjoy!

***End of Project***

## 21.12 True Studio Task

How would you like to make a rendering of your work that will look about as real as you can get it without actually building a real prototype and photographing it? With the True Studio Task, you can quickly see what your product will look like in different environments, under different lighting conditions, with different textures, and various perspectives. To proceed you select **View/Visualization/True Studio Task**. The system immediately applies some default values to the view.

**Before**

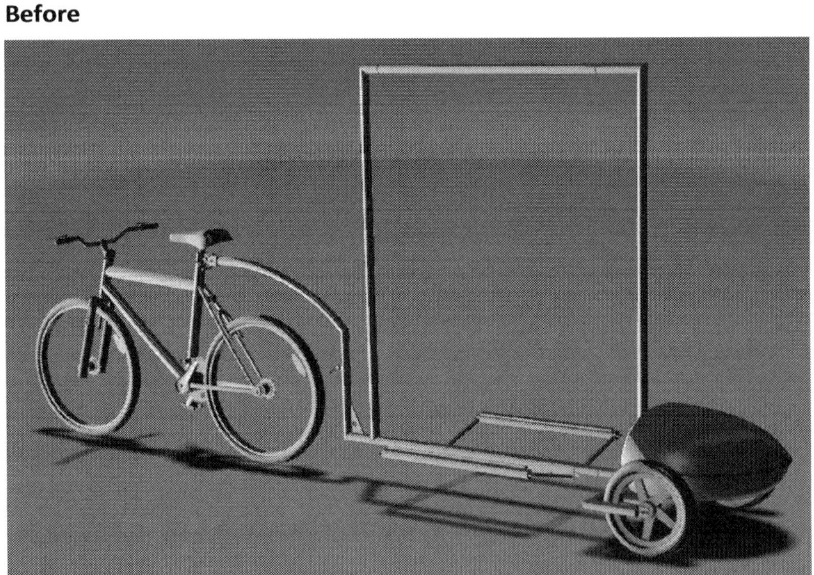

**After**

Perhaps the first thing to try is the "Systems Scenes" option. 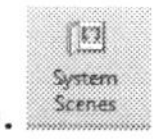

You get your choice of neat photo environments in which to showcase your geometry. For example, you can select "Congress Hall 1". Your image is shown on the plaza of what appears to be a museum.

Next you can edit the scene parameters using the "Scene Editor". 

For example. Try turning up the "Reflectivity of the floor" in the "Stage" tab.

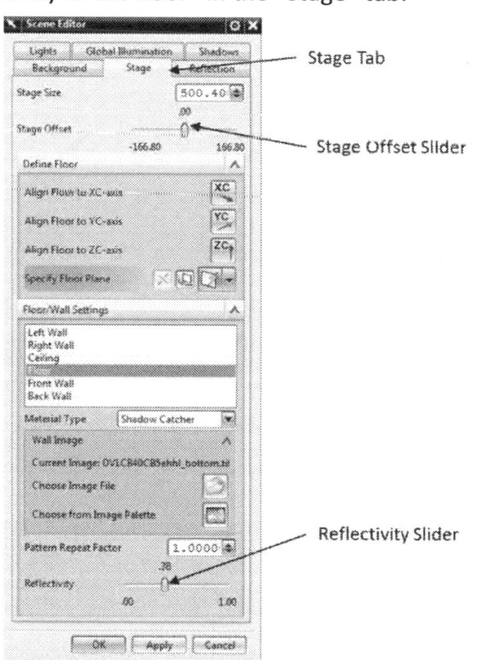

## 21.13 Introduction to Customer Defaults

It is very important to modify the default settings according to your need in order to work faster. You define your default settings by selecting ***File / Utilities / Customer Defaults***. Once you set this up, you don't have to go back and forth to modify individual files for the same thing. For example, if you work in Metric units you should set your **Default Units** as Millimeters. You can find this option in ***Customer Defaults / Gateway / General,*** as shown below.

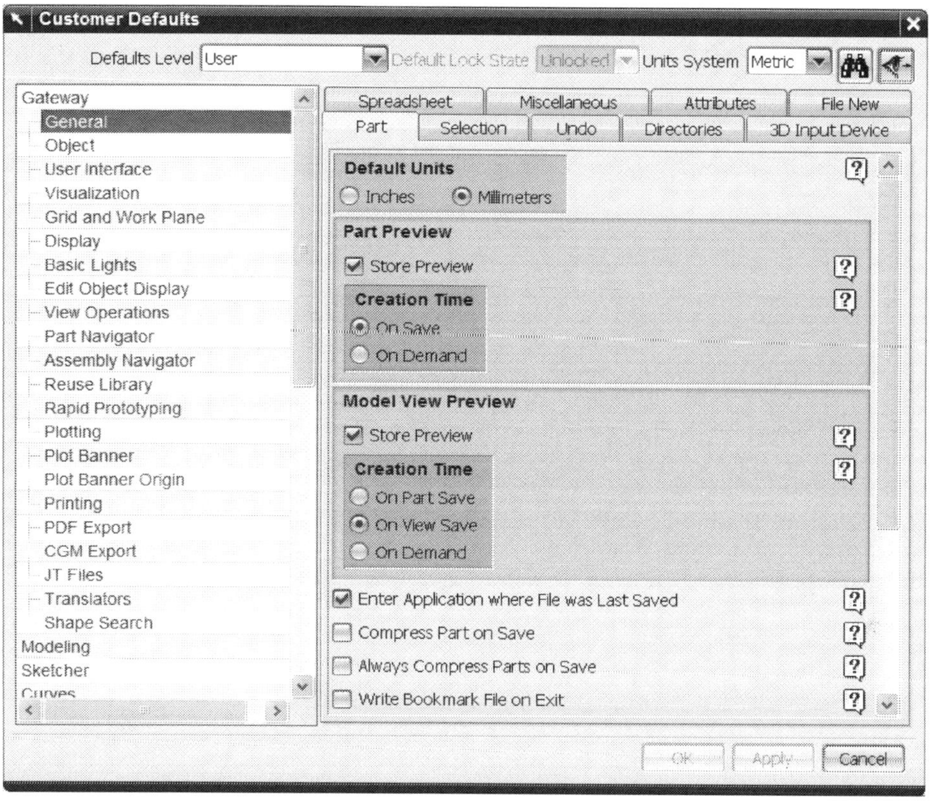

Explaining all the Customer Default options is out of the scope of this book. You can click on different menus and tabs to explore more options to set your defaults.

# 22. Practice Projects

## 22.1 Introduction

Sometimes, the best way to learn NX 8.5 is to be given a production job and a deadline. If you learn it this way, you will know many of the ins and outs of NX 8.5 – better than the people who wrote the software. This chapter focuses on the fine points of NX 8.5 by compiling a set of exercises that help you use what you have already learned and enable you to practice it all.

Most of the projects in this chapter rely on your creativity and overall new-found knowledge of the software. We have omitted detailed steps on purpose. The last section, 22.9 NX 8.5 Final Project, is the biggest challenge yet!

As always, if you have any questions, comments or feedback regarding these projects, do not hesitate to contact us at *info@designviz.com*.

## 22.2 Free Form Solid Using Swept

You need to build a drive curve in order to create a Swept feature (see the figure below).

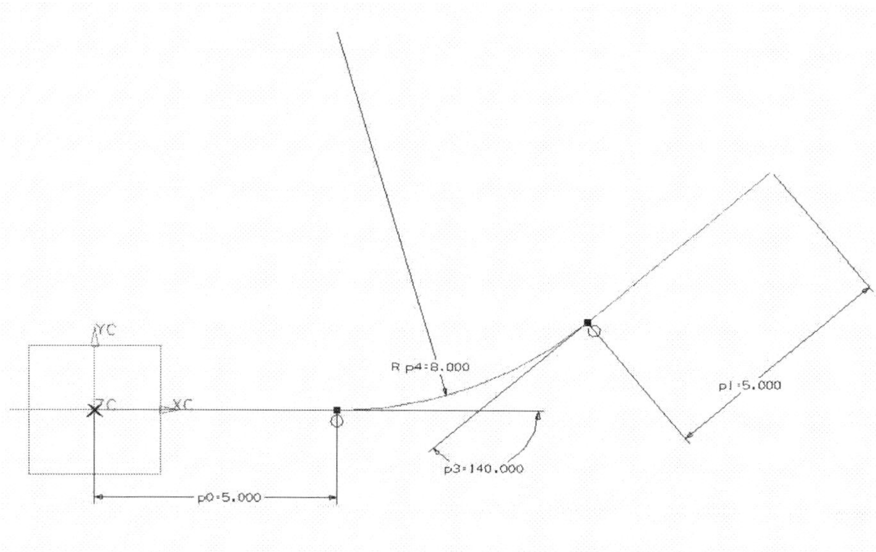


The dimensions of the basic drive curve sketch are shown in the figure above.
Several Datum Planes distributed along its length will enable you to place sections on it that define the shape. ***Hint:*** *Alternatively, you can use a sketch on a path.*

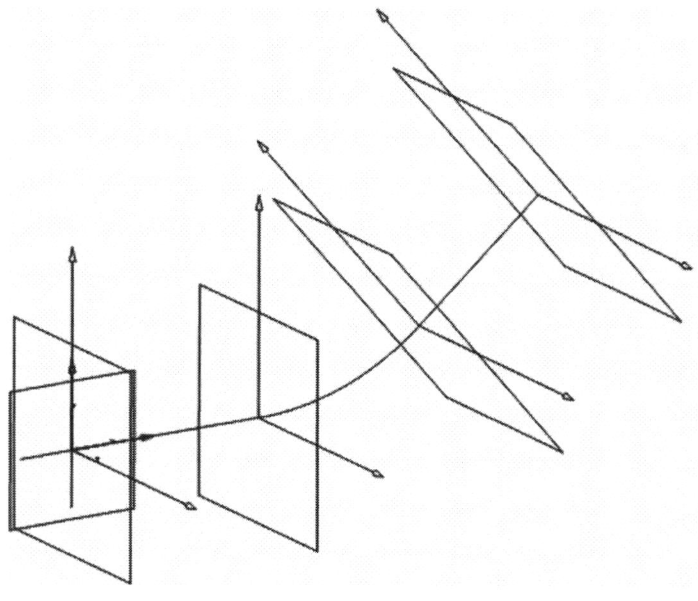

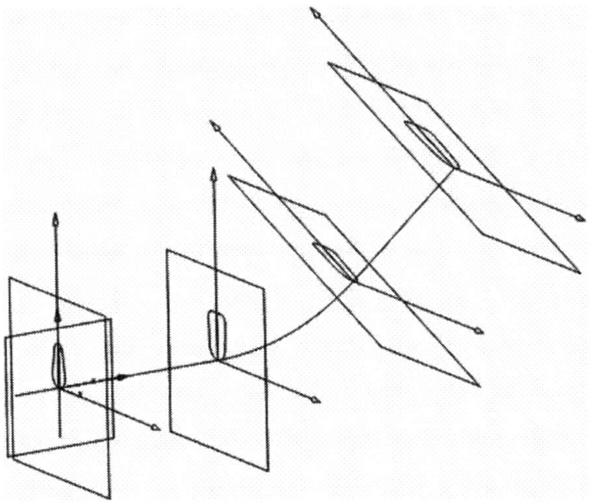

The dimensions of the sections are shown below.

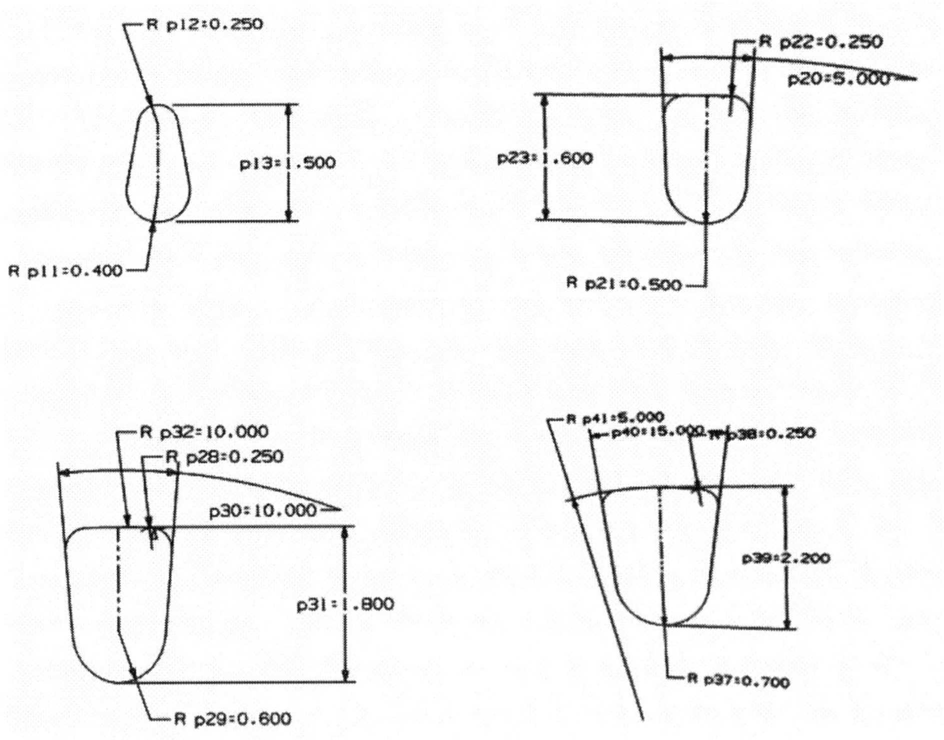


The resulting swept feature is shown below.

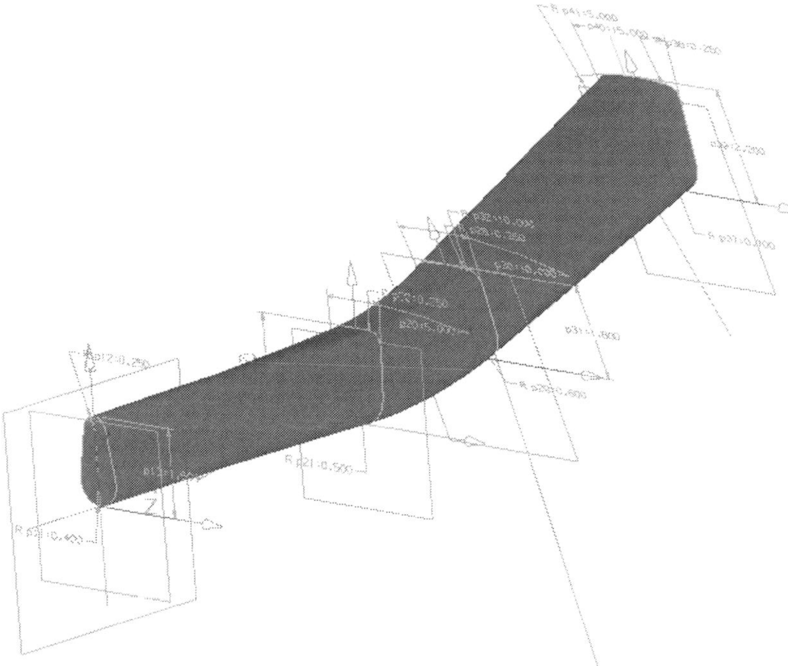

***End of Project***

## 22.3 The Product Housing

The housing of a typical consumer product is shown below. The model is fully parametric and features a curved parting line. All vertical walls have draft angles, and the model is ready for injection molding tooling.

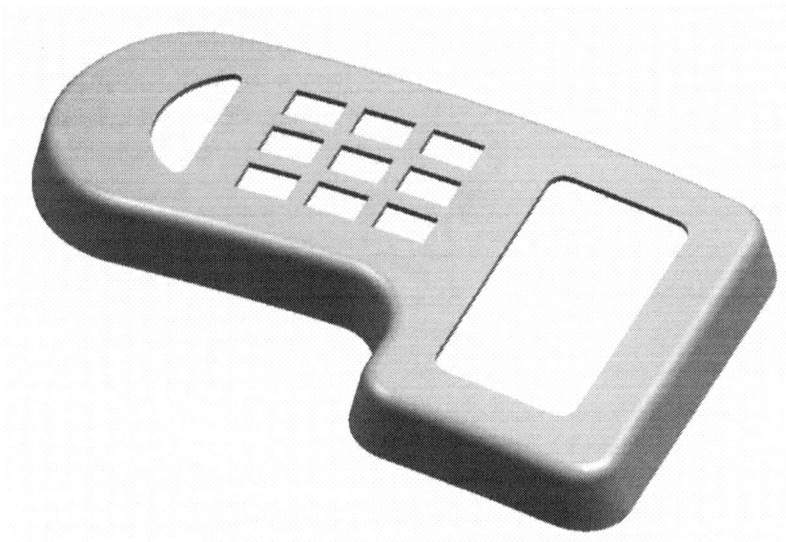

The Part Navigator (shown below) illustrates that the model is built from sketches, datum planes, extrudes, drafts, intersects and other features.

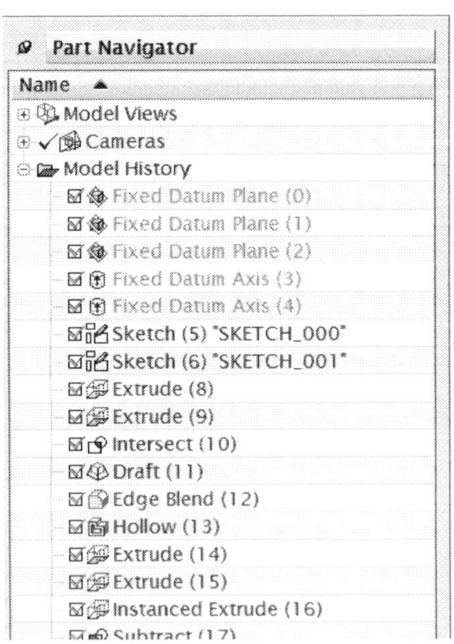

- To begin the project, create the three default datum planes and begin sketching on the X-Z plane. Create the sketch that is depicted in the figure below and center it.

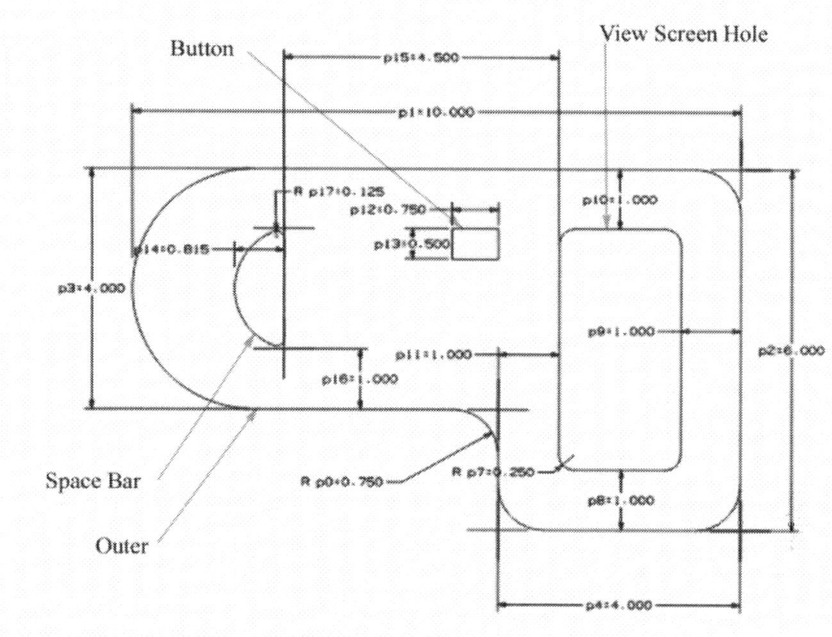


- Next, create the sketch depicted in the figure below on the X-Y plane.

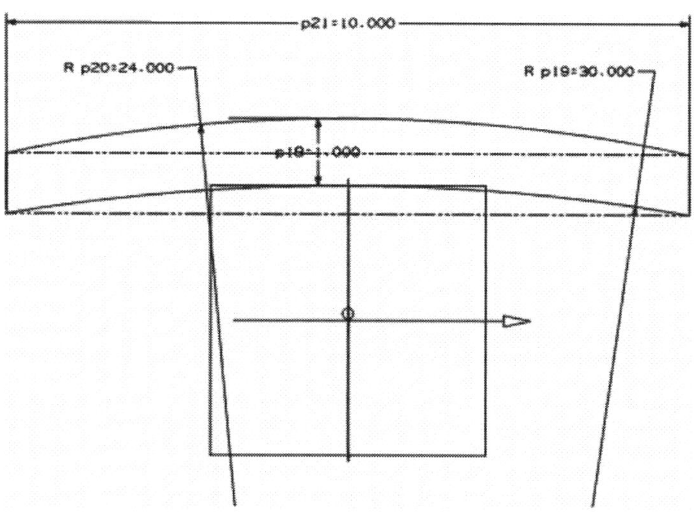


- Create an intersecting solid by extruding the two sketches as shown in the figure below. When you extrude the top view sketch, only use the outside contour. Do not include the button hole, the view screen or the space bar hole.

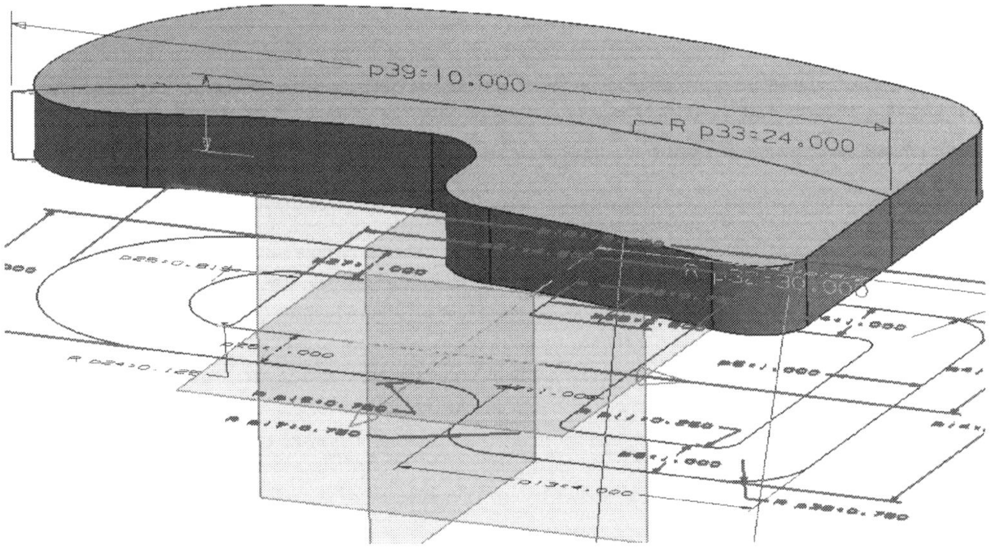


- Now you may add a draft to the solid from the bottom edge. The draft should be 12 degrees tilted in towards the Z-axis as shown in the figure below.

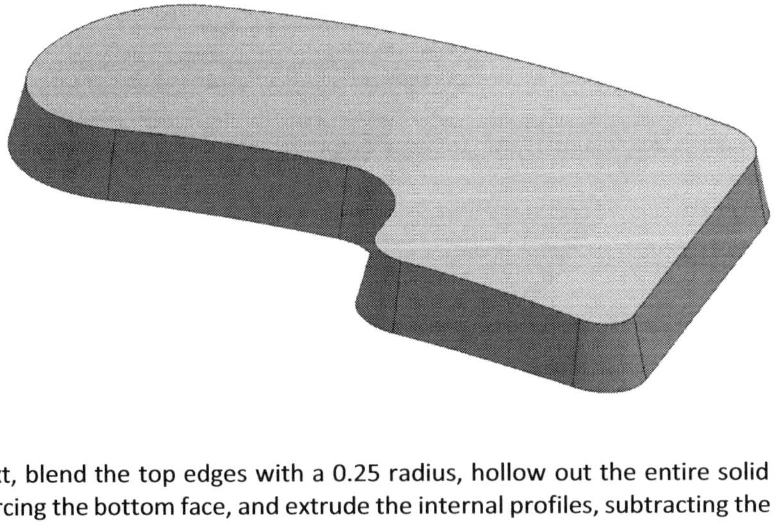

- Next, blend the top edges with a 0.25 radius, hollow out the entire solid (0.1), piercing the bottom face, and extrude the internal profiles, subtracting the holes as shown below.

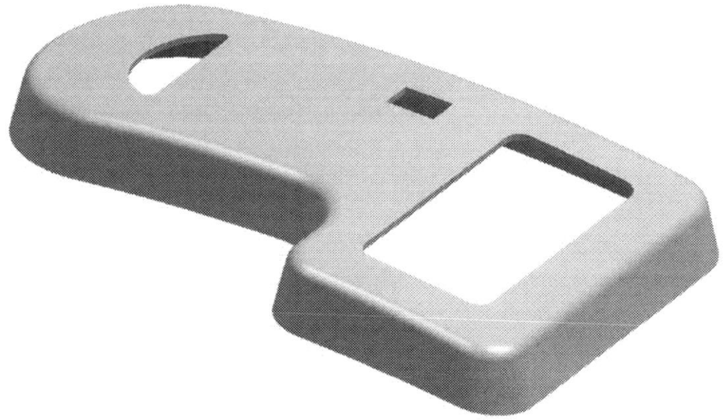

- Instance the button hole feature. The way the WCS is situated in the diagram above, you will need "Number in X = 3, X offset = -1, Number in Y = 3, Y offset = -0.75." (Note: The Instance function only provides X and Y values. If you need to change your WCS orientation temporarily to accommodate this command, go to ***Format / WCS***.)
- Compare your model with the figure below.

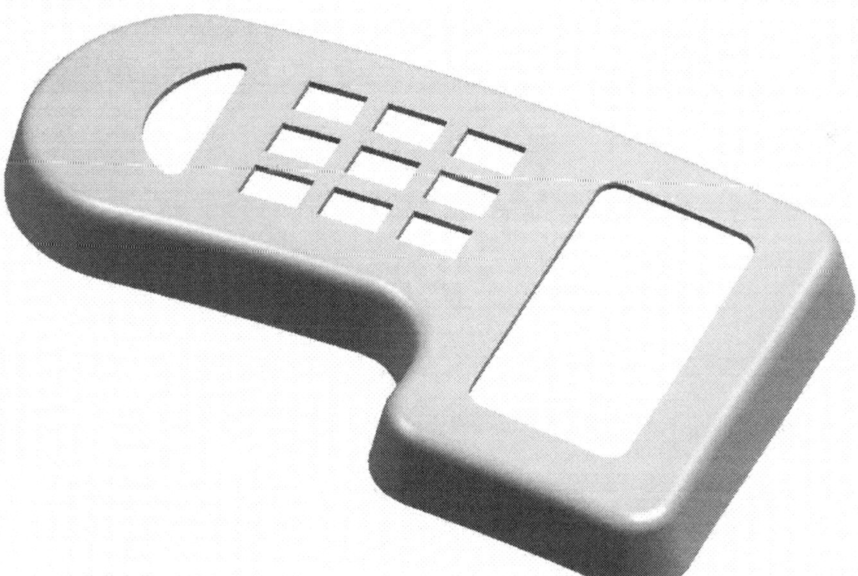

***End of Project***

## 22.4 Advanced Face Blend Features

The Face Blend feature is one of the most powerful tools available to perform difficult blends. In many situations where an Edge Blend will not work, a Face Blend will. And you can control Face Blend features in more ways. In the following example, you will apply several Face Blend features in various modeling configurations.

- Begin by creating the sketch shown in the figure below.

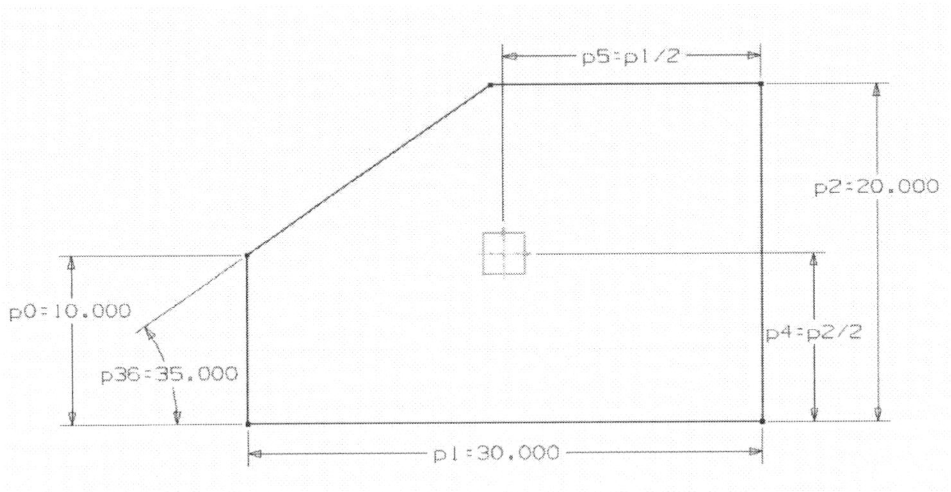


- Now extrude it from 0 to 40, and place a datum plane in the center as shown in the figure below.

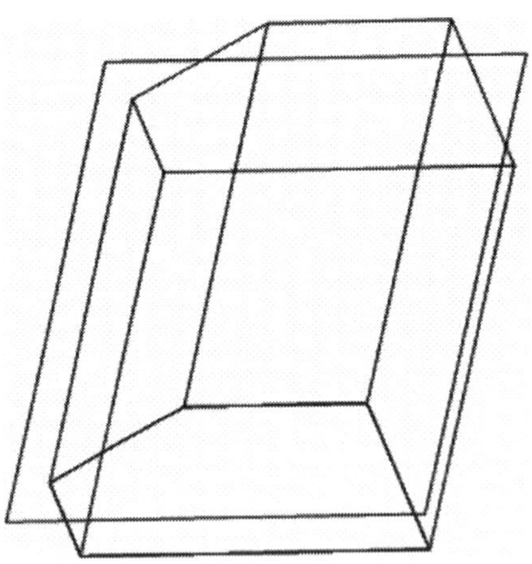

- Create a sketch on the center datum plane. Use the dimensions shown in the figure below.

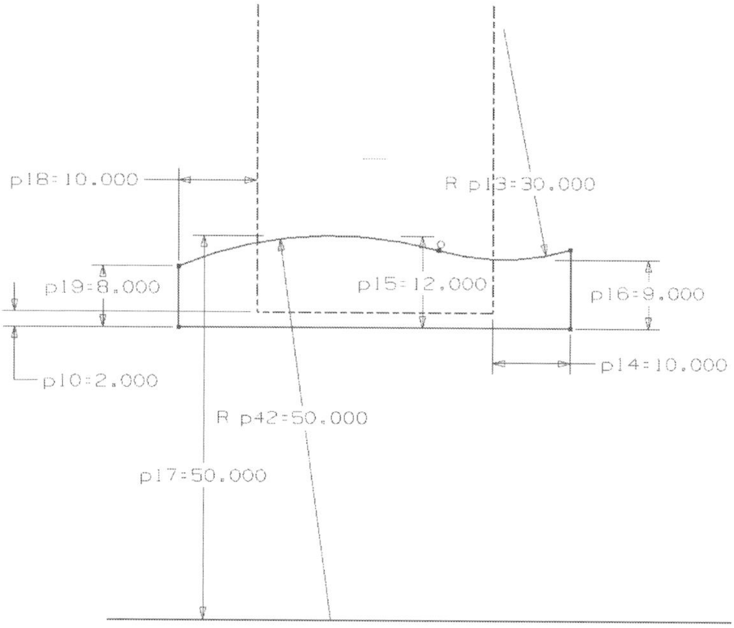


- Revolve the finished sketch about the centerline that is 50 units below the crest of the first arc. Revolve it from –30° to 30° to form the model shown in the figure below.

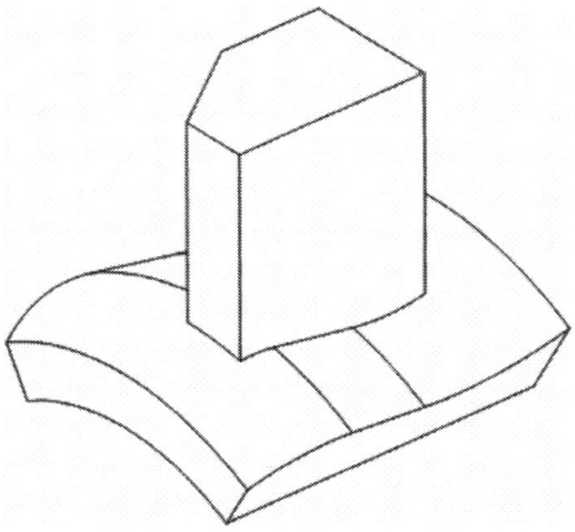

- Add 6 unit blends to the vertical corners of the extrusion as shown in the figure below.

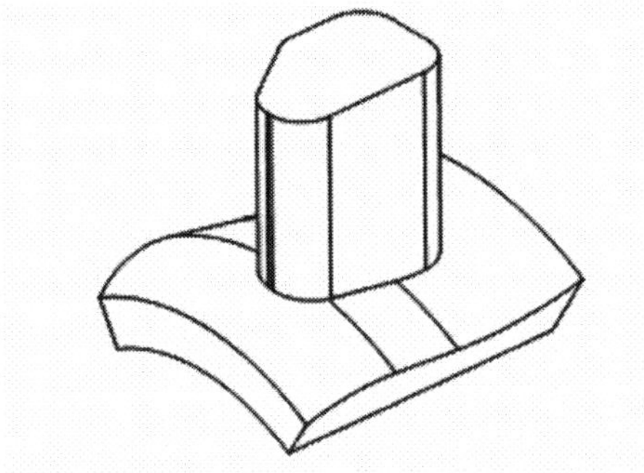

- The challenge is to create a face blend that begins at the point shown and whose radius follows the graph in the figure below. Notice how the graph begins at 10, then decreases to 1, and finally increases back to 10. This is a required condition if the blend is to go completely around the feature. You can accomplish this by creating the sketch shown below and by using the Law Controlled radius instead of a constant radius to control the blend. Select 'Define Law' and choose 'By Law Curve.' Follow the cues and select the Spine, the Law Curve, and the base line as shown in the figure below.

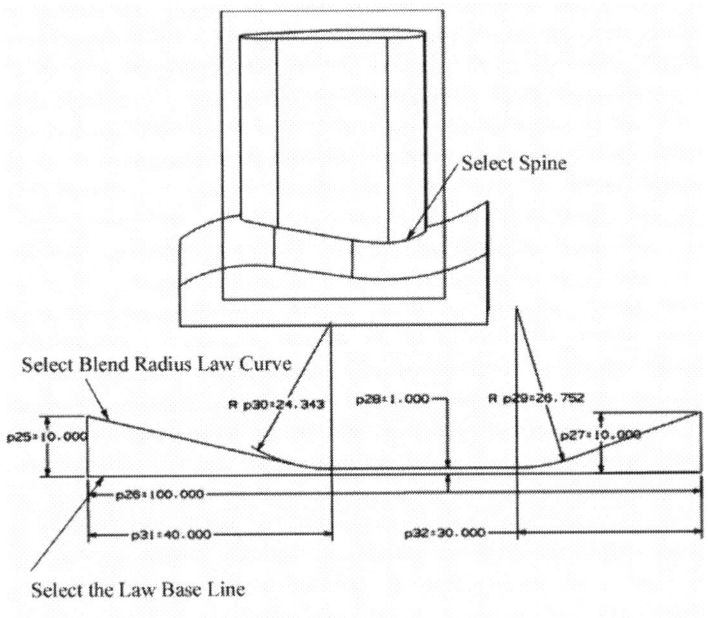


- Create the face to Face Blend.

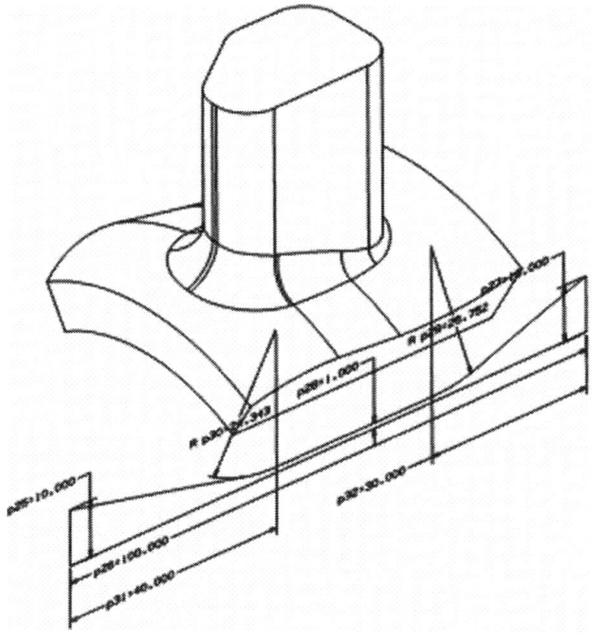


- What happens as you edit the sketch? Change the starting and ending values to 15 and the mid value to 5. You will automatically get a new blend. The starting value and ending value must always be the same.

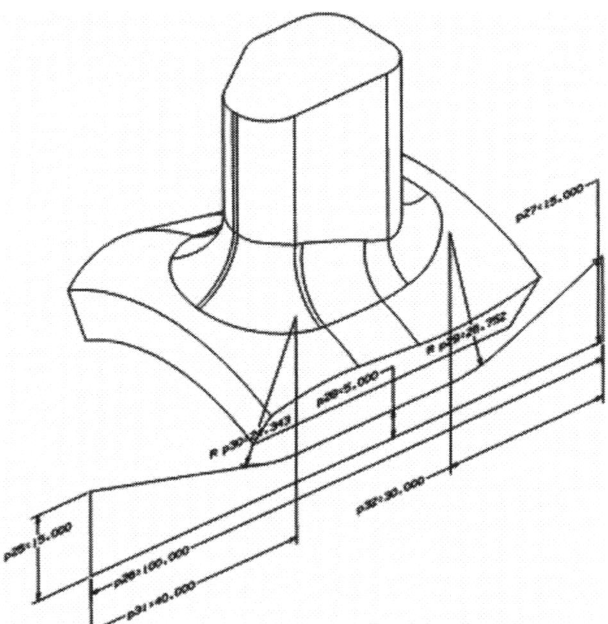


***End of Project***

## 22.5 Face Blending - Tangency Controlled

The Face Blend feature can be used with the tangency-controlled method as shown below.

- Begin by building the model shown in the figure below.

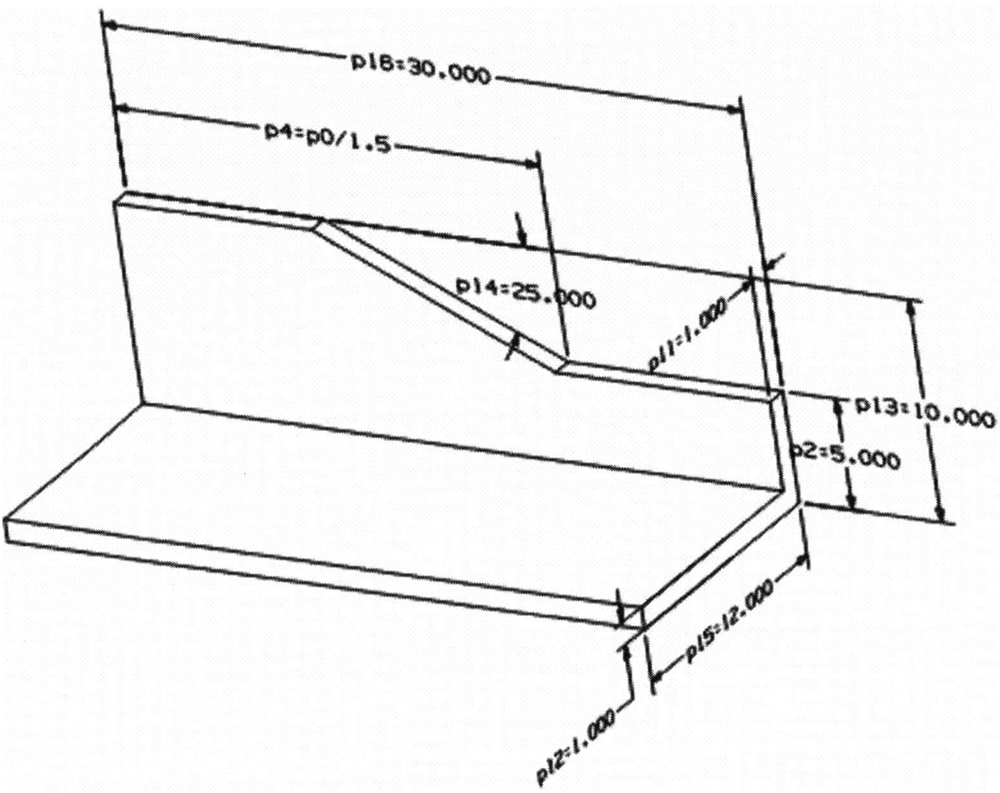


- Place an R=10 blend on the model that follows the downward slope of the edge to control tangency.

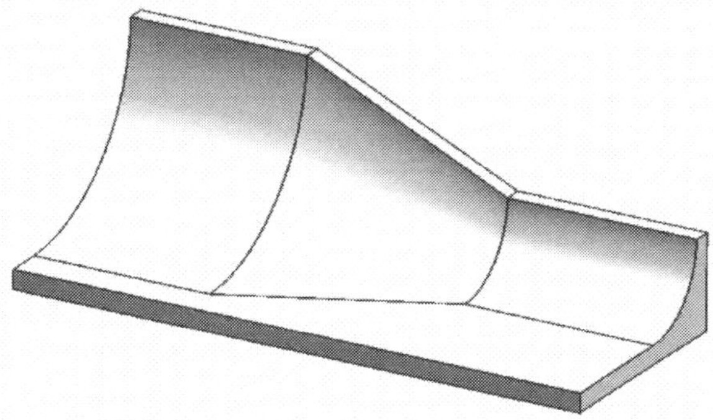

- Create a spline on the face of the model as shown and create a tangency-controlled blend as shown in the figure below.

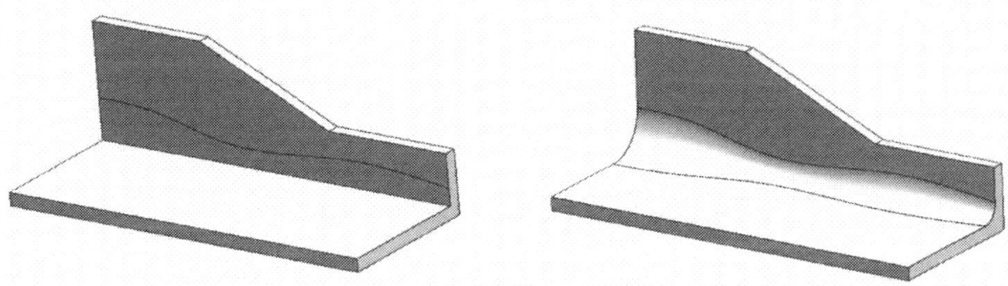

- Use the spline to trim the solid and place a cliff edge blend (radius 10) onto the face.

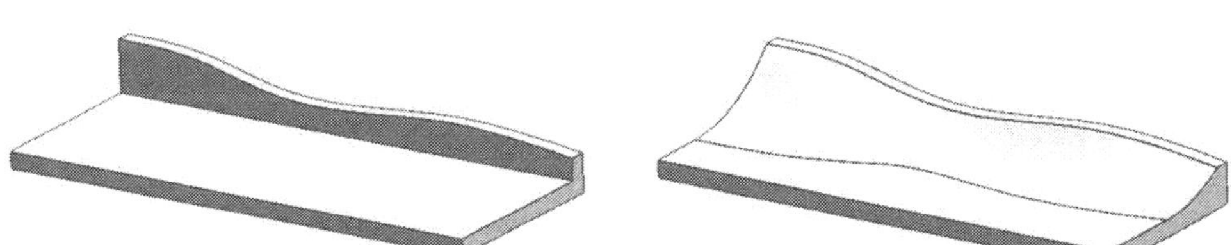

***End of Project***

## 22.6 A Bridge Surface using the Wave Geometry Linker

The Bridge Surface and Bridge Curve functions are very powerful in situations where you must blend geometries into each other. The functions sense the vector orientations of existing geometry and create smooth surfaces and curves. In this project, you will create four Bridge Curves, seal the open ends with the bounded plane command, and sew all sheets together to form a solid.

With the WAVE Geometry Linker, it is easy to link geometry from one component file into another component file. For example, you may design a bracket or transitional component to update when its interfacing components change their position.

Using the two techniques in concert produces a fabulous ability to place your design intent into a model so it automatically performs the design iterations.

- To proceed, create the sketch shown below and extrude it 5in. Be sure to pay attention to how the component is positioned relative to the global coordinate system. Use the specified number of curves shown in the figure (four – one semicircular arc and three straight lines.)

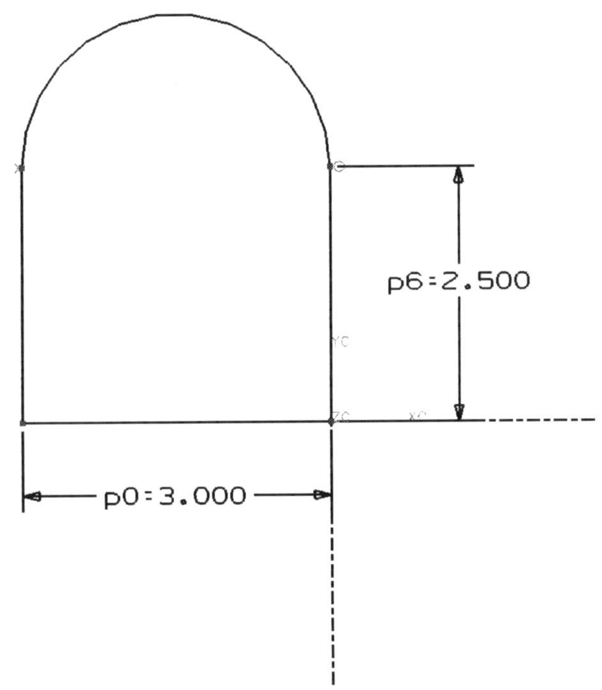


- Next, in a new part file, create the component shown below whose dimensions are shown in the next few diagrams. *Once again, be sure to use only the necessary curves – four total – two semicircular arcs and two straight lines.*

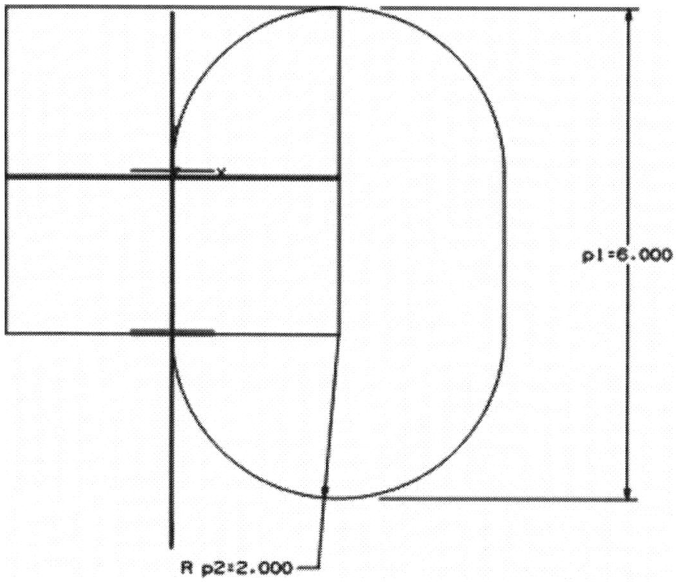


Notice the position of the global coordinate system as shown below.

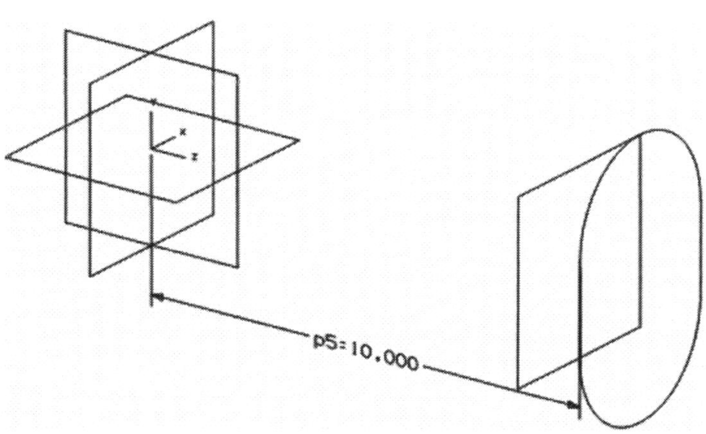


- The model is generated by creating a datum plane that is 10 inches away from the x-y plane, extruding it from 0 to 5, and hollowing it to 0.25, as shown below.

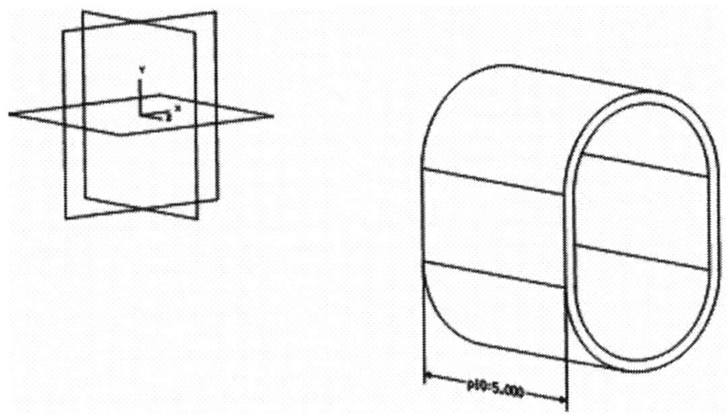

- Assemble the two components into one master assembly. Bring them in at the positions they were created using absolute coordinates as shown below.

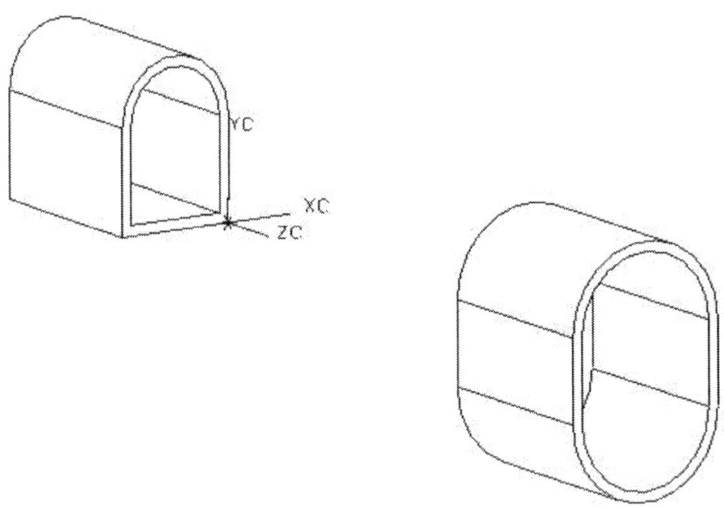


- Create and save a new component composed of both parts. Make this new assembly your work part.
- Now select the WAVE Geometry Linker (located at ***Insert / Associative Copy***) to associate the faces of one part with the corresponding faces on the other part. Select one side face on either part and the corresponding side face on the other part and select apply. *Do this for all four faces. Note: Be sure to associate outside faces with outside faces.*

- Create two bridge surfaces (Located at **Insert / Detail Feature / Bridge**) that link the two bodies together as shown in the figure below. Note: Be sure the corresponding bridge vectors on each part are pointing in the same direction.

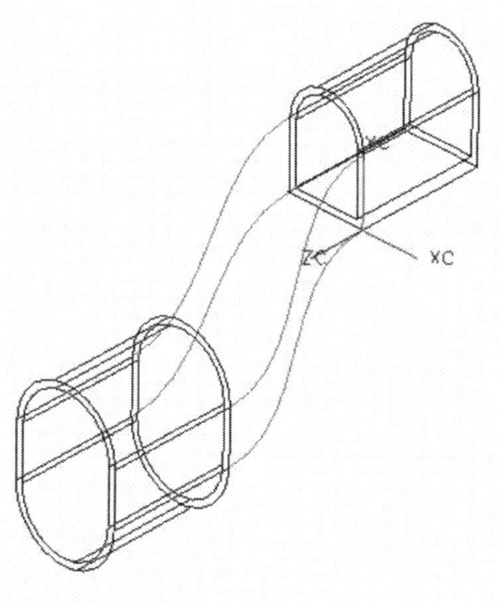


- Next, create the top and bottom bridge surfaces as shown below.
- Create two bounded plane surfaces to close both ends of the newly created sheets.
- Sew the six adjacent sheets together to form a solid interior and hollow it with wall thickness of 0.25 to match the thickness of the two original parts.

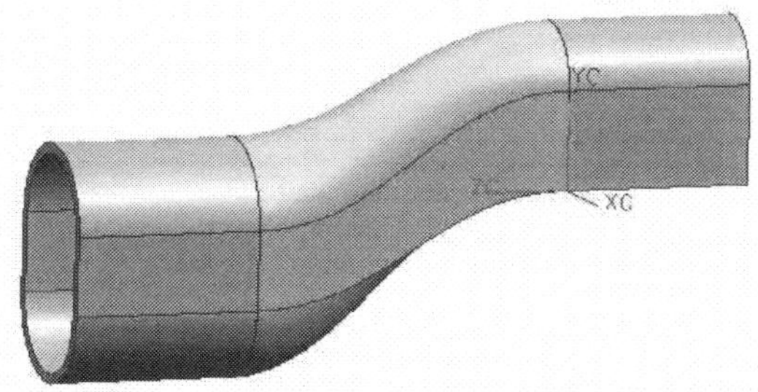


***End of Project***

## 22.7 Applying Shrinkage with Non-Uniform Scale

For those who make large castings and are involved with manufacturing processes that have shrinkages associated with them, the non-uniform shrink capability in NX 8.5 is especially interesting. The idea is this: in order to end up with accurate geometry, you must anticipate the shrinkage that the manufacturing process will produce. Then, you may place reverse distortions on the geometry. The following example shows this technique.

- Begin by building a hemisphere. You may do this using a sketch or a sphere primitive with a trim on the bottom along the XC-ZC plane. Make the radius **30in**. The result is shown below with grid lines for clarity.

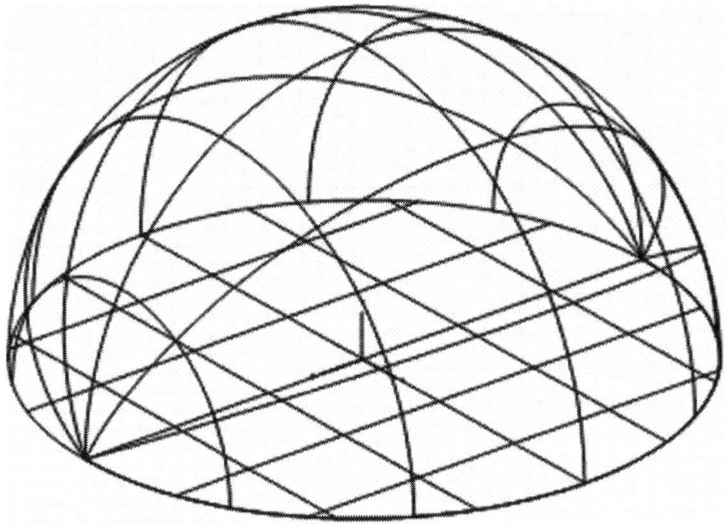

- The next step is to use the **Scale Body** function. You will find it in the ***Offset / Scale*** menu. Select **General** as your selection type.

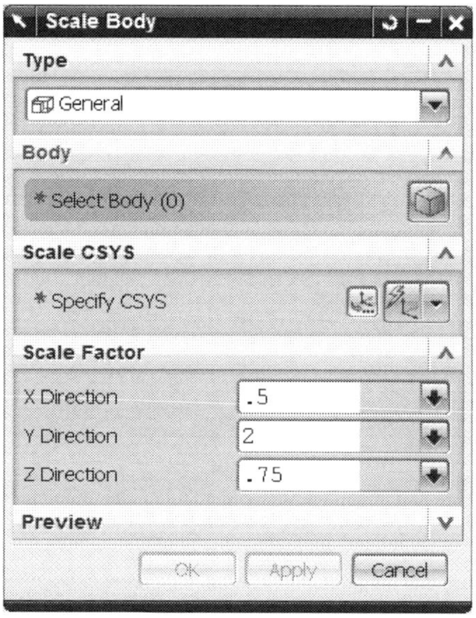


- Enter a scale of **0.5** in X direction, a scale of **2** in Y direction, and a scale of **0.75** in the Z direction, as shown below.

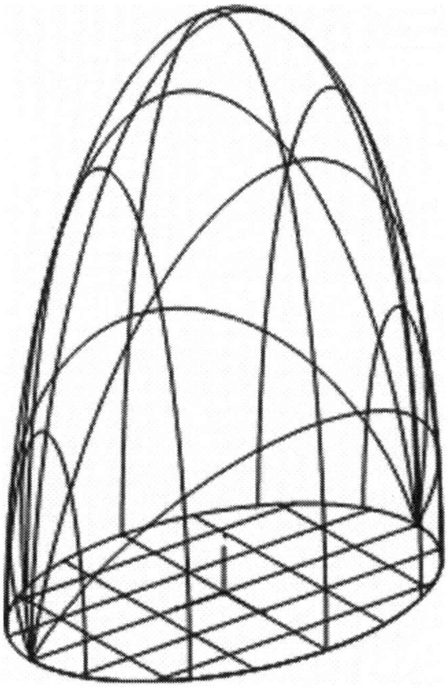

***End of Project***

## 22.8 Coil Springs

For those who make mechanisms of all types, there are occasions where a specialized spring is needed. In the following exercises, you will create a few different types of springs. You will create a simple garden-variety spring, a conical spring, and a spring where the last coil of each end is compressed. Knowledge of law curves is the easiest way to guarantee that you can model even the most difficult of springs. If the spring is easy, you can use the helix option. In ***Insert / Curve***, you can enter the appropriate dimensions, pitch, radius etc. and produce a spline. By using that spline to create the necessary wire diameter, you are good to go! However, with springs that are more difficult, you must use the more rigorous curve by equation method.

Create a right-handed coil spring with a 20 inch length, a pitch of 2, a radius of 2, and an 0.5 wire diameter.

- Begin by entering the following expressions: L=20 inches, p=2 (constant), r=1.5 inches, and wd=0.5 inches.
- Select Insert / Curves / Helix.

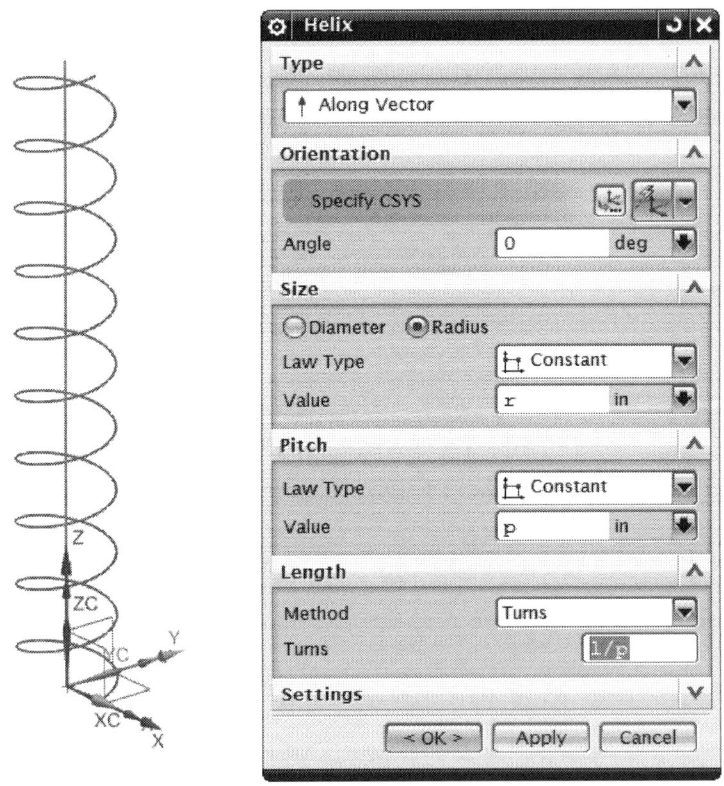


- Enter in the following: Number of Turns = L/p, Pitch = p, Radius = r.
- Choose **Apply**.

- Select Insert / Sweep / Tube.
- Enter Outer Diameter = wd and Inner Diameter = 0.

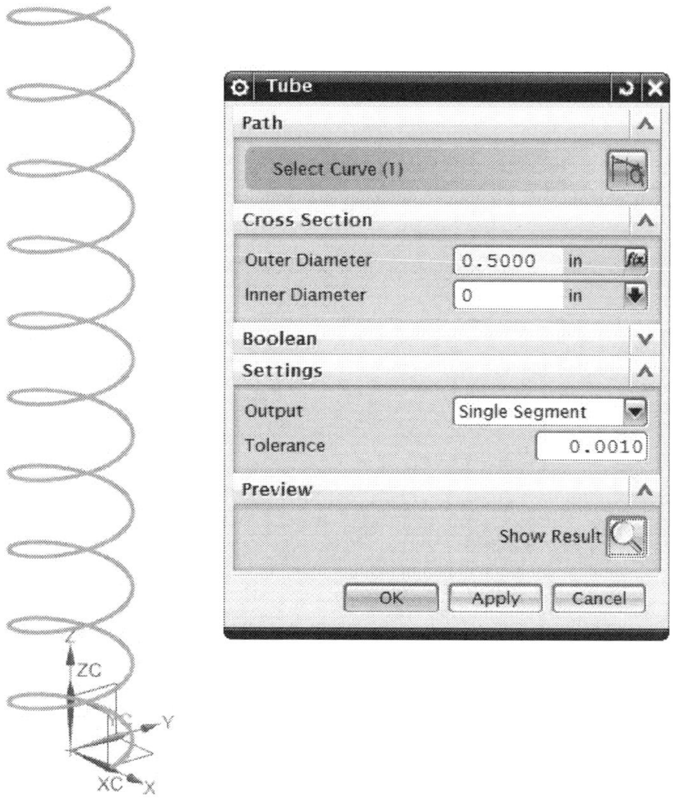


- Select the helix as the Guide String and choose OK.

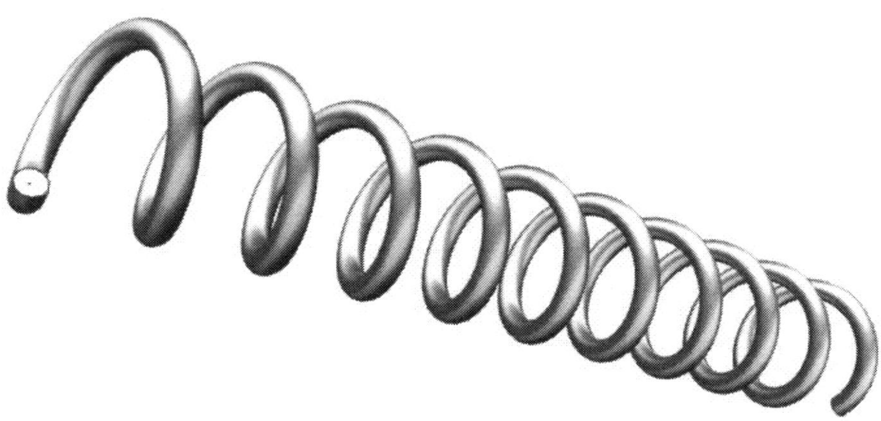

In order to do a more sophisticated spring, we may assign some kind of sophisticated shape to the radius. We can do this in a number of ways. In the next example, we will create a sketch that emulates its profile. Are you familiar with the baby buggies that they used to have in the fifties? The springs on those were always tapered on the ends with a fat middle. The figure below represents the radius as a function of its centerline length.

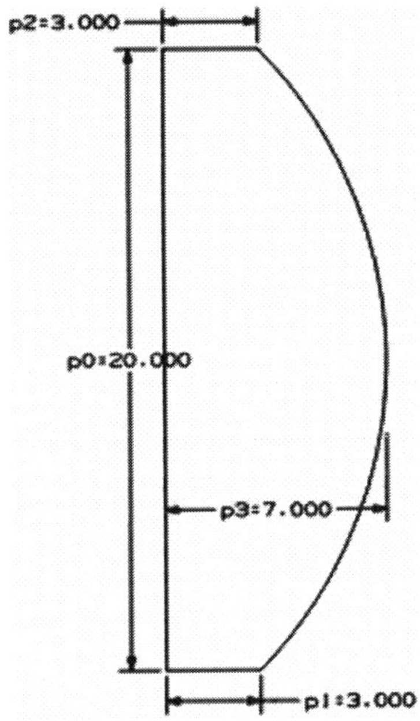


- Notice that the overall height is represented by p0.
- Create expressions that emulate the pitch (ph=1.5) and the wire diameter (wd=.5).
- Find the helix function and enter p0/ph for the number of turns.
- Enter ph for pitch.
- Select the 'Use Law' option for the radius. The following menu will appear.

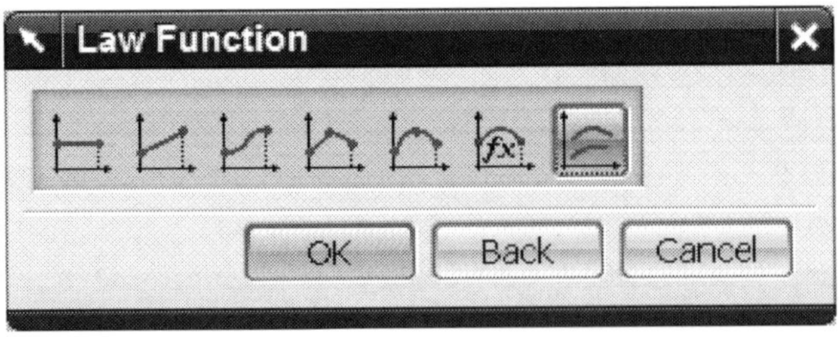


- Select the 'By Law Curve' option in the menu. Then select "curve" so that you can select only a portion of the sketch without having the entire sketch highlight.

- Select the bottom of the arc. Selecting on the bottom tells NX 8 that you want the initial radius to be equal to the distance between the arc and the vertical line where the arc starts.
- Choose OK and you will see an arrow that indicates the direction that NX 8 will march along the arc looking for radius values.
- Select OK again; this signifies that you are finished specifying the curves that NX 8 will determine values from.
- NX 8 will now prompt you to select the law base line. Select the bottom of the vertical line. You will get a second arrow. Both arrows should be pointing up.
- Rotate the WCS so that the ZC-axis is pointing straight up and Choose OK. You will get the following geometry.

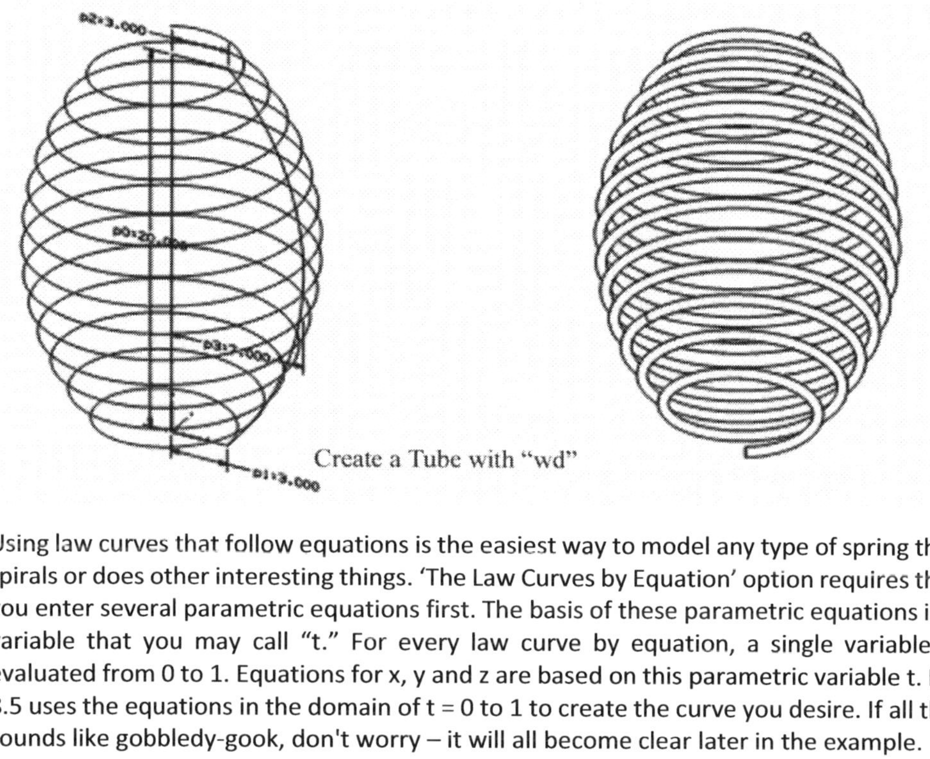

Create a Tube with "wd"

Using law curves that follow equations is the easiest way to model any type of spring that spirals or does other interesting things. 'The Law Curves by Equation' option requires that you enter several parametric equations first. The basis of these parametric equations is a variable that you may call "t." For every law curve by equation, a single variable is evaluated from 0 to 1. Equations for x, y and z are based on this parametric variable t. NX 8.5 uses the equations in the domain of t = 0 to 1 to create the curve you desire. If all this sounds like gobbledy-gook, don't worry – it will all become clear later in the example.

- Enter the following equations into the Expressions Editor:

t=1
radius=20+t*30
n=10
xt=radius*sin(t*n*360)
yt=0
zt=radius*cos(t*n*360)

- Now you are ready to create the law curve by equation. Select ***Insert / Curve / Law curve.***
- Select Law curve / By Equation.

- The system will prompt you to "Define X – Enter parameter expression." It will display the 'Law Curve' dialog box, as shown below. The following sequence allows you to use equations that you have already entered to define the law curve.

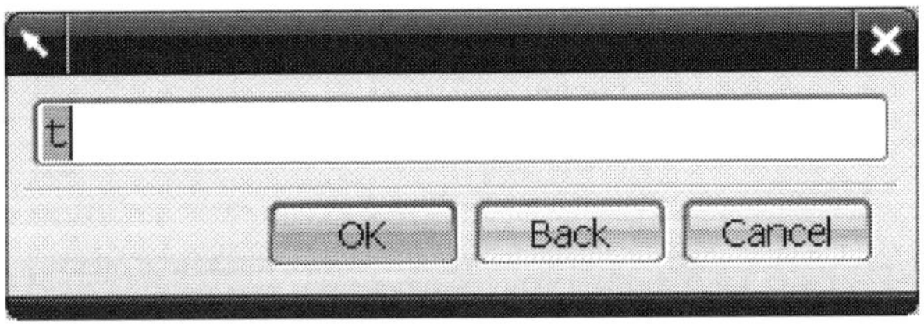


- Choose OK to accept "t."
- The system will then prompt you to "Define X – Enter function expression."
- Choose OK to accept "xt."
- Now select 'By Equation' and 'OK' to accept "t" and 'OK' to accept "yt."
- Now select 'By Equation' and 'OK' to accept "t" and 'OK' to accept "zt."
- Choose OK to finish.

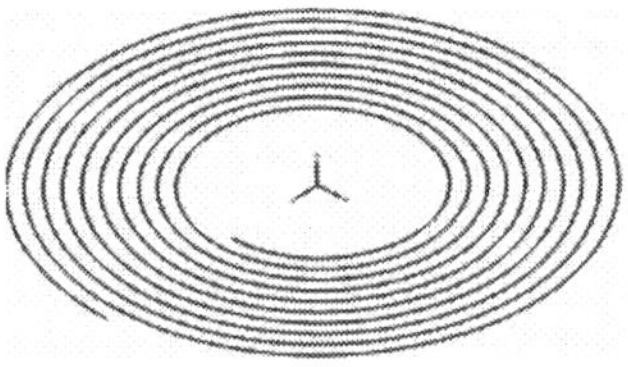

Now that you have created a flat coil by equations, imagine what would happen if you changed the equation that controls the Y axis. If yt equaled t*20 instead of zero, the spiral would begin at zero as t begins at zero, then by the time t=1, yt would equal 20.

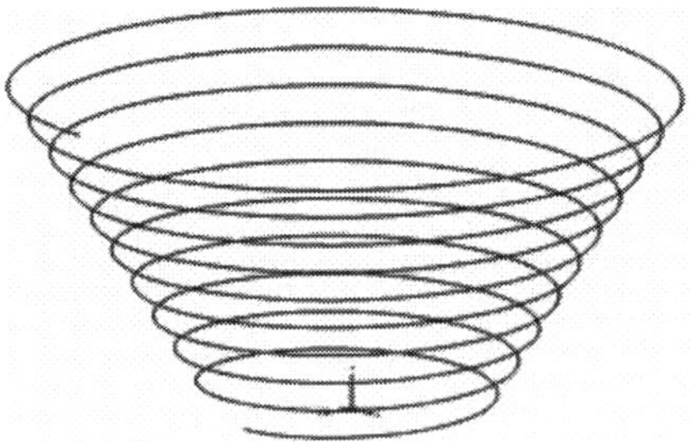

***End of Project***

## 22.9 NX 8.5 Final Project

In this project you will build five individual parts. These include TOP COVER, CIRCUIT BOARD, BOTTOM COVER, No.2 x 55 SS screw and the RUBBER FOOT. After making each part you will need to bring them into a separate assembly file to create the assembled part in the figure below.

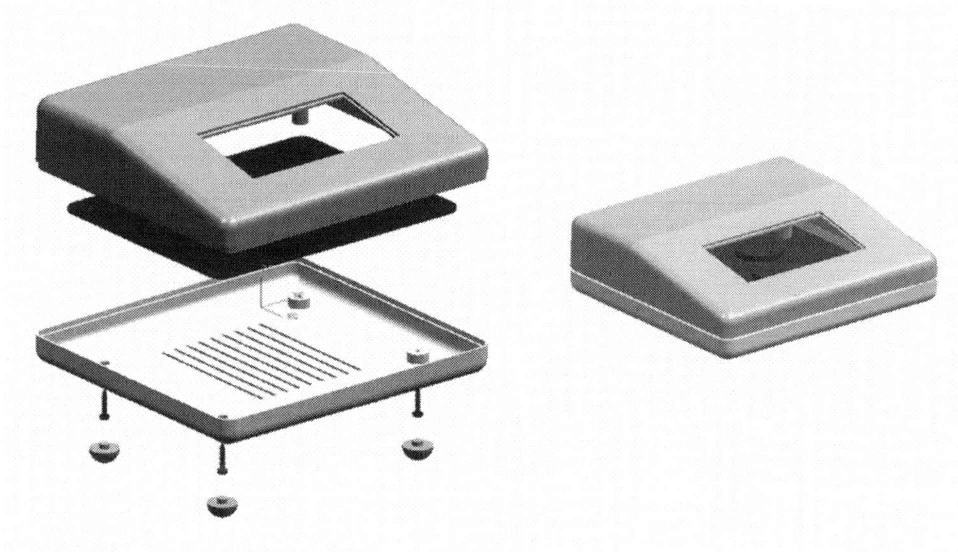

Finally, after the assembly is complete, make individual drawings for each part and one assembly drawing with an exploded view and a parts list (see the following figure). Create and use your drawing borders when creating the drawings. Remember the top cover and bottom cover pieces are made of injection molded ABS plastic; that is why a 1° taper (draft angle) is used all around the parts. The ICU placement on the circuit board is up to you as a designer.

You can use the following diagrams to create your part files:

1. ASSEMBLY
2. BOTTOM COVER
3. TOP COVER
4. CIRCUIT BOARD
5. RUBBER FOOT AND SCREW

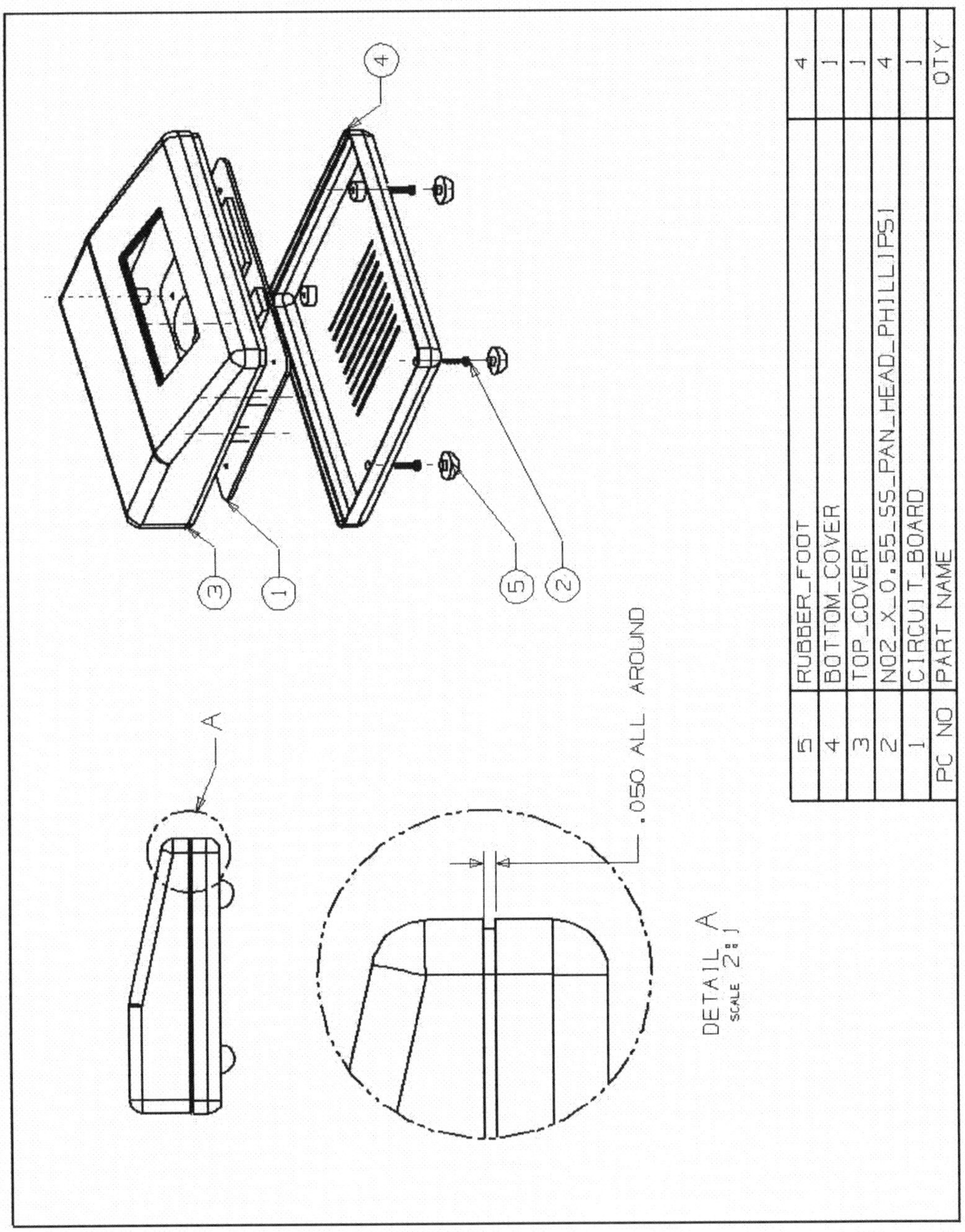
A
.050 ALL AROUND
DETAIL A
SCALE 2:1
1
2
3
4
5
PC NO | PART NAME | QTY
5 | RUBBER_FOOT | 4
4 | BOTTOM_COVER | 1
3 | TOP_COVER | 1
2 | N02_X_0.55_SS_PAN_HEAD_PHILLIPS1 | 4
1 | CIRCUIT_BOARD | 1

**(Assembly)**

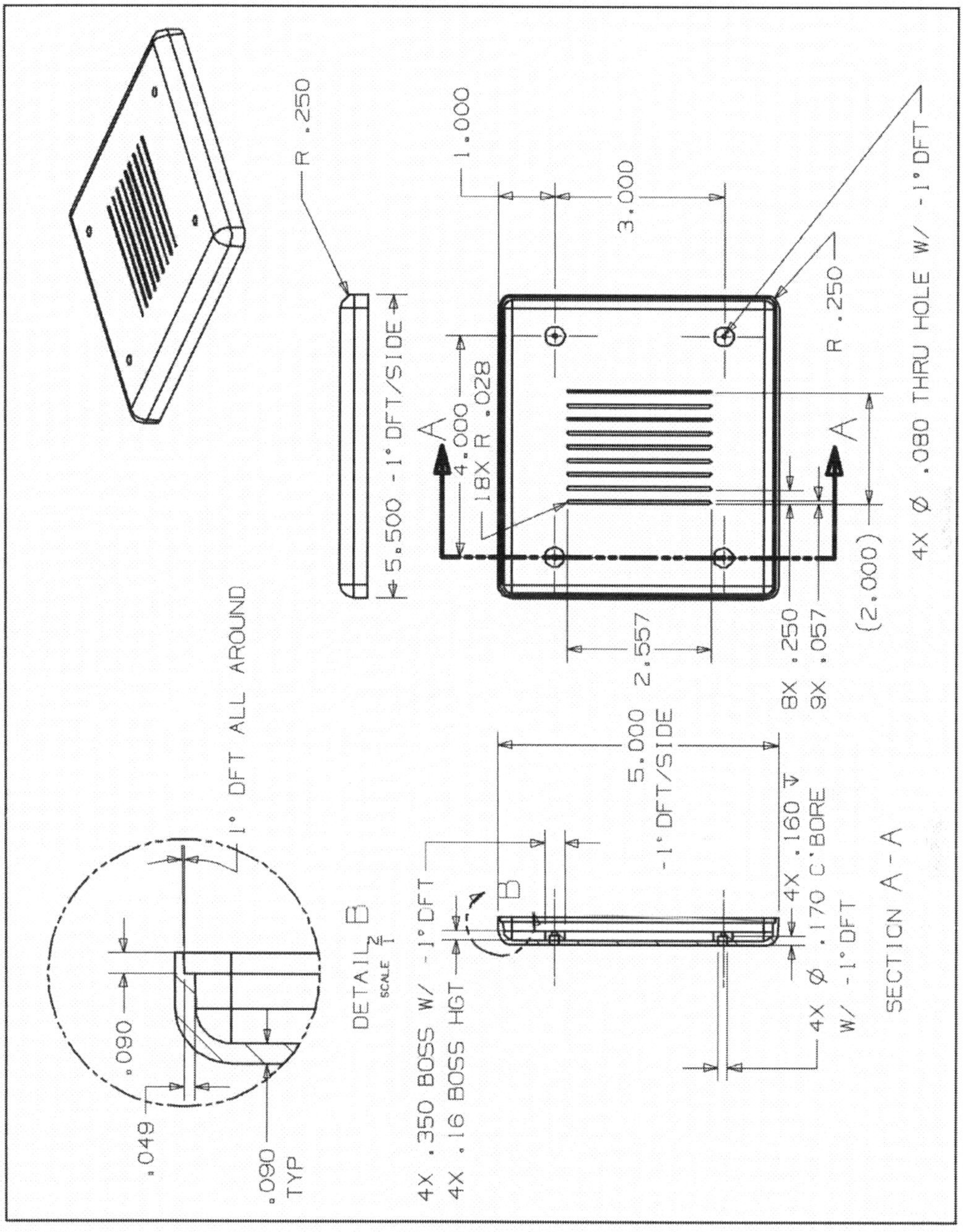


**(Bottom Cover)**

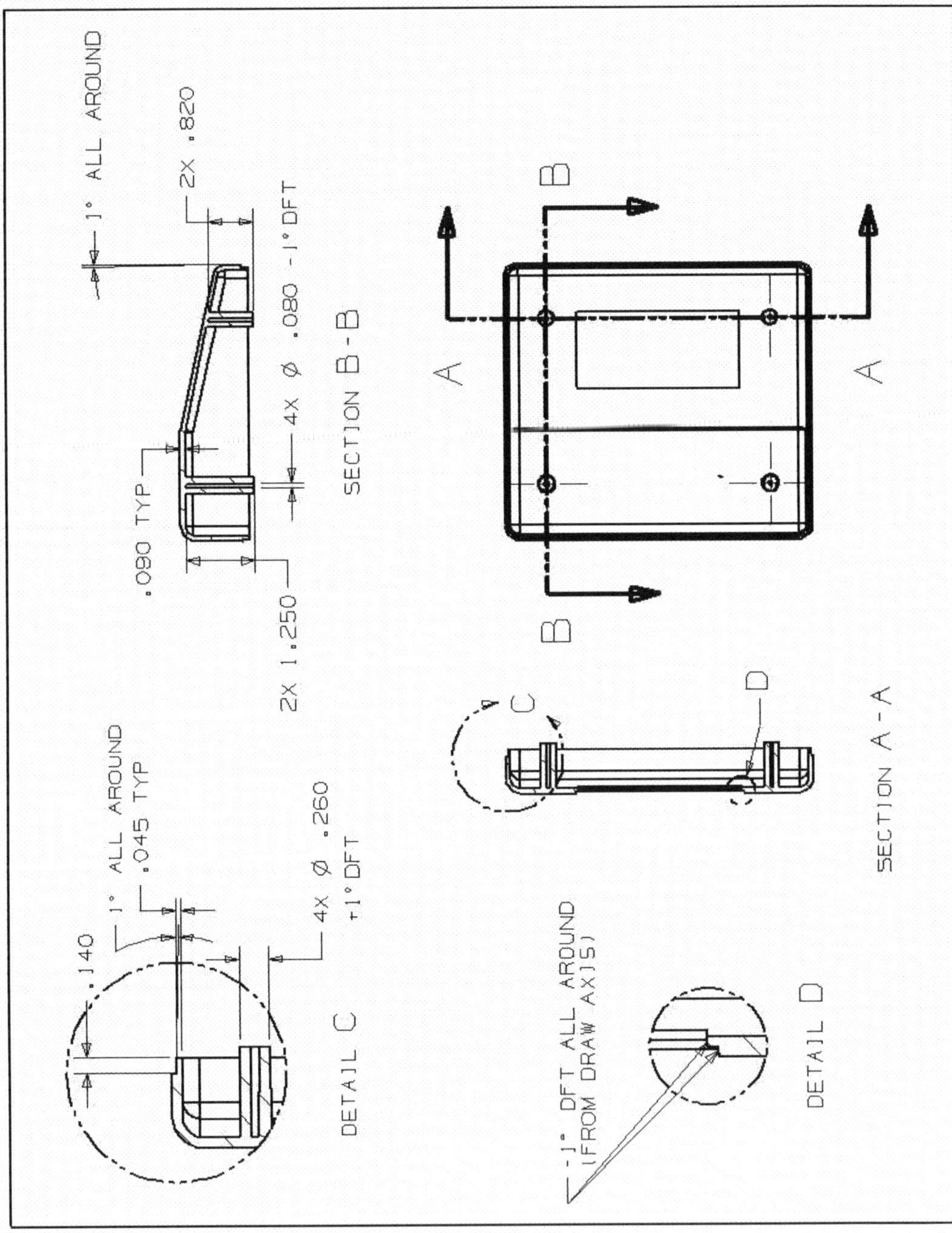


**(Top Cover Sheet 1)**

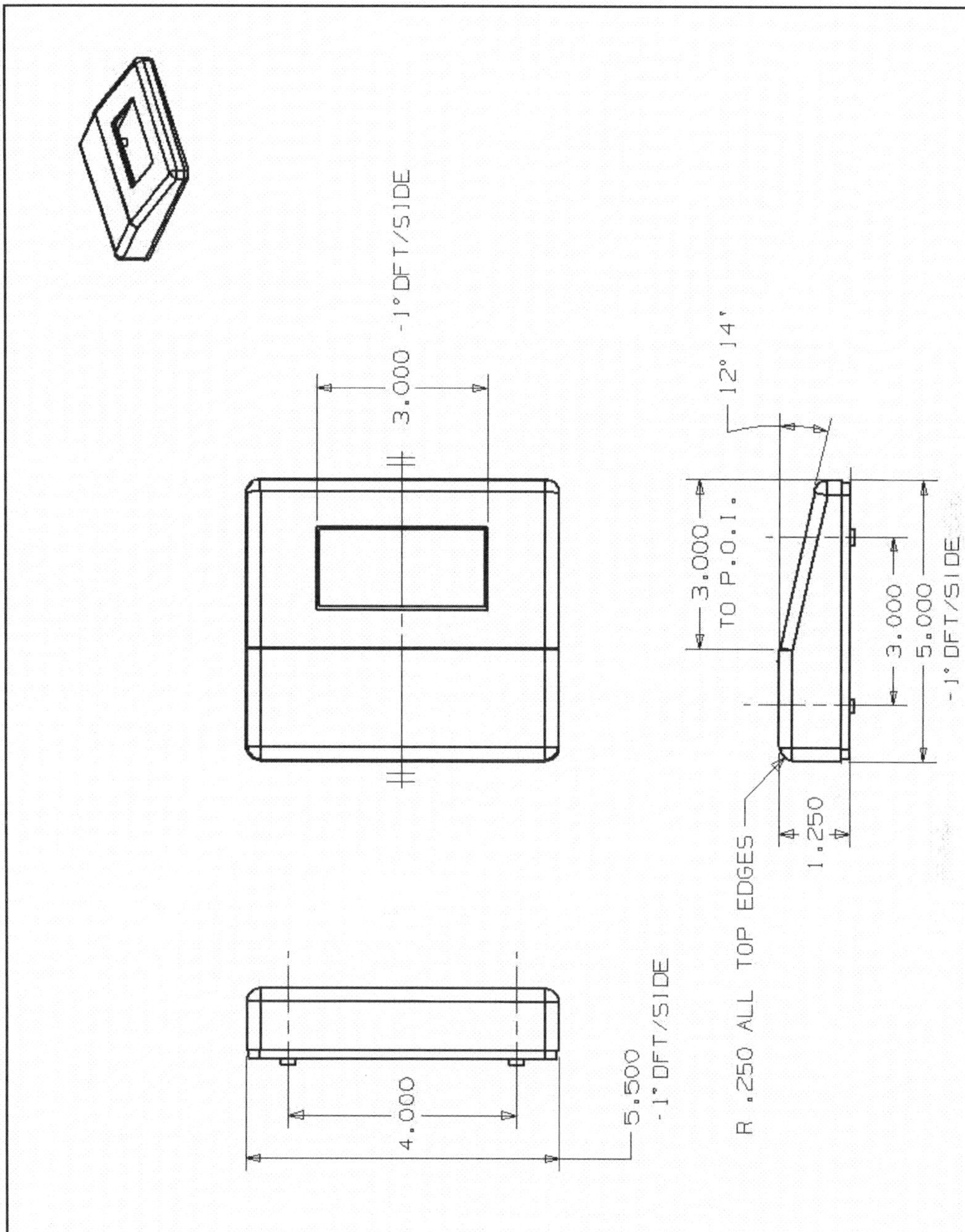


**(Top Cover Sheet 2)**

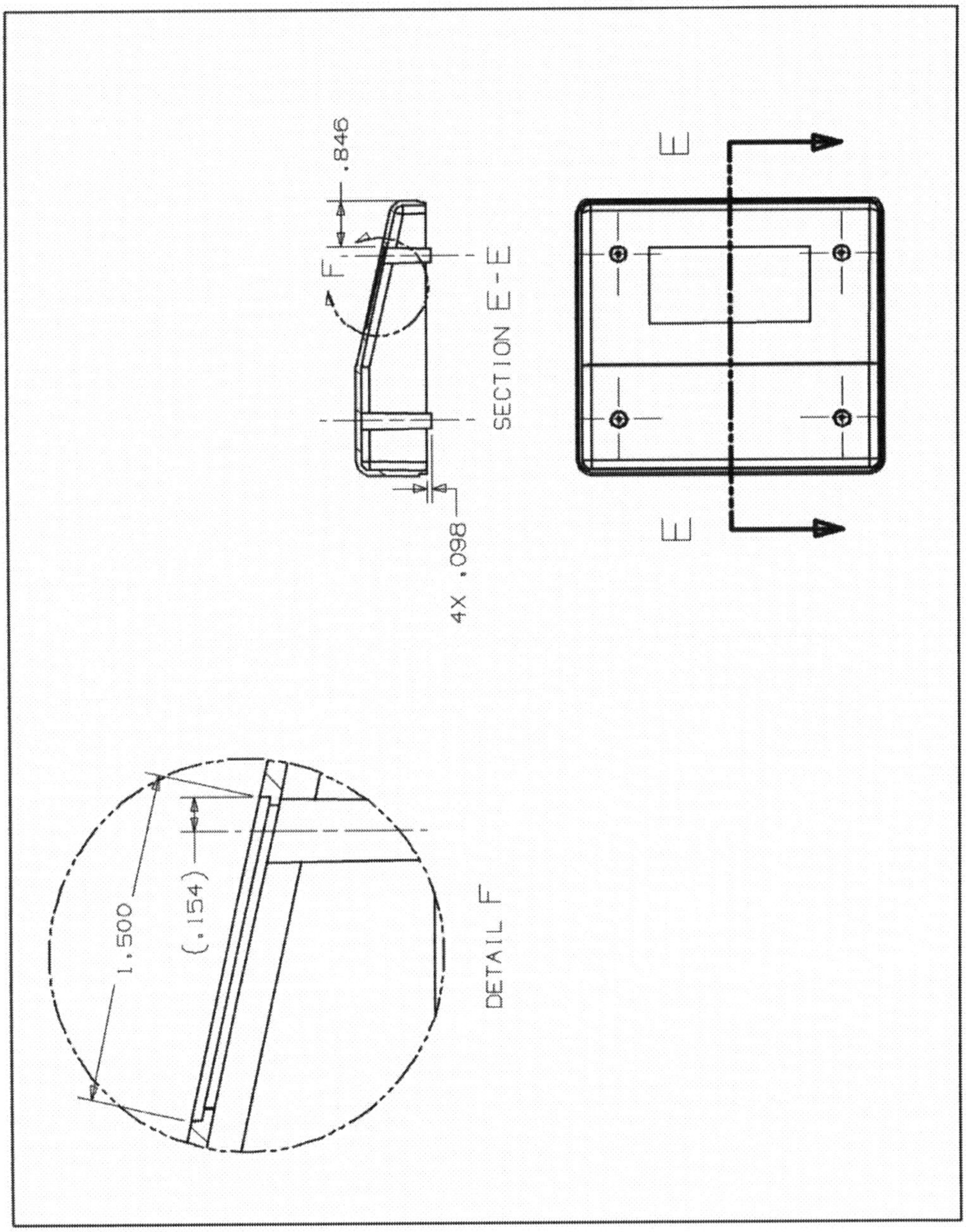


**(Top Cover Sheet 3)**

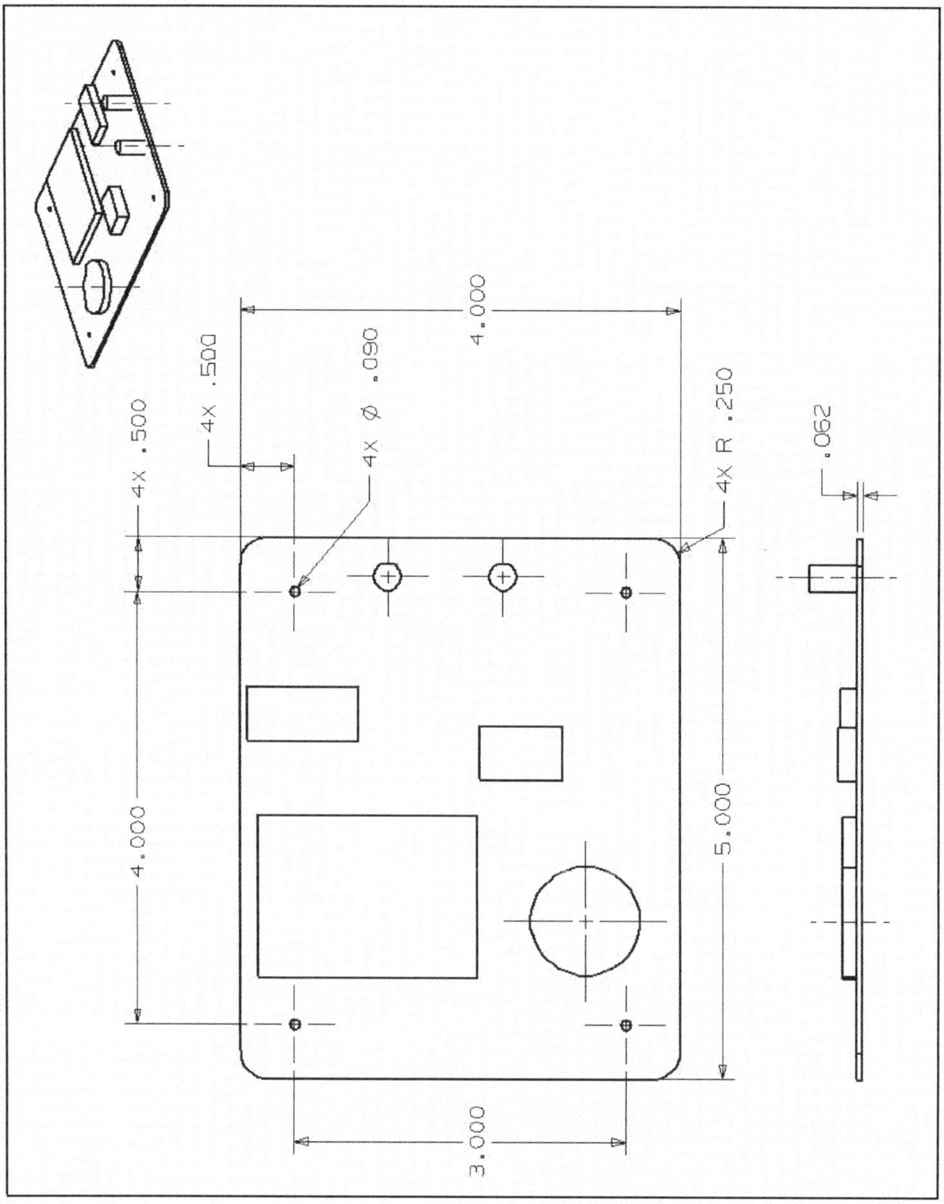


**(Circuit Board)**

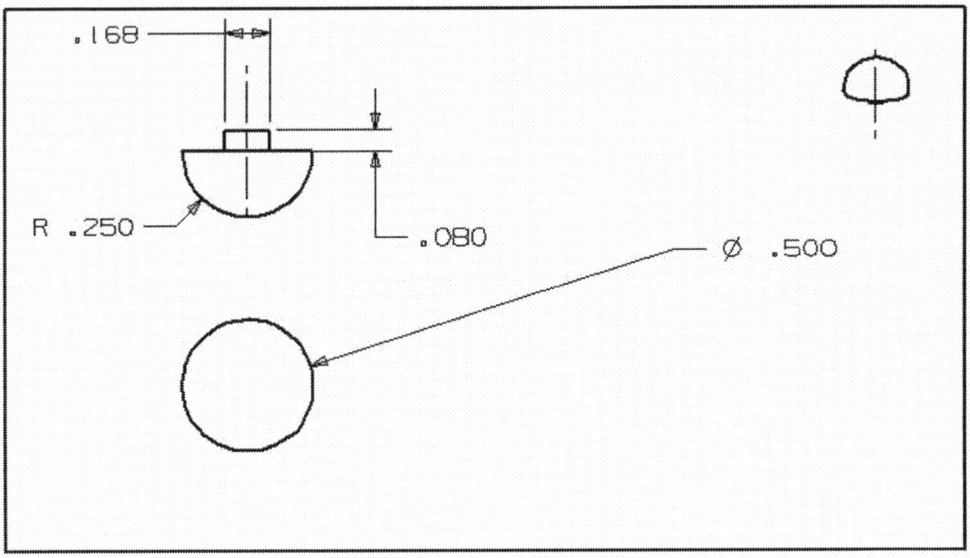


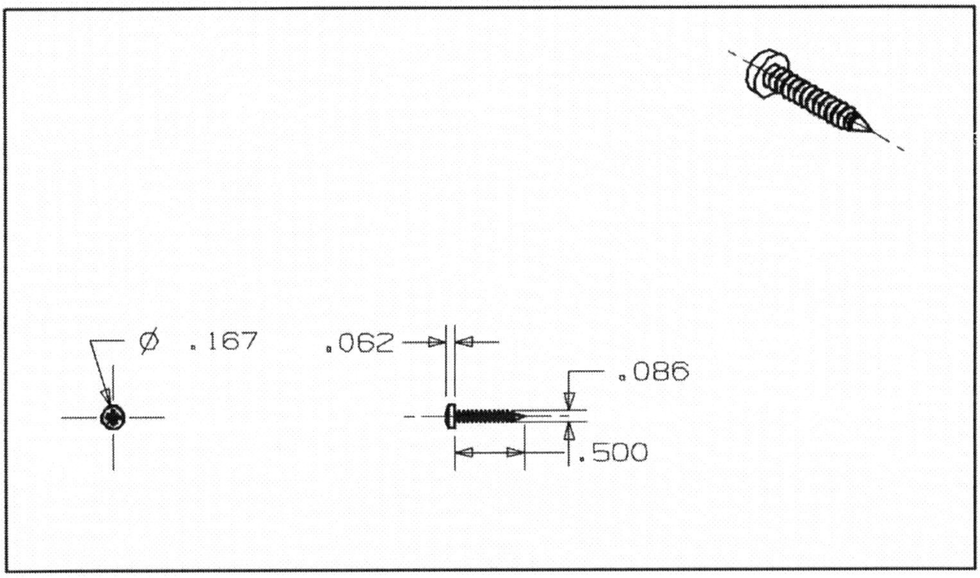


**(Rubber Foot and Screw)**

***End of Project***

# 23. Goodies

## 23.1 Introduction

Now that you've made your way through a mountain of NX 8.5 projects and verbiage, it's time for a reward. You've earned **goodies**!

At Design Visionaries, using computer aided design tools is a labor of love. We have created a few cool things in our spare time that we thought you might like to have. They are all on the web site *www.designviz.com* and nxtutorials.com. Below are a few examples.

## 23.2 Creative 3D Models

### *Detail: The Bee*

The bee is a combination of surfaces and solids. If you're wondering how we gave him his color scheme, it was not hard. We used the ***Insert / Trim / Divide Face*** command with curves that we had intersected onto his body from a series of datum planes. He is on the web site as dv_bee.zip. There you will find a Parasolid file and a part file.

## ***Detail: The DV_One Fish***

The dv_one_fish is made up of surfaces and solids. We used cool techniques like divide face and curved mesh surface. You will find him on the web site under dv_one_fish.zip.

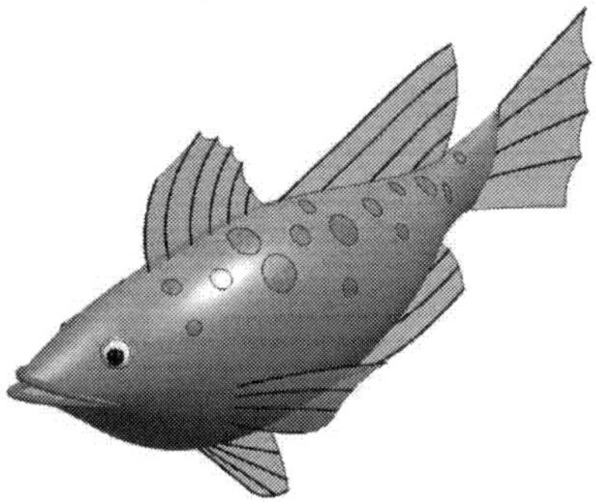

## ***Detail: DV_Fish***

These fish were created from the first fish. We sized him, reflected him, and generally placed him all over the place. He's available on the web site under dv_fishes.zip.

## ***Detail: DV_Flowers***

The flowers are a combination of surfaces and solids found on the website under flowers.zip.

## 23.3 Cool Stuff to Build

There's only one way to improve from here – build cool stuff and constantly push the envelope of what you can do. Here is a shape that you may try to create on your own. When you've given it a try, open the part file "Lantern v2.prt" and you can see how I did it.

## 23.4 Haven't Had Enough?

Visit www.designviz.com and access all this exiting content:

- Video tutorials from the authors of the Book –

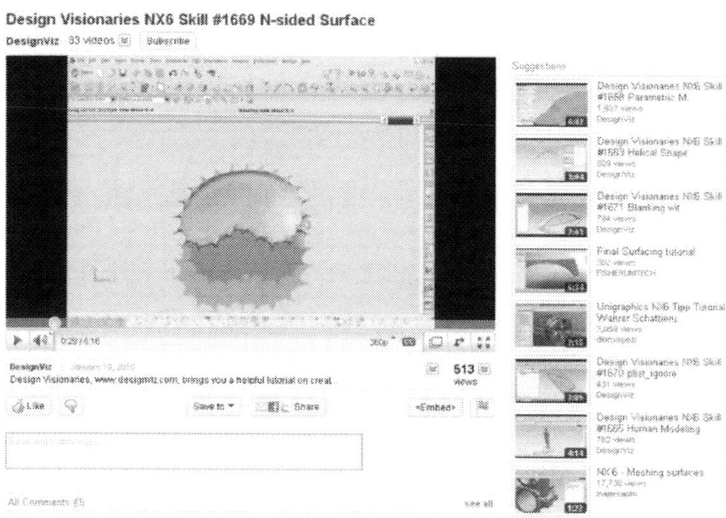


- Download even more amazing NX CAD models –

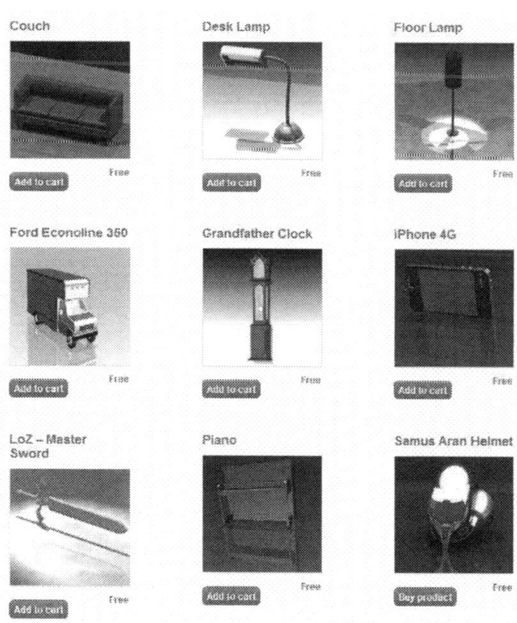

- Access to nxtutorials.com –

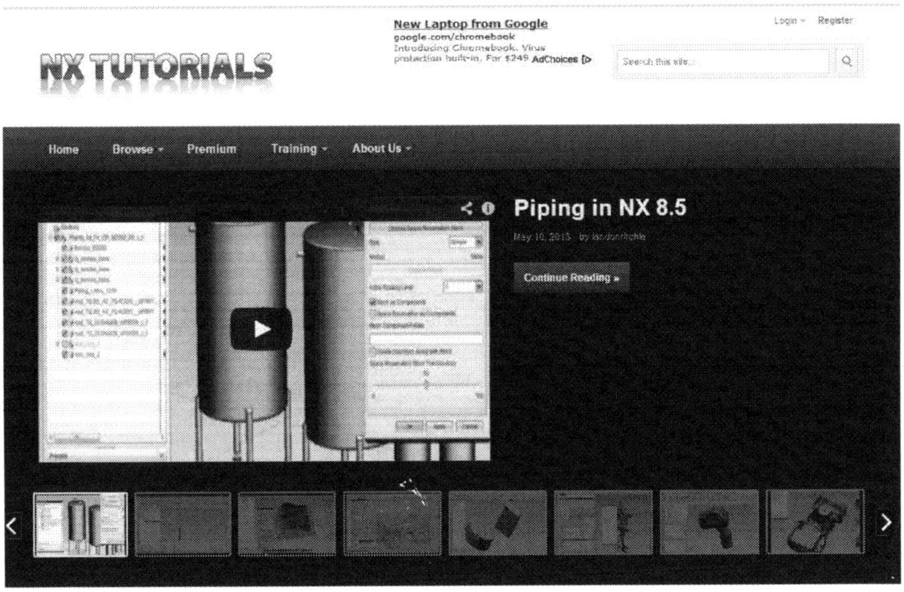


- Information on all our custom training services –

NX SOLID EDGE creo
TEAMCENTER SOLIDWORKS

# Glossary

|  |  |
|---|---|
| Absolute Coordinate System | A Cartesian coordinate system that is at the center of an infinite 3-D space where all entities reside. |
| Annotation | Text and/or symbols that you place on a drawing or solid model. |
| Annotation Editor | The annotation editor allows you to make changes to existing text entities and symbols. |
| Annotation Preferences | These govern the creation of text by predefining the characteristics of text and symbols. |
| Appended Text | Text that is added onto a dimension or note. |
| Arc | A circular curve or section of a circle. |
| Arrangements | This technique enables you to create, within one assembly, various alternate positions for components. You can move things around and save an "arrangement" for each new set of component positions. |
| Assembly | A collection of components that are arranged with geometric relationships, and, in many cases, involve mating conditions. |
| Assembly Cut | An assembly cut is a Boolean operation that can be performed on a component that is assembled into a larger assembly. The Boolean only exists in the component when it's in the assembly. Typically this technique is used when you are creating a component with some secondary machining operation. For example, two plates are welded together and subsequently drilled through with a hole. |
| Assembly Navigator | A tool that allows you to graphically view and manipulate the structure of an assembly. |

| | |
|---|---|
| Associative copy | A number of commands create geometry that references and has a parametric link to other geometry. |
| Auto Inherit | A function which allows NX to apply dimensions from a sketch onto a drawing. |
| Auto Ordinate Dimension | A function within ordinate dimensioning which automatically creates dimensions via use of a selection dialog. |
| Auto Placement | The term that describes how a dimension is automatically positioned when you create a sketch. |
| Auxiliary View | An orthographic projection that is taken at an arbitrary angle. |
| Axis Handles | Graphical interfaces (little yellow balls) that appear on a coordinate system which is used to manipulate it. |
| Basic Drawing Views | Eight standard views that are default views for NX; top view, side view etc. |
| Blanking | The temporary hiding of geometry. You can't blank an individual feature of a solid. |
| Blend | A surface created at the intersection of two surfaces. It is usually of circular cross section. |
| Block | A command used to create a solid by inputting length, width and height. |
| Boolean Operations | Boolean operations are the way you combine solid models. You can unite, subtract or intersect two or more solids. |
| Boss | A parametric cylinder that you can place on a planar surface. |
| Bridge curve | A fantastically useful function that creates a curve that flows smoothly from the end of one curve to the end of another. |

| | |
|---|---|
| Bridge surface | An extremely useful function that creates a surface that flows smoothly from the edge of one surface to the edge of another. |
| By Law Curve | Used when you are controlling the shape of entities with a curve that varies from one value to another. |
| Chamfer | An angled surface that is created at the intersection of two surfaces. |
| Circle | A planar entity that is drawn by default on the X-Y plane of the WCS. |
| Circular Instance Array | Used to make a parametric copy of features in a circular pattern. |
| Cliff Edge Blend | A blend that does not run out tangent to the surface that was used to define it. |
| Cloning | You are able to copy an entire assembly and rename each component according to naming rules that you set up using the clone command. |
| Coincident Constraint | Used to define one entity to be placed directly on top of another. |
| Collinear Line Constraints | To force two lines to be constrained by the same angle and direction. |
| Cone | A parametric primitive with angular cross-section. |
| Constrained | A sketch is fully constrained when it has no degree of freedom indicators. |
| Constrained Face | An entity that you can use to place rules or geometric relationships on surfaces. |
| Constraint | Geometric rules that you affix to various sketch entities. |

| | |
|---|---|
| Context control | Is an assembly related menu of commands. It allows you to set the work part, open parts by proximity, and a host of other useful commands. |
| Control point | A control point exists on certain entities like curves, lines and splines. For example, a circle has a control point at the 3, 6, 9 and 12 o'clock locations. |
| Counterbore | A cylindrical volume that is at the top of a hole with a larger diameter than the hole. |
| Countersink | A conical volume that exists at the top of a hole. |
| Cue Display | The bar at the top or the bottom of the display window that directs you through various modeling functions. |
| Curve | An entity family that includes lines, arcs, rectangles, splines, and even some text entities. |
| Cylinder | A parametric entity that is a cylindrical solid. |
| Datum Axis | A reference entity that is an infinitely long line used to position other geometry. |
| Datum CSYS | A datum entity that has three orthogonal planes and three axes. |
| Datum Plane | A reference entity that is infinitely wide and long that is used to position other geometry. You can also use it to trim solid geometry. |
| Deformable part | There are a number of components that take on different shapes when they are assembled than when they are stand-alone. For example, a spring that you stretch into place. These components can be created and defined with the "deformable part" technique. When you bring them into the assembly, you are prompted for the length, or whatever variable that you pre-specified, that you want for that particular instance. |

| | |
|---|---|
| Degree Of Freedom Indicator | A small yellow marker on a sketch entity that is used to show which parts are not fully constrained. |
| Detail Views | An enlarged partial view of a particular region of interest on a 2-D drawing. |
| Dimensions | Depicted numerical values that dictate length, angle etc. on a sketch or drawing. |
| Synchronous modeling | An incredibly powerful technique of creating geometric relationships between surface entities or other surface manipulations. |
| Divide face | A method of breaking single surfaces into multiple surface patches. It is commonly used when you want to apply different colors to different sections of existing solids. |
| Drafting | The application of dimensions, text and other fabrication or assembly instructions to the part file. |
| Drawings | The 2-D documents that are made from part geometry; the by-product of drafting. |
| Dynamic input boxes | An input menu that comes up when you are in the sketcher and other functions in which you input numerical values that control geometry. |
| Dynamic WCS | A coordinate system that you can graphically manipulate by handles. There are six handles for X, Y and Z translation and rotation, and a seventh for simultaneous translations in all directions. |
| Edge Blend | The surfaces that are created at the intersecting edge of two faces. |
| Ellipses | 2-D curves in the shape of elongated circles that are drawn by default on the x-y plane of the WCS. |

| | |
|---|---|
| Emboss | A command that allows you to create raised characters on an existing surface. It's very handy when you have to create labels and other raised geometry. |
| Expression | The mathematical parameters that capture the design intent of a cad model. They can be as simple as p0=2, or as complex as p1=sin (b) + 2*x^2. |
| Extrude | A command that is used to drive curve geometry in a direction to make a solid or sheet solid. |
| Face Blend | Used to create a curved surface in between two pre-existing surfaces. It is like an edge blend but is used when the edge is interrupted or for some other reason is not suitable for the edge blend function. |
| File management | The act of placing your part files in logical orders and locations. |
| Fix | A constraint that fully constrains a single sketch entity. |
| Fly-out | Describes the way the menu items on the resource bar appear and shift out when you move the mouse over them. |
| Form Features | Parametric entities that you build by filling out the pre-programmed option dialog boxes. |
| Fully constrain | To fully constrain a sketch is to take away all degree of freedom such that every entity is at a fixed location. |
| GD&T Symbols | The symbols on a drawing that control the tolerance scheme. They can indicate the design intent down to the last detail or acceptable limits of fit and/or form. |
| Geometric constraints | Used to control the relationships between sketch entities such as equal length lines, coincident points, or concentric arcs. |
| Graphics Window | The large display window where all the geometry is depicted. |

| | |
|---|---|
| Groove | A cylindrical volume or ring-shaped void that is created on a cylindrical face, such as one you might need for a snap-ring seat. |
| Hinge Line | An imaginary line that defines the direction of various types of section views. |
| Hole | The feature located under the heading of Design Feature. A parametric entity. |
| Horizontal Reference | Used to orient a sketch on a solid face or datum plane. |
| Hybrid | Hybrid Modeling is a combination of parametric and non-parametric entities. |
| ID Symbols | A drafting entity that is used to identify a component on a drawing - particularly for identifying the individual parts of a sketch. |
| Inferred Dimension | The icon that is smart enough to adjust to whatever entity the dimension is created on. |
| Inferred Point | NX's method of choosing the most likely point of interest. |
| Instance | Any one of a number of associative copies of features and bodies. |
| Intersect | Where two lines, a surface, and a solid or other pairs of entities meet. |
| Law | The law option is found in many geometry creation techniques. For example, when you make a helix, you can control the radius using a law. It can be a linear law, a cubic law, a law by equation, or a few others. For the example of a helix you can define a linear law that forces the helix to have a 3 inch radius at one end and varying linearly to a five inch radius at the other. |
| Law Curve | A curve whose definition is based on formulas or geometric relationships to other curves. |

| | |
|---|---|
| Layer | Any one of 256 huge volumes that you may place entities in, allowing compartmentalization of the model. |
| Layer Masking | The act of specifying which layer will be the work layer and which layers will be selectable, visible only, invisible, or work. |
| Leader | A linear indicator, usually an arrow, which connects annotations, dimensions, and symbols to an entity. |
| Line | The shortest distance between two entities. |
| Local Scale | The ability to enlarge or reduce specific portions of solid geometry. |
| Manual Placement | Choosing the location for a dimension's text curves yourself, in a non-automated fashion. |
| Mirror | A command in Instances that allows associative reflected copies of bodies or specific features. There is also a mirror command for curves. |
| Model Navigator | The graphical interface usually appearing on the left hand side of the graphics window that enables you to view and manipulate the feature list. Some people refer to it as the CSG tree (Constructive Solid Geometry tree). |
| Move Region | A synchronous modeling command that enables you to parametrically move surfaces in a variety of ways. |
| Non-Parametric | Entities that are no longer driven by numerical values of any sort. It is very common to bring geometry in from an IGES or Parasolid file and have it be non-parametric. |
| Numerate | To apply the correct numerical values to all dimensional constraints in a sketch. |
| Object Display | The actual state of an entity – it's color, transparency value, etc. |

| | |
|---|---|
| Offset Curves | Curves that are created at a set distance from another set. The offset is applied normal to the defining set at all locations. |
| Offset Face | An extremely useful command that allows you to move the face of a solid as if you applied a uniform coating everywhere. The offset is normal to the defining face and the adjacent faces are extended or trimmed to meet the face of the initial face. |
| Offset Region | A synchronous modeling technique that allows you to offset a collection of surfaces parametrically. |
| On curve | A method of placing a sketch on a plane that is perpendicular to a curve. |
| Open by proximity | An amazing technique that saves loads of time when you are working with a large assembly. You set your load options to "structure only," then you open the assembly – none of the parts will open, just the structure. Next you go to the assembly navigator and open the one part you will be working on and around. Next you use the "Open by Proximity" command to create a three dimensional boundary that will be around the part that you have opened. When you execute the command, just the components that touch that boundary will open. |
| Orient View | A command that allows you to set the directions of your view in a variety of ways. You may define vectors from entities, the WCS, directions, etc. |
| Orient View to Sketch | A method of getting the sketch view to rotate back to its original orientation. |
| Origin Tool | A tool that defines the origin in order to move annotations in relation to views, geometry and other annotations. |
| Orthographic | A direction of a view 90 degrees from an initial view. |
| Orthographic Views | Views that are projected 90 degrees from some base view. |

| | |
|---|---|
| Over-constrained | When a sketch has too many dimensions and/or geometric constraints. |
| Pad | A parametric solid that is united onto a solid face. |
| Pan | To move a view left, right, up, or down. |
| Parallel and Perpendicular Lines | Parallel lines run in the exact same angle, perpendicular lines run 90 degrees away from each other. |
| Parametric | Entities that are defined and controlled by numeric expressions. |
| Parent/Child Relationships | Parent entities are those that other entities or features rely on for their definition. Children are those that rely on the parents. If you created a model of a block with a hole in it, the hole would be the child of the block. |
| Part Families | A method of creating several associated and similar part files from one master part file. |
| Pattern Face | A command that allows you to make associative copies of face geometry. |
| Pockets | A parametric indentation made on a solid face which allows positional dimensioning during creation. |
| Point on Curve Constraint | Allows placing a point at a prescribed location along a line arc or spline. |
| Pop-Up Menus | Menus invoked with certain mouse button manipulations or clicks. |
| Primitive Features | "Design Features" as blocks, cylinders, spheres and cones. |

| | |
|---|---|
| Re-blend face | A synchronous modeling command that allows you to parametrically change the radius value of a blend above and beyond its original parametric definition. |
| Rectangular Instance Array | Allows you to parametrically create associative copies of a feature in a rectangular pattern. |
| Rectangular Pocket | A parametric feature that is subtracted from a planar face in the shape of a block which allows positional dimension during creation. |
| Reference Entities | Datum planes, datum axes, datum coordinate systems and points are all reference entities. |
| Refresh | To clear the screen of all artifacts from previous picks. |
| Reordering Features | To move a feature in the Constructive Geometry Tree. If you have a model that is made up of a block, then a hole, then a hollow, it will appear very different than a model that is defined by a block that has a hole in it and is then hollowed. |
| Replace Face | This command exists in synchronous modeling in the edit face menu. Replace face allows you to substitute one surface for another; adjacent faces of a solid are trimmed to meet the replaced face. |
| Replace view | Allows you to substitute one view with another in a layout. |
| Resize Face | A synchronous modeling command that allows you to redefine the parametric values of a conical or cylindrical face. |
| Resource Bar | The bar of functions on one side of the display where you can find the help, training, history and other resources. |
| Revolve | A command that allows you to create solid geometry by rotating 2-D curves about an axis. You can also revolve 3-D curves, but this is not as common. |

| | |
|---|---|
| Rotate | Used to turn a view about a defined axis or around a point. |
| Rotation Handles | Allows manipulation of a coordinate system. |
| Scale | To apply a magnification factor to a geometry about some reference point. |
| Section Views | Gives you a 2-D view of a 3-D model based on a cutting plane location. |
| Selectable | A type of layer in which non-work layers can have their entities modified. |
| Selection Ball | A small circle that indicates entities to be selected. When entities are within the circle they can be selected. |
| Selection Filter | The ability to specify that you will only select entities of certain types, such as lines or just arcs. |
| Sew | To unite surfaces into one sheet solid or one solid body. |
| Sew Tolerance | The numerical value used to make the connection of two surfaces possible. |
| Shade | To shade a solid or sheet is to render it as if light bounced off of it. |
| Shell | A tool to hollow out a body to any desired thickness. Called 'Hollow' in other programs. |
| Show and Hide | Formerly known as Blank / Unblank; allows you to temporarily display or un-display separate entities. |
| Sketch | A collection of lines, arc, splines and points along with dimensions and constraints that capture geometric shape and design intent. |

| **Sketch Constraints** | Geometric rules that you define for geometry within a sketch. |
|---|---|
| Sketch plane | A plane that defines the 3-D location of a sketch. |
| Slots | A parametric entity that is added and subtracted from a solid, positioned on a planar face. |
| Sphere | A primitive parametric entity found in the Design Feature menu. |
| Spline | Curves that are defined by a series of points. |
| Split | A command that allows you to cut a model in two using a datum plane or sheet, which renders the model non-parametric. |
| Status Display | A small window that tells you the state of a sketch – whether it is fully constrained, over-constrained, etc. |
| String Mode | The condition defining curve entities in a sketch; the end of the last entity is the beginning of the next. |
| Studio Spline | A spline that is created by points that you can manipulate as you create the spline. |
| Subtract | A command that allows you to take the volume defined by one set of features from the volume of another. |
| Suppress | To temporarily hide geometry from the Constructive Solid Geometry tree. |
| Sweep Along Guide | A method of creating geometry that requires a drive curve and a section curve. It produces a model that has a constant cross section like a molding. |
| Swept | Geometry created by a sweep along a guide, or a surface created by pushing section geometry along one or more drive curves. |

| | |
|---|---|
| Swept Free Form Features | A surface that is created by sweeping sections along drive curves. |
| Tangency Controlled Blend | A blend whose radius is controlled by a curve or edge. The distance that the curve is away from the intersection of surfaces defines the radius. |
| Tangent Constraint | A constraint that forces geometry in a sketch to be tangent. |
| Tangent Lines | A tangent line is one that runs in a direction that is 90 degrees away from normal at a specific location. |
| Taper | A command used to angle surfaces. An example is when you are creating ribs in an injection-molded or cast part; the surfaces must be tapered to release from the mold. |
| Taper Faces | Used when you want to angle an entire surface about a neutral plane relative to a vector. |
| Taper From Edges | Used when you must taper a face about one of its edges relative to a vector. |
| Text | There's a "text" command in the curves function that you can use to create extrudeable text of various fonts. |
| Through Curves | A surface that is created with multiple section curves. |
| Tip Angle | The angle of the cone at the bottom of a hole. |
| Toolbars | Menus all around the display window that contain the command icons. |
| Translucency | The property of surfaces and solids that allows you to see light that goes right through them. |
| Trim | Cutting back a curve, surface, or solid geometry with one entity or another. |

| | |
|---|---|
| Trim Curve | To cut a curve back with a point, line, arc, spline, or surface. |
| Tube | A command in which you submit a group of curves that are all tangent and define an inside and outside diameter. |
| Unblank | Allows you to un-hide temporarily hidden geometry. |
| Undo | Allows you to go back a step or series or steps in your command structure. |
| Unigraphics NX 8™ | The CAD program that allows you to define almost any geometry that you can imagine. |
| Unite | Adding solid geometry together and making it one solid. |
| Unsuppress | Allows you to make features active that you previously rendered inactive. |
| Updating the Drawing Display | You must update the drawing display when you've made a change to the underlying geometry that the drawing views are based on. |
| User-Defined Symbols | Any symbols that are not in the standard menu that you actually create yourself and save. |
| Utility Symbols | Indicators that you place on a drawing such as linear centerline, symmetric centerline, target point and others that assist in 2-D geometry definition. |
| Variable Radius Edge Blend | Edge blends that have more than one radius from one end to another. |
| Vector Constructor | Enables you to define the direction for various commands. |
| View Label | At the lower left hand corner of a view you can display a label that tells you the view's name. |

| | |
|---|---|
| Visible Only | When you can see the entities on a layer but they are un-selectable. |
| Wave Geometry linker | A fantastic command that allows you to make smart copies of geometry from one component to another. Found under *Insert / Associative copy*. |
| WCS | The work coordinate system used to place geometry, which is able to be moved into place in a variety of ways. |
| Wireframe | Geometry that is made up of only curves, such as lines, arcs, splines, and points. |
| Work Layer | The layer that any new geometry that you create goes on. There are only 256 layers that you can place geometry on. |
| Work Part | The part in the assembly in which the entities that you are creating reside. |
| Work Plane | The plane that is current; all 2-D geometry that you create resides on the current work plane. |
| XC-YC plane | Also known as the plane of the WCS and the default location for all new 2-D geometry. |
| Zoom | A command that allows you to bring your viewing closer or farther away. |

# INDEX


## 2

2D basic curves, 33

## A

Absolute Coordinate System, 27

Advanced Form Features, 131

Animation, 627

Annotation Editor, 501

Annotation Preferences, 492

Application, 20

Drafting, 453, 454, 457, 463, 464, 468, 471, 472, 477, 484, 491, 493, 504, 506, 511

modeling, 190, 308, 338, 340, 342, 346, 351, 354, 383, 425, 457, 472, 483, 511

**Arc**, 49

Assembly, 405, 406, 407, 408, 409, 410, 413, 414, 416, 417, 418, 419, 420, 421, 422, 423, 424, 425, 426, 427, 428, 430, 431, 432, 434, 452, 453, 660, 708

Bottom-Up Assembly, 405

Displayed Part, 20, 414

Exploded View, 432

Load Options, 405

Mating Conditions, 409, 411, 423

Reference Set, 406, 407, 408, 411, 427

Reposition, 417

Top-Down assembly design, 414

Work Part, 20

Assembly Navigator, 22, 423, 427, 708

Associative Copy, 246

Auto Placement', 77

# B

Bead, 545

Bee, 678

**Bend Edge**, 533

Blanking, 20

Blend, 15, 157, 180, 187, 241, 249, 328, 511, 512, 652, 656

Coincident Edge Blend, 163

Face Blend, 161

Tangency Controlled Blend, 165

Variable Radius Edge Blend, 160

Block, 211

Body Taper, 561

Boolean, 15, 127, 151, 220, 273, 274, 276, 279, 283

Intersect, 15, 274, 282, 300

Intersection, 279, 283

Subtract, 15, 151, 220, 275, 277

Unite, 15, 150, 249, 251, 275, 278, 312

boss, 147, 329, 330

Boss, 326

Bosses, 131, 606

Bridge, 259

bridge curve, 658

Bridge surface, 658

Butt, 549

By Law Curve, 667

# C

Chamfer, 15, 180

Changing Color, 417

Circular beads, 545

Class Selection, 53, 58

Closing Files, 31

Coincident Constraint, 79

Collinear, 280

Collinear constraint, 86

command finder, 26

Connected Curves, 275

constrained, 14

Constraint, 280

construction entities, 15

Consumer Product, 648

**Continuous Auto Dimensioning**, 62

**Contour**, 540

control point, 32

Convert to/from reference, 87

Convert To/From Reference, 87

Create, 277

CSYS, 27

**Cue Display**, 21

curve, 32

Curves, 14, 20, 32, 33, 40, 47, 53, 253, 271, 404, 664

Arcs, 32, 76, 78, 86, 378

Circles, 14, 32, 76, 77, 83, 271, 509

Control points, 14, 32, 33, 191, 197

Ellipses, 32

Helix, 50

Line, 14, 32, 34, 38, 64, 69, 77, 78, 81, 82, 83, 86, 257, 271, 309, 314, 378, 662

Offset Curves, 253, 257, 259, 262, 265

Splines, 32

String Mode, 257

# D

Datum axes, 64, 68

Datum Axes, 190

Datum Axis, 15, 190, 196, 200

Datum Plane, 15, 190, 191, 192, 193, 194, 195, 196, 197, 198, 199, 200, 415

Datum Planes, 190

Default Settings, 643

Delete, 76, 194, 195, 196, 197, 198, 327, 331, 423, 512, 708

Deleting, 53, 326, 327, 423

Deleting Features, 327

Dependencies Panel, 323

Design Features, 15

Detail Features, 20, 156

Die cast, 559

Dimensions, 14, 16, 62, 67, 174, 489

Appended Text, 489, 490, 506

Automatic Placement, 489, 490, 494

Inferred Dimension, 489

**Direct Mdeling**

**Replace Face**, 342, 343, 346

Direct Modeling, 332

**Offset Region**, 340

Pattern Face, 360, 361

Resize Face, 338

Direct Sketch, 62

Direct Sketching, 98

Draft, 15, 112, 174, 177, 250, 251, 258

Drafting, 453

Drafting Preferences, 491

Drawing

Annotation, 490, 492, 493, 501, 502, 504, 506, 507, 708

Annotation Editor, 490, 501, 502, 504, 506, 507, 708

Auxiliary Views, 479, 480

Basic Drawing Views, 468

Detail Views, 479, 481

Dimensions, 66, 69, 70, 77, 78, 130, 144, 145, 146, 148, 150, 222, 226, 275, 300, 303, 304, 305, 310, 317, 320, 340, 454, 455, 474, 475, 490, 493, 494, 500, 501, 503, 511, 512, 645, 646, 652, 658, 664

Drawing Parameters, 454

Drawing Symbols, 501, 508

GD&T Symbols, 501

ID Symbols, 508, 509, 511, 514

Leader, 490, 503

Lettering Sizes, 492

Line / Arrow Style, 492

notes, 454, 475, 490, 501, 502, 504, 506, 509

Orthographic, 479

Section Views, 480

Units, 489, 492, 493

User Symbols, 501

User-Defined Symbols, 509

Utility Symbols, 508, 511

Drawings, 453

Dumb, 47, see Non-Parametric

Dynamic Input Boxes, 67

# E

Edit, 325

Edit Parameters, 320

Exiting UG, 31

Expression, 302, 303, 304, 307, 311, 314, 315, 318, 319, 668, 707

Renaming variables, 314

Expressions, 309

Extract, 246

extrude, 16

Extrude, 15, 105, 106, 108, 109, 110, 111, 112, 114, 150, 151, 158, 167, 168, 171, 178, 181, 184, 207, 212, 216, 226, 235,

239, 249, 251, 275, 276, 281, 282, 283, 284, 285, 286, 288, 300, 303, 305, 318, 371, 376, 395, 396, 397, 519, 560, 572, 591, 649, 650, 658, 689

## F

Face Blend, 652, 654

Feature Operations, 15, 204, 273, 320

File Management, 29

Fish, 679

Flowers, 680

fly-out, 22

Form Features, 104, 131, 152, 375

Free Form Feature

Swept Free Form Features, 383

Free Form Features, 375

Through Curves, 376, 378, 387

free-form surfaces, 15

Full Miter, 549

fully constrain, 78

## G

GD&T, 454

Geometric constraints, 77

Colliniar Line Constraints, 86

Fix, 78, 79, 80

Tangent Constraint, 86

The Point on Curve Constraint, 83

Geometric Relations, 37

Parallel and Perpendicular Lines, 38

Tangent Lines, 33, 37

Goodies, 678

Graphics Window, 18, 19, 20, 21, 22, 23, 24, 27, 28, 30, 55, 57, 65, 192, 195, 326, 405, 410, 413, 414, 472, 487, 489, 494, 495, 496, 504, 505, 514, 706

Groove, 146, 148

Grooves, 131

## H

Heat Stake, 595

Helix, 218, see Curves

Hide, 55, 56, 240

Hinge Line, 479

history, 22

Hole, 131, 147, 150, 151, 187, 235, 249, 278, 339, 455, 483, 485, 486, 494, 495, 497, 498, 506, 513, 514, 649, 651

Counter bored, 131

Countersunk, 131

Holes, 131

Hollow, 207, 330

Horizontal is cool, 250

Hybrid. See Parametric, See Parametric, See Parametric

## I

Inferred, 68, 69

Injection molded, 559

Inset Flange, 535

Instance, 15, 221, 222, 236, 360, 651

Circular Instance Arrays, 225

Geometry, 237

Mirror, 234, 250, 251, 252

Rectangular Instance Array, 222

Instance Geometry Along a Path, 237

Intersect, 274, 279

***Isocline Curves***, 564

Isometric, 481

## J

Join, 262

## K

keyboard shortcuts, 622

## L

Law Curve, 654

Layer Masking, 468
Layers, 457, 468, 471, 472
  Invisible, 468, 472
  Layer Settings, 472
  Move to Layer, 472
  Selectable, 468, 472
  Work Layer, 472
Lips, 591

## M

Machinery, 549
Manual Placement, 77, 495
Master Modeling, 517
Match Taper, 562
Menu bar, 17, 29
Mirror, 234
mirror curve, 89
Mirror Feature and Mirror Body, 234
Mirror Sketch, 89
Model Navigator, 22, 320, 322, 327, 328, 330, 648, 708
Model Tree, 105
Mouse Button
  MB1, 23, 418, 421, 477, 484, 489, 494
  MB2, 18, 423
  MB3, 18, 28, 38

## N

Nominal Wall Thickness, 605
Non-Parametric, 13, 14, 16

## O

Object Display, 57, 58, 417, 430
Offset, 257, 287, 570
  Curve, 257
Offset Face, 15, 214, 248, 249, 707
Orient View, 22
Orient View to Sketch, 69
Origin Tool, 502
Orthographic Views, 470
over-constrained, 16

## P

pads, 131
Pan, 22
parallel, 37
Parallel Dimension, 73
Parallel Line. see Geometric Relations
Parameters, 325
Parametric, 13, 32, 481
parametric lines, 47
Parasolid, 678
Parent/Child Relationships, 328
Part Navigator, 32, 321, 322
Patch, 576
Patching, 576
Perimeter, 93
perimeter constraint, 93
perpendicular, 37
Perpendicular Dimension, 73
Perpendicular Line. see Geometric Relations
Pockets, 131
Point Constructor, 255
Point method, 37
Point on Curve, 48, 83, 84, 85, 86, 280, 309, 693
Primitive features, 16
  Block, 16, 125, 144, 147, 148, 211, 212, 213, 222, 328, 455, 457, 485, 486, 494, 513
  Cone, 16, 124
  Cylinder, 16, 127, 128, 131, 149, 151, 155, 199, 200, 207, 218, 219
  Sphere, 16, 124, 662
Primitive Features, 124
  Block, 16
Primitives, 124
Project Curves, 265

# Q

Quick-Pick. see Selection Confirmation

# R

Rectangular Pocket, 144
Reference Features, 15, 190
reference geometry, 15
reference line, 123
Refresh, 18
Relief, 535, 552
Reordering Features, 326, 328
Replace view, 22
Reposition, 27
Resource Bar, 22
Reverse Direction, 107
revolve, 16
Revolve, 105, 155, 276, 311, 312
Revolved, 121
Ribs, 579
Ribs and Gussets, 606
Rotate, 14, 22, 198, 203, 706
Rotate Point, 22
Rotation Handles, 28

# S

Scale, 662
Scale Label, 481
Select All, 57
Selection Ball, 24
Selection Confirmation, 24
Selection Filter, 23
Selection Order, 25
Selection Toolbar, 23
Sew, 287, 288
Sew Tolerance, 289
Sewing, 572
Shaded Face Edge, 628
Sheet Metal Feature / Bridge, 543
Shell, 207
Show, 56
**Show All**, 56
shrinkages, 662
Simple Miter, 549
Sink Marks, 608
Sketch, 14, 15, 16, 32, 62, 63, 64, 65, 67, 70, 71, 73, 76, 77, 78, 79, 80, 81, 82, 84, 103, 106, 108, 122, 123, 128, 129, 130, 150, 155, 180, 235, 248, 249, 250, 251, 275, 276, 281, 282, 300, 304, 305, 309, 310, 312, 314, 318, 375, 378, 379, 380, 381, 383, 385,415, 425, 489, 645, 649, 652, 653, 654, 655, 658, 662, 666, 667, 706
  Constrain, 14, 16, 65, 78, 405, 409, 410, 419, 421
  Geometric constraints, 14, 78, 300
  Geometric Constraints, 16
  On a Face, 204
Sketch Constraints, 69, 77, 80, 81, 85, 695
Sketch Environment, 60
**Sketch in Task Environment'**, 60
Sketch on Path, 60, 91
Sketch plane, 60, 64
Sketches, 14, 20, 60, 203
Slots, 131
Snap Points, 48
Snap-Fit, 601
***Solid Punch***, 554
Spline, 216, 253
Split, 204, 210, 213, 217
ssembly
  Work Part, 414, 415, 424, 425, 427, 430
**Status Display**, 21
Studio Spline, 95
Subdivide Face, 569
Subtract, 274, 275
Suppress, 327, 328, 423
Surface Texture, 605
Surfaces, 20

Sweep, 218
Sweep along Guide, 92
Sweep Along Guide, 218, 380, 382
Swept, 385, 386, 404, 645

## T

tangent, 37
Tangent Line. see Geometric Relations
Taper From Edges, 559
Text Curves, 116
The Degree Of Freedom Indicators, 78
Thicken, 217
Thicken Sheet, 581
Timestamp, 320
Toolbars, 19, 454
Tools, 623
Tracking Bar, 40
translucency, 429
Trim, 15, 43, 45, 46, 64, 210, 211, 212, 271, 313
Trim Curve, 43
Trim to Boundary, 573
Trimetric, 481
True Draft, 562
Tube, 218, 219, 246, 247, 248, 249, 665

## U

***UDF***, 603
Undo, 22
Unigraphics NX™, 13
Unite, 274, 275
Unselect, 24
Unsuppress, 327, 328, 423
Update Views, 472
Use law, 667

## V

Vector Constructor, 108, 109
View Label, 481

## W

wave geometry linker, 658
WCS, 27, 28, 30, 34, 40, 64, 124, 145, 191, 225, 433, 651, 667
  Axis Handles, 28
  Origin Handle, 28
  The Dynamic WCS, 27
Work Plane, 27

## X

XC-YC plane, 27

## Z

Z-bend, 543
Zoom, 14, 18, 22

# Test Answers

Chapter 3

1. True.
2. False. When trimming to shorten, select the part of the curve to be removed.
3. False. Offset curves will be parametric if the Associative Output radio button is turned on.

Chapter 4

1. False.
2. True.
3. False. You do not have to use the Class Selection menu. You can immediately begin picking geometry in the Graphics Window without applying any filters.

Chapter 5

1. True.
2. True.
3. False. Constraints that are duplicated will turn a different color. It is best to fix this problem before continuing the sketch process.
4. True.

Chapter 6

1. True.
2. False – positive tapers will angle inward toward the center of the solid.
3. False – it is common to rotate about an axis that is not on the section string.
4. False.
5. True. The direction reference is used to define the pocket's dimensions.
6. True. (They were non-parametric in NX3.)

Chapter 7

1. False. The 'Variable Blend' option is located under the 'Edge Blend' command.
2. True.

3. True.

Chapter 8

1. True.
2. True.
3. True.

Chapter 9

1. False. You have an option to define alternate thickness.
2. False. An 'Offset Face' feature can be positive or negative.
3. True.
4. False. You cannot 'Trim' an object if the result will break it into more than one piece.
5. True.

Chapter 10

1. False, the 'Spline' function allows you to create a non-parametric General Spline curve. It can be closed-loop but is not required to be.
2. True.
3. False. The result of the 'Project' command may be parametric or non-parametric.

Chapter 11

1. True.
2. False. It may be used for combining sheet or solid bodies.
3. True. It works with small gaps based on the tolerance you define.

Chapter 12

1. True.
2. False. An expression value can be just a number; however, it can also be a mathematical function, or even another variable.
3. False. You can create expressions before, during, or after the construction of the model.
4. False. An expression variable can be renamed at any time using Tools / Expression / Rename.
5. False. You can edit any expression value.

Chapter 13

1. False. If you know which parameters of a feature you would like to change, you can also change them in the expressions menu. You can also change them by right-clicking on the feature in the graphics window.
2. False. You can also delete a feature by right-clicking it in the Part Navigator and choosing 'Delete' from the pop-up menu.

Chapter 14

1. True
2. False
3. False. 'Instance' can be used only when the model is parametric, while pattern face can be used on any kind of dumb or smart geometry.

Chapter 15

1. True.
2. True.
3. True.

Chapter 16

1. False. When you create a new component in the assembly, it uses the units of the assembly part.
2. False. You can change or create mating conditions of two components by highlighting the component in the Assembly Navigator and selecting Edit / Mate.
3. True.

Chapter 17

1. True. You can select Edit / Views / Update Views to update the drawing to reflect any changes made to your model.
2. True. You can add text before or after the dimension using the Annotation Editor.
3. True. You can use Text Editor/ Relationships to add expressions, object attributes, and part attributes.

Chapter 18

1. True.

2. True.

3. True.

Chapter 19

1. True. In the Pre-NX 8.5 sheet metals tools found in the modeling environment, you had to use the separate Inset Flange tool.

2. False. You can Bend/Unbend a sheet metal hole, but you cannot Bend/Unbend a normal hole.

3. True. You can select options and define relief.

4. False. You only need a single line to define a Bead Feature.

Chapter 20

1. False.
2. False. They only update if you have selected the associative option.
3. True.

33550661R10395

Made in the USA
Lexington, KY
08 July 2014